Handbook of Quantitative Sustainable Finance

Handbook of Quantitative Sustainable Finance is an edited collection concerning the integration of sustainability and climate risk considerations into mathematical and quantitative finance. This comprehensive handbook provides a valuable resource for researchers, practitioners, policymakers, and students who are interested in understanding the practical role of quantitative techniques in delivering sustainable finance and investment.

The book is divided into four main parts: Risks and Regulation; Asset Pricing and Portfolio Management; Data, Measurement, and AI; and Product Design and Specific Markets. Although this structure offers a coherent, unifying structure to the book, each chapter has been written so as to be self-contained and useful to readers interested in any specific aspect of quantitative sustainable finance. Every chapter has been written by leading experts in their field, and offers a useful, authoritative window into the state of research and practice.

Features

- Numerous contributions from leading experts in sustainable finance
- Cutting edge analysis of recent technological advances in areas such as artificial intelligence
- Practical tools and ideas for working quants, as well as valuable material for academic study.

Peter Tankov is professor of quantitative finance at ENSAE, the French National School for Statistics and Economic Administration, having previously worked at Paris-Cite university and Ecole Polytechnique. He is a mathematician, specialist in applied probability and quantitative finance. He earned his PhD in applied mathematics from Ecole Polytechnique in 2004. His current research focuses on green and sustainable finance, where he aims to develop quantitative methodologies. Peter is the author of over 60 research articles on these and other topics and of the widely read book, Financial Modelling with Jump Processes. He is the recipient of the 2016 Best Young Researcher in Finance award of the Europlace Institute of Finance and the 2024 Louis Bachelier Prize of London Mathematical Society, SMAI, and Natixis Foundation. Professor Tankov is the scientific director of the Paris Agreement Research Commons foundation at Louis Bachelier Institute, where he leads data-oriented initiatives, and member of editorial boards of the main quantitative finance journals: Mathematical Finance and Finance and Stochastics.

Ruixun Zhang is an associate professor with tenure at Peking University. Ruixun earned a PhD in Applied Mathematics from MIT in 2015, and bachelor's degrees in Mathematics and Applied Mathematics, and Economics (double degree) from Peking University in 2011. Ruixun's research interests include green finance, machine learning, market microstructure, and evolutionary models of financial behavior. His research has appeared in leading journals such as *Proceedings of the National Academy of Sciences, Operations Research, Management Science*, and *Journal of the American Statistical Association.* His work has been recognized by the S&P Global Academic ESG Research Award, the International Centre for Pension Management (ICPM) Research Award, the Commodity and Energy Markets Association (CEMA) Questrom-CEMA Best Paper Prize, the SIAM Activity Group on Financial Mathematics and Engineering Conference Paper Prize, and the CFRI&CIRF-China Finance Review International Research Excellence Award. He serves on the editorial board of *Digital Finance* and *International Journal of Financial Engineering.*

Chapman & Hall/CRC Financial Mathematics Series

Recently Published Titles

Computational Methods in Finance, Second Edition
Ali Hirsa

Interest Rate Modeling
Theory and Practice, Third Edition
Lixin Wu

Data Science and Risk Analytics in Finance and Insurance
Tze Leung Lai and Haipeng Xing

Foundations of Quantitative Finance
Book VI: Densities, Transformed Distributions and Limit
Robert R. Reitano

Malliavin Calculus in Finance
Theory and Practice, Second Edition
Elisa Alòs and David Garcia Lorite

Arbitrage and Rational Decisions
Robert Nau

Equity Release Products
Radu S. Tunaru and Enoch B. Quaye

Quantitative Finance with Case Studies in Python
A Practical Guide to Investment Management, Trading and Financial Engineering
Chris Kelliher

XVA Analysis
Probabilistic, Risk Measure, and Machine Learning Issues
Stéphane Crépey

Risk Analysis in Finance and Insurance
Alexander Melnikov

For more information about this series please visit: https://www.routledge.com/Chapman-and-HallCRC-Financial-Mathematics-Series/book-series/CHFINANCMTH?srsltid=AfmBOornRwcwlEiKdCW6zJb5pugh5qIPk97qnp6be0Y-bhXBE82SHe9n

Handbook of Quantitative Sustainable Finance

Edited by
Peter Tankov and Ruixun Zhang

CRC Press
Taylor & Francis Group
Boca Raton London New York

CRC Press is an imprint of the
Taylor & Francis Group, an **informa** business
A CHAPMAN & HALL BOOK

Designed cover image: Authors

First edition published 2026
by CRC Press
2385 NW Executive Center Drive, Suite 320, Boca Raton FL 33431

and by CRC Press
4 Park Square, Milton Park, Abingdon, Oxon, OX14 4RN

CRC Press is an imprint of Taylor & Francis Group, LLC

ISBN: 978-1-032-62792-2 (hbk)
ISBN: 978-1-032-63624-5 (pbk)
ISBN: 978-1-032-63625-2 (ebk)

DOI: 10.1201/9781032636252

Typeset in CMR10 font
by KnowledgeWorks Global Ltd.

Publisher's note: This book has been prepared from camera-ready copy provided by the authors.

Contents

Chapter 1

Introduction: what can quantitative finance bring to sustainable finance?

Peter Tankov

CREST – ENSAE, Institut Polytechnique de Paris, France

Ruixun Zhang

Peking University, China

Quantitative finance is a collection of mathematical tools and methods, based on advanced probability and statistics, and used for solving a range of problems in the financial industry, such as pricing derivative products through dynamic hedging, systematic portfolio construction, quantitative risk management, and, more recently, algorithmic trading. Key milestones in the development of quantitative finance are the Markowitz portfolio theory (1952), which led to the development of modern portfolio optimization techniques based on a trade-off between risk and return, Black-Scholes option pricing formula (1973), which transformed the derivatives industry, and the Value at Risk, which was developed at JP Morgan in 1994 and became the cornerstone of the modern approach to financial risk management.

Sustainable finance involves taking environmental, social, and governance (ESG) considerations into account when making investment decisions, both in terms of maximizing positive impacts of investments on the environment and society and minimizing negative ones, and in terms of accounting for various risk sources related to them. It relies on a range of new data sources to quantify these characteristics. Nevertheless, sustainable investors also care about the financial performance of their assets; thus, risk and return analysis techniques are still relevant to them.

While the goal of this book is to present an overview of quantitative approaches in sustainable finance, not necessarily inspired by quantitative finance, in this introductory chapter, our aim is to compare the modeling approaches used in the two domains, emphasize the areas and applications where sustainable finance has something to learn from its quantitative cousin, and hopefully inspire more research and collaborations between scholars in related fields.

1.1 Modeling approaches

Models in quantitative finance are probabilistic representations of possible future evolutions of financial markets. They are used to price assets and financial products, quantify the risk of financial positions, design trading strategies, and determine provisions for unhedged risks. Although based on specific assumptions and simplifications of real-world phenomena, they allow us to structure decision making and mitigate financial risks if used wisely.

DOI: 10.1201/9781032636252-1

Quantitative models work best for short- to medium-term horizons—not too long so that stationarity can be assumed and forecasts remain reliable, and not too short so that market microstructure effects and other imperfections related to, say, discrete time, do not come into play.

A typical risk management framework for, say, an exotic options trading desk would consist of a pricing model, used to determine price quotes for options, strategies for hedging the risks that are well represented within the model, and provisions for the risks that are not well represented. In other words, the model may be too simple compared to reality, but if the trio model—hedging strategies—provisions is sufficiently conservative, it can still be successfully used to manage the trading desk's risks.

Many of the above problems also arise in sustainable finance. Although exotic structured products are less common, relevant problems include

- Pricing financial assets (stocks) by taking into account various ESG-related premia and market constraints;
- Pricing and hedging bonds exposed to climate and other nature-related risks;
- Finding a trade-off between financial and ESG performance in portfolio construction, and building net-zero and Paris-aligned portfolios.

These problems also require quantitative models that can be solved rigorously. The modeling is, however, complex and challenging due to the long-term nature of the risks involved and the often poor quality of data.

The long-term nature of climate and nature-related financial risks make it difficult to use so-called reduced-form models, which are common in quantitative finance. In reduced-form models, prices of fundamental assets, such as stocks, are modeled directly by exogenous stochastic processes, whose parameters are calibrated to data. By contrast, in structural models, prices are deduced from market fundamentals, and in equilibrium models, prices are obtained as a result of the interaction between market participants. Structural and equilibrium models are better suited for modeling long-term risks because of their capacity to adapt to changing market conditions. However, they require (much) more computational power to run as, for example, in equilibrium models, the decisions of individual agents must be described and taken into account.

Another reason for the complexity of models in sustainable finance comes from the need to model risk factors other than prices, such as climate and macroeconomic variables, which means that the model dimension (number of variables) is much larger than in a typical context in quantitative finance. Since these models describe different aspects of the future, such as climate, economy, and finance, they are known as integrated assessment models (IAM). As a result of this complexity, most IAMs are deterministic; that is, they do not include random noises and produce just a few simulations, or even a single one, whereas option prices in quantitative finance are typically computed using tens of thousands of simulations which explore the randomness of possible futures through "branching" scenarios (see Figure 1.1).

This complexity makes it difficult to analyze the robustness of models and quantify the impact of parameter uncertainty on the final results. Research on uncertainty quantification of climate-economy IAMs has only recently started to appear; see, for example, Cai and Lontzek (2019), where a variant of a rather classical IAM with stochastic risks that was solved in 8 hours on a Blue Waters supercomputer with 110,688 cores running in parallel. During the last 20 years, quantitative finance has developed efficient algorithms for solving high-dimensional stochastic optimization problems with much less computing power, using cleverly generated trajectories or neural networks. Perhaps, climate economics has something to learn here.

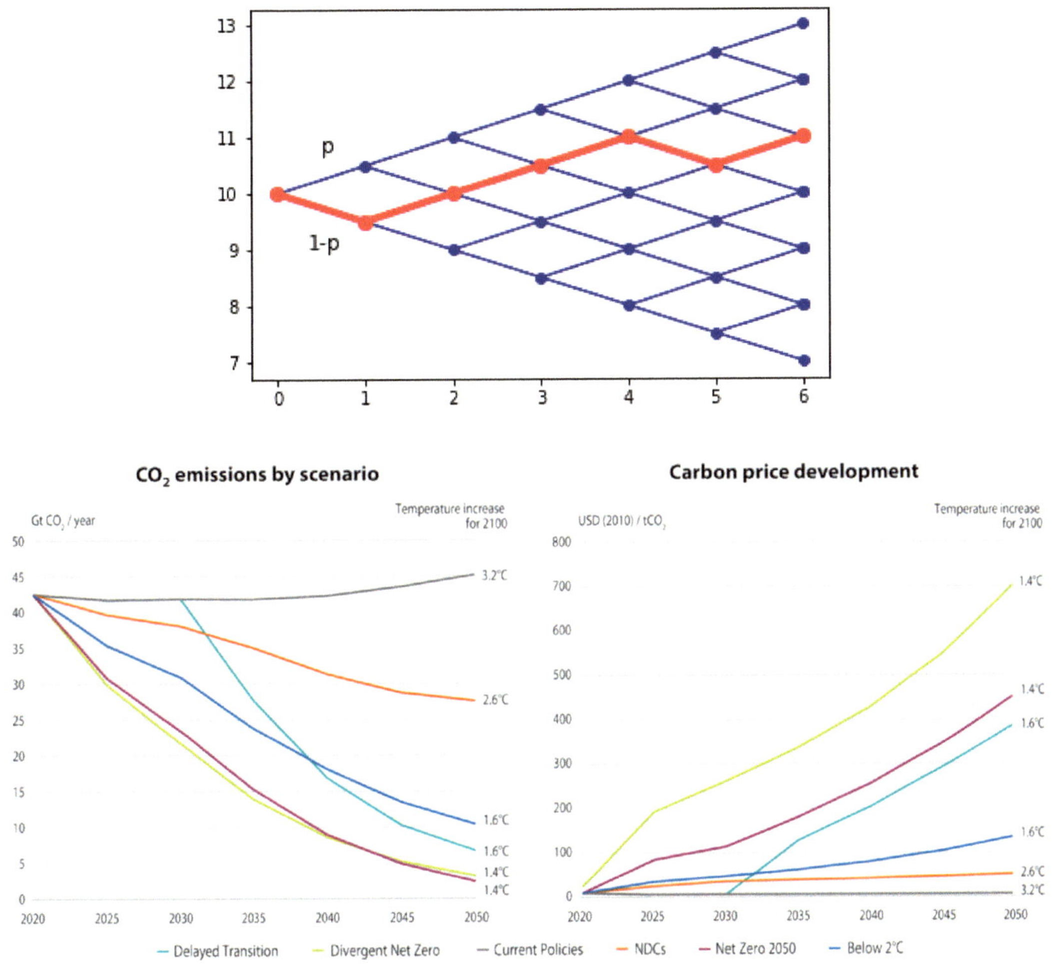

FIGURE 1.1: Top: a simple financial model for the evolution of an asset price: at every time step the trajectory branches up or down with specified probabilities. Source: author calculation. Bottom: evolution of CO_2 emissions and the carbon price in the scenarios from the NGFS database (Release 3): a small set of deterministic scenarios is used, with no possibility of branching at future dates. Source: NGFS.

1.2 Model governance and model validation

One important domain where sustainable finance has a lot to learn from quantitative finance is related to model governance and model validation. Models have been used in quantitative finance for at least 50 years, and sound principles have been established regarding the proper use of models and the management of model risk. For example, the famous SR 11-07 guideline of the Federal Reserve, published in 2011, defines the model as a combination of three components:

- Information input component, which delivers assumptions and data to the model;
- Processing component, which transforms inputs into estimates;
- Reporting component, which translates the estimates into useful business information.

In this framework, model risk is defined as the risk of either having a model that contains fundamental errors and may produce inaccurate outputs or having a correct model that is used incorrectly or inappropriately. According to the guideline, model risk should be managed through robust model development and implementation, a sound model validation process, and finally, a governance system that sets an effective framework with defined roles and responsibilities for clear communication of model limitations and assumptions, as well as the authority to restrict model usage. Tools for mitigating model risk include establishing limits on model use, monitoring model performance, adjusting or revising models over time, and supplementing model results with other analysis and information.

A key component of model governance in quantitative finance is a model validation framework, which includes the following steps:

- Evaluation of conceptual soundness that assesses the quality of the model design and construction;
- Ongoing monitoring, which must confirm that the model is appropriately implemented and used, and is performing as intended;
- Outcomes analysis, including back-testing that compares model outputs to the corresponding actual outcomes.

The documentation of model development and validation should be sufficiently detailed so that parties unfamiliar with a model can understand how the model operates, its limitations, and its key assumptions. These basic principles are further developed in the Fundamental Review of Trading Book guidance of the Basel Committee (2016).

In sustainable finance, the principles of sound model governance and validation are yet to be established. Currently, financial actors meet their needs through the use of off-the-shelf scenarios. This seems to be the simplest choice, but these scenarios are mostly based on black-box IAMs whose details are often not fully public, and whose validation, in consequence, is problematic. An alternative is to develop an in-house modeling capacity that would allow the financial actors to control the entire modeling chain and implement appropriate governance and validation procedures. This, however, in addition to the skills in financial and quantitative modeling, requires knowledge in macroeconomics, energy economics, and environmental sciences, which most financial institutions do not have at present. This can also lead to multiple models generating different conclusions, which is particularly dangerous in the highly politicized domain of climate finance. The answer is to intensify the dialogue between the financial institutions and various modeling communities, and to promote an interdisciplinary and open approach in model development along the lines of the guiding principles of model design and governance, established by the regulator for quantitative finance models.

1.3 Risk assessment and stress testing

The modern approach to risk assessment and banking supervision is arguably the most significant application of quantitative methods to finance. The principles of banking supervision were developed in the successive Basel accords, from the first one in 1988, dealing with the banking book only, to the third one, published in 2011 in the aftermath of the financial crisis, as well as additional regulations such as the Market Risk Amendment of 1996, which introduced the Value at Risk, and the Fundamental Review of Trading Book of 2016. Within the Basel Framework, supervisors monitor various regulatory ratios, the most important of which is the capital ratio, whose minimal allowed value is usually set to 8%:

$$\text{Capital ratio} := \frac{\text{Capital}}{\text{Risk-weighted assets}} \geq 8\%$$

Quantitative methods are used to calculate the capital (for certain illiquid assets) as well as the value of risk-weighted assets (risk measures such as Value at Risk, stressed Value at Risk, and the Expected Shortfall are used at this point). Everyday risk and capital adequacy assessments are thus based on probabilistic models of market evolution.

To complement this probabilistic perspective, supervisory prudential stress tests are conducted regularly to evaluate the banks' resilience to scenarios that may not be well represented by the probabilistic models, but are nevertheless relevant and sufficiently severe to warrant special attention (see Chapter 4). These stress tests are conducted by supervisors according to a specific schedule (for example, the ECB conducts them on a bi-annual basis), but banks are expected to implement internal stress tests at a higher frequency.

To put it simply, modern financial risk assessment is thus a combination of probabilistic approaches and stress testing to account for specific non-probabilistic scenarios.

In recent years, and especially since the climate-related financial risks came into the limelight with Mark Carney's speech "Tragedy of the Horizons" in 2015, supervisors started to develop frameworks for climate risk assessment, and the first "climate stress tests" were conducted by central banks starting from 2020 (see Chapter 4). These stress tests are similar to prudential stress tests in some aspects, but very different from them otherwise. Most importantly, while prudential stress tests use either historical scenarios corresponding to a crisis period or hypothetical scenarios where a specific risk factor is stressed, climate stress tests are typically based on prospective scenarios, produced by IAMs under specific climate and policy assumptions.

While the stress testing component of the financial risk management framework has been replicated for climate stress tests, it is not the case for the probabilistic component based on risk measures. This is to some extent justified by the long-term nature of climate risks and by the difficulty in assigning a probability to prospective scenarios. However, since climate stress testing is not a complementary tool aiming to quantify the impact of plausible but rare severe scenarios, but the primary method for assessing climate risks by exploring different likely outcomes, it is clearly not sufficient to use a small number of deterministic scenarios. Climate risk associated with, say, a disorderly transition scenario, strongly depends on the nature of this scenario, and a single scenario may miss important mechanisms of risk materialization. Since we are dealing with likely events whose nature is not perfectly known, it is important to explore a multitude of possible futures, and for this reason, a probabilistic approach is necessary. One possible path forward is given by the recent literature inspired by quantitative finance, which introduces probabilistic components into IAMs to quantify their uncertainty and generate a multitude of scenarios.

A second drawback of the current climate stress testing approaches is their inability to describe endogenous financial system dynamics: herd behavior, fire sales, and credit contagion, which can exacerbate financial crises triggered by external factors such as climate. The mechanisms behind these phenomena are relatively well understood in the quantitative finance literature, but climate stress tests are usually based on very simple models for the propagation of climate risks to the financial sector, which work at the level of a single asset and do not account for nonlinear interactions.

In Part I of this handbook, "Risks and Regulation," we have included four chapters broadly related to this topic. Chapter 2, contributed by Blasberg, Kiesel, and Stahl, discusses a risk management framework for transition risk. Chapter 3, contributed by Flora and Tankov, develops a real-options approach to evaluate assets and potential investment projects under dynamic climate transition scenario uncertainty. Chapter 4, contributed by Cartellier, provides an overview of climate stress testing and scenario analysis methodologies developed by central banks and scholars. Chapter 5, contributed by Grasselli and Milic, explores the potential role of central banks in combating climate change through the implementation of a green quantitative easing program.

1.4 Portfolio choice

A key objective of sustainable finance is to redirect financial flows away from activities with high negative impacts and toward activities compatible with sustainable development. One way to achieve this is to incorporate green or net-zero considerations into portfolio choice. Because passive and active investment strategies both account for a significant share of the investment management industry, it is necessary to develop both green indices for passive investment and green active investment strategies. Key issues here include (i) determining which investments are considered green or net zero and measuring their greenness, and (ii) understanding the investors' preference for greenness and constructing indices and strategies taking this preference into account.

The definition of green asset is provided, for example, by the European Taxonomy. However, assets that are not yet green but on the right track for net zero in 2050, and assets that are necessary for the transition, should also be funded. The notion of net-zero alignment is more difficult to define, as it involves not only the present activities of the company, but also its future trajectory and transition plan. Quantitative approaches are needed to measure the performance of the company beyond what the company itself wishes to report. For example, carbon emissions are reported on an annual basis, and the emissions for year N are usually published well into year $N+1$. Quantitative methods such as machine learning or satellite data analysis may be used to predict emissions (i) before they are available and at a high frequency, and (ii) for companies without published emission data. Quantitative methods (forecasting methodologies) may also be used to predict the future evolution of the company based on its past performance and the available data, to measure whether it is on track for net zero.

The second step, portfolio construction, is also based on quantitative approaches. One of the first and foundational results in quantitative finance is the Markowitz efficient frontier: if investors want to maximize portfolio return and minimize its risk, all efficient portfolios will draw a curve in the risk-return space, called the efficient frontier (see Figure 1.2). In addition, modern dynamic portfolio choice in a multi-period setting often uses stochastic control approaches, which originated with the celebrated Merton's portfolio problem.

When green, ESG, and net-zero considerations come into play, the classical theory of efficient frontier becomes insufficient, since the investors want, at the same time, to (i) maximize the portfolio's expected return, (ii) maximize its environmental performance, and (iii) minimize its risk. It is therefore essential to define an optimization criterion taking into account investor preferences in an appropriate way. Popular approaches include screening or re-weighting assets based on the ranking of a green measure, described in Chapter 8, and constructing portfolios with the highest utility or Sharpe ratio under a given green

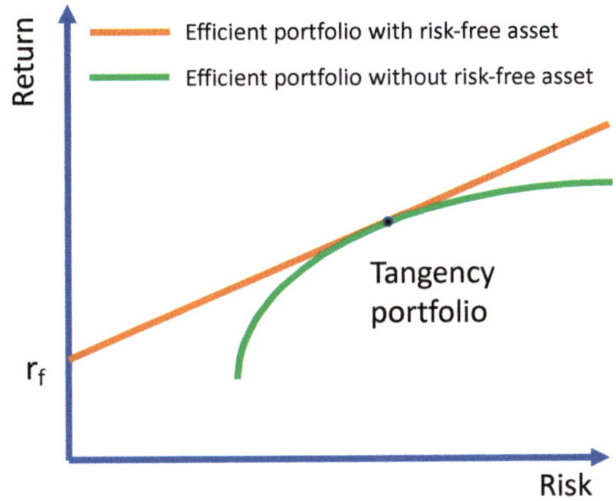

FIGURE 1.2: Markowitz' efficient frontier.

constraint, described in Chapter 9. Furthermore, investors' green preferences will also be reflected in equilibrium asset prices, described in Chapter 6.

An additional complication in green investing is that, though the ESG scores are observable, they provide uncertain information about the firm's environmental performance (see Chapter 14; (Avramov et al., 2022)). The same is true for reported carbon emissions, which are less subjective but only available with a delay and do not predict future environmental performance. Therefore, robust stochastic control methodologies may be used to build optimal portfolios in the presence of biases, errors, and uncertainty in the data.

In Part II of this handbook, "Asset Pricing and Portfolio Management," we have included five chapters broadly related to this topic. Chapter 6, contributed by Pástor, Stambaugh, and Taylor, shows that the high returns delivered by green assets in recent years reflect unexpectedly strong increases in environmental concerns, not high expected returns, a result motivated by their celebrated equilibrium model of sustainable investing (Pástor, Stambaugh, and Taylor, 2021). Chapter 7, contributed by De Angelis, Tankov, and Zerbib, shows how green investing spurs companies to mitigate their carbon emissions by raising the cost of capital of the most carbon-intensive companies. Chapters 8 and 9, both contributed by Lo and Zhang, discuss how to construct optimal sustainable portfolios and, in addition, propose a quantitative framework for assessing the financial impact of green and impact investing, including those constructed from green ranking and green constraints. Chapter 10, contributed by Roncalli, examines portfolio alignment in net-zero investing and distinguishes it from portfolio decarbonization.

1.5 Data and AI

Quantitative finance models are based on financial data. For example, in risk management, at least two types of data are used: risk management methodologies are based on historical time series data to compute risk measures and identify stress scenarios, and parameters of pricing and hedging models are calibrated to a snapshot of market data on any given day (such as listed option prices), to ensure the absence of arbitrage. Given the scarcity of historical data (only a single trajectory is available), generative modeling is often used, from well-established Monte Carlo methodologies to more recent AI-based methods such as generative adversarial networks (GAN) and diffusion models. In addition, it is also crucial to analyze the quality of data sources as part of the model validation and model risk management procedures.

In sustainable finance, there is much more data of poorer quality. Unlike financial data, which typically corresponds to prices of actual transactions recorded reliably, extra-financial data comes in a multitude of formats and can be acquired from different sources with varying degrees of reliability. This includes, for example:

- Numeric and textual data on environmental, social, and other impacts reported by companies and compiled by data providers with some quality checks, as well as the indices, ratings, and scores computed by data providers such as ESG scores;
- Textual data from newspapers;
- Data observed by physical measurements such as satellite images, temperature, carbon emissions, and weather conditions;
- Other asset-level data compiled by non-governmental organizations and data providers based on a variety of sources;
- Scenarios and projections produced by various organizations such as the Network for Greening the Financial System (NGFS), the Intergovernmental Panel on Climate Change (IPCC), the International Energy Agency (IEA), and the World Bank.

Many of these data sources are known to contain biases, and standards governing them are barely starting to emerge. Robust data governance and validation approaches must be established, and quantitative methods can be very helpful in this context. For example,

- Machine learning approaches can be used to estimate company impacts from asset-level data or satellite images, to double-check data reported by corporations;
- Natural language processing can be used to analyze news articles and/or company reports to detect greenwashing;
- Deep neural networks can be used to detect patterns in the data and forecast future impacts of extreme events and tipping points.

Given the multitude of data sources, the data quality issues, and the difficulty in accessing reliable data, especially for academic research, non-profit initiatives aiming to construct reliable and publicly available extra-financial databases have started to emerge in recent years. Examples include the Net-Zero Data Public Utility (NZPDU) for data on emissions and other company impacts, and Climate Trace for asset-level emission data.

In Part III of this handbook, "Data, Measurement, and AI," we have included four chapters broadly related to this topic. Chapter 11, contributed by Tantet and Tankov, presents a high-level overview of the data sources relevant for sustainable finance, and describes existing classification attempts and important open source initiatives. Chapter 12, contributed by Barreau, Fahmaoui, and Tankov, reviews existing machine learning models for estimating greenhouse gas emissions of companies using publicly available data, and develops a new interpretable approach, focusing on out-of-sample evaluation as well as on the data preprocessing steps such as outlier filtering, missing value imputation, and predictor selection. Chapter 13, contributed by Rui and Shan, presents firm-level measures of climate risk exposure, distinguishing between physical and transition risks, and demonstrates that these measures have significant implications for firm performance in capital markets and for predicting real economic outcomes. Chapter 14, contributed by Rigobon, discusses three key questions regarding the ESG scores, including the objective of their design and construction, how they are produced, and why they are needed.

1.6 Product design and specific markets

The transition to a sustainable economy requires reimagining traditional financial instruments and developing new markets that explicitly price environmental externalities. Product design in sustainable finance must align long-term climate objectives with short-term financial incentives, embedding environmental risks into pricing frameworks, and creating mechanisms that facilitate capital reallocation while maintaining market stability. Quantitative methods play a pivotal role in addressing these challenges by enabling the structural modeling of novel assets, optimizing regulatory interventions, and designing risk-sharing mechanisms tailored to sustainability goals.

A critical challenge lies in translating environmental constraints into financially viable instruments. For energy projects (Chapter 15), this involves modeling technology-specific operating leverage and endogenous investment thresholds under revenue uncertainty. In carbon markets (Chapter 17), it requires solving inverse problems where regulators dynamically calibrate allowance allocations to achieve emission targets while accounting for firms' abatement cost structures. Fixed income markets face their own paradox: while green bonds demonstrate measurable yield differentials (Chapter 16), the modest premium suggests investors prioritize financial fundamentals over environmental impact, requiring innovative coupon structures or regulatory incentives to deepen market penetration. Insurance solutions (Chapter 18) must bridge the gap between parametric triggers and complex sustainability outcomes.

The green bond market, in particular, despite rapid growth in recent years, still faces unique challenges, including defining the scope of eligible projects, ensuring transparency in the use of proceeds, and establishing credible certification standards. These challenges require sophisticated quantitative methods to measure and report the environmental impact of green investments.

One particularly promising area is the development of climate-linked derivatives, which allow firms and investors to hedge against climate-related risks. These products require robust models to quantify climate risks and to design effective hedging strategies that reflect the complexities of both physical and transition risks. The design of such derivatives must consider the interactions between climate policy, market prices, and real-world environmental outcomes, making them a unique and challenging area for quantitative finance.

The interplay between product innovation and market microstructure further complicates design choices. The integration of sustainability into financial markets introduces the potential for new types of market frictions, such as liquidity issues in green markets or mispricing of climate risk. Furthermore, regulatory frameworks are shaping the markets for sustainable products by establishing standards for what qualifies as a "green" investment and influencing investor preferences. These regulatory policies, along with market incentives such as tax breaks or subsidies for sustainable investments, play a crucial role in influencing both thc supply and demand for green financial products.

In Part IV of this handbook, "Product Design and Specific Markets," we have included four chapters that explore these issues in depth. Chapter 15, contributed by Detemple and Kitapbayev, studies investments in exclusive energy projects with different cost structures, revealing the effects of operating leverage on managerial policies, investment decisions, and values. Chapter 16, contributed by Zerbib, finds a small negative premium in the yield of green bonds, demonstrating a low impact of investors' pro-environmental preferences on bond prices, which does not represent, at this stage, a disincentive for investors to support the expansion of the green bond market. Chapter 17, contributed by Aïd and Biagini, deals with optimal dynamic carbon emission regulation of a set of firms and derives closed-form allocations that achieve emission targets with minimal market distortion. Chapter 18, contributed by Biffis and Hughes, analyzes wildfire-linked carbon offset policies in California's cap-and-trade system, and highlights the role of new insurance products designed to manage climate-related risks and incentivize sustainable business practices.

1.7 Sustainable finance: outlook

As we finish writing this introduction, sustainable finance is living through difficult days of *ESG backlash*. After the ESG investing boom of 2019–2021, the resurgence of inflation in 2022 triggered a lasting rise in interest rates, putting energy transition projects on hold. Facing diminishing ESG returns, some investors abandoned net zero and ESG investing in favor of the search for yield. At the same time, ESG got caught up in the political agenda in the US with some politicians labeling it as "woke investing." Some states, such as Texas and Florida, restricted the management of public funds using ESG factors in investment decisions. Following withdrawals of public money from their ESG funds, major US fund managers scaled back ESG priorities to avoid political and reputational risks. As a result, international net-zero initiatives were disbanded or reduced their ambitions. Examples include:

- Net-Zero Insurance Alliance was discontinued as of April 25, 2024;
- Net Zero Asset Managers initiative has been "in review" since Jan 13, 2025, following departures and threats of departure of major asset managers;
- Glasgow Financial Alliance for Net Zero (GFANZ), facing possible defections from US banks, issued a statement saying that coal finance targets were not binding.

In addition, climate reporting obligations are reduced or simplified: in the US, the SEC climate disclosure rule implementing TCFD recommendations was put on hold following court challenges, and the EU aims to simplify many of its corporate sustainability reporting requirements in the upcoming omnibus regulation.

There are valid reasons to criticize ESG investing. First, the ESG concept is too broad; it bundles together many things and often involves conflicting objectives. It is also prone to greenwashing risks due to measurement and data issues and investors' preference for warm-glow over impact. Finally, recent research has shown that the most common ESG strategies (exclusion, divestment) may not be effective in reducing emissions. To a certain degree, ESG is not quantitative and precise enough, whereas recent research in quantitative sustainable finance allows one to identify specific material risks related to climate and sustainability, and to understand how investors can achieve real-world impact. Thus, a shift from ESG investing to more modern concepts such as impact investing and transition planning, if it happens, will be much welcomed.

Many green technologies are now mature and do not require state support for large-scale deployment. According to the latest IEA report, the world now invests twice as much in clean energy as in fossil fuels (International Energy Agency, 2024). On the other hand, climate risks are now part of everyday life and few question their materiality. Thus, even though they may communicate less on these issues, companies will continue to address climate risks and invest in green technologies, and investors will continue to pressure them to do so. However, at times when the US government is removing information about climate change from its websites, the role of civil society, academia, and sustainable finance practitioners is becoming ever more important. It is up to you, the reader of this book, to convince the corporate and public decision makers that climate risks are indeed material, and that green investing can indeed be profitable when managed properly, and to provide them with quantitative tools to evaluate these risks and structure these investments.

1.8 Final remarks

The chapters in this handbook collectively demonstrate how quantitative finance can address the defining challenge of our era: aligning financial systems with planetary boundaries while maintaining economic stability and sustainability. Though the four parts form a cohesive narrative, each chapter stands on its own merits. Readers may navigate the collection based on their specific interests.

We extend our deepest gratitude to the contributors for sharing their cutting-edge research, and to you, the reader, for engaging with these critical issues. We thank two students, Minghao Liu and Binnan Wang, for excellent editorial assistance. May this volume serve as both an inspiration for future research and a practical toolkit for practitioners steering finance toward a sustainable future.

Thank you for joining us on this journey through quantitative sustainable finance.

References

Avramov, D., S. Cheng, A. Lioui, and A. Tarelli, 2022, Sustainable investing with ESG rating uncertainty, *Journal of Financial Economics* 145, 642–664.

Cai, Y., and T. S. Lontzek, 2019, The social cost of carbon with economic and climate risks, *Journal of Political Economy* 127, 2684–2734.

International Energy Agency, 2024, World Energy Investment, Technical report, IEA.

Pástor, L., R. F. Stambaugh, and L. A. Taylor, 2021, Sustainable investing in equilibrium, *Journal of Financial Economics* 142, 550–571.

Part I

Risks and Regulation

Chapter 2

Transition risks: a new frontier for risk management

Alexander Blasberg
University Duisburg-Essen, Germany

Rüdiger Kiesel
University Duisburg-Essen, Germany

Gerhard Stahl
HDI, Germany

This contribution establishes a risk management framework for transition risk. We start with a review of the definition given by the Network of Greening the Financial Systems (NGFS) and the Task Force on Climate-Related Financial Disclosures (TCFD). Next, we summarize the current literature on the topic, which by and large remains inconclusive. We explain possible reasons for differing findings on climate risk and propose a classification framework, which is based on a resilience rather than a capital-based risk management approach.

2.1 A first classification

Moving toward a more sustainable economy involves significant shifts in policies, laws, technologies, and market dynamics to tackle the challenges of climate change mitigation and adaptation. Depending on how swiftly and comprehensively these changes are implemented, organizations could face differing degrees of financial and reputational risks associated with transition challenges. The Task Force on Climate-Related Financial Disclosures (TCFD) provides examples of various types of transitional risks (Policy and Legal, Technology, Market, Reputation) which are summarized in Table 2.1.

Climate policy is in a state of constant evolution, generally aimed at either curbing activities that exacerbate the effects of climate change or enhancing resilience to its impacts. Examples include implementing carbon-pricing mechanisms to reduce greenhouse gas emissions, transitioning to cleaner energy sources, adopting energy-efficient technologies, promoting water conservation, and encouraging sustainable land-use practices. The financial implications and risks associated with these policies depend on their nature and timing.

Additionally, legal risk is growing, evidenced by increasing climate-related litigation involving property owners, municipalities, states, insurers, shareholders, and advocacy groups. Such litigation often revolves around allegations of inadequate mitigation and adaptation

DOI: 10.1201/9781032636252-2

efforts, as well as insufficient disclosure of climate-related financial risks. As the costs of climate-related losses mount, the likelihood of litigation risk escalating also increases. On the other hand, mitigating climate-related risk, for example by establishing low-emission portfolios, has also led to lawsuits for neglecting investors' interests.

TABLE 2.1: Examples of transition risks and their potential financial impacts. According to TCFD (2017)

Climate-related risks	*Potential financial impact*
Policy and Legal	
Increasing GHG emission prices	Increasing operational (compliance, insurance, ..) costs
Enhanced emission-reporting obligations	Write-offs, asset impairment stranded assets
Technology	
Substitution of products and with lower emission variants	Write-offs and stranded assets Reduced demand for products/services
Unsuccessful investment in new technology	R & D expenditures in alternative and new technologies
Market	
Changing customer behavior	Reduced demand due to shift in consumer preferences
Uncertainty in market signals	Change in revenue mix
Increased cost of raw materials	Increased production costs Abrupt changes in energy prices Re-pricing of assets (e.g., fossil fuel reserves)

Technological advancements or innovations that aid the shift to a low-carbon, energy-efficient economy can profoundly affect organizations. For instance, the emergence and utilization of technologies such as renewable energy sources, battery storage, energy efficiency, and carbon capture and storage will influence the competitiveness of almost all organizations, as well as their production and distribution costs, and ultimately the demand for their products and services. As new technologies replace old systems and disrupt parts of the current economic framework, this process of "creative destruction" will produce both winners and losers. However, a significant uncertainty in evaluating technology risk is the timing of the development and implementation of these technologies.

The impact of climate change on markets is varied and complex, but one major way is the manifestation of shifts in the supply and demand for certain commodities, products, and services as climate-related risks and opportunities are increasingly considered.

How and to what extent market participants adjust asset valuations due to climate transition risk is the subject of ongoing theoretical and empirical research. The aim of this chapter is to contribute to this debate by proposing an alternative classification framework.

2.2 The possible effect of transition risk on other standard risk types

The complex interplay of transition risk with other risk categories (e.g., market risk, credit risk, etc.) raises the question of whether transition risk can be considered a systemic risk vis-a-vis being a latent part of other financial risks. In the latter case, the issue of double-counting may occur, which can lead to substantial problems concerning adequate risk capital. In the following, we will review the literature on how transition risk is incorporated in market and credit risk categories.[1] We will particularly focus on understanding to which extent transition risk is accounted for and which regions or sectors seem to increasingly be affected.

2.2.1 Market risk

The extent to which and if at all transition risk is priced in equities is still an open research question. While some scholars find that green stocks exhibit higher returns than brown stocks, others find the opposite effect or no effect at all.

Görgen et al. (2020), who investigate carbon risk in global equity prices, find that while brown firms have higher average returns and decreasing greenness leads to lower announcement returns, there is no evidence of a carbon risk premium due to opposing price movements and unpriced cash-flow changes. In contrast, Kacperczyk and Bolton (2020) find a widespread carbon premium, with higher stock returns for companies with higher carbon emissions across all sectors and continents, driven by both direct and indirect emissions. They also note a recent increase in the premium along with widespread divestment by institutional investors, particularly targeting companies from non-local markets. Bauer et al. (2022) report mixed evidence. They examine the relative equity pricing of green versus brown companies and find that green stocks in the G7 have generally outperformed brown stocks over the past decade, but during the early 2022 energy crisis, brown stocks yielded higher returns, highlighting conflicting evidence in previous research on this issue. Trinks et al. (2022) find that the transition to lower-carbon production systems increases regulatory market risks for high-emitting firms, resulting in a distinct positive impact of carbon intensity on the cost of equity capital (CoE), with a higher carbon intensity correlating with a CoE premium, especially in high-emitting sectors, EU countries, and firms under carbon pricing regulation, suggesting that reducing carbon emissions can be an effective risk mitigation strategy. Zhang (2023) highlight that the carbon premium is not observed in the U.S. or internationally after accounting for data release lag and value weighting, with less carbon-intensive firms delivering higher returns in the U.S. and the carbon return being zero on average internationally, though higher in countries more exposed to climate change after controlling for in-sample green shocks. Bolton and Kacperczyk (2023) find that higher stock returns are associated with higher levels and growth rates of carbon emissions across sectors and countries, with higher carbon premia found in less economically developed countries, those with larger energy sectors, less inclusive political systems, and stricter climate policies. Chan and Wan (2024) investigate the impact of firms' net-zero commitments on the carbon risk premium across 1,100 listed firms globally, revealing that such declarations can either increase or decrease the premium based on firms' readiness for transition, while institutional investors tend to divest from high-emitting firms, thereby influencing stock market dynamics related to carbon risk. Reboredo and Ugolini (2022) examine how climate transition risk impacts the financial performance and stock pricing of publicly traded European and U.S. firms. They report that firms with lower transition risk perform better financially, especially in Europe, and that stocks less exposed to transition risk provide higher returns to investors, although market reaction to these risks diminished after COP21.

Focusing on the U.S. market, Bolton and Kacperczyk (2021) find that firms with higher carbon emissions earn higher stock returns, a carbon premium unexplained by known risk

[1] There also exists a substantial literature on other risk categories, but we focus on market and credit risk as these are typically treated first.

factors, and observe institutional investors using exclusionary screening based on emission intensity in certain industries, suggesting investors seek compensation for carbon emission risk exposure. Nag, Chakrabarty, and Basu (2021) discover that investor anticipation of impending carbon regulations has led to a significant transition risk premium among large-cap stocks in the U.S. market. They use a transition risk measure to model firms' maximum exposure and explore trade-offs between premia and risks. Sautner et al. (2023) estimate an evolving risk premium for firm-level climate change exposure among S&P 500 stocks, finding an insignificant unconditional risk premium overall but a positive premium before the financial crisis and a steady increase afterward, with forward-looking proxies showing a maximum premium of 0.5%–1% annually between 2011 and 2014, declining since 2015 and primarily driven by uncertainty about climate-related opportunities. Michelfelder and Pilotte (2023) investigate pricing of direct greenhouse gas (GHG) emissions, particularly focusing on electricity producers post-Paris Accord, finding significant impacts on beta, alpha, and predictive power of emissions metrics for returns. This highlights the robust pricing of carbon risk in the period following the Paris Agreement.

For Europe, Alessi, Ossola, and Panzica (2021) identify a negative greenium in stock returns, where both environmental performance and transparency are priced by the market, and suggest that a stressed scenario favoring greener firms could result in global losses, highlighting the need for climate stress tests for significant financial institutions. Santi and Moretti (2021) find that investor worries about climate change significantly impact the pricing of carbon-intensive and low-emission stocks, with a notable carbon risk premium only emerging when investor concerns are high. Bua et al. (2024) examine transition risk pricing in European markets under European Union (EU) climate regulations aiming for net-zero emissions, finding significant risk premia post-2015 using novel daily indicators constructed through text analysis, and demonstrating that firm-level transition-risk exposures affect stock returns differently for green and brown stocks, thus highlighting economic implications and investor demands for compensation related to transition risks. Bernardini et al. (2021) observe that the decarbonization process has challenged traditional value-creation models for fossil fuel-reliant electricity companies, negatively impacting their finances and posing risks to institutional investors, while analyses show that investing in low-carbon companies during this transition has yielded higher returns without increasing overall risk. Witkowski, Adamczyk, and Franek (2021) assess the impact of the EU Emissions Trading Scheme (EU ETS) on the carbon risk premium, focusing on energy-intensive sectors and employing a dynamic criterion to classify firms into clean and dirty portfolios, revealing a statistically significant positive carbon premium from 2003 to 2012, a negative premium from 2013 to 2015, and no statistically significant premium from 2016 to 2019, challenging the stability of the carbon premium over time and highlighting the limitations of static portfolio approaches in such analysis.

With a focus on particular countries, Oestreich and Tsiakas (2015) find that during the initial years of the EU ETS, German firms receiving free carbon emission allowances significantly outperformed those that did not, indicating a "carbon premium" driven by increased cash flows from the allowances, and that a carbon risk factor partly explains the variation in stock returns, with high-emission firms showing higher expected returns due to greater exposure to carbon risk. For South Korea, Kim, An, and Kim (2015) observe that higher carbon intensity increases the cost of equity capital for firms, regardless of voluntary sustainability reporting, but the impact is lower for firms in high-emission sectors, implying that effective GHG management can reduce the cost of capital and enhance firm value. Lemma et al. (2019)'s study of South African firms reveals that voluntary carbon disclosure is linked to a lower cost of capital and that firms with higher carbon risk tend to provide higher quality disclosures to avoid negative market reactions, although the market generally demands higher returns from companies in carbon-intensive sectors. Ni and Sun (2023) explore how corporate ESG performance impacts stock returns in China, finding that firms with lower ESG scores showed higher returns post-Paris Agreement announcement and even more so after President Xi's carbon neutrality pledge, indicating investors are rewarded for ESG-related risks. Particularly, state-owned firms with larger market capitalizations and stronger financial and operational performance seem to be favored. Using the Shenzhen pilot as a case study, Wen, Wu, and Gong (2020) employ a difference-in-differences method to analyze how environmental regulations on carbon emissions impact companies' stock returns, finding that the establishment of China's carbon emissions trading market positively influences the excess returns of companies engaged in carbon emission allowances trading, with an increasing carbon premium observed post-establishment and a significant positive coefficient for the carbon risk factor due to higher carbon exposures among participating firms.

In the global corporate bond market, Duan, Li, and Wen (2023) examine how carbon risk is priced, finding that bonds from more carbon-intensive firms yield lower returns contrary to the "carbon risk premium" hypothesis, with evidence suggesting investor under reaction to factors like carbon intensity predictability, firm cash-flow news, creditworthiness, and environmental incidents. Dumrose and Höck (2023) explore how global climate mitigation efforts impact companies' business prospects, examining the relationship between carbon risk and credit risk, emphasizing the role of carbon risk management and regulatory factors, finding significant effects on credit risk in a global sample of corporate bonds, with implications for policymakers, investors, and companies, particularly those with substantial carbon risk exposure. Mastouri, Mendiratta, and Giese (2022) underscore the financial significance of climate-change risks for developed-market corporate-bond portfolios, highlighting that while current asset pricing and bond spreads do not fully reflect climate policies, these factors can materially impact firm asset values, potentially posing significant downside risks for bondholders and creditors, thereby urging greater investor and policymaker involvement in accelerating the transition to a low-carbon economy.

Seltzer, Starks, and Zhu (2022) analyze how investor concerns regarding climate and environmental regulatory risks influence corporate bond risk assessment and pricing in the U.S. market, finding that firms with weaker environmental profiles or higher carbon footprints tend to receive wider yield spreads, especially in states with stringent regulatory enforcement; furthermore, using the Paris Agreement as a regulatory shock, it demonstrates that climate regulatory risks directly impact bond credit ratings, yield spreads, and institutional ownership composition. Lichtenberger, Braga, and Semmler (2022) investigate the impact of green bonds as a significant financing tool for climate change mitigation, utilizing dynamic portfolio models and econometric analyses to demonstrate that green bonds generally exhibit lower volatility and superior Sharpe ratios compared to non-green bonds, suggesting they can stabilize portfolio returns and mitigate risks associated with oil price fluctuations and business cycles, thereby encouraging policymakers to promote these instruments for accelerating the transition to a low-carbon economy. Bats, Bua, and Kapp (2023) explore how climate regulations in the EU impact climate risk premiums in euro area corporate bond markets, finding that transition risk premiums are small but still evolving, underscoring the increasing investor interest in climate-resilient assets.

Moving beyond the equity and bond market, Ilhan, Sautner, and Vilkov (2020) demonstrate that climate policy uncertainty is reflected in the option market, where U.S. firms with higher carbon intensity face higher costs for downside protection during periods of heightened public concern about climate change, which decreased following the election of a climate change skeptic President Trump.

2.2.2 Credit risk

The identification of climate transition risk in the context of credit risk is an ongoing research topic. Typically, the impact is examined through a multitude of various channels such as credit ratings, default probabilities, credit default swaps (CDS) or cost of debt (loans).

Credit ratings.

Using a sample of international firms, Carbone et al. (2021) find that high GHG emissions increase firm credit ratings, while disclosing emissions and setting ambitious emission reduction targets lower credit ratings. These effects are more pronounced for European firms post-Paris Agreement, which again highlights the importance of climate-related disclosures and strategies in managing transition risks. Focusing on the global energy sector, Chodnicka-Jaworska (2022) examines the significant impact of environmental, social, and governance (ESG) measures on credit ratings during the COVID-19 crisis. They find that ESG factors were more influential during the crisis but their importance diminishes over time due to limited non-financial reporting, regulatory gaps, and the large, often government-supported nature of energy companies, particularly in less developed countries.

For the U.S., Safiullah, Kabir, and Miah (2021) report that high firm-level carbon emissions negatively impact credit ratings of firms, with this effect being linked to increased cash flow uncertainty. Similarly, Shahrour, Arouri, and Rao (2024) identify a negative relationship between firms' carbon emissions scores and credit risk overall, but a positive relationship in the Industrials sector, highlighting the need for companies to integrate climate risk into their risk management strategies and offering important insights for portfolio managers, socially responsible investors, and policymakers. Using voluntary disclosure of carbon emissions information, Lee (2022) discover a positive impact on credit ratings for non-financial companies in South Korea, especially when managed by competent managers.

Default risk.

Using model-implied default probabilities, Bouchet and Le Guenedal (2020) examine the impact of setting an effective carbon price on corporate credit risk. They find that the energy, materials, and utilities sectors are most affected, with the risk appearing sooner and more variably for utilities. However, a carbon price scenario limiting global warming to two degrees Celsius has no significant impact on global credit risk, suggesting it is achievable without significant financial losses. Capasso, Gianfrate, and Spinelli (2020) provide evidence that higher carbon emissions and carbon intensity are associated with a higher risk of corporate default, as measured by a reduced distance-to-default, especially following policy shifts such as the Paris Agreement. This indicates that climate risks significantly impact the creditworthiness of corporate loans and bonds and should be a key consideration for financial regulators and policymakers. In a similar fashion, Kabir et al. (2021) find that carbon emissions significantly increase firms' default risk, as measured by a reduced distance-to-default, with environmental commitments and green initiatives mitigating this effect, while environmental controversies exacerbate it. They identify return on assets and cash flow volatility as key channels, and suggest that firm-level emissions should be considered in policy decisions.

Ramos-García, López-Martín, and Arguedas-Sanz (2023) find that climate factors significantly affect default risk for companies listed on the STOXX Europe 600 exchange, with notable differences before and after the Paris Agreement. Thus, traditional financial factors alone cannot explain these differences, and the importance of incorporating climate change risks and opportunities labels into credit risk evaluation is highlighted. Guo et al. (2024) develop a credit risk assessment model for Chinese listed companies that integrates macroeconomic indicators, ESG scores, and industry variables to analyze the impact of carbon risk on default probabilities, revealing a negative relationship between high ESG scores and default risks and highlighting the significance of environmental factors in carbon finance credit risk analysis. Thornton, Altunbaş, and Nobile (2024) link default rates on bank loans of Italian firms to the GHG emissions of their sectors and investigates the impact of the Paris Agreement, finding that higher emissions correlate with higher default rates, and the Paris Agreement led to a significant increase in default rates across nearly all sectors, indicating early costs associated with transitioning to a lower carbon economy are already affecting Italian firms negatively. Chherawala (2024) develops a climate-adjusted expected default frequency metric to quantify transition risk for carbon-intensive Indian steel companies, revealing that extreme shocks from global climate policies can substantially heighten firm-level credit risk, with implications for internal bank risk assessments and regulatory climate stress tests.

Credit default swaps (CDS).

In the realm of CDS, Blasberg, Kiesel, and Taschini (2021) create a forward-looking market-implied carbon risk factor and demonstrate that stringent carbon regulations significantly increase firms' credit protection costs in North America and Europe, particularly under explicit carbon pricing regimes and for firms with high direct emissions, with impacts varying by sector and leading to higher anticipated costs for short-term transitions. Using natural language processing to quantify regulatory climate risk disclosures, Kölbel et al. (2022) find that disclosing transition risks increases CDS spreads of U.S. firms, with significant impacts observed after the Paris Agreement and Trump's election, aligning with theoretical predictions. In a similar vein, Zhang and Zhao (2022) observe a significant positive relationship between carbon emissions and CDS spreads, with higher emissions raising CDS term structures due to long-term environmental policy concerns, especially for financially constrained firms, thus contributing to the understanding of asset pricing implications of stricter ESG policies. Zhang, Lio, and Wang (2023) reveal that direct carbon emissions increase CDS spreads for U.S. companies, while indirect emissions are less impactful, with a positive correlation between carbon risk and CDS term structure indicating long-term credit risk concerns, and findings remain robust post-Paris Agreement, highlighting channels like sustainability awareness and green transition in pricing carbon risk in the credit market.

Christ, Hertel, and Kocian (2022) identify that high corporate sustainability performance (CSP), particularly in environmental and social dimensions, is associated with lower credit default swap (CDS) spreads for European firms, with country- and industry-specific risk premiums for climate-related performance, and that global climate policy events, unlike firm-specific ESG controversies, generally affect CDS spreads without nuanced investor reactions, with results robust across various measures and operationalizations of CSP. Similarly, Vozian and Costola (2023) examine major European firms' climate-related transition indicators and their impact on CDS-implied credit risk, revealing heightened transition risk pricing, particularly for carbon-intensive sectors, since the Paris Agreement, and highlights challenges in valuing transition risk management efforts and predicting allowance prices in the EU ETS market. Based on Japanese data, Okimoto and Takaoka (2024) find that carbon emissions influence corporate CDS spreads with time-varying effects, particularly in sectors where transitioning to carbon-free energy is simpler and cheaper, and high emissions steepen the CDS spread curve; investor pressure notably reduces emissions, especially for investment-grade firms, supporting the "investor awareness" hypothesis across sectors and credit qualities.

Cost of debt.

Focusing on the cost of debt, Kleimeier and Viehs (2018) observe that international firms that voluntarily disclose their carbon emissions receive more favorable lending conditions with lower loan spreads, while higher carbon emissions lead to higher loan spreads, indicating that environmental risks, rather than investor preferences, drive these spread premia for both norms-constrained and unconstrained lenders. On a similar note, Ehlers, Packer, and de Greiff (2022) show that since the Paris Agreement, banks have applied a significant "carbon premium" to syndicated loans based on borrowers' CO_2 emission intensity, indicating broader industry impacts beyond stranded asset risks, though the risk pricing remains relatively low considering potential material risks, and notably, only scope 1 emissions are priced, not the overall carbon footprint, with no discernible difference in pricing by "green" banks compared to others.

In the eurozone, Palea and Drogo (2020) find that lenders have increasingly factored carbon emissions into the cost of debt financing, with the Paris Agreement marking a turning point where even less polluting sectors faced higher spreads, highlighting the effectiveness of public policies and climate-related disclosures in influencing lender decisions and mitigating corporate carbon risk. Likewise, Pizzutilo et al. (2020) explore the link between carbon risk and the cost of debt among Eurostoxx 600 Index companies, revealing a positive relationship and addressing a gap in the literature concerning the economic and financial impacts of environmental risk amid increasing global attention to climate change.

For the Australian market, Jung, Herbohn, and Clarkson (2018) uncover that firms' exposure to carbon-related risks affects their cost of financing, with a positive association between higher carbon risk and increased cost of debt for firms not responding to the Carbon Disclosure Project (CDP) survey, but this penalty is negated for firms demonstrating carbon risk awareness, emphasizing the importance of sustainability strategies in lending decisions, and highlighting the debt market's consideration of both historical emissions and future carbon performance. Lee and Choi (2021) investigate how firms' carbon risk management practices in South Korea affect their cost of debt capital. They find that effective management reduces costs, reflecting the impact of government carbon policies and contributing to both theoretical insights and practical strategies for improving carbon risk management.

2.2.3 Incorporation of transition risk by credit agencies

Financial actors, regulators and societies rely on credit ratings to assess the creditworthiness of debt instruments, corporates and sovereigns. Thus, it is interesting to study how credit agencies treat climate risk. As pointed out in the previous section climate risk does affect the usual factors, such as default probabilities, distance-to-default, or CDS spreads, credit agencies use to asses credit risk. Thus, one may argue that what started as climate risk may eventually transform into credit risk due to the assessment of credit agencies. In the following, we will outline how the three major credit rating agencies, Standard & Poor's (S&P), Moody's and Fitch, integrate transition risk into their credit rating formation process. We also discuss the current challenges that go along with it.

S&P incorporates transition risk into its credit ratings by utilizing scenario analysis and a battery of tools, also known as Climate Credit Analytics, to assess the financial impacts of regulatory changes, market dynamics, and technological advancements related to climate change. This includes evaluating the potential effects of carbon pricing and other policies on companies' financial stability, particularly in sectors sensitive to energy transition. They integrate these assessments into their overall credit rating process, examining both immediate and long-term risks, and conduct detailed sector-specific analyses to provide nuanced credit assessments. Thus, S&P mainly considers how regulatory shifts and market demand changing to low-carbon solutions affect companies' operations and strategies, see Dawuda (2020), Global (2021), Global (2022). Moody's accounts for transition risk by utilizing the Climate-Adjusted Expected Default Frequency (EDF) framework, which assesses the impact of climate policies and technological changes on a company's default risk. They conduct scenario analysis and stress testing to simulate different regulatory and market conditions, evaluating how these might impact financial performance. The agency also employs Carbon Transition Indicators (CTIs) to systematically assess how companies manage their carbon footprints and adapt to regulatory changes (Ratings (2022)). Additionally, Moody's integrates ESG factors into its credit ratings, using ESG scores to reflect the material impact of transition risks (Ratings (2024)).

Fitch incorporates transition risk through its Climate Vulnerability Scores, which evaluate the long-term creditworthiness impact of transition risks based on scenarios such as the UN Principles for Responsible Investments "Inevitable Policy Response Forecast Policy Scenario." Fitch conducts sector-specific analysis to identify how transition risks affect different industries, particularly those highly exposed to climate change, such as oil and gas, utilities, and chemicals. The agency uses forward-looking assessments to project future financial performance under various transition pathways and integrates these assessments into its credit rating process to provide investors with a clear understanding of long-term risks (Institute (2022), Fitch Ratings (2022)).

Credit rating agencies assert they have always considered climate-related risks when identifiable, relevant, and material. However, the assessment is mainly based on regulatory (carbon certificate prices, carbon taxes or subsidies) and customer preference changes. Furthermore, they still face challenges due to inconsistent and poor-quality data. Also, transparency around the methodologies for incorporating climate risks into ratings and their influence on final ratings is still lacking. Additionally, credit ratings typically assess short-term creditworthiness, which may not fully account for the longer-term impacts of climate change (NGFS (2022)).

Beyond the agency-specific issues in the credit rating formation process, there also exist significant differences between multiple agencies. Using an analytical framework, Breitenstein, Ciummo, and Walch (2022) compare agencies' definitions, methodologies, assessment models, data usage, and disclosure practices and find significant variations across agencies and asset classes. Based on these results, they highlight three key areas for improvement in climate-related disclosures: the specificity of climate change risk definitions, transparency of the models and methods used to estimate climate risk exposure, and the clarity regarding the impact magnitude of material climate change risk on credit ratings.

2.2.4 The classification of risks

Table 2.2 summarizes the findings of the above sections.

Clearly, evidence of the impact of climate risk is mixed. However, there is a tendency for findings of strong impacts. As a consequence, the pricing of climate risk is not done appropriately by the markets with obvious consequences for the risk management process.

We now discuss possible reasons for the mixed evidence of climate risk in the context of market risk. Standard asset pricing arguments predict that green assets should have lower expected returns than brown assets because investors sacrifice financial returns for environmental benefits and see green assets as a better hedge against climate change. However, empirical studies only find inconclusive evidence on the relation between green characteristics and equity returns. This may be due to imprecise definitions in the context of climate risks. As Reisinger et al. (2024) point out, even basic Greenhouse Gas classifications are improperly used. In addition, ESG ratings, such as the MSCI ESG Ratings, also lack precise definitions and may not be sufficiently correlated, see Berg, Koelbel, and Rigobon (2020) for details. Also, the use of either efficient market theory employing factor models or investment-style analysis with the construction of green-brown portfolios may affect the results. Finally, realized returns, where typically indices for market sentiment from printed

TABLE 2.2: Overview on impact of climate risk

Transition risk Premium	*Market risk*
Strong Evidence	Alessi, Ossola, and Panzica (2021), Kacperczyk and Bolton (2020), Bolton and Kacperczyk (2021), Bolton and Kacperczyk (2023), Ilhan, Sautner, and Vilkov (2020), Nag, Chakrabarty, and Basu (2021), Mastouri, Mendiratta, and Giese (2022), Seltzer, Starks, and Zhu (2022), Trinks et al. (2022), Dumrose and Höck (2023), Michelfelder and Pilotte (2023), Ni and Sun (2023),
Weak Evidence	Görgen et al. (2020), Bernardini et al. (2021) Duan, Li, and Wen (2023), Bauer et al. (2022), Reboredo and Ugolini (2022)
Mixed Evidence	Santi and Moretti (2021), Witkowski, Adamczyk, and Franek (2021), Lichtenberger, Braga, and Semmler (2022), Sautner et al. (2023) Zhang (2023), Chan and Wan (2024).
Transition risk Premium	*Credit risk*
Strong Evidence	Capasso, Gianfrate, and Spinelli (2020), Palea and Drogo (2020), Pizzutilo et al. (2020), Blasberg, Kiesel, and Taschini (2021), Lee and Choi (2021), Kölbel et al. (2022), Carbone et al. (2021), Kabir et al. (2021) Zhang and Zhao (2022), Ramos-García, López-Martín, and Arguedas-Sanz (2023), Guo et al. (2024), Thornton, Altunbaş, and Nobile (2024).
Weak Evidence	
Mixed Evidence	Bouchet and Le Guenedal (2020), Ehlers, Packer, and de Greiff (2022), Zhang, Lio, and Wang (2023), Shahrour, Arouri, and Rao (2024).

media or online news feeds are constructed, will generate different results than expected returns (see Ardia et al. (2022) for further discussion and analysis).

To make our point more clearly, we now use specific examples to illustrate the differences in the classification of firms and the risk metric. Kacperczyk and Bolton (2020) find *strong evidence* and use total absolute emissions (low emissions are classified as green) in their analysis. However, they particularly point out that emission intensity is not a risk factor. For assessing risk, they rely on traditional efficient capital markets theory, which applies a standard risk and expected return tradeoff to green and brown firms. They report that brown firms face a higher risk and thus investors require a higher return. Alternatively, the line of arguments from an investment point of view, sees potential for higher average returns of green firms, or in their terminology, a low-carbon alpha (and not primarily a reduced riskiness). However, increased investors' demand for green assets will tend to bid up prices of green firms relative to brown ones, reducing the green expected returns and resulting in a carbon risk premium.

As an example for *Weak Evidence* Bauer et al. (2022) examine the relative equity pricing of green versus brown companies with absolute emissions used for classification. They find that green stocks have generally outperformed brown stocks over the past decade. However, during the early 2022 energy crisis, brown stocks yielded higher returns. Their analysis is based on panel regression with various factor models and portfolio construction approaches.

Finally, as an example for *Mixed Evidence* Sautner et al. (2023) estimate an evolving risk premium for firm-level climate change exposure among S&P 500 stocks. For classification, attention to climate risk paid by market participants, measured by the proportion of earnings calls devoted to climate risk, is used.

Let us close this section with a few additional points that complement the above findings. Using carbon intensity for classification may lead to differing results depending on the choice of the denominator, as pointed out in Bajic, Hellmich, and Kiesel (2023), who also provide an analysis of the impact of the time lag in measurement of emissions and economic indicators (see also Zhang (2023)). Also, the time period considered is of high importance. Carbon risk may diffuse in different risk categories over time—most notably credit risk.

In light of the discussion above, it is obvious that the standard classification and risk management approaches are not adequate for climate risk. In the next section, we outline an alternative approach.[2].

2.3 From the viable system model to the risk landscape

The treatment of climate risks within a risk management process is anchored in risk identification. Compared with many other risks, e.g., market risks, climate risks are insofar special, however, as any human activity will impact climate risks in one way or another. Technically speaking, climate risk stands for a so-called *background risk*, Tsanakas (2008). Hence, climate risks co-move and confound with many other variables, risks and activities. That our life is embedded in climate risks will also cause an availability bias in the perception of climate risk.

An adequate communication of climate risks forms the basis for decision-making. Unfortunately, climate risks are not directly observable in assets or liabilities, hence the need to filter them out in order to avoid e.g., double counting. The following example considers *stranded assets* in order to highlight this point.

Example 1 (Stranded Assets) *The energy sector, including coal, oil, gas reserves, and fossil fuel power generators, already gave well-known examples of stranded assets caused by climate risks. In general, stranded assets are considered as those that lose value or turn into liabilities before the end of their expected economic life. In the context of the energy sector, this means that fossil fuels will not be extracted (they remain stranded in the ground) or power plants may be prematurely switched off.*

Assume a corporation faces such risks in its business model. Of course, rating agencies would timely take measures like a negative outlook or even a downgrade. This points to a characteristic of background variables: the confounding with other variables. In our case with credit risk, with the risk of a double-counting effect, if the climate risk component is not filtered out properly.

Example 1 is in so far of additional importance as stranded assets are the motivation underpinning many regulatory initiatives for climate risks inaugurated by central banks. It highlights furthermore the emergence of operational risks related to a naive understanding of climate risks.

We pick up in greater detail the implications that come along with the fact that climate risk admits the character of a background variable. Note that background variables are endogenous, not attributable to individual decisions, of systemic nature, complex and non-hedgeable. In Daníelsson (2022) such attributes motivate the *Illusion of Control*, the imperative for corporates not to count on regulatory solutions w.r.t. climate risk, but to implement an agile and resilient risk management framework.

The long time horizon is a well-known stylized fact (perhaps the most important risk parameter) of climate risks, hence the question of timing w.r.t. decisions is an important component of the management of climate risks. This is related in various ways to the theory of *weak signals*, see Ansoff (1975). This approach emphasizes the importance of filtering out

[2]For additional discussion see Kiesel and Stahl (2022) and Kiesel and Stahl (2023).

important information about emerging risks at an early stage, see Gomez and Lambertz (2023) for an up-to-date overview. Thus, an insight-oriented information gathering process, which leads to new rules, comes into play. This philosophical perspective motivates the need for a clear categorization w.r.t. climate risks from an epistemological point of view.

The development of adequate models requires descriptions of a phenomenon in time and space. We focus here on the aspect of operational risks a corporation may face by tackling climate risks. We use the Viable System Model (VSM), a cybernetic model introduced in Beer (1959) in order to describe how companies are structured and how they adapt to a changing environment, to develop a model to analyze transition risks.

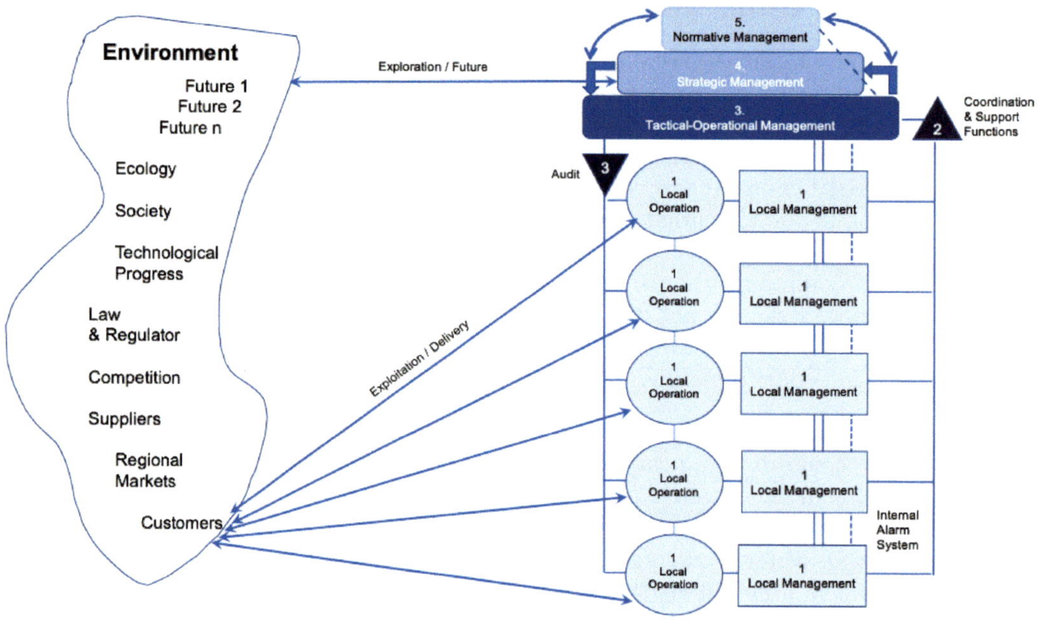

FIGURE 2.1: Structures of a VSM.

The changes w.r.t. the environment, given in the left part of figure 2.1, will create risks or uncertainties for the company with its participations or assets given on the right side. Financial supervisors require financial institutions (such as banks or insurers) to hold capital in order to buffer these risks. Initiatives like Solvency II or Basle III give important examples. Furthermore, rating agencies follow a similar, yet less bureaucratic approach.

We will use Solvency II (S-II) as a blueprint, because a market consistent balance sheet—the so-called Solvency II balance sheet—defines the underlying variable of interest. The capital requirements are determined from a related forecast distribution. Due to the fact that the S-II balance sheet is a proxy of the IFRS balance sheet, the generalizations of our exposition are at hand. We yield roughly:

$$\text{IFRS} \sim \text{S-II} + \text{Goodwill} \tag{2.1}$$

So, in terms of the time dimension, the balance sheet evaluates a company with its own funds, OF_t at time t as the difference between the value $\nu(A_t)$ of the assets minus the value of the liabilities, $\nu(L_t)$:

$$OF_t = \nu(A_t) - \nu(L_t). \tag{2.2}$$

For the space component, we use homogeneous risk categories

$$\mathfrak{R}_i \text{ for i} = 1,....,\text{r} \tag{2.3}$$

in order to describe a risk situation, denoted by $\mathfrak{R}$. *Epistomologically ideal*, risk categories yield an exhaustive and separating setting, i.e.:

$$\mathfrak{R} = \bigcup_{i=1}^{r} \mathfrak{R}_i \text{ and } \mathfrak{R}_i \cap \mathfrak{R}_j = \emptyset. \tag{2.4}$$

Typically, prices allow for bridging the gap between a balance sheet and the risk categories. Let

$$\pi(X) = \mathbb{E}(X \mid \mathcal{F}_t) \tag{2.5}$$

be the price of a financial instrument X, say. The filtration $\mathcal{F}_t$ is assumed to be generated by the time series of the random vector of risk resp. price factors $\mathbf{R} = (R_1, \ldots, R_n)$. This allows to represent (2.5) by a price function

$$\pi(X) = f(\mathbf{R}) = f(R_1, \ldots, R_n). \tag{2.6}$$

The changes of $\pi(X)$, i.e., gains or losses, are thus explainable with the risk factor $\mathbf{R}$. This allows for reformulating the S-II balance sheet, which characterizes the *paradigm* of Solvency II. More formally:

$$\nu(\mathbf{A}_t) - \nu(\mathbf{L}_t) \approx \pi(\mathbf{A}_t) - \pi(\mathbf{L}_t), \tag{2.7}$$

where $\mathbf{A}_t = \sum_{i=1}^{a} A_t^i$ denotes the portfolio of the insurer's assets (analog $\mathbf{L}_t$).

Risk categories allow a helicopter view on the risk situation, i.e., *Occam's Razor* comes into play. Compared to risk factors, risk categories provide *generic terms*, which reduce the dimension of the universe of risk factors to a low number of risk categories. This simplifies the expression of risk preferences for stakeholders as risk categories $\mathfrak{R}_i$ may be linked to *utility functions* (preferences), which eases communication at various levels. In addition, frameworks for risk charges (from regulatory bodies, rating agencies, management boards) express risk preferences in terms of risk categories. This explains why (2.4) is so crucial, in order to avoid *double counting* as a consequence of unintended redundancy.

The following example, quoted from the Basel document, Basel Committee on Banking Supervision (2009), p.17, shows that in the case of operational risks the separation in (2.4) can even for risks under Pillar I not be taken for granted:

Example 2 (Regulatory Risk related to Operational Risk) *Some losses [i.e., $\Delta\nu(X)$] are clearly the result of operational risk. For others, it is less clear whether they should be classified as operational risk or credit, market or strategic risk. Still, in other cases, it may be appropriate to allocate an individual loss partially to operational risk and partially to credit or some other risk category. These classification issues are broadly described as* ***boundary issues****.*

Note that operational risks can be seen as a background variable. The original *decision* to take risks follows a business strategy. Operations are merely a vehicle (which cannot be circumvented) in order to bring the business strategy into reality. This causal relation implies dependencies and causes the boundary issue. However, in contrast to climate risk, operational risk is of *endogenous* nature. This implies that the decomposition related to the boundary issue for climate risk is much more complicated or even more complex (due to unexpected feedbacks).

Let us now have a closer look at the category of (physical and transition) climate risk (denoted by $\mathfrak{C}$).

Example 3 (Stylized facts of $\mathfrak{C}$) *We recall from Kiesel and Stahl (2022), Kiesel and Stahl (2023), and Hellmich and Kiesel (2021), a number of stylized facts and relate them to risk categories.*

- $\mathfrak{C}$ *is not adequately priced. Hence, tight price boundaries are lacking, which results in model risk or in the case of insured climate risks (such as natural catastrophes) in addition to business risk.*
- $\mathfrak{C}$ *may relate to a long time horizon, which generates strategic and business risk.*
- $\mathfrak{C}$ *carries regulatory risks as various supervisors (such as the ECB, rating agencies, national supervisors, ...) have different reaction times and/or different interests. Furthermore, changing governments may reverse policies.*
- $\mathfrak{C}$ *is prone to boundary issues, which implies operational risk (compare example 2).*
- $\mathfrak{C}$ *is affected by social behavior, which by its feedback effects and complexity (see the VUCA discussion below) generates systemic risk.*

Thus, climate change is a background variable that interacts with nearly all variables of interest.

It is worth mentioning that the risk categories considered in the example above differ fundamentally w.r.t. their role in a risk-based regulation, but also in the perception of risk. Some are considered merely by qualitative methods (business risk, strategic risk, reputational risk) others however are related to quantitative methods (natural catastrophes, market risk, ..). By and large the borderline is given by the criterion of *hedgeability*, which requires a sufficient good pricing, hence (2.5) to work. An approximate understanding of hedgeability may be deduced from the axiom of translation invariance for risk functionals

$$\varrho(X + A) = \varrho(X) - a, \tag{2.8}$$

where the position in cash, A may be interpreted as a replication portfolio of X (which allows a hedge). Of course, the practice knows a number of risks, where regulators require capital requirements, and full hedges do not exist. However, translation invariance is at least approximately valid and not in question. In order to show that the current categorization for climate risk is not epistemologically optimal, we put the insights from example 3 in a more formal setting and denote those risks for which no tight price boundaries exist by $\mathfrak{S}_j$. These categories complement $\mathfrak{R}_i$, where good price bounds are assumed.

With Example 3 we lifted *boundary risk* at another level, due to

$$\mathfrak{C} \;\cap\; \mathfrak{R}_i \neq \emptyset \tag{2.9}$$

$$\mathfrak{C} \;\cap\; \mathfrak{S}_j \neq \emptyset. \tag{2.10}$$

Whereas boundary risk in example 2 leads to a problem of decomposition (unknown knows), climate risk is related to known or unknown unknowns. In risk-based regulations, the $\mathfrak{R}_i$ structure is used to define the regulatory capital requirements, whereas $\mathfrak{S}_j$ is used to address issues of qualitative risk management, i.e., requirements to manage risks beyond hedging and/or capital requirements. Important examples include the improvement of (business) processes, e.g., underwriting effectiveness, financial market alpha, etc. This is mirrored by the interplay of the relations (2.4)–(2.7). The following example will elaborate on the relation between climate risk $\mathfrak{C}$, and the risks $\mathfrak{S}_j$, in greater detail.

Example 4 *[The new Challenge - VSM revisited] As in Gomez and Lambertz (2023), see pp 24, a firm is understood through the VSM as a* ***socio-technical construct****. These hybrid processes, which are of physical, business and social nature. Their synthesis determines the corporates'* ***boundaries****, denoted by $\mathcal{B}_C$. These characterize the companies' beliefs (their purpose) and structure. These boundaries influence the perception of risk and, in particular, the $\mathfrak{S}_j$. The associated longer time horizons for $\mathfrak{S}_j$ compared to $\mathfrak{R}_i$ are one important reason.*

The firm-specific nature of the boundaries $\mathcal{B}_C$ implies that the associated risk factors $\mathfrak{S}$ depend conditionally on the particular firm C. This explains, on one hand, why $\mathfrak{R}$ and $\mathfrak{S}$ are treated so differently by regulators. On the other hand, the attribution process, i.e., the determination of responsibility and causality, e.g., by means of a root cause analysis, depends very much on the firm's own philosophy.

Example 4 shows that $\mathcal{B}_C$ requires another materiality level compared to the boundary issue encountered in example 2. A lack of understanding of the importance $\mathcal{B}_C$ might bias the perception of risk. This could create a huge exposure to operational risk that might harm the company existentially. Thus, a firm-specific (knowledge) process has to be implemented in order to identify weak signals w.r.t. changes concerning events related to $\mathfrak{S}$.

As pointed out above, for risk under $\mathfrak{C}\cap\mathfrak{S}$, a resilient risk strategy should be put in place. Such a strategy could quantify risks related to $\mathfrak{S}\cap\mathfrak{C}$ via a generic model for the balance sheet (2.2), as in Korn and Stahl (2023), in combination with a system dynamics approach, see Mella (2021). This gives a flexible starting point, which allows for incorporating the available information.[3] An implementation which is practically tractable will, on the one hand, rely on techniques borrowed from *System Dynamics* and, on the other hand, should use *Bayesian Belief Networks.* This flexible approach will enrich current techniques applied by companies, which are dominated by those approaches that capture causality and diversification. In addition, the extended approach will also treat issues related to knowledge and feedback.

In this context Taleb and Sandis (2013) and Taleb (2018) start from an approach characterized by a triad consisting of an economic, mathematical and philosophical (or say normative; or in managerial parlance: culture) components. In addition, it is argued to harmonize knowledge and uncertainty, value-based behavior with the tackling of complex situations. This is in line with the *model in a wide sense*, see Kiesel and Stahl (2022), Kiesel and Stahl (2023) and Kiesel and Stahl (2024). The latter sketches the changes w.r.t. required skill profile of actuaries that will cope with the challenges ahead. The VSM, see Figure 2.1, locates the normative value at the top of its pyramid in the value management. A short overview of the impact of current changes about social values (common value approaches etc..) in the context of the VSM approach is given in Gomez, Meynhardt, and Lambertz (2019).

We paraphrase one of Peter Drucker's famous quotes (Culture Eats Strategy for Breakfast) by *Culture Eats Methods for Breakfast.* In other words, the understanding of the culture's role in the interplay of quantitative methods and change management is paramount. This roots in the fact that tail risks (longer time horizons: distresses, downgrades...) are very induced by systemic risk, which in turn cannot be captured without value-based, subjective approaches.

Figure 2.2 compares, by means of two incomplete lists of buzzwords, the current (merely quantitative) approach of risk management, whose direction was implemented and constantly improved (both by practitioners and academics) from the mid-nineties on. Currently, a discussion is taking place on how to cope with the challenges in a VUCA world. We advocate a hybrid approach, which we call *Risk Informed Leadership*, which combines traditional

[3]CorporateMetrics, see Metricsgroup (1999), contains various hints on how to build bottom-up stochastic models for balance sheets.

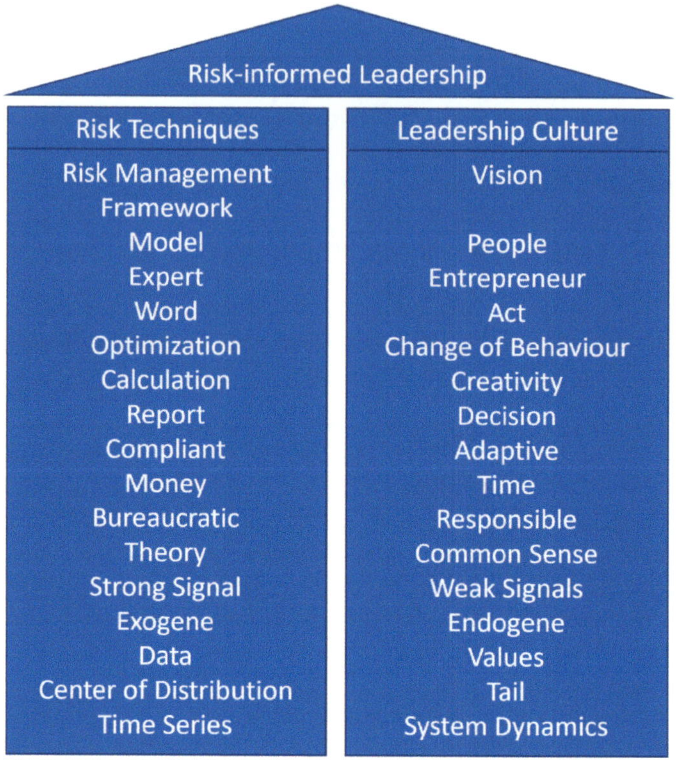

FIGURE 2.2: The Risk Informed Leadership Approach as hybrid of a VUCA adapting culture (Leadership Culture) and current risk management practice (Risk Techniques), further details are found in Kiesel and Stahl (2022) and Kiesel and Stahl (2024) and the literature cited there.

actuarial and risk management techniques, seen as information inputs for workshops or focus groups, which bring in the leaders' view about future developments and express by this way the leaders' risk maturity. In general, the output of this process will deliver scenarios with associated probabilities. The latter might be fed in smoothly in a System Dynamic model approach—as a shortcut for a fully fledged control theoretic framework.

2.4 Conclusions

This contribution discusses the difficulties a traditional price-based risk management framework faces in the context of climate transition risk. We review the current literature on the extent to which climate risk is affecting market and credit risk to illustrate the fact that climate risk is not appropriately priced. Besides, we describe the different approaches used in the empirical literature and to what extent the difference in approaches may lead to different outcomes. Finally, we outline a risk management approach that is based on risk-informed leadership in a viable system model.

References

Alessi, L., E. Ossola, and R. Panzica, 2021, What greenium matters in the stock market? the role of greenhouse gas emissions and environmental disclosures, *Journal of Financial Stability* 54, 100869.

Ansoff, H. I., 1975, Managing strategic surprise by response to weak signals, *California Management Review* 18, 21–33.

Ardia, D., K. Bluteau, K. Boudt, and K. Inghelbrecht, 2022, Climate change concerns and the performance of green versus brown stocks, *Management Science* 69, 7607–7632.

Bajic, A., M. Hellmich, and R. Kiesel, 2023, Handle with care: challenges and opportunities of using company-level emissions data for assessing financial risks from climate change, *Journal of Climate Finance; accepted.*

Basel Committee on Banking Supervision, 2009, Observed range of practice in key elements of advanced measurement approaches (AMA).

Bats, J., G. Bua, and D. Kapp, 2023, Physical and transition risk premiums in euro area corporate bond markets, Working Papers 761, DNB.

Bauer, M. D., D. Huber, G. D. Rudebusch, and O. Wilms, 2022, Where is the carbon premium? global performance of green and brown stocks, *Journal of Climate Finance* 1, 100006.

Beer, S., 1959, *Cybernetics and Management* (The English Universities Press Ltd.).

Berg, F., J. F. Koelbel, and R. Rigobon, 2020, Aggregate confusion: The divergence of esg ratings, *Review of Finance* 26, 1315–1344.

Bernardini, E., J. D. Giampaolo, I. Faiella, and R. Poli, 2021, The impact of carbon risk on stock returns: evidence from the European electric utilities, *Journal of Sustainable Finance & Investment* 11, 1–26.

Blasberg, A., R. Kiesel, and L. Taschini, 2021, Carbon default swap – disentangling the exposure to carbon risk through cds, *SSRN Electronic Journal.*

Bolton, P., and M. Kacperczyk, 2021, Do investors care about carbon risk?, *Journal of Financial Economics* 142, 517–549.

Bolton, P., and M. Kacperczyk, 2023, Global pricing of carbon-transition risk, *The Journal of Finance* 78, 3677–3754.

Bouchet, V., and T. Le Guenedal, 2020, Credit risk sensitivity to carbon price, *SSRN Electronic Journal.*

Breitenstein, M., S. Ciummo, and F. Walch, 2022, Disclosure of climate change risk in credit ratings, Technical report, European Central Bank.

Bua, G., D. Kapp, F. Ramella, and L. Rognone, 2024, Transition versus physical climate risk pricing in European financial markets: a text-based approach, *The European Journal of Finance* 30, 2076–2110.

Capasso, G., G. Gianfrate, and M. Spinelli, 2020, Climate change and credit risk, *Journal of Cleaner Production* 266, 121634.

Carbone, S., M. Giuzio, S. Kapadia, J. S. Krämer, K. Nyholm, and K. Vozian, 2021, The low-carbon transition, climate commitments and firm credit risk, Working Paper Series 2631, European Central Bank.

Chan, K., and W. Wan, 2024, The double-edged sword of net-zero commitment declaration on the carbon risk premium, *SSRN Electronic Journal*.

Chherawala, T., 2024, Impact of climate transition risk on firm-level credit risk: The case of India's steel companies, Working paper, National Institute of Bank Management.

Chodnicka-Jaworska, P., 2022, Environmental, social, and governance impact on energy sector default risk—long-term issuer credit ratings perspective, *Frontiers in Energy Research* 10, 817679.

Christ, J., T. Hertel, and J. Kocian, 2022, Corporate sustainability performance and the cost of debt – an analysis of the impact of country- and industry-specific climate risk exposures, *SSRN Electronic Journal*.

Daníelsson, J., 2022, *The illusion of control: why financial crises happen, and what we can (and can't) do about it* (Yale University Press).

Dawuda, S., 2020, Linking Climate Transition Risks and Credit Risks, https://www.spglobal.com/marketintelligence/en/news-insights/blog/linking-climate-transition-risks-and-credit-risks, Accessed: 2024-08-01.

Duan, T., F. W. Li, and Q. Wen, 2023, Is carbon risk priced in the cross section of corporate bond returns?, *Journal of Financial and Quantitative Analysis* 60, 1–35.

Dumrose, M., and A. Höck, 2023, Corporate carbon-risk and credit-risk: The impact of carbon-risk exposure and management on credit spreads in different regulatory environments, *Finance Research Letters* 51, 103414.

Ehlers, T., F. Packer, and K. de Greiff, 2022, The pricing of carbon risk in syndicated loans: Which risks are priced and why?, *Journal of Banking & Finance* 136, 106180.

Fitch Ratings, 2022, Transportation – Long-Term Climate Vulnerability Scores, Technical report, Fitch Ratings.

Global, S., 2021, General Criteria: Environmental, Social, And Governance Principles In Credit Ratings, https://www.spglobal.com/ratings/en/research/articles/211010-general-criteria-environmental-social-and-governance-principles-in-credit-ratings-12085396, Accessed: 2024-08-01.

Global, S., 2022, ESG Brief: Incorporating Climate Transition Risk In U.S. States Credit Ratings, https://www.spglobal.com/ratings/en/research/articles/220322-esg-brief-incorporating-climate-transition-risk-in-u-s-states-credit-ratings-12313215, Accessed: 2024-08-01.

Gomez, P., and M. Lambertz, 2023, *Leading by Weak Signals: Using Small Data to Master Complexity*, volume 5 (Walter de Gruyter GmbH & Co KG).

Gomez, P., T. Meynhardt, and M. Lambertz, 2019, *Verantwortungsvoll führen in einer komplexen Welt: Denkmuster-Werkzeuge-Praxisbeispiele* (Haupt Verlag).

Guo, C., X. Wang, S. Chen, and D. Xia, 2024, Credit risk assessment of carbon assets: Evidence from Chinese listed firms, *Proceedings of the 3rd International Conference on Big Data Economy and Digital Management (BDEDM 2024)*, Ningbo, China, January 12–14, 2024, EAI, https://doi.org/10.4108/eai.12-1-2024.2347205.

Görgen, M., A. Jacob, M. Nerlinger, R. Riordan, M. Rohleder, and M. Wilkens, 2020, Carbon risk, *SSRN Electronic Journal*.

Hellmich, M., and R. Kiesel, 2021, *Carbon Finance: A Risk Management View* (World Scientific).

Ilhan, E., Z. Sautner, and G. Vilkov, 2020, Carbon Tail Risk, *The Review of Financial Studies* 34, 1540–1571.

Institute, G. A., 2022, Fitch Incorporates Low-Carbon Transition Risks Into Credit Scores, `https://www.ga-institute.com/index.php?id=444&L=1&tx_ttnews%5Btt_news%5D=4347`, Accessed: 2024-08-01.

Jung, J., K. Herbohn, and P. Clarkson, 2018, Carbon Risk, Carbon Risk Awareness and the Cost of Debt Financing, *Journal of Business Ethics* 150, 1151–1171.

Kabir, M. N., S. Rahman, M. A. Rahman, and M. Anwar, 2021, Carbon emissions and default risk: International evidence from firm-level data, *Economic Modelling* 103, 105617.

Kacperczyk, M., and P. Bolton, 2020, Carbon Premium around the World, CEPR Discussion Papers 14567, C.E.P.R. Discussion Papers.

Kiesel, R., and G. Stahl, 2022, Prolegomenon for managing climate risk, *Available at SSRN 4243603*.

Kiesel, R., and G. Stahl, 2023, An uncertainty-based risk management framework for climate change risk, *Annals of Actuarial Science* 17, 420–437.

Kiesel, R., and G. Stahl, 2024, Actuaries of the 6th kind—individuals with qualities.

Kim, Y.-B., H. T. An, and J. D. Kim, 2015, The effect of carbon risk on the cost of equity capital, *Journal of Cleaner Production* 93, 279–287.

Kleimeier, S., and M. Viehs, 2018, Carbon Disclosure, Emission Levels, and the Cost of Debt, *SSRN Electronic Journal*.

Korn, R., and G. Stahl, 2023, Insights from backtesting under solvency ii-overdose or implicit change of paradigm?, *Available at SSRN 4643633*.

Kölbel, J. F., M. Leippold, J. Rillaerts, and Q. Wang, 2022, Ask BERT: How Regulatory Disclosure of Transition and Physical Climate Risks Affects the CDS Term Structure*, *Journal of Financial Econometrics* 22, 30–69.

Lee, J., 2022, Voluntary disclosure of carbon emissions information, managerial ability, and credit ratings, *Sustainability* 14, 7504.

Lee, S.-Y., and D.-K. Choi, 2021, Does corporate carbon risk management mitigate the cost of debt capital? evidence from South Korean climate change policies, *Emerging Markets Finance and Trade* 57, 2138–2151.

Lemma, T. T., M. Feedman, M. Mlilo, and J. D. Park, 2019, Corporate carbon risk, voluntary disclosure, and cost of capital: South African evidence, *Business Strategy and the Environment* 28, 111–126.

Lichtenberger, A., J. P. Braga, and W. Semmler, 2022, Green bonds for the transition to a low-carbon economy, *Econometrics* 10.

Mastouri, A., R. Mendiratta, and G. Giese, 2022, Corporate bonds and climate-change risk, *The Journal of Portfolio Management* 48, 78.

Mella, P., 2021, *Magic Ring* (Springer).

Metricsgroup, R., 1999, Corporate metrics, Technical report, RiskMetrics Group.

Michelfelder, R., and E. Pilotte, 2023, U.S. electric utility greenhouse gas emissions and carbon risk premiums, *SSRN Electronic Journal*.

Nag, S., S. Chakrabarty, and S. Basu, 2021, From carbon-transition premium to carbon-transition risk. *arXiv preprint arXiv:2107.06518.*

NGFS, 2022, Credit Ratings and Climate Change – Challenges for Central Bank Operations, Technical report, Network for Greening the Financial System.

Ni, Y., and Y. Sun, 2023, Environmental, social, and governance premium in Chinese stock markets, *Global Finance Journal* 55, 100811.

Oestreich, A. M., and I. Tsiakas, 2015, Carbon emissions and stock returns: Evidence from the EU emissions trading scheme, *Journal of Banking & Finance* 58, 294–308.

Okimoto, T., and S. Takaoka, 2024, Credit default swaps and corporate carbon emissions in Japan, *Energy Economics* 133, 107504.

Palea, V., and F. Drogo, 2020, Carbon emissions and the cost of debt in the eurozone: The role of public policies, climate-related disclosure and corporate governance, *Business Strategy and the Environment* 29, 2953–2972.

Pizzutilo, F., M. Mariani, A. Caragnano, and M. Zito, 2020, Dealing with carbon risk and the cost of debt: Evidence from the European market, *International Journal of Financial Studies* 8, 61.

Ramos-García, D., C. López-Martín, and R. Arguedas-Sanz, 2023, Climate transition risk in determining credit risk: evidence from firms listed on the STOXX Europe 600 index, *Empirical Economics* 65, 2091–2114.

Ratings, M., 2022, Integration of ESG, `https://ratings.moodys.io/integration-of-esg-into-credit-risk`, Accessed: 2024-08-01.

Ratings, M., 2024, Moody's on Climate for Banking: Navigating climate risk for resilience and growth, `https://climate.moodys.com/banking`, Accessed: 2024-08-01.

Reboredo, J. C., and A. Ugolini, 2022, Climate transition risk, profitability and stock prices, *International Review of Financial Analysis* 83, 102271.

Reisinger et al., 2024, Science-based targets miss the mark, *communication: earth & environment* available at https://doi.org/10.1038/s43247-024-01535-z.

Safiullah, M., M. N. Kabir, and M. D. Miah, 2021, Carbon emissions and credit ratings, *Energy Economics* 100, 105330.

Santi, C., and A. Moretti, 2021, Carbon risk premium and worries about climate change, *SSRN Electronic Journal*.

Sautner, Z., L. van Lent, G. Vilkov, and R. Zhang, 2023, Pricing climate change exposure, *Management Science* 69, 7540–7561.

Seltzer, L. H., L. Starks, and Q. Zhu, 2022, Climate regulatory risk and corporate bonds, Working Paper 29994, National Bureau of Economic Research.

Shahrour, M. H., M. Arouri, and S. Rao, 2024, Linking climate risk to credit risk: Evidence from sectorial analysis, *SSRN Electronic Journal*.

Taleb, N. N., and C. Sandis, 2013, The skin in the game heuristic for protection against tail events, *arXiv preprint arXiv:1308.0958*.

Taleb, N. N., 2018, *Skin in the game: Hidden asymmetries in daily life* (Random House).

TCFD, 2017, Recommendations of the Task Force on Climate-related Financial Disclosures, Technical report, Task Force on Climate-related Financial Disclosures.

Thornton, J., Y. Altunbaş, and L. Nobile, 2024, Credit risk and firm carbon emissions: A dynamic regression discontinuity approach, *SSRN Electronic Journal*.

Trinks, A., G. Ibikunle, M. Mulder, and B. Scholtens, 2022, Carbon intensity and the cost of equity capital, *The Energy Journal* 43, 181–214.

Tsanakas, A., 2008, Risk measurement in the presence of background risk, *Insurance: Mathematics and Economics* 42, 520–528.

Vozian, K., and M. Costola, 2023, Pricing climate transition risk: Evidence from European corporate CDS, SAFE Working Paper Series 387, Leibniz Institute for Financial Research SAFE.

Wen, F., N. Wu, and X. Gong, 2020, China's carbon emissions trading and stock returns, *Energy Economics* 86, 104627.

Witkowski, P., A. Adamczyk, and S. Franek, 2021, Does carbon risk matter? evidence of carbon premium in EU energy-intensive companies, *Energies* 14, 1855.

Zhang, S., 2023, Carbon premium: Is it there?, *SSRN Electronic Journal*.

Zhang, Y., Y. Lio, and H. Wang, 2023, How credit default swap market measures carbon risk, *Environmental Science and Pollution Research* 30, 82696—-82716.

Zhang, Z., and R. Zhao, 2022, Carbon emission and credit default swaps, *Finance Research Letters* 50, 103286.

Chapter 3

Green investment and asset stranding under transition scenario uncertainty

Maria Flora

Capital Fund Management, France

Peter Tankov

CREST – ENSAE, Institut Polytechnique de Paris, France

Risks and opportunities related to environmental transition are usually evaluated through the use of scenarios, produced and maintained by international bodies such as the International Energy Agency. This approach assumes perfect knowledge of the scenario by the agent, but in reality, scenario uncertainty is an important obstacle for making optimal investment or divestment decisions. This chapter develops a real-options approach to evaluate assets and potential investment projects under dynamic climate transition scenario uncertainty. It uses off-the-shelf Integrated Assessment Model (IAM) scenarios, assuming that the economic agent acquires the information about the scenario progressively by observing a signal, such as the carbon price or the greenhouse gas emissions. The problem of valuing an investment is formulated as an American option pricing problem, where the optimal exercise time corresponds to the time of entering a potential investment project or the time of selling a potentially stranded asset. To illustrate our approach, this chapter employs representative scenarios from the scenario database of the Network for Greening the Financial System to two energy-related examples: the divestment decision from a coal-fired power plant without Carbon Capture and Storage (CCS) technology and the potential investment into a green coal-fired power plant with CCS. In both cases, the real option value is very sensitive to scenario uncertainty: the value of the coal-fired power plant is reduced by 25% and that of the green coal investment project is reduced by 7% when the agent deduces the scenario by observing carbon emissions, compared to the setting when the true scenario is known. It is also shown that scenario uncertainty can lead to considerable delays in the implementation of green investment projects, emphasizing the importance of timely and precise climate policy information.

This chapter is a reprint from an article originally published in Energy Economics. Please cite this work as follows:
Flora, M., & Tankov, P. (2023). Green investment and asset stranding under transition scenario uncertainty. Energy Economics, 124, 106773.

We thank Frédéric Ghersi, Jean-Charles Hourcade and Stéphane Voisin for numerous insightful discussions. This research was supported by ADEME (Agency for Ecological Transition) in the context of SECRAET project, and by the FIME (Finance for Energy Markets) research initiative of the Institut Europlace de Finance.

DOI: 10.1201/9781032636252-3

3.1 Introduction

As the global climate is changing, the need for a major decarbonization of the energy system has become evident (Teske, 2019; Bogdanov et al., 2019), while climate change impacts are expected throughout the energy system itself (Stanton, Dessai, and Paavola, 2016; Cronin, Anandarajah, and Dessens, 2018). While there is little doubt that the low-carbon transition will lead to profound changes in the energy system in the years and decades to come, it is difficult to predict the exact nature of these changes and the pace of the transition. Faced with such uncertainty, the scenario approach has emerged as a means to provide decision-makers with the tools to optimize their actions. Produced with the help of integrated assessment models, transition scenarios are published by international bodies such as the IEA (International Energy Agency), IPCC (Intergovernmental Panel on Climate Change), NGFS (Network for Greening the Financial System), IIASA (International Institute for Applied System Analysis), and used by the economic agents to understand the possible futures they need to prepare for. In particular, the NGFS maintains a database of six transition scenarios, described in detail in Section 3.3.[1] These scenarios are used by financial institutions to conduct climate stress tests: for example, by comparing the value of a bank's portfolio under orderly and disorderly transition scenario, one may evaluate the risk of disorderly transition.

The scenario approach, however, suffers from a number of drawbacks from the point of view of risk management and asset pricing. In most existing approaches the scenario is assumed to be given and known to the agent, thus the influence of the agent's actions on the scenario, and the imperfect knowledge of the scenario by the agent are not taken into account. Yet, owners of energy assets with a risk of stranding, or of construction permits for green energy projects, make their decisions to sell the asset, or to build the plant, without the perfect knowledge of the scenario to come. Instead, they evaluate the prospects of a given asset / investment based on partial information about the state of the energy transition.

In this paper we therefore extend the real options (RO) approach to take into account (i) the transition scenario uncertainty and (ii) the progressive discovery of information about the transition scenario by the investors. Traditional RO theory already acknowledges the value of waiting and postponing the investment decision in favor of flexibility and in view of acquiring more information on the evolution of the underlying stochastic variables affecting the project value. However, the sources of uncertainty in a RO model usually stem from the evolution of asset prices or cost variables specific to the project. We consider an additional layer of uncertainty, that is the uncertainty stemming from the energy transition scenario, potentially affecting the distribution of all stochastic variables. Then, we introduce an active Bayesian learning component to this framework. Namely, the agents continuously update their beliefs regarding the likelihood of being in a certain climate scenario, and evaluate the projects accordingly.

Our approach does not require developing a new integrated assessment model, but instead works with any set of scenarios, provided they contain the required variables to determine the cash flow stream of the energy asset. To incorporate scenario uncertainty, we then add to the scenario database a stochastic model describing progressive information discovery and progressive updates of posterior scenario probabilities in the context of Bayesian learning.

Our paper contributes to different strands of literature. The literature on RO in the context of energy project valuation is vast. Indeed, RO is a prominent approach to evaluate

[1] This chapter uses Release 3 of NGFS scenarios; starting from Release 4, the number of scenarios was increased to 7.

capital investments under uncertainty and irreversibility, and energy projects provide a natural field of application given their relatively high capital costs and the multiple sources of uncertainty relative to commodity prices and future electricity demand and supply. Siddiqui and Fleten (2010) evaluate how a firm may proceed with staged commercialization and deployment of competing alternative energy technologies, and find that the option of investing in such projects increases the value of the firm. Fuss et al. (2012) analyze the impact of uncertainty for deriving the optimal portfolio of energy technologies for a profit-maximizing investor. Boomsma, Meade, and Fleten (2012) analyze investment decisions in renewable energy under policy interventions, and find that a feed-in tariff leads to earlier investment. Abadie, Chamorro, and González-Eguino (2011) employ a binomial lattice model to compute the value of the option to abandon a coal-fired power plant; Laurikka and Koljonen (2006) analyze how the uncertainty related to an emission trading scheme affects the value of an option to invest in a coal power plant; Flora and Vargiolu (2020) use a least-squares Monte Carlo approach to solve the optimization problem of decision making in case of a power producer who is considering switching from a carbon-intensive technology to a renewable one under a carbon price floor; Hach and Spinler (2016) assess the effectiveness of capacity payments in promoting gas-fired generation investments under different degrees of feed-in tariff. Detemple and Kitapbayev (2020) (Chapter 15 of this book) develop a real options model where a firm seeking to build a new power plant has the exclusive choice between two technologies, namely wind and gas.

A related literature is also that of climate-related stranded assets. McGlade and Ekins (2015) employ an IAM and find that a third of oil reserves, half of gas reserves and over 80 per cent of current coal reserves must remain underground to maintain the global temperature rise compared to pre-industrial level below 2° C. A more recent study (Welsby et al., 2021) finds that the 1.5° C scenario requires nearly 60 per cent of oil and fossil methane gas, and 90 per cent of coal to remain unextracted. Mercure et al. (2018) estimate, with an integrated global economy-environment simulation model, the discounted global wealth loss from stranded fossil fuel assets. Rozenberg, Vogt-Schilb, and Hallegatte (2020) analyze the impact of alternative policy instruments on costs and dynamics of transition from polluting to clean capital, and study their implications for asset stranding. Van der Ploeg and Rezai (2020) study the determinants of asset stranding in the fossil-fuel industry. Mo, Cui, and Duan (2021) focus on the case of China, and show that carbon pricing increases the risk for newly built coal power plants to become stranded.

Finally, our paper also relates to the literature on active learning. In this context, closest to our paper is Dalby et al. (2018). They study policy uncertainty in the form of an unexpected downward adjustment of a fixed feed-in tariff (FIT) scheme, with a learning perspective. Their agent expects an adverse transition between two regimes of fixed FIT, and has the option to invest in a green energy project. Our work differs in multiple respects. First, their model pertains to renewable energy policy uncertainty rather than broader transition scenario uncertainty. Second, it is specific to renewable energy (RE) project valuation, while we develop a flexible approach that can be used for several potential applications in decision making analysis. Third, they consider a single policy revision, with two possible regimes: either the change in the FIT payment has not yet occurred (this is the starting point), or it has occurred, and in such a case the value of the option to invest in the green energy project becomes zero. In contrast, our model can include several different scenarios, and climate uncertainty shapes the trajectories of all state variables.

The paper is structured as follows. In Section 3.2, we present our model of an agent learning the likelihood of a global transition scenario from a climate-related signal, for the purpose of decision-making. In Section 3.3, we describe an empirical application of our approach, and analyze the sensitivity of our results to a set of parameters. Section 3.4 concludes.

3.2 Model and scenario uncertainty with Bayesian learning

3.2.1 Description of the model

We consider a discrete time model, where the integer-valued variable t denotes time measured in years. In the context of long-term investment/divestment decisions, it seems reasonable to assume that the agent may revise her investment/divestment strategy once a year. A risk-neutral and profit-maximizing economic agent (owner of a power-generating asset or of the potential investment project), facing both revenue risk and scenario uncertainty, has the option to either sell/decommission the asset or to invest into the project at a future date τ. The revenues of the asset prior to closure / of the project after investment are determined by future values of risk factors (fuel prices, electricity price, carbon price) whose evolution is stochastic and whose distribution depends on the realized transition scenario. We assume there are N scenarios corresponding to different climate, economic and policy assumptions. The true scenario is not known to the agent ex ante, however, the agent observes a signal (e.g., global CO_2 emissions), which contains noisy information about the scenario, and allows the agent to progressively update her posterior probability of realization for each scenario. For example, if the emissions decrease at a steady rate, the agent will assume that an orderly transition scenario is more likely than a delayed transition scenario.

Under each scenario, consider a number K of stochastic risk factors. We assume that the value $P_{k,t}$ of the risk factor k at time t under the scenario i follows an autoregressive dynamics with scenario-dependent mean $\mu^i_{k,t}$. To write the risk factor dynamics in matrix form, we denote by $\mathbf{P}_t$ the K-dimensional vector of risk factors, by $\boldsymbol{\mu}^i_t$ the vector of scenario-dependent means, by $\boldsymbol{\Phi}$ the $(K \times K)$ matrix of mean reversion rates, and by $\boldsymbol{\sigma}$ the Cholesky decomposition of the instantaneous variance-covariance matrix of the risk factors. Thus, in matrix form we have

$$\mathbf{P}_t = \widetilde{\mathbf{P}}_t + \boldsymbol{\mu}^i_t, \quad \widetilde{\mathbf{P}}_t = \boldsymbol{\Phi}\widetilde{\mathbf{P}}_{t-1} + \boldsymbol{\sigma}\boldsymbol{\varepsilon}_t, \quad t > 0, \quad \widetilde{\mathbf{P}}_0 = 0. \tag{3.1}$$

under scenario i, where $(\boldsymbol{\varepsilon}_t)$ is a sequence of i.i.d. K-dimensional standard normal vectors.

3.2.2 Bayesian updates

As mentioned above, the agent does not know the true scenario i, but observes a noisy signal y_t, and infers the likelihood of being in scenario i based on this signal. Ideally, the signal should be a (scenario-dependent) variable that is highly affected by or that has a high correlation with the scenario. Thus, the signal will "reveal" to the agent, with error, the state i. For example, the signal the agent relies on could be the price of carbon in the region where the production asset is located or the agent plans to invest; or the total emissions of greenhouse gases. Let us assume the signal is normally distributed with mean $\mu^i_{y,t}$ and standard deviation σ_y, that is

$$y_t = \mu^i_{y,t} + \sigma_y \eta_t\,, \qquad \text{with} \quad \eta_t \sim N(0,1) \text{ i.i.d.} \tag{3.2}$$

We assume that, given the scenario, the signal is independent from the risk factors, or, in other words, η_t is independent from ε^k_t for all k. This does not imply that the signal is completely independent from the risk factors, which is not a realistic assumption, but only that their dependence is explained by the scenario. For example, if the signal corresponds to the carbon price and the risk factor is the electricity generation, then in a net zero scenario we expect both of them to rise, but, conditionally on the scenario, the noises of these two processes may be independent.

At every time step, the agent updates her prior knowledge of the state to obtain a posterior probability of each state i. Since our aim is to understand the role of emissions/carbon price as a signal informing economic agents about scenario uncertainty, we assume that the updates are only based on the value of the signal, but not on the values of the risk factors P_t.

The probability π_t^i of being in scenario i at time t based on the signal y_t is

$$\pi_t^i = \mathbb{P}[I = i|\mathcal{F}_t], \quad \mathcal{F}_t = \sigma(y_s, s \leq t), \tag{3.3}$$

where $\{\mathcal{F}_t : t \geq 0\}$ is the filtration generated by the observable process $\{y_t : t \geq 0\}$. In particular, the Bayesian update of π_t^i at each time step t is

$$\pi_t^i = \mathbb{P}[I = i|y_t, \mathcal{F}_{t-1}] = \frac{\mathbb{P}[I = i, y_t \in dy|\mathcal{F}_{t-1}]}{\mathbb{P}[y_t \in dy|\mathcal{F}_{t-1}]} = \pi_{t-1}^i \frac{\mathbb{P}[y_t \in dy|I = i, \mathcal{F}_{t-1}]}{\mathbb{P}[y_t \in dy|\mathcal{F}_{t-1}]}, \tag{3.4}$$

where we use the notation $\mathbb{P}[y_t \in dy|\mathcal{F}_{t-1}]$ as a short-hand for the density of y_t given $\mathcal{F}_{t-1}$, and similarly for other notation in the above equation.

The *unnormalized posterior probability* of being in scenario i is then given by

$$\hat{\pi}_t^i = \hat{\pi}_{t-1}^i e^{-\frac{\left(y_t - \mu_{y,t}^i\right)^2}{2\sigma_y^2}}, \quad \hat{\pi}_0^i = \pi_0^i, \tag{3.5}$$

and the normalized probability is

$$\pi_t^i = \frac{\hat{\pi}_t^i}{\sum_i \hat{\pi}_t^i}. \tag{3.6}$$

Provided that the scenarios offer a range of sufficiently diversified trends $\mu_{y,t}^i$ for the signal, the lower the standard deviation of the signal σ_y, the sooner the agent's belief on the likelihood of being in a certain scenario i will converge to either 0 or 1.

In real option theory, the decision making analysis of the agent is similar to the pricing of an American option. To enforce our assumption that the scenario information is extracted solely from the signal, we suppose that the pricing procedure uses only the random part of the risk factors $\widetilde{\mathbf{P}}$, but not the scenario-dependent part. In practice this means that the agent uses a stochastic model for, e.g., the difference between the observed monthly local price and a slowly varying global price index, and uses the scenario information extracted from the signal to project the slowly varying price index.

The value function (value of the asset) at date t is related, through the dynamic programming principle, to the value function at date $t+1$, and computed by backward induction. This approach relies on the specification of a dynamics for the underlying stochastic processes, that is, we need to determine the joint dynamics of the (scenario-independent random parts of the) risk factors, the signal, and the scenario probabilities π_t^i, or, in other words, the rule of updating $\widetilde{\mathbf{P}}_t$, y_t and π_t^i from $\widetilde{\mathbf{P}}_{t-1}$, y_{t-1} and π_{t-1}^i. The rule for updating the probabilities is given by Equations (3.5–3.6) and that for the risk factors is given by Equation (3.1). For updating the signal, we compute the conditional law of y_t given $\mathcal{F}_{t-1}$:

$$\mathbb{P}\left[y_t \in B|I = i\right] = \frac{1}{\sqrt{2\pi\sigma_y^2}} \int_B e^{-\frac{\left(z - \mu_{y,t}^i\right)^2}{2\sigma_y^2}} \mathrm{d}z\,, \tag{3.7}$$

Thus, conditionally on the information available to the agent, the risk factors and the signal are distributed as a multivariate Gaussian mixture model.

Since the update rule for y_t does not depend on y_{t-1}, the process obtained with this update rule is a Markov process, as shown by the following proposition.

Proposition 1 *Let $(U_t)_{t=0,1,\dots}$ be a sequence of independent random variables with uniform distribution on the interval $[0,1]$, independent from the sequences (ε_t) and (η_t). Define the sequence*

$$
\begin{aligned}
\widetilde{\mathbf{P}}_t &= \boldsymbol{\sigma}\boldsymbol{\varepsilon}_t + \boldsymbol{\Phi}\widetilde{\mathbf{P}}_{t-1},\\
I_t &= \min\{i=1,\dots,N : \sum_{j=1}^{i} \tilde{\pi}^j_{t-1} \geq U_t\},\\
\tilde{y}_t &= \sigma_y\eta_t + \sum_{i=1}^{N} \mathbf{1}_{I_t=i}\,\mu^i_{y,t}\\
\tilde{\pi}^i_t &= \frac{\tilde{\pi}^i_{t-1} e^{-\frac{\left(\tilde{y}_t-\mu^i_{y,t}\right)^2}{2\sigma_y^2}}}{\sum_{j=1}^N \tilde{\pi}^j_{t-1} e^{-\frac{\left(\tilde{y}_t-\mu^j_{y,t}\right)^2}{2\sigma_y^2}}}, \quad i=1,\dots,N,
\end{aligned}
$$

for $t=1,2,\dots,$ with initial conditions $\widetilde{\mathbf{P}}_0 = 0$ and $\tilde{\pi}^i_0 = \pi^i_0$, $i=1,\dots,N$.

Then the process $(\widetilde{\mathbf{P}}_t, \tilde{y}_t,\ \tilde{\boldsymbol{\pi}}_t)$ has the same law as the process $(\widetilde{\mathbf{P}}_t, y_t,\ \boldsymbol{\pi}_t)$, where y and $\boldsymbol{\pi}$ are defined by

$$
y_t = \sigma_y\eta_t + \sum_{i=1}^{N} \mathbf{1}_{I=i}\,\mu^i_{y,t}, \quad \pi^i_t = \mathbb{P}[I=i|\mathcal{F}_t], \quad \mathcal{F}_t = \sigma(y_s, s\leq t),
$$

where

$$
I = \min\{i=1,\dots,N : \sum_{j=1}^{i} \pi^j_0 \geq U\}
$$

and U is an independent random variable with uniform distribution on $[0,1]$.

Proof 1 *Since the risk factors $\widetilde{\mathbf{P}}$ are independent from the scenario, it is sufficient to consider the law of the process $(\tilde{y}_t,\ \tilde{\boldsymbol{\pi}}_t)$. Moreover, we have shown by Bayes formula that the conditional probabilities $\boldsymbol{\pi}_t$ satisfy the same update rules (as function of the signal) as the probabilities $\tilde{\boldsymbol{\pi}}$ defined in the proposition. Thus it is sufficient to show that the signal $\tilde{y}$ defined in the proposition has the same law as the signal y. Further, for any $i=1,\dots,N$, any $t>0$, and any bounded measurable f,*

$$
\begin{aligned}
\mathbb{E}[f(\tilde{y}_t)\tilde{\pi}^i_t \,|\, \tilde{\boldsymbol{\pi}}_{t-1}] &= \mathbb{E}\left[\left.\frac{f(\tilde{y}_t)\tilde{\pi}^i_{t-1} e^{-\frac{\left(\tilde{y}_t-\mu^i_{y,t}\right)^2}{2\sigma_y^2}}}{\sum_{j=1}^N \tilde{\pi}^j_{t-1} e^{-\frac{\left(\tilde{y}_t-\mu^j_{y,t}\right)^2}{2\sigma_y^2}}}\right| \tilde{\boldsymbol{\pi}}_{t-1}\right]\\
&= \mathbb{E}\left[\left.\sum_{k=1}^N \frac{f(\sigma_y\eta_t+\mu^k_{y,t})\tilde{\pi}^k_{t-1}\tilde{\pi}^i_{t-1} e^{-\frac{\left(\sigma_y\eta_t+\mu^k_{y,t}-\mu^i_{y,t}\right)^2}{2\sigma_y^2}}}{\sum_{j=1}^N \tilde{\pi}^j_{t-1} e^{-\frac{\left(\sigma_y\eta_t+\mu^k_{y,t}-\mu^j_{y,t}\right)^2}{2\sigma_y^2}}}\right|\tilde{\boldsymbol{\pi}}_{t-1}\right]\\
&= \frac{1}{\sigma_y\sqrt{2\pi}}\int_{\mathbb{R}} \sum_{k=1}^N \frac{f(y+\mu^k_{y,t})\tilde{\pi}^k_{t-1}\tilde{\pi}^i_{t-1} e^{-\frac{\left(y+\mu^k_{y,t}-\mu^i_{y,t}\right)^2}{2\sigma_y^2}}}{\sum_{j=1}^N \tilde{\pi}^j_{t-1} e^{-\frac{\left(y+\mu^k_{y,t}-\mu^j_{y,t}\right)^2}{2\sigma_y^2}}} e^{-\frac{y^2}{2\sigma_y^2}}\,dy\\
&= \frac{1}{\sigma_y\sqrt{2\pi}}\int_{\mathbb{R}} \frac{f(y)\tilde{\pi}^i_{t-1} e^{-\frac{\left(y-\mu^i_{y,t}\right)^2}{2\sigma_y^2}}}{\sum_{j=1}^N \tilde{\pi}^j_{t-1} e^{-\frac{\left(y-\mu^j_{y,t}\right)^2}{2\sigma_y^2}}} \sum_{k=1}^N \tilde{\pi}^k_{t-1} e^{-\frac{(y-\mu^k_{y,t})^2}{2\sigma_y^2}}\,dy\\
&= \frac{\tilde{\pi}^i_{t-1}}{\sigma_y\sqrt{2\pi}}\int_{\mathbb{R}} f(y) e^{-\frac{\left(y-\mu^i_{y,t}\right)^2}{2\sigma_y^2}}\,dy = \tilde{\pi}^i_{t-1}\mathbb{E}[f(\sigma_y\eta_t+\mu^i_{y,t})].
\end{aligned}
$$

Fix an integer $T < \infty$ and consider bounded measurable functions $f_1, \dots, f_T$. Then, using repeatedly the previous formula,

$$
\begin{aligned}
\mathbb{E}[f_T(\tilde{y}_T) \dots f_1(\tilde{y}_1)] &= \mathbb{E}[\mathbb{E}[f_T(\tilde{y}_T) \,|\, \tilde{y}_{T-1} \dots \tilde{y}_1]\, f_{T-1}(\tilde{y}_{T-1}) \dots f_1(\tilde{y}_1)] \\
&= \sum_{i=1}^{N} \mathbb{E}[f_T(\sigma_y \eta_T + \mu^i_{y,T})]\, \mathbb{E}\left[\tilde{\pi}^i_{T-1} f_{T-1}(\tilde{y}_{T-1}) \dots f_1(\tilde{y}_1)\right] \\
&= \sum_{i=1}^{N} \mathbb{E}[f_T(\sigma_y \eta_T + \mu^i_{y,T})]\, \mathbb{E}[f_T(\sigma_y \eta_{T-1}) + \mu^i_{y,T-1})]\, \mathbb{E}\left[\tilde{\pi}^i_{T-2} f_{T-2}(\tilde{y}_{T-2}) \dots f_1(\tilde{y}_1)\right] \\
&= \dots = \sum_{i=1}^{N} \pi^i_0 \prod_{t=1}^{T} \mathbb{E}[f_t(\sigma_y \eta_t + \mu^i_{y,t})] = \mathbb{E}\left[\prod_{t=1}^{T} f_t(y_t)\right],
\end{aligned}
$$

which finishes the proof.

3.2.3 Optimal project valuation under scenario uncertainty

Our model is general enough to be applicable in many contexts. In this section, we will focus on two types of investment decisions: (1) an optimal exit problem, where the agent owns a carbon-intensive plant and is considering decommissioning the plant; and (2) an optimal entry problem, where the agent has the option to invest in a green energy project. These problems can be both modeled as an American option pricing problem, and solved numerically by a Least Squares Monte Carlo (LSMC) approach.

In our model, the agent's incentive to delay the investment decision lies not only in the opportunity to wait for future price information, but also in that of learning about the macroeconomic scenario with greater accuracy. The agent has indeed two main sources of uncertainty: the one stemming from the fluctuations of risk factor values $\widetilde{\mathbf{P}}_t$, and the transition scenario uncertainty. This is the main difference with respect to standard real options models, where the value of the underlying only depends on stochastic variables that are commodity prices or other asset prices.

We assume that the agent makes the decision based on the information from the observable risk factor values, and the information about the posterior scenario probabilities deduced from the signal. We thus consider the Markov process $(\mathbf{X}_t := (\widetilde{\mathbf{P}}_t, \boldsymbol{\pi}_t))_{t=0,1,\dots}$ defined in the previous section, and denote by $\mathbb{G} = (\mathcal{G}_t)_{t=0,1,\dots}$ the natural filtration of this process.

1. **Optimal exit**
   The problem of an agent who is considering selling (or decommissioning) a potentially stranded asset can be written as the following optimal stopping problem:

$$
\sup_{\tau \in \mathcal{T}} \mathbb{E}\left[\sum_{t=1}^{\tau} \beta^t h(\mathbf{P}_t) - \beta^\tau K(\tau)\right], \tag{3.8}
$$

   where $\mathcal{T}$ is the set of $\mathbb{G}$-stopping times with values in $0, 1, 2, \dots, T$, $h(\mathbf{P}_t)$ is the Profit&Loss function of year t, β is a discount factor that accounts both for the time value of money and for the riskiness of the investment, and K is the cost of decommissioning the plant or, when negative, the price at which the plant is sold. At each point in time until the asset's lifetime T, the agent has to decide whether to continue operating the plant or not.

2. **Optimal entry**
   The problem of an agent who is considering a potential investment before time T_0 in a green energy project with lifetime T can be written as the following optimal stopping problem:

$$\sup_{\tau \in \mathcal{T}} \mathbb{E}\left[\sum_{t=\tau}^{\tau+T} \beta^t h(\mathbf{P}_t) - \beta^\tau K(\tau)\right], \tag{3.9}$$

   where $\mathcal{T}$ is the set of $\mathbb{G}$-stopping times with values in $0, 1, 2, \ldots, T_0$, $h(\mathbf{P}_t)$ is the profit and loss (P&L) function of year t, β is a discount factor, and K is the capital cost needed to undertake the project. At each point in time, the agent has to decide whether to exercise the real option, or to postpone the decision until more precise information about the potential profitability of the project becomes available.

In what follows, to simplify our presentation, we will focus on case (1), that is the optimal exit problem, with the understanding that our setting fully addresses case (2) as well. Introduce the value function

$$\begin{aligned} V(t, \widetilde{\mathbf{P}}, \boldsymbol{\pi}) &:= \sup_{\tau \in \mathcal{T}_t} \mathbb{E}\left[\sum_{s=t+1}^{\tau} \beta^{s-t} h(\mathbf{P}_s) - \beta^{\tau - t} K(\tau) \,\middle|\, (\widetilde{\mathbf{P}}_t, \boldsymbol{\pi}_t) = (\widetilde{\mathbf{P}}, \boldsymbol{\pi})\right] \\ &= \sup_{\tau \in \mathcal{T}_t} \mathbb{E}\left[\sum_{s=t+1}^{\tau} \beta^{s-t} \sum_{j=1}^{N} \pi_s^j \, h(\widetilde{\mathbf{P}}_s + \boldsymbol{\mu}_s^j) - \beta^{\tau - t} K(\tau) \,\middle|\, (\widetilde{\mathbf{P}}_t, \boldsymbol{\pi}_t) = (\widetilde{\mathbf{P}}, \boldsymbol{\pi})\right], \end{aligned}$$

where $\mathcal{T}_t$ is the set of $\mathbb{G}$-stopping times with values in $t, \ldots, T$. By standard backward induction argument, it can be shown that the value function satisfies the following dynamic programming principle:

$$V(t, \widetilde{\mathbf{P}}, \boldsymbol{\pi}) := \max\left\{-K(t), \beta \mathbb{E}\left[h(\mathbf{P}_{t+1}) + V(t+1, \mathbf{P}_{t+1}, \boldsymbol{\pi}_{t+1}) \,\middle|\, (\widetilde{\mathbf{P}}_t, \boldsymbol{\pi}_t) = (\widetilde{\mathbf{P}}, \boldsymbol{\pi})\right]\right\} \tag{3.10}$$

$$= \max\{-K(t), CV\left(t, \widetilde{\mathbf{P}}_t, \boldsymbol{\pi}_t\right)\}, \tag{3.11}$$

where CV is the so called continuation value:

$$CV\left(t, \widetilde{\mathbf{P}}_t, \boldsymbol{\pi}_t\right) = \beta \mathbb{E}\left[h(\mathbf{P}_{t+1}) + V(t+1, \widetilde{\mathbf{P}}_{t+1}, \boldsymbol{\pi}_{t+1}) \,|\, \widetilde{\mathbf{P}}_t, \boldsymbol{\pi}_t\right]. \tag{3.12}$$

As mentioned, Eq. (3.10) can be solved numerically by LSMC, that couples backward oriented dynamic programming techniques with forward oriented simulation techniques. The LSMC algorithm works by backward induction, and at each point in time it compares the convenience of immediate exercise with that of delaying the decision. As outlined in Longstaff and Schwartz (2001), the continuation value at each possible exercise point can be estimated from a least squares cross-sectional regression using the simulated paths. The algorithm (see Algorithm 1) then returns both the value of the real option V_t and the optimal exercise time τ. In the algorithm, $\widetilde{V}_t$ denotes an auxiliary process whose conditional expectation equals the continuation value.

As the algorithm shows, a crucial part in the LSMC procedure is to use the cross-sectional information in the simulation to estimate the expectation on future cash flows. In all the empirical applications that follow, we employ a quadratic specification to regress the discounted value of the payoff at future dates over the simulated state variables. Specifically, our least-squares specification with which we cross-sectionally regress the continuation values of the different simulated paths j at time t is the following:

$$CV_{t,j} = \alpha_t + \boldsymbol{\vartheta}_t f\left(\boldsymbol{\pi}_{t,j}, \widetilde{\mathbf{P}}_{t,j}\right) + \varepsilon_{t,j}, \tag{3.13}$$

Algorithm 1: Least Squares Monte Carlo

Simulate N_{sim} realizations of the true scenario I and corresponding trajectories of $X_t = [\pi_t^1, \ldots, \pi_t^N, \widetilde{P}_t^1, \ldots, \widetilde{P}_t^K]$, for $t = 1, \ldots, T$.
For each trajectory, set $\tilde{V}_T = h(\widetilde{\mathbf{P}}_T + \boldsymbol{\mu}_T^I) - K(T)$;
for $t = T-1 : -1 : 1$ **do**
  Perform a polynomial regression of $Y_t = \beta \tilde{V}_{t+1}$ on X_t;
  Use the result to estimate continuation value CV_t on each trajectory
  if $CV_t + K(t) < 0$ **then**
    exercise is optimal on this trajectory;
    $V_t = -K(t)$ and $\tilde{V}_t = h(\widetilde{\mathbf{P}}_t + \boldsymbol{\mu}_t^I) - K(t)$;
  else
    continuation is optimal on this trajectory;
    $V_t = CV_t$ and $\tilde{V}_t = h(\widetilde{\mathbf{P}}_t + \boldsymbol{\mu}_t^I) + CV_t$
  end
end
$V_0 = \frac{\beta}{N_{sim}} \sum \tilde{V}_1$.

where $f(\cdot)$ is a second-order polynomial function. A sensitivity analysis with respect to the choice of the basis functions has been performed and shows that the standard quadratic basis provides a good trade-off in terms of improving numerical accuracy and reducing the risk of overfitting in our context.

3.3 Empirical application

3.3.1 Climate scenarios

Integrated assessment models (IAMs) encompassing feedbacks between the global economy, the energy system and the climate system, are the convenient tool to analyze the economic impacts of climate change and climate change mitigation measures. IAMs are used to generate scenarios of evolution of the economy consistent with given climate objectives, based on a set of assumptions.

In this section, we illustrate how our method can be applied to model transition scenario uncertainty. We employ scenario data from an IAM in the NGFS scenario database[2], namely REMIND-MAGPIE 3.0-4.4. This model is a global multi-regional general equilibrium model with a rather detailed representation of the energy system, belonging to the class of intertemporal optimization models with perfect foresight.

The NGFS scenario database includes 6 alternative scenarios produced with REMIND 3.0 model:

- **Current Policies:** existing climate policies remain in place, and there is no strengthening of ambition level of these policies;

[2]Release 3.4, available at `https://data.ene.iiasa.ac.at/ngfs/`

- **Nationally Determined Contributions (NDCs):** currently pledged unconditional NDCs are implemented fully, and respective targets on energy and emissions in 2025 and 2030 are reached in all countries;
- **Delayed Transition (Disorderly):** there is a "fossil recovery" from 2020 to 2030; thus this scenario follows the trajectory of the Current Policies scenario until 2030. Only thereafter countries with a clear commitment to a specific net-zero policy target at the end of 2020 are assumed to meet the target, representing regional fragmentation. Regionally fragmented carbon prices converge to a global price near 2070 to keep the 67-percentile of warming below 2°C in 2100, which also allows for temporary overshoot;
- **Below 2°C:** this scenario assumes that optimal carbon prices in line with the long-term targets are implemented immediately after 2020 and keeps the 67-percentile of warming below 2°C throughout the 21st century;
- **Divergent Net Zero (Disorderly):** optimal carbon prices in line with the long-term targets are implemented immediately after 2020 to bring the median temperature below 1.5°C in 2100, after a limited temporary overshoot. Policy pressure and mitigation efforts are unevenly distributed across sectors;
- **Net Zero 2050:** global CO_2 emissions are at net-zero in 2050. Furthermore, countries with a clear commitment to a specific net-zero policy target at the end of 2020 are assumed to meet this target.

From top to bottom, the scenarios display a range of levels of stringency of the climate policy that underlies scenario assumptions. Thus, each scenario entails different paths for the macroeconomic variables in the model, and overall they provide a comprehensive overview of the possible climate states.

3.3.2 Optimal exit from a carbon-intensive power plant

As a first empirical application, we consider an agent who owns a pulverized coal power plant with combined cycle, without Carbon Capture and Storage (CCS) technology, and wants to know when it is economically optimal to decommission (or sell) the plant. We assume the cost of decommissioning the plant corresponds to a fraction k of the capital cost of building a new coal power plant, C_C^c. When k is negative, the agent is selling the plant at price $k\, C_C^c$. In the results showcased in Section 3.3.2.1, unless specified otherwise, we assume $k = 0$, *i.e.,* there are no decommissioning costs. A sensitivity analysis for this parameter is shown is Section 3.3.2.2. The plant is a price taker, and supplies every year a quantity of electricity which depends on the plant's utilization rate (a variable available in each scenario) but not on energy prices. We assume it has a nominal capacity $W = 1000\,\text{MW}$ and it is located in Germany. We further suppose that the plant has a residual lifetime of 30 years, thus, assuming $t_0 = 2020$, we restrict our investment valuation framework to years 2020-2050 of the scenario dataset. Finally, we assume a risk-adjusted discount factor $\beta = e^{-r\Delta t}$, with $r = 1\%$ unless specified otherwise. Because it is not equipped with a CCS filter, the plant has to buy some emission allowances in every period, to comply with an emission trading scheme (ETS). Moreover, in every period, it incurs some operation and maintenance (O&M) costs, both fixed (C_F^c) and proportional to the production output (C_V^c). The production output of the plant depends on the plant utilization rate R_U^c, which equals the ratio of the yearly net electricity production Q^c to the total installed capacity T^c for this specific technology in the selected region. The P&L function of year t of the coal plant, with the appropriate conversion factors, is thus equal to

$$h_t^c = W \cdot R_{Ut}^c \cdot 365.25 \cdot 24 \left(P_{Et} - \frac{P_{Ft}}{R_C^c} - R_E^c P_{Ct} - C_V^c \right) - W \cdot C_F^c, \tag{3.14}$$

where

$$R_{Ut}^c = \text{Utilization rate} = \frac{Q_t^c}{T_t^c \cdot 365.25 \cdot 24}. \tag{3.15}$$

The variables used and their corresponding unit of measure are listed in Table 3.1 (we omit the superscript c, indicating the coal technology, to avoid clutter).

TABLE 3.1: Variables used in the plant evaluation

Variables	*Name*	*Units*
Capital costs	C_C	USD2010/kW
O&M costs, fixed	C_F	USD2010/MW/year
O&M costs, variable	C_V	USD2010/MWh
Fuel price	P_F	USD2010/MWh
Conversion efficiency	R_C	per cent
Carbon price	P_C	USD2010/t CO_2
Electricity price	P_E	USD2010/MWh
Net electricity production	Q	GWh
Total installed capacity	T	GW
Emission rate	R_E	t CO_2/MWh
Plant rated power	W	MW

Some of the variables in (3.14) are more likely to have a high variability, and thus we model them as stochastic, according to (3.1). Such risk factors, in this example application, are

$$\mathbf{P} = \begin{pmatrix} P_E \\ P_F \\ P_C \end{pmatrix}.$$

As mentioned, their mean values $\mu_{:,t}^i$ depend on the scenario i of the chosen scenario dataset, and are given by the corresponding i-th scenario path for that variable. Figure 3.1 plots the evolution of the mean value $\mu_{k,t}^i$ of each risk factor together with a sample simulation of the three risk factors. All other variables in Eq. (3.14) except for the risk factors are modeled as deterministic, and their time-dependent values are extracted from the scenario database, when available. Moreover, when possible (for coal, carbon and electricity price), the variables are taken from the downscaled NGFS data for the specific region, otherwise (for capital costs and utilization rate) the variables for the EU-28 region of the original REMIND 3.0 model are used. For example, the right bottom graph in Figure 3.1 shows the evolution of the coal utilization rate in the six scenarios; it is clear that the variability of this quantity is much lower than that of the risk factors. In addition, we assume constant fixed costs C_F^c at 58,000 EUR/MW/year (see ACIL (2014)), variable costs C_V^c at 2.6 EUR/MWh, conversion efficiency R_C^c of 48% and emission rate R_E^c of 0.71 tonCO_2 per MWh of electricity generated (see Metz et al. (2007, Sec. 4.4.3.1)).

The parameters $\mathbf{\Phi}$ and $\boldsymbol{\sigma}$ in (3.1) are estimated from historical data. Specifically, we use front-month futures data at the monthly frequency from January 2015 to October 2021 of ICE Rotterdam Coal and of ICE EUA (carbon allowances traded in the EU ETS). For electricity, we employ the average hourly price in the day-ahead market for electricity in Germany traded in the EEX, at the monthly frequency. All data were retrieved using the

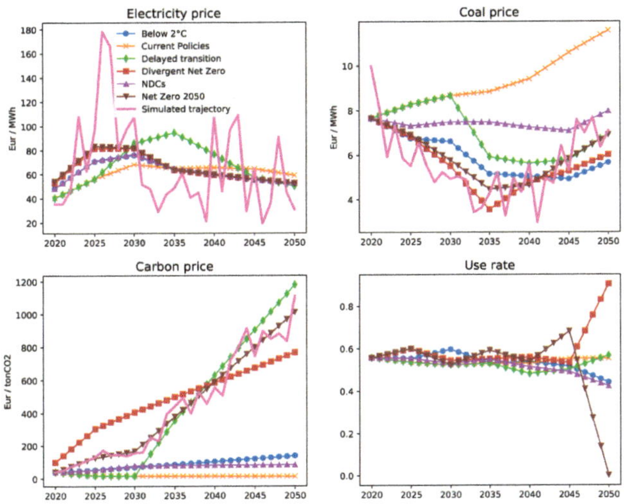

FIGURE 3.1: Evolution of the mean value of the three risk factors and of the coal utilization rate in the different scenarios produced with the REMIND 3.0 IAM, together with a sample simulation of the three risk factors.

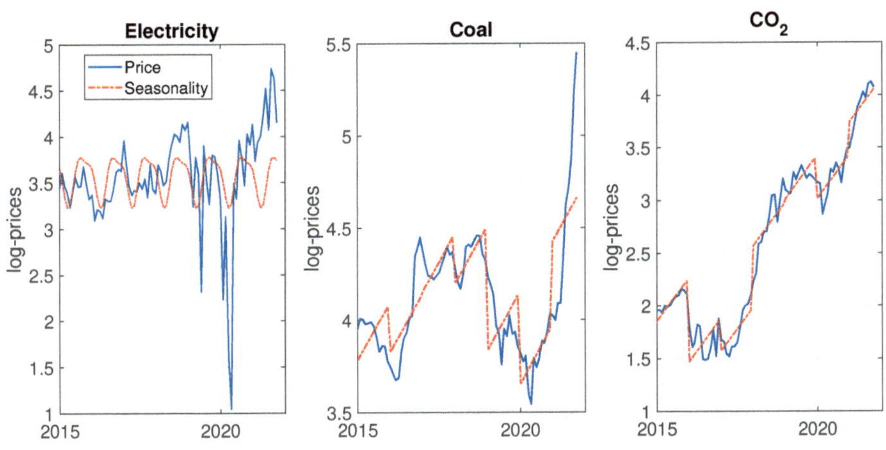

FIGURE 3.2: From left to right, EEX day-ahead electricity log-prices in Germany, ICE Rotterdam front-month futures log-prices, ICE EUA front-month futures log-prices (blue lines), and their relative de-trending (red dotted lines).

Macrobond database. Because seasonality is a peculiar stylized fact of electricity prices, we first de-seasonalize our monthly log-price electricity sample with a sinusoidal function accounting for half-yearly and annual seasonality. As for carbon and coal log-prices, we de-trend them by subtracting the annual mean level and a linear trend (see Figure 3.2). We then estimate a multivariate autoregressive model by maximum likelihood using (de-trended) log-prices. Table 3.2 shows the resulting estimates for the mean-reversion rates and for the volatility parameters. In the empirical application that follows, we will only

TABLE 3.2: Monthly parameter estimates of (3.1) obtained by estimating a multivariate autoregressive model on de-trended prices of electricity, coal, and carbon. Standard errors are in parenthesis. $^{*} = p < 0.1$, $^{**} = p < 0.05$, $^{***} = p < 0.01$

		E	F	C
Φ	E	0.552***	0.023	0.019
		(0.095)	(0.031)	(0.028)
	F	−0.232	0.779***	−0.093
		(0.261)	(0.085)	(0.077)
	C	0.173	0.119	0.382***
		(0.354)	(0.116)	(0.104)
Σ	E	0.174***	0.015**	0.007***
		(0.027)	(0.007)	(0.003)
	F	0.015**	0.018***	0.004**
		(0.007)	(0.006)	(0.002)
	C	0.007***	0.004**	0.015***
		(0.003)	(0.002)	(0.002)

employ estimates that are significant at the 95% level, i.e., all entries in the matrix Σ, and all diagonal entries in Φ.

To update her beliefs about the transition scenario, the agent needs to choose which signal to rely on. In this illustration, we assume that the agent chooses the greenhouse gases (GHG) emission level from energy production in the region where the plant is located (Germany). To estimate the volatility of such a signal y, we use a yearly time series from the Macrobond data set, spanning almost 50 years (from 1970 to 2018), and we fit a Gaussian model. The estimated annual signal volatility parameter is $\hat{\sigma}_y = 108.75$ million tons of CO_2 equivalent. We then employ this value to simulate the signal as in Proposition 1. On the bottom graph, Figure 3.3 shows a simulated sample path for the signal y_t (solid line), when using the scenarios from the REMIND 3.0 model. The REMIND 3.0 values of the mean GHG emissions path in each scenario are represented by the dashed lines. On the top graph, the figure shows the corresponding evolution of the relative conditional probabilities π_t^i, for $i \in \{1, \dots, 6\}$, computed as in Eq. (3.6) (where the marks indicating the scenario are the same in both graphs). In this case, the conditional probability of the scenario "Below 2° C" approaches one, while the others converge to zero;

All simulation examples below use $N_{sim} = 200,000$ simulated trajectories.

3.3.2.1 Numerical results

First, before moving on to analyze the impact of scenario uncertainty revealed by the signal, we compute the value of the project in each individual scenario (assumed known by the agent), the expected optimal exit date for each individual scenario, and the expected value of the project if it operates for the entire lifetime of 30 years with no possibility of exit. This information is shown in Table 3.3. We see that there are three classes of scenarios: in the least constrained scenario (current policies) it is optimal to operate the plant until the end of its lifetime; in the three intermediate scenarios (below 2° C, delayed transition and NDCs), it is optimal to stop after 10-11 years, and finally in the most constrained scenarios (divergent net zero and net zero 2050) immediate exit is optimal. Stopping at the optimal time leads to a positive value for the plant in all but the most constrained scenarios (where the plant has a small negative value because stopping is only possible after one year of operation). On the other hand, operating the plant until the end of its lifetime has a

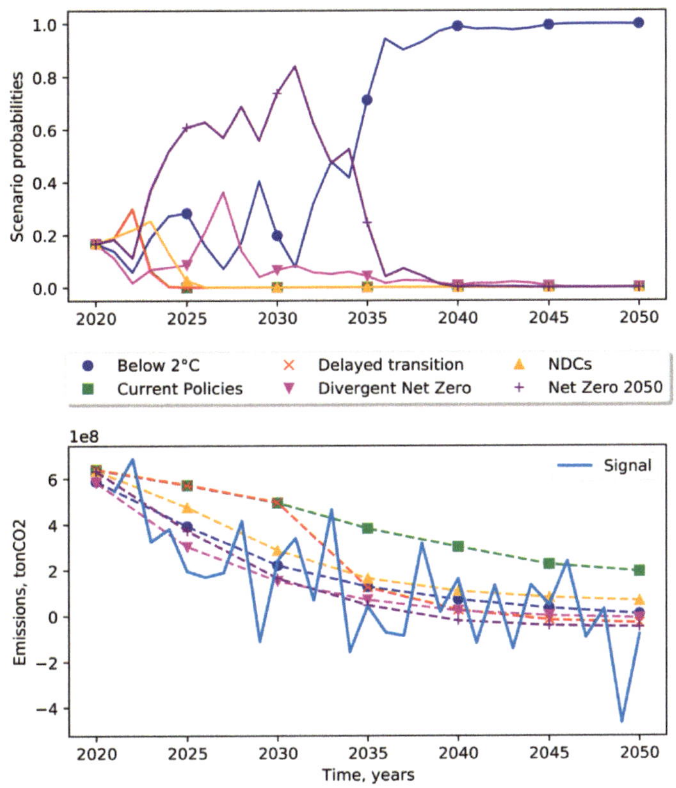

FIGURE 3.3: Bottom graph: a simulated path for the signal y_t (solid line), together with the GHG emission trend evolution in each of the six scenarios according to REMIND 3.0 data. Top graph: conditional probabilities π_t^i, for $i \in \{1, \ldots, 6\}$ estimated from the observations of the signal. The marks are the same in both graphs and indicate the specific scenario. For illustration purposes, here the signal volatility was taken equal to 2 times the estimated value.

negative net present value in all but one scenario and entails a particularly massive loss in three of them. Determining the optimal exit time precisely is thus very important in this context.

We now proceed to analyze the impact of the scenario uncertainty. Different scenario datasets may differ not only in the nature of the base assumptions underlying the scenario specifications, but also in the number of scenarios itself. The NGFS REMIND 3.0 dataset offers 6 different scenarios, as discussed in Section 3.3.1. The ones that matter the most in the context of this empirical application are the ones related to the emissions trajectory, since the signal the agent relies on is closely related to this variable. The span of the paths of this variable in different scenarios represents in a sense the extent to which the agent can decode the state of the system and form a belief about possible implications. It is thus important that the agent has a number of sufficiently diversified scenarios available. To see this, first, let us consider the case when an agent has only two available emission scenarios, and let us suppose that the two are very far apart in their range of values for the emissions

TABLE 3.3: Project values with and without possibility of exit, and optimal exit times in the absence of scenario uncertainty

Scenario	*Below 2°C*	*Current policies*	*Delayed transition*	*Divergent net zero*	*NDCs*	*Net zero 2050*
Project value, MEur	426	2777	992	−315	254	−15
Project value, MEur, no exit	−1942	2777	−28319	−40080	−1191	−26937
Optimal exit time	11.02	30	11	1	9.21	1

variable. If scenarios have a range of values much wider than the variability in the signal, the signal, albeit noisy, will likely immediately identify the state of the system, and climate uncertainty would immediately resolve. In this case, the learning process of the agent would end soon, and it would thus not affect much the value of the real option.

Our results confirm this intuition. Indeed, Figure 3.4, left graph, shows the sensitivity in the results of our model to the value of the volatility of the signal in the case of two divergent scenarios, "Current Policies" and "Net Zero 2050". We let the signal volatility vary in a range $[0; 3\,\hat{\sigma}_y]$. As Figure 3.3 shows, these two scenarios entail very different trajectories for the GHG emissions. When the signal volatility is in a range of 0 to 0.5 times the estimated $\hat{\sigma}_y$, which is not sufficient to cover the range of values spanned by the two scenarios, the value of the project and the optimal exercise time change at a slow pace, and it is only for higher values of signal volatility that the project value starts to decline quickly due to information loss. We remark also that the expected optimal exit time increases as function of signal volatility. To understand this effect, one may examine Figure 3.5, left graphs, which show the distribution of optimal exit times for each of the two scenarios. We see that when the signal volatility is high (lower panel), the "Current policies" scenario is stable, meaning that when this scenario is the true one, the agent is unlikely to make wrong decision and stop early. On the other hand, the "Net Zero 2050" scenario is unstable: when this scenarios the true one and the signal is very noisy, the agent may spend a long time learning the true scenario before finally deciding to stop.

Figure 3.4, right graph, shows the results of our model when all six available scenarios from REMIND 3.0 are included. In this case, scenarios span a wide range of possible trajectories for the GHG emissions, so that the learning process of the agent is more important in her decision making. Indeed, now the value of the real option declines quickly as the signal volatility increases and the information available to the agent becomes more noisy. To understand the decreasing profile of the optimal exercise time, once again we examine the distribution of stopping times in each scenario in Figure 3.5, right graphs. We see that while the "Current policies" scenario remains quite stable, the intermediate scenarios such as "NDCs" and "Below 2° C" are less so: when one of this scenarios is the true one, and the signal is very noisy, the agent may mistakingly think that the true scenario is one of the two most constrained ones ("Net Zero 2050" or "Divergent Net Zero") and stop early. Indeed, by looking at Figure 3.3 it is clear that these four scenarios are hard to distinguish in the beginning. We then conclude that while the price of the real option always decreases with increasing scenario uncertainty, the expected waiting time until exit may both increase and decrease, depending on the nature of the underlying scenarios.

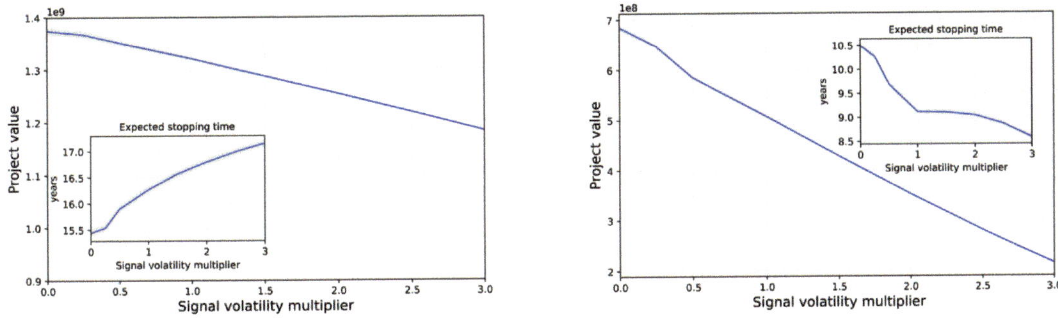

FIGURE 3.4: Optimal exit problem. Sensitivity of the project value (in EUR) to the volatility of the signal, σ_y. In the figure, σ_y varies in a range $[0; 3\,\hat{\sigma}_y]$. In the left panel, only scenarios "Current Policies" and "Net Zero 2050" from the REMIND 3.0 are included in the model. In the right panel, all scenarios are included. The shaded gray area represents the 95% confidence interval (sometimes invisible because it is too narrow). The inset plot shows the sensitivity of the average waiting time before divestment τ to the volatility of the signal σ_y, with a 95% confidence interval.

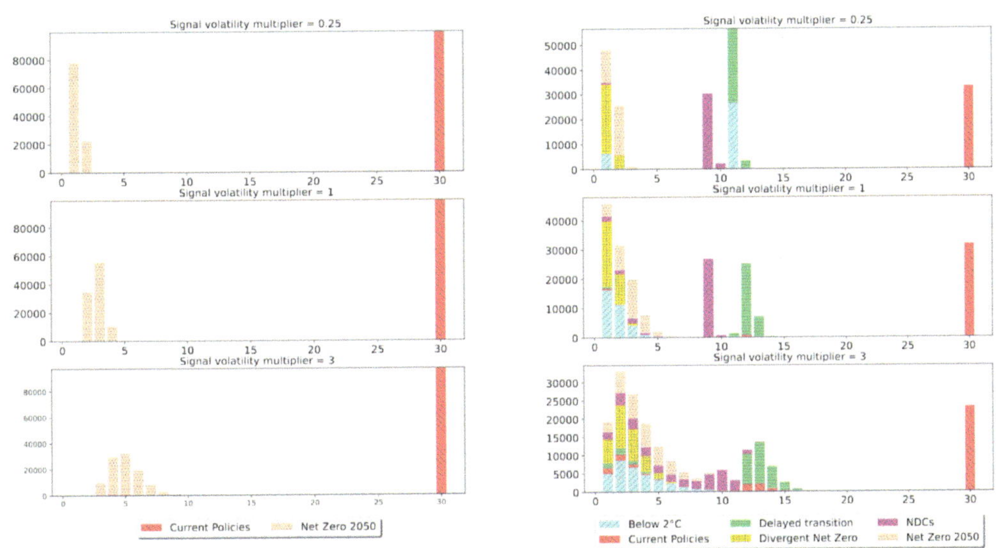

FIGURE 3.5: Optimal divestment timing under scenario uncertainty: distribution of the optimal exit times according to the true scenario. Top row: $\sigma_y = 0.25\,\hat{\sigma}_y$, middle row: $\sigma_y = \hat{\sigma}_y$, bottom row: $\sigma_y = 3\,\hat{\sigma}_y$. In the left graphs, only scenarios "Current Policies" and "Net Zero 2050" are included. In the right graphs, all scenarios are included.

In applications of our model, we recommend using a sufficiently large number of sufficiently diverging scenarios, so as to describe the range of possible futures in a realistic way.

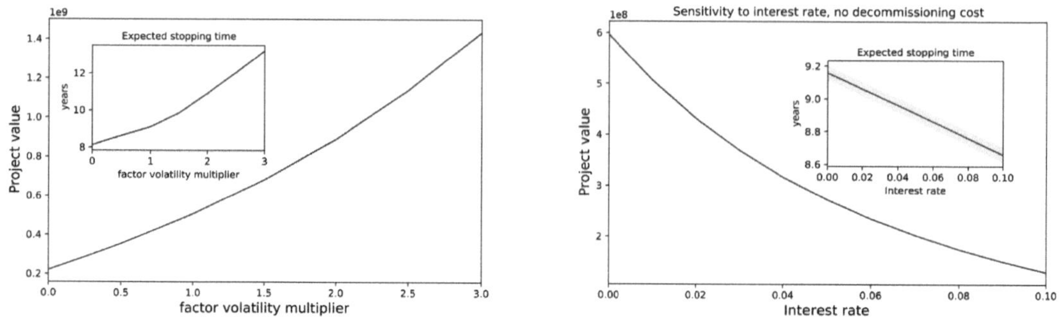

FIGURE 3.6: Optimal exit problem. Sensitivity of the project value (in EUR) to the risk factors volatility, Σ (left panel) and to the discount rate r (right panel). In the left panel, Σ varies in a range $[\frac{1}{4}; 5\,\hat{\Sigma}]$, and the project value is computed with a risk-adjusted discount rate $r = 1\%$. All scenarios from REMIND 3.0 dataset are included. The shaded gray area represents the 95% confidence interval. The inset plot shows the sensitivity of the average waiting time before divestment τ, with a 95% confidence interval.

3.3.2.2 Sensitivity analysis

A crucial parameter in our model is the discount rate reflecting the riskiness of the investment project. Thus, we perform a sensitivity analysis to assess the extent to which our results are affected by the choice of this parameter. Economic intuition commands that lower discount factors β, and thus higher discount rates r, will lead to underweighting future cash flows, and thus to lower values of the project. Indeed, the right panel of Figure 3.6 shows that the project value declines as r increases in the range $[1\%; 8\%]$. The inset panel in the figure also shows a mild decrease in the optimal stopping time. This figure also shows the sensitivity to another set of parameters, that is the estimated risk-factors' variance-covariance matrix Σ (left panel). Here, we multiply the estimated variance-covariance matrix by a factor that varies in a range $[\frac{1}{4}; 5]$. When risk factors are more volatile, the flexibility given by owning the option is more valuable, and the RO value is higher. The uncertainty related to the risk factors leads to postponing the decision, and thus to a longer optimal waiting time.

To better understand the impact of the interest rate on the optimal stopping time and at the same time study the impact of decommissioning costs or selling revenues on the real option value, we plot in Figure 3.7 the sensitivity of the project value and optimal stopping time to the interest rate in the presence of such costs and revenues. In the presence of a positive decommissioning cost (negative cash flow, left graph), the optimal stopping time is increasing as function of the interest rate: for higher rates the agent prefers to wait longer before paying the decommissioning cost, so that its impact is reduced by discounting. In the presence of a decommissioning premium (right graph), the stopping time is decreasing as function of the interest rate: for higher rates the agent prefers to stop earlier to benefit from the decommissioning premium before its value is reduced by discounting. The case when there are no decommissioning costs and no premium is an intermediate one, when there are both positive and negative cash flows and none of them clearly dominates the other. For this reason, in Figure 3.6, right graph, we observe only mild dependence of the optimal exit time on the discount rate.

This analysis shows how, by providing a decommissioning premium to carbon-intensive power plants, or by helping them reduce the decommissioning costs, the regulator may accelerate the low-carbon transition of the electrical energy industry.

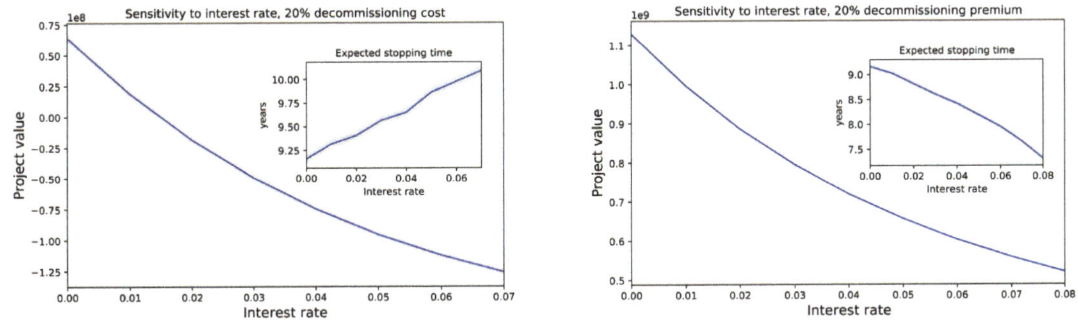

FIGURE 3.7: Sensitivity of the project value (in EUR) to the discount rate. Left decommissioning cost equal to 20% of the capital costs. Right: decommissioning premium (selling revenue) equal to 20% of the capital costs.

3.3.3 Optimal entry in a green investment project

As a second empirical application, we consider an agent who has the option to invest in a combined cycle coal-fired power plant with Carbon Capture and Storage (CCS) technology, and wants to know when it is economically optimal to exercise the RO. The plant is again a price taker, and will supply electricity inelastically. We assume it has a nominal capacity $W = 1000\,\mathrm{MW}$ and is located in Germany. We further assume the plant will have a lifespan of $T = 50$ years, and we assume $t_0 = 2020$. We set the maturity of the RO to be $T_0 = 2050$. Finally, we assume a risk-adjusted discount factor $\beta = e^{-r\Delta t}$, with a reference discount rate of $r = 3\%$. The value of interest rate is higher than in the previous example since the project value is very sensitive to the interest rate in the present case, and the low value of 1% leads to unrealistic results. Sensitivity to the interest rate parameter is analyzed in the following paragraph. The problem the agent needs to solve is now that of Eq. (3.9). Exercising the option entails a stream of revenues and costs starting from the exercise time τ throughout the plant lifetime, T. The strike price of the RO corresponds to the capital cost of building a new power plant, C_C^{gc} (with gc standing for green coal). Namely, the payoff of the RO at time τ is

$$\max\left(0\,;\mathbb{E}\left[\sum_{t=\tau}^{\tau+T}\beta^{t-\tau}h^{gc}(\mathbf{P}_t)\right]-C_C^{gc}(\tau)\right).$$

If the agent exercises the option, and thus builds the power plant, in every period t she will incur some output-dependent operation and maintenance (O&M) costs, both fixed (C_F^{gc}) and proportional to the production output (C_V^{gc}). The production output of the plant depends on the plant's utilization rate R_U^{gc}. The P&L function of year t of the plant, with the proper conversion factors, is thus equal to

$$h_t^{gc} = W\cdot R_{U\,t}^{gc}\cdot 365.25\cdot 24\left(P_{Et}-\frac{P_{Ft}}{R_C^{gc}}-C_V^{gc}\right)-W\cdot C_F^{gc}\,. \tag{3.16}$$

We refer the reader to Table 3.1 for the explanation of the remaining variables in Eq. (3.16), as well as for the variables' units of measure.

The risk factors in this example application, that we model according to (3.1), are now two:

$$\mathbf{P}=\begin{pmatrix}P_E\\P_F\end{pmatrix}.$$

TABLE 3.4: Value of the investment opportunity and optimal exit times in the absence of scenario uncertainty

Scenario	*Below 2°C*	*Current policies*	*Delayed transition*	*Divergent net zero*	*NDCs*	*Net zero 2050*
Project value, MEur	747	0	1810	1454	177	1390
Optimal investment time	3	∞	6.02	2	2.99	1.37

Indeed, since the plant is a green investment, it is not required to comply with an ETS, and thus it does not have any carbon-related costs.[3] As before, the risk factors' mean values $\mu^i_{\cdot,t}$ depend on the scenario i of the chosen scenario dataset, and are given by the corresponding i-th scenario path for that variable. The random parts of the risk factors are modeled as in the previous example, and we use the same estimated parameter values.

In this case, the agent chooses to use as a signal the price of the EUA. To estimate the volatility of such a signal y, we employ the EEX EUA front-month futures detrended monthly log-prices and we fit a Gaussian model. The estimated monthly signal volatility parameter is $\hat{\sigma}_y = 0.317$. We then employ this value to simulate the signal as in Eq. (3.2). For comparison, we also plot the impact of the scenario uncertainty using GHG emissions as the signal (the same one as in the first example).

All variables in Eq. (3.16) except for the risk factors are modeled as deterministic, and they follow a trend depending on the availability of scenario trajectories for each one of them. In this case, the REMIND 3.0 model includes scenario paths for the risk factors and for the capital cost C^b_C. However, the total installed capacity of coal-fired power plants and the electricity production from coal is not available after 2050. For this reason, we use a fixed utilization rate $R^{gc}_U = 85\%$ International Energy Agency (2020). Following the same source, the O&M costs of a coal-fired power plant with CCS are summarized as variable costs of 19.34 USD/MWh. The conversion efficiency is assumed to be $R_C = 48\%$ as in the previous example.

3.3.3.1 Numerical results

First, before analyzing the impact of scenario uncertainty, we compute the value of the project and the optimal investment time in each individual scenario, assumed known by the agent. This information is shown in Table 3.4. In the "Current policies" scenario, due to relatively high coal prices and low electricity prices, it is never optimal to invest in a CCS coal power plant, and the project value is zero. In other scenarios, investment happens relatively quickly, except for the delayed transition scenario, where it is optimal to wait for some time to allow the coal prices to go down.

We now proceed to analyze the impact of scenario uncertainty. Figure 3.8 shows the sensitivity of both the RO value and the optimal exercise time to the value of the volatility of the signal for the entry problem, when all six available scenarios from REMIND 3.0 are included. We let the signal volatility vary in a range between 10% and 300% of the estimated value. The left graph uses the EUA emission allowance price as the signal, and the right graph uses the GHG emissions from energy production in Germany. The inset plots show

[3]In reality, modern CCS technology can only capture about 90% of the emissions, but we neglect the residual emissions in our stylized model since our aim is to illustrate the case of a green investment.

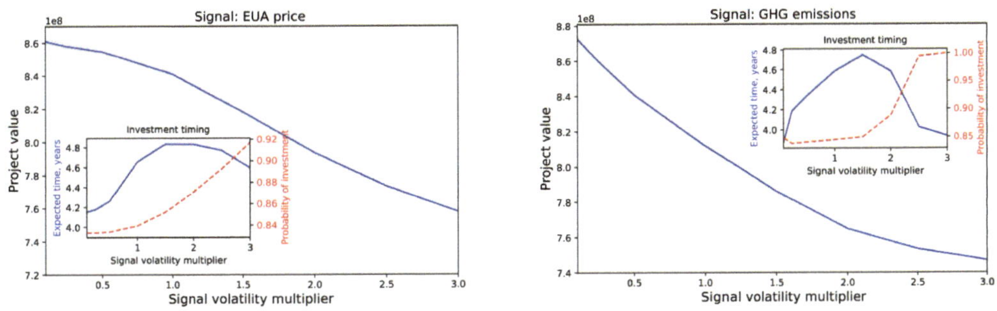

FIGURE 3.8: Optimal entry problem. Sensitivity of the project value (in EUR) to the volatility of the signal, σ_y. In the figure, σ_y varies in a range $[0.1, 3\,\hat{\sigma}_y]$. In the left graph, the EUA emission allowance price is used as the signal. In the right graph, the carbon emissions from energy production in Germany are used. The inset plot shows the sensitivity of the average optimal waiting time before investment, conditional on the investment taking place (solid line, left scale), and the sensitivity of the probability of the investment (dotted line, right scale), to the volatility of the signal σ_y.

the expected stopping time at which the investment takes place, and the probability of the investment. In this problem, on some trajectories it is never optimal to invest, and these trajectories are not taking into account in the computation of the expected stopping time.

Both signals have similar impact on the project value and the investment strategy. In particular, the optimal investment time is first increasing, as the agent needs to wait longer to acquire information, and then starts to decrease, as the signal becomes too noisy and its information value drops. The probability of investment increases with signal volatility, as it becomes more difficult for the agent to precisely detect the only scenario in which investment is not optimal (current policies). Finally, the GHG emissions appear to have a higher information value for the agent than the EUA emission allowance price, as the project value drops faster in the latter case.

To better understand the optimal investment timing in this case, we plot in Figure 3.9 the distribution of optimal investment times according to true scenario, when EUA emission allowance price is used as the signal. For higher volatility values, investment happens earlier in the delayed transition scenario, and in some cases even in the current policies scenario (where it is not optimal to invest under full information), because the low information content of the signal does not allow the agent to distinguish these scenarios from the ones where early investment is optimal.

Figure 3.10 shows the sensitivity of the RO value, of the average optimal investment time conditional on the investment taking place, and of the investment probability, to the volatility of the risk factors Σ (left panel) and the discount rate r (right panel). We let r vary in a range $[1\%; 6\%]$, and we multiply the estimated variance-covariance matrix by a factor that varies in a range $[\frac{1}{4}; 5]$. As in the optimal exit problem, the RO value is increasing in the volatility of the risk factors and decreasing in the discount rate. However, here, the average optimal exercise time is decreasing in the volatility of the risk factors. Thus, the higher the variability relative of the future cash flows, the earlier it is convenient to invest in the green energy project. The probability of investment is sharply decreasing as function of the interest rate: the construction of a green coal power plant requires a large up-front payment, and for high interest rates, none of the NGFS scenarios promize sufficiently high cash flows to compensate this after discounting.

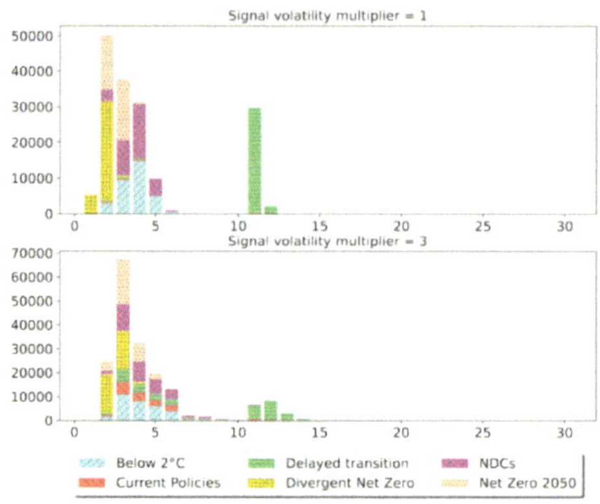

FIGURE 3.9: Optimal investment timing under scenario uncertainty: distribution of the optimal investment times according to the true scenario. Top: $\sigma_y = \hat{\sigma}_y$, bottom: $\sigma_y = 3\hat{\sigma}_y$. EUA emission allowance price is used as the signal.

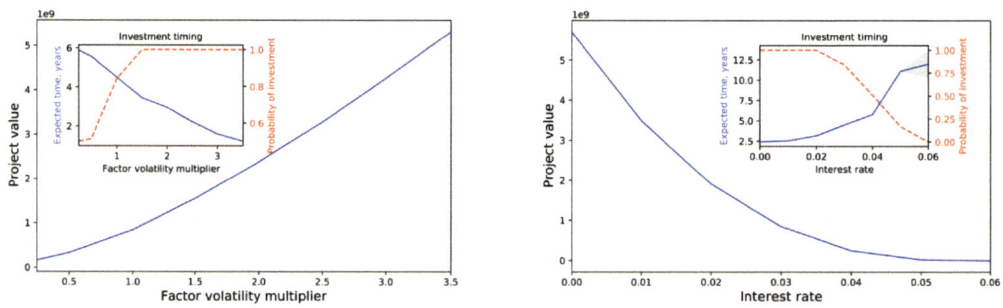

FIGURE 3.10: Optimal entry problem. Sensitivity of the project value (in EUR) to the risk factors volatility, Σ (left panel) and to the discount rate r (right panel). In the left panel, Σ varies in a range $[\frac{1}{4}; 5\,\hat{\Sigma}]$. The inset plot shows the sensitivity of the average optimal waiting time before investment, conditional on the investment taking place (solid line, left scale), and the sensitivity of the probability of the investment (dotted line, right scale). The confidence interval is large for high interest rate because investment only happens on a small number of trajectories.

3.4 Conclusion

In this paper, we present a new strategy for evaluating investment projects, by combining standard real options techniques with a macroeconomic approach for climate transition analysis. In our model, the agent observes a noisy climate-related signal, and forms a belief relative to the likelihood of the possible current macroeconomic climate scenarios. The agent then bases her entry or exit decisions on the posterior probability of being in a certain macroeconomic scenario. The agent's learning about the level of climate risk, via Bayesian updating, then plays an active role in the decision making process.

We showcase the potential of our strategy using public available scenario data from the NGFS scenario database, which are representative of a range of several different environmental transition pathways. Each of them is associated to a certain level of climate policy stringency, from low (existing climate policies remain in place, with no further effort to mitigate climate change), to high (there is a clear commitment to a specific net-zero policy target that results in CO_2 emissions to be at net-zero in 2050).

These scenarios are employed to analyze the value of real options in two energy-related examples: the divestment decision from a coal-fired power plant without Carbon Capture and Storage (CCS) technology and the potential investment into a green coal-fired power plant with CCS. In both cases, we find that the real option value is very sensitive to scenario uncertainty: the value of the coal-fired power plant is reduced by 25% and that of the green coal investment project is reduced by 7% when the agent deduces the scenario by observing carbon emissions, compared to the setting when the true scenario is known. We also find that scenario uncertainty can lead to considerable delays in the implementation of green investment projects. These results underscore the importance of reliable and detailed climate scenarios and indicate that taking into account scenario uncertainty and its progressive resolution is essential for precise valuation of energy projects and for determining the optimal investment timing.

The main limitation of our study is that it is based on a stylized model of power plant operation and uses low-granularity scenario data from the NGFS database: some variables are only available for the entire EU-28 region and the time step is generally 5 years before 2050 and 10 years after this date. Features like higher time granularity and increased geographical diversity would add precision to our results. Finally, for the purpose of energy project valuation, it would be important to have an IAM that includes energy producer-specific variables, such as the wholesale price of electricity at the primary level, rather than at the secondary one (which also includes dispatching costs).

References

Abadie, L. M., M. Chamorro, and M. González-Eguino, 2011, Optimal abandonment of EU coal-fired stations, *The Energy Journal* 32, 175–207.

ACIL, 2014, Fuel and technology cost review – report to Australian energy market operator, Technical report, ACIL.

Bogdanov, D., J. Farfan, K. Sadovskaia, A. Aghahosseini, M. Child, A. Gulagi, A. S. Oyewo, L. Barbosa de Souza Noel Simas, and C. Breyer, 2019, Radical transformation pathway towards sustainable electricity via evolutionary steps, *Nature Communications* 10, 1–16.

Boomsma, T. K., N. Meade, and S.-E. Fleten, 2012, Renewable energy investments under different support schemes: A real options approach, *European Journal of Operational Research* 220, 225–237.

Cronin, J., G. Anandarajah, and O. Dessens, 2018, Climate change impacts on the energy system: a review of trends and gaps, *Climatic Change* 151, 79–93.

Dalby, P. A., G. R. Gillerhaugen, V. Hagspiel, T. Leth-Olsen, and J. J. Thijssen, 2018, Green investment under policy uncertainty and Bayesian learning, *Energy* 161, 1262–1281.

Detemple, J., and Y. Kitapbayev, 2020, The value of green energy: Optimal investment in mutually exclusive projects and operating leverage, *The Review of Financial Studies* 33, 3307–3347.

Flora, M., and T. Vargiolu, 2020, Price dynamics in the European Union Emissions Trading System and evaluation of its ability to boost emission-related investment decisions, *European Journal of Operational Research* 280, 383–394.

Fuss, S., J. Szolgayová, N. Khabarov, and M. Obersteiner, 2012, Renewables and climate change mitigation: Irreversible energy investment under uncertainty and portfolio effects, *Energy Policy* 40, 59–68.

Hach, D., and S. Spinler, 2016, Capacity payment impact on gas-fired generation investments under rising renewable feed-in—a real options analysis, *Energy Economics* 53, 270–280.

International Energy Agency, 2020, Projected costs of generating electricity, Technical report, IEA.

Laurikka, H., and T. Koljonen, 2006, Emissions trading and investment decisions in the power sector—a case study in Finland, *Energy Policy* 34, 1063–1074.

Longstaff, F. A., and E. S. Schwartz, 2001, Valuing American options by simulation: a simple least-squares approach, *The Review of Financial Studies* 14, 113–147.

McGlade, C., and P. Ekins, 2015, The geographical distribution of fossil fuels unused when limiting global warming to 2°C, *Nature* 517, 187–190.

Mercure, J.-F., H. Pollitt, J. E. Viñuales, N. R. Edwards, P. B. Holden, U. Chewpreecha, P. Salas, I. Sognnaes, A. Lam, and F. Knobloch, 2018, Macroeconomic impact of stranded fossil fuel assets, *Nature Climate Change* 8, 588–593.

Metz, B., O. Davidson, P. Bosch, R. Dave, and L. Meyer, eds., 2007, *Contribution of Working Group III to the Fourth Assessment Report of the Intergovernmental Panel on Climate Change* (Cambridge University Press, Cambridge, United Kingdom and New York, NY, USA).

Mo, J., L. Cui, and H. Duan, 2021, Quantifying the implied risk for newly-built coal plant to become stranded asset by carbon pricing, *Energy Economics* 99, 105286.

Rozenberg, J., A. Vogt-Schilb, and S. Hallegatte, 2020, Instrument choice and stranded assets in the transition to clean capital, *Journal of Environmental Economics and Management* 100, 102183.

Siddiqui, A., and S.-E. Fleten, 2010, How to proceed with competing alternative energy technologies: A real options analysis, *Energy Economics* 32, 817–830.

Stanton, M. C. B., S. Dessai, and J. Paavola, 2016, A systematic review of the impacts of climate variability and change on electricity systems in Europe, *Energy* 109, 1148–1159.

Teske, S., 2019, *Achieving the Paris climate agreement goals: global and regional 100% renewable energy scenarios with non-energy GHG pathways for +1.5° C and +2° C* (Springer Nature).

Van der Ploeg, F., and A. Rezai, 2020, The risk of policy tipping and stranded carbon assets, *Journal of Environmental Economics and Management* 100, 102258.

Welsby, D., J. Price, S. Pye, and P. Ekins, 2021, Unextractable fossil fuels in a 1.5°C world, *Nature* 597, 230–234.

Chapter 4

Climate stress testing

Fanny Cartellier
University of Zurich, Switzerland

Climate stress testing and climate scenario analysis have been developed in recent years to shed light on the exposure and vulnerability of the financial system to climate-related risks. This chapter offers a methodological review of the various climate scenario exercises that have been carried out by central banks and designed by scholars. It answers the following question: are existing methodologies well-suited to assess the extent to which climate-related risks may induce financial risks and impair financial stability? We identify blind spots in the coverage of climate-related risks transmission channels to the financial system, such as liquidity risk. We also discuss the extent to which deep uncertainty regarding climate-related risks is taken into account in these methodologies, and stress the importance of exploring its consequences further. We advocate complementing exploratory scenario analyses with the use of properly designed adverse scenarios in order to quantify tail events and prevent extreme losses. The development of these methodologies has been an important first step in raising financial institutions' awareness of climate-related risks, and has fostered the development of tools and best practices to assess them.

4.1 Introduction

Since Mark Carney's famous speech on "Breaking the tragedy of the horizon" (Carney, 2015), awareness that climate change and the transition to a low-carbon economy may be sources of significant financial risk has grown among central banks and an increasing number of financial players (Bolton and Kacperczyk, 2021; Krüger, Sautner, and Starks, 2020; Network of Central Banks and Supervisors for Greening the Financial System, 2018, 2019). The financial risks associated with climate change are generally classified in two main categories (Task Force on Climate-Related Financial Disclosures, 2017). Physical risks encompass the financial risks caused by the physical consequences of climate change, such as the increased frequency and intensity of extreme weather events or sea-level rise, which can damage economic and human capital. Transition risks represent the financial risks caused by the transition to a low-carbon economy, for example through the rise in carbon prices, which may impair the profitability of companies relying on carbon-intensive activities.

More and more central banks and supervisors undertake the assessment of these risks as part of their financial stability mandate (Financial Stability Board and Network of Central Banks and Supervisors for Greening the Financial System, 2022). However, their assessment raises a number of significant methodological challenges (Bolton et al., 2020). Climate change and the low-carbon transition disrupt historical trends. This diminishes the relevance of the statistical use of past observed data to assess these risks, and instead calls

DOI: 10.1201/9781032636252-4

for an approach that is fundamentally forward-looking to measure them. The drivers of these risks and the ways they channel through the economy to transform into financial risks is complex, and new to financial risk analysis. Finally, the extent of these risks is deeply uncertain. This uncertainty holds under two dimensions: scientifically, the extent and consequences of climate change for a given amount of greenhouse gas (GHG) emissions are extremely complex to model, which makes them hard to predict (Intergovernmental Panel on Climate Change, 2021); and fundamentally, there is uncertainty about the future trajectory of GHG emissions, as it will depend on complex (geo)political and societal evolutions (Intergovernmental Panel on Climate Change, 2022).

The difficulty to measure and forecast the extent of these risks leads to information opacity about firms' and financial institutions' exposure to climate-related risks, and casts doubt on the ability of current prices to integrate them in their full extent (Acharya et al., 2023; Campiglio et al., 2023; Daumas, 2023). The possibility of these risks being misjudged by financial players means that they could be a source of financial instability (Bolton et al., 2020; Financial Stability Board, 2020; Litterman et al., 2020; Semieniuk et al., 2021).

Climate stress tests and scenario analyses have been developed as an answer to these measurement challenges. They are scenario-based exercises designed to estimate the impact on financial institutions of various scenarios of climate-related risks' materialization. Inspired by the practice of financial stress testing, they adapt some of its features to the specificity of climate-related risks. For example, scenario horizons can be much longer in climate scenario-based exercises. Moreover, climate or climate-economy models are often included to model the materialization and propagation of climate-related risks in the economy.

After the pioneering scenario-based exercises by Battiston et al. (2017), the Dutch National Bank (Vermeulen et al., 2021) and the French central bank (Autorité de Contrôle Prudentiel et de Résolution (ACPR), 2020), climate stress tests and scenario analyses were undertaken by a number of financial authorities, fostered by the creation of the Network of Central Banks and Supervisors for Greening the Financial System (NGFS), which published a number of reference scenarios (Network of Central Banks and Supervisors for Greening the Financial System, 2020, 2021, 2022b, 2023c). In 2022, a joint synthesis report by the Financial Stability Board (FSB) and the NGFS reported that 35 climate scenario analysis exercises had already been completed by financial authorities throughout the world, and 19 more were in progress (Financial Stability Board and Network of Central Banks and Supervisors for Greening the Financial System, 2022). In parallel, academic researchers studied specific methodological steps of climate stress tests and analyzed the resilience of the financial system to climate-related risks (Battiston et al., 2019; Roncoroni et al., 2021; Reinders, Schoenmaker, and Van Dijk, 2020; Nguyen et al., 2020; Bikakis, 2020; Mandel, 2021; Mandel et al., 2021; Gourdel and Sydow, 2023; Bressan et al., 2024; Le Guenedal, Drobinski, and Tankov, 2022).

This chapter provides an overview of climate stress testing and scenario analysis methodologies developed by central banks and scholars. In particular, it questions the ability of these methodologies to assess the full scope and extent of climate-related risks to the financial system, in order to reliably test the stability of the financial system to those risks. It is organized as follows. Section 4.2 introduces the categories of physical and transition risks, and discusses the channels through which they can affect financial institutions, as well as the threat that they pose to financial stability. Section 4.3 sets out the methodological challenges involved in assessing these risks to the financial system, and introduces scenario-based climate exercises by showing how they are conceptually relevant to answer these challenges. Section 4.4 reviews climate stress testing methodologies developed by central banks and scholars, and discusses their ability to assess the full scope and extent of climate-related risks to the financial system. Section 4.5 concludes.

4.2 Climate-related risks to the financial system

Climate-related risks channel through the economy in a complex way, and can reach financial institutions through various categories of financial risk. In this section, we first define climate-related risks to the financial system and then discuss how they can pose a threat to the financial stability.

4.2.1 Physical and transition risks

Climate change is the source of two main types of new risks for the financial system: physical risks and transition risks (Task Force on Climate-Related Financial Disclosures, 2017; Network of Central Banks and Supervisors for Greening the Financial System, 2019). Physical risks arise from the material consequences of climate change. They can be caused by two types of drivers, which affect the economy at different time scales. Acute physical risks, such as extreme climate events (droughts, floods, wildfires) of which the intensity and frequency are increasing with climate change, can cause sudden shocks to the productivity of human, physical and natural capital in affected areas. This can affect the solvency of debtors (companies, individuals) located in these areas, as well as devalue their assets (housing and companies' productive tangible assets), and trigger liquidity crises through the sudden withdrawal of their deposits (Brei, Mohan, and Strobl, 2019). These are all risks for financial players (banks, investors) holding securities or financial assets exposed to these counterparties. Chronic physical risks, such as rising sea levels and ocean acidification, can have slower-onset effects, such as the gradual devaluation of seaside dwellings or a decline in fishing activities. But they can also cause abrupt changes in agents' anticipations, leading to sudden revaluations of certain assets, for example when some information is revealed about the future exposure of a given territory. The effects of acute and chronic physical risks are already present today, and they are expected to continue to intensify, becoming very significant in the second half of the century (Intergovernmental Panel on Climate Change, 2021).

Transition risks are financial risks caused by the process of moving toward a low-carbon economy, which affect the profitability of carbon-intensive and GHG-intensive economic activities. Hence, they impact companies that are high carbon-emitters or which depend on emission-intensive activities through their insertion in value chains (Cahen-Fourot et al., 2020, 2023; Adenot et al., 2022). There are four main drivers of the low-carbon transition that can cause these risks to emerge (Task Force on Climate-Related Financial Disclosures, 2017): (i) *climate policies* encouraging a shift to a low-carbon economy, including carbon taxes and subsidies for investment in renewable energies for example, (ii) *technological innovations*, which make renewable energies more affordable, improve the energy efficiency of production and transport systems, and allow capturing and storing carbon, (iii) *changes in market sentiment*, which reflects investors' anticipations and preferences regarding the course of the low-carbon transition, and which influences their investment decisions, and (iv) *changes in consumer preferences*, which determine the relative demand for products relying on carbon-intensive activities. Some transition risks are unpredictable and can materialize at short, mid-term and long-term horizons: the sudden implementation of stringent climate policies could trigger short-term transition risks. However transition risks are likely to materialize until 2050, as this is the horizon at which the net-zero target should be reached to meet the Paris agreement goals (Network of Central Banks and Supervisors for Greening the Financial System, 2020, 2023c).

TABLE 4.1: Transmission channels of physical risks to financial institutions. Source: Basel Committee (2021) and additions by the author

	Microeconomic channels and direct impacts	*Macroeconomic channels and indirect impacts*
Credit risk	Reduction in debtors' wealth, revenue or loan collateral (households, corporates, public institutions)	Decline in labor productivity, output reduction; increase in sovereign spreads or defaults; downgrade of (local) public institutions in vulnerable areas
Market risk	Uncertainty about physical risk materialization: mispricing, increased volatility, increased difficulty of hedging	Same as microeconomic channels, which can spread through value chains and macroeconomic variables
Liquidity risk	Increase in funding cost due to exposure to physical risk; increased demand for liquidity following a natural disaster	Rush on liquidity following important natural disasters
Operational risk	Disruption of transportation facilities, telecommunications or market infrastructures; damages to tangible fixed assets	Same as microeconomic channels, which can spread through value chains and macroeconomic variables
Underwriting risk (for insurers)	Higher than expected increase in claims on damaged insured assets (for non-life insurers), higher than expected mortality or morbidity rates	No specific macroeconomic channel identified
Underwriting risk (for financial institutions)	Decrease in insurance coverage leading to increased financial institutions' exposure to physical risks	No specific macroeconomic channel identified

Climate-related risks can have an impact on financial institutions through roughly all standard Basel financial risk categories (Basel Committee, 2021): credit, market, liquidity, operational and underwriting risks. The way these risk categories can be affected by physical and transition risks is summarized in Tables 4.1 and 4.2, respectively.[1]

4.2.2 Climate-related risks and financial stability

Concerns have been raised both among central banks and in the academic literature about the possibility that physical and transition risks could be a threat to financial stability (Carney, 2015; Network of Central Banks and Supervisors for Greening the Financial

[1] The operational risk is reported here to raise awareness about its existence. However, it is not considered by any climate scenario-based exercise mentioned in this paper, as it requires an assessment framework very different from the other types of risk channels. Therefore, it will not be considered further in this chapter.

TABLE 4.2: Transmission channels of transition risks to financial institutions. Source: Basel Committee (2021) and additions by the author

	Microeconomic channels	*Macroeconomic channels*
Credit risk	Decrease in creditworthiness of corporates vulnerable to transition risk; higher litigation costs for corporates and governments	Changes in production value chains structure leading to output reduction (decrease in creditworthiness of households, corporates and governments)
Market risk	Uncertainty about corporates exposure to transition risk: mispricing, risk of abrupt repricing events (ex: stranded assets), increased volatility	The spread of microeconomic market risk channels through value chains and macroeconomic variables
Liquidity risk (asset liquidity)	Assets becoming illiquid (ex: stranded assets)	Spillover effects between companies of same sectors or through value chains
Liquidity risk (funding cost)	Increase in funding cost due to exposure to transition risk or uncertainty related to this exposure	Cascading increases in funding costs due to dependence on carbon intensive activities in upstream value chains

System, 2019; Alogoskoufis et al., 2021a; Emambakhsh et al., 2022; Bolton et al., 2020; Semieniuk et al., 2021; Campiglio et al., 2023; Daumas, 2023). In the context of climate change, there is a risk of widespread mispricing in financial markets due to the radical uncertainty of climate-related risks (Financial Stability Board, 2020; Litterman et al., 2020). The threat of abrupt repricing events related to changes in expectations of investors should therefore be taken seriously. Such events may be triggered by election outcomes, technological breakthroughs, new research findings about the speed and nature of physical climate impacts, or the occurrence of major natural catastrophes (Litterman et al., 2020).

These phenomena may have an impact on the economy through feedback loops between the financial system and the real economy (Financial Stability Board, 2020). For example, there may be a widespread increase in risk premia due to the opacity of exposure to climate-related risks (Financial Stability Board, 2020), or banks may have to restore their solvency after a huge climate physical shock, which would lead to a reduction in lending, decreasing banks' support to the real economy. Moreover, the increasing probability and intensity of weather-related natural catastrophes could lead to a reduction in insurance coverage of these events, transferring this risk to economic agents. Climate-related risks can also lead to a self-reinforcing deterioration of prospects for sovereigns and their national financial institutions through the sovereign-bank nexus (Battiston et al., 2019).

4.3 Climate stress testing and scenario analysis

Assessing climate-related risks to the financial system comes with a significant number of challenges. Climate stress testing and climate scenario analysis are scenario-based methodologies that were developed in the last years to address these measurement issues, inspired by financial stress testing.

This section presents the methodological challenges raised by this assessment. It then outlines the principles of financial stress testing. Finally, it introduces climate stress testing and climate scenario analysis, with an emphasis on the way they address the specificities of climate-related risks.

4.3.1 Challenges raised by the assessment of climate-related risks

Climate-related risks raise a number of challenges when it comes to their assessment (Bolton et al., 2020).

Disruption of historical trends First, they induce a disruption in many historical trends. Climate change leads to long-term and even irreversible changes in environmental and meteorological conditions around the world, with a nonlinear evolution and various tipping points. Climate change mitigation requires a structural change of the economy, with a shift from high-carbon to low-carbon activities. These two aspects cause major disruptions in historical trends, reducing the relevance of past data to forecast future outcomes. Instead, it calls for the use of a fundamental forward-looking approach, based on sophisticated modeling frameworks.

Deep uncertainty Second, the extent of climate-related risks and the way they will affect the economy are deeply uncertain. They are both subject to scientific and political uncertainty (Intergovernmental Panel on Climate Change, 2021, 2022). This hinders the estimation of probability distributions of climate-related risks, as they are conditional on many uncertain events. However, these risks can be extremely severe, and it is highly likely that some combination of these risks will materialize in the future.

Long-term horizons Third, a significant part of these risks will materialize and will be most severe at a long-term horizon. Assessing them therefore requires significant projection efforts.

Complex and nonlinear channels Finally, climate-related risks are channeled to the economy through complex and non-linear mechanisms. They affect all agents of the economy, and all types of financial assets. For example, both physical and transition risks can affect companies that are not directly exposed to them through value chains (Cahen-Fourot et al., 2020).

4.3.2 Financial stress testing

These features of climate-related risks make many of the current risk measurement methods unsuitable for assessing these risks. For example, many such methods are backward-looking, based on probabilistic estimations and historical data (Institute for Climate Economics, 2019). This leads Bolton et al. (2020) to call for an "epistemological break"[2] with respect to financial risk management. Among the available risk assessment methods, stress testing distinguishes itself by its scenario-based approach, which does not attribute probabilities to narratives of risk materialization. This feature makes stress testing compatible with the deep uncertainty of climate-related risks. Moreover, stress testing has the advantage to be a very flexible tool (Borio, Drehmann, and Tsatsaronis, 2014).

[2]This concept comes from Bachelard Bachelard (1938).

Financial stress testing is a scenario-based exercise in which the resilience of a single financial institution (microprudential) or of a financial system (macroprudential) to a number of adverse scenarios is tested. The financial outcomes under a stress scenario are compared to those obtained under a baseline scenario, usually following "business-as-usual" assumptions. Methodologies to build stress scenarios vary across the type of risk that is tested, but can, for example, include macrofinancial modeling in order to account for macrofinancial feedback in macroprudential stress tests (Aikman et al., 2023). The horizon of scenarios in financial stress tests typically ranges from very short-term to business cycles of three to five years. While financial stress testing is mostly used as a regulatory tool for banks and insurers, it can also include other types of financial institutions in macroprudential system-wide exercises, such as investment funds (Aikman et al., 2023).

Financial stress testing has been widely used in recent years to assess the resilience of financial institutions to various types of economic and financial risks. It became an essential tool of both microprudential and macroprudential regulatory toolkits in the wake of the Global Financial Crisis (GFC, 2007-2009) (Farmer et al., 2022, Chapter 3). It can be noted that macroprudential system-wide exercises were mostly developed after the GFC (Aikman et al., 2023). Financial stress testing is used by financial supervisors to test the capital buffers of banks and insurers, and to formulate individual capital requirements for the largest of these financial institutions, as they are key to global financial stability (Bank for International Settlements, 2017; Baudino et al., 2018; European Insurance and Occupational Pensions Authority, 2019).

4.3.3 Climate stress testing and scenario analysis

Climate scenario-based exercises, an expression that will be used to designate both climate stress tests and climate scenario analyses, have been developed to assess the financial impacts of climate-related risks and the resilience of financial institutions to their materialization. The main challenge in the development of these methodologies has been to adapt financial stress testing methodologies to the specific characteristics of climate-related risks.

The recent history of the development of climate scenario-based exercises can be summarized as follows. The beginnings of climate stress testing are rooted in the practice of stress testing for systemic risk, which developed in the wake of the GFC.[3] Drawing on this practice, Battiston et al. (2017) offered the first example of climate stress testing. Their paper assesses the impact of various climate policy scenarios on banks' balance sheets in the Euro area, accounting for second-round effects in financial networks. A number of other academic publications contributed to the research on climate stress testing methodology. A few examples are Mandel et al. (2021), who assess the impact of flood risk on aggregate financial sectors of many countries, emphasizing the role of international exposures as a risk transmission channel, and Jung, Engle, and Berner (2021), who develop a market-based stress testing methodology, by assessing the sensitivity of bank equity values to an aggregate measure of climate risk.[4]

In parallel, financial authorities quickly jumped on the bandwagon, and the adoption of this tool to assess climate-related risks to the financial system soon became widespread

[3]See Battiston and Caldarelli (2013); Acemoglu, Ozdaglar, and Tahbaz-Salehi (2015); Hurd et al. (2016) for examples of systemic risk stress tests, and Vodenska et al. (2021) for a comprehensive overview of this practice.

[4]A more exhaustive list would, to the best of my knowledge, be the following: Roncoroni et al. (2021); Battiston et al. (2019); Bressan et al. (2024); Gourdel and Sydow (2023); Nguyen et al. (2020); Mandel (2021); Le Guenedal, Drobinski, and Tankov (2022); Desnos et al. (2023); Reinders, Schoenmaker, and Van Dijk (2020); Baer et al. (2022); Bikakis (2020); Grippa, Mann et al. (2020); Caloia and Jansen (2021); Mandel et al. (2021); Jung, Engle, and Berner (2021).

among supervisors. A pioneering top-down climate stress testing exercise was carried out by the Dutch National Bank (DNB) in 2019, assessing the impact of adverse transition scenarios on 80 Dutch financial institutions (Vermeulen et al., 2021). In particular, it took into account macroeconomic feedback and industry-specific consequences of these shocks, within a 5-year scenario horizon.

The creation of the Network of Central Banks and Supervisors for Greening the Financial System (NGFS) gave a strong boost to the adoption of climate stress testing methodologies by central banks. It was created at the Paris "One Planet Summit" in 2017, with the stated aim to develop the analysis and management of climate-related risks (Network of Central Banks and Supervisors for Greening the Financial System, 2018). It developed reference scenarios for the use of central banks and the industry. They were built thanks to a suite of climate-economy models (Integrated Assessment Model, IAM), and with long scenario horizons of 30 years (Network of Central Banks and Supervisors for Greening the Financial System, 2020, 2021, 2022b, 2023c). A number of central banks quickly followed suit by conducting their own climate scenario-based exercises, mostly relying on the NGFS scenarios (Financial Stability Board and Network of Central Banks and Supervisors for Greening the Financial System, 2022). Among them, the French central bank (Autorité de Contrôle Prudentiel et de Résolution (ACPR), 2020; Allen et al., 2020), the Bank of England (Bank of England, 2019) and the European Central Bank (ECB) (European Central Bank, 2022) can be cited as other inspiring precursors.

Climate scenario-based exercises have adapted the following features of financial stress testing to the specificities of climate-related risks.

- The scenario horizon can be very long, up to 30 years for transition risks (Allen et al., 2020; Network of Central Banks and Supervisors for Greening the Financial System, 2020, 2023c) and even 60 years for physical risks (Mandel et al., 2021). However, there exist also exercises with short-term horizon, to account for the sudden adoption of stringent climate policies (Vermeulen et al., 2021; Reinders, Schoenmaker, and Van Dijk, 2020; European Central Bank, 2022; Network of Central Banks and Supervisors for Greening the Financial System, 2023b) and the advent of extreme climate events (Caloia and Jansen, 2021; Gourdel and Sydow, 2023).

- The scenario modeling can include climate and climate-economy models to represent physical and transition risks to the economy. Climate modeling allows projecting the amplitude of specific physical risk drivers under various temperature scenarios (Mandel, 2021; Le Guenedal, Drobinski, and Tankov, 2022). Additionally, special effort must be made to understand how their materialization can impact the valuation of financial assets. For transition risks, Integrated Assessment Models (IAMs) are used to estimate carbon tax trajectories that would be compatible with the Paris agreement temperature target of "well below" 2°C under various technological assumptions (Allen et al., 2020; Network of Central Banks and Supervisors for Greening the Financial System, 2020, 2021, 2022b, 2023c). Given their overall complexity, IAMs cannot afford a high-level sectoral granularity and macroeconomic complexity. Therefore, in the French central bank exercise and in the NGFS scenarios, they are complemented by macroeconomic models and input-output sectoral models to understand, for example, the detailed impact of the implementation of a carbon tax in the various sectors of the economy and through value chains.

- The geographical and sectoral granularity of scenario output variables can be very high, as climate-related risks can affect companies very differently, even when they are in the same country, the same region, or the same industry. The exposure of financial institutions' counterparties to physical risks mostly depends on the location of their physical assets and on their dependence to other exposed companies through their insertion in value chains. Having data with enough granularity and precision about physical assets' locations can be very challenging, but is essential for a good assessment of companies' exposure to physical risks (Bressan et al., 2024). The exposure and vulnerability to transition risks vary both across industries, which are more or less carbon-intensive due to their activities, but also between companies within the same industry (Battiston et al., 2017), based primarily on their carbon emissions of scopes 1, 2, and 3,[5] but also on their size, financial health, the diversity of their activities, and whether they have a forward-looking strategy for carbon emissions abatement (Monnin, 2018; Baer et al., 2022). This explains the choice, in some climate scenario analyses, to derive transition scenario output variables at the company level (Allen et al., 2020; De Guindos, 2021). However, having data for all these variables is very challenging too (Nguyen et al., 2020; Baer et al., 2022).

- The choice of a baseline scenario for climate-scenario based exercises is not as straightforward as it is in other financial stress tests, as it is certain that some combination of climate-related risks will materialize in the future (Allen et al., 2020). Hence, a baseline "business-as-usual" scenario that would include neither physical nor transition risk is unrealistic. Different choices of baseline scenarios have therefore been made. The NGFS, along with the French central bank exercise of 2020 (Autorité de Contrôle Prudentiel et de Résolution (ACPR), 2020), encourages the use of the "orderly transition" scenario as a baseline scenario, arguing that it is the ideal scenario in the sense that it minimizes climate-related risks, and that it should be a collective target, in line with the Paris agreement. Other exercises have made the choice to build an artificial "business-as-usual" scenario without any climate-related risks as a reference. The Bank of England did this in its long-term exercise (Bank of England, 2021), and this is also more often used in short-term exercises, under the reasonable assumption that most climate-related risks will not (significantly) materialize within a short timespan (Vermeulen et al., 2021; Jung, Engle, and Berner, 2021; Battiston et al., 2017; Nguyen et al., 2020).

Terminology Two terminologies have been used to designate these exercises: climate stress testing and climate scenario analyses. However, no clear difference exists between the two, and they are sometimes used interchangeably (Acharya et al., 2023; Carè, 2023). It seems that the term "climate scenario analysis" has more frequently been employed by supervisors using the long-term NGFS scenarios in their exercises (Financial Stability Board and Network of Central Banks and Supervisors for Greening the Financial System, 2022), while "climate stress testing" is more often used by scholars or for exercises including short-term stress scenarios (Battiston et al., 2017; Vermeulen et al., 2021). An attempt to draw a line between these two types of exercises could be made by looking at their respective objectives. Climate scenario analyses have exploratory objectives, aiming at better understanding climate-related risks and the way that they are channeled through the economy to financial institutions, along with the objective to raise awareness about climate-related risks and foster their management. As for climate stress tests, they more specifically aim at assessing the resilience of financial institutions to tail climate-related risks. Though it is interesting to have these nuances in mind, the second type of objective was also included in most of central banks' scenario analysis objectives (Financial Stability Board and Network

[5]Scope 1 emissions refer to the direct carbon emissions of a company, arising from its production process. Scope 2 emissions represent emissions indirectly related to its production process, through energy consumption for example. Scope 3 emissions include all other indirect emissions that are necessary for the company's production upward and downward its value chain, including the life cycle of the company's output.

of Central Banks and Supervisors for Greening the Financial System, 2022). Moreover, as these scenario analyses are undertaken by financial supervisors in the frame of their financial stability mandates, they also deserve to be included in this discussion about their ability to test for the resilience of the financial system to climate-related risks.[6]

4.4 Overview of methodologies

A number of climate scenario-based exercises were developed in recent years, answering various objectives. Many financial authorities conducted climate scenario analyses as part of their supervisory framework, motivated by their financial stability mandate. One of their primary aims was to understand and assess climate-related risks to the financial system, and to identify financial system vulnerabilities to those risks, as surveyed in a joint report by the FSB and the NGFS (Financial Stability Board and Network of Central Banks and Supervisors for Greening the Financial System, 2022). However, the understanding and management of these risks are still at their infancy, preventing these scenario analyses to capture them properly at this stage. In particular, major data gaps have to be overcome (Network of Central Banks and Supervisors for Greening the Financial System, 2022a). Hence, these first exercises were also conducted with the purpose to raise awareness about these risks and improve capabilities for their assessment and management. For the same reason, no climate scenario analysis mentioned by the synthesis report of the FSB and the NGFS (Financial Stability Board and Network of Central Banks and Supervisors for Greening the Financial System, 2022) has led to any microprudential or macroprudential answer at this stage, such as a change in capital requirements.

Academic researchers have also developed various top-down climate stress testing methodologies, generally contributing to one or two key steps of the exercise. For example, Nguyen et al. (2020) assess the impact of transition risks on U.S. banks' portfolios of syndicated loans, based on machine learning predictions of corporate carbon footprints, to compensate for missing data associated to each loan. Le Guenedal, Drobinski, and Tankov (2022) develop a statistical methodology to project future cyclone tracks under various climate change scenarios and estimate their financial impact.

This section provides a representative overview, if not exhaustive, of the methodologies that have been developed both by scholars and financial supervisors. In particular, it discusses their ability to provide an in-depth assessment of climate-related risks to financial stability. As physical and transition risks raise different challenges in the construction of scenario-based exercises, this section starts with methodologies for transition risks, and follows with methodologies for physical risks. Before moving to these two categories, a few common patterns of these exercises can be outlined and commented.

Scope Scenario-based exercises mostly focus on assessing banks' exposure and vulnerability to climate-related risks. The report by the FSB and the NGFS (Financial Stability Board and Network of Central Banks and Supervisors for Greening the Financial System, 2022) mentions that one third of the surveyed exercises, conducted by financial authorities, also involved insurers, and that a few of them also included investment funds, such as the joint system-wide exercise by the ECB and the European Systemic Risk Board (ESRB) in

[6]In its fourth scenario release, the NGFS publishes the results of a survey that shows that a significant percentage of users of the NGFS scenarios use them "To better understand how climate risk could impact financial stability" (Network of Central Banks and Supervisors for Greening the Financial System, 2023c).

2022 (European Central Bank and European Systemic Risk Board, 2022). A few academic papers also include other financial institutions in their assessment, such as Mandel et al. (2021) who look at aggregate national financial exposures, and Roncoroni et al. (2021) and Gourdel and Sydow (2023) who explicitly develop a framework involving banks and/or investment funds to understand the interplay between these two types of institutions and the transmission of climate-related risks across financial networks.

Coverage of financial risk transmission channels Climate scenario-based exercises, both conducted by financial authorities and developed by academic researchers, have mostly provided an assessment of credit and market risk (Financial Stability Board and Network of Central Banks and Supervisors for Greening the Financial System, 2022; Acharya et al., 2023). Liquidity risk, instead, is almost never modeled in these exercises. Notable exceptions are Gourdel and Sydow (2023) and Roncoroni et al. (2019) who model shocks in asset liquidity due to transition risks, for example caused by fire sales. Their results show that the asset liquidity risk could be significant. Moreover, natural catastrophes could have a significant impact on banks' funding liquidity due to sudden deposit withdrawal (Brei, Mohan, and Strobl, 2019), calling for an assessment of this risk in scenario-based exercises on physical risks.

Second-round effects A few climate scenario-based exercises take into account second-round effects in financial networks (Battiston et al., 2017; Roncoroni et al., 2021; Gourdel and Sydow, 2023; Bressan et al., 2024; Daumas, 2024; European Central Bank and European Systemic Risk Board, 2022). Their results show that second-round effects can significantly increase the impact of climate-related risks on financial institutions, and hence that these effects should be included in climate stress tests to have a complete picture of climate-related risks to the financial system.

4.4.1 Methodologies for transition risks

Scenario-based methodologies for transition risks can be divided into two main frameworks. The "long-term framework" comprises exercises conducted by financial authorities with a long-term horizon. These exercises, mostly based on the NGFS scenarios with a 2050 horizon, are generated using a sophisticated modeling framework (Network of Central Banks and Supervisors for Greening the Financial System, 2020, 2021, 2022b, 2023c). The "short-term framework" covers more heterogeneous exercises, with scenarios up to a 5-year horizon, that are developed both by financial authorities and scholars. They generally have a lighter modeling framework, sometimes without macroeconomic feedback.

The main characteristics and challenges of long-term and short-term frameworks for transition risks can be described as follows.

- The long-term framework has been developed based on the long-term scenarios provided by the NGFS. The NGFS scenarios, ranging from a number of five to seven in each release, are defined across four categories of narratives of climate-related risks materialization with a 2050 horizon (Network of Central Banks and Supervisors for Greening the Financial System, 2020, 2021, 2022b, 2023c). Scenarios with strong climate change mitigation can be either "orderly" or "disorderly," representing the level of transition risks caused by a low-carbon transition achieving Paris agreement mitigation goals. Scenarios with low climate change mitigation are either "hot-house world" scenarios in which there is little transition risk, or the "too-little, too-late" scenarios in which high transition and physical risks coexist. Most of the NGFS scenarios that have been released so far belong to the two categories of scenarios with successful climate change mitigation. More "hot-house world" scenarios have been developed in the latest releases, and only one "too little, too late" scenario was introduced for the first time in the last release (Network of Central Banks and Supervisors for Greening the Financial System, 2023c). Each scenario is aligned with a certain temperature target reached at the end of the century. Transition scenarios are generated using an integrated assessment model under various assumptions about climate policy and technological change (Network of Central Banks and Supervisors for Greening the Financial System, 2020). Output variables describing the scenario are calculated every five years, covering the modeled period. Many central bank exercises, taking as inputs the transition NGFS scenarios, also added a sectoral input-output model to increase the sectoral granularity and account for risk transmission across value chains, such as the French central bank's exercise (Allen et al., 2020; Financial Stability Board and Network of Central Banks and Supervisors for Greening the Financial System, 2022).

- The short-term framework comprises exercises with short scenario horizons, up to 5 years. They rarely, or only very approximately, correspond to any temperature goal, as they cover a much shorter period of the low-carbon transition. This frees the scenario design from the constraint to use a climate-economy model, which is a heavy and complex modeling tool, and allows for very flexible scenario design. Scenario narratives are often simpler than in the long-term framework, based on the description of one or two shocks (Reinders, Schoenmaker, and Van Dijk, 2020; Gourdel and Sydow, 2023), with sometimes the use of macroeconomic modeling to derive macroeconomic feedback (Vermeulen et al., 2021; European Central Bank, 2022). A few exercises also derive short-term shocks from existing databases of long-term transition scenarios (Battiston et al., 2017; Roncoroni et al., 2021; Nguyen et al., 2020).

Scenario adversity The long-term modeling framework based on NGFS scenarios tends to smooth out the impact of transition risks, as business cycles dynamics are not modeled by the IAMs that are employed (Hourcade et al., 2021; Jacquetin, 2021). Hence, disorderly transition scenarios do not include any economic recession, but instead translate into a lower growth compared to the orderly scenario (Autorité de Contrôle Prudentiel et de Résolution, 2021). The impact of sudden and abrupt climate policies is therefore minimized. Long-term scenarios should rather be interpreted as exploratory scenarios (Allen et al., 2020; Bank of England, 2021). To better represent the adversity of transition scenarios, the long-term exploratory framework therefore needs to be complemented by short-term, more adverse scenarios. The interest of this approach has recently been acknowledged by the NGFS, which intends to add adverse short-term scenarios to its scenario database (Network of Central Banks and Supervisors for Greening the Financial System, 2023b). For now, some central banks have complemented their use of NGFS scenarios with short-term in-house stress scenarios (European Central Bank, 2022; Financial Stability Board and Network of Central Banks and Supervisors for Greening the Financial System, 2022).

Harmonization vs. flexibility One of the goals of the NGFS reference scenario database is to ensure comparability of scenario analyses across countries. In practice, the methodologies relying on NGFS scenarios are indeed quite harmonized, though some differences exist across countries (Financial Stability Board and Network of Central Banks and Supervisors for Greening the Financial System, 2022). However, relying on a small set of exploratory scenarios has a number of drawbacks. Since the development of these scenarios requires a heavy modeling framework, there is an important path-dependency in their design, as illustrated by the development of the NGFS scenarios, where each new release builds upon the previous phase. These scenarios are not necessarily well suited to specific features of each country, requiring some adjustments, which reduce the overall comparability of the results. Moreover, having few very prominent scenarios creates the risk of reducing the field of vision of financial players. In contrast, the flexibility of short-term frameworks allows to be creative about future outcomes and react to new challenges.

Dynamic balance sheets Due to the 30-year horizon of its scenarios, the long-term framework requires describing the evolution of balance sheets during the modeled period. Most of the exercises covered by the FSB and NGFS report (Financial Stability Board and Network of Central Banks and Supervisors for Greening the Financial System, 2022) have used static balance sheets, both due to technical and data limitations preventing the projection of balance sheet evolutions, but also as a more conservative assumption. Allowing for portfolio reallocations sounds more realistic, but can also play down the financial losses in each scenario, as financial institutions can mitigate their exposure accordingly. As most of these scenario analyses involve a large number of financial institutions in a given country, dynamic balance sheets also raise the risk of incoherence between banks' sectoral portfolio reallocation and the scenario's projection of the economy. Hence, when implementing the dynamic balance sheet assumption, the French central bank exercise included an a posteriori adjustment of the banks' portfolio reallocation projections to each scenario projections of sectoral outputs (Autorité de Contrôle Prudentiel et de Résolution, 2021).

Number of scenarios Most scenario-based exercises use a relatively small number of scenarios corresponding to different narratives. However, a few climate stress tests have been conducted with a higher number of scenarios. Daumas (2024) has made the deliberate choice to use the full range of scenarios available in the NGFS scenario database, corresponding to three sets of six scenario narratives generated by three different IAMs (Network of Central Banks and Supervisors for Greening the Financial System, 2021, 2022b). This allows emphasizing the model-dependence of the scenario design, and showing how climate stress test outcomes can vary across models for a given narrative. A few climate stress tests designed by academic researchers use an even larger scenario database, comprising a large number of scenario narratives, with scenarios being generated by alternative models (Battiston et al., 2017; Roncoroni et al., 2021; Nguyen et al., 2020; Desnos et al., 2023). The long-term scenarios are used to build plausible transition shocks, which are then pooled together to mimic a representative distribution of transition shocks. This allows deriving risk measures such as value at risk. This approach relies on the ability to overcome the deep uncertainty of transition risks, as it involves attributing probabilities to shock scenarios. However, the derived distributions remain subject to the assumptions and biases underlying the generation of the scenario database.

Coverage of transition risk drivers The four drivers of the low-carbon transition, identified in Section 4.2, are unequally covered by scenario-based exercises for transition risks. The most represented driver is the climate policy driver, through the raise in the carbon price or carbon taxes, that is mentioned in about all scenario narratives. This driver is often accompanied by assumptions about technological change, especially for long-term scenarios, as done in the NGFS scenarios (Network of Central Banks and Supervisors for Greening the Financial System, 2023c). Very few narratives include changes in market sentiment. A notable exception is Gourdel and Sydow (2023), who account for the risk of fire sales due to abrupt changes in climate anticipations of investment funds. As for changes in consumer sentiment, this driver is rarely discussed or included in the main assumptions of a transition scenario narrative. However, it seems to be included in some long-term transition scenarios, which make assumptions about consumption in their macroeconomic modeling framework (Financial Stability Board and Network of Central Banks and Supervisors for Greening the Financial System, 2022).

4.4.2 Methodologies for physical risks

Climate scenario-based exercises that include physical risks arrived after the development of the first climate stress tests for transition risks, facing some specific difficulties related to the modeling of physical risks. A major difficulty related to physical risks is the need of climate modeling to project the extent of acute and chronic risks in different scenarios of temperature increase. This requires the cooperation of the financial community with climate scientists, which represents an obstacle (Bouchet, Dayan, and Contoux, 2022). Moreover, current knowledge in climate science does not allow downscaling locally the precise consequences of climate change for every type of acute and chronic risks (Fiedler et al., 2021). For example, projecting the hazards related to climate change is easier for temperature variations or precipitation levels than for cyclones, as these phenomena are better understood by climate scientists (Pinchot et al., 2021).

Therefore, climate stress tests for physical risks that were conducted by scholars either rely on hypothetical scenarios that do not relate the shocks to a given level of climate change (Caloia and Jansen, 2021; Gourdel and Sydow, 2023), or focus on a specific kind of hazard that is particularly relevant and for which climate models provide usable projections such as flood risk (Mandel et al., 2021; Mandel, 2021; Bikakis, 2020). A notable exception is the efforts made by Le Guenedal, Drobinski, and Tankov (2022) to develop their own cyclone generation algorithm to quantify the financial implications of tropical cyclone-related physical risks implied by climate change, leading to a publication in a geoscience journal. Another challenge is to trace the exact location of companies' physical assets, and the share of companies' revenues that they generate, to properly derive the financial implications of climate hazards, as emphasized by the physical risk stress test conducted by Bressan et al. (2024) on Mexican financial institutions.

Facing these modeling difficulties, central banks have adopted various strategies. For its climate pilot exercise, the French central bank collaborated with the Caisse Centrale de Réassurance, using projections of the French weather forecaster, Météo France, to assess the property damages caused by natural disasters. In its initial scenario releases, the NGFS only covered chronic physical risks (Network of Central Banks and Supervisors for Greening the Financial System, 2020, 2021). Acute physical risks were included in the later releases (Network of Central Banks and Supervisors for Greening the Financial System, 2022b, 2023c).

Due to these technical obstacles, scenario-based exercises for physical risks are still far from providing a mature picture of how physical risks can affect financial institutions. However, interesting exercises have already been released, allowing us to draw a first picture of the specific methodological challenges that are to be addressed, and to outline a range of solutions to them.

Time horizon Physical risks will become increasingly significant throughout the 21st century (Intergovernmental Panel on Climate Change, 2021). This makes the choice of the scenario horizon particularly tricky. The French central bank makes the choice to limit its climate projections at 2050. This leads to a lower-bound estimation of physical risks that financial institutions will face throughout the century. The Bank of England makes a different choice (Bank of England, 2021): climate outcomes that are expected between 2050 and 2080 in the absence of further mitigation actions are assumed to instead take place between 2020 and 2050. This amounts to assessing what would happen for financial institutions if they did not change their balance sheet exposure to physical risks until 2050.

Mandel et al. (2021) assess the impact of physical risks projected at the years 2030, 2050 and 2080 under different temperature scenarios on current balance sheets.

Balance sheet assumptions It is rare to see dynamic balance sheet assumptions in scenario-based exercises assessing the impact of physical risks on financial institutions. Though it seems unrealistic to believe that balance sheets' exposure and vulnerability to physical risks would remain the same in 2050 or 2080 as today, the tools to understand how to project their evolution with respect to physical risks are lacking. Moreover, assessing directly the impact of physical risks at such long-term horizons on current balance sheets still provides a useful transparent upper bound for the estimation of these risks.

Assumptions about adaptation In the same way as assumptions about mitigation are critical when assessing the impact of transition risks on financial institutions, assumptions about adaptation are a key element when assessing the impact of physical risks on financial institutions. Adaptation efforts can reduce both the exposure and the vulnerability of financial institutions' counterparties to physical risks. However, it is rare to see explicit assumptions about adaptation in scenario-based exercises for physical risks. A noticeable exception is Mandel et al. (2021) who mention assumptions about adaptation efforts: scenarios either assume that investments are made so that protection levels of infrastructures remain constant, or that such investments are not made, in which case protection levels deteriorate in time. This lack of attention to adaptation efforts can be directly related to the lack of interest and attention for adaptation investments, which are a crucial aspect of the fight against climate change. For example, worst-case scenarios could use the hypothesis that infrastructures deteriorate in time. Exploratory scenarios with various adaptation assumptions could be made to assess the trade-off between adaptation costs and physical risks in terms of financial stability, as the European central bank has done to assess the trade-off between mitigation costs and no climate action by balancing transition risks and physical risks (Alogoskoufis et al., 2021b).

Second-round effects In addition to second-round effects related to financial contagion that are also relevant for transition risks, physical risks can lead to a quite specific second-round effect, that is from insurance companies to banks. Physical risks can lead insurance companies to lower their coverage or raise their premia. Banks are exposed to these underwriting risks, as they finance companies that are insured against climate-change-related hazards. As for financial contagion, Mandel et al. (2021) show the importance of taking into account international inter-exposures of financial institutions. The financial institutions of the countries in which physical risks are the highest will be highly exposed to these physical risks; and high international interconnectedness of financial institutions may lead to important financial contagion.

4.5 Conclusion

Climate stress testing and scenario analysis have been used by both central banks and academic researchers to assess climate-related risks to the financial system. Significant work has been done to develop methodologies addressing the specificities of climate-related risks. In particular, the NGFS has facilitated the implementation of climate scenario-based exercises by financial authorities thanks to its long-term reference scenarios.

Assessing climate-related risks to the financial system raises significant challenges. Taken together, climate scenario-based methodologies that were developed by financial authorities and academic researchers offer solutions to most of these challenges. To identify precisely the financial system's vulnerabilities to climate-related risks, the main remaining obstacles are: data gaps regarding exposure and vulnerabilities to climate-related risks of the counterparties of financial institutions, the state of the art in climate science, which limits the precision of physical risk projections, and the lack of understanding and projection capabilities of adaptation efforts to understand how the infrastructures' vulnerabilities to physical risks will evolve.

The methodologies reviewed in this chapter also include a number of blind spots. The way climate-related risks can channel into liquidity risk is very little modeled, and would deserve more consideration. Similarly, the risk of sudden repricing events, associated with changes in market sentiment, has been overlooked by most of the exercises, and more work would need to be done to understand the extent of this risk and assess its consequences for financial institutions. Finally, a key factor contributing to the adversity of climate-related risks is their compounded nature, as the accumulation of these risks can progressively undermine the stability of the financial system (Acharya et al., 2023). Most of these flaws have been acknowledged by the NGFS and central banks, and work is being done in the direction of correcting them (Network of Central Banks and Supervisors for Greening the Financial System, 2023b, 2022a, 2023a; Financial Stability Board and Network of Central Banks and Supervisors for Greening the Financial System, 2022).

By improving our understanding of financial system's vulnerabilities to climate-related risks, these climate-scenario based exercises have played an important role in raising awareness about climate-related risks among financial authorities, financial institutions, and even non-financial corporations. Financial authorities that have already conducted climate scenario analyses, and financial institutions that were involved in those exercises, report progress in their understanding, assessment and management of climate-related risks (Financial Stability Board and Network of Central Banks and Supervisors for Greening the Financial System, 2022; Autorité de Contrôle Prudentiel et de Résolution, 2021). Moreover, recent evidence suggest that participating in such exercises changes the financing behavior of banks, leading them to charge higher interest rates for high-carbon emitters, boost lending for green purposes and foster their borrowers' efforts to improve their environmental performance (Fuchs et al., 2024). Hence, climate scenario-based exercises conducted by financial authorities may also play a role in redirecting financial flows to the financing of the low-carbon transition, as pledged in the second article of the Paris agreement.

References

Acemoglu, D., A. Ozdaglar, and A. Tahbaz-Salehi, 2015, Systemic risk and stability in financial networks, *American Economic Review* 105, 564–608.

Acharya, V. V., R. Berner, R. Engle, H. Jung, J. Stroebel, X. Zeng, and Y. Zhao, 2023, Climate stress testing, *Annual Review of Financial Economics* 15, 291–326.

Adenot, T., M. Briere, P. Counathe, M. Jouanneau, T. Le Berthe, and T. Le Guenedal, 2022, Cascading effects of carbon price through the value chain: Impact on firm's valuation, *Available at SSRN 4043923*.

Aikman, D., D. Beale, A. Brinley-Codd, A.-C. Hüser, G. Covi, and C. Lepore, 2023, Macroprudential stress test models: A survey, Working Paper 23/173, International Monetary Fund.

Allen, T., S. Dees, V. Chouard, L. Clerc, A. de Gaye, A. Devulder, S. Diot, N. Lisack, F. Pegoraro, M. Rabate, et al., 2020, Climate-related scenarios for financial stability assessment: an application to France, Working Paper 774, Banque de France.

Alogoskoufis, S., S. Carbone, W. Coussens, S. Fahr, M. Giuzio, F. Kuik, L. Parisi, D. Salakhova, and M. Spaggiari, 2021a, Climate-related risks to financial stability, *Financial Stability Review* 1.

Alogoskoufis, S., N. Dunz, T. Emambakhsh, T. Hennig, M. Kaijser, C. Kouratzoglou, M. A. Muñoz, L. Parisi, and C. Salleo, 2021b, ECB economy-wide climate stress test: Methodology and results, ECB Occasional Paper 281, European Central Bank.

Autorité de Contrôle Prudentiel et de Résolution, 2021, A first assessment of financial risks stemming from climate change: The main results of the 2020 climate pilot exercise, Tech Report.

Autorité de Contrôle Prudentiel et de Résolution (ACPR), 2020, Pilot exercise on climate-related risks, Technical Report.

Bachelard, G., 1938, *La formation de l'esprit scientifique : contribution à une psychanalyse de la connaissance* (Vrin).

Baer, M., B. Caldecott, J. Kastl, A. M. Kleinnijenhuis, and N. Ranger, 2022, TRISK - A climate stress test for transition risk, *Available at SSRN 4254114*.

Bank for International Settlements, 2017, Supervisory and bank stress testing: range of practices, Tech Report.

Bank of England, 2019, Financial stability report, Tech Report.

Bank of England, 2021, Key elements of the 2021 biennial exploratory scenario: Financial risks from climate change, Tech Report.

Basel Committee, 2021, Climate-related risk drivers and their transmission channels, Tech Report.

Battiston, S., and G. Caldarelli, 2013, Systemic risk in financial networks, *Journal of Financial Management, Markets and Institutions* 1, 129–154.

Battiston, S., P. Jakubik, I. Monasterolo, K. Riahi, and B. van Ruijven, 2019, Climate risk assessment of the sovereign bond portfolio of European insurers, In *Financial Stability Report*, 69–89.

Battiston, S., A. Mandel, I. Monasterolo, F. Schütze, and G. Visentin, 2017, A climate stress-test of the financial system, *Nature Climate Change* 7, 283–288.

Baudino, P., R. Goetschmann, J. Henry, K. Taniguchi, and W. Zhu, 2018, Stress-testing banks - A comparative analysis, Technical report, Financial Stability Institute.

Bikakis, A., 2020, Climate change, flood risk and mortgages in the UK: A scenario analysis, *New School Economic Review* 10.

Bolton, P., M. Despres, L. A. P. Da Silva, F. Samama, R. Svartzman, et al., 2020, The green swan, *BIS Books*.

Bolton, P., and M. T. Kacperczyk, 2021, Do investors care about carbon risk?, *Journal of Financial Economics* 142, 517–549.

Borio, C., M. Drehmann, and K. Tsatsaronis, 2014, Stress-testing macro stress testing: does it live up to expectations?, *Journal of Financial Stability* 12, 3–15.

Bouchet, V., H. Dayan, and C. Contoux, 2022, Finance and climate science: worlds apart?, *Journal of Risk Research* 25, 176–197.

Brei, M., P. Mohan, and E. Strobl, 2019, The impact of natural disasters on the banking sector: Evidence from hurricane strikes in the Caribbean, *The Quarterly Review of Economics and Finance* 72, 232–239.

Bressan, G., A. Duranović, I. Monasterolo, and S. Battiston, 2024, Asset-level assessment of climate physical risk matters for adaptation finance, *Nature Communications* 15, 5371.

Cahen-Fourot, L., E. Campiglio, L. Daumas, M. G. Miess, and A. Yardley, 2023, Stranding ahoy? Heterogeneous transition beliefs and capital investment choices, *Journal of Economic Behavior & Organization* 216, 535–567.

Cahen-Fourot, L., E. Campiglio, E. Dawkins, A. Godin, and E. Kemp-Benedict, 2020, Looking for the inverted pyramid: An application using input-output networks, *Ecological Economics* 169, 106554.

Caloia, F., and D.-J. Jansen, 2021, Flood risk and financial stability: Evidence from a stress test for the Netherlands, Working Paper 2021/730, De Nederlandsche Bank.

Campiglio, E., L. Daumas, P. Monnin, and A. von Jagow, 2023, Climate-related risks in financial assets, *Journal of Economic Surveys* 37, 950–992.

Carè, R., 2023, Climate-related financial risks: exploring the known and charting the future, *Current Opinion in Environmental Sustainability* 65, 101385.

Carney, M., 2015, Breaking the tragedy of the horizon – Climate change and financial stability, *Speech Given at Lloyd's of London* 29, 220–230.

Daumas, L., 2023, Financial stability, stranded assets and the low-carbon transition – A critical review of the theoretical and applied literatures, *Journal of Economic Surveys* 38, 601–716.

Daumas, L., 2024, A climate stress-test for transition risks with a stock-flow consistent model, *Available at SSRN 4913244*.

De Guindos, L., 2021, Shining a light on climate risks: the ECB's economy-wide climate stress test, ECB blog, March 18, European Central Bank.

Desnos, B., T. Le Guenedal, P. Morais, and T. Roncalli, 2023, From climate stress testing to climate value-at-risk: A stochastic approach, *Available at SSRN 4497124*.

Emambakhsh, T., M. Giuzio, L. Mingarelli, D. Salakhova, and M. Spaggiari, 2022, Climate-related risks to financial stability, *Financial Stability Review* 1.

European Central Bank, 2022, 2022 climate risk stress test, Tech Report.

European Central Bank and European Systemic Risk Board, 2022, The macroprudential challenge of climate change, Tech Report.

European Insurance and Occupational Pensions Authority, 2019, Methodological principles of insurance stress testing, Tech Report.

Farmer, J. D., A. M. Kleinnijenhuis, T. Schuermann, and T. Wetzer, 2022, *Handbook of Financial Stress Testing* (Cambridge University Press).

Fiedler, T., A. J. Pitman, K. Mackenzie, N. Wood, C. Jakob, and S. E. Perkins-Kirkpatrick, 2021, Business risk and the emergence of climate analytics, *Nature Climate Change* 11, 87–94.

Financial Stability Board, 2020, The implications of climate change for financial stability, Tech Report.

Financial Stability Board, and Network of Central Banks and Supervisors for Greening the Financial System, 2022, Climate scenario analysis by jurisdictions: Initial findings and lessons, Tech Report.

Fuchs, L., H. Nguyen, T. Nguyen, and K. Schaeck, 2024, Climate stress test, bank lending, and the transition to the carbon-neutral economy, *IWH Discussion Papers* 9/2024.

Gourdel, R., and M. Sydow, 2023, Non-banks contagion and the uneven mitigation of climate risk, *International Review of Financial Analysis* 89, 102739.

Grippa, P., S. Mann, et al., 2020, Climate-related stress testing: Transition risks in Norway, Working Paper WP/20/232, International Monetary Fund.

Hourcade, J.-C., P. Tankov, S. Voisin, F. Ghersi, and J. Lefèvre, 2021, Les modèles intégrés économie-climat : quels usages pour quelles décisions ?, *Opinions & Débats* 23, 16–42.

Hurd, T. R., et al., 2016, *Contagion!: Systemic Risk in Financial Networks*, volume 42 (Springer).

Institute for Climate Economics, 2019, Towards an alternative approach in finance to climate risks: taking uncertainties fully into account, Tech Report.

Intergovernmental Panel on Climate Change, 2021, *Climate Change 2021: The Physical Science Basis. Contribution of Working Group I to the Sixth Assessment Report of the Intergovernmental Panel on Climate Change* (Cambridge University Press, Cambridge, United Kingdom and New York, NY, USA).

Intergovernmental Panel on Climate Change, 2022, *Climate Change 2022: Mitigation of Climate Change. Contribution of Working Group III to the Sixth Assessment Report of the Intergovernmental Panel on Climate Change.* (Cambridge University Press, Cambridge, United Kingdom and New York, NY, USA).

Jacquetin, F., 2021, Stress-tests climatiques par scénarios : De l'analyse des risques à la modélisation, Document de travail, ADEME.

Jung, H., R. F. Engle, and R. B. Berner, 2021, Climate stress testing, Working Paper 977, Federal Reserve Bank of New York, New York, NY.

Krüger, P., Z. Sautner, and L. Starks, 2020, The importance of climate risk for institutional investors, *Review of Financial Studies* 33, 1067–1111.

Le Guenedal, T., P. Drobinski, and P. Tankov, 2022, Cyclone generation algorithm including a THERmodynamic module for integrated national damage assessment (CATHERINA 1.0) compatible with Coupled Model Intercomparison Project (CMIP) climate data, *Geoscientific Model Development* 15, 8001–8039.

Litterman, R., C. E. Anderson, N. Bullard, B. Caldecott, M. L. Cheung, J. T. Colas, R. Coviello, P. W. Davidson, J. Dukes, H. P. Duteil, et al., 2020, Managing climate risk in the U.S. financial system, Technical report, U.S. Commodity Futures Trading Commission.

Mandel, A., 2021, Impacts of extreme weather events on mortgage risks and their evolution under climate change: A case study on Florida, *Available at SSRN 3929927*.

Mandel, A., T. Tiggeloven, D. Lincke, E. Koks, P. Ward, and J. Hinkel, 2021, Risks on global financial stability induced by climate change: the case of flood risks, *Climatic Change* 166, 1–24.

Monnin, P., 2018, Integrating climate risks into credit risk assessment – Current methodologies and the case of central banks corporate bond purchases., Discussion note, Council on Economic Policies.

Network of Central Banks and Supervisors for Greening the Financial System, 2018, First progress report, Tech Report.

Network of Central Banks and Supervisors for Greening the Financial System, 2019, A call for action – Climate change as a source of financial risk, Tech Report.

Network of Central Banks and Supervisors for Greening the Financial System, 2020, NGFS climate scenarios for central banks and supervisors - Phase I, Tech Report.

Network of Central Banks and Supervisors for Greening the Financial System, 2021, NGFS climate scenarios for central banks and supervisors - Phase II, Tech Report.

Network of Central Banks and Supervisors for Greening the Financial System, 2022a, Final report on bridging data gaps, Tech Report.

Network of Central Banks and Supervisors for Greening the Financial System, 2022b, NGFS climate scenarios for central banks and supervisors - Phase III, Tech Report.

Network of Central Banks and Supervisors for Greening the Financial System, 2023a, Compound risks: Implications for physical climate scenario analysis, Tech Report.

Network of Central Banks and Supervisors for Greening the Financial System, 2023b, Conceptual note on short-term climate scenarios, Tech Report.

Network of Central Banks and Supervisors for Greening the Financial System, 2023c, NGFS climate scenarios for central banks and supervisors - Phase IV, Tech Report.

Nguyen, Q., I. Diaz-Rainey, D. Kuruppuarachchi, M. McCarten, and E. K. Tan, 2020, Climate transition risk in U.S. loan portfolios: Are all banks the same?, *Available at SSRN 3766592*.

Pinchot, A., L. Zhou, G. Christianson, J. McClamrock, I. Sato, et al., 2021, Assessing physical risks from climate change: do companies and financial organizations have sufficient guidance, *World Resources Institute* 1–44.

Reinders, H. J., D. Schoenmaker, and M. A. Van Dijk, 2020, A finance approach to climate stress testing, *Available at SSRN 3573107*.

Roncoroni, A., S. Battiston, M. D'Errico, G. Hałaj, and C. Kok, 2019, Interconnected banks and systemically important exposures, Working Paper 2331, European Central Bank.

Roncoroni, A., S. Battiston, L. O. Escobar-Farfán, and S. Martinez-Jaramillo, 2021, Climate risk and financial stability in the network of banks and investment funds, *Journal of Financial Stability* 54, 100870.

Semieniuk, G., E. Campiglio, J.-F. Mercure, U. Volz, and N. R. Edwards, 2021, Low-carbon transition risks for finance, *Wiley Interdisciplinary Reviews: Climate Change* 12, e678.

Task Force on Climate-Related Financial Disclosures, 2017, Recommendations of the Task Force on Climate-related Financial Disclosures, Tech Report.

Vermeulen, R., E. Schets, M. Lohuis, B. Kölbl, D.-J. Jansen, and W. Heeringa, 2021, The heat is on: A framework for measuring financial stress under disruptive energy transition scenarios, *Ecological Economics* 190, 107205.

Vodenska, I., H. Aoyama, A. P. Becker, Y. Fujiwara, H. Iyetomi, and E. Lungu, 2021, From stress testing to systemic stress testing: the importance of macroprudential regulation, *Journal of Financial Stability* 52, 100803.

Chapter 5

Combating climate change with green quantitative easing

Matheus Grasselli

McMaster University, Canada

Luka Milic

Toronto Metropolitan University, Canada

This chapter explores the potential role of central banks in combating climate change through the implementation of a green quantitative easing (QE) program. By incorporating a central banking sector into a stock-flow consistent integrated climate-economic model, we analyze the effectiveness of green QE, wherein central banks purchase green bonds to encourage investments in low-carbon production processes. Through numerical simulations, our key findings suggest that green QE, individually or in conjunction with targeted carbon taxes and abatement subsidies, can stabilize debt levels for firms, foster green investment, and mitigate the adverse effects of climate change. These findings are also robust with respect to changes in policy parameters, initial conditions, and reaction functions of economic agents, demonstrating that green QE significantly increases the chances of achieving a low-carbon economy while also reducing long-term financial instability. This analysis provides valuable insights into how central banks can be pivotal in steering economies toward sustainability while maintaining financial stability in the face of climate-related challenges.

5.1 Introduction

Most central banks now recognize climate change as a major source of risk for the financial system. The Network for Greening the Financial System - a voluntary coalition that expanded from the initial eight members in 2017 to currently over 140 members - issued a call to action in 2019 with a number of recommendations for central banks, supervisors, regulators, and policy makers focusing primarily on measuring and monitoring climate-related financial risk, including consistent disclosure, knowledge sharing, and addressing data gaps (NGFS (Network of Central Banks and Supervisors for Greening the Financial System), 2019). One of the recommendations calls for integrating sustainability factors into a central bank's own portfolio management, that is to say "exploring the interaction between climate change and central banks' mandates (beyond financial stability) and the effects of climate-related risks on the monetary policy frameworks, paying due regard to their respective legal mandates (see page 5 in NGFS (Network of Central Banks and Supervisors for Greening the Financial System) (2019)).

DOI: 10.1201/9781032636252-5

As argued in Campiglio (2016), a key potential role for central consists in helping to increase the amount of credit available to green investment, which continues to be much lower than what is estimated to be needed for achieving the goals of the Paris Agreement[1]. Although the focus in Campiglio (2016) is on regulatory changes that would favor credit expansion toward green investment, such as differentiated reserve and capital requirements for banks providing loans to sustainable projects, other authors have proposed more direct forms of intervention, for example through green quantitative easing (Dafermos, Nikolaidi, and Galanis, 2018), that is to say, large scale purchase of debt instruments, such as green bonds, issued with the explicit purpose of financing investment in low-carbon activities. Once considered unconventional, in the short span between the Global Financial Crisis of 2008 and the COVID-19 shock in early 2020 quantitative easing went from a last-resort measure to a mainstream policy, with central banks in all major economies engaging in purchasing a vast array of debt instruments, including previously unthinkable ones, such as corporate bonds (Grasselli, 2022). Against this backdrop of expansion of the role of central banks in highly financialized economies, authors such as Guttmann (2022) predict that quantitative easing programs will soon include purchases of green bonds with a goal to increase financial flows to environmentally friendly projects.

The purpose of this Chapter is to build a mathematical model to assess the effectiveness of such green quantitative easing programs. We use as our baseline the integrated climate-economic model proposed in Bovari, Giraud, and Mc Isaac (2018) and known as GEMMS (General Monetary and Multisectoral Macrodynamics for the Ecological Shift)[2], which combines a stock-flow consistent economic module based on Grasselli and Lima (2012) and Grasselli and Nguyen-Huu (2015) with a climate module adapted from Nordhaus (2018). As further analyzed in Bovari, Lecuyer, and Isaac (2018), a core result of GEMMS is the existence of a fundamental tension between limiting the amount of damages caused by climate change through reduction in carbon emissions and keeping the debt levels of the private sector at manageable levels. This delicate dance of debt and damages arises because reduction in emissions is achieved through abatement costs, which in turn decrease the profits of firms. One way to alleviate this tension is through the type of government subsidy explored in Bovari, Lecuyer, and Isaac (2018), which the authors show to reduce the debt burden of firms, although their analysis omits the corresponding effect on public debt levels. Even conceding that sovereign governments issuing debt in their own currency have more financial flexibility than private firms, it is nevertheless important to track the effect that climate subsidies have on public finances, which is one of the modifications that we introduce in our model. The other, much more consequential modification consists of adding to the economic module a central bank sector pursuing green quantitative easing exactly in the way described in Guttmann (2022), namely by creating a "green bond spread" between the interest rate on conventional loans and the yield on green bonds. The corresponding lower cost of borrowing by issuing such green bonds allows for an alternative way to lessen the debt burden of firms without increasing the debt of the government sector.

The remainder of the Chapter is organized as follows. Section 5.2 describes the economic module in stock-flow consistent form, that is to say specifying the relevant sectors of the economy, their balance sheets, transactions and flow of funds, as well as the behavior of each sector. Importantly, we introduce in this section the key extensions of the model in Bovari, Giraud, and Mc Isaac (2018): (i) a differentiation between conventional and green capital, (ii) the issuance of green bonds by firms as an additional source of funding, (iii) the decisions by firms to invest in green or conventional capital as a function of the green

[1]For a comprehensive and relatively recent overview of climate finance, see the interactive tool at https://www.climatepolicyinitiative.org/publication/global-landscape-of-climate-finance-2021/

[2]See https://www.afd.fr/en/actualites/gemmes-application-understanding-effects-climate-change-economy.

bond spread, and (iv) the targeting of this spread by the central bank as a function of desired emissions. Section 5.3 describes the climate module, mainly for completeness, as it is essentially the same as in Bovari, Giraud, and Mc Isaac (2018), with the exception of the way damages are allocated to capital. Section 5.4 derives the reduced-form dynamical system satisfied by the state variables of the model asymptotically once the transition to a zero-emissions economy is completed, from which one can analyzed the impact of climate damages on the asymptotic equilibrium of the system.

Section 5.5 contains the main results of the Chapter, obtained through numerical simulation of the full set of dynamic equations in the model. In Section 5.5.2, we show the trajectories of key variables in the model under different policy scenarios, with the main takeaway being that the green quantitative easing program is effective in stabilizing the debt ratio of the firms sector with or without the presence of subsidies provided by the government sector. In Section 5.5.3, we conduct a sensitivity analysis with respect to some of the key parameters of the model, namely the government subsidy parameter, the reaction of the firms to changes in the green spread and the reaction of the central bank to changes in observed emissions, with the key result being that the level of subsidy required for the economy to converge to a good equilibrium decreases as the willingness of the central bank to engage in green quantitative easing increases. This is followed in Section 5.5.4 by a complementary analysis of the basin of attraction of the interior equilibrium of the model for different values of initial conditions for some of the key state variables. In accordance with the previous results, here we observe that the green quantitative easing program considerably increases the set of initial conditions leading to a good equilibrium. Lastly, Section 5.5.5 presents Monte Carlo simulations of several state variables under random sampling of the parameters. Focusing on the same two variables explored in Bovari, Lecuyer, and Isaac (2018), namely the temperature anomaly caused by climate change and the debt level of firms, our conclusion is that green quantitative easing unequivocally increases the chances of keeping both variables under acceptable thresholds, while also reducing the overall level of debt for the government sector. Section 5.6 offers concluding remarks.

5.2 The economic module

5.2.1 Accounting framework

We begin by first adopting the Stock-Flow Consistency (SFC) approach from Godley and Lavoie (2007). For our purposes, we divide the simplified economy into five sectors, namely households, firms, banks, government and a central bank sector. Accordingly, we create matrices representing the balance sheet, transactions, and flow of funds for all monetary exchanges between the various sectors of the closed economy.

The balance sheet matrix in Table 5.1 records the *stocks* (i.e., levels), expressed in units of currency, for example USD, of all assets and liabilities within our economic system. More specifically, a positive entry indicates an asset of the sector, while a liability is recorded as a negative entry. It is clear from the table that, apart from a mere accounting device, the balance sheet matrix already embodies many of the simplifying assumptions in the model, notably: (i) the households sector holds bank deposits M_h as a monetary asset and also owns the firms and banks sector through private equity $\mathcal{E}_h$, (ii) the firms sector holds all the productive capital $K = K_C + K_G$ in the economy in addition to bank deposits M_f and borrows through bank loans L_f and the issuance of green bonds Γ, (iii) banks make loans,

TABLE 5.1: Balance sheet matrix

	Households	*Firms*	*Banks*	*Government*	*Central bank*	*Sum*
Conventional capital		pK_C				pK_C
Green capital		pK_G				pK_G
Deposits	M_h	M_f	$-M$			0
Loans		$-L_f$	L_f			0
Green bonds		$-\Gamma$			Γ	0
Government bonds				$-B$	B	0
Reserves			R		$-R$	0
Equities (households, firms, banks)	$\mathcal{E}_h$	$-\mathcal{E}_f$	$-\mathcal{E}_b$			0
Equities (government, central bank)				$\mathcal{E}_g$	$-\mathcal{E}_{cb}$	0
Sum (net worth)	X_h	$X_f = 0$	$X_b = 0$	X_g	$X_{cb} = 0$	pK

accept deposits, and hold central bank reserves R, (iv) the government issues bonds B, and also owns the central bank through equity $\mathcal{E}_g$, and finally (v) the central bank acts as a buffer for the entire economy by purchasing government bills and green bonds, while providing reserves for the banks sector. Observe that our balance sheet matrix is an extension of that shown in Table 1 in Bovari, Giraud, and Mc Isaac (2018) through the addition of a stylized government and central bank, as further explained below. Accordingly, from the column sums in the balance sheet matrix, we obtain the following

$$X_h = M_h + \mathcal{E}_h \qquad \text{(net worth of households)}, \tag{5.1}$$

$$\mathcal{E}_f = pK_C + pK_G + M_f - L_f - \Gamma \qquad \text{(equity of firms)}, \tag{5.2}$$

$$\mathcal{E}_b = L_f - M + R \qquad \text{(equity of banks)}, \tag{5.3}$$

$$X_g = \mathcal{E}_g - B \qquad \text{(net worth of the government)}, \tag{5.4}$$

$$\mathcal{E}_{cb} = \Gamma + B - R \qquad \text{(equity of the central bank)}, \tag{5.5}$$

from which is follows that the total net worth for the entire economy is

$$X = X_h + X_f + X_b + X_g + X_{cb} = pK_C + pK_G = pK, \tag{5.6}$$

that is to say, the nominal value of the total capital stock, as expected.

The transactions matrix in Table 5.2 accounts for all monetary *flows* between sectors, expressed in units of currency per time, for example USD/year. We highlight that a positive sign indicates any transaction involving an incoming monetary flow (i.e., a source of funds), whereas a negative sign denotes a transaction involving an outgoing flow (i.e., a use of funds). Note that the current account registers any monetary payments received or made by the productive sector, while the capital account shows how the investment in real assets is funded. Comparing again with Table 1 in Bovari, Giraud, and Mc Isaac (2018), we see that the extensions in our model consist of: (i) differentiating between conventional investment I_C and green invest I_G and corresponding depreciation, (ii) explicit accounting for the transactions of the Government sector (e.g., in Bovari, Giraud, and Mc Isaac (2018) the carbon taxes paid by the firms were received by Households), (iii) introduction of abatement subsidies pS^C as in Bovari, Lecuyer, and Isaac (2018), and (iv) interest on green bonds – all of which will be explained in detail in what follows.

Finally, the balance sheet is integrated with the transaction flows through the flow of funds matrix in Table 5.3, which logs the change in stocks resulting from all the monetary transactions occurring at each period of time.

With the closed economic system constructed, we can begin to appreciate the added realism of our updated framework. For instance, even before introducing any behavioral equations, we see that households no longer implement fiscal policy and are now indeed the

TABLE 5.2: Transactions matrix

	Households	*Firms*		*Banks*	*Government*	*Central bank*	*Sum*
Consumption	$-pC$	pC					0
Conventional investment		pI_C	$-pI_C$				0
Green investment		pI_G	$-pI_G$				0
Government spending		pG			$-pG$		0
Accointing memo [GDP]		$[pY]$					
Wages	W	$-W$					0
Conventional capital depreciation		$-\bar{\delta}_D pK_C$	$\bar{\delta}_D pK_C$				0
Green capital depreciation		$-\bar{\delta}_D pK_G$	$\bar{\delta}_D pK_G$				0
Carbon taxes		$-pT^C$			pT^C		0
Non-carbon taxes	$-pT_h$	$-pT_f$			pT		0
Abatement subsidies		pS^C			$-pS^C$		0
Interest on loans		$-r_L L_f$		$r_L L_f$			0
Interest on deposits	$r_M M_h$	$r_M M_f$		$-r_M M$			0
Interest on bonds					$-r_B B$	$r_B B$	0
Interest on reserves				$r_R R$		$-r_R R$	0
Interest on green bonds		$-r_\Gamma \Gamma$				$r_\Gamma \Gamma$	0
Banks' dividends	Π_b			$-\Pi_b$			0
Firms' dividends	Π_d	$-\Pi_d$					0
Central bank profits					Π_{cb}	$-\Pi_{cb}$	0
Column sum (financial balance)	S_h	Π_r	$-pI + \delta_D pK$	S_b	S_g	0	0

TABLE 5.3: Flow of funds matrix

	Households	*firms*	*Banks*	*Government*	*Central bank*	*Sum*
Change in conventional capital		$p\dot{K}_C$				$p\dot{K}_G$
Change in green capital		$p\dot{K}_G$				$p\dot{K}_G$
Change in deposits	$\dot{M}_h$	$\dot{M}_f$	$-\dot{M}$			0
Change in loans		$-\dot{L}_f$	$\dot{L}_f$			0
Change in bonds				$-\dot{B}$	$\dot{B}$	0
Change in reserves			$\dot{R}$		$-\dot{R}$	0
Change in green bonds		$-\dot{\Gamma}$			$\dot{\Gamma}$	0
Column sum	S_h	Π_r	S_b	S_g	0	$p\dot{K}$
Change in equities	$\dot{\mathcal{E}}_h$	$-\dot{\mathcal{E}}_f = -(\Pi_r + \dot{p}K)$	$-\mathcal{E}_b = -S_b$			0
Change in net worth	$\dot{X}_h = S_h + \mathcal{E}_h$	$\dot{X}_f = 0$	$\dot{X}_b = 0$	$\dot{X}_g = S_g$	$\dot{X}_{cb} = 0$	$p\dot{K} + \dot{p}K$

passive sector in the economic system, as was presumed in Bovari, Giraud, and Mc Isaac (2018). Equivalently, the government and central bank sectors have a clear function and can independently implement fiscal and monetary policy instead of operating under an aggregated public sector, as seen in Bovari, Lecuyer, and Isaac (2018). Consequently, this improvement allows us to accurately follow the interactions between the private, banking, government and central banking sectors that would occur in reality.

5.2.2 Sector behavior

We next explore the set of behavioral assumptions that describe the evolution of the continuous-time dynamics for the macroeconomic monetary system. In what follows, we denote a constant parameter by a bar above the letter corresponding to it, for example $\bar{\nu}$ for the constant capital-to-output, whereas any other quantities without a bar are understood to be time-dependent variables. As usual, $\dot{x}$ denotes the time derivative of x.

Production and Sales

We postulate that firms produce a yearly real output Y_0 consisting of a homogeneous good using two different methods of production: one based on conventional capital K_C and another based on green capital K_G, as proposed in Dafermos, Nikolaidi, and Galanis (2018). Namely, we formalize the two distinct production processes as

$$Y_C^0 = \frac{K_C}{\bar{\nu}}, \quad Y_G^0 = \frac{K_G}{\bar{\nu}}, \quad Y^0 = Y_C^0 + Y_G^0, \tag{5.7}$$

where $\bar{\nu} > 0$ is the constant capital-to-output ratio, assumed for simplicity to be the same across the two methods of production, although this can easily be relaxed. Observe that we use $K = K_C + K_G$ to denote the total stock of capital, so that the ratio of green capital to total can be defined as

$$\varkappa = \frac{K_G}{K}. \tag{5.8}$$

We interpret green capital as generating output using renewable resources, such as wind, solar or hydro, whereas conventional capital produces goods using non-renewable energies such as fossil fuels. Accordingly, green production has a lower emission intensity than conventional production and releases less industrial emissions into the atmosphere. In essence, greater adoption of green capital leads to a lower-carbon economy, as will be explained in more detail below.

Because the production process releases industrial emissions E_{ind} into the atmosphere, in an effort to reduce CO_2 emissions, the public sector will levy a carbon tax on the productive sector. Hence, to alleviate some of the tax burdens, the private sector may elect to divert a fraction A of production to various abatement activities aimed at lowering their emission rate. Furthermore, a fraction $\mathbf{D}^Y$ of the remaining output may be destroyed as a consequence of climate change and rising temperatures, as argued in Nordhaus (2014). Therefore, the portion of production that is available for sales is

$$Y = (1 - \mathbf{D}^Y)(1 - A)Y^0. \tag{5.9}$$

The Labor Market and Inflation

The private sector is assumed to follow a minimally rational behavior, where firms will minimize their costs by hiring the required amount of labor at full capacity, such that

$$Y^0 = \frac{K}{\bar{\nu}} = aL \implies L = \frac{Y^0}{a} = \frac{K}{a\bar{\nu}}, \tag{5.10}$$

where L refers to the total employed labor and a represents the labor productivity, which grows exogenously at a constant rate of

$$\bar{\alpha} = \frac{\dot{a}}{a} \geq 0. \tag{5.11}$$

Likewise, we can define the endogenously determined employment rate as

$$\lambda = \frac{L}{N}. \tag{5.12}$$

for an exogenously defined global workforce N satisfying

$$g_N := \frac{\dot{N}}{N} = \bar{\delta}_N \left(1 - \frac{N}{\bar{N}_{max}}\right), \tag{5.13}$$

where $\bar{N}_{max}$ denotes the upper bound of the world's labor force and $\bar{\delta}_N$ represents the rate of convergence toward $\bar{N}_{max}$.

Finally, the wage-price dynamics follow Grasselli and Nguyen-Huu (2018) as the wage dynamics follow the short-run Phillips curve expressed by

$$\dot{\mathrm{w}} = \mathrm{w}(\Phi(\lambda) - \bar{m}i), \tag{5.14}$$

where w is the money wage per capita, $\bar{m}$ denotes the money illusion parameter and $\Phi(\cdot)$ is an increasing real-valued function with range in $[0, 1)$, which we choose to be

$$\Phi(\lambda) = \bar{\phi}_0 + \bar{\phi}_1\lambda. \tag{5.15}$$

Generally, (5.14) describes the phenomenon where wages will rise at an increasing rate as the economy reaches full employment, since workers gain bargaining power due to the lack of competition, with the constant parameter $\bar{m}$ specifying the degree to which workers consider inflation when bargaining for wages. Naturally, these dynamics leads to higher inflation rate, which we can capture through

$$i(\omega) = \frac{\dot{p}}{p} = \bar{\eta}(\bar{\xi}\omega - 1), \tag{5.16}$$

where $p \geq 0$ is the price level and $\bar{\eta}$ is the speed of convergence toward its long-run value, which is set at a constant mark-up $\bar{\xi} > 0$, multiplied by the wage share

$$\omega = \frac{\mathrm{w}L}{pY}. \tag{5.17}$$

Firms Sector Financing

Using Table 5.2 we can define the private sector's nominal profit before dividends as the nominal output minus the cost of production, that is

$$\Pi = pY - \mathrm{w}L - \delta_{\mathbf{D}}pK - r_L L_f + r_M M_f - r_\Gamma \Gamma + p(S^C - T^C). \tag{5.18}$$

We see that the total cost of production is determined by: (i) the wage bill $\mathrm{w}L$, (ii) the capital depreciation $\delta_{\mathbf{D}}pK$, where $\delta_{\mathbf{D}} = \bar{\delta} + \mathbf{D}^K$ is the capital depreciation rate obtained as a constant rate $\bar{\delta}$ plus damages to capital $\mathbf{D}^K$, (iii) the interest payment $r_L L_f$ on bank loans and the interest receipt $r_M M_f$ on deposits at instantaneous interest rates r_L, r_M, (iv) the interest payment $r_\Gamma \Gamma$ on green bonds, where r_Γ is instantaneous green bond yield (to be defined below), Γ is the dollar amount of green bonds issued, and (v) the net transfers $(S^C - T^C)$ from the government to the productive sector.

Accordingly, we can define the profit ratio, π as:

$$\pi = \frac{\Pi}{pY}. \tag{5.19}$$

We assume that firms distribute a fraction of sales to Households according to the following function of the profit ratio:

$$\Delta(\pi) = \min\left(\bar{\Delta}_{max}, \max\left(0, \bar{\Delta}_0 + \bar{\Delta}_1\pi\right)\right). \tag{5.20}$$

Thus, the retained earnings of the corporate sector Π_r are described by

$$\Pi_r = \Pi - \Pi_d, \tag{5.21}$$

where, $\Pi_d = \Delta(\pi)pY$.

In addition to paying out dividends, the productive sector will allocate a percentage of output destined for re-investment. As in Dafermos, Nikolaidi, and Galanis (2018), we formalize the private sector's investment process in two stages. During the first phase, firms decide their aggregate demand for investment I, which is driven by the profit share and captures the risk appetite of the productive sector

$$I = \kappa(\pi)Y, \tag{5.22}$$

where $\kappa(\cdot)$ is an increasing function depending on the profit ratio π as follows

$$\kappa(\pi) = \min\left(\bar{\kappa}_{max}, \max\left(0, \bar{\kappa}_0 + \bar{\kappa}_1\pi\right)\right). \tag{5.23}$$

In the second stage, firms determine the amount of green investment I_G as a proportion β of the overall desired investment, namely

$$I_G = \beta(\delta_\Gamma)I, \tag{5.24}$$

where $\beta(\cdot)$ is an increasing function that depends on the green bond spread

$$\delta_\Gamma = r_L - r_\Gamma \tag{5.25}$$

between the interest rate on conventional loans r_L and the yield on green bonds r_Γ, which we assume to be of the form

$$\beta(\delta_\Gamma) = \min\left(\bar{\beta}_{max}, \max\left(\bar{\beta}_{min}, \bar{\beta}_0 + \bar{\beta}_1\delta_\Gamma\right)\right). \tag{5.26}$$

Naturally, the demand for conventional investment I_C is then determined to be the residual amount

$$I_C = I - I_G. \tag{5.27}$$

Correspondingly, we differentiate between the capital accumulation funded by green investment and conventional investment according to

$$\dot{K}_G = I_G - \delta_{\mathbf{D}}K_G, \qquad \dot{K}_C = I_C - \delta_{\mathbf{D}}K_C \tag{5.28}$$

Because current profits may not be enough to fund the private sector's gross demand for investment, firms obtain external funding either by issuing green bonds or by borrowing from banks. Staying in line with Dafermos, Nikolaidi, and Galanis (2018), we assume that green bonds are solely used to finance green investment, whereas loans are used to finance the remaining conventional investment. That is, green bonds evolve according to

$$\dot{\Gamma} = pI_G - \delta_{\mathbf{D}}pK_G + r_\Gamma\Gamma - \kappa_\Gamma\Gamma, \tag{5.29}$$

where κ_Γ represents the repayment rate on green bonds, whereas conventional loans evolve according to

$$\dot{L_f} = pI_C - \delta_{\mathbf{D}}pK_C + r_L L_f - \kappa_L L_f, \tag{5.30}$$

where κ_L is the repayment rate on conventional loans. Note that for both green bonds and conventional loans, we set the repayment rate to be equal to the interest rate on the respective debt instrument. That is, $\kappa_\Gamma = r_\Gamma$ and $\kappa_L = r_L$.

Moreover, we postulate that private-sector deposits evolve according to

$$\dot{M_f} = pY - \mathrm{w}L - \delta_{\mathbf{D}}pK + r_M M_f - \kappa_L L_f - \kappa_\Gamma\Gamma + p(S^C - T^C) - \Pi_d. \tag{5.31}$$

Thus, we can express the change in total debts by the private sector as

$$\dot{D} = (\dot{L_f} + \dot{\Gamma}) - \dot{M_f} = pI - \delta_{\mathbf{D}}pK - \Pi_r = p\dot{K} - \Pi_r, \tag{5.32}$$

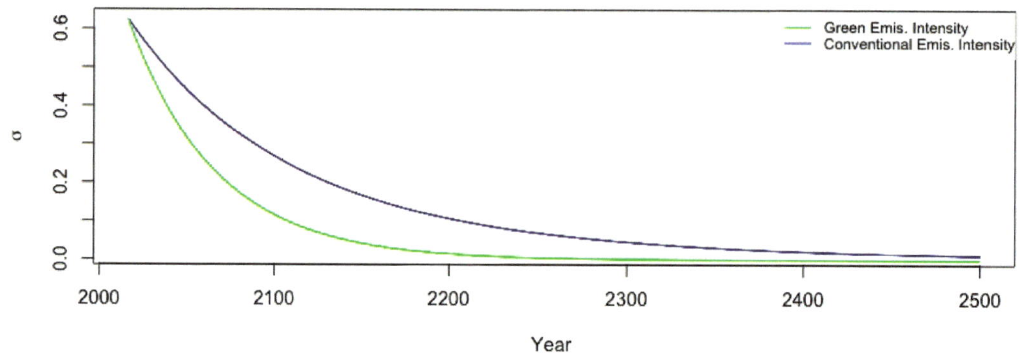

FIGURE 5.1: Exogenous dynamics for conventional and green emission intensity.

in agreement with the column sum for firms in Table 5.3.

Finally, debt-to-output, loan-to-output, green bond-to-output and firm's deposits-to-output ratios can be defined as

$$d = \frac{D}{pY}, \quad \ell_f = \frac{L_f}{pY}, \quad \gamma = \frac{\Gamma}{pY}, \quad m_f = \frac{M_f}{pY}, \tag{5.33}$$

respectively.

Emissions, Taxation and Abatement Decisions

We assume that production releases industrial emissions E_{ind} according to

$$E_{ind} = \left(Y_C^0\sigma_C + Y_G^0\sigma_G\right)(1-n). \tag{5.34}$$

where $n \in [0,1]$ denotes the endogenously determined emission reduction rate (to be defined) while σ_C and σ_G represents the emission intensities of conventional and green production, which are assumed to evolve exogenously according to

$$\frac{\dot{\sigma_C}}{\sigma_C} = g_{\sigma_C}, \qquad \frac{\dot{g_{\sigma_C}}}{g_{\sigma_C}} = \bar{\delta}_{g_{\sigma_C}}, \tag{5.35}$$

$$\frac{\dot{\sigma_G}}{\sigma_G} = g_{\sigma_G}, \qquad \frac{\dot{g_{\sigma_G}}}{g_{\sigma_G}} = \bar{\delta}_{g_{\sigma_G}}. \tag{5.36}$$

Moreover, we posit that green technologies become more efficient over time and reach economies of scale such that conventional capital also emits lower emissions after sufficient time has passed (see Figure 5.1 for an example). We capture this mathematically by setting $g_{\sigma_G}(0) < g_{\sigma_C}(0) < 0$, while also maintaining $\bar{\delta}_{g_{\sigma_C}} < \bar{\delta}_{g_{\sigma_G}} < 0$ for the parameters controlling the rate of decrease.

As mentioned above, we assume that the public authorities levy a carbon tax T^C on emissions expressed as

$$T^C := p_c E_{ind} = p_c\left(Y_C^0\sigma_C + Y_G^0\sigma_G\right)(1-n). \tag{5.37}$$

where p_c denotes the inflation-adjusted price per ton of CO_2-e. Naturally, to minimize the burden of a carbon tax, the productive sector will elect to reduce emissions as much as possible, subject to the associated abatement costs, which we assume to depend on the chosen emission reduction rate as follows:

$$AY^0 = \frac{\left(\sigma_C Y_C^0 + \sigma_G Y_G^0\right) p_{BS} n^\theta}{\theta}. \tag{5.38}$$

Here p_{BS} is the price of back-stop technology, assumed to evolve as

$$\frac{\dot{p}_{BS}}{p_{BS}} = \bar{\delta}_{p_{BS}} < 0, \tag{5.39}$$

and θ describes the convexity of the cost of abatement. Observe that by adopting different carbon intensities σ_C and σ_G for conventional and green production in (5.34) and (5.38) we depart from Bovari, Giraud, and Mc Isaac (2018) and Bovari, Lecuyer, and Isaac (2018), in the sense that investment in green capital has a direct effect on the abatement processes. Moreover, we additionally assume that

$$\theta = \varkappa\bar{\theta}_G + (1 - \varkappa)\bar{\theta}_C > 0, \tag{5.40}$$

where $\varkappa$ is the ratio of green capital to total capital defined in (5.8). Note that $\bar{\theta}_G > \bar{\theta}_C$ implies that green abatement is a cheaper substitute compared to conventional abatement.

Furthermore, the government authorities may bankroll a constant fraction $\bar{s}_a$ of abatement costs, such that firms receive an inflation-adjusted subsidy of the form

$$S^C = \bar{s}_a AY^0. \tag{5.41}$$

Thus, we can define the net transfer from the public to the private sector as having a real value of $(S^C - T^C)$.

Accordingly, firms must choose the emission reduction rate that minimizes the abatement cost with a carbon tax and subsidy

$$\min_{n\in[0,1]} AY^0 + T^C - S^C, \tag{5.42}$$

which leads to the optimal emissions reduction rate of the form

$$n^* = \min\left\{\left(\frac{p_c}{(1-\bar{s}_a)p_{BS}}\right)^{\frac{1}{\theta-1}}; 1\right\}. \tag{5.43}$$

In the expression above, we capped the maximum emissions reduction rate at one, but this can also be relaxed by allowing the possibility of net negative emissions, for example through advances in carbon extraction technologies. Accordingly, the carbon price, which plays the role of a control variable for the government sector, is modeled as

$$p_c = \max\left(p_{BS}, \widetilde{p}_C\right), \tag{5.44}$$

where

$$\frac{\dot{\widetilde{p}}_C}{\widetilde{p}_C} = \bar{a}_{p_C} + \frac{\bar{b}_{p_C}}{t} \tag{5.45}$$

is the functional form adopted in Bovari, Giraud, and Mc Isaac (2018) for the increasing carbon price until equaling the price of the backstop technology, from which point onwards both decrease exponentially at the rate $\bar{\delta}_{p_{BS}}$. Alternative functional forms for the price of carbon can be adopted to simulate other policy scenarios of interest.

Government Policies, Central Bank and Banking Sector

We assume the government has two roles within our economic framework: collecting a carbon tax and subsidizing climate-related efforts. Thus, apart from those fiscal mechanisms, the government runs a balanced budget in the sense that remaining government spending is always equal to non-carbon taxation, or $pG = pT$. Hence, to bankroll climate subsidies, the government will increase its liabilities through the issuance of bonds as follows:

$$\dot{B} = p(S^C - T^C) + r_B B - \Pi_{cb}, \tag{5.46}$$

in accordance with the Government column sums in Tables 5.2 and 5.3. Accordingly, we can define the public debt ratio as

$$b = \frac{B}{pY}. \tag{5.47}$$

Next, we assume that the central bank will adjust the short-term interest rate, r_B, according to the Taylor rule

$$r_B(\omega) = \max\{0, \bar{r}^* + i(\omega) + \bar{\phi}_1(i(\omega) - \bar{i}^*)\}, \tag{5.48}$$

where $\bar{r}^*$ is the long-term interest rate, $\bar{i}^*$ is the inflation rate target (for example, 2%), and $\bar{\phi} > 0$ is a parameter that controls the magnitude of the central bank's response to inflation. We further assume that the interest rate on bonds, deposits and reserves are all equal to the short-term interest rate, that is $r_B = r_M = r_R$. Moreover, the interest rate on conventional bank loans is assumed to be the short-term interest rate plus a constant spread $\bar{\delta}_L$, that is

$$r_L = r_B + \bar{\delta}_L, \tag{5.49}$$

where $\bar{\delta}_L$ is assumed to be constant.

The introduction of green financing creates an innovative role for the central bank within our economic framework. Specifically, controlling the yield on green bonds r_Γ according to

$$r_\Gamma = \max\{r_B, r_L - \delta_\Gamma\}, \tag{5.50}$$

where δ_Γ represents the spread between the bank loans rate r_L and the yield on green bonds, which is assumed to vary according to

$$\delta_\Gamma = \bar{\phi}_2 \frac{(\sigma_C Y_C^0 + \sigma_G Y_G^0)}{E_T}, \tag{5.51}$$

where $\bar{\phi}_2$ denotes the magnitude of the central bank's response to rising pre-abatement emissions and E_T is an exogenous function representing the target emission pathway calibrated to the IPCC (2022) emissions pathway which limits warming to 2°C with 67% likelihood, namely

$$E_T = \max\left\{\bar{E}_{min}, \bar{E}_0 + \bar{E}_1 t\right\}. \tag{5.52}$$

In essence, (5.51) indicates that the central bank considers firms' pre-abatement emissions when controlling yields. Thus, as emissions prior to any abatement efforts grow, the central bank will intervene and increase the spread between the interest rate on bank loans and the yield on green bonds, causing the green borrowing rate to decline. Naturally, this will encourage the private sector to allocate a more significant portion of funds toward energy-efficient production because of the form of green investment assumed in (5.24). On the other hand, if pre-abatement emissions fall, the central bank will reduce the spread between rates, resulting in firms becoming indifferent between green and conventional investment.

However, instead of explicitly establishing a bond market, as in Dafermos, Nikolaidi, and Galanis (2018), we implicitly assume the central bank will purchase the necessary amount of green bonds to ensure the relationship $r_L > r_\Gamma$ holds. Moreover, we postulate that the central bank will engage in this type of *green QE* until bond yields reach the overnight rate r_B.

Moving along, the central bank reserves now evolve according to

$$\dot{R} = p(S^C - T^C) + r_R R - r_\Gamma \Gamma + \dot{\Gamma} \tag{5.53}$$

and we can represent the reserve-to-output ratio as

$$\rho = \frac{R}{pY}, \tag{5.54}$$

Consequently, we see the full effects of the green QE program as the net worth of the central bank decreases whenever the central bank accumulates green bonds, whereas the net worth rises as the reserve bank tapers its green QE policy.

5.3 The climate module

Next, we examine the climate component of our integrated assessment model and the damage function that couples the economic and climate modules. The emissions and climate dynamics are a continuous-time adaptation of the famous DICE model from Nordhaus (2018), reproduced directly from Bovari, Giraud, and Mc Isaac (2018) and Bovari, Lecuyer, and Isaac (2018), with the exception of the damages to capital in (5.65).

5.3.1 Emissions

First, global CO_2-e emissions are assumed to be of the form

$$E = E_{ind} + E_{land}, \tag{5.55}$$

that is, the sum of industrial emissions E_{ind} as defined in (5.34) and land-use emissions E_{land} assumed to evolve according to

$$\frac{\dot{E}_{land}}{E_{land}} = \bar{\delta}_{E_{land}} < 0. \tag{5.56}$$

5.3.2 The carbon cycle, radiative forcing and temperature

The carbon cycle is described using a three-layer model where emissions can accumulate in the atmosphere (CO_2^{AT}), the upper ocean and the biosphere (CO_2^{UP}) and the deep ocean (CO_2^{LO}). Thus, given global CO_2-e emissions E, we can depict the change in carbon concentration in each layer as

$$\begin{pmatrix} \dot{CO}_2^{AT} \\ \dot{CO}_2^{UP} \\ \dot{CO}_2^{LO} \end{pmatrix} = \begin{pmatrix} E \\ 0 \\ 0 \end{pmatrix} + \begin{pmatrix} \bar{\phi}_{12} & \bar{\phi}_{12}\frac{\bar{C}^{AT}}{\bar{C}^{UP}} & 0 \\ \bar{\phi}_{12} & -\bar{\phi}_{12}\frac{\bar{C}^{AT}}{\bar{C}^{UP}} - \bar{\phi}_{23} & \bar{\phi}_{23}\frac{\bar{C}^{UP}}{\bar{C}^{LO}} \\ 0 & \bar{\phi}_{23} & \bar{\phi}_{23}\frac{\bar{C}^{UP}}{\bar{C}^{LO}} \end{pmatrix} \begin{pmatrix} CO_2^{AT} \\ CO_2^{UP} \\ CO_2^{LO} \end{pmatrix}, \tag{5.57}$$

where $\bar{\phi}_{12}, \bar{\phi}_{23}$ are diffusion parameters and the ratios

$$\bar{C}_i^j = \frac{\bar{C}_{pre}^j}{\bar{C}_{pre}^i}, \tag{5.58}$$

are obtained from estimated pre-industrial carbon concentration levels $\bar{C}_{pre}^i$, with $i, j \in \{AT, UP, LO\}$,

A consequence from the accumulation of greenhouse gases is a rise in radiative forcing F, which is determined to be the sum of the residual forcing F_{exo} and industrial forcing F_{ind}, that is

$$F = F_{ind} + F_{exo}, \tag{5.59}$$

where F_{ind} is modeled by

$$F_{ind} = \frac{\bar{F}_{2\times CO_2}}{log(2)} \left(\frac{CO_2^{AT}}{\bar{C}_{pre}^{AT}} \right). \tag{5.60}$$

Observe that the constant $\bar{F}_{2\times CO_2}$ represents the increase in radiative forcing resulting from a doubling of the preindustrial CO_2-e concentration in the atmosphere. We assume that F_{exo} grows exogenously and linearly until the year 2100 and remains constant afterwards, namely

$$F_{exo} = \min \left\{ \bar{F}_{max}, \bar{F}_0 + \bar{F}_1 t \right\}. \tag{5.61}$$

Finally, the rise in radiative forcing results in a corresponding change in the average temperature deviation. This phenomena is described by the following two equations

$$\bar{C}\dot{T} = F - \frac{\bar{F}_{2\times CO_2}}{\bar{S}} T - \bar{h}(T - T_0), \tag{5.62}$$

$$\bar{C}_0 \dot{T}_0 = \bar{h}(T - T_0), \tag{5.63}$$

where, T represents the mean temperature deviations in the atmosphere, biosphere and upper ocean layer, and $\bar{c}$ denotes the heat capacity of the three layers, respectively. Similarly, T_0 denotes the deep ocean average temperature anomaly, while $\bar{c}_0$ describes the heat capacity of the layer. In both expressions, $\bar{h}$ signifies the heat exchange between layers, while $\bar{S}$ is the equilibrium climate sensitivity.

5.3.3 Environmental damages

Lastly, the environmental damage function quantifies the economic loss due to climate change for a specified temperature anomaly, T, and can be expressed as

$$\mathbf{D} = 1 - \frac{1}{1 + \bar{\pi}_1 T + \bar{\pi}_2 T^2 + \bar{\pi}_3 T^{\bar{\zeta}}}, \tag{5.64}$$

where $\bar{\pi}_1, \bar{\pi}_2, \bar{\pi}_3$ and $\bar{\zeta}$ are parameters.

Furthermore, Dietz and Stern (2015) posit that increasing temperatures will affect not only the real output but capital growth, as well. However, we depart from their damage specifications and postulate that damages from climate change are not a trade-off between output and capital stock. Instead, total output will face the full force of climate damages, with capital stock suffering a fraction of the impacts as well. In other words, we allocate a portion of total damages, $\bar{f}_K \in [0, 1]$, to the stock of capital $\mathbf{D}^K$ according to

$$\mathbf{D}^K = \bar{f}_K \mathbf{D}, \tag{5.65}$$

while the damage to output $\mathbf{D}^Y$ is expressed as

$$\mathbf{D}^Y = \mathbf{D}. \tag{5.66}$$

5.4 Post-transition reduced-form dynamical system

The model discussed above forms a 23-dimensional dynamical system consisting of differential equations for the following 17 state variables for the economic module

$$(a, N, \mathrm{w}, p, K_C, K_G, \Gamma, L_f, M_f, \sigma_C, g_{\sigma_C}, \sigma_G, g_{\sigma_G}, p_{BS}, p_C, B, R)\,, \tag{5.67}$$

coupled with differential equations for the following 6 state variables for the climate module

$$\left(E_{land}, \mathrm{CO}_2^{AT}, \mathrm{CO}_2^{UP}, \mathrm{CO}_2^{LO}, T, T^{LO}\right). \tag{5.68}$$

Compared with the 16-dimensional system proposed in Bovari, Giraud, and Mc Isaac (2018), we see that our economic module has separate dynamics for K_C and K_G instead of for the total capital $K = K_C + K_G$, as well as separate dynamics for the three financing components for firms, namely loans L_f, deposits M_f and green bonds Γ instead of the net debt $D = L_f + \Gamma - M_f$ (or simply loans minus deposits in the case of Bovari, Giraud, and Mc Isaac (2018)), in addition to the two state variables B and R corresponding to the Government and central bank sectors. Similarly, our climate module has separate dynamics for the emissions intensities of conventional and green production (as well as their growth rates), instead of a single emission intensity for the economy, resulting in 7 additional state variables in total.

As in Bovari, Giraud, and Mc Isaac (2018), the climate and macroeconomic modules are coupled through the summation of industrial and land emissions (5.55), along with the damage function defined in (5.64)-(5.66).

To further our analysis and stay in line with Bovari, Giraud, and Mc Isaac (2018), we consider a time in the distant future when the transition toward net-zero emissions has already been completed. Accordingly, with $E = 0$ in (5.57), carbon concentrations eventually stabilizes near their equilibrium values, resulting in an equilibrium for the industrial forcing F_{ind} in (5.60). Additionally, since the residual forcing F_{exo} grows exogenously and linearly until the year 2100, where it ultimately plateaus, we know that the upper limit is reached at every asymptotic equilibrium, so that total forcing F also approaches an equilibrium value. Consequently, the temperature anomalies in (5.62)-(5.63) also stabilize. Using the equilibrium atmospheric temperature anomaly T_{eq}, we find that damage to output and capital also converge to asymptotic limits $\mathbf{D}^Y_{eq}$ and $\mathbf{D}^K_{eq}$, respectively. Lastly, the portion of abatement undertaken by the productive sector tends toward $A_{eq} = 0$, as the emissions intensities σ_G and σ_C both go to zero. We can then insert these asymptotic damages back into the economic module and consider its effect on the equilibrium values for the economic variables.

We can first note that the growth rate of total output evolves according to

$$g(\pi) = \frac{\dot{Y}}{Y} = \frac{(1 - \mathbf{D}^Y_{eq})\kappa(\pi)}{\bar{\nu}} - \delta_{eq}. \tag{5.69}$$

where $\delta_{eq} = \bar{\delta} + \mathbf{D}^K_{eq}$ is the equilibrium depreciation rate. Next, we can assume that the population has reached its carrying capacity $\bar{N}_{max}$, so that $g_N = 0$ in (5.13). Finally, since pre-abatement emissions are also zero in this asymptotic regime, we see from (5.51) that $\delta_\Gamma = 0$, so that $r_\Gamma = r_L$ and $\beta(\delta_\Gamma) = \bar{\beta}_0$, meaning the green investment converges to a constant proportion of overall desired investment by firms, leading to a constant ratio $\varkappa_{eq}$ of green capital to total capital.

Using the definitions in (5.12), (5.17) and (5.33), it is straightforward to show that, post-transition to a zero-emissions economy, the main economic model reduces to

$$\begin{cases} \dot{\omega} = \omega\left[\Phi(\lambda) - i(\omega) - \bar{\alpha}\right] \\ \dot{\lambda} = \lambda\left[g(\pi) - \bar{\alpha}\right] \\ \dot{\ell_f} = \kappa(\pi)(1-\bar{\beta}_0) - \frac{\delta_{eq}\bar{\nu}(1-\varkappa_{eq})}{(1-\mathbf{D}^Y_{eq})} - \ell_f[i(\omega)+g(\pi)] \\ \dot{\gamma} = \kappa(\pi)\bar{\beta}_0 - \frac{\delta_{eq}\bar{\nu}\varkappa_{eq}}{(1-\mathbf{D}^Y_{eq})} - \gamma[i(\omega)+g(\pi)] \\ \dot{m_f} = 1-\omega - \frac{\delta_{eq}\bar{\nu}}{(1-\mathbf{D}^Y_{eq})} - (r_B(\omega)+\bar{\delta}_L)(\ell_f+\gamma) - \Delta(\pi) \\ \qquad +m_f(r_B(\omega) - i(\omega) - g(\pi)) \end{cases}, \tag{5.70}$$

where the asymptotic variable π is defined as

$$\pi = 1 - \omega - (r_B(\omega)+\bar{\delta}_L)(\ell_f+\gamma) + r_B(\omega)m_f - \frac{\delta_{eq}\bar{\nu}}{(1-\mathbf{D}^Y_{eq})} \tag{5.71}$$

and the functions Φ, i, g, κ, r_B and Δ are defined in (5.15), (5.16), (5.69), (5.23), (5.48), and (5.20), respectively.

We also obtain a secondary differential system containing the state variables (b, ρ) as follows

$$\begin{cases} \dot{b} = r_B(\omega)\rho - (r_B(\omega)+\bar{\delta}_L)\gamma - b(i(\omega)+g(\pi)) \\ \dot{\rho} = \kappa(\pi)\bar{\beta}_0 - (r_B(\omega)+\bar{\delta}_L)\gamma \\ \qquad +\rho\left(r_B(\omega) - i(\omega) + g(\pi)\right)) - \frac{\delta_{eq}\bar{\nu}\varkappa_{eq}}{(1-D^Y_{eq})} \end{cases}. \tag{5.72}$$

Starting from the second equation in (5.70), we observe that the equilibrium growth rate must satisfy

$$g_{eq} = \bar{\alpha}, \tag{5.73}$$

which describes the balanced growth path or "Solovian" equilibrium, observed in Bovari, Giraud, and Mc Isaac (2018). Consequently, we can use (5.69) to determine the equilibrium profit rate as

$$\pi_{eq} = \kappa^{-1}\left[\frac{\bar{\nu}(\bar{\alpha}+\delta_{eq})}{1-\mathbf{D}^Y_{eq}}\right]. \tag{5.74}$$

Similarly, we can use the first expression in (5.70) to establish that the equilibrium employment must satisfy

$$\lambda_{eq} = \Phi^{-1}(\bar{\alpha}+i_{eq}) = \Phi^{-1}\left(\bar{\alpha}+\bar{\eta}(\bar{\xi}\omega_{eq}-1)\right). \tag{5.75}$$

Similarly, it follows from the other three equations in (5.70)) that the equilibrium loans, green bonds and deposit ratio must satisfy

$$\ell_{f_{eq}} = \frac{\kappa(\pi_{eq})(1-\bar{\beta}_0) + \frac{\delta_{eq}\bar{\nu}(1-\varkappa_{eq})}{(1-\mathbf{D}^Y_{eq})}}{\bar{\eta}(\bar{\xi}\omega_{eq}-1)+\bar{\alpha}}, \tag{5.76}$$

$$\gamma_{eq} = \frac{\kappa(\pi_{eq})\bar{\beta}_0 + \frac{\delta_{eq}\bar{\nu}\varkappa_{eq}}{(1-\mathbf{D}^Y_{eq})}}{\bar{\eta}(\bar{\xi}\omega_{eq}-1)+\bar{\alpha}}, \tag{5.77}$$

$$m_{f_{eq}} = \frac{\omega_{eq} - \frac{\delta_{eq}\bar{\nu}}{(1-\mathbf{D}^Y_{eq})} + (r_B(\omega_{eq})+\bar{\delta}_L)(\ell_{f_{eq}}+\gamma_{eq}) + \Delta(\pi_{eq}) - 1}{r_B(\omega_{eq})+\bar{\delta}_L - \bar{\eta}(\bar{\xi}\omega_{eq}-1) - \bar{\alpha}}. \tag{5.78}$$

Finally, we use (5.71) to obtain that the equilibrium wage must satisfy

$$\omega_{eq} = 1 - \pi_{eq} - (r_B(\omega_{eq}) + \bar{\delta}_L)(\ell_{f_{eq}} + \gamma_{eq}) + r_B(\omega_{eq})m_{f_{eq}} - \frac{\delta_{eq}\bar{\nu}}{(1 - \mathbf{D}^Y_{eq})}. \tag{5.79}$$

That is to say, an equilibrium for (5.70) can be found as a solution to the nonlinear systems of equations (5.75)–(5.79).

Once this is found, the equilibrium values for the public debt and reserve ratios are then given by

$$b_{eq} = \frac{r_B(\omega_{eq})\rho_{eq} - (r_B(\omega_{eq} + \bar{\delta}_L)\gamma_{eq}}{\bar{\eta}(\bar{\xi}\omega_{eq} - 1) + \bar{\alpha}}, \tag{5.80}$$

$$\rho_{eq} = \frac{\frac{\delta_{eq}\bar{\nu}\varkappa_{eq}}{(1-\mathbf{D}^Y_{eq})} + (r_B(\omega_{eq} + \bar{\delta}_L)\gamma_{eq} - \bar{\beta}_0\kappa(\pi_{eq})}{r_B(\omega_{eq} - \bar{\eta}(\bar{\xi}\omega_{eq} - 1)\bar{\alpha}}. \tag{5.81}$$

We can use this characterization of the post-transition equilibrium to understand the destabilizing effects of a rising equilibrium temperature and climate damages. More specifically, when looking at (5.74), we see that an increase in either damage to capital or damage to output forces an increase in investment in order to maintain the economy on the balanced growth path. Naturally, this raises the equilibrium profit share and, by extension, the asymptotic loans share. Additionally, climate damages reduce deposits, causing the overall debt share to increase. Ultimately this combined rise in profit share and debt-to-output results in a reduction in the wage share. As expected, this behavior is magnified with higher temperature anomalies where beyond a particular average temperature, the corresponding debt share tends toward infinity while the wage share and employment rate vanishes to zero. In other words, we observe the bad equilibrium implying an economic collapse.

In the subsequent section, this phenomenon will be demonstrated numerically. We will see that incorporating climate damages causes the economy to converge toward an unstable growth path characterized by an infinite debt ratio and collapsing wage share and employment rate.[3] Hence, to avoid an economic collapse, an aggressive combination of climate policies, such as a carbon tax, abatement subsidies, and a green QE program designed to incentivize a shift toward renewable energies, must be adopted to reverse the economy back toward a stable growth path.

5.5 Numerical analysis

In this section, we begin by introducing the four policy scenarios that will form the basis of our numerical analysis. Subsequently, we explore the transitional dynamics of our macroeconomic model to gain insight into the economic shock of climate change and the impact of incorporating various fiscal and monetary policies. Afterwards, we divide our sensitivity analysis into a medium and long-term comparison to examine the effectiveness of a green QE program over time. Finally, we perform Monte Carlo simulations for our respective policy scenarios to account for the uncertainty inherent in the underlying parameters. Note that all numerical results were obtained by simulating the model in `R` (R Core Team, 2024) using the `deSolve` (Karline Soetaert, Thomas Petzoldt, and R. Woodrow Setzer, 2010) package and `lsoda` integration technique.

[3]See Bovari, Giraud, and Mc Isaac (2018) for numerical illustration of the bifurcation from the Solovian steady state as a result of rising temperatures and climate damages.

TABLE 5.4: Summary of policy scenarios considered

Scenario	*No policy*	*Strong subsidy*	*Strong subsidy with green QE*	*Weak subsidy with green QE*
Carbon Tax	No	High	High	High
Subsidy	No	64%	64%	10%
Green QE	No	No	Yes	Yes

5.5.1 Policy scenarios

For our numerical analysis, we will be comparing four different policy scenarios. In the spirit of Bovari, Giraud, and Mc Isaac (2018), we will first examine the "business-as-usual" *No Policy scenario*, where we introduce the Weitzman damage function from Weitzman (2012) with a damage to capital allocation of $\bar{f}_K = 1/3$. Here, state intervention is minimal as the government implements an inefficient carbon tax with a growth rate of 2% per year. This scenario aligns with Nordhaus (2018) and acts as a benchmark for our remaining scenarios. Next, in the *Strong subsidy scenario*, we consider the identical damage specifications but include a government subsidy on abatement of 64% along with an aggressive carbon tax according to the High-Level Commission on Carbon Prices (2017), namely a carbon tax of US\$80/tCO_2 by 2020 increasing linearly to US\$100/tCO_2 by 2030. Beyond 2030, we assume the price of carbon continues to grow linearly with a rate of change of 2% per year. As a natural next step, we also consider the *Strong subsidy scenario with green QE* where the central bank implements a green asset purchasing program in addition to the *Strong subsidy scenario*. Lastly, to highlight the efficacy of the green QE program, we analyze the *Weak subsidy scenario with green QE*, where the public sector only provides a 10% subsidy on abatement, while the central bank maintains its green QE program. We summarize the four scenarios in Table 5.4 for quick reference.

5.5.2 Transitional dynamics

Figures 5.2 and 5.3 depicts some of the key economic and climate-related trajectories for our four scenarios.

Beginning with the *No policy scenario*, we see that the economy initially remains on its balanced growth path as the growth rate is approximately equal to the exogenously set labor productivity of 2%.[4] However, by the year 2049, the increasing emissions and temperature levels beyond the 2°C threshold causes damage to output and capital to grow. Absent any climate policies, this behavior continues, and the damage to GDP significantly reduces the growth rate. Consequently, the debt ratio of the productive sector reaches the threshold of $\bar{\nu} = 2.7$, beyond which the net debt level of the sector exceeds the total stock of capital, meaning that the equity of the sector reaches zero according to (5.2), a situation that we associate with potential generalized defaults in the economy. As expected, by the end of the century, we see the beginning of a collapsing wage share and employment rate, as well.

Shifting the focus to the *Strong subsidy scenario*, it becomes clear that imposing a carbon tax on production immediately depresses the growth rate and inflates the debt levels of the productive sector, due to increase abatement costs for firms. However, the presence of a 64% subsidy on abatement costs allows firms to reduce emissions quickly and remain on a balanced growth path. In fact, under this policy combination, the economy reaches net-zero emissions by 2043. This is much earlier than the target year of 2050, which many countries

[4]As in Bovari, Giraud, and Mc Isaac (2018), absent climate damages, our dynamical system admits a "Solovian" equilibrium or balanced growth path, where the real growth rate stabilizes such that $g_{eq} = \frac{\dot{Y}}{Y} = \bar{\alpha}$.

FIGURE 5.2: Economic trajectories for the *No policy scenario* (red line), *Strong subsidy scenario* (blue line), *Strong subsidy with Green QE scenario* (green line) and *Weak subsidy with Green QE scenario* (orange line). The dotted black line in the debt share plot indicates the 2.7 debt-to-output threshold.

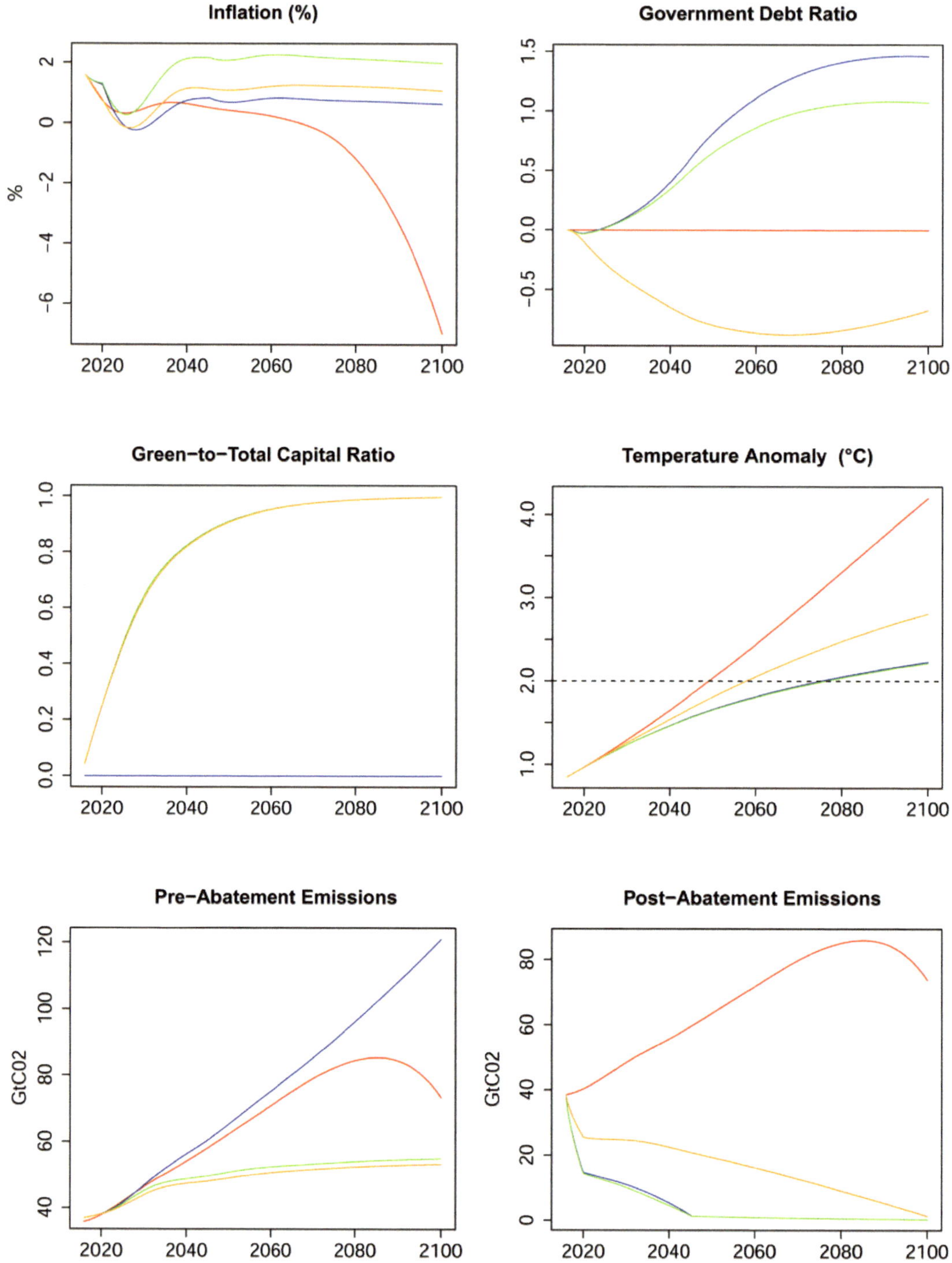

FIGURE 5.3: Supplementary economic and climate trajectories for the *No policy scenario* (red line), *Strong subsidy scenario* (blue line), *Strong subsidy with Green QE scenario* (green line) and *Weak subsidy with Green QE scenario* (orange line). The dotted black line in the temperature anomaly plot indicates the 2°C target set by the Paris Climate Agreement.

that signed the Paris Agreement have set as the deadline to achieve carbon neutrality. Note that the high subsidy rate causes the government debt ratio to reach 1.46 by the year 2100. Furthermore, the aggressive climate policies implemented by the public sector merely delays the global temperature from increasing beyond the 2°C ceiling in comparison to the *No policy scenario.* Nevertheless, the end-of-century temperature anomaly is restricted to 2.2°C in this scenario, which is a substantial improvement from the 4.2°C temperature levels observed in the *No policy scenario.*

Now, despite successfully returning the economy to its balanced growth path and reducing the effects of climate change, the policy combination observed in the *Strong subsidy scenario* places an unprecedented burden and expectation on the government sector. But when we introduce a green asset purchasing program undertaken by the central bank, the economic and climate outlook improves even further, as illustrated in the *Strong subsidy with green QE.* More specifically, the favorable spread between bank loans and green bonds not only decreases the private sector's debt levels but also improves the firms' short-term profit share and entices the sector to seek more investment in green technologies. Naturally, the subsequent spike in production and the growth rate increases the employment rate, wage share and inflation. Additionally, the rapid adoption of green capital, along with the increase in central bank profits from green QE, leads to a corresponding decrease in the government debt ratio, as well. Following the successful transition toward net-zero emissions we see the growth rate and the economy as a whole beginning to stabilize toward a balanced growth path. Looking at the climate trajectories, as the productive sector increasingly shifts production toward renewable energies, there is an analogous decrease in the pre-abatement emission, which reduces post-abatement CO_2 emissions even quicker, represented by the area under the curve being smaller. Moreover, the temperature anomaly only marginally decreases compared to the *Strong subsidy scenario.*

Finally, in the *Weak subsidy scenario with green QE,* we see that without a significant subsidy on abatement, firms take full advantage of the central banks' green bond-buying program and willingly adopt green technologies causing the economy to return to its stable growth path and observe comparable debt levels, wage share and employment rates to the *Strong subsidy scenario.* Additionally, since green QE acts as the main climate policy driver, the government observes a surplus due to the low subsidy rate on abatement and accumulating revenue from the tax on carbon. On the climate side, the lack of aggressive emissions reduction by the private sector leads to net-zero emissions being achieved only at the end of the century. Naturally, this leads to a higher temperature anomaly of 2.8°C by the year 2100.

As evident in the transitional dynamics, the introduction of green bonds creates another channel through which public authorities can reduce the debt share of the productive sector via a green asset purchasing program. Moreover, by maintaining a wide spread between the bank loan rate and the yield on green bonds, the central bank incentivizes the adoption of more green technologies that are conducive to a lower-carbon economy. Naturally, this reduces the equilibrium temperature anomaly and the long-term climate damages absorbed by the private sector. Furthermore, the burden of the government sector to maintain the economy on a stable growth path is reduced since both fiscal and monetary authorities can now effectively share the responsibility of addressing the climate crisis. In the subsequent section, we will turn to the model's medium and long-term behavior to highlight the flexibility of the various policy specifications that successfully maintain the economy on its balanced growth path.

5.5.3 Sensitivity analysis

Within our economic framework, the productive sector can issue green bonds to finance its renewable production processes, while the central bank can ease financial conditions and propel a shift toward more sustainable energies by implementing a green QE program. At the same time, the government sector can levy a carbon tax along with minimizing the cost of abatement efforts through the implementation of a subsidy.

Thus, to examine the interactions and trade-offs between each of these three mechanisms, we perform a sensitivity analysis by varying the subsidy rate, $\bar{s}_a \in [0, 0.99]$, the magnitude of the central bank's response to rising pre-abatement emissions, $\bar{\phi}_2 \in [0, 0.05]$, and slope of the $\beta(\cdot)$ function, $\bar{\beta}_1 \in [0, 0.99]$. Broadly speaking, $\bar{\beta}_1$ captures the speed with which the productive sector embraces green technologies. Additionally, we extend the scope of potential economic outcomes by comparing the parameter ranges with and without a carbon tax. Note that the parameter values are selected from a sequence of 15 equally spaced points. Furthermore, all long-term analyses seen below are simulated up to the year 2500 to guarantee the dynamical system has reached equilibrium and to account for the long-term effects of climate change on the macroeconomic system.

Following Bolker, Grasselli, and Holmes (2021), the economic outcomes of each simulation were classified according to three possibilities:

(i) a "good" outcome if $0.4 \leq \omega \leq 1, 0.4 \leq \lambda \leq 1$ and $d \leq 2.7$, which resembles the Solovian equilibrium discussed above, where we observe non-zero wage share, employment rate, and finite debt levels;

(ii) an "outside bounds" outcome if either $\lambda > 1$ or $\omega > 1$; otherwise

(iii) a "bad" outcome that is synonymous with a vanishing wage share and employment rate, along with infinite debt levels.

The results of the long-term sensitivity analysis are pictured in Figure 5.4. Evidently, in the absence of an effective carbon pricing scheme, no amount of green debt instruments, government or central bank intervention prevents the economy from converging toward the climate-induced bad equilibrium, as can be seen in Figure 5.4a, while with a carbon tax in effect, the amount of good economic outcomes drastically improves, as shown in Figure 5.4b. Furthermore, we see an inverse relationship emerge between the magnitude of the central bank's green QE policy and the government sector's subsidy rate. More specifically, as the central bank becomes increasingly willing to maintain a large spread between bank loans and the yield on green bonds, the subsidy parameter required for the economy to converge to the good equilibrium decreases. In fact, beyond $\bar{\beta}_1 \geq 0.2829$ and $\bar{\phi}_2 \geq 0.0071$, the government does not need to provide any abatement subsidies as the rapidly falling emission intensity due to the adoption of renewable energies limits abatement costs for the productive sector.

For further insights, we also performed a medium-term sensitivity analysis by solely varying $\bar{\phi}_2$ in the *Weak subsidy with green QE scenario* and *Strong subsidy with green QE scenario* to examine the different pathways of the debt share and temperature anomaly over time, as seen in Figures 5.5 and 5.6. Confirming the long-term results, we see in these figures that debt-to-output ratio significantly decreases, along with an analogous decline in the temperature anomaly, for increasingly aggressive QE policies. Moreover, Figures 5.5a and 5.6a illustrates a bifurcation with respect to the parameter $\bar{\phi}_2$, in the sense that the system transitions from explosive to finite debt levels as $\bar{\phi}_2$ increases.

Interestingly, as the subsidy on abatement grows, the effectiveness of the green QE program diminishes. More specifically, there is a maximum $\bar{\phi}_2$ value where the debt-to-output ratio no longer decreases. Looking at both Figures 5.5a and 5.6a, we see that the

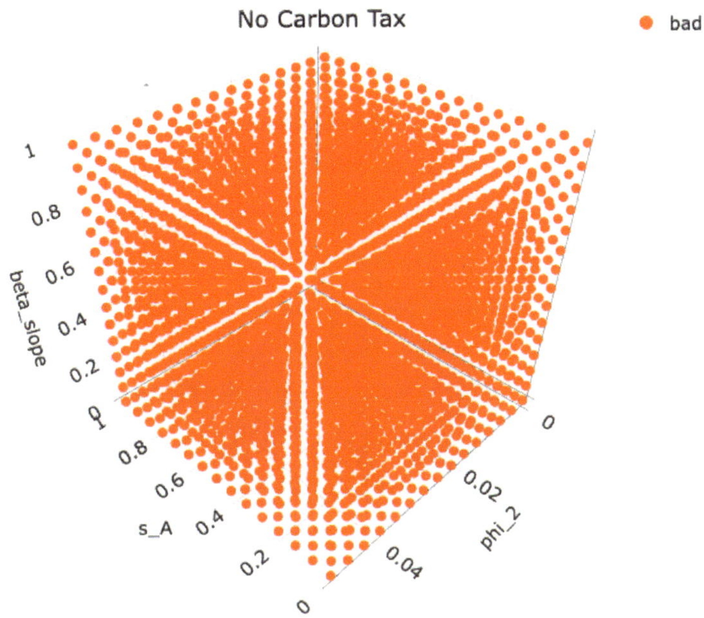

(a)

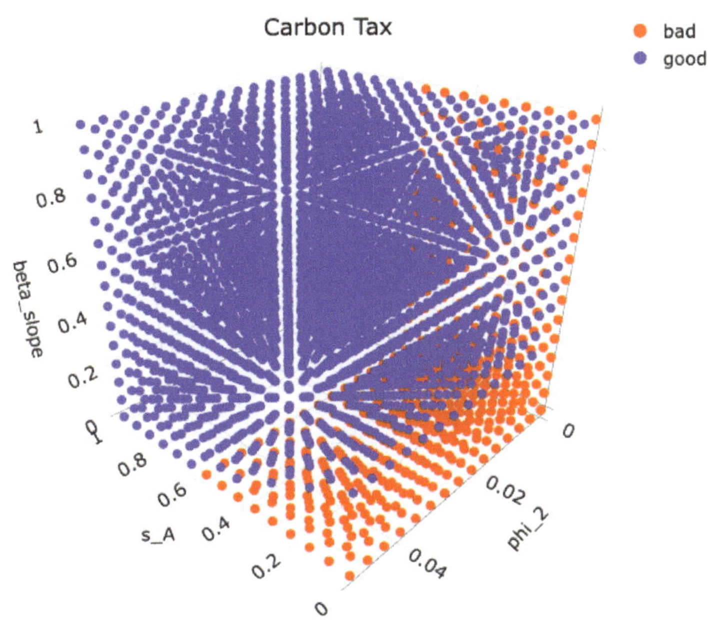

(b)

FIGURE 5.4: Economic outcomes with and without a carbon tax for a range of abatement subsidies $\bar{s}_a$, the magnitude of central banks response to rising emissions $\bar{\phi}_2$, and the slope of the $\beta(\cdot)$ function $\bar{\beta}_1$.

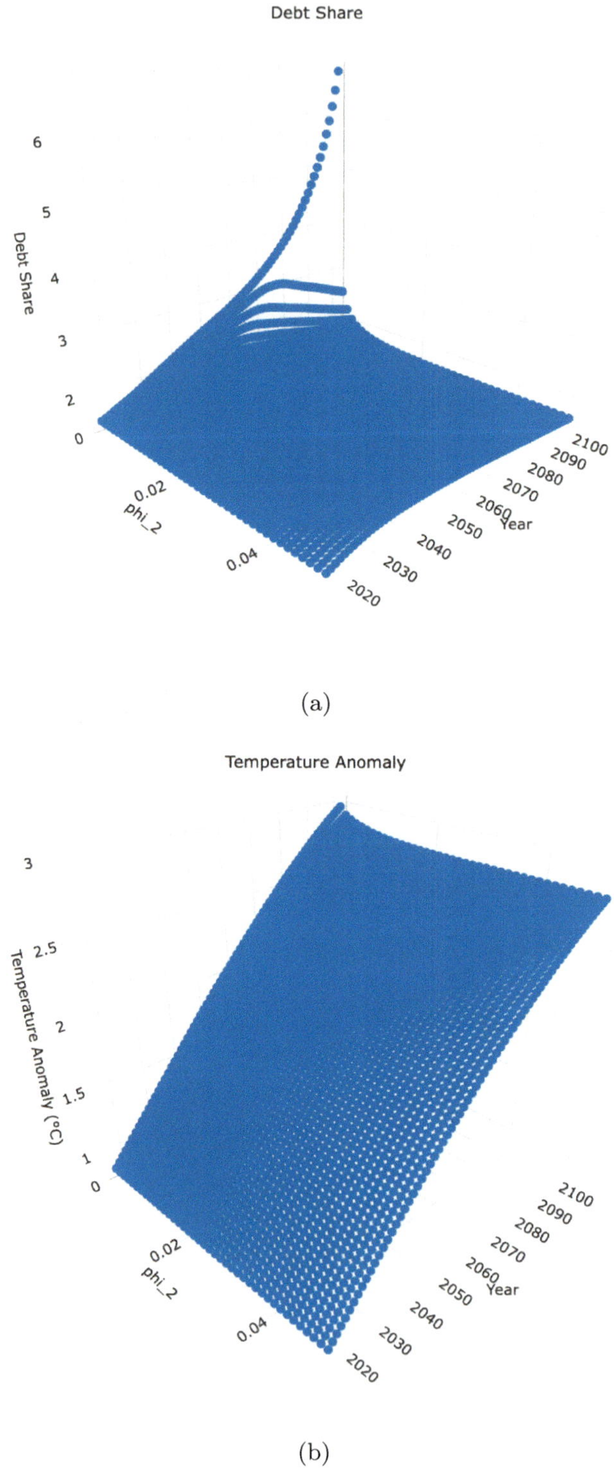

FIGURE 5.5: Time plots of the private sector debt share and temperature anomaly for *Weak subsidy scenario with green QE* for a range of values for the central bank response parameter $\bar{\phi}_2$.

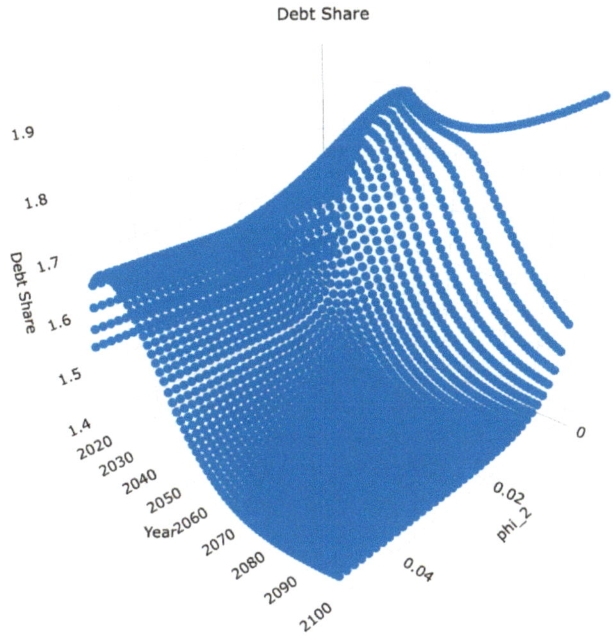

(a)

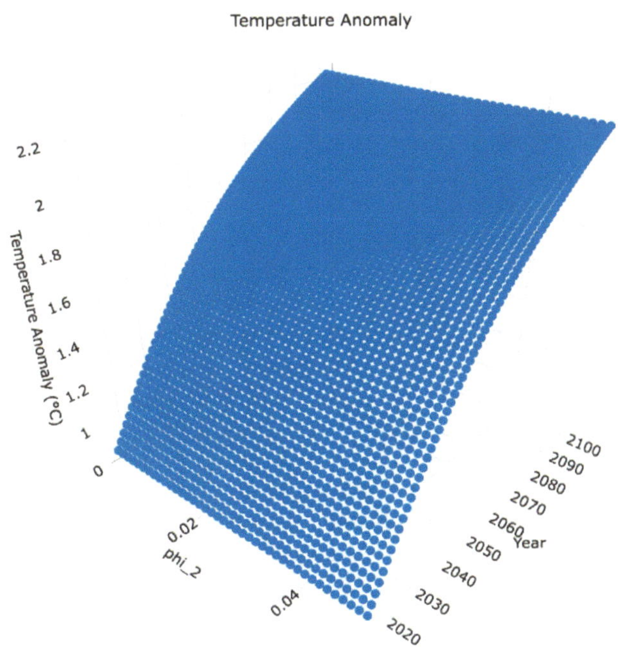

(b)

FIGURE 5.6: Time plots of the private sector debt share and temperature anomaly for *Strong subsidy scenario with green QE* for a range of values for the central bank response parameter $\bar{\phi}_2$.

end values of the debt share remain constant in the year 2100 beyond $\bar{\phi}_2 \geq 0.03$. This is due to the spread between the rate on bank loans and the yield on green bonds reaching the lower bound, $r_\Gamma = r_B$. Beyond this limit, we would undoubtedly observe further decreases the debt share, but it would be more realistic for the government sector to increase the subsidy rate than allow the central bank to promote green bond yields below the policy rate. Thus, in the presence of a subsidy, the central bank gains little from adopting an ultra-loose monetary policy beyond a certain maximum spread.

Hence, an explicit central bank QE policy designed to promote renewable energies in the production process significantly relieves the government's financial burden to bankroll abatement subsidies. Nevertheless, we must stress that the correct policy combination is not dependent only on the long-term economic and climate outcome. Depending on our initial economic conditions, we may observe very different long-term equilibria. Not to mention that imposing a carbon tax raises the productive sector's short-term debt levels and can accelerate the convergence toward an economic collapse. Thus, policymakers must account for the present monetary circumstances when setting appropriate environmental policies. For this, we turn to the basin of attraction analysis.

5.5.4 Basin of attraction

To perform our basin of attraction analysis, we classify economic outcomes for a range of initial conditions concerning the wage share, employment rate and debt share for each of our four scenarios, as these were the variables considered in Bovari, Giraud, and Mc Isaac (2018). Note that we make no alterations to the outcome classification from the previous section. The "initial set" for the state variables, $(\omega, \lambda, d) \in [0.2 : 0.99]^2 \times [0.1 : 2.7]$, is borrowed directly from Bovari, Giraud, and Mc Isaac (2018) and corresponds to a reasonable range of initial conditions where we can observe both favorable and unfavorable conditions. By favorable we simply mean an economy where both the initial wage share and employment rate are high, whereas the initial debt level is low.

Looking at Figure 5.7a, we see that no matter how favorable the initial conditions, the impact of climate change in the absence of any meaningful policies causes the economy to converge toward the bad equilibrium. Conversely, when incorporating an aggressive carbon pricing scheme and a 64% subsidy on abatement, we see considerably more good economic outcomes as the convergence set is approximately $(\omega, \lambda, d) \in [0.2 : 0.99] \times [0.6 : 0.99] \times [0.1 : 2.14]$, as shown in Figure 5.7b. In other words, we see that in the presence of climate damages, it is economies with a high initial wage share and employment rates, along with low debt levels, that can withstand the price associated with the carbon tax. Nonetheless, this still places an unprecedented burden on the government sector to alleviate the abatement costs and return these economies to their stable growth path.

On the other hand, when we introduce a green asset purchasing program undertaken by the central bank, we see the range of desirable economic outcomes expand even further as the set of initial conditions that converge toward the Solovian equilibrium increases to $(\omega, \lambda, d) \in [0.2 : 0.99] \times [0.2 : 0.99] \times [0.1 : 2.32]$, as shown in Figure 5.8a. This improvement is a direct consequence of the central bank's bond-buying program, which not only lowers the private sector's financial burden in the near term but also incentivizes firms to adopt a cleaner and less emission-intensive production process, which minimizes the long-term effects of climate change.

Lastly, if the government reduces its subsidy on abatement costs to 10% while the central bank maintains its green-bond buying program, the set of initial values that converge toward the good equilibrium is $(\omega, \lambda, d) \in [0.2 : 0.99] \times [0.48 : 0.99] \times [0.1 : 2.14]$, as shown in Figure 5.8b. This region is a noticeably larger basin of attraction compared to the *Strong subsidy*

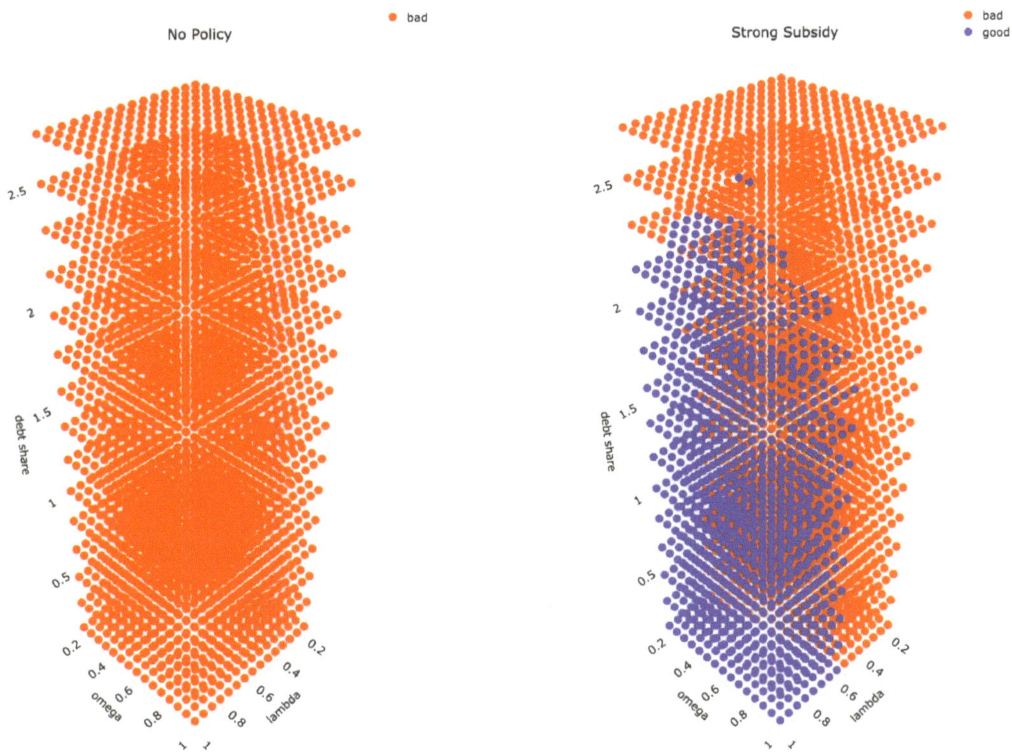

(a) (b)

FIGURE 5.7: Basin of attraction results for the *No policy scenario* and *Strong subsidy scenario* for a range of initial conditions of the wage share ω, employment rate λ, and debt share d.

scenario and highlights the efficacy of the central bank's QE program in alleviating the fiscal responsibility of the public sector in managing the risks associated with climate change.

Now, we must emphasize that our results up to this point are deterministic in the sense that they only show the dynamics and economic outcomes for a very particular set of calibrated parameters. Thus, to consider the inherent uncertainty of the economic and climate parameters, we will perform a final sensitivity analysis for our medium-term dynamics in the form of Monte Carlo simulations. Doing so will strengthen our deterministic results and allow us to examine the number of simulations that remain below the 2.7 debt-to-output and 2°C temperature thresholds, respectively.

5.5.5 Parameter uncertainty

For the Monte Carlo simulations, we stay in line with Bovari, Lecuyer, and Isaac (2018) and vary the labor productivity $\bar{a}$ in (5.11), the equilibrium climate sensitivity $\bar{S}$ in (5.62), and the concentration of preindustrial CO_2-e in the biosphere and upper ocean $\bar{C}^{UP}_{pre}$ in (5.57). All of these parameters have been widely studied in climate literature, and estimates for each probability density function (PDF) could be found. The PDFs, in our case, are taken directly from Bovari, Lecuyer, and Isaac (2018) and Nordhaus (2018).

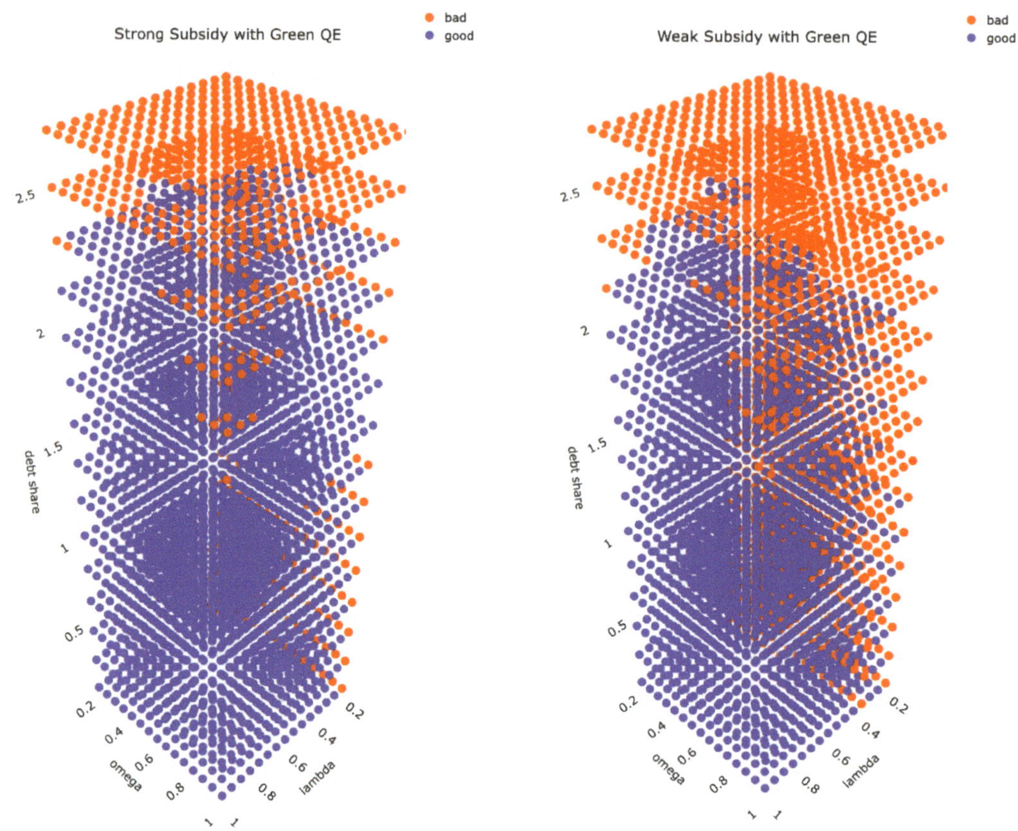

(a) (b)

FIGURE 5.8: Basin of attraction results for the *Strong subsidy scenario with green QE* and *Weak subsidy scenario with green QE* for a range of initial conditions of the wage share ω, employment rate λ, and debt share d.

Specifically, we assume that the growth rate of labor productivity is normally distributed with a mean of 0.0206 and a standard deviation of 0.0112. Likewise, we vary the equilibrium climate sensitivity parameter according to a log-normal distribution with a mean of 1.107 and a standard deviation of 0.264. This parameter is vital in dictating how resistant the environment is toward an increase in CO_2 emissions. Finally, we assume that the size of the carbon reservoir follows a log-normal distribution with mean 5.8855763 and standard deviation 0.2512867. Note, we sample the PDFs for each parameter 1000 times and run each simulation from 2016-2100.

Figure 5.9 depicts the trajectories of our Monte Carlo simulations for each policy scenario. Note, we omitted the *Weak subsidy with green QE scenario* from our results and subsequent analysis to reduce clutter in our graphical representation. Additionally, the transitional dynamics for each scenario do not change from the deterministic results observed in Section 5.5.2. Thus, we move straight into discussing the probability of staying below the critical temperature and debt thresholds.

In the *No policy scenario*, the temperature anomaly by 2100 has a median value of 4.09°C and stays below the 2°C target of the Paris Agreement in only 0.1% of the simulations. Given

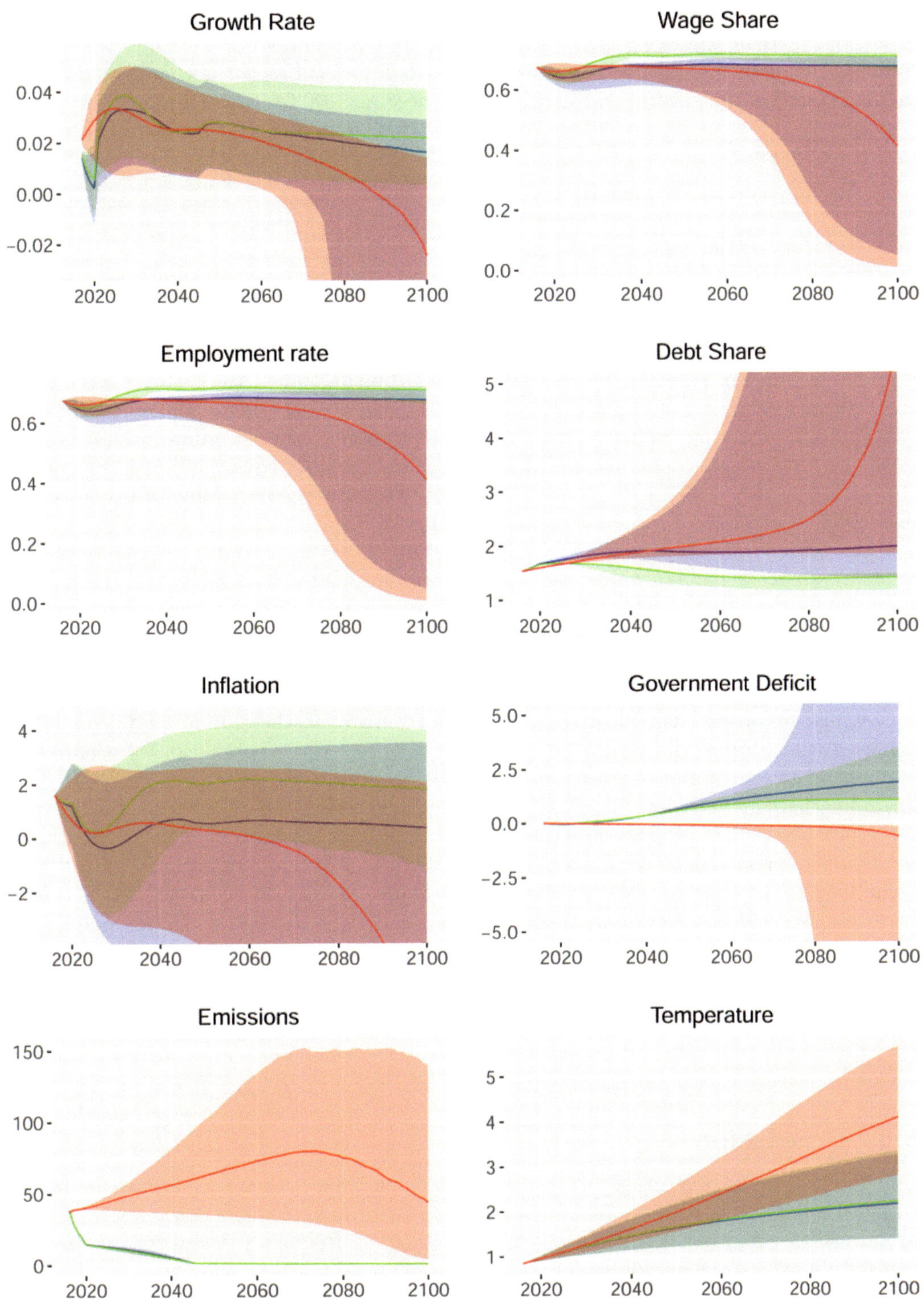

FIGURE 5.9: Median values and [0.05; 0.95] probability interval for the *No policy scenario* (red line), *Strong subsidy scenario* (blue line), *Strong subsidy with Green QE scenario* (green line).

TABLE 5.5: Median values of the temperature anomaly T and percentage of runs with $T < 2°C$, median values for private debt ratio and percentage of the runs with $d < 2.7$, followed by the percentage of the runs with both T and d below their thresholds, and the median values of the government debt ratio in the year 2100 for each climate policy scenario

Scenario	T(°C)	$T < 2°C$	d	$d < 2.7$	$T < 2°C$ *and* $d < 2.7$	g
No Policy	4.09	0.1%	5.83	31.2%	0.1%	−0.63
Strong Policy	2.15	38.7%	1.99	70.1%	28.8%	1.87
Strong Policy with Green QE	2.21	36.4%	1.40	100.0%	36.4%	1.05

that no carbon tax is levied by the government, by the end of the century we only start to observe the impacts of climate damages on the private debt ratio, which show a median value of 5.83 and stays below the 2.7 threshold in 31.2% of the simulations. Accordingly, both the temperature anomaly and debt share stay simultaneously below their respective thresholds in only 0.1% of the simulations. Note that the median value of the government debt ratio is -0.63.

The outlook drastically improves in the *Strong policy scenario* as the temperature anomaly shows a median value of 2.15 and remains under 2°C in 38.7% of the simulations, while the private debt ratio has a median value of 1.99 and remains below the 2.7 threshold in 70.1% of runs. Accordingly, the temperature and debt trajectories remain simultaneously below their thresholds in 28.8% of simulations. As expected, the implementation of government subsidies causes the median debt ratio to increase to 1.87 by the year 2100.

Finally, we observe the most significant improvement in the climate and economic outcomes in the *Strong policy with green QE scenario*, as the combination of a carbon tax, abatement subsidies and a green bond purchasing program results. Whereas the temperature anomaly stays below the Paris Agreement targets in 36.4% of runs with a median value of 2.21 (that is, slightly worse than in the previous scenario if taken in isolation), we see that the private debt ratio avoids its threshold in 100% of simulations, with a median value of 1.4. Once again, this highlights the effectiveness of the central bank's green QE program. Accordingly, the temperature anomaly and debt share stay simultaneously below their thresholds in 36.4% of the runs under this policy. Moreover, the median value of the government debt ratio in this scenario is 1.05, again indicating a marked improvement from carbon taxes and a strong subsidy alone. We summarize the results in Table 5.5.

5.6 Conclusion

Climate change presents a formidable challenge for the human race. Whereas the future physical effects of a warmer planet have been established beyond controversy, and some are already being observed with increasingly alarming frequency, there remains much uncertainty on the transmission channels through financial markets and the general economy. Moreover, the combination of externalities, mismatch of time horizons, and truly global nature of the risks involved, make it prone to be subject to market failures and lack of coordination in the potential responses by the several agents involved. It is exactly in these circumstances, which resembles the components of the worst financial crises, that intervention from central banks become of paramount importance.

In this Chapter, we proposed a stock-flow consistent climate-economy integrated model with the purpose of exploring the consequences of a particular type of intervention, namely large scale purchase by central banks of debt securities issued by firms with the purpose of funding projects in the low-carbon economy, that is to say, green quantitative easing. Through simulations of the model, we established that such program can be effective in keeping the global temperature increase within manageable levels, while also achieving sustainable private and public debt level. These results are robust with respect to changes in policy parameters, initial conditions and reaction functions of economic agents, as well as uncertainty with respect to underlying climate parameters.

We conclude by stressing that implementing a green quantitative easing program should not be viewed simply as an augmentation of the current central bank mandate. Despite their widespread deployment since the 2008 crisis, quantitative easing programs aimed at government bills, mortgage-based securities and other forms of debt instruments were designed to be a policy action to stabilize financial markets once the central bank's usual monetary toolkit had been exhausted, that is to say, temporary by nature and only use in extraordinary circumstances. On the other hand, the type of green quantitative easing explored here would require a much longer commitment spanning multiple decades and implemented in parallel to usual monetary policy. Therefore, instituting a green bond purchasing program would entail a redefining, or at the very least, a re-examination of the role the central bank play and how climate change factors into the tenets of financial stability, price stability and full employment.

5.7 Appendix

TABLE 5.6: Initial conditions for state variables

Symbol	*Description*	*Equation*	*Value*
a	Labor productivity	(5.11)	18.1924
N	Workforce, in billions	(5.13)	4.83
w	Wage rate	(5.14)	10.5014
p	Composite good price level	(5.16)	1
K_C	Conventional capital	(5.28)	53.5532
K_G	Green capital	(5.28)	6.3980
Γ	Green Bonds, in trillions USD	(5.29)	3.630964
L_f	Conventional corporate loans, in trillions USD	(5.30)	83.65741
M_f	Corporate deposits, in trillions USD	(5.31)	0.124
σ_C	Conventional emissions reduction rate	(5.35)	0.03
g_{σ_C}	Growth rate of the conventional emissions intensity of the economy	(5.35)	−0.0105
σ_G	Green emissions reduction rate	(5.36)	0.03
g_{σ_G}	Growth rate of the green emissions intensity of the economy	(5.36)	−0.02
p_{BS}	Price level of backstop technology, in 2016 USD	(5.39)	547.22
p_C	Price of carbon, in 2016 USD	(5.44)	0.721
B	Government bonds issued, in trillions USD	(5.46)	0
R	Central bank reserves in trillions USD	(5.53)	0
E_{land}	Exogenous land use CO_2-e emissions, in Gt CO_2-e	(5.56)	2.6
CO_2^{AT}	CO_2-e concentration in the atmosphere layer, in Gt C	(5.57)	851
CO_2^{UP}	CO_2-e concentration in the biosphere and upper ocean layer, in Gt C	(5.57)	460
CO_2^{LO}	CO_2-e concentration in the lower ocean layer, in Gt C	(5.57)	1740
T	Temperature anomaly, in °C	(5.62)	0.85
T_0	Temperature anomaly in lower ocean layer, in °C	(5.63)	0.0068

TABLE 5.7: Initial conditions for auxiliary variables

Symbol	*Description*	*Equation*	*Value*
$\varkappa$	Ratio of green capital to total capital	(5.8)	0.04
n	Emission reduction rate	(5.43)	0.03

TABLE 5.8: Parameters values for model

Symbol	*Parameter description*	*Equation*	*Value*
$\bar{\nu}$	Capital-to-output ratio	(5.7)	2.7
$\bar{\alpha}$	Productivity of growth rate	(5.11)	0.02
$\bar{\delta}_N$	Growth rate of workforce	(5.13)	.0305
$\bar{N}_{max}$	Upper limit of workforce dynamics in billions	(5.13)	7.056
$\bar{m}$	Money illusion parameter	(5.14)	0
$\bar{\phi}_0$	Constant of short-term Phillips curve, $\Phi(\cdot)$	(5.15)	$-.292$
$\bar{\phi}_1$	Slope of short-term Phillips curve, $\Phi(\cdot)$	(5.15)	.4629
$\bar{\eta}$	Relaxation parameter for inflation	(5.16)	0.192
$\bar{\xi}$	Markup constant	(5.16)	1.875
$\bar{\delta}$	Depreciation rate of capital	(5.18)	0.04
$\bar{\Delta}_0$	Constant of the dividend function, $\Delta(\cdot)$	(5.20)	0.138
$\bar{\Delta}_1$	Slope of the dividend function, $\Delta(\cdot)$	(5.20)	0.4729
$\bar{\Delta}_{max}$	Maximum of the dividend function, $\Delta(\cdot)$	(5.20)	0.3
$\bar{\kappa}_0$	Constant of the investment function, $\kappa(\cdot)$	(5.23)	0.0318
$\bar{\kappa}_1$	Slope of the investment function, $\kappa(\cdot)$	(5.23)	0.575
$\bar{\kappa}_{max}$	Maximum of the investment function	(5.23)	0.3
$\bar{\kappa}_{min}$	Minimum of the investment function	(5.23)	0
$\bar{\beta}_0$	Constant of the portion of desired green investment function, $\beta(\cdot)$	(5.26)	0.052825
$\bar{\beta}_1$	Slope of the portion of desired green investment function, $\beta(\cdot)$	(5.26)	0.349616
$\bar{\beta}_{max}$	Maximum of the green investment function	(5.26)	1
$\bar{\beta}_{min}$	Minimum of the green investment function	(5.26)	0.05
$\bar{\delta}_{g_{\sigma_C}}$	Variation rate of the growth of conventional emission intensity	(5.35)	-0.001
$\bar{\delta}_{g_{\sigma_G}}$	Variation rate of the growth of green emission intensity	(5.35)	$-.00001$
$\bar{\delta}_{p_{BS}}$	Growth rate of the price of backstop technology	(5.39)	-0.0051
$\bar{\theta}_C$	Conventional abatement cost	(5.40)	2.6
$\bar{\theta}_G$	Green abatement cost	(5.40)	2.8
$\bar{s}_a$	Subsidy parameter (strong policy, weak policy)	(5.41)	(0.64, 0.10)
$\bar{a}_{pC}$	Growth rate of carbon price (no policy)	(5.44)	0.02
$\bar{b}_{pC}$	Time-varying component of growth rate of carbon price (no policy)	(5.44)	0
$\bar{r}^*$	Long-term interest rate	(5.48)	0.01
$\bar{\phi}_1$	Magnitude of reactivity of the monetary policy	(5.48)	0.5
$\bar{i}^*$	Interest rate targeted by the central bank	(5.48)	0.02
$\bar{\delta}_L$	Prime rate spread on bank loans	(5.49)	0.03
$\bar{\phi}_2$	Magnitude of central banks response to a change in emissions	(5.51)	0.03
$\bar{E}_0$	Initial amount of CO_2 emissions released by the productive sector	(5.52)	38.45
$\bar{E}_{min}$	Minimum/target amount of CO_2 emissions to be released by the productive sector	(5.52)	0.00001
$\bar{E}_1$	Speed of decrease (slope) of emission pathway function	(5.52)	-0.6638889
$\bar{\delta}_{E_{land}}$	Growth rate of land use change CO_2-e emissions	(5.56)	-0.022
$\bar{C}^{AT}_{pre}$	Preind. concentration of CO_2 in the atmosphere layer, in Gt C	(5.57)	588
$\bar{C}^{UP}_{pre}$	Preind. concentration of CO_2 in the biosphere/upper ocean layer, in Gt C	(5.57)	360
$\bar{C}^{LO}_{pre}$	Preind. concentration of CO_2 in the lower ocean layer, in Gt C	(5.57)	1720
$\bar{\phi}_{12}$	Transfer coefficient for carbon from AT to UP	(5.57)	0.0239069
$\bar{\phi}_{23}$	Transfer coefficient for carbon from UP to LO	(5.57)	0.0013409
$\bar{F}_{2\times CO_2}$	Change in radiative forcing from a doubling of preind CO_2, in W/m^2	(5.60)	3.6813
$\bar{F}_0$	Initial value of exogenous radiative forcing	(5.61)	0.5
$\bar{F}_{max}$	Value of exogenous radiative forcing in the year 2100	(5.61)	1
$\bar{F}_1$	Slope of the exogenous radiative forcing function	(5.61)	0.00595
$\bar{C}$	Heat capacity of the atmosphere, biosphere and upper ocean, in SI	(5.62)	1.02
$\bar{C}_0$	Heat capacity of the deeper ocean, in SI	(5.63)	3.52
$\bar{h}$	Heat exchange coefficient between temperature layers, in SI	(5.62)	0.0176
$\bar{\pi}_1$	Damage function parameter, in $1/°C$	(5.64)	0
$\bar{\pi}_2$	Damage function parameter, in $1/°C^2$	(5.64)	0.00236
$\bar{\pi}_3$	Damage function parameter from Weitzman (2012), in $1/°C^{\bar{\zeta}}$	(5.64)	.00000507
$\bar{\zeta}$	Damage function parameter from Weitzman (2012)	(5.64)	6.754

References

Bolker, B. M., M. R. Grasselli, and E. Holmes, 2021, Sensitivity analysis of an integrated climate-economic model, *SIAM Journal on Financial Mathematics* 12, SC44–SC57.

Bovari, E., G. Giraud, and F. Mc Isaac, 2018, Coping With Collapse: A Stock-Flow Consistent Monetary Macrodynamics of Global Warming, *Ecological Economics* 147, 383–398.

Bovari, E., O. Lecuyer, and F. M. Isaac, 2018, Debt and damages: What are the chances of staying under the 2° C warming threshold?, *International Economics* 155, 92–108.

Campiglio, E., 2016, Beyond carbon pricing: The role of banking and monetary policy in financing the transition to a low-carbon economy, *Ecological Economics* 121, 220–230.

Dafermos, Y., M. Nikolaidi, and G. Galanis, 2018, Climate change, financial stability and monetary policy, *Ecological Economics* 152, 219–234.

Dietz, S., and N. Stern, 2015, Endogenous growth, convexity of damage and climate risk: how Nordhaus' framework supports deep cuts in carbon emissions, *The Economic Journal* 125, 574–620.

Godley, W., and M. Lavoie, 2007, Monetary economics. An integrated approach to credit, money, income, production and wealth (Springer).

Grasselli, M. R., and B. C. Lima, 2012, An analysis of the Keen model for credit expansion, asset price bubbles and financial fragility, *Mathematics and Financial Economics* 6, 191–210.

Grasselli, M. R., 2022, Monetary policy responses to Covid-19: A comparison with the 2008 crisis and implications for the future of central banking, *Review of Political Economy* 34, 420–445.

Grasselli, M. R., and A. Nguyen-Huu, 2015, Inflation and speculation in a dynamic macroeconomic model, *Journal of Risk and Financial Management* 8, 285–310.

Grasselli, M. R., and A. Nguyen-Huu, 2018, Inventory growth cycles with debt-financed investment, *Structural Change and Economic Dynamics* 44, 1–13.

Guttmann, R., 2022, Central banks and the zero-carbon transition: an institutional challenge, in L.-P. Rochon, S. Kappes, and G. Vallet, eds., *Central Banking, Monetary Policy and the Environment*, Chapters, chapter 1, 21–53 (Edward Elgar Publishing).

IPCC, 2022, Summary for policymakers, in P. Shukla, J. Skea, R. Slade, A. A. Khourdajie, R. van Diemen, D. McCollum, M. Pathak, S. Some, P. Vyas, R. Fradera, M. Belkacemi, A. Hasija, G. Lisboa, S. Luz, and J. Malley, eds., *Climate Change 2022: Mitigation of Climate Change. Contribution of Working Group III to the Sixth Assessment Report of the Intergovernmental Panel on Climate Change* (Cambridge University Press, Cambridge, UK and New York, NY, USA).

Karline Soetaert, Thomas Petzoldt, and R. Woodrow Setzer, 2010, Solving differential equations in R: Package deSolve, *Journal of Statistical Software* 33, 1–25.

NGFS (Network of Central Banks and Supervisors for Greening the Financial System), 2019, A call for action: Climate change as a source of financial risk., Technical report, NGFS Secretariat, Paris.

Nordhaus, W., 2014, Estimates of the social cost of carbon: concepts and results from the DICE-2013R model and alternative approaches, *Journal of the Association of Environmental and Resource Economists* 1, 273–312.

Nordhaus, W., 2018, Projections and uncertainties about climate change in an era of minimal climate policies, *American Economic Journal: Economic Policy* 10, 333–360.

R Core Team, 2024, *R: A Language and Environment for Statistical Computing*, R Foundation for Statistical Computing, Vienna, Austria.

Weitzman, M. L., 2012, GHG targets as insurance against catastrophic climate damages, *Journal of Public Economic Theory* 14, 221–244.

Part II

Asset Pricing and Portfolio Management

Chapter 6

Dissecting green returns

Ľuboš Pástor
University of Chicago and NBER, USA; CEPR UK; and National Bank of Slovakia

Robert F. Stambaugh
University of Pennsylvania and NBER, USA

Lucian A. Taylor
*University of Pennsylvania, USA**

Green assets delivered high returns in recent years. This performance reflects unexpectedly strong increases in environmental concerns, not high expected returns. German green bonds outperformed their higher-yielding non-green twins as the "greenium" widened, and U.S. green stocks outperformed brown as climate concerns strengthened. Despite that outperformance, we estimate lower expected returns for green stocks than for brown, consistent with theory. We estimate expected returns in two ways: ex ante, using implied costs of capital, and ex post, using realized returns purged of shocks from climate concerns and earnings. A theoretically motivated green factor explains much of value stocks' recent underperformance.

This chapter is a reprint of an article originally published in the Journal of Financial Economics. Please cite this work as follows:
Pástor, Ľ., Stambaugh, R.F. and Taylor, L.A., 2022. Dissecting green returns. Journal of Financial Economics, 146(2), pp.403-424.

* The views in this chapter are the responsibility of the authors, not the institutions they are affiliated with. We are grateful for comments from Alex Edmans, Stefano Giglio, Mark Hulbert, Ralph Koijen, Stefan Nagel, Dimitris Papanikolaou, Zacharias Sautner, Lee Seltzer, Kent Smetters, David Zerbib, Lu Zhang; conference participants at the 2022 AFA meetings, 2022 Frankfurt 1st Sustainability Standards Watchers Conference, 2022 Jackson Hole Finance Conference, 2022 Luxembourg Sustainable Finance Conference, 2022 Mayo Center/ICI Virtual Asset Management Symposium, 2022 NBER Asset Pricing meeting, 2021 NBER Conference on Measuring and Reporting Corporate Carbon Footprints and Climate Risk Exposure, 2021 Australasian Banking and Finance Conference, 2021 Brazilian Finance Society Meeting, 2021 KPMG Annual M&A and Economic Forum, 2021 Q Group Fall Seminar, 2021 QES Global Quant and Macro Investing Conference; and seminar participants at the Bank of Italy, Erste Bank Asset Management, Jacobs Levy Center webinar, RCEA webinar, U.S. Department of the Treasury, and the following universities: Chicago, Colorado, CUHK, Drexel, Georgetown, LBS, Massachusetts at Amherst, McGill, Peking, Pennsylvania, South Carolina, Southern California, Southern Methodist, Texas at Austin, Texas at Dallas, Utah, Warwick, Washington, Zurich. We thank Livia Amato and Anna-Theresa Helmke for excellent research assistance. This research was funded in part by the Fama-Miller Center for Research in Finance at the University of Chicago Booth School of Business.

Corresponding author email: luket@wharton.upenn.edu.

DOI: 10.1201/9781032636252-6

6.1 Introduction

The growth of sustainable investing is a leading trend in the investment industry over the past decade. Sustainable investing applies environmental, social, and governance (ESG) criteria, with environmental concerns playing the leading role. For example, 88% of the clients of BlackRock, the world's largest asset manager, rank environment as "the priority most in focus" among ESG criteria (BlackRock, 2020). Investments considered environmentally friendly are often referred to as "green," with "brown" denoting the opposite.

Investors often cite improved returns as a top motivation for applying ESG criteria.[1] Moreover, asset managers often market sustainable investment products as offering superior risk-adjusted returns.[2] Past performance is a popular marketing tool, and indeed a number of studies report superior historical returns to sustainable strategies (e.g., Edmans, 2011; Nagy, Kassam, and Lee, 2016; In, Park, and Monk, 2019). Of course, as the SEC generally requires of all marketed funds, managers must warn that past performance does not necessarily predict future performance. In this study we show why investors would be especially well advised to heed that warning when investing in green assets.

What does the past performance of green assets imply about their future performance? We address this question empirically, guided by the equilibrium model of Pástor, Stambaugh, and Taylor (2021, henceforth PST). The PST model predicts that green assets have lower *expected* returns than brown, for two reasons: investors have green tastes, and greener assets are a better hedge against climate risk. Greener assets' lower expected returns thus reflect both a taste premium and a risk premium. PST also explain, however, that green assets can have higher *realized* returns while agents' demands shift unexpectedly in the green direction. This wedge between expected and realized returns is central to our study. PST identify two ways green demands can shift. First, investors' demand for green assets can increase, directly driving up green asset prices. Second, consumers' demand for green products can strengthen—for example, due to environmental regulations—driving up green firms' profits and thus their stock prices. Similarly, investors' demand for brown assets or consumers' demand for brown products can decrease, again making green stocks outperform.

Our analysis focuses primarily on the U.S. stock market. Using environmental ratings from MSCI, a leading provider of ESG ratings, we assign greenness measures to individual stocks. Our sample begins in November 2012, when MSCI's data coverage increased sharply, and ends in December 2020. Over this period, the value-weighted portfolio of stocks in the top third of greenness outperformed the bottom third by a cumulative return difference of 174%. This return spread, which we denote as GMB (green-minus-brown), has a monthly Sharpe ratio of 0.33, larger than the stock market's Sharpe ratio during this bull-market period. In short, green stocks strongly outperformed brown in recent years.

Should green stocks' recent outperformance lead one to expect high green returns going forward? No, we argue. That outperformance likely reflects an unanticipated increase in environmental concerns. We reach this conclusion after computing a measure of concerns about climate change, using the media index constructed by Ardia et al. (2023). We observe a strong increase in climate concerns during the last decade, with the level of our measure nearly doubling. We find that shocks to climate concerns exhibit a significant positive

[1] Improved returns is the first- or second-ranked motivation for ESG investing in surveys of investors by BlackRock (2020), Paribas (2019), and Schroders (2020). In addition, in the BNP Paribas survey, 60% of respondents expect their ESG portfolios to outperform over the next five years.

[2] For example, BlackRock believes that "integrating sustainability can help investors build more resilient portfolios and achieve better long-term, risk-adjusted returns" (Fink, 2021). According to State Street, "ESG is a source of alpha that leads to positive portfolio performance" (Lester, He, and McKnett, 2018).

relation to GMB. Green stocks thus tend to outperform brown when there is bad news about climate change, consistent with green stocks being better hedges against climate shocks.

We compute an ex post estimate of GMB's expected return by purging unanticipated shocks from its average realized return. If we set the climate shocks to zero, GMB's counterfactual performance becomes essentially flat. That is, green stocks would not have outperformed brown without strengthened climate concerns. In fact, they would have underperformed had there been no surprises to either climate concerns or earnings of green versus brown firms. If we zero out both climate and earnings shocks, GMB's counterfactual performance becomes slightly negative, indicating a negative expected return for GMB.

Our empirical explanation of green stocks' outperformance accords with the PST model. During a period when climate concerns strengthen sufficiently and unexpectedly, GMB delivers a positive return, as investors demand greener stocks or customers demand greener products. Outperformance caused by the strengthening of investor concerns is followed by lower expected performance of GMB going forward. That is, a shift in GMB's expected future performance relates inversely to its realized performance.

An inverse relation between realized returns and shifts in expected returns is not new in the stock return literature.[3] With stocks, a challenge to documenting this relation is that expected stock returns are generally hard to estimate. With bonds, however, we can see the relation more clearly. The inverse relation between a bond's realized return and the change in its yield to maturity is well understood, and the yield provides direct information about expected return, especially for buy-and-hold investors.

The case of German "twin" bonds illustrates this inverse relation in the context of climate concerns. Since 2020, the German government has issued green bonds, along with virtually identical non-green twins. The green bonds trade at lower yields, indicating lower expected returns compared to non-green bonds. The yield spread between the green and non-green twins, known as the "greenium," reflects investors' willingness to accept a lower return in exchange for holding assets more aligned with their environmental values. Since issuance, the 10-year greenium experienced almost a four-fold widening, possibly due to growing climate concerns. As a result, the green bond outperformed its non-green twin by a significant margin over the same period. However, this outperformance does not imply green outperformance going forward. Rather the opposite is clearly true, given the now wider greenium.

We define an equity analogue to the greenium, the "equity greenium," as the difference between the expected returns of green and brown stocks, i.e., GMB's expected return. Given the difficulty in estimating expected stock returns, the equity greenium cannot be measured as precisely as the greenium for bonds. Complementing our ex post approach, we estimate the equity greenium ex ante by the difference between the implied costs of capital of green and brown stocks. We find a consistently negative equity greenium throughout our sample. This evidence lends further support to our argument that the outperformance of green stocks in our sample period was unexpected. The equity greenium widens in the second half of our sample, consistent with strengthening investor demands for green assets.

Our main results relating climate shocks to stock returns rely on the time series of GMB. We also conduct a parallel analysis by running panel regressions on individual stocks, leading to several findings. First, there is a positive cross-sectional relation between a stock's greenness and its return. Second, that relation disappears when we interact greenness with climate-concern shocks, revealing that these shocks fully account for the superior performance of green stocks during the sample period. In fact, the relation becomes slightly negative when we add earnings shocks as controls. These results echo our time-series evidence: despite having lower expected returns, green stocks outperform brown due to positive surprises over the sample period. Finally, industry-level greenness, as opposed to within-

[3]For example, this inverse relation figures prominently in empirical analyses of the equity premium by Fama and French (2002) and Pástor and Stambaugh (2001, 2009).

industry differences in greenness, largely accounts for the superior performance of green stocks as well as the importance of climate shocks in explaining that performance.

We find that small stocks react to climate news with a delay. In panel regressions of individual stock returns on greenness interacted with climate shocks, previous-month shocks enter more strongly than current-month shocks, indicating a delayed reaction for some stocks. There is no significant delay in the response of GMB, whose long and short legs are value-weighted, to climate shocks. But when we replicate GMB's construction separately within the large- and small-cap segments, we find that small-cap GMB responds mostly to previous-month climate shocks, whereas large-cap GMB responds mostly to same-month shocks. At a weekly frequency, large-cap GMB reacts more strongly than small-cap GMB to climate shocks in the current and previous week, whereas small-cap GMB reacts more strongly to shocks at longer lags, especially the three-week lag. This evidence suggests that smaller stocks react more slowly to climate news, consistent with prior evidence that small stocks react more slowly to news in general (Lo and MacKinlay, 1990). Our evidence of a delay complements that of Hong, Li, and Xu (2019). They also find that stock prices are slow to react to climate-change risks, but they look at different assets (stocks in food industries across countries) and different climate shocks (trends in the risks of drought).

Green stocks' recent outperformance helps us understand the poor performance of value stocks in the 2010s, the worst decade on record for the HML factor of Fama and French (1993). We leverage PST's theoretical result that assets are priced by a two-factor model, where the factors are the market portfolio and the ESG factor. Focusing on the "E" part of ESG, we construct a "green factor" by following PST's procedure. The green factor is the return on a portfolio that goes long green and short brown stocks, where the stocks are weighted by their greenness. We find that the two-factor model explains much of HML's recent underperformance. From November 2012 through December 2020, HML's monthly CAPM alpha is a marginally significant −71 bps, whereas HML's two-factor alpha is an insignificant −15 bps. In contrast, the green factor's alpha with respect to the Fama-French three-factor model is a significant 38 bps. The green factor and HML are negatively correlated, as value stocks are more often brown than green. Insofar as recent average performance, however, the two-factor model explains HML's underperformance better than the three-factor model explains the green factor's outperformance. The two-factor model can also explain the momentum strategy's positive performance over the same period: momentum's monthly CAPM alpha is 66 bps, whereas its two-factor alpha is −6 bps.

Our study relates to a large empirical literature investigating returns on green versus brown assets. One set of studies examine returns on an ex ante basis, using proxies for expected future returns. In the bond market, for example, Baker et al. (2022), Zerbib (2019), and Larcker and Watts (2020) analyze yields on green bonds versus brown. In the stock market, Chava (2014) and El Ghoul et al. (2011) compare implied costs of capital estimated for green firms versus brown. Most of these studies find lower ex ante returns on green assets, consistent with theory. A second, larger set of studies examine returns on an ex post basis, measuring realized green-versus-brown returns, generally for stocks. Examples include Garvey, Iyer, and Nash (2018), In, Park, and Monk (2019), Bolton and Kacperczyk (2021, 2023), Görgen et al. (2020), Hsu, Li, and Tsou (2023), and Aswani, Raghunandan, and Rajgopal (2024). We examine returns both ex ante and ex post, focusing on the distinction between expected and realized returns, in the spirit of Elton (1999). We show why high green returns realized in recent years are likely to be misleading predictors of the future.

Our evidence on how climate shocks affect realized returns also relates to studies investigating the pricing of climate risk. Recent work examines that pricing in equities (e.g., Bolton and Kacperczyk, 2021, 2023;Hsu, Li, and Tsou, 2023; Faccini, Matin, and Skiadopoulos, 2023), corporate bonds (Huynh and Xia, 2021; Seltzer, Starks, and Zhu, 2021); municipal bonds (Painter, 2020; Goldsmith-Pinkham et al., 2023), options (Ilhan, Sautner, and Vilkov, 2021), and real estate (Bernstein, Gustafson, and Lewis, 2019; Baldauf, Garlappi, and Yannelis, 2020; Giglio et al., 2021). Engle et al. (2020) develop a procedure to dynamically hedge climate risk with the help of mimicking portfolios and textual analysis of news sources. Krueger, Sautner, and Starks (2020) document the importance of climate risk in a survey of institutional investors. For a survey of the climate finance literature, see Giglio, Kelly, and Stroebel (2021).

Our empirical analysis is guided by the theoretical model of PST, in which investors' tastes for green assets play a key role. Other models featuring tastes for green assets can be found in Fama and French (2007), Baker et al. (2022), Pedersen, Fitzgibbons, and Pomorski (2021), Avramov et al. (2022), and Zerbib (2022). PST's model assumes that markets are efficient, so that if green firms are expected to be more profitable than brown in the future, this difference is reflected in current prices. Pedersen, Fitzgibbons, and Pomorski show that if some investors anticipate this greater profitability before market prices respond, those investors expect higher returns on green assets. While our analysis is motivated primarily by PST's efficient-market perspective, we do find some evidence of slow price response, as noted earlier.

Our results have important implications for research and practice. They underline the danger in using recent average returns to estimate expected returns. In particular, the recent outperformance of green assets does not imply high green returns going forward. In fact, if the outperformance resulted from increased demands by ESG investors, then green assets' expected returns are lower today than a decade ago. In the same spirit, value stocks' recent underperformance is less likely to continue, because value stocks tend to be brown and growth stocks green. From the corporate finance perspective, our findings imply that greener firms have lower costs of capital than their recent stock performance might suggest. This is good news for ESG investors, because one way they exert social impact is by decreasing green firms' cost of capital (e.g., Heinkel, Kraus, and Zechner, 2001; PST).

This chapter is organized as follows. Section 6.2 highlights the gap between expected and realized returns in the context of German twin bonds. Section 6.3 describes how we measure greenness for U.S. stocks. Section 6.4 compares the realized performance of green versus brown stocks. Section 6.5 implements two approaches to estimating the expected return on the green-minus-brown stock portfolio. Section 6.6 documents the delayed reaction of stock prices to climate news. Section 6.7 discusses how we construct the green factor and explore its role in pricing value and momentum. Section 6.8 concludes.

6.2 German twin bonds

This chapter emphasizes the difference between expected and realized returns on green assets. In this section, we illustrate this difference for bonds. Bonds' expected returns are tightly linked to yields to maturity, which are easily observable.

Since 2020, the government of Germany, the largest European economy, has been issuing green securities to finance its transition toward a low-carbon, sustainable economy.[4] The first green security, a 10-year bond, was issued in September 2020 in the amount of 6.5 billion euros. The second green security, a 5-year note, followed two months later in the amount of 5 billion euros. Both securities have zero coupon rates. Germany plans to issue at least one green security per year. We refer to these securities as "green bonds."

[4] For more details, see `https://www.deutsche-finanzagentur.de/en/institutional-investors/federal-securities/green-federal-securities/`.

Each green bond is issued with the same characteristics as an existing conventional bond issued by the German government. Besides having the same issuer, the two bonds have the same maturity date, the same coupon rate, and the same coupon payment dates. This pairing creates "twin" bonds, which offer identical streams of cash flows with identical credit risk but different greenness. This clean twin pairing makes German government bonds uniquely well suited for our purposes. By comparing the prices of twin bonds, we can gain some insight into the value assigned to greenness by bond market investors.

Even though the twin bonds are paired very carefully, some differences between them remain. First, the issuance date of the green bond always comes after the initial issuance date of the conventional bond. For example, the green bond issued in September 2020 has a conventional twin issued in June 2020. Second, conventional bonds tend to be issued at larger volumes than their green twins. For example, in 2020, the issuance of conventional bonds was almost five times larger than that of the corresponding green bonds. Conventional bonds could thus in principle be more liquid than their green twins. However, the German Finance Agency has committed to play an active role in the secondary market for green bonds to make their liquidity comparable to that of conventional bonds.

We obtain daily data on the first pair of twin bonds, downloading the end-of-day bond prices and mid-yields to maturity for the 10-year green bond (ISIN DE0001030708) and the 10-year non-green bond (DE0001102507) from Bloomberg. We download data since the first date of trading for the green bond, which is September 8, 2020, through November 17, 2021. Over this time period, the two bonds' annual yields fluctuate between −67 and −15 bps. We plot these yields in Panel A of Figure 6.1 and show their means in Panel A of Table 6.1.[5]

Panel B of Figure 6.1 plots the difference between the yields of green and non-green bonds, also known as the greenium (e.g., Larcker and Watts, 2020). The greenium is always negative, averaging −4.6 bps and ranging mostly between −7 and −2 bps per year.[6] Therefore, for investors holding the bonds to maturity, the green bond always has a lower expected return than the non-green bond. This evidence is in line with theories predicting that green assets offer lower expected returns than non-green assets (e.g., PST).[7]

Given the lower yield of the green bond, one would expect it to deliver a lower return than its conventional twin. Instead, the green bond delivered a higher return in our sample. We calculate bond returns as daily percentage changes in bond prices and report them in Panel B of Table 6.1. The full-sample cumulative returns are negative, −1.53% for the green bond and −1.90% for the non-green bond, due to a rise in yields between September 2020 and November 2021. More interesting, the green bond outperforms its non-green twin over this time period, as shown by Panel C of Figure 6.1. The figure plots cumulative returns on a long-short portfolio, which goes long the green bond and short the non-green bond. The portfolio's average daily return of 0.12 bps is statistically significant ($t = 2.19$), and its cumulative return of 37 bps is substantial relative to German government bond yields.

[5]In the Appendix, we plot the counterpart of Figure 6.1 for the second pair of twin bonds (five-year bonds), which was first issued in November 2020. The results are similar to those presented in Figure 6.1. We prioritize the first twin pair due to its longer history. The Appendix is on the authors' websites.

[6]These greenium values are close to those estimated by prior studies in different settings. For example, Baker et al. (2022) estimate a greenium of about −6 bps in a sample of over 2,000 U.S. municipal and corporate green bonds, whereas Zerbib (2019) estimates −2 bps in a sample of over 1,000 supranational, sub-sovereign and agency, municipal, corporate, financial, and covered green bonds.

[7]This conclusion is reinforced by liquidity considerations. As noted earlier, the non-green bond has been issued at larger volume than its green twin. If this volume difference makes the conventional bond more liquid despite the aforementioned efforts of the German Finance Agency, then the resulting liquidity premium pushes the greenium up, and the expected return penalty associated with greenness is even larger.

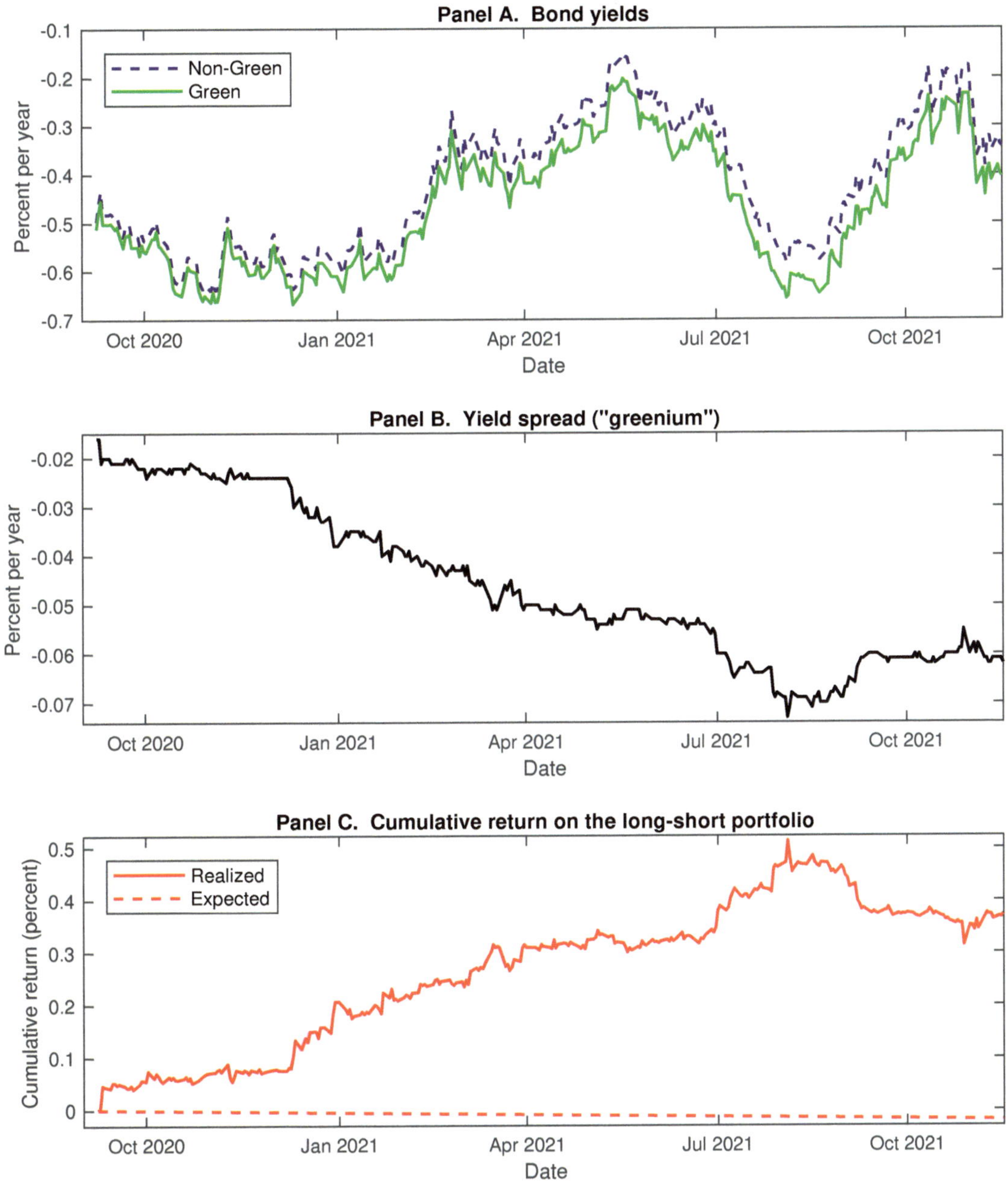

FIGURE 6.1: German twin bonds. Panel A plots the daily time series of annual yields on the German government's 10-year green bond and its non-green twin. Panel B plots the "greenium," the yield difference between the green bond and its non-green twin. Panel C plots the performance of a portfolio that goes long the 10-year green bond and short its non-green twin. The solid line plots this long-short portfolio's daily cumulative realized return. The dashed line plots the expected cumulative return as of the first day of trading of the green bond (September 8, 2020), absent a subsequent change in the greenium, which was −1.6 bps on that day.

TABLE 6.1: German government bond yields and returns.
Panel A reports the yields to maturity of the 10-year German government green bond (column 1), 10-year German government conventional bond (column 2), and their difference (column 3), all in basis points per year. Average yields are computed over the full sample period of September 8, 2020 to November 17, 2021. The yields on the first and last days of this period are also reported. Panel B reports the realized returns on the same green bond (column 1), same conventional bond (column 2), and their difference (column 3). Average returns are in basis points per day. Cumulative returns are in percent over the full sample period. The t-statistics, which are shown in parentheses, are adjusted for any statistically significant autocorrelation in the underlying series

	Green bond	*Non-green bond*	*Difference*
Panel A. Yields to maturity (basis points per year)			
Average	−46.72	−42.09	−4.63
	(−13.53)	(−10.90)	(−6.19)
First day	−51.20	−49.60	−1.60
Last day	−40.60	−34.40	−6.20
Panel B. Realized returns			
Average	−0.47	−0.59	0.12
	(−0.35)	(−0.44)	(2.19)
Cumulative	−1.53	−1.90	0.37

Importantly, the positive average return of the long-short portfolio does not imply that the portfolio's expected return is positive. On the contrary, we know with certainty that the portfolio's expected return is negative if the bonds are held to maturity. For example, on September 8, 2020, the green bond's yield was −51.2 bps per year, whereas the non-green bond's yield was −49.6 bps. Therefore, if both bonds are held to maturity, the green bond delivers a return 1.6 bps lower than the non-green bond. The green bond's expected return is lower also if the bonds are not held to maturity under a variety of plausible conditions, such as changes in the greenium being unpredictable. That condition is likely to hold, especially in efficient, or near-efficient, markets. Under that condition, the green bond's expected return is lower at the beginning of the sample, and the expected return of the long-short portfolio is negative. The cumulative value of this expected return is plotted by the dashed line in Panel C of Figure 6.1, which is gently downward-sloping.

How can we reconcile the higher realized return of the green bond with its lower expected return? The answer is that that the greenium in Panel B grows increasingly negative between September 2020 and November 2021, deepening from −1.6 to −6.2 bps. This deepening is responsible for the outperformance of the long-short portfolio in Panel C. In the language of PST, if investors' tastes shift toward green assets, they push up the price of the green bond relative to the non-green bond. However, the green bond's outperformance is temporary, as it comes entirely at the expense of the bond's future return. Investors buying the bonds on September 8, 2020 and holding them to maturity expected to earn 1.6 bps less from the green bond, but those buying on November 17, 2021 expected to earn 6.2 bps less.

Investors' tastes for green assets could plausibly have shifted unexpectedly due to heightened concerns about climate change. Those concerns are likely to have risen in July 2021 when Germany, along with several other European countries, experienced catastrophic floods caused by heavy rainfall that followed unprecedented heat waves. Germany experienced around 200 fatalities in those floods, which were the deadliest natural disaster in the country in almost six decades. Consistent with a shift in investors' tastes toward green German bonds, the greenium widened from −5.5 bps on June 29 to −7.3 bps on August 4, before easing back to −6 bps by mid-September. These changes suggest a possible link between investors' tastes and climate concerns.[8] We further explore this link later in the chapter.

The case study of German twin bonds illustrates how shifts in expected return drive a wedge between returns expected ex ante and those realized ex post. Even though the green bond's realized return is higher than that of the non-green bond, the green bond's expected long-term return is demonstrably lower. In other words, the expected return of the long-short portfolio is negative even though the portfolio's average realized return is positive and significant. Unexpected events often happen, and one of them was likely the outperformance of the German green bond in the first 14 months of its existence.

6.3 Measuring stocks' greenness

While the German bond example is clean, it is essentially a case study. In this section, we begin our main analysis, which examines U.S. stocks. Focusing on stocks allows us to examine the role of greenness in a larger asset universe over a longer time period.

We compute stock-level environmental scores based on MSCI ESG Ratings data, a successor to the MSCI KLD data used in many academic studies. Our data have a number of advantages. According to Eccles and Stroehle (2018), MSCI is the world's largest provider of ESG ratings. The MSCI ESG Ratings data are used by more than 1,700 clients, including pension funds, asset managers, consultants, advisers, banks, and insurers.[9] MSCI covers more firms than other ESG raters, such as Asset4, KLD, RobescoSAM, Sustainalytics, and Vigeo Eiris (Berg, Koelbel, and Rigobon, 2022). Berg et al. (2021) find that MSCI's ESG scores are the least noisy among the eight ESG data vendors they consider. MSCI generates its ratings based on a variety of sources and updates those ratings at least annually. MSCI's ESG research unit employs more than 200 analysts and incorporates artificial intelligence, machine learning, and natural language processing into its methodology.

The availability of industry-unadjusted granular data is another advantage of the MSCI data. With industry adjustment, a heavily polluting firm is classified as green if it pollutes less than other firms in its heavily polluting industry. Without industry adjustment, such a firm is classified as very brown. In principle, either classification could be more relevant for green-versus-brown effects on investor and consumer demands. The MSCI data allow us to explore that issue. MSCI's composite ESG rating is industry-adjusted, as are ratings from other leading providers, whereas MSCI's granular data allow us to compute a greenness measure that is not industry adjusted. We conduct our primary analyses using the latter all-in measure. This approach seems reasonable, as we later show that the effects we identify are strongly associated with industry-level greenness.

[8] According to the September 2021 ARD-DeutschlandTREND survey, 33% of Germans view climate as the first or second most important problem facing Germany, ahead of immigration (22%), the coronavirus (18%), and social injustice (16%). In the pre-flood June 2021 survey, the proportion favoring climate was lower, 28%, indicating a substantial shift in Germans' climate concerns in the summer of 2021.

[9] See `https://www.msci.com/our-solutions/esg-investing`, as of May 2021. In addition, MSCI has been voted 'Best firm for SRI research' in the Extel & SRI Connect Independent Research in Responsible Investment Survey in each year from 2015 through 2019 (`https://www.msci.com/zh/esg-ratings`).

We use the MSCI variables "Environmental pillar score" (E_score) and "Environmental pillar weight" (E_weight). E_score is a number between 0 and 10 measuring the firm's weighted-average score across 13 environmental issues related to climate change, natural resources, pollution and waste, and environmental opportunities. These scores are designed to measure a company's resilience to long-term environmental risks. E_weight, which is typically constant across firms in the same industry, is a number between 0 and 100 measuring the importance of environmental issues relative to social and governance issues.[10]

We compute the unadjusted greenness score of firm i at the beginning of month t as

$$G_{i,t-1} = -(10 - E\_score_{i,t-1}) \times E\_weight_{i,t-1}/100 \ , \tag{6.1}$$

where $E\_score_{i,t-1}$ and $E\_weight_{i,t-1}$ are from company i's most recent MSCI ratings date before month t, looking back no more than 12 months. The quantity $10 - E\_score_{i,t-1}$ measures how far the company is from a perfect environment score of 10. The product $(10 - E\_score_{i,t-1}) \times E\_weight_{i,t-1}$ measures how brown the firm is, specifically, the interaction of how badly the firm scores on environmental issues and how large the environmental impacts are for the industry's typical firm (i.e., $E\_weight_{i,t-1}$). The initial minus sign converts the measure from brownness to greenness.

Including E_weight in Eq. (6.1) is important for capturing a company's greenness. For example, in 2019, Exxon Mobil and Best Buy had similar E_score values: 4.2 and 4.1, respectively. If we only used E_score, we would judge these companies to be similarly green. But E_weight is 48 for Exxon and only 11 for Best Buy, reflecting that oil and gas companies have larger environmental impacts than consumer retail companies. Exxon and Best Buy end up with $G_{i,t} = -2.78$ and -0.65, respectively, indicating that Best Buy is much greener than Exxon. Similar to us, MSCI uses the interaction of E_score and E_weight when computing firms' composite ESG ratings.[11]

The environmental score we use in our analysis is

$$g_{i,t} = G_{i,t} - \overline{G}_t \ , \tag{6.2}$$

where $\overline{G}_t$ is the value-weighted average of $G_{i,t}$ across all firms i. Since we subtract $\overline{G}_t$, $g_{i,t}$ measures the company's greenness relative to the market portfolio, as in PST. If w_t and g_t denote the vectors containing stocks' market weights and $g_{i,t}$ values in month t, then

$$w_t' g_t = 0 \ , \tag{6.3}$$

a condition imposed by PST.

We compute $g_{i,t}$ in the sample of stocks with non-missing MSCI data and CRSP share codes of 10 or 11. We merge CRSP and MSCI by using a combination of CUSIP, ticker, and company name. Our sample extends from November 2012 to December 2020. We begin in November 2012 because MSCI's coverage increases dramatically in October 2012, when MSCI began covering small U.S. stocks.[12] Figure 6.2 plots the number of U.S. stocks with

[10] MSCI's E, S, and G weights sum to 100. According to MSCI, "The weightings take into account both the contribution of the industry, relative to all other industries, to the negative or positive impact on the environment or society; and the timeline within which we expect that risk or opportunity for companies in the industry to materialize...." We follow MSCI in using the GICS sub-industry classification.

[11] MSCI's composite ESG rating is based on their "Weighted Average Key Issue" score, which equals $[E\_score \times E\_weight + S\_score \times S\_weight + G\_score \times G\_weight]/100$, where S and G refer to social and governance. So if MSCI used a formula like Eq. (6.1) to compute greenness not just on environmental but also on social and governance dimensions, then we could express MSCI's composite ESG score as 10 plus the sum of E, S, and G greenness.

[12] Before October 2012, MSCI covered only the largest 1,500 companies in the MSCI World Index, plus large companies in the UK and Australia MSCI indexes. In October 2012 MSCI added many smaller U.S. firms when it began covering also the MSCI U.S. Investible Market Index.

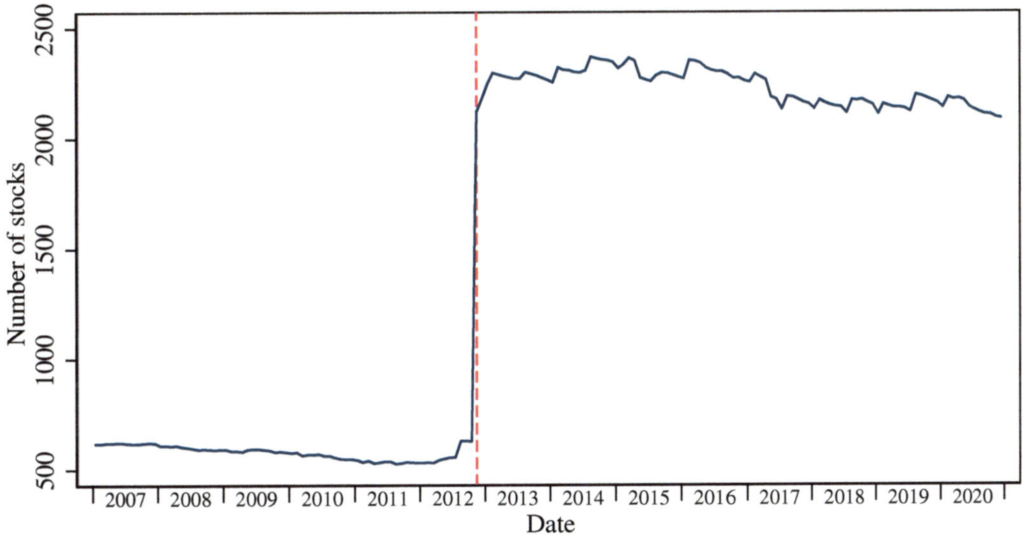

FIGURE 6.2: MSCI coverage. The figure plots the number of stocks in our sample with non-missing MSCI environmental scores at the beginning of the month. The dashed red line is at November 2012, where our sample begins. MSCI expanded its coverage in October 2012. We begin our sample in November 2012, as we require lagged environmental scores.

non-missing lagged MSCI ratings. This number increases sharply in November 2012, from roughly 500 to over 2,000. Our purchased MSCI data end in March 2020, but we extend our sample through December 2020 by looking back up to 12 months when computing $G_{i,t-1}$.

Table 6.2 shows industries ranked by their equal-weighted average $g_{i,t}$ scores at the end of 2019. The lowest-ranked industries include chemicals, oil and gas exploration and production, steel, mining (including coal), paper and forest products, and marine transport. It is reassuring that these industries, which are generally viewed as having negative environmental impacts, appear at the bottom of our ranking.

Among the 64 industries considered in Table 6.2, only 20 have positive values of average $g_{i,t}$ at the end of 2019. This fact may appear at odds with our assumption that the average value of $g_{i,t}$ across all stocks is zero. However, our assumption pertains to the market-value-weighted average (see Eq. (6.3)). While the equal-weighted average of $g_{i,t}$ at the end of 2019 is -0.33, the value-weighted average is indeed zero, by construction. The value-weighted average exceeds the equal-weighted one because greener firms tend to be larger.

6.4 Realized green stock returns

Green stocks strongly outperformed brown in recent years. Figure 6.3 displays the performance of green and brown stocks from November 2012 to December 2020. The solid line, representing green stocks, plots the cumulative value-weighted return on the portfolio of stocks with greenness scores in the top third. The dashed line, representing brown stocks, plots the corresponding return for stocks with scores in the bottom third. We see that green stocks strongly outperformed brown in the 2010s, with a cumulative return difference of 174% over our 8.2-year sample period. The monthly return difference, which we denote

TABLE 6.2: Industries ranked by environmental scores.
Average g is the environmental score averaged across firms within each MSCI industry at the end of 2019. MSCI uses the GICS sub-industry classification

Rank	*MSCI industry*	*Avg. g*	*Rank*	*MSCI industry*	*Avg. g*
1	Asset Management & Custody Banks	0.87	33	Textiles, Apparel & Luxury Goods	−0.50
2	Professional Services	0.85	34	Auto Components	−0.51
3	Telecommunication Services	0.84	35	Property & Casualty Insurance	−0.51
4	Consumer Finance	0.84	36	Casinos & Gaming	−0.54
5	Health Care Equipment & Supplies	0.84	37	Real Estate Development...	−0.55
6	Health Care Providers & Services	0.83	38	Semiconductors...	−0.66
7	Life & Health Insurance	0.76	39	Electrical Equipment	−0.75
8	Interactive Media & Services	0.74	40	Construction & Farm Machinery...	−0.76
9	Diversified Financials	0.73	41	Tobacco	−0.89
10	Media & Entertainment	0.70	42	Trading Companies & Distributors	−0.99
11	Diversified Consumer Services	0.61	43	Industrial Machinery	−1.04
12	Biotechnology	0.57	44	Containers & Packaging	−1.09
13	Pharmaceuticals	0.49	45	Energy Equipment & Services	−1.16
14	Multi-Line Insurance & Brokerage	0.40	46	Real Estate Management & Services	−1.20
15	Investment Banking & Brokerage	0.39	47	Airlines	−1.21
16	Banks	0.35	48	Hotels & Travel	−1.57
17	Restaurants	0.31	49	Building Products	−1.62
18	Construction & Engineering	0.13	50	Utilities	−1.90
19	Aerospace & Defense	0.10	51	Integrated Oil & Gas	−2.01
20	Commercial Services & Supplies	0.07	52	Food Products	−2.02
21	Air Freight & Logistics	−0.06	53	Beverages	−2.04
22	Household Durables	−0.12	54	Metals and Mining, Precious	−2.19
23	Software & Services	−0.13	55	Oil & Gas Refining, Marketing...	−2.52
24	Electronic Equipment, Instruments...	−0.17	56	Construction Materials	−2.56
25	Leisure Products	−0.17	57	Specialty Chemicals	−2.82
26	Automobiles	−0.22	58	Marine Transport	−2.83
27	Retail - Food & Staples	−0.25	59	Paper & Forest Products	−2.93
28	Retail - Consumer Discretionary	−0.27	60	Metals and Mining, Non-Precious	−2.95
29	Road & Rail Transport	−0.30	61	Steel	−2.96
30	Household & Personal Products	−0.30	62	Oil & Gas Exploration & Production	−3.01
31	Industrial Conglomerates	−0.36	63	Diversified Chemicals	−3.21
32	Technology Hardware, Storage...	−0.39	64	Commodity Chemicals	−3.78

GMB (green-minus-brown), averaged 65 bps per month (t-statistic: 3.23), as reported in the first column of Table 6.3. The monthly Sharpe ratio of GMB is 0.33, larger than even the market portfolio's Sharpe ratio of 0.30 over the same period.

This strong performance of GMB cannot be explained by exposures to return factors prominent in the asset pricing literature. The remaining columns of Table 6.3 report results of regressing GMB on various factors, including those in the three- and five-factor models of Fama and French (1993, 2015), the momentum factor (UMD) as constructed by those authors, the traded liquidity factor of Pástor and Stambaugh (2003), and the factors of Hou, Xue, and Zhang (2015) and Hou et al. (2021). In all cases, GMB's alpha (regression constant) is economically and statistically significant, ranging from 47 to 71 bps per month, with t-statistics between 1.99 and 2.91.

GMB's lowest alpha in Table 6.3 occurs in column 4, where we adjust for the three Fama-French factors and momentum. Its exposures to SMB, HML, and UMD indicate that GMB tilts toward large stocks, growth stocks, and recent winners. Net of those exposures, the alpha of GMB is 47 bps per month ($t = 2.14$).

At first sight, these results appear at odds with those of Bolton and Kacperczyk (2021), who find that stocks of firms with higher carbon emissions earn higher risk-adjusted returns. However, Bolton and Kacperczyk's sample period, 2005 to 2017, is substantially different from ours. Moreover, the sign of the carbon premium depends on how exactly carbon emissions are measured. Bolton and Kacperczyk find a positive carbon premium associated with total emissions, but not with emission intensity (i.e., emissions per unit of sales). Görgen et al. (2020) find an insignificantly negative carbon premium when they combine multiple carbon-emission-related measures and use a sample period closer to ours, 2010 to 2017. Finally, carbon emissions are only one of 13 firm characteristics that enter MSCI's environmental scores, which we use to determine firm greenness. For example, MSCI also considers

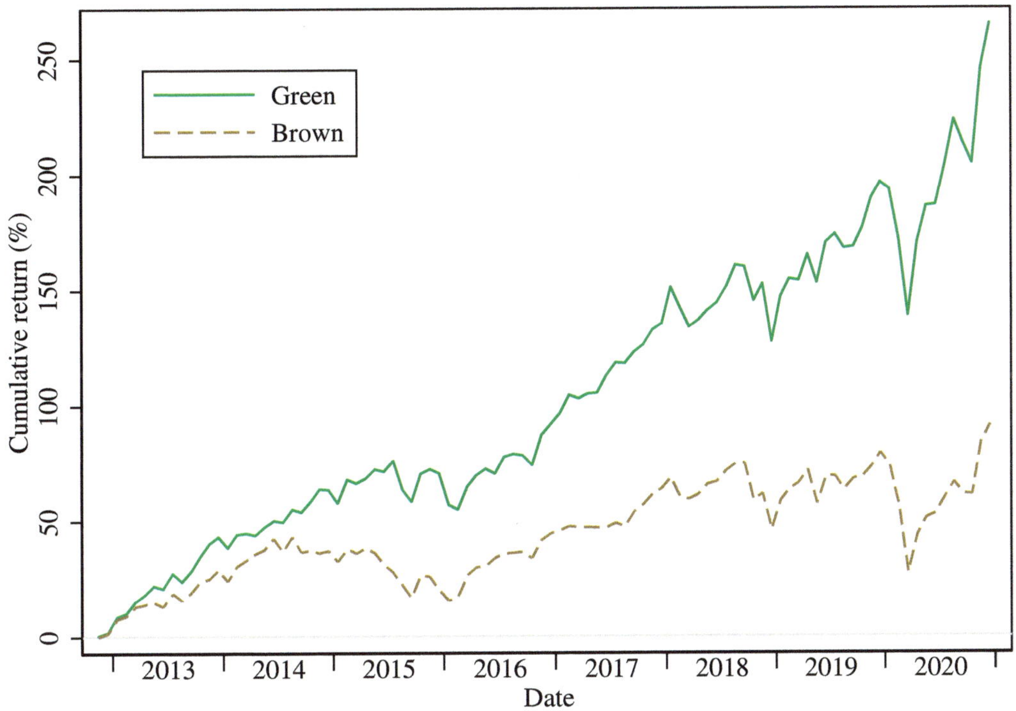

FIGURE 6.3: Returns on value-weighted green and brown portfolios. This figure plots the green and brown portfolios' cumulative returns. The values of the green and brown lines at the end of 2020 are 264.9 and 91.3, implying green stocks outperformed brown by $264.9 - 91.3 = 174$ percentage points over this period.

the firm's handling of land use, water stress, raw material sourcing, toxic waste, and opportunities in clean tech, green building, and renewable energy.

6.5 The equity greenium

We next explore the equity analog to the bond greenium analyzed in Section 6.2. The equity greenium captures the difference in expected returns on green versus brown stocks. For concreteness, we define the equity greenium as the expected return on the GMB spread. Expected stock returns are not directly observable, so the equity greenium must be estimated. This section presents two approaches to the estimation.

One approach uses ex ante data while the other uses ex post data. The first approach estimates each stock's expected return as its implied cost of capital (ICC), which is the discount rate that equates the stock's current price to the present value of expected future cash flows, with the latter estimated using data available when the price is observed. With this ex ante approach, we construct the expected GMB return from the underlying stocks' ICCs. The second approach estimates the expected GMB return as the average ex post return purged of unanticipated shocks to quantities affecting the return. To identify those

TABLE 6.3: GMB performance.
We estimate monthly time-series regressions using data from November 2012 to December 2020. The dependent variable is the difference between the returns on the green and brown portfolios (GMB). Mkt-Rf is the excess market return. SMB and HML are the size and value factors of Fama and French (1993). UMD is the momentum factor of Carhart (1997). LIQ is the traded liquidity factor of Pástor and Stambaugh (2003). RMW and CMA are the profitability and investment factors of Fama and French (2015). ME, I/A, and Roe are the size, investment, and profitability factors of Hou, Xue, and Zhang (2015). Eg is the expected-growth factor of Hou et al. (2021). Returns are in percent per month. Robust t-statistics are in parentheses

	(1)	(2)	(3)	(4)	(5)	(6)	(7)	(8)
Constant	0.65	0.71	0.50	0.47	0.50	0.50	0.55	0.49
	(3.23)	(2.91)	(2.23)	(2.14)	(2.25)	(2.38)	(2.28)	(1.99)
Mkt-RF		−0.05	0.02	0.05	0.01	0.04	−0.00	0.01
		(−0.78)	(0.32)	(0.87)	(0.21)	(0.77)	(−0.05)	(0.23)
SMB			−0.14	−0.11	−0.16	−0.26		
			(−1.49)	(−1.23)	(−1.56)	(−2.59)		
HML			−0.26	−0.18	−0.26	−0.21		
			(−3.36)	(−1.99)	(−3.26)	(−2.60)		
UMD				0.13				
				(2.00)				
LIQ					0.04			
					(0.60)			
RMW						−0.39		
						(−2.90)		
CMA						−0.10		
						(−0.60)		
ME							−0.15	−0.13
							(−1.48)	(−1.28)
I/A							−0.30	−0.25
							(−2.21)	(−1.59)
Roe							0.09	0.02
							(0.99)	(0.20)
Eg								0.12
								(0.67)
Observations	98	98	98	98	98	98	98	98
R^2	0.00	0.01	0.19	0.22	0.19	0.26	0.13	0.14

shocks, we follow PST in noting that GMB's realized performance can be positive in periods of unanticipated increases in demands for green firms' products and stocks (or decreases in demands for brown firms' products and stocks). These demands can change for various reasons, but a likely leading source is increased concerns about climate change. We use climate-concern shocks and earnings-news shocks when pursuing the second approach.

As we detail below, the ex ante and ex post approaches deliver similar negative estimates of the expected GMB return. These estimates contrast sharply with GMB's strongly positive realized performance. Later in the section we show that our main conclusions about expected versus realized returns are robust along various dimensions, such as including additional shocks and examining returns at the individual stock level. We also show that our results are driven more by industry-level greenness than within-industry greenness.

6.5.1 ICC estimates of the equity greenium

Our first approach to estimating the equity greenium, using ex ante data, computes the ICC for each stock-month. The ICC combines data on market prices and forecasted cash flows to infer a discount rate using the standard discounted-cash-flow formula. We follow the approach of Hou, Van Dijk, and Zhang (2012), which builds on the classic framework of Gebhardt, Lee, and Swaminathan (2001) but replaces analysts' earnings forecasts with regression-based forecasts. Lee, So, and Wang (2021) compare ICC methods used in a number of finance studies. We choose the method they find produces the most precise expected return estimates in the cross section. The Appendix provides further details.

Panel A of Figure 6.4 plots the time series of the ICCs of the green and brown portfolios, the long and short legs of GMB. Each portfolio's ICC is the market-weighted average of its stocks' ICCs. During our sample period, the green portfolio's ICC declines from 7.6% to 4.9% per year, whereas the brown portfolio's ICC falls from 8.8% to 6.8% per year. Importantly, at each point in time, the green portfolio's ICC is below that of the brown portfolio, indicating a consistently negative equity greenium, i.e., lower expected return on green stocks versus brown. Panel B plots the difference between the two ICCs. This difference, which is the GMB portfolio's ICC, ranges from −0.4% to −2.4%, with an average of −1.4% per year. To the extent that the ICC is a good proxy for expected stock return, this evidence of a consistently negative equity greenium supports our argument that GMB's strong performance in our sample period was unexpected.

Additional insight into the equity greenium comes from a panel regression of a stock's ICC on the stock's greenness, with month fixed effects. This regression produces a highly significant negative slope estimate ($t = -11.90$), again consistent with a negative equity greenium. When we add the interaction of greenness with a time trend to the right-hand side of the regression, both greenness and this interaction command highly significant slope estimates, with t-statistics of about −5.5. (We tabulate the results in the Appendix.) These estimates imply not only that the equity greenium is negative but also that it widened over our sample period. Consistent with the latter, in Panel B of Figure 6.4 the ICC of GMB declines from −1.2% to −1.9% per year during our sample period, albeit far from monotonically. The greenium's decline is especially steep from 2017 to 2020. This evidence of a widening equity greenium is consistent with investors' demand for green assets strengthening during our sample period.

6.5.2 Inferring expected return from past realizations

Our second approach to estimating the equity greenium addresses the general problem of inferring an asset's unconditional expected return, $\mu = E\{r_t\}$, using ex post data. One option is to use the asset's sample average return, $\bar{r}$, as the estimate of μ. Another approach, which we follow, is to exploit the additional information in the contemporaneous history of another variable, x_t, that is correlated with the return and for which $E\{x_t\} = 0$. For example, as in our setting, x_t can be an unanticipated change in climate concerns. In the regression,

$$r_t = a + bx_t + \epsilon_t\,, \tag{6.4}$$

$a = \mu$ because x_t has zero mean ex ante. Therefore, we can estimate μ by the sample estimate of a. This estimate is given by the OLS intercept $\hat{a} = \bar{r} - \hat{b}\bar{x}$, where $\hat{b}$ is the OLS

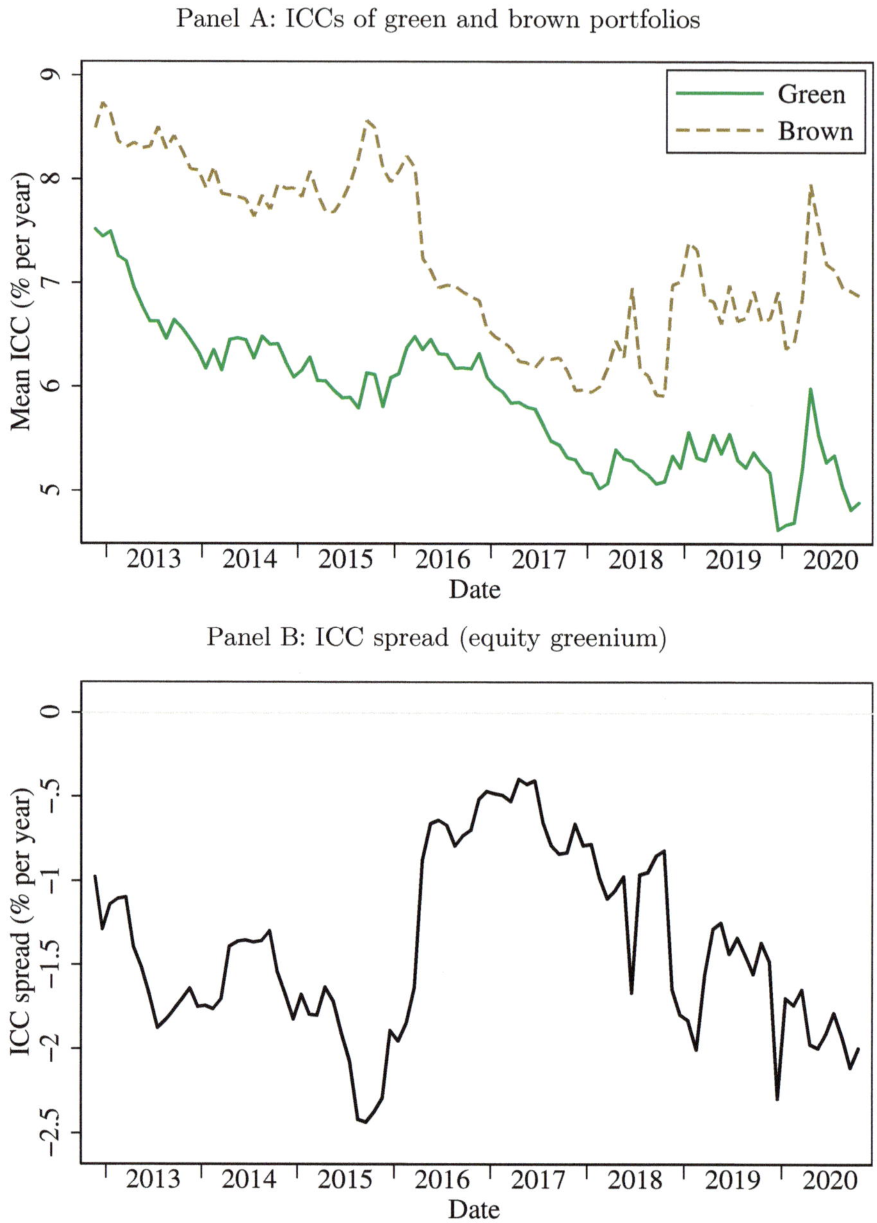

FIGURE 6.4: Implied costs of capital. Panel A plots the ICCs of green and brown portfolios, computed as value-weighted averages of annual ICCs of stocks within each portfolio. Panel B plots the green-minus-brown difference between the ICCs from Panel A.

slope and $\bar{x}$ is the sample average of x_t. We thus have two alternative estimators of μ:

$$\text{Estimator 1:} \quad \bar{r} \tag{6.5}$$

$$\text{Estimator 2:} \quad \hat{a} = \bar{r} - \hat{b}\bar{x}\,. \tag{6.6}$$

To obtain more insight into the second estimator, let x_t be signed such that $b > 0$. Suppose the realizations of x_t exceed their expectation on average, so that $\bar{x} > 0$. As a result, $\bar{r}$ overstates μ by $b\bar{x}$ on average. This overstatement is essentially removed by the second estimator, $\hat{a}$, which reduces $\bar{r}$ by $\hat{b}\bar{x}$. Similarly, when instead $\bar{x} < 0$, one expects $\bar{r}$ to understate μ, and $\hat{a}$ essentially adds back the understatement. In general, with $\bar{x} \neq 0$, the regression intercept removes the associated ex post distortion in $\bar{r}$. The same argument applies if x_t is a vector of variables whose sample means differ from their zero ex ante means.

To illustrate quantitatively how $\bar{r}$ and $\hat{a}$ can provide different inferences, we analyze a setting that corresponds roughly to our regressions presented later, just simplified to the above case of one explanatory variable, x_t. Specifically, we set T equal to 68 months, the sample length in our regressions, and we assume that the regression in Eq. (6.4) has an R-squared of 20%, which is broadly representative of our estimated regressions. We also assume that the ϵ_t's are normally distributed, independently and identically across months, and that the monthly return, r_t, has a standard deviation of 2%, matching that of GMB. Finally, we set that spread's expected return, μ, equal to -10 bps per month, which is representative of both the -11.6 bps implied by the earlier ICC estimate (-1.4% per year on average) as well as the estimates we obtain later using $\hat{a}$.

Figure 6.5 displays comparisons of $\bar{r}$ and $\hat{a}$ as estimators of μ. Panel A shows the probability that an estimate of μ is positive, i.e., has the wrong sign. The probability is conditioned on the magnitude of $\bar{x}$, which we express on the horizontal axis in terms of $t_{\bar{x}}$, the t-statistic for $\bar{x}$.[13] Regardless of $t_{\bar{x}}$, if μ is estimated by $\hat{a}$, the probability of getting the wrong sign is about 0.33. For $t_{\bar{x}} = 0$, that value is also the probability of getting the wrong sign when estimating μ by $\bar{r}$, but the probability in this latter case rises quickly as $t_{\bar{x}}$ increases, to the extent that getting the wrong sign becomes very likely when $\bar{x}$ is strongly significant. Panel B shows the probability that each estimate of μ not only gets the wrong sign but is also statistically significant at the two-tailed 5% level. If μ is estimated by $\hat{a}$, this probability is consistently less than 1%. If μ is instead estimated by $\bar{r}$, the probability grows quickly as $t_{\bar{x}}$ increases. For example, if $t_{\bar{x}} = 4$, there is a 25% probability of having $\bar{r}$ be statistically significant with the wrong sign. Overall, for samples in which $\bar{x}$ departs significantly from its zero mean, the advantage of using $\hat{a}$ rather than $\bar{r}$ to estimate μ seems clear.

6.5.3 Measuring shocks to climate concerns and earnings

To implement the above approach that uses $\hat{a}$, we must specify x_t. We generalize the latter to be a vector of shocks having two sources. First, climate concerns are likely to play a key role in boosting demands by consumers for green firms' products as well as demands by investors for green firms' stocks (and reducing demands for brown firms' products and stocks). Therefore, news regarding climate concerns serves as one source of return shocks in x_t.[14] Second, while the product-demand channel for climate news impacts returns via expectations of firms' earnings, non-climate information also impacts earnings expectations and thus returns. We therefore include earnings news directly as another source of return shocks in x_t. Next, we describe how we measure both sources of shocks.

Climate concerns

We measure concerns about climate change with the Media Climate Change Concerns index (MCCC) of Ardia et al. (2023). This index, which is available from January 2003 through June 2018, is constructed by using data from eight major U.S. newspapers. It captures the number of climate news stories each day as well as their negativity and focus on risk. For each news article discussing climate change, Ardia et al. compute a "concern"

[13] The probabilities in Figure 6.5 are derived in the Appendix.

[14] We do not take a stand on whether customers' and investors' responses to climate news reflect genuine concerns about climate or just virtue signaling. Either way, asset prices can be affected.

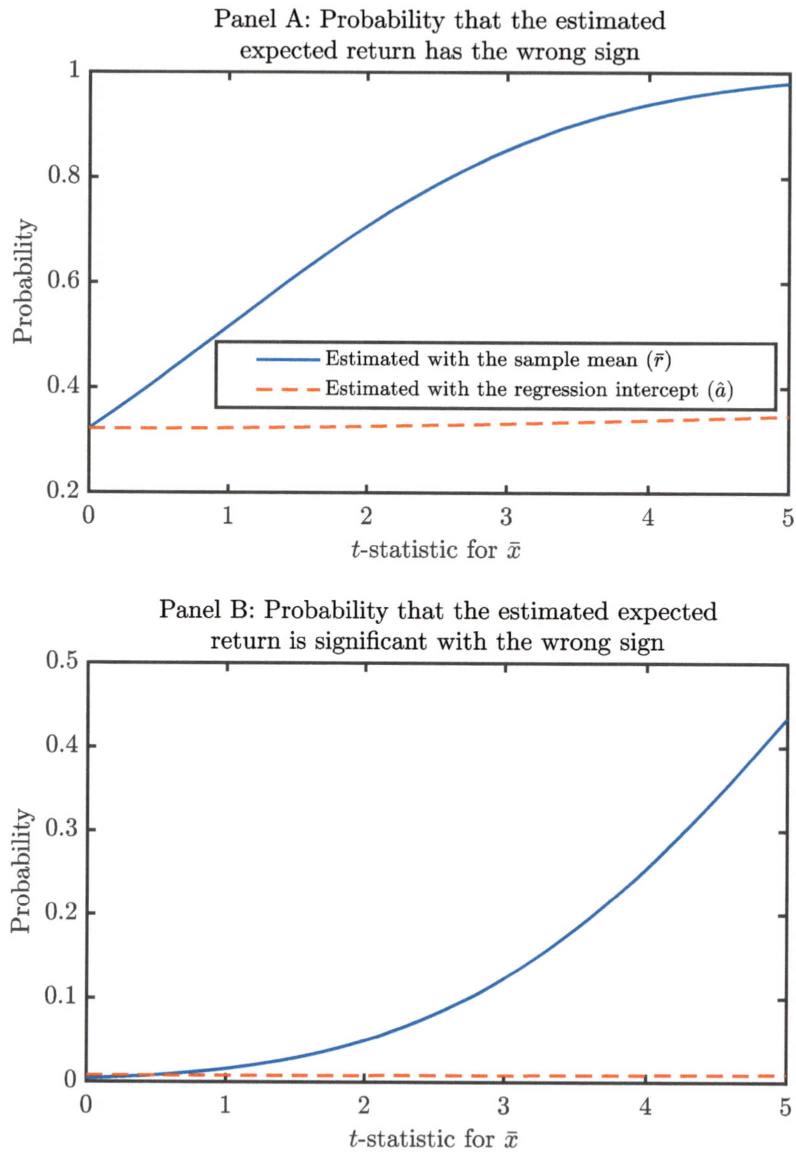

FIGURE 6.5: Comparing estimators of expected return. The figure displays the indicated probabilities when the number of observations equals 68, the regression R-squared is 0.02, and the monthly return has a true mean (μ) of -10 basis points and a standard deviation of 2%. In Panel B, statistical significance is at the two-tailed 5% level.

measure that interacts two quantities: the fraction of total words related to risk and the scaled difference between negative and positive words. They aggregate this measure to the newspaper-day level by adding the concern values across stories. Next, they aggregate to the day level by averaging across newspapers, after adjusting for heterogeneity across newspapers. Finally, they take the square root of this daily measure because, as they put it, "One concerning article about climate change may increase concerns, but 20 concerning articles are unlikely to increase concerns 20 times more."

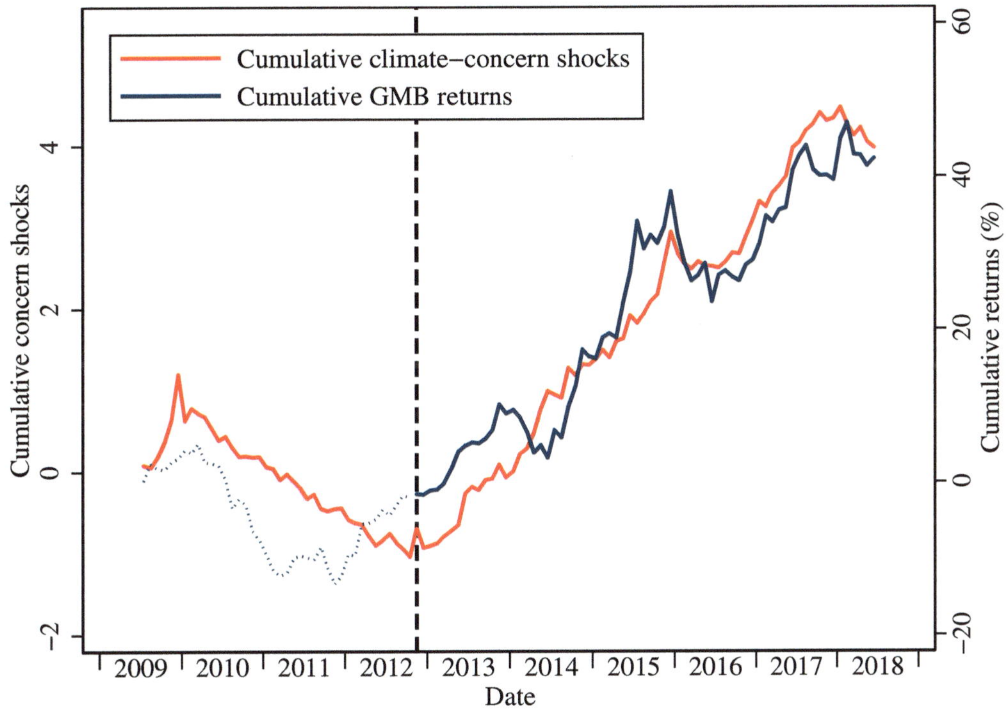

FIGURE 6.6: Climate concerns and GMB. Climate-concern shocks are prediction errors from rolling AR(1) models fitted to the monthly MCCC index. The dashed vertical line is at November 2012, where our sample begins. Before November 2012, the GMB return, shown as a dotted line, is constructed using a much smaller number of stocks (recall Figure 6.2).

Following Ardia et al. (2023), we measure shocks to climate concerns as prediction errors from AR(1) models applied to the underlying MCCC index. To compute the prediction error in month t, we estimate an AR(1) model using the 36 months of MCCC data ending in month $t-1$ (including data before November 2012), and we set the prediction error to month t's realization of MCCC minus the AR(1) model's prediction.

Figure 6.6 plots the cumulative shocks to climate concerns over the ten-year period between July 2009 and June 2018. We begin the plot immediately after the financial-crisis-induced recession, which ended in June 2009. The cumulative shocks trend down initially but then trend up sharply from 2013 through 2017, before dipping slightly in 2018.[15] GMB's performance, also plotted in Figure 6.6, looks strikingly similar. It performs strongly in 2013 through 2018, cumulatively returning over 40%, whereas its pre-2013 performance is negative. Of course, GMB's performance before November 2012 is only approximate because it is computed based on a sample of firms that is much smaller and biased toward large-capitalization firms (recall Figure 6.2). We plot GMB's imprecise earlier performance for comparison purposes, but we do not use it in any of our analysis.

[15]Sautner et al. (2023) provide independent evidence that climate concerns strengthen after 2012. They measure firms' climate change exposures by the extent to which climate change topics are discussed in firms' earnings calls, finding a sharp increase in climate change exposure between 2013 and 2018.

We include month t's climate-concern shock, which we denote as ΔC_t, in x_t. We also include in x_t the previous month's shock, ΔC_{t-1}, given our evidence of delayed stock price reactions to climate news. (That evidence is analyzed later in Section 6.6.)

Earnings news

Next, we include in x_t two measures of earnings news constructed using data from CRSP and I/B/E/S. The first measure is based on the idea that a large portion of earnings news occurs on days when firms make earnings-related announcements (Beyer et al., 2010). We consider two types of announcements: those of quarterly earnings and voluntary forward guidance regarding future earnings. We compute stock returns in excess of the market during the three-trading-day windows centered on these announcement dates. We add the excess returns across unique events within a given stock-quarter. For about 70% of observations, no summation is required because the forward-guidance date coincides with the earnings-announcement date. We find that our announcement-return measure explains about 20% of the variance of quarterly stock-level returns (see the Appendix).

Our second measure captures news about long-term earnings. Such news can arrive gradually over time, in between the quarterly announcements. This second measure uses data on analysts' forecasts of each firm's long-run earnings growth rate. For firm i and quarter t, the measure equals the earliest mean analyst forecast in quarter $t + 1$ minus the latest mean forecast in quarter $t - 1$. Using forecasts from quarters $t - 1$ and $t + 1$ helps to capture all news arriving in quarter t. The measure may also include a small amount of information that arrives in quarters $t - 1$ or $t + 1$, but those inclusions are innocuous since they should not help explain returns in quarter t. We winsorize this measure at the 1% level. We find that this measure is significantly related to quarterly stock-level returns but explains less than 1% of their variance (see the Appendix).

Since GMB is a spread between portfolio returns, we need to convert our firm-level earnings measures into the appropriate portfolio-level quantities to be included in x_t. We do so following GMB's construction, each month computing value-weighted averages of the firm-level measures within GMB's green and brown legs and then taking the difference.

Measuring the part of returns coming from earnings news is known to be difficult, and our measures surely miss important earnings news. Our first measure misses short-term news that arrives outside the three-day announcement windows. One limitation of our second measure is that analysts' forecasts can differ from investors' forecasts. In addition, analysts' long-term forecasts are only three- to five-year forecasts, so the second measure also misses changes in expectations of earnings that lie more than five years in the future.

Changes in expectations of distant future earnings seem especially likely to arise from shocks to climate concerns. For example, the meteoric rise of Tesla's stock price in 2020 may have been caused in part by climate-driven revisions to forecasts of electric vehicle sales at horizons longer than five years. Such climate-driven shocks to earnings, and thus to returns, can nevertheless be captured by our climate news measure, ΔC_t, which is included in x_t. In general, the return shocks captured by our specification of x_t can reflect changes in earnings expectations, either climate-driven, and thus captured by ΔC_t, or non-climate-driven, and thus captured by the earnings news measures.

6.5.4 Estimates of the equity greenium using past realizations

The first two columns of Table 6.4 report results from regressions of GMB returns on x_t, both with and without the earnings variables included in x_t. Including those variables raises the R-squared from 0.14 to 0.25. The same-month climate shock, ΔC_t, and the earnings announcement return both enter significantly with their expected positive signs (t-statistics:

TABLE 6.4: Sources of GMB performance.
We estimate monthly time-series regressions using data from November 2012 through June 2018. The dependent variable is the GMB return in columns 1–2 and GMB's Fama-French three-factor alpha in columns 3–4. We estimate alphas in a time-series regression of GMB on the Fama-French factors. We set each month's alpha equal to the regression's intercept plus residual. Both returns are in percent per month. "Δ Climate concerns" is the prediction error from rolling AR(1) models applied to the MCCC index. The two earnings news measures, "Earnings announcement returns" and "Δ Earnings forecasts," are described in Section 6.5.5. They correspond to the quarter that contains the given month. Robust t-statistics are in parentheses

	Dependent variable			
Independent variable	*GMB return*		*GMB alpha*	
Δ Climate concerns (same month)	4.08	3.75	3.95	3.44
	(2.70)	(2.69)	(2.79)	(2.70)
Δ Climate concerns (prev. month)	2.98	2.86	2.64	2.33
	(1.86)	(1.77)	(1.97)	(1.82)
Earnings announcement returns		0.77		0.63
		(2.64)		(2.31)
Δ Earnings forecasts		6.93		14.16
		(0.44)		(0.96)
Constant	0.05	−0.04	−0.10	−0.15
	(0.20)	(−0.15)	(−0.41)	(−0.66)
Observations	68	68	68	68
R^2	0.14	0.25	0.14	0.26

2.69 and 2.64). The positive ΔC_t coefficient supports the prediction that an increase in climate concerns is worse news for brown stocks than green stocks. This conclusion, based on monthly returns, echoes the conclusion reached by Ardia et al. (2023) at the daily frequency. The previous month's climate shock, ΔC_{t-1} also enters positively and is marginally significant (t-statistic: 1.77). This result, which suggests delayed stock price response to climate news, emerges more strongly among smaller stocks, as we show in Section 6.6. The only variable falling well short of statistical significance is the change in analysts' long-term forecasts, although its coefficient does have the expected positive sign.

The key quantity of interest, the equity greenium, is estimated by the regression intercept $\hat{a}$. With the earnings variables included in x_t, the estimated equity greenium is −4 bps per month. Recall that the ICC estimate is about −12 bps per month. Both the ex ante and ex post approaches thus suggest a negative equity greenium whose magnitude is modest, at least in comparison to the 65 bps for GMB's realized average return.

As noted earlier, GMB tilts toward large growth stocks. Size and growth effects are not driving our results, however. The remaining columns of Table 6.4, labeled "GMB alpha," repeat the above regressions with the dependent variable redefined as the GMB return adjusted by the three factors of Fama and French (1993). To construct that return, we take the intercept plus the residual from the time-series regression of GMB on the factors. The resulting slope coefficients are all quite similar to their counterparts in the first two columns. The intercept when the earnings variables are included shifts downward somewhat, from −4 bps to −15 bps.

What if there had been no climate-concern shocks or other shocks to green-versus-brown earnings? Panel A of Figure 6.7 compares GMB's realized performance to its counterfactual performance in the absence of climate and earnings shocks. Using the regression estimated in column 2 of Table 6.4, we compute the counterfactual monthly GMB return as the regression intercept, $\hat{a}$, plus the estimated residual. (Equivalently, the counterfactual equals the realized value minus the regressors times their respective coefficients.) The dashed line plots the cumulative counterfactual return, and the solid line shows the cumulative realized return. We also plot a 95% confidence interval around the counterfactual, recognizing that the regression coefficients are estimated with error. To compute this interval, we repeatedly draw regression coefficients from their estimated sampling distribution, use those coefficients to compute simulated counterfactual returns, and plot the simulated returns' 95% confidence intervals. Panel B of Figure 6.7 repeats the same analysis using the GMB alpha and the regression estimated in column 4 of Table 6.4.

Both panels of Figure 6.7 deliver the striking message that, absent shocks to climate concerns and earnings, GMB's performance is slightly downward-trending, reflecting the negative intercepts in the second and fourth columns of Table 6.4. Moreover, GMB's counterfactual performance is reliably below its realized performance, as the latter lies well outside the 95% confidence interval in both panels.

The sharp contrast between the realized and counterfactual performance in Figure 6.7 reflects the difference between $\bar{r}$ and $\hat{a}$, the two estimators in Eqs. (6.5) and (6.6). The main source of this difference is that the climate-concern shock, ΔC_t, had average realizations that were unexpectedly high during the sample period. Note in column 1 of Table 6.4 that when controlling for just climate-concern shocks, $\hat{a}$ is merely 5 bps, compared to 65 bps for GMB's average return, $\bar{r}$. The t-statistic for the average of ΔC_t is 4.01. Recall from Figure 6.5 that with such an outcome in the single-variable version of x_t, getting a misleading estimate of the equity greenium is much more likely when using the average realized performance, $\bar{r}$, than when using the average counterfactual performance, $\hat{a}$. In essence, given that a substantial portion of the increase in climate concerns was unanticipated, so too was GMB's significant positive performance. Accordingly, that performance should not lead one to infer that the expected return on green stocks is higher than brown.

Given the high realized average of ΔC_t, one might question whether its non-zero values were truly unanticipated. An alternative story could be that positive shocks early in the sample period led investors to anticipate subsequent increases in the climate-concern index. In considering that story, first recall that we use a moving 36-month estimation window when computing ΔC_t as the prediction error from an AR(1) model. The AR(1) model's intercept absorbs the recent level and trend in the climate index. Second, an equal split of the sample period gives results contrary to the above anticipation story. When we estimate the regressions in Table 6.4 separately in both subperiods, the climate-concern shocks actually enter somewhat more strongly in the second half (results are in the Appendix). If ΔC_t had become anticipated later in the sample period, then returns should have reacted to ΔC_t less strongly in the second subperiod, not more strongly.

6.5.5 ESG flows and assets

Increased climate concerns can impact green-versus-brown stock returns not only through expected earnings (via product demands) but also by impacting investors' desires to hold green stocks rather than brown. As perhaps the most prominent recent trend in the investment management industry, sustainable investing has experienced rapid growth. At the same time, however, the dollar amounts reallocated by sustainable investing thus far, especially in the U.S., appear to be fairly small relative to aggregates. Consider the universe

Panel A: GMB

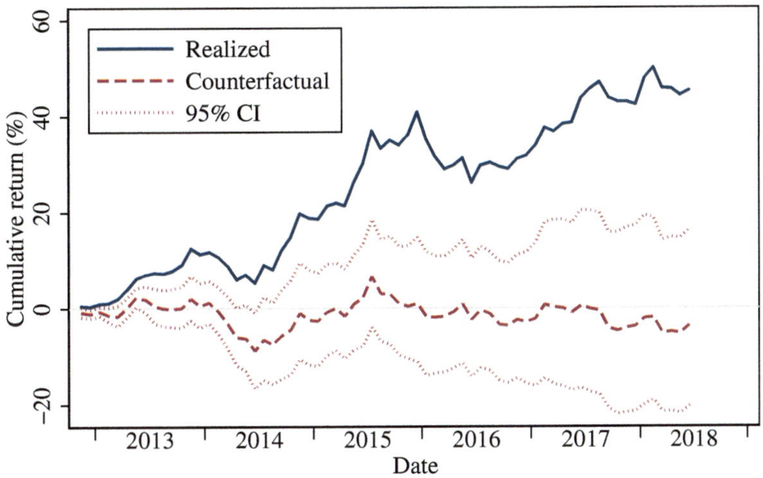

Panel B: GMB alpha

FIGURE 6.7: Counterfactual GMB performance. The solid lines shows realized cumulative, compounded returns on GMB (Panel A) and GMB's Fama-French three-factor alpha (Panel B). Alphas are computed as in Table 6.4. The dashed lines shows the returns' counterfactual counterparts computed from columns 2 and 4, respectively, of Table 6.4. The counterfactual monthly return equals the realized return minus the regressors times their respective regression slopes. Dotted lines indicate the counterfactual's 95% confidence interval. We compute confidence intervals using the following steps: (1) Estimate the regression from column 2 of Table 6.4 and store the estimated coefficients and their covariance matrix. (2) Repeat the following steps (2a)–(2c) 500 times: (2a) draw a new coefficient vector from a normal distribution with mean and variance saved in step (1); (2b) use the new coefficient to compute each period's counterfactual return; (2c) compute and store cumulative counterfactual returns. (3) Each month, compute the 2.5th and 97.5th percentiles of the counterfactual cumulative returns stored in step (2c).

of U.S. mutual funds and ETFs, for example. In 2020 its assets totaled about \$29 trillion, but sustainable funds' assets accounted for only \$230 billion, less than 1% of the total.[16]

When sustainable investing's asset share is small, so too is the likely effect of that investment on expected stock returns. In PST's calibrated version of their model, a small value for the fraction of the market's total assets owned by ESG-conscious investors (λ in their setting) implies a small effect on expected return. Berk and Binsbergen (2021) show that the effects of ESG divestment on expected return are quite small in both theoretical and empirical settings where the fraction of assets being divested is small.

Important to remember, though, is that the magnitude of the equity greenium does not depend only on such taste-related investment effects. Green stocks' expected returns can also reflect those stocks' greater ability to hedge against adverse climate news. Evidence of such ability appears in our Table 6.4 in the form of a significantly positive relation between GMB and ΔC_t, as well as in prior studies mentioned earlier. All investors can be willing to pay for that climate-hedging property of green stocks, whether or not some investors reallocate due to the warm glow (anguish) they get from holding green (brown) stocks. If climate-hedging demand increased during our sample period, this is yet another source of increased investor demand, and hence unexpected returns, for green assets.

Given that the asset footprint of ESG investing is still fairly small, one might reasonably surmise that ESG investing did not exert significant effects on GMB's realized returns. Nevertheless, some exploration of such potential effects seems warranted, especially given evidence that stock prices can respond significantly to seemingly small demand shifts (e.g., Koijen and Yogo, 2019; Gabaix and Koijen, 2021). In looking for ESG investing effects, we also examine green and brown returns separately, because such effects are more likely to be evident in brown stocks. For example, Nofsinger, Sulaeman, and Varma (2019) find that institutions are more likely to underweight stocks with negative environmental and social indicators than they are to overweight stocks scoring positively on those dimensions. The experimental evidence of Humphrey et al. (2021) shows the strengths of green versus brown preferences exhibit a similar asymmetry.

We construct two variables to investigate effects of ESG investing. The first uses flows into sustainable funds as a proxy for shifts in investor demand for green assets. From Morningstar's *2021 Sustainable Funds U.S. Landscape Report*, we obtain data on quarterly total flows into U.S. sustainable funds.[17] We scale these flows, which we refer to as "ESG flows," by the average total market capitalization of CRSP stocks during the quarter. ESG flows increased dramatically in 2013–2020, especially beginning in 2019.

The second investing variable uses sustainable funds' lagged total assets (AUM) as a proxy for the level of investors' ESG tastes. This variable is motivated by PST's theoretical result that expected green-minus-brown returns depend negatively on the average strength of ESG tastes (see Eq. (33) in PST), and the size of the ESG industry depends positively on those tastes (see Figure 5 in PST). We compute sustainable fund AUM from the previously mentioned Morningstar report, as detailed in the Appendix. We scale ESG AUM by the total market capitalization of CRSP stocks.

Columns 1 and 2 of Table 6.5 report results from regressions of GMB returns on the two investing variables and the previous climate and earnings variables. In columns 3 and 4, the dependent variable is the return on the green leg, and in columns 5 and 6, the brown leg of the GMB spread. Reverse causation is a potential concern when regressing returns

[16]Sources: Morningstar's *2021 Sustainable Funds U.S. Landscape Report* and the Investment Company Institute's *2021 Investment Company Fact Book*.

[17]The data combine active and passive funds, equity and bond funds, open-end funds, and ETFs. Morningstar defines a sustainable fund as follows: "For a fund to be included in the sustainable funds universe, it must hold itself out to be a sustainable investment. In other words, ESG concerns must be central to its investment process and the fund's intent should be apparent from a simple reading of its prospectus...."

TABLE 6.5: The roles of ESG flows and assets.
This table builds on Table 6.4 by adding controls for ESG flows and assets. "ESG flows" equals the quarter's dollar flow into ESG funds scaled by the average total CRSP market capitalization during the quarter that contains the given month, times 1000. In the specifications that drop Δ Climate concerns, the sample extends through September 2020. We instrument for contemporaneous ESG flow by using its previous-quarter value. The first-stage t-statistic for lagged flows is 3.50 in the shorter samples and 5.68 in the longer samples. We do not tabulate R^2 because it is difficult to interpret in an IV setting. "ESG assets" equals total AUM in ESG funds scaled by the total CRSP market capitalization and measured at the beginning of the quarter containing the given month, times 1000. The dependent variable is the GMB return in columns 1–2 and the market-hedged return on GMB's green (brown) leg in columns 3–4 (5–6), all in percent per month. We compute the market-hedged portfolio returns by replacing individual stock returns with $\tilde{r}_t^e$, the market-adjusted return defined in Section 6.7.1. Remaining details are the same as in Table 6.4

	Dependent variable					
Independent variable	*GMB return*		*Green leg*		*Brown leg*	
Δ Climate concerns (same month)	4.02		1.93		−1.90	
	(2.74)		(2.78)		(−1.69)	
Δ Climate concerns (prev. month)	3.30		0.42		−3.04	
	(2.13)		(0.67)		(−2.62)	
Earnings announcement returns	0.84	0.88	0.20	0.22	−0.63	−0.60
	(3.06)	(2.69)	(1.61)	(1.54)	(−2.88)	(−2.51)
Δ Earnings forecasts	9.97	2.00	4.56	5.20	−6.70	3.23
	(0.70)	(0.15)	(0.82)	(1.06)	(−0.59)	(0.31)
ESG flows	32.96	9.00	7.30	1.91	−29.69	−11.99
	(1.51)	(1.42)	(0.95)	(0.54)	(−1.60)	(−2.55)
ESG assets	−0.56	−0.74	−0.27	−0.27	0.61	0.84
	(−0.82)	(−1.13)	(−0.89)	(−0.81)	(1.08)	(1.78)
Constant	−0.34	1.80	0.18	0.64	0.03	−1.75
	(−0.28)	(1.37)	(0.28)	(1.02)	(0.03)	(−1.80)
Observations	68	95	68	95	68	95

on contemporaneous flows. Instead of flows (or shifts in investors' ESG demands) causing returns, flows could be chasing recent returns within the same period. We address this potential endogeneity by instrumenting for same-quarter ESG flow using its previous-quarter value and estimating the regression by two-stage least squares. The exclusion restriction plausibly holds, because flows cannot chase future realized returns. We find large first-stage t-statistics, indicating that the relevance condition holds and there is no problem with weak instruments.

The coefficients on ESG flows and assets in Table 6.5 all have their predicted signs, whether or not climate concerns are included in the regression. That is, ESG flows enter positively for the GMB spread and its green leg but negatively for the brown leg, whereas ESG assets enter negatively for the GMB spread and its green leg but positively for the brown leg. For the GMB spread and its green leg, none of the above coefficients are statistically significant. This insignificance could be related to the fact, noted above, that ESG investment during this period is still relatively small. For the brown leg, however, when climate concerns are excluded from the regression, ESG flows enter with a t-statistic of

−2.55, and ESG assets get a marginally significant t-statistic of 1.78. These stronger effects of ESG investing on the brown leg are consistent with the asymmetry noted earlier. When climate concerns are included in the regression, though, the brown leg's coefficients on ESG flows and assets also become insignificant.[18] A reasonable interpretation is that the effects of ESG investing on brown stocks' returns are driven largely by climate concerns. Overall, the results in Table 6.5 justify having excluded ESG flows and assets from our primary regression analyses in Table 6.4.

6.5.6 Adding other shocks

As explained earlier, our measure of climate concerns builds on that of Ardia et al. (2023). Those authors in turn acknowledge the prior work of Engle et al. (2020), who construct two media-based measures of climate concerns. Ardia et al. (2023) discuss those alternative measures and explain that their measure adds risk as another component of climate concerns. We rely on that more recent measure, but we also examine the robustness of our results to including the Engle et al. measures. We find that doing so does not change our conclusions. We augment the independent variables in column 2 of Table 6.4 by including climate-concern shocks based on both Engle et al. measures. One of their measures enters significantly, whereas the Ardia et al. measure always enters positively and significantly, either for the current or previous month. When we add the one significant measure from Engle et al. to the right-hand side of the regression in column 2 of Table 6.4, we obtain the same conclusions: the counterfactual GMB return slopes down slightly. The plot is in the Appendix.

Besides their MCCC index, Ardia et al. (2023) also construct sub-indices capturing eight themes related to climate change: agreement and summit, agricultural impact, disaster, environmental impact, financial and regulation, research, societal impact, and "other." To see which themes correlate most closely with GMB returns, we first compute the ΔC_t series for each of the eight sub-indices and then regress GMB on both ΔC_t and its lag, analogous to our analysis for the MCCC index. For each of the eight themes, we find positive slope estimates on both ΔC_t and its lag. At least one of those measures is statistically significant for five of the eight themes. (We tabulate the results in the Appendix.) Therefore, the results in Table 6.4 are not driven by any single type of climate concerns.

The three themes that deliver the largest R-squared in the above regressions are agreement and summit ($R^2 = 0.15$), societal impact ($R^2 = 0.12$), and financial and regulation ($R^2 = 0.12$). Ardia et al. (2023) find these three themes are also closely related to the returns on their green-minus-brown portfolio, which is constructed differently from ours. Moreover, these are the three most discussed themes in the media, according to Ardia et al. The theme that delivers the lowest R-squared is disaster ($R^2 = 0.02$). GMB returns are thus more closely associated with climate-related policy news than with news about disasters.

We also examine the robustness of our results to the inclusion of oil price shocks, which have clear environmental relevance, and long-term bond returns, which could be related to differences in duration between green and brown stocks. We measure oil price shocks as the monthly change in the expected "front month" value of oil, derived from oil futures contracts.[19] We take the long-term bond return to be the return on the 30-year U.S. Treasury bond. When we add both variables to the right-hand side of the regression in column 2 of Table 6.4, the counterfactual performance of GMB is again essentially flat. See the Appendix.

[18]When we adjust the GMB spread for the three Fama-French factors, all the coefficients on ESG flows and assets retain the same signs, but none of them are statistically significant. See the Appendix for details.

[19]We thank Erik Gilje for providing these data.

TABLE 6.6: Greenness and individual stock returns.
This table shows results from panel regressions in which the dependent variable is stock i's percent return in month t. $g_{i,t-1}$ is the stock's lagged greenness. ΔC_t is month t's change in aggregate climate concerns, computed as the prediction error from a rolling AR(1) model applied to the MCCC index. "[Earnings announcement ret.]$_{i,t}$" is the stock's sum of the three-trading-day excess percent returns (stock minus market) around earnings announcements and management earnings forecasts (if available) during the quarter containing month t. "[Δ Earnings forecast]$_{i,t}$" is the change in analysts' mean long-term earnings growth rate forecast for stock i during the quarter containing month t. The last column adds controls (untabulated) for the log of lagged market equity, log of lagged book-to-market ratio, and the stock's return from months $t-12$ through $t-2$, following Lewellen (2015). The sample begins in November 2012. All regressions include month fixed effects, cluster by month, and use robust standard errors

	(1)	(2)	(3)	(4)
$g_{i,t-1}$	0.21	0.00	−0.02	−0.04
	(2.24)	(0.02)	(−0.23)	(−0.41)
$g_{i,t-1} \times \Delta C_t$		0.83	0.81	0.72
		(1.42)	(1.59)	(1.28)
$g_{i,t-1} \times \Delta C_{t-1}$		1.70	1.54	1.65
		(2.66)	(2.78)	(2.68)
[Earnings announcement ret.]$_{i,t}$			0.32	0.32
			(13.28)	(12.38)
[Δ Earnings forecast]$_{i,t}$			5.89	5.91
			(5.02)	(4.58)
Observations	218,208	153,884	133,290	114,355
R^2	0.18	0.11	0.18	0.19
Additional controls	No	No	No	Yes

6.5.7 Greenness and individual stock returns

All of our empirical analysis thus far is based on the time series of green-versus-brown portfolio returns. To show that our conclusions do not hinge solely on portfolio returns, we next run panel regressions using individual stocks.

Table 6.6 reports regressions of individual stock returns in month t on various regressors. All regressions include time fixed effects and therefore capture cross-sectional variation in returns. We begin in column 1 with a single regressor, the stock's greenness, $g_{i,t-1}$. The remaining columns add regressors that capture shocks to climate concerns and earnings. The climate-concern shocks are interacted with the stock's greenness, and the earnings variables are the firm-level constituents of the earlier portfolio-level versions used in Table 6.4. The last column includes additional stock-specific variables as controls: log of market equity, log of book-to-market, and return from months $t-12$ through $t-2$.

When greenness is the only regressor, it has a significantly positive relation to return (column 1), consistent with the outperformance of green stocks reflected in GMB. The coefficient on greenness becomes negative when the regressors include the climate-concern and earnings variables (columns 3 and 4), consistent with GMB's negative expected return estimate given by the intercept in column 2 of Table 6.4. Thus, consistent with the GMB results, the relation between greenness and return flips from strongly positive to modestly negative when controlling for shocks to returns from climate concerns and earnings.

The coefficients on the climate-concern variables indicate that green stocks outperform when climate concerns increase, consistent with that same conclusion delivered by the regressions for GMB in Table 6.4. The timing is somewhat different, however, in that the lagged climate-concern shock now enters more strongly than the contemporaneous shock. In Table 6.6, the coefficient on $g_{i,t-1}$ interacted with ΔC_t is positive but insignificant, while the coefficient on the interaction with ΔC_{t-1} is larger and significant. In Section 6.6, we further analyze the delayed reaction of stock prices to climate news.

6.5.8 Industry greenness

Our analysis thus far is based on $g_{i,t}$, a measure of the firm's total greenness that reflects two components: the greenness of the firm's industry and the relative greenness of the firm within its industry. How do each of those components contribute to our results? To investigate this question, we decompose $g_{i,t}$ as

$$g_{i,t} = gAcross_{i,t} + gWithin_{i,t}, \tag{6.7}$$

with $gAcross_{i,t}$ equal to the average $g_{i,t}$ of all firms within the same industry as stock i in month t, and $gWithin_{i,t} = g_{i,t} - gAcross_{i,t}$.

Figure 6.8 displays the original GMB analyzed earlier as well as an alternative GMB series constructed the same way but with $gWithin_{i,t}$ replacing $g_{i,t-1}$. We see that the cumulative performance of this alternative, industry-adjusted GMB is much weaker than the original. While the original GMB's average return is positive and highly significant ($t = 3.23$; see column 1 of Table 6.3), the average return of the industry-adjusted GMB is four times smaller and insignificant ($t = 0.99$; see the Appendix). Therefore, the original GMB's performance owes much to industry-level greenness.

The technology industry, especially "big tech," has delivered high stock returns in recent years. However, our results are not driven by big tech. To show this, we compute monthly returns on the value-weighted portfolio of "FAANG" stocks, which include Meta (formerly Facebook), Amazon, Apple, Netflix, and Alphabet (formerly Google). The FAANG portfolio's return is not significantly related to either the original GMB or changes in climate concerns. The Appendix shows the details.

The importance of industry greenness is also evident in individual stock returns. Table 6.7 reports the same regressions as Table 6.6, except that each independent variable containing $g_{i,t}$ is replaced by two variables, one for each component in Eq. (6.7). We first see that, as with GMB, the superior performance of green stocks relative to brown is largely attributable to industry greenness. In column 1 of Table 6.7, the coefficient on $gAcross_{i,t}$, industry greenness, is 3.6 times the coefficient on $gWithin_{i,t}$, within-industry greenness; the former coefficient is statistically significant ($t = 2.14$), whereas the latter is not ($t = 1.11$).

Industry greenness continues to play the dominant role in Table 6.7's remaining columns, which analyze the sources of green stocks' outperformance. Recall that a key result in Table 6.6 is the significantly positive coefficient on $g_{i,t-1}$ interacted with month $t-1$'s climate-concern shock. When the latter shock is instead interacted with industry greenness ($gAcross_{i,t}$), the coefficient on that variable is significantly positive in each of columns 2 through 4. In contrast, when the same climate shock is interacted with within-industry greenness ($gWithin_{i,t}$), the coefficient is insignificant throughout. Therefore, we conclude that industry greenness is the key component of a firm's greenness, capturing both the superior past performance of green stocks as well as the climate-shock source of that performance.

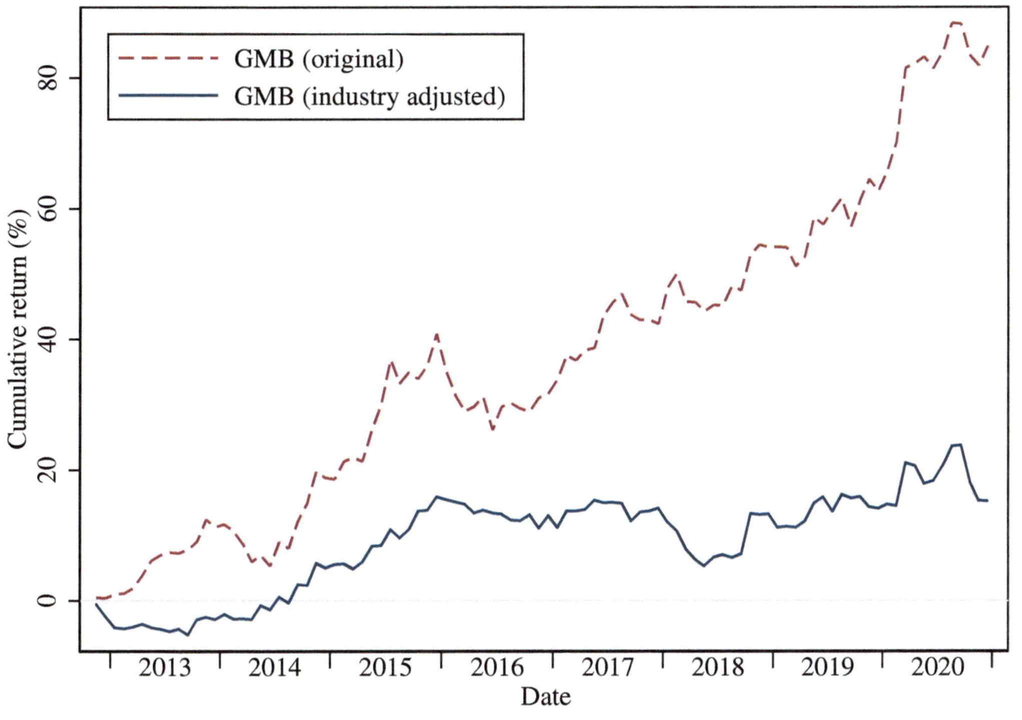

FIGURE 6.8: Effect of industry adjustment. The dashed line plots the cumulative return on the original GMB (green-minus-brown) portfolio constructed with firms' total greenness (i.e., not industry-adjusted). The solid line plots the cumulative return on an industry-adjusted GMB portfolio, which is constructed using g scores demeaned at the industry $\times$ month level.

6.6 Delayed stock price reaction to climate news

In this section, we take a closer look at the timing of the strong positive relation between green-versus-brown returns and the shock to climate concerns, ΔC. As shown in the previous section, the GMB return is strongly related to ΔC in the current month, whereas ΔC in the previous month enters more strongly in the panel regression using individual stocks. We conjecture this difference relates to stocks' market capitalization. The long- and short-leg portfolios of GMB are value-weighted, making GMB most representative of the largest stocks. The panel regression is instead representative of a typical stock, which is substantially smaller than the largest stocks.

To investigate the role of size, we replicate GMB's construction separately within the large- and small-cap segments. Small (large) stocks are those in the bottom (top) quartile of market capitalization based on NYSE breakpoints. As in the original GMB, we continue value-weighting stocks within each GMB spread's green and brown legs. We then regress these GMB returns on the current and lagged month's ΔC. The first row of Table 6.8 reports the results. For large-cap GMB, same-month ΔC is significant ($t = 2.46$), whereas previous-month ΔC is not ($t = 1.74$). In contrast, for small-cap GMB, same-month ΔC is insignificant ($t = 1.23$), whereas previous-month ΔC is strongly significant ($t = 2.99$), and its coefficient significantly exceeds its large-cap counterpart ($t = 2.35$). In sum, while

TABLE 6.7: Greenness and individual stock returns: Effects within and across industries. This table repeats the regressions in Table 6.6, except that we decompose g into $gAcross$ and $gWithin$, representing across- and within-industry variation. We define $gAcross$ as the average of g within the industry×month, and $gWithin = g - gAcross$, so that $g = gAcross + gWithin$

	(1)	(2)	(3)	(4)
$gAcross_{i,t-1}$	0.25	−0.00	−0.02	−0.05
	(2.14)	(−0.01)	(−0.18)	(−0.39)
$gWithin_{i,t-1}$	0.07	0.02	−0.02	−0.01
	(1.11)	(0.28)	(−0.27)	(−0.11)
$gAcross_{i,t-1} \times \Delta C_t$		1.08	1.05	0.94
		(1.49)	(1.66)	(1.33)
$gWithin_{i,t-1} \times \Delta C_t$		−0.13	−0.08	−0.09
		(−0.28)	(−0.17)	(−0.16)
$gAcross_{i,t-1} \times \Delta C_{t-1}$		2.01	1.86	1.94
		(2.58)	(2.74)	(2.57)
$gWithin_{i,t-1} \times \Delta C_{t-1}$		0.49	0.34	0.56
		(1.05)	(0.70)	(1.00)
$[\text{Earnings announcement ret.}]_{i,t}$			0.32	0.32
			(13.28)	(12.38)
$[\Delta\ \text{Earnings forecast}]_{i,t}$			5.85	5.88
			(5.01)	(4.57)
Observations	218,208	153,884	133,290	114,355
R^2	0.18	0.11	0.18	0.19
Additional controls	No	No	No	Yes

both large-cap and small-cap GMB exhibit strong positive reactions to ΔC, the reaction is significantly more delayed in the small-cap segment.

We bring sharper focus to the timing of this size-related difference in reactions by looking at a weekly frequency. We construct the weekly ΔC as the prediction error from an AR(1) model estimated using the previous three years of observations of the weekly $MCCC$, computed as the within-week average of the daily $MCCC$ values. We also compute the weekly difference in returns between large-cap and small-cap GMB. Then we regress that return difference on ΔC lagged each of τ weeks, for $\tau = 0, \ldots, 7$. Figure 6.9 displays the estimated coefficients along with their 95% confidence intervals. The plot reveals that large-cap GMB reacts more strongly than small-cap GMB to ΔC in the current and most recent week, with the difference being statistically significant for the current week. In contrast, small-cap GMB reacts more strongly at longer lags, significantly so at the three-week lag.

The apparent slower reaction of small stocks to climate news is consistent with prior evidence that small stocks react more slowly in general. For example, Lo and MacKinlay (1990) show that returns on small stocks generally lag those of large stocks. Also, it is well known that small stocks are less liquid and less covered by analysts and media, potentially making them more susceptible to underreaction. A large literature attributes numerous

TABLE 6.8: Stock size and the response to climate news.
This table shows results from six time-series regressions of monthly portfolio percent returns on ΔC, the change in climate concerns, from the same and previous months. The first row shows results for two value-weighted GMB portfolios, the second shows results for two green portfolios, and the third shows results for two brown portfolios. The green and brown portfolios form the two legs of GMB. For each portfolio, we form one version using small-cap stocks and a second version using large-cap stocks. Small (large) stocks are those in the bottom (top) quartile of market capitalization, using monthly unconditional NYSE breakpoints. The column "Lg.−Sm." shows the difference between the large- and small-cap portfolios' coefficients. The green and brown portfolios' returns are market-adjusted, meaning we subtract from each stock's excess return the stock's estimated market beta times the excess market return. Each regression has 68 observations and uses data from November 2011 through June 2018. Robust t-statistics are in parentheses

	ΔC (*same month*)			ΔC (*prev. month*)		
Portfolio	*Small*	*Large*	*Lg.−Sm.*	*Small*	*Large*	*Lg.−Sm.*
GMB	2.83	3.91	1.08	7.49	2.79	−4.70
	(1.23)	(2.46)	(0.56)	(2.99)	(1.74)	(−2.35)
Green	0.03	2.27	2.24	−0.14	0.62	0.75
	(0.01)	(3.19)	(0.75)	(−0.06)	(0.83)	(0.28)
Brown	−1.81	−1.39	0.42	−8.49	−2.35	6.14
	(−0.61)	(−1.26)	(0.15)	(−2.71)	(−2.10)	(2.29)

return anomalies to underreaction, and Chen et al. (2023) find those anomalies are stronger among firms with lower media coverage.

Underreaction of stock prices to climate news may not be limited to small stocks. Recall from the first row of Table 6.8 that large-cap GMB's coefficient on the previous month's ΔC borders on significance ($t = 1.74$). Stronger evidence of large-cap underreaction emerges when we examine GMB's green and brown legs separately in the remaining rows of Table 6.8. For large stocks, both the green and brown legs exhibit significant reactions to ΔC in the expected directions, i.e., positive for green and negative for brown. However, the green leg's significant reaction occurs for the same month ($t = 3.19$), whereas the brown leg's significant reaction occurs for the previous month ($t = -2.10$). The other t-statistics for the large-cap legs are insignificant.

Separating the green and brown legs also reveals that the effect of climate news on small stocks is limited to brown stocks. The green leg's coefficients on both the same-month and previous-month ΔC are virtually zero ($t = 0.01$ and -0.06, respectively), whereas the brown leg exhibits negative reactions to both ΔC's. Only the previous-month ΔC is statistically significant, not surprisingly, given the similar result for the small-cap GMB spread.

6.7 The green factor

PST introduce an ESG factor and show that, along with the market factor, the ESG factor prices assets in equilibrium. Motivated by that insight, we use stocks' greenness scores to construct a green factor, thereby continuing our focus on the prominent role of "E" in ESG investing. In this section, we explain the green factor's construction and show that

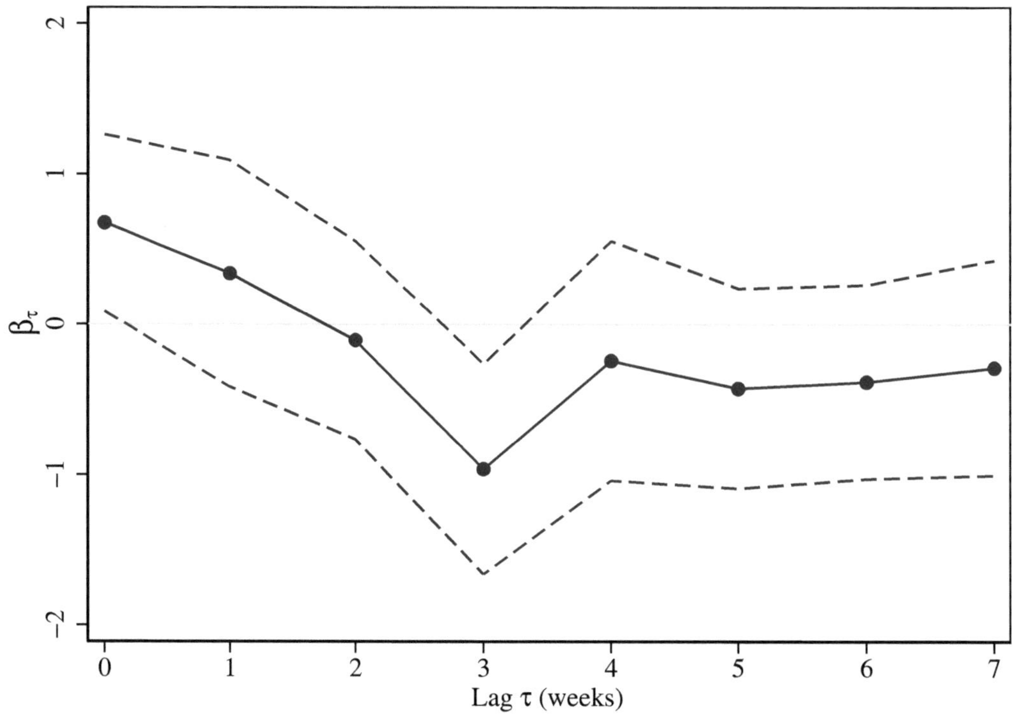

FIGURE 6.9: Weekly response of GMB to climate news: Large versus small stocks. This figure plots the coefficients β_τ from a regression of large-minus-small GMB weekly percent returns on weekly shocks to climate concerns lagged by τ weeks, for $\tau = 0, \ldots, 7$ weeks. The large-minus-small GMB portfolio is defined in Table 6.8. Weekly shocks to climate concerns are prediction errors from rolling AR(1) models fitted to the weekly MCCC index. Dashed lines indicate 95% confidence intervals.

it helps explain the recent underperformance of value stocks. We also find that the green factor, appropriately scaled, is empirically similar to GMB analyzed above.

6.7.1 Constructing the green factor

We apply the PST methodology to construct the green factor. PST show that the factor's realizations can be estimated month by month by running cross-sectional regressions of market-adjusted excess stock returns on the stocks' greenness, with no intercept. The slope from one such regression, which represents the green factor's realization in month t, is given in Eq. (34) of PST as

$$\hat{f}_{gt} = \frac{g'_{t-1}\tilde{r}^e_t}{g'_{t-1}g_{t-1}}\,, \tag{6.8}$$

where $\tilde{r}^e_t \equiv \tilde{r}_t - \beta_{m,t-1}\tilde{r}_{mt}$ is the vector of stocks' market-adjusted excess returns. Specifically, $\tilde{r}_t$ is the vector of stocks' returns in excess of the risk-free rate, $\tilde{r}_{mt}$ is the market return in excess of the risk-free rate, and $\beta_{m,t-1}$ is the vector of stocks' market betas, which we estimate from rolling monthly regressions of individual stocks' excess returns on excess market returns using up to 60 months (and no less than 36 months) of data ending in month t.

Eq. (6.8) implies that $\hat{f}_{gt}$, a linear combination of the elements of $\tilde{r}_t^e$, is the market-hedged excess return on a portfolio containing long positions in green stocks (with positive $g_{i,t-1}$'s) and short positions in brown stocks (with negative $g_{i,t-1}$'s). In common terminology, the green factor is therefore the return on a "zero-cost" long-short factor. The green factor, however, differs in both motivation and construction from typical zero-cost factors in the finance literature. Motivation for the latter factors is often empirical, whereas PST derive the green factor theoretically, showing that assets are priced in equilibrium by two factors, the market portfolio and the green factor. The construction of many zero-cost factors can be somewhat arbitrary, with stocks in the long and short legs often weighted by market cap (e.g., Fama and French, 1993, 2015). In contrast, the analytically derived green factor weights each stock by its greenness, with green stocks receiving positive weights and brown stocks negative weights. Also, the typical factor's long and short returns are not market-hedged, whereas the green factor is constructed with market-hedged excess returns.

In addition to market hedging and weighting stocks by greenness, the green factor's construction differs from that of the typical zero-cost factor in another technical respect. The typical factor is a difference between two unlevered rates of return: the return on the long leg minus the return on the short leg. The green factor is generally not a return difference with the same simple form, at least not quite. Specifically, we can rewrite Eq. (6.8) as $\hat{f}_{gt} = \hat{f}_{green,t} - \hat{f}_{brown,t}$, with the contribution from green stocks being $\hat{f}_{green,t} = g_{t-1}^{\prime +}\tilde{r}_t^{e+}/(g_{t-1}^{\prime}g_{t-1})$, where g_{t-1}^{+} contains positive values of g_{t-1} and $\tilde{r}_t^{e+}$ contains those stocks' excess returns. Similarly, the contribution to $\hat{f}_{gt}$ from brown stocks is $\hat{f}_{brown,t} = -g_{t-1}^{\prime -}\tilde{r}_t^{e-}/(g_{t-1}^{\prime}g_{t-1})$, where g_{t-1}^{-} contains negative values of g_{t-1} and $\tilde{r}_t^{e-}$ contains those stocks' excess returns. Both $\hat{f}_{green,t}$ and $\hat{f}_{brown,t}$ are market-hedged excess returns on portfolios, but generally those portfolios have differing degrees of implied leverage, because the sum of the elements of $g_{t-1}^{\prime +}$ does not necessarily equal minus the sum of the elements of $g_{t-1}^{\prime -}$. In our data, for example, the latter sum's magnitude is about 1.9 times the former sum, on average. Moreover, neither of those sums generally equals $g_{t-1}^{\prime}g_{t-1}$ in magnitude, meaning that neither $\hat{f}_{green,t}$ nor $\hat{f}_{brown,t}$ is the unlevered excess return on the market-hedged portfolio of its underlying stocks. Note finally that g_{t-1} is meaningfully defined only up to multiplication by a positive scalar, whose value is irrelevant to satisfying the condition in Eq. (6.3). The right-hand side of Eq. (6.8) can be multiplied by any positive scalar without affecting the green factor's pricing ability. We choose the scalar to achieve a desired volatility of the green factor, as explained next.[20]

Recall that GMB is the green-versus-brown return spread analyzed earlier. We scale the green factor to have its monthly volatility match that of GMB over the sample period, 1.99%. Figure 6.10 plots the green factor's cumulative return (solid line) along with that of the cumulative GMB return (dashed line). The plotted lines exhibit strong similarities in both cumulative performance and fluctuations. The monthly Sharpe ratios over the period are similar, 0.29 for the green factor versus 0.33 for GMB, and the monthly correlation between the green factor and GMB is 0.72. Therefore, despite the various differences in construction between GMB, a typical zero-cost return, and our green factor, the latter is well viewed empirically as a green-versus-brown return difference over the sample period.[21]

[20]Note that the green factor's greenness always equals one. Following PST, a portfolio with market-adjusted excess return $x_{t-1}^{\prime}\tilde{r}_t^e$ has greenness equal to $x_{t-1}^{\prime}g_{t-1}$. The green factor in Eq. (6.8) has $x_{t-1} = (1/g_{t-1}^{\prime}g_{t-1})g_{t-1}$, so the factor's greenness equals $(1/g_{t-1}^{\prime}g_{t-1})g_{t-1}^{\prime}g_{t-1} = 1$.

[21]This result seems somewhat similar to an observation made independently by Lioui and Tarelli (2022).

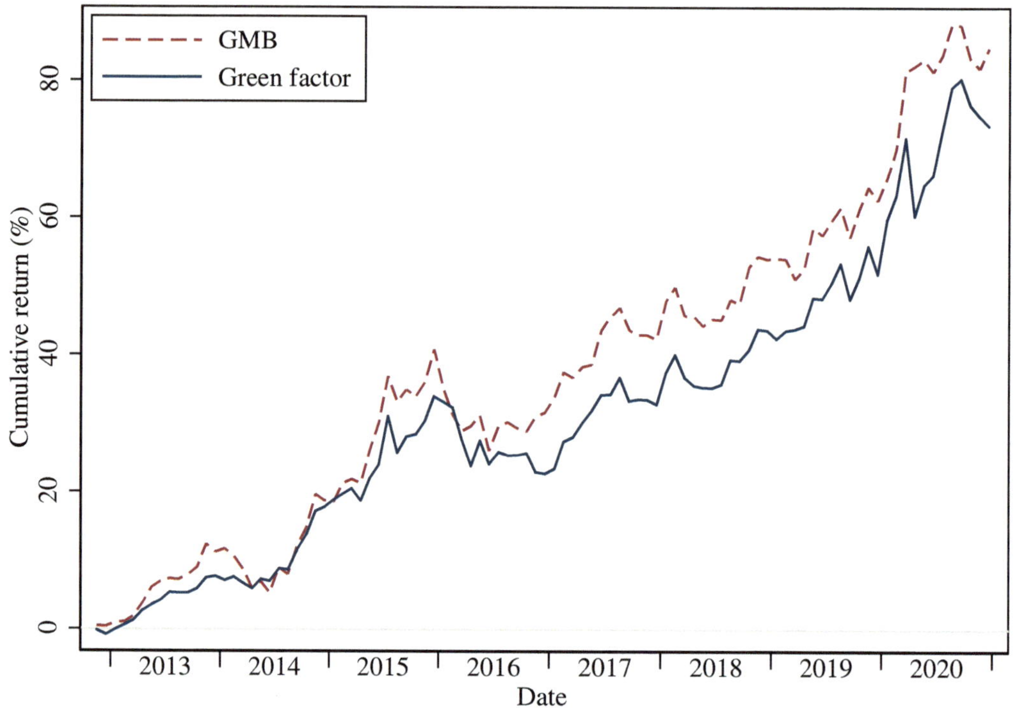

FIGURE 6.10: Green factor. This figure compares the cumulative returns of the green factor (solid line) and GMB (dashed line).

6.7.2 Value and momentum

During our sample period, the market-adjusted monthly alphas of HML and UMD are −71 bps and 66 bps, respectively, with t-statistics of −1.93 and 1.92, as shown in columns 1 and 3 of Table 6.9. GMB's significant exposures to value and momentum, noted earlier, prompt us to ask a performance question in the reverse direction: To what extent can the strong performance of green stocks relative to brown account for the last decade's historic underperformance of value, or for the positive performance of momentum?

To address this question, we turn to PST's two-factor model, in which the factors are the market portfolio and the green factor. HML's and UMD's alphas with respect to the two-factor model, which are shown in columns 2 and 4 of Table 6.9, are much smaller in magnitude than with just market adjustment. HML's alpha becomes −15 bps instead of −71 bps; UMD's alpha becomes −6 bps instead of 66 bps. The t-statistics shrink to −0.50 and −0.22. These results show that nearly 80% of HML's negative alpha, and all of UMD's positive alpha, disappear after controlling for the green factor's strong performance. Recognizing the brown nature of value stocks, and the green nature of growth stocks, thus helps us understand why the value strategy experienced its worst decade ever in the 2010s.

The green factor also explains about two thirds of the underperformance of an industry-neutral HML factor.[22] This factor's monthly CAPM alpha in our sample period is −66 bps ($t = -2.69$), but the alpha drops to −23 bps ($t = -1.37$) when we add the green factor to the regression. As with the original HML, the industry-neutral HML has a significantly negative loading on the green factor. See the Appendix for details.

[22]We thank Peter Hecht of AQR for providing this factor's returns.

TABLE 6.9: Explaining value and momentum with the green factor.
We estimate monthly time-series regressions of either HML (in columns 1 and 2) or UMD (in columns 3 and 4) on the excess market return and the green factor by using data from November 2012 to December 2020. Returns are in percent per month. Robust t-statistics are in parentheses

	Value		*Momentum*	
Constant	−0.71	−0.15	0.66	−0.06
	(−1.93)	(−0.50)	(1.92)	(−0.22)
Mkt-RF	0.14	0.07	−0.37	−0.27
	(1.18)	(0.70)	(−3.75)	(−3.14)
Green factor		−0.80		1.05
		(−4.55)		(6.18)
Observations	98	98	98	98
R^2	0.04	0.35	0.17	0.49

Recall that PST's two-factor model includes the green factor, not GMB, as the second factor alongside the market. When we depart from the model and replace the green factor with GMB, the two-factor alphas of both HML and UMD move farther away from zero: HML's alpha becomes −32 bps instead of −15 bps, and UMD's alpha becomes 21 bps instead of −6 bps. A full table is in the Appendix. The green factor thus performs better than GMB in explaining value and momentum over the past decade.

While exposure to the green factor explains most or all of HML and UMD, the reverse is not true. The green factor's strong performance over the last decade survives controlling for HML and UMD exposures. When we rerun the regression reported in column 4 of Table 6.3, replacing GMB with the green factor, $\hat{f}_{gt}$, we find a positive and significant alpha of 34 bps per month ($t = 2.46$). Details are in the Appendix.

6.8 Conclusion

Realized returns are a popular proxy for expected returns in the empirical asset pricing literature. However, high realized returns do not always indicate high expected returns, especially if realized over a relatively short period. We offer the salient example of high returns on green assets over the past decade. We show that these high returns were unexpected, reflecting news about environmental concerns rather than high expected returns. After constructing a green-minus-brown portfolio from U.S. stock data, we show that the portfolio's recent outperformance vanishes after removing the effects of unexpected increases in climate concerns. Another proxy for the portfolio's expected return, its implied cost of capital, is also consistently negative. A two-factor asset pricing model featuring a theoretically motivated green factor absorbs much of the historic underperformance of value stocks in the 2010s. Finally, our evidence suggests that small stocks underreact to climate news.

Our results contain a warning for investigations of the pricing of climate risk. We find that green stocks typically outperform brown when climate concerns increase. This result echoes similar findings by Choi, Gao, and Jiang (2020), Engle et al. (2020), and Ardia et al. (2023). Equilibrium expected returns of stocks that are better hedges against adverse climate shocks include a negative hedging premium if the representative investor is averse to such shocks (e.g., PST). Empirically confirming a climate risk premium, however, must confront the large unanticipated positive component of green stock returns during the last decade. Without accounting for those unexpectedly high returns on stocks that appear to be good climate hedges, one could be led astray. That is, one could infer that stocks providing better climate hedging have higher expected returns, not lower as theory predicts.

We use two approaches to estimate the green-minus-brown portfolio's expected return, which we label the equity greenium. The first approach, the implied cost of capital, has been applied by prior studies to different data. The second approach, which removes unanticipated shocks from the realized average return, seems novel. Future research can apply this latter approach to estimate expected returns in other settings.

Data Availability

We have uploaded our code and a portion of our data to Mendeley Data. The remaining data are proprietary.

Supplementary material

Supplementary material associated with this chapter can be found, in the online version, at `doi.org/10.1016/j.jfineco.2022.07.007`.

References

Ardia, D., K. Bluteau, K. Boudt, and K. Inghelbrecht, 2023, Climate change concerns and the performance of green vs. brown stocks, *Management Science* 69, 7607–7632.

Aswani, J., A. Raghunandan, and S. Rajgopal, 2024, Are carbon emissions associated with stock returns?, *Review of Finance* 28, 75–106.

Avramov, D., S. Cheng, A. Lioui, and A. Tarelli, 2022, Sustainable investing with esg rating uncertainty, *Journal of Financial Economics* 145, 642–664.

Baker, M., D. Bergstresser, G. Serafeim, and J. Wurgler, 2022, The pricing and ownership of us green bonds, *Annual Review of Financial Economics* 14, 415–437.

Baldauf, M., L. Garlappi, and C. Yannelis, 2020, Does climate change affect real estate prices? only if you believe in it, *The Review of Financial Studies* 33, 1256–1295.

Berg, F., J. F. Koelbel, and R. Rigobon, 2022, Aggregate confusion: The divergence of esg ratings, *Review of Finance* 26, 1315–1344.

Berg, F., J. Kölbel, A. Pavlova, and R. Rigobon, 2021, Esg confusion and stock returns: tackling the problem of noise, Unpublished working paper.

Berk, J., and J. Binsbergen, 2021, The impact of impact investing, Unpublished working paper.

Bernstein, A., M. T. Gustafson, and R. Lewis, 2019, Disaster on the horizon: The price effect of sea level rise, *Journal of Financial Economics* 134, 253–272.

Beyer, A., D. A. Cohen, T. Z. Lys, and B. R. Walther, 2010, The financial reporting environment: Review of the recent literature, *Journal of Accounting and Economics* 50, 296–343.

BlackRock, 2020, Global sustainable investing survey, Technical report, BlackRock.

Bolton, P., and M. Kacperczyk, 2021, Do investors care about carbon risk?, *Journal of Financial Economics* 142, 517–549.

Bolton, P., and M. Kacperczyk, 2023, Global pricing of carbon-transition risk, *The Journal of Finance* 78, 3677–3754.

Carhart, M. M., 1997, On persistence in mutual fund performance, *The Journal of Finance* 52, 57–82.

Chava, S., 2014, Environmental externalities and cost of capital, *Management Science* 60, 2223–2247.

Chen, X., W. He, L. Tao, and J. Yu, 2023, Attention and underreaction-related anomalies, *Management Science* 69, 636–659.

Choi, D., Z. Gao, and W. Jiang, 2020, Attention to global warming, *The Review of Financial Studies* 33, 1112–1145.

Eccles, R. G., and J. C. Stroehle, 2018, Exploring social origins in the construction of esg measures, *Available at SSRN 3212685*.

Edmans, A., 2011, Does the stock market fully value intangibles? employee satisfaction and equity prices, *Journal of Financial Economics* 101, 621–640.

El Ghoul, S., O. Guedhami, C. C. Kwok, and D. R. Mishra, 2011, Does corporate social responsibility affect the cost of capital?, *Journal of Banking & Finance* 35, 2388–2406.

Elton, E. J., 1999, Presidential address: expected return, realized return, and asset pricing tests, *The Journal of Finance* 54, 1199–1220.

Engle, R. F., S. Giglio, B. Kelly, H. Lee, and J. Stroebel, 2020, Hedging climate change news, *The Review of Financial Studies* 33, 1184–1216.

Faccini, R., R. Matin, and G. Skiadopoulos, 2023, Dissecting climate risks: Are they reflected in stock prices?, *Journal of Banking & Finance* 155, 106948.

Fama, E. F., and K. R. French, 1993, Common risk factors in the returns on stocks and bonds, *Journal of Financial Economics* 33, 3–56.

Fama, E. F., and K. R. French, 2002, The equity premium, *The Journal of Finance* 57, 637–659.

Fama, E. F., and K. R. French, 2007, Disagreement, tastes, and asset prices, *Journal of Financial Economics* 83, 667–689.

Fama, E. F., and K. R. French, 2015, A five-factor asset pricing model, *Journal of Financial Economics* 116, 1–22.

Fink, L., 2021, Blackrock ceo's annual letter to shareholders, *https://www.blackrock.com/corporate/investor-relations/larry-fink-chairmans-letter*.

Gabaix, X., and R. S. Koijen, 2021, In search of the origins of financial fluctuations: The inelastic markets hypothesis, Technical report, National Bureau of Economic Research.

Garvey, G. T., M. Iyer, and J. Nash, 2018, Carbon footprint and productivity: does the" e" in esg capture efficiency as well as environment?, *Journal of Investment Management* 16, 59–69.

Gebhardt, W. R., C. M. Lee, and B. Swaminathan, 2001, Toward an implied cost of capital, *Journal of accounting research* 39, 135–176.

Giglio, S., B. Kelly, and J. Stroebel, 2021, Climate finance, *Annual Review of Financial Economics* 13, 15–36.

Giglio, S., M. Maggiori, K. Rao, J. Stroebel, and A. Weber, 2021, Climate change and long-run discount rates: Evidence from real estate, *The Review of Financial Studies* 34, 3527–3571.

Goldsmith-Pinkham, P., M. T. Gustafson, R. C. Lewis, and M. Schwert, 2023, Sea-level rise exposure and municipal bond yields, *The Review of Financial Studies* 36, 4588–4635.

Görgen, M., A. Jacob, M. Nerlinger, R. Riordan, M. Rohleder, and M. Wilkens, 2020, Carbon risk, *Available at SSRN 2930897*.

Heinkel, R., A. Kraus, and J. Zechner, 2001, The effect of green investment on corporate behavior, *Journal of Financial and Quantitative Analysis* 36, 431–449.

Hong, H., F. W. Li, and J. Xu, 2019, Climate risks and market efficiency, *Journal of Econometrics* 208, 265–281.

Hou, K., H. Mo, C. Xue, and L. Zhang, 2021, An augmented q-factor model with expected growth, *Review of Finance* 25, 1–41.

Hou, K., M. A. Van Dijk, and Y. Zhang, 2012, The implied cost of capital: A new approach, *Journal of Accounting and Economics* 53, 504–526.

Hou, K., C. Xue, and L. Zhang, 2015, Digesting anomalies: An investment approach, *The Review of Financial Studies* 28, 650–705.

Hsu, P.-H., K. Li, and C.-Y. Tsou, 2023, The pollution premium, *The Journal of Finance* 78, 1343–1392.

Humphrey, J., S. Kogan, J. Sagi, and L. Starks, 2021, The asymmetry in responsible investing preferences, Technical report, National Bureau of Economic Research.

Huynh, T. D., and Y. Xia, 2021, Climate change news risk and corporate bond returns, *Journal of Financial and Quantitative Analysis* 56, 1985–2009.

Ilhan, E., Z. Sautner, and G. Vilkov, 2021, Carbon tail risk, *The Review of Financial Studies* 34, 1540–1571.

In, S., K. Park, and A. Monk, 2019, Is 'being green' rewarded in the market? an empirical investigation of decarbonization and stock returns, Unpublished working paper.

Koijen, R. S., and M. Yogo, 2019, A demand system approach to asset pricing, *Journal of Political Economy* 127, 1475–1515.

Krueger, P., Z. Sautner, and L. T. Starks, 2020, The importance of climate risks for institutional investors, *The Review of Financial Studies* 33, 1067–1111.

Larcker, D. F., and E. M. Watts, 2020, Where's the greenium?, *Journal of Accounting and Economics* 69, 101312.

Lee, C. M., E. C. So, and C. C. Wang, 2021, Evaluating firm-level expected-return proxies: implications for estimating treatment effects, *The Review of Financial Studies* 34, 1907–1951.

Lester, A., C. He, and C. McKnett, 2018, Harnessing esg as an alpha source in active quantitative equities, *State Street Global Advisors White Paper*.

Lewellen, J., 2015, The cross-section of expected stock returns, *Critical Finance Review* 4, 1–44.

Lo, A. W., and A. C. MacKinlay, 1990, When are contrarian profits due to stock market overreaction?, *The Review of Financial Studies* 3, 175–205.

Nagy, Z., A. Kassam, and L.-E. Lee, 2016, Can esg add alpha? an analysis of esg tilt and momentum strategies, *The Journal of Investing* 25, 113–124.

Nofsinger, J. R., J. Sulaeman, and A. Varma, 2019, Institutional investors and corporate social responsibility, *Journal of Corporate Finance* 58, 700–725.

Painter, M., 2020, An inconvenient cost: The effects of climate change on municipal bonds, *Journal of Financial Economics* 135, 468–482.

Paribas, B., 2019, The esg global survey, Technical report, BNP Paribas.

Pástor, L., and R. F. Stambaugh, 2001, The equity premium and structural breaks, *The Journal of Finance* 56, 1207–1239.

Pástor, L., and R. F. Stambaugh, 2003, Liquidity risk and expected stock returns, *Journal of Political Economy* 111, 642–685.

Pástor, L., and R. F. Stambaugh, 2009, Predictive systems: Living with imperfect predictors, *The Journal of Finance* 64, 1583–1628.

Pástor, L., R. F. Stambaugh, and L. A. Taylor, 2021, Sustainable investing in equilibrium, *Journal of Financial Economics* 142, 550–571.

Pedersen, L. H., S. Fitzgibbons, and L. Pomorski, 2021, Responsible investing: The esg-efficient frontier, *Journal of Financial Economics* 142, 572–597.

Schroders, 2020, Global investor study, Technical report, Schroders.

Seltzer, L., L. Starks, and Q. Zhu, 2021, Climate regulatory risks and corporate bonds (working paper), *Nanyang Business School*.

Zerbib, O. D., 2019, The effect of pro-environmental preferences on bond prices: Evidence from green bonds, *Journal of Banking & Finance* 98, 39–60.

Zerbib, O. D., 2022, A sustainable capital asset pricing model (s-capm): Evidence from environmental integration and sin stock exclusion, *Review of Finance* 26, 1345–1388.

Chapter 7

Climate impact investing

Tiziano De Angelis

University of Turin, Italy

Peter Tankov

CREST – ENSAE, Institut Polytechnique de Paris, France

Olivier David Zerbib

CREST – ENSAE, Institut Polytechnique de Paris, France

This paper shows how green investing spurs companies to mitigate their carbon emissions by raising the cost of capital of the most carbon-intensive companies. Companies' emissions decrease when the wealth share of green investors and their sensitivity to climate externalities increase. We show that the impact of green investors primarily governs companies' long-run emissions. Companies are further incentivized to reduce their emissions when green investors anticipate tighter climate regulations and climate-related technological innovations. However, heightened uncertainty regarding future climate risks alleviates green investors' pressure on the cost of capital of companies and pushes them to increase their emissions. Calibrated on United States data, our model suggests that, albeit effective, the impact of green investors remains limited given their current wealth share and practices.

This chapter is a reprint of an article originally published in Management Science. Please cite this work as follows:
De Angelis, T., Tankov, P., & Zerbib, O. D. (2023). Climate impact investing. Management Science, 69(12), 7669-7692.

We thank Marco Ceccarelli, Patricia Crifo, Joost Driessen, Caroline Flammer, Jesse Grabowski, Ying Jiao, Sonia Jimenez Garces, Rüdiger Kiesel, Frank de Jong, Lionel Melin, Martin Oehmke, Christian Robert, Bert Scholtens, Luca Taschini, Dimitri Vayanos, as well as participants at the Bachelier Finance Society One World Seminar, the seminairs of CREST - Ecole Polytechnique, the University of Zurich, the Universität Duisburg-Essen, the University of Edinburgh, the 2020 GRASFI Annual Meeting and the 2020 PRI Academic Network Conference, for their valuable comments and suggestions. We thank ISS for kindly providing us with data on shareholder proposals. We gratefully acknowledge the financial support from the Europlace Institute of Finance. T. De Angelis was also supported by EPSRC Grant EP/R021201/1 and P. Tankov was supported by the FIME (Finance for Energy Markets) research initiative of the Institut Europlace de Finance.

DOI: 10.1201/9781032636252-7

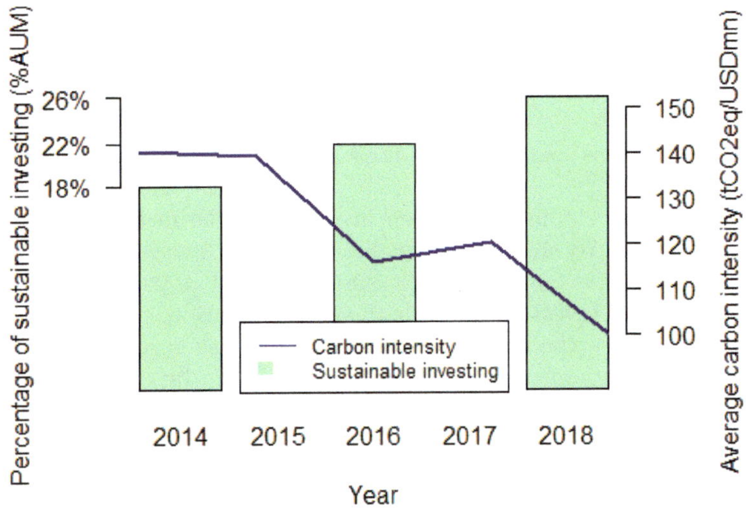

FIGURE 7.1: **Percentage of sustainable investments and average carbon intensity of the AMEX, NASDAQ, and NYSE stocks.** This figure presents the evolution of the proportion of sustainable investing relative to total managed assets over time, according to the Global Sustainable Investment Alliance (2018), as compared to the average carbon intensity of AMEX, NASDAQ and NYSE companies provided by S&P-Trucost between 2014 and 2018. The carbon intensity corresponds to the direct (scope 1 and 2) and indirect (upstream scope 3) greenhouse gas emissions of the companies, expressed in tCO_2e per million dollars of revenue generated.

7.1 Introduction

From 2014 to 2018, sustainable investments grew from 18% to 26% of the total assets under management (AUM) in the United States (U.S.) (US SIF, 2018) while, over the same period, the average carbon intensity of the companies listed on the National Association of Securities Dealers Automated Quotations (NASDAQ), American Stock Exchange (AMEX), and New York Stock Exchange (NYSE) decreased from 140 tons of CO_2 equivalent per million dollars of revenue (tCO_2e/USDmn) to 100 tCO_2e/USDmn (Figure 7.1).[1] The downward trend in corporate greenhouse gas intensity may be driven by several factors, such as the reductions in the costs of green technologies, tighter climate regulations, consumer pressure for more sustainable practices, and pressure exerted by green investors.[2] The two main channels through which green investors can have an impact on companies' practices are portfolio screening and shareholder engagement. Through portfolio climate screening, by underweighing or excluding the most carbon-intensive[3] companies from their investment scope, green investors increase these companies' cost of capital (Heinkel, Kraus,

---

[1]The carbon intensity of a company is defined as its emission rate relative to its revenue over one year. This metric is expressed in terms of tons of equivalent carbon dioxide per million dollars.

[2]Green investing is a form of socially responsible investing aimed at contributing to environmental objectives, mostly reducing greenhouse gas emissions by internalizing climate externalities.

[3]We refer to *carbon-intensive* companies and companies with high *greenhouse gas emissions* interchangeably since carbon dioxide is the main gas contributing to global warming. In the United States (U.S.), it

and Zechner, 2001; Pastor, Stambaugh, and Taylor, 2021; Pedersen, Fitzgibbons, and Pomorski, 2021; Zerbib, 2021) and can push them to reform. We focus on the specific channel of climate screening (referred to as *green investing* hereinafter) and address the issue of impact investing by answering the following questions: does green investing push companies to reduce their greenhouse gas emissions? If so, what are the factors that lead companies to mitigate their emissions? And how do these factors affect the *dynamics* of companies' emissions?

We show that the development of green investing—both in terms of the proportion of AUM and the sensitivity to climate externalities of green investors—pushes companies to reduce their greenhouse gas emissions by raising their cost of capital. By internalizing the negative impact of green investors on their financial valuation, companies are incentivized to pay a price to mitigate their emissions by adopting less carbon-intensive technologies, thereby lowering their cost of capital. These incentives are further strengthened when investors anticipate tighter climate regulations, climate-related technological innovations, and when they account for the negative impact of other companies' emissions on the company under consideration. However, in a sufficiently large or diversified market, investors' uncertainty regarding future climate risks reduces the incentives for companies to mitigate their emissions.

We develop a dynamic equilibrium model populated by heterogeneous investors and companies. We model two different groups of investors with constant absolute risk aversion (CARA investors). Both groups determine their optimal allocation by maximizing their expected wealth at a given terminal date, but they differ in their climate beliefs. Of the two groups of investors, one is a group of *green investors* and the other one of *regular investors*. Because green investors are aware of environmental risks, their magnitude, and their timing, or better equipped to anticipate them, they internalize the expected financial impact of future climate externalities of the companies in which they invest, while regular investors do not. These externalities correspond, for example, to greenhouse gas emissions and reflect companies' exposure to various climate transition risks, such as the rise in carbon price (Jakob and Hilaire, 2015) or changes in consumer preferences. While climate externalities will have a negative financial impact for the most carbon-intensive companies, they may positively impact the greenest companies benefiting from their favorable position. In the first version of our model, green investors internalize *deterministic* climate externalities.

These investors invest in n companies, each with a different marginal cost of reducing greenhouse gas emissions (referred to as *marginal abatement cost* hereinafter). Corporate managers are stock-value optimizers, who balance the benefit of mitigating greenhouse emissions, thus attracting green investors, against the cost of implementing these reforms. To represent the fact that a company reforms its environmental practices over a long period of time, at the initial date, $t = 0$, each company chooses a deterministic greenhouse gas emission schedule up to a final date T to maximize its expected discounted future market value. We allow companies to have their own climate beliefs. In addition, each company accounts for the strategies adopted by all other companies, hence reducing the companies' problem to a nonzero-sum game. This framework notably differs from standard heterogeneous belief models because the choice of each company's emission schedule directly affects the parameter on which investors disagree—companies' climate externalities.

We obtain a tractable formula of the equilibrium asset prices and show that they include an *externality premium*. This premium increases with the future *financial impact of climate externalities* (for simplicity, referred to as *climate externalities* hereinafter) internalized by green investors, which can be either positive or negative, and scales in proportion to the

---

accounted for more than 80% of the total emissions in 2018: `https://www.epa.gov/ghgemissions/overview-greenhouse-gases`.

relative wealth of green investors. Therefore, all else being equal, the asset price of a carbon-intensive company (also referred to as a *brown* company hereinafter) will be lower than that of a company with a low carbon footprint (also referred to as a *green* company hereinafter). Consequently, the expected returns increase when the climate externalities are negative and decrease when they are positive.

We characterize companies' optimal emission schedules in a general setup and derive an explicit solution for the case when climate externalities are measured as a decreasing quadratic function of the company's emissions. In equilibrium, emissions decrease with the proportion of assets managed by green investors, their sensitivity to climate externalities (also referred to as *climate sensitivity* hereinafter), as well as companies' climate sensitivity, but increase with the marginal abatement cost. In particular, we show that investors' climate sensitivity mainly drives long-term emissions whereas companies' climate sensitivity predominantly drives short-term emissions. Therefore, when the average climate sensitivity of investors is lower than that of the companies, optimal emissions decrease over time. In addition, corporate emissions are a convex function of time, with the degree of convexity increasing in the rate of time preference. We calibrate the model on the AMEX, NASDAQ, and NYSE stocks between 2004 and 2018 using the carbon intensity of companies as a proxy for their emissions. We then simulate emissions' mitigation in several scenarios by considering an electrical equipment manufacturing company. Consistent with the small effect of divestment on companies' cost of capital estimated by Berk and van Binsbergen (2021), our simulations suggest that, albeit effective, green investors' impact is still limited: an electrical equipment manufacturing company reduces its emissions by an average of 1% per year over a 20-year period when green investments account for 25% of the total AUM in the economy. When either green investments account for 50% of the AUM or when green investors' climate sensitivity doubles, this company reduces its emissions by an average of 3% per year over the same period.

We also show that tighter climate regulations as well as climate-related technological innovations, when anticipated and internalized by green investors, increase the pressure on companies to further reduce their emissions. Using previous research, we recalibrate the marginal abatement cost and the climate sensitivities accounting for the effects of technological change and expected regulatory action, respectively. Our model estimates that emissions decline 2.2 times faster when green investors anticipate regulatory tightening. We also find that emissions decline 3.6 times faster when they anticipate climate-related technological change. Finally, when both changes are anticipated, declines are 4 times faster. In addition, we illustrate the effect of strategic interactions between companies by showing that when green investors internalize the negative financial impact of the economy's average emissions, companies are further incentivized to curb their emission schedules.

We next extend our model to the case where green investors also internalize *uncertainty* about future climate externalities. Climate risks, such as a rise in carbon price or the occurrence of natural disasters, usually have non-Gaussian fat-tailed distributions (Weitzman, 2009; Barnett, Brock, and Hansen, 2020). Therefore, we model future climate risks internalized by green investors as a stochastic jump process (Poisson process). We give a tractable expression of optimal portfolio allocations, asset prices, expected returns and emission schedule in equilibrium. We show that, in a sufficiently large or diversified market, uncertainty about future climate risks leads green investors to lower the risk of their portfolios by reducing their tilt toward green assets, shifting portfolio allocations away from green assets and into brown assets. Consequently, climate uncertainty reduces the magnitude of the externality premium on expected returns. As a result, the incentive for companies to reform is also reduced, leading them to increase their optimal emission schedule in equilibrium.

The results of this paper are of interest to both investors and policymakers. For investors, we identify three major implications. First, the findings show that investors can increase their impact on companies by raising their environmental requirements, for example by restricting the range of companies in which they invest or by significantly underweighing the most carbon-intensive companies. However, the small simulated effect suggests that other impact strategies, such as shareholder engagement, might be more efficient at this stage, in line with the conclusions of Broccardo, Hart, and Zingales (2020). Indeed, as long as the wealth share of sustainable investors is low, the assets they exclude will be purchased by regular investors whose wealth is large enough to dampen the effect on the cost of capital and thus the impact on corporate practices. Second, to increase the climate impact of their asset allocations, investors also have a key role to play as shareholders in pressing companies to increase transparency about future climate-related risks and raise their environmental standards. Third, impact investing is financially beneficial if investors favor companies that are on a pathway toward reducing their climate footprints. Investors can also benefit from financial gains by investing in green companies for which information on their climate footprints is still poorly available.

From the public authorities' viewpoint, the results of this paper have four implications. First, they highlight their role in supporting the development of green investments. In particular, they suggest policymakers should set rigorous standards for environmental impact assessments and disclosure to foster and increase impact investing. This is consistent with the recommendations of the on Sustainable Finance (2018) and the Commission (2018)'s Action Plan, which led to the recent development of a green taxonomy and an official standard for green bonds. Second, these results emphasize the importance of access to information regarding companies' climate footprints, which enables green investors to internalize climate externalities as accurately as possible, thereby maximizing their impact on companies. Third, these results clarify the effect of climate regulations and their predictability on the adjustment of investors' beliefs: a tighter climate regulation is amplified by the adjustment of green investors' expectations, which increases their pressure on companies' cost of capital, thereby forcing them to further reduce their emissions. Fourth, our analysis shows the importance of low-cost access to greener technological solutions as a lever for companies to mitigate their climate impact. Specifically, industries for which green alternatives are limited, such as cement or aircraft, face a structural barrier to which an increase in R&D is an essential response. In addition, enabling investors to better forecast and internalize the likelihood of future technological innovations increases their impact on corporate emissions.

Related literature. This paper contributes to the emerging literature on asset pricing and impact investing in sustainable finance. First, from an asset pricing perspective, we clarify the relationship between the development of sustainable investing[4] and asset returns. Building on the seminal paper by Heinkel, Kraus, and Zechner (2001), three recent papers by Pastor, Stambaugh, and Taylor (2021), Pedersen, Fitzgibbons, and Pomorski (2021) and Zerbib (2021) study this relationship using a single-period model. These papers show that the stock returns of the brownest companies are increased by a positive premium. Bolton and Kacperczyk (2021), Ardia et al. (2021) and Pastor, Stalbaugh, and Taylor (2021) empirically support this finding.[5] Avramov et al. (2021b) extend this modeling framework with dynamic

[4]Sustainable investing can be motivated by pecuniary or non-pecuniary motives (Krüger, Sautner, and Starks, 2020). Riedl and Smeets (2017) and Hartzmark and Sussman (2020) highlight the positive effect of sustainable preferences on sustainable fund flows. Pro-social and pro-environmental preferences also impact asset returns since they induce an increase in the return on sin stocks (Hong and Kacperczyk, 2009), a decrease in the return on impact funds (Barber, Morse, and Yasuda, 2021) and a decrease in the return on bonds (Baker et al., 2018; Zerbib, 2019).

[5]It is worth noting that the global empirical literature on the effects of Environment, Social and Governance (ESG) integration on asset returns is mixed: some authors highlight the negative impact of ESG performance on asset returns, while others suggest a positive relationship or find no significant impact. For negative impacts, see Brammer, Brooks, and Pavelin (2006), Renneboog, Ter Horst, and Zhang (2008),

preferences for sustainability. We contribute to this literature by characterizing the dynamic of asset prices and expected returns when green investors internalize non-Gaussian climate uncertainty. Consistent with Avramov et al. (2021a) who study the effect of uncertainty in a Gaussian setup, we show that, in the presence of non-Gaussian uncertainty about future climate risks, the expected return gap between brown and green assets narrows as green investors diversify their exposure to mitigate their risk.

We also contribute to the emerging literature on impact investing. In a seminal study, by constructing a single-period model in which green investors have the ability to exclude the most polluting companies, Heinkel, Kraus, and Zechner (2001) show that such companies are pushed to reform because exclusionary screening negatively impacts their valuations. Chowdhry, Davies, and Waters (2018) study the optimal contracting for a company that cannot commit to social objectives and show that impact investors must hold a large enough financial claim to incentivize the company to internalize social externalities. Oehmke and Opp (2019) show how sustainable investors can enable a scale increase of clean production by internalizing social costs in the presence of financing constraints. The authors emphasize the importance of sustainable investors unconditionally caring for sustainability issues (i.e., irrespective of whether they are investors in the firm) and coordinating among themselves. They also show that non-sustainable investors and sustainable investors can jointly achieve higher surplus than either investor type alone. When markets are subject to search friction, Landier and Lovo (2020) show that the presence of an ESG fund forces companies to partially internalize externalities. In contrast, by analyzing venture capital funds, Barber, Morse, and Yasuda (2021) suggest instead that the pressure exerted by sustainable investors on companies is tied to their willingness to pay to invest in impact funds. Finally, through an asset pricing model, Pastor, Stambaugh, and Taylor (2021) show that green investors produce positive social impact by shifting investment toward green firms and making firms greener. We also contribute to the literature on impact investing from an asset pricing perspective along three different avenues. First, we characterize the *dynamics* of companies' optimal emissions and show that increases in either the share of green investors or their climate sensitivity pushes companies to cut their emissions by affecting their cost of capital. Notably, long-term emission dynamics are governed by green investors' beliefs, while short-term dynamics are driven by companies' beliefs. Second, when investors internalize future climate-related regulations and technological innovations, they increase their pressure on companies' cost of capital, thereby incentivizing them to further reduce their emissions. Finally, we show that green investors' uncertainty about future climate risks leads companies to increase their emissions.

The remainder of this paper is structured as follows. Section 2 introduces an economy populated by greenhouse gas emitting companies and investors with heterogeneous beliefs. Section 3 describes the equilibrium pricing equations and companies' emission schedules when green investors internalize deterministic climate externalities. Section 4 extends the model to non-Gaussian stochastic climate externalities. Section 5 concludes the paper. Proofs of the mathematical results are presented in detail in the Appendix.

Sharfman and Fernando (2008), ElGhoul et al. (2011), Chava (2014), Barber, Morse, and Yasuda (2021), and Hsu, Li, and Tsou (2019). For positive impacts, see Derwall et al. (2005), Statman and Glushkov (2009), Edmans (2011), Eccles, Ioannou, and Serafeim (2014), Krüger (2015) and Statman and Glushkov (2016). Finally, Bauer, Koedijk, and Otten (2005), Galema, Plantinga, and Scholtens (2008) and Trinks et al. (2018) find no significant impact.

7.2 A simple economy with greenhouse gas emitting companies and heterogeneous beliefs

We develop a simple model with heterogeneous beliefs in which climate externalities are internalized by green investors in a deterministic way. We introduce the dynamics of the assets available in the market and heterogeneous beliefs about climate externalities of three types of agents—a group of regular investors, a group of green investors and n companies. We then present the investors' and companies' optimization programs.

7.2.1 Securities market

In this section, we consider a financial market consisting of n risky stocks and a risk-free asset, which is assumed to be free of arbitrage and complete. The risk-free asset is in perfectly elastic supply and we assume that the risk-free rate is zero without loss of generality.[6] Each stock $i \in \{1, \ldots, n\}$ is in positive net supply of one unit and is a claim on a single liquidating dividend D_T^i at horizon T. The terminal dividend of the i-th asset, D_T^i, is broken down into three terms: (i) the cost of an environmental reform, decided by the company at time $t = 0$ and implemented between $t = 0$ and $t = T$, with an associated greenhouse gas emission schedule; (ii) the initial dividend forecast at time $t = 0$ net of the cost of the environmental reform; (iii) the entire cash flow news sequence between $t = 0$ and $t = T$. The values of the first two terms are public information at time $t = 0$ and their sum is also referred to as the initial dividend forecast. We isolate the cost of environmental reform from the initial dividend forecast because it is a parameter of interest in our model. This cost is known at the initial state because a company's decision to reduce its emissions is usually made over a long period of time. For example, the transformation of a generating fleet by an electric utility, or the development of a line of electric vehicles by a car manufacturer are the result of long-term decisions known to the public well in advance of their completion. Therefore, the vector of terminal dividends for all assets, $D_T \in \mathbb{R}^n$, reads

$$D_T = \underbrace{\underbrace{\int_0^T c_t(\psi_t - \psi_b)dt}_{\text{Cost of reform}} + \underbrace{d}_{\substack{\text{Initial dividend forecast} \\ \text{net of cost of reform}}}}_{\text{Initial dividend forecast}} + \underbrace{\int_0^T \sigma_t dB_t}_{\text{Cash flow news}} . \tag{7.1}$$

In the first term of the above expression, representing the cost of an environmental reform, ψ_t is the vector of greenhouse gas emissions per unit of time of the companies at date t.[7] We refer to greenhouse gas emissions for simplicity, but ψ can be seen as a measure of relative emissions compared to a level of production (e.g., carbon intensity). The constant $\psi_b \in \mathbb{R}_+^n$ is the vector of initial (or *baseline*) emissions, and c_t is a diagonal matrix with elements (c_t^i), which correspond to the marginal abatement costs for each company. Thus, by reducing its emissions by x, the i-th company reduces its terminal dividend by $c_t^i x$ per unit of time t. The process $(\psi_t^i)_{t\in[0,T]}$, which is also referred to as the emission schedule of the i-th company, is determined at time $t = 0$: it is deterministic and does not depend on the future cash flow news so as to reflect the long-term nature of companies' reform strategies. The second term of the expression, d, is a constant vector corresponding to the initial dividend forecast under a reference probability measure net of the cost of the environmental reform. In the third

[6] As stressed by Atmaz and Basak (2018), the interest rate can be taken as exogenous since consumption occurs only at time T, i.e., there is no intermediate consumption.

[7] Formally, $\psi \in F([0, T], \mathbb{R}_+^n)$, where $F([0, T], \mathbb{R}_+^n)$ is the set of Borel-measurable functions of $[0, T]$ in $\mathbb{R}_+^n$.

term, $\sigma_t dB_t$ ($t \in [0,T]$) is the sequence of cash flow news, where $(B_s)_{s\in[0,T]}$ is a standard n-dimensional Brownian motion defined on a probability space $(\Omega, \mathcal{F}, \mathbb{P})$ equipped with a filtration $(\mathcal{F}_s)_{s\in[0,T]}$. We refer to $\mathbb{P}$ as the *reference probability measure* and we will later introduce other probability measures for the companies and the investors, which reflect their different beliefs. For each $s \in [0,T]$, σ_s is a deterministic, $n \times n$, invertible matrix.

Denoting by $(p_t)_{t\in[0,T]}$ the equilibrium assets price process in $\mathbb{R}^n$, we assume $p_T = D_T$. We also denote the dividend forecast under the reference probability measure at $t \in [0,T]$ by

$$D_t = \mathbb{E}[D_T|\mathcal{F}_t] = \int_0^T c_t(\psi_t - \psi_b)dt + d + \int_0^t \sigma_s dB_s, \tag{7.2}$$

and in particular,

$$D_0 = \mathbb{E}[D_T|\mathcal{F}_0] = \int_0^T c_t(\psi_t - \psi_b)dt + d.$$

This Gaussian continuous-time specification of the dividend dynamics is consistent with previous literature on models with heterogeneous beliefs that study investors' reaction to good and bad news (Veronesi, 1999), excess confidence (Scheinkman and Xiong, 2003) and extrapolation bias (Barberis et al., 2015).[8] We choose a setup with Gaussian dividends and prices because we are after explicit formulae for equilibrium prices, which companies use to endogenously determine their prospective greenhouse gas emissions.

7.2.2 Investors' and companies' beliefs

The market is populated by two types of investors, *regular* and *green*, who have different expectations regarding companies' future cash flow news. Regular investors only consider the information related to the flow of financial news, in addition to the initial dividend forecast. Therefore, under their probability measure $\mathbb{P}^r$, conditional on the information in t, they account for the past cash flow news, $\int_0^t \sigma_s dB_s$, which is known at time t, but the expectation of the future cash flow news, $\int_t^T \sigma_s dB_s$, is zero as B_s is a Brownian motion. Denoting by $\mathbb{E}_t^r$ this conditional expectation,

$$\mathbb{E}_t^r(D_T) = D_t, \tag{7.3}$$

that is, the dividend forecast of the regular investors coincides with the dividend forecast under the reference probability measure. From the point of view of the properties of the cash flow news, $\sigma_s dB_s$, there is no difference between the measures $\mathbb{P}$ and $\mathbb{P}^r$, and we can simply assume $\mathbb{P} = \mathbb{P}^r$. However, it should be noted that $\mathbb{P}$ is a technical device and, as such, we make no assumptions about the realistic nature of this measure. The expectations of regular investors are thus not necessarily consistent with the realized events.

In contrast, compared to regular investors, green investors are more aware of environmental risks, their magnitude, and their timing, or better equipped to anticipate them. Therefore, they internalize the financial impact of the expected climate externalities of the companies in which they invest. These externalities correspond, for example, to greenhouse gas emissions and reflect companies' exposure to various climate transition risks, such as the rise in carbon price (Jakob and Hilaire, 2015) or changes in consumer preferences. While climate externalities will have a negative financial impact for the brownest companies, they may positively impact the greenest companies. In our first model specification (Sections 7.2 and 7.3), we assume that green investors have a perfect anticipation of the future climate

[8]Other articles on heterogeneous beliefs adopt this same setup in discrete time such as Hong and Stein (1999), Barberis and Shleifer (2003) and Barberis et al. (2018).

externalities. As a result, in addition to accounting for the cash flow news and the initial dividend forecast as regular investors do, green investors internalize, under their probability measure, the financial impact of future climate externalities at date $t \in [0, T]$. The latter is expressed by

$$\int_t^T \theta_s(\psi_s)ds. \tag{7.4}$$

Here, $\theta_s(\psi_s) \in \mathbb{R}^n$ is the vector of the *financial impact of climate externalities* (for convenience, referred to as *climate externalities*) at date $s \in [0, T]$. Naturally, we assume that θ_s^i is a decreasing function of ψ_s^i so that higher emissions of the i-th company correspond to stronger negative financial impact on the i-th asset. However, in a general case, θ_s^i depends on the emissions of all companies because more global emissions increases financial risks for each company. This is why θ_s^i is a function of the whole vector ψ_s. Moreover, θ_s^i is a function of time to reflect growing climate sensitivity of green investors, or their anticipation of stronger regulatory pressure. As a consequence, green investors internalize their climate beliefs regarding the i-th company by attributing a fundamental value to the i-th stock at time t that is higher (if $\int_t^T \theta_s^i(\psi_s)ds$ is positive) or lower (if $\int_t^T \theta_s^i(\psi_s)ds$ is negative) than the value of the dividend forecast (see Equation (7.2)). Denoting by $\mathbb{E}_t^g$ the expectation of the green investors conditional on the information at time t, we have

$$\mathbb{E}_t^g(D_T) = D_t + \int_t^T \theta_s(\psi_s)ds, \tag{7.5}$$

that is, the dividend forecast of the green investors equals the forecast under the reference measure augmented by the financial impact of future climate externalities internalized by these investors. It is worth noticing that the variable D_t is constructed from the actual realization of the past cash flow news between 0 and t, which is a known quantity to investors at time t. However, from a probabilistic point of view, the stochastic process D_t under the probability $\mathbb{P}^g$ includes an additional drift $\theta_s(\psi_s)ds$ associated with the beliefs of green investors.[9]

Alongside the two types of investors, we also introduce the productive sector by modeling the views of the companies about the n assets available on the market. As in Oehmke and Opp (2019), corporate managers (referred to as *companies* hereinafter) also have subjective beliefs about the financial impact of climate externalities on the dividend dynamics of each of the n companies. We denote by $\theta_s^c(\psi_s)$ the vector of the climate externalities internalized by all companies. Denoting by $\mathbb{E}_t^c$ the expectation of the companies conditional on the information in t, we have

$$\mathbb{E}_t^c(D_T) = D_t + \int_t^T \theta_s^c(\psi_s)ds. \tag{7.6}$$

Similarly to Equation (7.5), the dividend forecast of the companies equals the forecast under the reference measure augmented by the financial impact of future climate externalities internalized by the companies. Here again, the variable D_t is constructed from the actual realization of the past cash flow news between 0 and t, which is known to market participants at time t. However, from a probabilistic point of view, the dynamic of the stochastic

[9]Formally, the probability measure of green investors is constructed through a change of measure. The Radon-Nikodym density that connects the two probability measures $\mathbb{P}^g$ and $\mathbb{P}^r$ is $Z_T = e^{\int_0^T \lambda_s^\top dB_s - \frac{1}{2}\int_0^T \|\lambda_s\|^2 ds}$, where $\lambda_t := \sigma_t^{-1}\theta_t(\psi_t)$. Under $\mathbb{P}^g$, the dynamics of the stochastic process D_t reads $D_t = \int_0^T c_s(\psi_s - \psi_b)ds + d + \int_0^t \sigma_s dB_s^g + \int_0^t \theta_s(\psi_s)ds$, where B^g is a Brownian motion under $\mathbb{P}^g$.

process D_t under the probability $\mathbb{P}^c$ includes an additional drift $\theta_s^c(\psi_s)$ associated with the companies' beliefs.[10]

7.2.3 Investors' preferences and optimization

Regular and green investors are assumed to have CARA preferences. Subject to their budget constraints, investors maximize the expected exponential utility of their terminal wealth[11] W_T, which reads

$$\mathbb{E}^j(1 - e^{-\gamma^j W_T^j}), \quad \gamma^j > 0, \quad j \in \{r, g\},$$

where the superscripts r and g refer to the regular and green investors, respectively, and γ^js are their absolute risk aversions. The wealth processes follow the dynamics

$$W_t^r = w^r + \int_0^t (N_s^r)^\top dp_s, \qquad W_t^g = w^g + \int_0^t (N_s^g)^\top dp_s, \tag{7.7}$$

where N_t^r and N_t^g are quantities of assets held by the regular and green investors, respectively, at time t, and prices $(p_t)_{t\in[0,T]}$ are determined by the market clearing condition. The initial wealth levels of regular and green investors are denoted by w^r and w^g, respectively, and the symbol $\top$ stands for the transposition operator.

In what follows, we denote by γ^* the global risk aversion, defined by $\frac{1}{\gamma^*} = \frac{1}{\gamma^r} + \frac{1}{\gamma^g}$, and set $\alpha = \frac{\gamma^r}{\gamma^r+\gamma^g}$ and $1 - \alpha = \frac{\gamma^g}{\gamma^r+\gamma^g}$. To simplify the interpretation of the impact of green and regular investors' wealth on the variables in equilibrium, and without losing generality, we assume that green and regular investors have equal relative risk aversions; that is, $\gamma^R = \gamma^g w^g = \gamma^r w^r$, where γ^R denotes the relative risk aversion. In this case, α is the proportion of the green investors' initial wealth at $t = 0$, and $1-\alpha$ is that of the regular investors; that is, $\alpha = \frac{w^g}{w^g+w^r}$ and $1 - \alpha = \frac{w^r}{w^g+w^r}$.

7.2.4 Companies' utility and optimization

A company's decision to implement reforms to mitigate its climate footprint is made over a sufficiently long time horizon. Therefore, at $t = 0$, the i-th company chooses its emission schedule, $(\psi_t^i)_{t\in[0,T]}$, up to the horizon T so as to maximize its future valuation. Consequently, in our setup, we endogenize companies' emissions through their market values: on the one hand, investors allocate their wealth according to companies' emission schedules,[12] thereby impacting companies' market values; on the other hand, companies take into account their market values to determine their emission schedules. We denote by ρ the rate of time preference and by ψ^{-i} the vector of emission schedules of companies other than the i-th company. The market value of the i-th company's asset at time t is denoted by $p_t^i(\psi^i, \psi^{-i})$ to reflect its dependence on the vector ψ of all companies' emission schedules. The companies have a linear utility and risk neutral preferences (Lambrecht and Myers, 2017; van Binsbergen and Opp, 2019). Therefore, at $t = 0$, the i-th company chooses $(\psi_t^i)_{t\in[0,T]}$ so as to maximize the following objective function:

$$\mathcal{J}^i(\psi^i, \psi^{-i}) = \mathbb{E}^c \left[\int_0^T e^{-\rho t} p_t^i(\psi^i, \psi^{-i}) dt \right]. \tag{7.8}$$

[10] In the same way as presented in footnote 9, the companies' measure is constructed through a similar change of measure where $\theta_t(\psi_t)$ is replaced by $\theta_t^c(\psi_t)$.

[11] As Atmaz and Basak (2018) point out, investors' preferences are based on their wealth at the terminal date rather than on intermediate dates, which would have led to endogenizing the interest rate in equilibrium.

[12] Green investors internalize climate externalities and all investors account for the costs of environmental reforms.

Maximizing the sum of the market values over the entire period is consistent with Pastor, Stambaugh, and Taylor (2021) as well as recent studies on Chief Executive Officers' (CEO) compensation plans. Larcker and Tayan (2019), for example, report that "stock-based performance awards have replaced stock options as the most prevalent form of equity-based pay." In addition, CEOs are generally required to hold their companies' stocks. Managers are therefore directly interested in the valuation of their company's stock price at each date, which endogenizes the financial impact of the company's emission schedule. This optimization program is also in line with the approach of Heinkel, Kraus, and Zechner (2001) in the context of a multi-period model where the company's climate impact is endogenized.

The optimal emission schedule, ψ^*, corresponds to a Nash equilibrium in which each company $i \in \{1, ..., n\}$ determines its own emissions, $\psi^{i,*}$, in $t = 0$, so that

$$\mathcal{J}^i(\psi^{*,i}, \psi^{*,-i}) \geq \mathcal{J}^i(\psi^i, \psi^{*,-i}), \quad \text{for all } \psi^i \in F([0,T], \mathbb{R}_+). \tag{7.9}$$

Table 7.1 summarizes the preferences and optimization programs of the different players and their interactions in the economy we model.

TABLE 7.1: **Summary of agents' actions.** This table summarizes the optimization programs of each agent as well as their interactions between $t = 0$ and $t = T$

Date	*Agent*	*Choose*	*Given*
At $t = 0$	Companies	Their deterministic emission schedule, ψ, from 0 to T	- Their expected market value between 0 and T - The cost of reducing their emissions between 0 and T
$\forall t \in [0, T]$	Regular investors	Their asset allocation, N^r	- The observed cash flow news between 0 and t, and the expected cash flow news between t and T - The cost of reducing companies' emissions between 0 and T
$\forall t \in [0, T]$	Green investors	Their asset allocation, N^g	- The observed cash flow news between 0 and t, and the expected cash flow news between t and T - The cost of reducing companies' emissions between 0 and T - The expected financial impact of companies' emissions between t and T

7.3 Equilibrium in the presence of greenhouse gas emitting companies and heterogeneous beliefs

This section presents equilibrium asset prices and returns in the model developed in Section 7.2. The optimal portfolio allocations of regular and green investors are found explicitly. We derive the optimal dynamics of companies' emissions, which we render tractable by assuming that climate externalities are quadratic. Finally, we analyze the effects of regulatory changes, technological changes, and companies' interactions on companies' optimal emission schedules.

7.3.1 Equilibrium stock prices and returns

In equilibrium, investors choose their allocations, which maximize their expected utility. Equilibrium prices are determined such that the market clears. Denoting $\Sigma_t = \sigma_t^\top \sigma_t$, and letting $\mathbf{1}$ be the vector of ones of length n, prop 2 gives the equilibrium prices and allocations.

Proposition 2 *Given an emission schedule $(\psi_t)_{t\in[0,T]}$, asset prices in equilibrium read*

$$p_t = D_t - \int_t^T \mu_s ds \quad \textit{with} \quad \mu_t = \gamma^* \Sigma_t \mathbf{1} - \alpha\theta_t(\psi_t), \tag{7.10}$$

where $-\alpha\theta_t(\psi_t)$ is the externality premium. The optimal number of shares for the regular and green investors are

$$N_t^r = (1-\alpha)\left(\mathbf{1} - \frac{1}{\gamma^g}\Sigma_t^{-1}\theta_t(\psi_t)\right) \quad \textit{and} \quad N_t^g = \alpha\left(\mathbf{1} + \frac{1}{\gamma^r}\Sigma_t^{-1}\theta_t(\psi_t)\right), \tag{7.11}$$

respectively.

The different beliefs of green investors introduce an *externality premium*, which is an additional drift in the price dynamics. When future climate externalities are negative (i.e., the emissions are high), the price is adjusted downward proportionally to the fraction of the initial wealth held by the green investors, α. Conversely, when future externalities are positive (i.e., the emissions are low), green investors bid up the price, which is adjusted upwards. However, the cost of environmental reform arises in the dividend forecast, D_t (see Equation (7.2)), and impacts the price in the opposite direction: a reduction (increase) in emissions thus lowers (increases) the price. Therefore, the net effect of a change in emissions on the price depends on the intensity with which green investors internalize climate externalities (through θ_t^i) and the cost of reform (through c_t^i).

The effect of heterogeneous beliefs on climate externalities can also be analyzed in terms of expected dollar returns (referred to as *expected returns* hereinafter), $\mathbb{E}(dp_t) = \mu_t dt$.

Since θ_t^i is a decreasing function of ψ_t^i, expected returns increase with companies' emissions. The externality premium on asset returns can be positive ($\theta_t^i(\psi) < 0$) or negative ($\theta_t^i(\psi) > 0$). This result is supported by extensive empirical evidence, including Renneboog, Ter Horst, and Zhang (2008), Sharfman and Fernando (2008), Chava (2014), Barber, Morse, and Yasuda (2021), Bolton and Kacperczyk (2021) and Hsu, Li, and Tsou (2019). It is also consistent with the theoretical works of Pastor, Stambaugh, and Taylor (2021), Pedersen, Fitzgibbons, and Pomorski (2021) and Zerbib (2021), who show, through a single-period model, that expected returns increase along with a company's climate impact as green investors require a higher cost of capital.

The number of shares purchased by investors is also adjusted by the climate externalities. Green investors overweigh assets with the higher positive externalities and underweigh or short assets with the higher negative externalities. Regular investors have a symmetrical allocation by providing liquidity to green investors. This result is consistent with optimal allocations in disagreement models where some investors have an optimistic market view and others a pessimistic one (Osambela, 2015; Atmaz and Basak, 2018): the risk is transferred from pessimists to optimists who increase their holding of the asset under consideration.

7.3.2 Equilibrium emission schedules

At the initial date, companies choose their optimal emission schedules by maximizing their expected market values between times 0 and T.

Proposition 3 *The i-th company's optimal emission schedule, ψ^i, given a vector ψ^{-i} of all other companies' emissions, is the one that maximizes for all $t \in [0,T]$*

$$\beta_t^c \theta_t^{c,i}(\psi_t) + \alpha\beta_t\theta_t^i(\psi_t) + c_t^i\psi_t^i, \tag{7.12}$$

where

$$\beta_t^c = \frac{e^{-\rho t} - e^{-\rho T}}{1 - e^{-\rho T}} \quad \text{and} \quad \beta_t = \frac{1 - e^{-\rho t}}{1 - e^{-\rho T}}.$$

At each time t, the i-th company maximizes the sum of three terms. The first and second terms measure the financial benefits associated to two climate externality premia: one endogenized by the company ($\theta_t^{c,i}(\psi_t)$) and the other endogenized by the green investors ($\alpha\theta_t^i(\psi_t)$), both adjusted by suitable time factors (β_t^c and β_t, respectively). The third one, $c_t^i\psi_t^i$, accounts for the financial benefits obtained by not reducing its emissions. The optimal emission schedule of a company is a trade-off between the positive effect of reducing its emissions—especially, the positive effect on its cost of capital through $\alpha\beta_t\theta_t^i(\psi_t)$—and the cost of reform to achieve the target emission schedule.[13]

Research in environmental economics consensually suggests the use of a convex specification to model the economic damage associated with environmental risks (Dietz and Stern (2015); Burke, Hsiang, and Miguel (2015); Burke, Davis, and Diffenbaugh (2018)). Following Nordhaus (2014), who argues that "damages can be reasonably well approximated by a quadratic function of temperature change", we use a quadratic climate damage function to model the economic impact associated with climate change. We assume that climate externalities are quadratic in carbon emissions and consider a first simple case where the climate externalities of a given company, θ_t^i, depend only on its own emissions, ψ_t^i.[14] In this case, Proposition 3 yields a simple solution detailed in Corollary 1.

Corollary 1 *Assuming $\theta_t^i(x) = \kappa_{0,t} - \frac{\kappa_t}{2}x^2$ and $\theta_t^{c,i}(x) = \kappa_{0,t}^c - \frac{\kappa_t^c}{2}x^2$, for $x \geq 0$, where κ_t, κ_t^c, $\kappa_{0,t}$ and $\kappa_{0,t}^c$ are positive deterministic functions of time,[15] the i-th company's optimal emission schedule reads*

$$\psi_t^{*,i} = \frac{c_t^i}{\beta_t^c\kappa_t^c + \alpha\beta_t\kappa_t} \tag{7.13}$$

Emissions decrease with respect to the proportion of wealth held by green investors, α, and when green investors' and companies' sensitivities to climate externalities increase. Indeed, κ_t and κ_t^c measure the sensitivity with which green investors and companies, respectively, internalize climate externalities at time t (also referred to as *climate sensitivity*). Thus, green investors can increase their impact on companies by raising their climate sensitivity, for example by restricting the range of companies in which they invest or by significantly underweighing the most carbon-intensive companies. It should be noted that even

---

[13]The companies' optimization problem in the reduced form of (7.12) is obtained by writing the equilibrium price (7.10) under the companies' probability measure $\mathbb{P}^c$:

$$p_t = D_t - \int_t^T \mu_s ds = \int_0^T c_t(\psi_t - \psi_b)dt + d + \int_0^t \sigma_s dB_s^c + \int_0^t \theta_s^c(\psi_s)ds - \gamma^* \int_t^T \Sigma_s \mathbf{1} ds + \alpha \int_t^T \theta_s(\psi_s)ds,$$

where B^c is a Brownian motion under the measure $\mathbb{P}^c$. Substituting this expression for the price into Equation (7.8) leads to the result of Proposition 3 after an integration by parts.

[14]The financial impact of climate externalities, θ_t^i, represents the opposite of a damage function (θ_t^i decreases as carbon emissions increase), so it is concave. Other types of climate externalities functions can be considered, such as an exponential function (Barnett, Brock, and Hansen, 2020). However, in the case of our model, the exponential specification does not lead to tractable solutions.

[15]For simplicity we assume that κ_t, κ_t^c, $\kappa_{0,t}$ and $\kappa_{0,t}^c$ are the same for all companies but the generalization to different constants is straightforward.

if the company does not internalize climate externalities ($\kappa_t^c = 0$), green investors' beliefs and the threat they pose to a company's market value are sufficient to prompt a company to reduce its climate impact. In such a case, the optimal emission schedule simplifies to:

$$\psi_t^{*,i} = \frac{c_t^i}{\alpha \beta_t \kappa_t}. \tag{7.14}$$

As expected, the emissions of the i-th company decrease when the marginal abatement cost, c_t^i, decreases. In the special case where the marginal abatement cost is zero, the company cuts its emissions to zero. The marginal abatement cost is a company specific factor that plays an important role in the greening dynamics of the economy. R&D in industries where green alternatives are still limited (e.g., cement, aviation) is therefore a key tool to support and accelerate the ecological transition.

Irrespective of technological and regulatory changes, which will be analyzed in Section 7.3.3, the emission schedule is not necessarily constant over time. The effect of investors' beliefs, $\alpha\kappa_t$, which changes with $\beta_t \simeq \frac{t}{T}$ (for small ρ), grows over time. In the long run, this effect becomes prominent over the effect of companies' beliefs, κ_t^c, which changes with $\beta_t^c \simeq 1 - \frac{t}{T}$ (for small ρ) and fades over time. This effect is explained by the following reasoning: green investors make equilibrium prices depend on their long-term beliefs (Equation (7.10)) by internalizing their perceived future climate risks in their investment decisions; therefore, since companies maximize their expected future valuations, they are pushed to adjust their long-term climate footprint according to green investors' beliefs. More precisely and from an analytical standpoint, this dynamic is driven by the fact that the asset prices at time t under the companies' probability measure depend on companies' climate beliefs between 0 and t ($\int_0^t \theta_s^c(\psi_s)ds$) and on green investors' climate beliefs between t and T through the externality premium ($\alpha \int_t^T \theta_s(\psi_s)ds$) as detailed in footnote 13. Therefore, by maximizing the expectation of the price's integral between 0 and T, companies give more weight to their beliefs at the beginning of the period and to green investors' beliefs at the end of the period. Consequently, when the climate sensitivity of companies is lower than the climate sensitivity of the average investor, $\kappa^c < \alpha\kappa$,[16] they emit more at the beginning of the period and less at the end of the period because the pressure exerted by investors is more pronounced over the long run. Conversely, the emission schedule increases over time if companies have a higher climate sensitivity than that of the average investor. In the limiting case where companies' climate beliefs are equal to investors' average climate beliefs, $\kappa^c = \alpha\kappa$, the emission schedule is constant over time. In short, companies' climate sensitivity, κ^c, mainly drives short-term emissions, whereas green investors' proportion of wealth, α, and their climate sensitivity, κ, mainly drive long-term emissions.[17] This result captures two main routes available to green

[16] $\alpha\kappa$ is the climate sensitivity of the average investor because green investors' climate sensitivity, κ, is weighted by their proportion of wealth, α, and regular investors have zero climate sensitivity.

[17] Two special cases arise: when κ^c is zero (that is, companies do not internalize climate externalities), emissions tend to infinity close to time $t = 0$ (because β_t tends to 0); similarly, when $\alpha\kappa$ is zero (that is, there are no green investors), emissions tend to infinity close to time $t = T$ (because β_t^c tends to 0). This effect is due to the fact that, in order to obtain a tractable and interpretable solution for the emission schedule, we have opted for a constant marginal abatement cost for each date t, which is profitable to companies when their emissions are higher than the initial level, ψ_b (for example, because they are exempt from the cost of infrastructure maintenance). Therefore, companies benefit from letting their emissions grow when there are no incentives to reduce them (in $t = 0$ when κ^c is zero, and in $t = T$ when $\alpha\kappa$ is zero). In the paper, we focus on this simplified model because (i) the main objective is to analyze the impact of green investors-induced incentives on companies and (ii) the model allows us to obtain tractable formulae that explain the pressures exerted by investors and companies on greenhouse gas emissions. However, in the Internet Appendix to the original article, we study a less tractable version of the model in which companies have zero marginal gain from letting their emissions grow above the initial level, ψ_b. In that framework the emissions' dynamics is similar to the one observed here but for an important difference: companies never increase their emissions above the initial level, ψ_b, irrespective of whether α, κ or κ^c are zero.

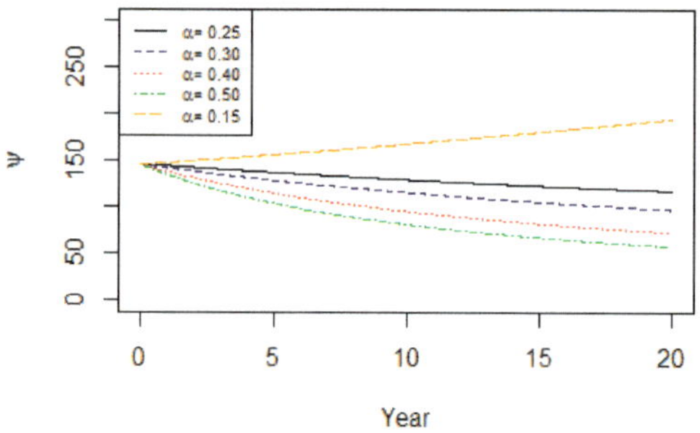

(a) ψ_t with different values for α

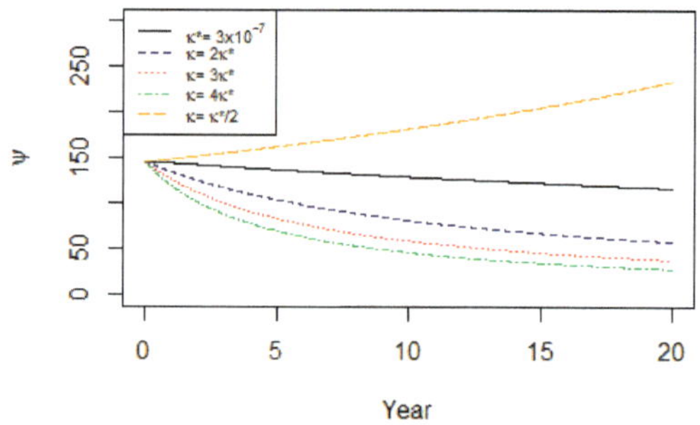

(b) ψ_t with different values for κ

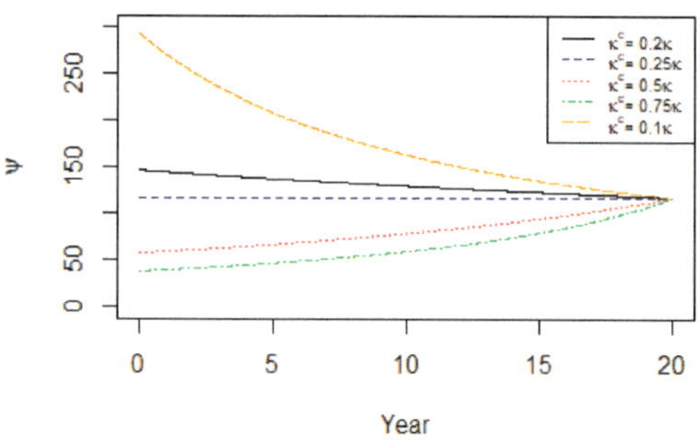

(c) ψ_t with different values for κ^c

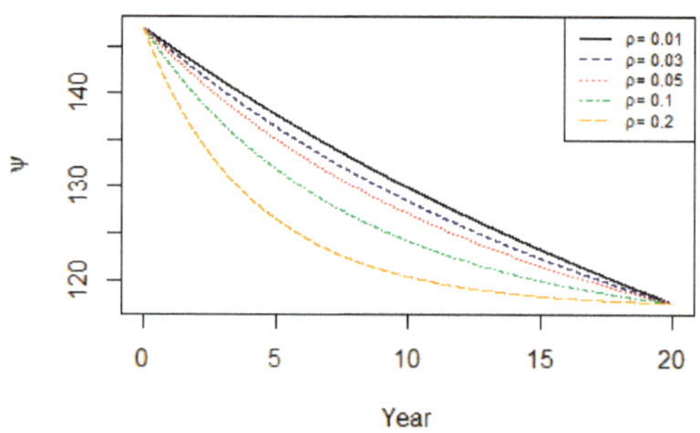

(d) ψ_t with different values for ρ

FIGURE 7.2: **Emission schedules.** This figure shows the optimal emission schedules, ψ_t, according to several values of the proportion of green investors (α, sub-figure (a)), the green investors' climate sensitivity (κ, sub-figure (b)), the companies' climate sensitivity (κ^c, sub-figure (c)), and the rate of time preference (ρ, sub-figure (d)). The parameters are calibrated according to the values estimated in Appendix B: $\psi_b = 147$, $\alpha = 0.25$, $\rho = 0.01$, $\kappa = 3 \times 10^{-7}$, $\kappa^c = 6 \times 10^{-8}$, $c^{\text{elec}} = 8 \times 10^{-6}$.

investors to impact corporate decisions. First, green investors have the ability to reduce the long-term emission target of the companies in which they invest by internalizing climate externalities in their investment decisions. Second, they can also contribute to reducing companies' short-term emissions by pushing them to internalize climate externalities, for example through shareholder engagement (Dimson, Karakaş, and Li, 2015; Broccardo, Hart, and Zingales, 2020). Aligning corporate objectives with climate issues can be achieved via incentive mechanisms in managers' compensation schemes as suggested by Edmans et al. (2012), Varas (2018) and Aggarwal, Dizon-Ross, and Zucker (2020).

Figure 7.2 shows several optimal emission schedules for an electrical equipment company as a function of α, κ, κ^c, and ρ. The calibration is detailed in Appendix B. It is reasonable to assume a lag in a company's responses to pressure from investors. To represent this delay, we use as an example the case where the company internalizes climate externalities with slightly less sensitivity than the average investor, $\kappa^c < \alpha\kappa$, and thus where emissions decrease over time. With these parameter values, when 25% of total AUM are managed by green investors, the company reduces its emissions by only 1% per year on average. This magnitude is coherent with the small effect of divestment on companies' cost of capital estimated by Berk and van Binsbergen (2021). The emission reduction reaches 3% per year on average when green investments account for 50% of total AUM, or when green investors' climate sensitivity doubles. The decrease in emissions is convex in time because of the dynamics of the time factors β_t and β_t^c: when ρ is small, the emission schedule has a hyperbolic temporal dynamic of the form $1/t$. This convexity increases with the rate of time preference, ρ, which accelerates the substitution effect between the impact of the company's beliefs (through β_t^c) and the impact of green investors' beliefs (through β_t) on the optimal emission schedule.[18] In the present example where the average investor is more climate sensitive than the company, $\kappa^c < \alpha\kappa$, green investors' pressure on long-term emissions occurs earlier and accelerates the emissions decline, thereby increasing the convexity of the schedule. Therefore, in cases where executives have weak incentives to reduce their companies' climate footprints, a strong preference for the present—for example, through short-term objectives in compensation schemes (Bolton, Scheinkman, and Xiong, 2006; Marinovic and Varas, 2019)—might mitigate their adverse impact on the companies' optimal emission schedules. Conversely, a company with low climate requirements and a low preference for the present will emit more greenhouse gases at the optimum.[19]

As in Pastor, Stambaugh, and Taylor (2021), this model extends the work of Heinkel, Kraus, and Zechner (2001) by (i) endogenizing the climate impacts of companies and (ii) allowing them to choose among a continuum of climate impacts, in contrast to Heinkel, Kraus, and Zechner (2001) where companies reform in a binary way (from *brown* to *green*). Compared to Pastor, Stambaugh, and Taylor (2021), in this first approach where externalities are deterministic, we develop a dynamic model that allows us to characterize the *dynamics* of companies' climate footprints (Section 7.3.2) as well as to study the dynamic impact on companies' climate footprints of (i) technological changes (Section 7.3.3), (ii) regulatory changes (Section 7.3.3), and (iii) interaction effects between companies' emissions (Section 7.3.4).

7.3.3 Technological and regulatory changes

The anticipation of technological changes and more demanding climate regulations by green investors can further push companies to reduce their climate footprints. We develop the analysis of this mechanism in this section.

Technological changes. Technological changes allowing companies to reduce their climate footprints can take three major forms: the use of new machines that improve energy efficiency, that is, the ratio of energy use over emissions; end-of-pipe innovations, such as carbon capture technologies for utilities, which reduce emissions without modifying the production process; and process innovations, which offer alternative production processes reducing the use of fossil fuels. Although the effect on the marginal abatement cost curve is not unequivocal (Amir, Germain, and Van Steenberghe, 2008; Bauman, Lee, and Seeley, 2008),[20] technological breakthroughs generally induce a decrease in the marginal

[18]The Internet Appendix (Figure IA.1.) shows the dynamics of β_t^c and β_t for different rates of time preference, ρ.

[19]If companies optimize over an infinite horizon, the economic insights remain the same. Indeed, in equilibrium, the optimal allocations, prices and emission schedules converge to a well-defined limit when T tends to $+\infty$. Specifically, regarding optimal emissions, the direct consequence of an infinite horizon optimization is that the effect of green investors' beliefs on climate externalities, which impact long-term emissions, substitutes more slowly for the effect of companies' beliefs. For example, for a rate of time preference of 0.01, the two effects have the same weight after 69 years ($\beta_{69} = \beta_{69}^c = 0.5$).

[20]The decrease in the marginal abatement cost is consensual for end-of-pipe innovations; for process innovations, the decrease in the marginal abatement cost is favored by a strong substitutability between the two factors of production (energy and capital).

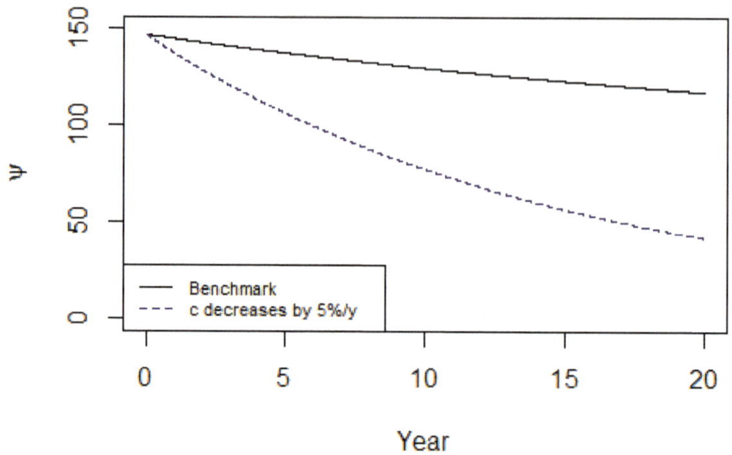

FIGURE 7.3: **Emission schedule with technological change.** This figure shows the optimal emission schedules, ψ_t, without technological change (Benchmark) and with technological change for which the marginal abatement cost decreases by 5% per year. The parameters are calibrated according to the values estimated in Appendix B: $\psi_b = 147$, $\alpha = 0.25$, $\rho = 0.01$, $\kappa = 3 \times 10^{-7}$, $\kappa^c = 6 \times 10^{-8}$, $c^{\text{elec}} = 8 \times 10^{-6}$ (Benchmark).

abatement cost (Milliman and Prince, 1989; Palmer, Oates, and Portney, 1995; Jaffe, Newell, and Stavins, 2002).

Mekaroonreung and Johnson (2014) estimate the effect of technological change on nitrogen oxides (NO_x) marginal abatement costs of U.S. coal power plants in 2000–2008 by analyzing 325 boilers operating in 134 bituminous coal power plants. They find that technological change reduced the NO_x marginal cost by 28.3% in 2000–2004 and 26.5% in 2004–2008. Based on the order of magnitude of their result and using the parameters calibrated in Appendix B, we simulate the effect of technological changes that would reduce the marginal cost of carbon intensity abatement by 5% per year (Figure 7.3). Such technological changes, when anticipated by companies and investors, push companies to multiply the pace of emissions reduction by a factor of 3.6 (from 1% to 3.6% per year on average). Compared to the situation where no technological change is anticipated, the carbon intensity is reduced by 40% after 10 years and by 64% after 20 years.

This result not only underscores the importance of R&D, particularly in sectors where marginal abatement costs are high, but also the need to enable agents to forecast and internalize a likely path of future technological change. Even if, by definition, the occurrence of climate-related innovations is unpredictable, it is possible to anticipate a vigorous dynamic of technological change when R&D is largely supported by public and private funding; the development of renewable energy infrastructures over the last 20 years as well as ongoing research on energy storage[21] or carbon capture and storage[22] are insightful examples.

[21] https://ec.europa.eu/energy/sites/ener/files/documents/batstorm_d10_roadmap.pdf

[22] https://www.energy.gov/fe/science-innovation/carbon-capture-and-storage-research/carbon-capture-rd

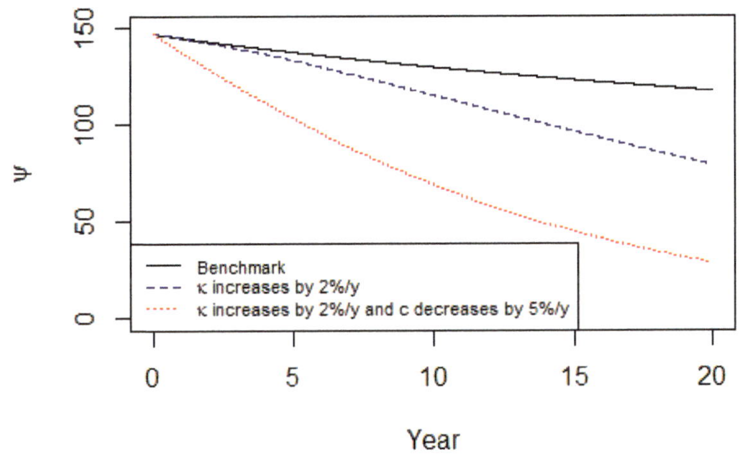

FIGURE 7.4: **Emission schedule with regulatory change.** This figure shows the optimal emission schedules, ψ_t, without regulatory change (Benchmark) and with regulatory change for which (i) green investors' climate sensitivity, κ_t, increases by 2% per year, and (ii) green investors' climate sensitivity, κ_t, increases by 2% per year and the marginal abatement cost, c^{elec}, decreases by 5% per year. The parameters are calibrated according to the values estimated in Appendix B: $\psi_b = 147$, $\alpha = 0.25$, $\rho = 0.01$, $\kappa = 3 \times 10^{-7}$ (Benchmark), $\kappa^c = 6 \times 10^{-8}$, $c^{\text{elec}} = 8 \times 10^{-6}$ (Benchmark).

Regulatory changes. Tightening climate regulations can take two main forms: the introduction of more demanding standards or an increase in the price of carbon, whether through taxes or pollution permits. For green investors, such regulatory changes raise the future financial risks of the brownest companies, specifically the transition risks. Therefore, when investors anticipate regulatory tightening, given the same level of emissions, ψ_t, they adjust their expected financial impact of climate externalities, $\theta_t(\psi_t)$, by increasing their climate sensitivity, κ_t. However, tighter regulations can have an additional effect: as suggested by Porter and van der Linde (1995), they also encourage companies to innovate and may lower the marginal abatement cost. Porter's hypothesis has been supported by empirical evidence, such as the introduction of sulfur emission standards in India (Sugathan et al., 2018) and a carbon emissions permit trading system in China (Xian et al., 2020).

To calibrate the effect of tighter regulation on the climate sensitivity of green investors, we use the carbon price trajectory. Although carbon prices vary from USD 1/tCO_2e to USD 119/tCO_2e in different jurisdictions, the High-Level Commission on Carbon Prices estimates that carbon prices of USD 40-80/tCO_2 by 2020 and USD 50-100/tCO_2 by 2030 are required to reduce emissions in line with the temperature goals of the Paris Agreement (Bank, 2020). Consistent with this trajectory, we therefore consider an increase of κ_t by 2% per year.

Figure 7.4 shows the effects of tighter regulation on companies' optimal emission schedule, using the parameters calibrated in Appendix B. In addition, we also show the effect of a simultaneous policy change and technological change by including a 5% annual drop in marginal abatement cost along with the 2% annual increase in climate sensitivity.

When investors anticipate regulatory tightening, they push companies to multiply the rate of emissions reduction by a factor of 2.2 (2.2% per year on average). This rate is multiplied by four when investors and companies account for technological changes. Compared to the situation without regulatory changes, the carbon intensity decreases by 11% (47%) after 10 years and by 33% (76%) after 20 years without (with) technological change. Here again, the anticipation of future regulatory tightening encourages companies to launch projects that emit less greenhouse gases and to adopt a more ambitious emission schedule, partly as a result of the increased pressure exerted by investors on their cost of capital. Therefore, by announcing plans for more stringent climate standards or the future carbon price trajectory early enough, governments are sending a signal not only to companies but also to investors, which further strengthens the climate impact they have on companies.

7.3.4 Interaction effects

In Corollary 1 we have made the simplifying assumption that investors and companies internalize the financial impact of the i-th company's climate externalities as a function of its own emissions only, $\theta_t^i(\psi_t^i)$. However, a company can be financially impacted by the emissions of other companies in the economy. For example, the risk of tightening climate regulations is likely to be greater if the companies in the same geographical area have a large climate footprint. Proposition 4 shows that when agents internalize the negative impact of the economy's average emissions through an interaction effect, companies decrease their optimal emission schedules compared to the case with no such interaction.

Proposition 4 *Let $\psi_t^{*,i} = c_t^i(\beta_t^c \kappa_t^c + \alpha\beta_t\kappa_t)^{-1}$ be the i-th company's optimal emission schedule without interaction effect (Corollary 1), $\bar{\psi}_t^* = \frac{1}{n}\sum_{j=1}^n \psi_t^{*,j}$ the average optimal emissions of the n companies without interaction effect, and $\varepsilon \geq 0$ an elasticity parameter such that*

$$\theta_t^i(\psi) = \kappa_{0,t} - \frac{\kappa_t}{2}\left[(\psi_t^i)^2 + \varepsilon\, \psi_t^i\, \bar{\psi}_t\right] \quad \text{and} \quad \theta_t^{c,i}(\psi) = \kappa_{0,t}^c - \frac{\kappa_t^c}{2}\left[(\psi_t^i)^2 + \varepsilon\, \psi_t^i\, \bar{\psi}_t\right],$$

where $\bar{\psi}_t = \frac{1}{n}\sum_{j=1}^n \psi_t^j$, and $\kappa_{0,t}$, $\kappa_{0,t}^c$, κ_t and κ_t^c are deterministic functions of time. Then, we have the following results:

(i) Companies lower their optimal emissions compared to the case without interaction effect (that is, when $\varepsilon = 0$), and the i-th company's optimal emission schedule reads

$$\psi_t^{*,\varepsilon,i} = \psi_t^{*,i} - \frac{\varepsilon}{2n+\varepsilon}\left(\psi_t^{*,i} + \frac{2n^2}{2n+(n+1)\varepsilon}\bar{\psi}_t^*\right). \tag{7.15}$$

(ii) When the number of companies is large and much larger than the elasticity parameter, i.e., $n \gg \varepsilon \geq 0$, the optimal emission schedule is approximated by

$$\psi_t^{*,\varepsilon,i} \simeq \psi_t^{*,i} - \frac{\varepsilon}{\varepsilon+2}\bar{\psi}_t^*. \tag{7.16}$$

In the above proposition, we show that when investors and companies internalize the negative financial impact of the economy's average emissions, companies are further incentivized to curb their optimal emission schedules. Let us take a simple example where a company has an optimal carbon intensity of 150 tCO_2e per million dollars of revenue generated (tCO_2e/USDmn), while the average optimal carbon intensity of the $n = 1000$ companies in the economy is 100 tCO_2e/USDmn in the absence of interaction effect. If the agents internalize the impact of the economy's average emissions with an elasticity of $\varepsilon = 0.5$, the optimal emissions decrease to 130 tCO_2e/USDmn, resulting in a 13% reduction. The downward adjustment of a company's emission schedule in an economy with a sufficiently large number of companies is proportional to $\frac{\varepsilon}{\varepsilon+2}$ times the average emissions of the economy. Therefore, in the particular case where companies give equal weight to their emissions and to the average emissions of the economy ($\varepsilon = 1$), they reduce their optimal emission schedules by one third of the average emissions of the economy. As expected, in the absence of interaction effect ($\varepsilon = 0$), the optimal emission schedule is maximal and it equals $\psi_t^{*,i} = c_t^i(\beta_t^c \kappa_t^c + \alpha \beta_t \kappa_t)^{-1}$.

7.4 Equilibrium with climate uncertainty

We extend the model presented in Section 7.2 to the case where the climate externalities are internalized by green investors as a non-Gaussian stochastic process. We characterize the optimal emission schedule under the new setup, and we show that uncertainty about future climate externalities reduces the incentive for companies to lower their emissions.

7.4.1 Climate uncertainty

The internalization of deterministic climate externalities is an imperfect approach. Barnett, Brock, and Hansen (2020) note that "given historical evidence alone it is likely to be challenging to extrapolate climate impacts on a world scale to ranges which many economies have yet to experience. Both richer dynamics and alternative nonlinearities may well be essential features of the damages that we experience in the future due to global warming." Indeed, climate risks are characterized by fat tails (Weitzman, 2009, 2011) and abrupt changes beyond tipping points (Alley et al., 2003; Lontzek et al., 2015; Cai et al., 2015) that will severely impact the world economy (Dietz, 2011).

We therefore extend our model to the case where green investors internalize uncertainty about climate-related financial risks. As climate-related financial risks are not Gaussian and occur in jerks and turns, we describe the arrival of these risks by a Poisson process. On the same filtered probability space, $(\Omega, \mathcal{F}, (\mathcal{F}_t)_{t\geq 0}, \mathbb{P})$, we define a time-homogeneous Poisson process $\mathcal{N} := (\mathcal{N}_t)_{t\in[0,T]}$ (counter of the shocks), whose intensity is denoted by λ. To ensure comparability with the deterministic case, we make the average effect of jumps in climate externalities equal to the average effect of deterministic linear externalities over a given period: we denote by $\mathcal{N}_t^\lambda := \lambda^{-1}\mathcal{N}_t$ the normalized version of $\mathcal{N}$, so that $\mathbb{E}[\mathcal{N}_t^\lambda] = t$ for any λ. Therefore, the only factor governing uncertainty is the intensity parameter, λ, which modulates the strength of climate uncertainty. When λ is small, climate uncertainty is strong: the shocks are rare but large; when λ is large, climate uncertainty is low: the shocks are frequent but small.

7.4.2 Investors' and companies' beliefs

As before, we assume that regular investors do not internalize the financial impact of climate externalities. Therefore, according to regular investors beliefs, the vectors of terminal dividends, D_T, and dividend forecast, $\mathbb{E}_t^r(D_T)$, are still given by Equations (7.1) and (7.3), respectively. However, in contrast to Sections 2 and 3, green investors internalize climate externalities by taking into account their uncertainty. Although the expression of the expected terminal dividends, $\mathbb{E}_t^g(D_T)$, given in Equation (7.6) continues to hold, it offers an average picture of green investors' beliefs and, therefore, does not allow us to specify the dynamics of jumps introduced in this section. To account properly for the jump process, we need to slightly modify the definition of the terminal dividends as perceived by the green investors. Specifically, green investors assume that the vector of terminal dividends contains an additional factor that depends on the climate externalities of the companies, $\theta^s(\psi^s)$, and the Poisson process, $\mathcal{N}^\lambda$:

$$D_T = \int_0^T c_t(\psi_t - \psi_b)dt + d + \int_0^T \sigma_t dB_t + \int_0^T \theta_s(\psi_s)d\mathcal{N}_s^\lambda, \tag{7.17}$$

where now B is a standard Brownian motion under all probability measures that we consider (that is, under the reference measure $\mathbb{P} = \mathbb{P}^r$ and under the measures of the green investors $\mathbb{P}^g$ and of the companies $\mathbb{P}^c$).

This framework generalizes the first model where climate externalities were deterministic. Indeed, when the intensity, λ, increases, the number of shocks increases and their size decreases; in the limiting case where λ tends to $+\infty$, the uncertainty disappears, and we recover the setting of Section 7.2:

$$\lim_{\lambda \to \infty} \int_0^T \theta_s(\psi_s)d\mathcal{N}_s^\lambda = \int_0^T \theta_s(\psi_s)ds.$$

Like in the case where the externalities were deterministic, the variable D_t is constructed from the actual realization of the past cash flow news between 0 and t, which is known at time t. However, from a probabilistic point of view, the dynamic of the stochastic process D_t under green investors' beliefs (probability measure $\mathbb{P}^g$) includes a jump process corresponding to the internalization of the uncertainty about future climate externalities:

$$D_t = \int_0^T c_t(\psi_t - \psi_b)dt + d + \int_0^t \sigma_t dB_t + \int_0^t \theta_s(\psi_s)d\mathcal{N}_s^\lambda.$$

Similarly, under companies' beliefs (probability measure $\mathbb{P}^c$) , we define the dynamics of the terminal dividend using $\theta_s^c(\psi_s)$ as

$$D_T = \int_0^T c_t(\psi_t - \psi_b)dt + d + \int_0^T \sigma_t dB_t + \int_0^T \theta_s^c(\psi_s)d\mathcal{N}_s^\lambda,$$

and the dynamic of the stochastic process D_t as

$$D_t = \int_0^T c_t(\psi_t - \psi_b)dt + d + \int_0^t \sigma_t dB_t + \int_0^t \theta_s^c(\psi_s)d\mathcal{N}_s^\lambda.$$

7.4.3 Equilibrium stock prices and returns

The optimization framework and notation remain similar to those of the first model. The *equilibrium price process* is denoted by $(p_t)_{t\in[0,T]}$, and it is assumed that $p_T = D_T$. In equilibrium, investors choose their allocation to maximize the expected exponential utility of their terminal wealth (Equations (7.7)), and equilibrium prices are determined by the market clearing condition. Proposition 5 gives the equilibrium price and allocations.

Proposition 5 *Given an emission schedule $(\psi_t)_{t\in[0,T]}$, the asset price in equilibrium reads*

$$p_t = D_t - \int_t^T \mu_s ds \quad \textit{with} \quad \mu_t = \gamma^* \Sigma_t \mathbf{1} - \alpha y_\lambda \theta_t(\psi_t), \tag{7.18}$$

and the optimal number of shares for the regular and green investors are

$$N_t^{r,\lambda} = (1-\alpha)\left(\mathbf{1} - \frac{y_\lambda}{\gamma^g}\Sigma_t^{-1}\theta_t(\psi_t)\right) \quad and \quad N_t^{g,\lambda} = \alpha\left(\mathbf{1} + \frac{y_\lambda}{\gamma^r}\Sigma_t^{-1}\theta_t(\psi_t)\right), \tag{7.19}$$

respectively, where $y_\lambda > 0$ *is obtained by solving the one-dimensional equation*

$$y_\lambda = \exp\left\{-\frac{\alpha\gamma^g}{\lambda}\left(\mathbf{1}^\top\theta_t(\psi_t) + \frac{y_\lambda}{\gamma^r}\theta_t(\psi_t)^\top\Sigma_t^{-1}\theta_t(\psi_t)\right)\right\}. \tag{7.20}$$

When green investors internalize uncertainty about climate externalities, an additional factor, y_λ, arises in the equilibrium allocations, $N_t^{r,\lambda}$ and $N_t^{g,\lambda}$, the equilibrium price, p_t, and in particular the expected return, $\mu_t dt$. Notice that y_λ is determined at each time t by Equation (7.20). Thus, it should be denoted by $y_\lambda(t)$. We choose the simpler notation y_λ as no confusion shall arise. The equation for y_λ admits a unique solution.[23] prop 6 explains the effect of y_λ on the optimal allocations and expected returns depending on the properties of green investors' optimal portfolio *without* uncertainty, N_t^g.

Proposition 6 *Fix an emission schedule* ψ_t. *Recall that* $N_t^g = \alpha\left(\mathbf{1} + \frac{1}{\gamma^r}\Sigma_t^{-1}\theta_t(\psi_t)\right)$ *is the optimal portfolio of green investors with deterministic climate externalities, and* $N_t^{g,\lambda} = \alpha\left(\mathbf{1} + \frac{y_\lambda}{\gamma^r}\Sigma_t^{-1}\theta_t(\psi_t)\right)$ *is the optimal portfolio of green investors when the uncertainty level of climate externalities equals* λ. *Then,*

(i) When the level of climate uncertainty increases, green investors decrease the market risk of their portfolio, i.e., $\lambda \mapsto (N_t^{g,\lambda})^\top\Sigma_t N_t^{g,\lambda}$ *is increasing in* λ.

(ii) If the green investors' portfolio in the absence of climate uncertainty has positive climate externalities, meaning that $\theta_t(\psi_t)^\top N_t^g > 0$, *then, the function* $\lambda \mapsto y_\lambda$ *is positive, monotonically increasing and satisfies* $\lim_{\lambda\to\infty} y_\lambda = 1$. *Consequently, for a given level of emission schedule,* ψ_t, *the increase in climate uncertainty (*λ *decreases)*

- *pushes green investors to reallocate their wealth toward brown assets;*
- *decreases (increases) the expected returns,* $\mu_t dt$, *of the brown (green) companies.*

(iii) If the green investors' portfolio in the absence of climate uncertainty has negative climate externalities, meaning that $\theta_t(\psi_t)^\top N_t^g < 0$, *then, the function* $\lambda \mapsto y_\lambda$ *is positive, monotonically decreasing and satisfies* $\lim_{\lambda\to\infty} y_\lambda = 1$. *Consequently, for a given level of emission schedule,* ψ_t, *the increase in climate uncertainty (*λ *decreases)*

- *pushes green investors to reallocate their wealth toward green assets;*
- *increases (decreases) the expected returns,* $\mu_t dt$, *of the brown (green) companies.*

(iv) If the green investors' portfolio in the absence of climate uncertainty has neutral climate externalities, meaning that $\theta_t(\psi_t)^\top N_t^g = 0$, *then* $y_\lambda = 1$ *for all* $\lambda > 0$. *Consequently, green investors' portfolio and companies' expected returns do not depend on the level of climate uncertainty.*

Uncertainty about future climate externalities is an additional source of risk for green investors who, as a result, reduce the overall risk of their portfolio. However, the adjustment of their asset allocation depends on the climate externalities of their optimal portfolio without uncertainty, $\theta_t(\psi_t)^\top N_t^g$. Lemma 1 further elaborates on the case in which the green investors' optimal portfolio without climate uncertainty is green, that is, $\theta_t(\psi_t)^\top N_t^g > 0$.

[23] Existence and uniqueness of y_λ are given as part of the proof of the next proposition.

Lemma 1 *Fix an emission schedule* ψ_t. *Assume that the sum of the companies' climate externalities is not too negative, such that* $\mathbf{1}^\top \theta_t(\psi_t) > -\frac{1}{\gamma^r}\theta_t(\psi_t)^\top \Sigma_t^{-1}\theta_t(\psi_t)$. *Then, the green investors' portfolio in the absence of climate uncertainty has positive climate externalities, that is,* $\theta_t(\psi_t)^\top N_t^g > 0$.

The situation described by Lemma 1 is the most common.[24] It requires that the sum of the climate externalities of the companies in the market is greater than $-\frac{1}{\gamma^r}\theta_t(\psi_t)^\top \Sigma_t^{-1}\theta_t(\psi_t)$, which is always negative. Typically, this occurs when the market is not excessively composed of brown assets, for example, when it is sufficiently diversified in terms of companies' climate externalities, that is, when the market is formed by companies with positive and negative $\theta_t^i(\psi_t)$. Moreover, the condition is more easily verified for large markets because the quadratic correction, $\frac{1}{\gamma^r}\theta_t(\psi_t)^\top \Sigma_t^{-1}\theta_t(\psi_t)$, becomes larger. Pastor, Stambaugh, and Taylor (2021) assume that the ESG externalities of the market portfolio are zero—here $\mathbf{1}^\top \theta_t(\psi_t) = 0$—which is a sufficient condition for Lemma 1 to be valid. Furthermore, Zerbib (2021) validates the Pastor, Stambaugh, and Taylor (2021) hypothesis by estimating the climate externalities of the market portfolio in the U.S. between 2007 and 2019 at a value close to zero.

In the main case described by Lemma 1 where the market is not excessively brown, green investors are able to build a green portfolio in the absence of uncertainty. Therefore, $y_\lambda \in [0, 1[$, which means that uncertainty about future climate externalities pushes green investors to lower the risk of their portfolio by reducing their exposure toward green assets. The size of this effect scales with the degree of uncertainty: the larger the uncertainty, the more green investors decrease their allocation in green assets and increase their allocation in brown assets. Consequently, for a given level of emission schedule, ψ_t, climate uncertainty reduces the effect of the externality premium, $-\alpha y_\lambda \theta_t(\psi_t)$, on expected returns, which lowers the cost of capital of brown companies and increases the cost of capital of green companies. This result is consistent with that of Avramov et al. (2021a) who show that gaussian uncertainty weakens the negative relationship between ESG and financial performances.

In less common cases, green investors fail to construct a green optimal portfolio in the absence of uncertainty. This situation is possible, for example, when all or much of the economy is brown. Therefore, $y_\lambda > 1$, which indicates that the increase in climate uncertainty pushes green investors to invest more in green assets and divest from brown assets to diversify their allocation and mitigate their risk. Consequently, for a given level of emission schedule, ψ_t, climate uncertainty amplifies the effect of the externality premium, $-\alpha y_\lambda \theta_t(\psi_t)$, on expected returns, which reduces the cost of capital of green companies and increases that of brown companies.

The following section presents the companies' optimal emission schedules when they account for the climate uncertainty internalized by green investors.

7.4.4 Equilibrium emission schedule

Companies fix their emission schedules at the initial date by maximizing their future market values as in the deterministic case (Equation (7.8)). The situation here is more complex, however, because of the appearance of the new parameter y_λ in the price vector (see Proposition 5). Indeed, for a fixed value of y_λ, the optimal emissions schedule of the i-th company is the one that maximizes for all $t \in [0, T]$ the expression:

$$\beta_t^c \theta_t^{c,i}(\psi_t) + \alpha \beta_t y_\lambda \theta_t^i(\psi_t) + c_t^i \psi_t^i, \tag{7.21}$$

by the same arguments as the ones used in Proposition 3 for deterministic externalities. However, this shows that the equilibrium emissions schedule depends on the choice of y_λ,

[24] The proof of the lemma is an immediate consequence of the expression for N_t^g (see Equation (7.11)).

which will affect the investors' portfolio allocations and asset prices in Proposition 5, creating a feedback effect. In order to maintain tractability in our model we make an assumption of a large market that allows us to partially decouple the optimization problems for investors and companies. Then we derive equilibrium emission schedules in the next proposition.[25]

Proposition 7 *Let the functions of climate externalities, θ_t^i and $\theta_t^{c,i}$, be defined as in Corollary 1. Assuming that the market is large enough so that the aggregated climate externalities of the market are not strongly affected by a change in the climate externalities of each company, the optimal emission schedule for the i-th company is*

$$\psi_t^{i,*}(y_\lambda^*) = \frac{c_t^i}{\alpha\beta_t\kappa_t y_\lambda^* + \beta_t^c\kappa_t^c}, \tag{7.22}$$

where y_λ^* is a solution of the fixed-point *equation*

$$y = \exp\left\{-\frac{\alpha\gamma^g}{\lambda}\left[\mathbf{1}^\top\theta_t(\psi_t(y)) + \frac{\alpha y}{\gamma^r}\theta_t(\psi_t(y))^\top\Sigma_t^{-1}\theta_t(\psi_t(y))\right]\right\}. \tag{7.23}$$

As climate uncertainty adds a multiplicative factor y_λ to the externality premium, it also adds a multiplicative factor y_λ^* to the company's optimization program (Equation (7.12)) on the factor driven by green investors' beliefs, $\alpha\beta_t y_\lambda^*\theta_t^i(\psi_t)$. As a result, the optimal emission schedule in the presence of uncertainty (Equation (7.22)) is adjusted by y_λ^* in its denominator. Proposition 8 clarifies the behavior of y_λ^* depending on λ.

Proposition 8 *Under the same assumptions as in Proposition 7, the following holds true:*

(i) Existence: For all $\lambda \in (0,\infty)$, Equation (7.23) admits at least one positive solution.

(ii) Uniqueness: If, for all $i \in \{1,\ldots,n\}$, and all $t \in [0,T]$,

$$\inf_{y>0}\sum_{j=1}^{n}(\Sigma_t^{-1})_{ij}\, y\,\theta_t^j(\psi_t^j(y)) \geq -\frac{\gamma^r}{\alpha} \tag{7.24}$$

the solution of Equation (7.23) is unique. In particular, in the asymptotic regime where uncertainty is low ($\lambda \to \infty$), the solution is unique.

(iii) Assume that the condition (ii) is satisfied and denote by y_λ^* the unique solution of Equation (7.23). Let $\psi_t^{*,0,i} = c_t^i(\beta_t^c\kappa_t^c + \alpha\beta_t\kappa_t)^{-1}$ be the equilibrium emission schedule without climate uncertainty (given by (7.13)), and $N_t^g = \alpha\left(\mathbf{1} + \frac{1}{\gamma^r}\Sigma_t^{-1}\theta_t(\psi_t^{*,0})\right)$ be the green investors' corresponding optimal allocation without climate uncertainty (given by (7.11)).

- *If the green investors' portfolio in the absence of climate uncertainty has positive climate externalities, meaning that $\theta_t(\psi_t^{*,0})^\top N_t^g > 0$, then, the function $\lambda \mapsto y_\lambda^*$ is positive, monotonically increasing and satisfies $\lim_{\lambda\to\infty} y_\lambda^* = 1$. Consequently, an increase in climate uncertainty (λ decreases) increases the companies' optimal emission schedule, ψ_t^*, given by (7.22).*

[25] If the market is large enough, we can assume that the aggregate quantities are not strongly affected by each individual company, and therefore, the i-th company may maximize its optimization functional by assuming that y_λ is constant. This assumption is standard, for example, in mean-field game optimization problems. At the mathematical level, our assumption says that each company solves (7.21) for a fixed value of y_λ, thus obtaining $\psi_t^i(y_\lambda)$. Plugging this expression back into (7.20), we arrive at a fixed point equation for y_λ. This equation is still one-dimensional but different from Equation (7.20) due to the additional dependence of ψ_t on y_λ.

- *If the green investors' portfolio in the absence of climate uncertainty has negative climate externalities, meaning that* $\theta_t(\psi_t^{*,0})^\top N_t^g < 0$*, then, the function* $\lambda \mapsto y_\lambda^*$ *is positive, monotonically decreasing and satisfies* $\lim_{\lambda\to\infty} y_\lambda^* = 1$*. Consequently, an increase in climate uncertainty (*λ *decreases) decreases the companies' optimal emission schedule,* ψ_t^**, given by* (7.22)*.*
- *If the green investors' portfolio in the absence of climate uncertainty has neutral climate externalities, meaning that* $\theta_t(\psi_t^{*,0})^\top N_t^g = 0$*, then, the function* $\lambda \mapsto y_\lambda^*$ *is constant and equal to* 1*. Consequently, climate uncertainty does not affect the companies' optimal emission schedule,* ψ_t^**, given by* (7.22)*.*

In the most common case described by Lemma 1 where green investors have a green optimal portfolio in the absence of climate uncertainty, $\theta_t(\psi_t^{*,0})^\top N_t^g > 0$, increasing uncertainty (λ decreases) decreases y_λ below 1. This effect dampens the contribution of green investors' beliefs, $\alpha\beta_t y_\lambda \theta_t^i(\psi_t)$, in the companies' optimization program (Equation (7.12)). Indeed, by mitigating the externality premium on expected returns (Proposition 6), green investors reduce the incentive for companies to decrease their emissions at unchanged abatement cost, $c_t^i\psi_t^i$. Therefore, all companies increase their optimal emission schedules in the presence of climate uncertainty compared to the situation without uncertainty.

In the case where green investors have a brown optimal portfolio in the absence of climate uncertainty, $\theta_t(\psi_t^{*,0})^\top N_t^g < 0$, increasing climate uncertainty (λ decreases) increases y_λ above 1 and thus amplifies the contribution of green investors' beliefs, $\alpha\beta_t y_\lambda \theta_t^i(\psi_t)$, in the companies' optimization program. By strengthening the incentive for companies to reduce their emissions, green investors push companies to reduce their emissions compared to the situation without uncertainty.

Figure 7.5 shows optimal emission schedules according to different levels of uncertainty: the cases where information on climate risks (or the materialization of climate risks) becomes available on average annually, every five years, every ten years, and every twenty years are displayed. When green investors internalize these levels of uncertainty, companies increase their emissions by 0.4%, 2%, 4%, and 8.5%, respectively, over a 20-year horizon compared to the case where climate risks are perfectly known.

This result underscores the value of increasing the transparency of companies' climate impacts as well as improving the forecasting of climate-related financial risks. It also emphasizes the importance of predictability of public policies in favor of climate transition, notably, the carbon price upward trajectory. Transparency and predictability are key pillars for a better integration of climate-related financial risks in green investors' asset allocation, which provides incentives for companies to better internalize their climate externalities and thus reduce their climate footprints more rapidly.

7.5 Conclusion

In this paper we show how green investing impacts companies' practices by increasing their cost of capital. Companies are pushed to internalize their climate externalities and thereby reduce their greenhouse gas emissions. Green investors' impact is further strengthened when they anticipate tighter climate regulations, technological advances, and when they account for the negative financial impact of the economy's average emissions. However, uncertainty about climate risks pushes green investors to diversify their asset allocation, thereby reducing the incentive for companies to mitigate their climate footprints.

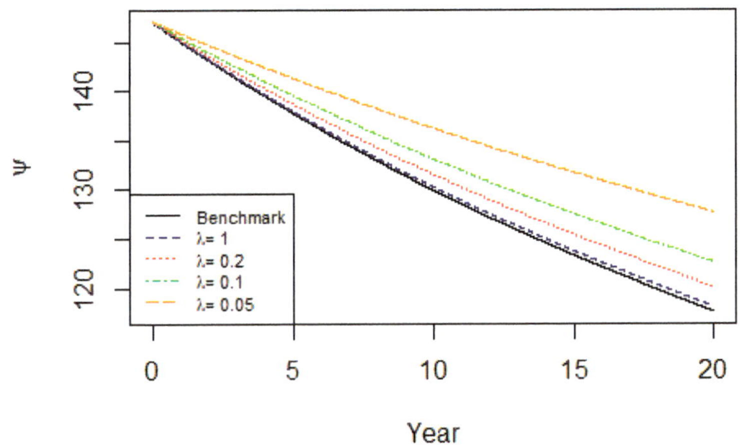

FIGURE 7.5: **Emission schedule with uncertainty.** This figure shows the optimal emission schedules with uncertainty of an electrical equipment company with an initial carbon intensity of $\psi_b^{\text{elec}} = 147$ $\text{tCO}_2\text{e/USDmn}$. This company operates in a market with two companies: the second company is a coal company with an initial carbon intensity of $\psi_b^{\text{coal}} = 555$ $\text{tCO}_2\text{e/USDmn}$. The correlation between the assets of these two companies is 50%. We consider different levels of uncertainty through λ. The parameters are calibrated according to the values estimated in Appendix B: $\alpha = 0.25$, $\rho = 0.01$, $\kappa = 3 \times 10^{-7}$, $\kappa_0 = 0.047$, $\kappa^c = 6 \times 10^{-8}$, $c^{\text{elec}} = 8 \times 10^{-6}$, $\gamma_r = 0.1$.

The results of this paper suggest that investors can increase their impact on companies by raising their environmental requirements as well as by pressing companies to increase transparency and their environmental standards. In addition, impact investing is financially beneficial if investors favor companies that are on a pathway toward reducing their climate footprints or green companies for which information on their climate footprints is still poorly available. From the viewpoint of public authorities, this study emphasizes the importance of developing a regulatory framework that supports the development of green investing and encourages the transparency of information on companies' climate footprints. These actions are naturally compatible with the strengthening of climate regulation and support for climate-related technological innovation, which, when anticipated by green investors, enhance the pressure the latter exert on companies to cut their emissions.

A natural avenue for future research would be to estimate the impact of green investing based on the equilibrium equations of this model. However, we acknowledge that rigorous identification is challenging for at least two major reasons. First, the proportion of green investors, their sensitivities to climate risks, and their perception of climate uncertainty must be approximated. Using green fund holdings is an ambitious approach because it is necessary to identify funds that truly have green practices (Raghunandan and Rajgopal, 2021; Yang, 2021) and to use a dynamic fund list that is not subject to survivor bias. Second, it is crucial to control for the other effects that may impact companies' practices, such as shareholder engagement, global environmental policy stringency, and climate-related technological innovation.

Besides, impact investing may go beyond climate screening, for example, by favoring brown companies that are inclined to green up quickly or small green companies that would benefit from financial support to grow (Green and Roth, 2020; Heeb and Kölbel, 2020). Therefore, future research could also analyze the impact of these new forms of investment on corporate practices, including their ability to further reduce the aggregate emissions of an economy. In addition, the impact of climate screening could be empirically compared to that of shareholder engagement, which Broccardo, Hart, and Zingales (2020) find more effective in reducing the environmental footprint of companies. Finally, a line of theoretical research could introduce the ability for companies to reform dynamically in response to the stochastic dynamics of cash flow news and investors' portfolio allocations.

Appendix A: Proofs

In this appendix we collect proofs and some supporting mathematical materials, needed to justify rigorously our claims.

Proof of Proposition 2

Since the market is assumed to be free of arbitrage and complete, there exists a unique *state price density* ξ_T, i.e., a positive $\mathcal{F}_T$-measurable integrable random variable such that the market price at time t of every contingent claim with terminal value X_T, satisfying $\mathbb{E}[\xi_T|X_T|] < \infty$, is given by

$$\xi_t^{-1}\mathbb{E}[\xi_T X_T|\mathcal{F}_t], \tag{7.25}$$

where $\xi_t := \mathbb{E}[\xi_T|\mathcal{F}_t] = \mathbb{E}_t[\xi_T]$. In particular, since the interest rate is zero, $\mathbb{E}[\xi_T] = 1$. It is worth recalling that $\mathbb{P} = \mathbb{P}^r$ and that $(B_t)_{t\in[0,T]}$ is a Brownian motion under this measure.

The optimization problems of the two investors read:

$$\min_{W_T^r\in\mathcal{A}_T} \mathbb{E}^r\left[e^{-\gamma^r W_T^r}\right], \qquad \min_{W_T^g\in\mathcal{A}_T} \mathbb{E}^g\left[Z_T e^{-\gamma^g W_T^g}\right], \tag{7.26}$$

subject to the budget constraints

$$\mathbb{E}[\xi_T W_T^r] = w^r, \qquad \mathbb{E}[\xi_T W_T^g] = w^g, \tag{7.27}$$

where $w^r > 0$ and $w^g > 0$ are the initial wealth of the regular and green investor, respectively. Both investors use the real-world probability measure for pricing but every investor uses her subjective measure for computing the utility function. Here we consider admissible controls from the class

$$\mathcal{A}_T := \{X \in \mathcal{F}_T : \mathbb{E}^r[\xi_T|X|] < \infty\}$$

and denote by Z_T the Radon-Nikodym density that connects the two probability measures $\mathbb{P}^g$ and $\mathbb{P}^r$. More precisely, recalling (7.3) and (7.5), we have

$$Z_T = e^{\int_0^T \lambda_s^\top dB_s - \frac{1}{2}\int_0^T \|\lambda_s\|^2 ds}, \tag{7.28}$$

where we set $\lambda_t := \sigma_t^{-1}\theta(\psi_t)$, to simplify the notation, and $\|\cdot\|$ is the Euclidean norm in $\mathbb{R}^n$.

The optimization problem is over the set of all admissible contingent claims, but we shall see later that the optimal claims will be attainable. Moreover, we assume that

$$\mathbb{E}^r[\xi_T|\log\xi_T|] < \infty \quad \text{and} \quad \mathbb{E}^r[\xi_T|\log Z_T|]. \tag{7.29}$$

This assumption will be checked a posteriori for the equilibrium state price density.

By the standard Lagrange multiplier argument, the solutions to problems (7.26)-(7.27) are given by

$$W_T^r = w^r - \frac{1}{\gamma^r}\log\xi_T + \frac{1}{\gamma^r}\mathbb{E}^r\left[\xi_T\log\xi_T\right], \qquad W_T^g = w^g - \frac{1}{\gamma^g}\log\frac{\xi_T}{Z_T} + \frac{1}{\gamma^g}\mathbb{E}^r\left[\xi_T\log\frac{\xi_T}{Z_T}\right]. \tag{7.30}$$

The equilibrium state price density ξ_T is found from the market clearing condition

$$W_T^r + W_T^g = \mathbf{1}^\top D_T + K,$$

where K is a constant that allows the market to clear since the bond supply is endogenous. Recall that the interest rate and the initial wealth are exogenous.

Substituting the formulas for W_T^r and W_T^g, yields

$$\xi_T = c\exp\left(-\gamma^*\mathbf{1}^\top D_T + \frac{\gamma^*}{\gamma^g}\log Z_T\right)$$

for some constant c, where we recall $\frac{1}{\gamma^*} = \frac{1}{\gamma^r} + \frac{1}{\gamma^g}$. Note that since D_T and $\log Z_T$ are Gaussian, our a priori assumptions (7.29) are satisfied.

We can now use the fact that $\mathbb{E}^r[\xi_T] = 1$ to conclude that:

$$\xi_T = \frac{\exp\left(-\gamma^*\mathbf{1}^\top D_T + \frac{\gamma^*}{\gamma^g}\log Z_T\right)}{\mathbb{E}^r\left[\exp\left(-\gamma^*\mathbf{1}^\top D_T + \frac{\gamma^*}{\gamma^g}\log Z_T\right)\right]}.$$

Substituting the explicit formulae for D_T and Z_T (see (7.1) and (7.28)) and using that

$$\int_0^T\left(-\gamma^*\mathbf{1}^\top\sigma_t + \frac{\gamma^*}{\gamma^g}\lambda_t^\top\right)dB_t$$

is normally distributed with zero mean and variance

$$\int_0^T\left\|-\gamma^*\mathbf{1}^\top\sigma_t + \frac{\gamma^*}{\gamma^g}\lambda_t^\top\right\|^2 dt,$$

because $(\sigma_t)_{t\in[0,T]}$ and $(\lambda_t)_{t\in[0,T]}$ are deterministic, we have:

$$\xi_T = \mathcal{E}\left(\int_0^\cdot\left\{-\gamma^*\mathbf{1}^\top\sigma_t + \frac{\gamma^*}{\gamma^g}\lambda_t^\top\right\}dB_t\right)_T. \tag{7.31}$$

Here $\mathcal{E}$ denotes the stochastic exponential, i.e., for any adapted square integrable process $X \in \mathbb{R}^n$,

$$\mathcal{E}\left(\int_0^\cdot X_s dB_s\right)_t = \exp\left(\int_0^t X_s dB_s - \frac{1}{2}\int_0^t \|X_s\|^2 ds\right).$$

From (7.31) and (7.28) we can easily verify that (7.29) holds, since (σ_t) and (λ_t) are deterministic.

Using the no-arbitrage pricing rule (7.25), the vector of equilibrium prices is then given by

$$p_t = \xi_t^{-1}\mathbb{E}_t^r[\xi_T D_T] = D_0 + \int_0^t \sigma_s dB_s + \mathbb{E}_t^{\mathbb{Q}}\left[\int_t^T \sigma_s dB_s\right],$$

where $\mathbb{Q}$ is the risk-neutral measure defined by

$$\left.\frac{d\mathbb{Q}}{d\mathbb{P}^r}\right|_{\mathcal{F}_T} = \xi_T.$$

Under $\mathbb{Q}$, the process

$$\widetilde{B}_t = B_t - \int_0^t \left\{-\gamma^*\sigma_s^\top \mathbf{1} + \frac{\gamma^*}{\gamma^g}\lambda_s\right\} ds$$

is a standard Brownian motion. Hence, the equilibrium prices are computed as follows.

$$\begin{aligned} p_t &= \xi_t^{-1}\mathbb{E}_t^r[\xi_T D_T] \\ &= D_0 + \int_0^t \sigma_s dB_s + \int_t^T \sigma_s \left\{-\gamma^*\sigma_s^\top \mathbf{1} + \frac{\gamma^*}{\gamma^g}\lambda_s\right\} ds \\ &= D_t + \int_t^T \{-\gamma^*\Sigma_s \mathbf{1} + \alpha\theta_s(\psi_s)\}\, ds, \end{aligned} \tag{7.32}$$

with

$$D_t = D_0 + \int_0^t \sigma_s dB_s, \quad \Sigma_t = \sigma_t\sigma_t^\top, \qquad \theta_t(\psi_t) = \sigma_t\lambda_t, \quad \text{and} \quad \alpha = \frac{\gamma^r}{\gamma^r + \gamma^g}.$$

This completes the proof of (7.10).

Next we determine the number of shares that each investor holds in her portfolio. The values of the investors' portfolios are determined through the no-arbitrage pricing rule (7.25). In particular, we have

$$\begin{aligned} W_t^r &= \xi_t^{-1}\mathbb{E}_t^r[\xi_T W_T^r] \\ &= w^r - \frac{1}{\gamma^r}\mathbb{E}_t^r\left[\frac{\xi_T}{\xi_t}\left(\log\frac{\xi_T}{\xi_t} + \log\xi_t\right)\right] + \frac{1}{\gamma^r}\mathbb{E}^r\left[\xi_t\left(\frac{\xi_T}{\xi_t}\log\frac{\xi_T}{\xi_t}\right) + \left(\frac{\xi_T}{\xi_t}\right)\xi_t\log\xi_t\right], \end{aligned}$$

by simple algebraic manipulations. Then, using that ξ_T/ξ_t is independent of $\mathcal{F}_t$ (hence of ξ_t) and that $\mathbb{E}^r[\xi_t] = \mathbb{E}^r[\xi_T] = \mathbb{E}_t^r[\xi_T/\xi_t] = 1$ we obtain the wealth at time t of the regular investor

$$W_t^r = w^r - \frac{1}{\gamma^r}\log\xi_t + \frac{1}{\gamma^r}\mathbb{E}^r\left[\xi_t\log\xi_t\right]. \tag{7.33}$$

By construction $W_t^r = \mathbb{E}^{\mathbb{Q}}[W_T^r|\mathcal{F}_t]$, hence it is a $\mathbb{Q}$-martingale. Moreover, by (7.33) we see that the only stochastic term in the dynamics of (W_t^r) is $-1/\gamma^r\log\xi_t$. Then, using

$$\xi_t = \mathcal{E}\left(\int_0^\cdot \left\{-\gamma^*\mathbf{1}^\top\sigma_s + \frac{\gamma^*}{\gamma^g}\lambda_s^\top\right\} dB_s\right)_t,$$

we can conclude that, under the measure $\mathbb{Q}$, the process (W_t^r) has martingale dynamics

$$W_t^r = w^r + (1-\alpha)\int_0^t \left\{\mathbf{1}^\top\sigma_s - \frac{1}{\gamma^g}\lambda_s^\top\right\} d\widetilde{B}_s.$$

The price derived in (7.32), on the other hand, has martingale dynamics under the measure $\mathbb{Q}$ given by

$$p_t = p_0 + \int_0^t \sigma_s d\widetilde{B}_s,$$

where

$$p_0 = D_0 + \int_0^T (-\gamma^* \Sigma_s \mathbf{1} + \alpha\theta_s(\psi_s))\, ds.$$

It follows that the optimal claim for the investor is replicable by a self-financing portfolio whose value can be written as follows:

$$\begin{aligned} W_t^r &= w^r + (1-\alpha) \int_0^t \left\{ \mathbf{1}^\top \sigma_s - \frac{1}{\gamma^g} \lambda_s^\top \right\} \sigma_s^{-1} dp_s \\ &= w^r + (1-\alpha) \int_0^t \left\{ \mathbf{1}^\top - \frac{1}{\gamma^g} \theta_s(\psi_s)^\top \Sigma_s^{-1} \right\} dp_s. \end{aligned}$$

We conclude that the vector of quantities of shares held by the regular investor at time t is given by

$$N_t^r = (1-\alpha) \left\{ \mathbf{1} - \frac{1}{\gamma^g} \Sigma_t^{-1} \theta_t(\psi_t) \right\},$$

while that of the green investor is given by

$$N_t^g = \alpha \left\{ \mathbf{1} + \frac{1}{\gamma^r} \Sigma_t^{-1} \theta_t(\psi_t) \right\}.$$

The latter can be obtained by the former and the market clearing condition. Alternatively, the risk-neutral pricing principle and calculations analogous to the ones above allow us to deduce that

$$W_t^g = \xi_t^{-1} \mathbb{E}_t^r [\xi_T W_T^g] = w^g + \alpha \int_0^t \left\{ \mathbf{1}^\top + \frac{1}{\gamma^r} \theta_t(\psi_s)^\top \Sigma_s^{-1} \right\} dp_s$$

from the formula in (7.30). Hence, the expression of N_t^g follows.

Proof of Proposition 3

Recalling (7.6), the measure $\mathbb{P}^c$ has density with respect to the measure $\mathbb{P}^r$ given by

$$Z_T^c = e^{\int_0^T (\lambda_s^c)^\top dW_s - \frac{1}{2} \int_0^T \|\lambda_s^c\|^2 ds},$$

where $\lambda_t^c := \sigma_t^{-1} \theta^c(\psi_t)$.

Using (7.32) and Girsanov theorem, the vector of expected equilibrium prices under the measure $\mathbb{P}^c$ reads

$$\mathbb{E}^c(p_t) = d + \int_0^T c_t(\psi_t - \psi_b) dt + \int_0^t \theta_t^c(\psi_s) ds + \alpha \int_t^T \theta_t(\psi_s) - \gamma^* \int_t^T \Sigma_s \mathbf{1}\, ds.$$

Then, the profit function of the i-th company reads

$$\begin{aligned} \mathcal{J}^i(\psi^i, \psi^{-i}) = \int_0^T e^{-\rho t} \Big(d &+ \int_0^T c_s^i(\psi_s^i - \psi_b^i) ds + \int_0^t \theta_s^{c,i}(\psi_s) ds \\ &+ \alpha \int_t^T \theta_s^i(\psi_s) ds - \gamma^* \int_t^T [\Sigma_s \mathbf{1}]_i ds \Big) dt, \end{aligned}$$

where $[\Sigma_s \mathbf{1}]_i$ is the i-th coordinate of the vector $\Sigma_s \mathbf{1}$.

Maximizing $\mathcal{J}^i(\psi^i, \psi^{-i})$ over ψ^i is equivalent to maximizing

$$\tilde{\mathcal{J}}^i(\psi^i, \psi^{-i}) = \int_0^T e^{-\rho t} \left(\int_0^T c_s^i \psi_s^i ds + \int_0^t \theta_s^{c,i}(\psi_s) ds + \alpha \int_t^T \theta_s^i(\psi_s) ds \right) dt.$$

Applying integration by parts to the integral with respect to 'dt' we have

$$\tilde{\mathcal{J}}^i(\psi^i, \psi^{-i}) = \int_0^T \left(\frac{e^{-\rho t} - e^{-\rho T}}{\rho} \theta_t^{c,i}(\psi_t) + \alpha \frac{1 - e^{-\rho t}}{\rho} \theta_t^i(\psi_t) + c_t^i \frac{1 - e^{-\rho T}}{\rho} \psi_t^i \right) dt. \quad (7.34)$$

The problem reduces to maximizing the integrand above along the entire trajectory of $(\psi_t^i)_{t \in [0,T]}$. That is

$$\max_{\psi_t^i} \left(\frac{e^{-\rho t} - e^{-\rho T}}{\rho} \theta_t^{c,i}(\psi_t) + \alpha \frac{1 - e^{-\rho t}}{\rho} \theta_t^i(\psi_t) + c_t^i \frac{1 - e^{-\rho T}}{\rho} \psi_t^i \right),$$

and the claim follows (see (7.12)).

Proof of Proposition 4

Recall the optimization problem (7.12). Let us denote by F_i the function that the i-th company needs to maximize:

$$F_i(\psi_t^i, \psi_t^{-i}) := \beta^c(t) \theta_t^{c,i}(\psi_t) + \alpha \beta(t) \theta_t^i(\psi_t) + c_t^i \psi_t^i$$

Since $\psi \mapsto F_i(\psi, \varphi)$ is concave for each φ, it is enough to impose first order conditions:

$$\beta^c(t) \partial_{\psi^i} \theta_t^{c,i}(\psi_t) + \alpha \beta(t) \partial_{\psi^i} \theta_t^i(\psi_t) + c_t^i = 0$$

for all $i = 1, 2, \ldots n$. This leads to

$$\psi_t^{*,\varepsilon,i} = \psi_t^{*,i} - \frac{\varepsilon}{2n} \left(\sum_{j=1}^n \psi_t^{*,\varepsilon,j} + \psi_t^{*,\varepsilon,i} \right), \quad (7.35)$$

where $\psi^{*,i} := c_t^i [\beta^c(t) \kappa^c + \alpha \beta(t) \kappa]^{-1}$ is the solution for $\varepsilon = 0$ (i.e., without interaction).

Taking sums over $i = 1, 2, \ldots n$ on both sides of the equation, we have

$$\sum_{i=1}^n \psi_t^{*,\varepsilon,i} = \sum_{i=1}^n \psi_t^{*,i} - \frac{\varepsilon}{2} \sum_{i=1}^n \psi_t^{*,\varepsilon,i} - \frac{\varepsilon}{2n} \sum_{i=1}^n \psi_t^{*,\varepsilon,i},$$

which gives,

$$\sum_{i=1}^n \psi_t^{*,\varepsilon,i} = \frac{2n}{2n + (n+1)\varepsilon} \sum_{i=1}^n \psi_t^{*,i}.$$

Substituting $\sum_{i=1}^n \psi_t^{*,\varepsilon,i}$ back into Equation (7.35), we get

$$\psi_t^{*,\varepsilon,i} = \psi_t^{*,i} - \frac{\varepsilon}{2n + \varepsilon} \left(\psi_t^{*,i} + \frac{2n^2}{2n + (n+1)\varepsilon} \bar{\psi}_t^* \right),$$

where $\bar{\psi}_t^* = \frac{1}{n} \sum_{j=1}^n \psi_t^{*,j}$. The term in brackets in the expression above is positive, hence proving (i).

In addition, it is clear that

$$\frac{\varepsilon}{2n + \varepsilon} \left(\psi_t^{*,i} + \frac{2n^2}{2n + (n+1)\varepsilon} \bar{\psi}_t^* \right) \underset{n \to +\infty}{\sim} \frac{\varepsilon}{\varepsilon + 2} \bar{\psi}_t^*, \quad (7.36)$$

which proves (ii).

Proof of Proposition 5 and 6

The standard approach to the problem, via dynamic programming, requires us to introduce the value processes for the two agents:

$$V_t^r = \min_{N \in \mathcal{A}_{t,T}^r} \mathbb{E}_t^r \left[\exp\left(-\gamma^r W_T^r\right)\right], \qquad V_t^g = \min_{N \in \mathcal{A}_{t,T}^g} \mathbb{E}_t^g \left[\exp\left(-\gamma^g W_T^g\right)\right],$$

where, for $t \leq T$ and $j \in \{r, g\}$, we define

$$\mathcal{A}_{t,T}^j := \{(N_s^\lambda)_{t \leq s \leq T} : N^\lambda \text{ is } \mathbb{R}^n\text{-valued, } (\mathcal{F}_s)_{t \leq s \leq T}\text{-adapted and } \mathbb{P}^j\text{-square integrable}\}$$

and $\mathbb{P}^j$-square integrable means

$$\mathbb{E}^j \left[\int_0^T |N_t^\lambda|^2 dt\right] < +\infty.$$

Moreover, we assume that the equilibrium price has the following dynamics.

$$p_t = p_0 + \int_0^t \mu_s ds + \int_0^t \sigma_s dB_s + \int_0^t \theta_s(\psi_s) d\mathcal{N}_s^\lambda \tag{7.37}$$

under the probability $\mathbb{P}^g$ of the green investors and

$$p_t = p_0 + \int_0^t \mu_s ds + \int_0^t \sigma_s dB_s \tag{7.38}$$

under the probability $\mathbb{P}^r$ of the regular investors, where μ is deterministic and must be found in equilibrium. We shall show a posteriori that an equilibrium price process of this form can indeed be found.

Following a well-known ansatz we expect

$$V_t^r = \exp\left(-\gamma^r W_t^r + Q_t^r\right), \qquad V_t^g = \exp\left(-\gamma^g W_t^g + Q_t^g\right),$$

where Q^r and Q^g are absolutely continuous deterministic processes with

$$dQ_t^r = q_t^r dt \quad \text{and} \quad dQ_t^g = q_t^g dt.$$

Applying the Itô's formula for jump processes to V^g under the green investor measure yields

$$\begin{aligned}
dV_t^g &= V_{t-}^g \left(-\gamma^g \, dW_t^g + q_t^g \, dt + \frac{(\gamma^g)^2}{2} \, d[W^g]_t^c + \left(e^{-\gamma \, \Delta W_t^g} - 1 + \gamma \, \Delta W_t^g\right)\right) \\
&= V_{t-}^g \Big(-\gamma^g \, (N_t^{g,\lambda})^\top \, dp_t + q_t^g \, dt + \frac{(\gamma^g)^2}{2} \, (N_t^{g,\lambda})^\top \, d[p]_t^c \, N_t^{g,\lambda} \\
&\qquad + \left(e^{-\gamma \, (N_{t-}^{g,\lambda})^\top \Delta p_t} - 1 + \gamma^g \, (N_{t-}^{g,\lambda})^\top \Delta p_t\right) \Big) \\
&= V_{t-}^g \Big(-\gamma^g \, (N_t^{g,\lambda})^\top \mu_t + q_t^g + \frac{(\gamma^g)^2}{2} \, (N_t^{g,\lambda})^\top \sigma_t \sigma_t^\top \, N_t^{g,\lambda} \\
&\qquad + \lambda \left(e^{-\frac{\gamma^g}{\lambda} \, (N_t^{r,\lambda})^\top \, \theta_t(\psi_t)} - 1\right) \Big) dt + M_t.
\end{aligned}$$

where (M_t) is a $\mathbb{P}^g$-martingale on $[0,T]$ and $[W^g]^c$ is the continuous part of the quadratic variation of the process W^g. Since V^g must be a martingale along the trajectory of the optimal process $(N_t^{g,\lambda})$ and a submartingale along every trajectory, we conclude that the drift term in 'dV_t^g' must be non-negative and

$$\min_{N_t^\lambda}\left(-\gamma^g(N_t^\lambda)^\top\mu_t+q_t^g+\frac{(\gamma^g)^2}{2}(N_t^\lambda)^\top\Sigma_tN_t^\lambda+\lambda(e^{-\frac{\gamma^g}{\lambda}(N_t^\lambda)^\top\theta_t(\psi_t)}-1)\right)=0 \tag{7.39}$$

for each $t\in[0,T]$. Since Σ is nondegenerate, the function to be minimized is strictly convex and coercive (i.e., it tends to $+\infty$ as $\|N_t^\lambda\|\to\infty$), thus the unique minimum is always attained. With a slight abuse of notation, we denote the minimizer of (7.39) (which does not depend on q_t) by $N_t^{g,\lambda}$, as this will be the number of assets held by the green investors. By imposing first order conditions we have that $N_t^{g,\lambda}$ must be the unique solution of

$$\mu_t-\gamma^g\Sigma_tN_t^\lambda+e^{-\frac{\gamma^g}{\lambda}(N_t^\lambda)^\top\theta_t(\psi_t)}\theta_t(\psi_t)=0.$$

By the same logic, the regular investors use the measure $\mathbb{P}^r$ to compute the dynamic 'dV_t^r' and find the optimal quantity of assets. In particular, the optimal quantity $N_t^{r,\lambda}$ is the minimizer of

$$\min_{N_t^\lambda}\left(-\gamma^r(N_t^\lambda)^\top\mu_t+q_t^r+\frac{(\gamma^r)^2}{2}(N_t^\lambda)^\top\Sigma_tN_t^\lambda\right)=0.$$

Since the regular investors do not take into account the climate uncertainty, there is no jump term in this formula, and the minimizer is given explicitly by

$$N_t^{r,\lambda}=\frac{1}{\gamma^r}\Sigma_t^{-1}\mu_t.$$

The market clearing condition therefore allows to compute $(\mu,N^{r,\lambda},N^{g,\lambda})$ by solving the following system of equations:

$$\begin{aligned}&N_t^{r,\lambda}=\frac{1}{\gamma^r}\Sigma_t^{-1}\mu_t;\\&\mu_t-\gamma^g\Sigma_tN_t^{g,\lambda}+e^{-\frac{\gamma^g}{\lambda}(N_t^{g,\lambda})^\top\theta_t(\psi_t)}\theta_t(\psi_t)=0;\\&N_t^{r,\lambda}+N_t^{g,\lambda}=\mathbf{1}.\end{aligned} \tag{7.40}$$

Substituting μ_t from the second equation into the first one, allows to eliminate it, obtaining the following equation:

$$\gamma^r\Sigma_t(\mathbf{1}-N_t^{g,\lambda})-\gamma^g\Sigma_tN_t^{g,\lambda}+e^{-\frac{\gamma^g}{\lambda}(N_t^{g,\lambda})^\top\theta_t(\psi_t)}\theta_t(\psi_t)=0. \tag{7.41}$$

The left-hand side of this equation coincides with the gradient of the strictly convex, differentiable and coercive function

$$f(N):=-\gamma^r\mathbf{1}^\top\Sigma_tN+\frac{\gamma^r+\gamma^g}{2}N^\top\Sigma_tN+\frac{\lambda}{\gamma^g}e^{-\frac{\gamma^g}{\lambda}N^\top\theta_t(\psi_t)},$$

which proves existence and uniqueness of the solution of (7.41). Let us write $y_\lambda=e^{-\frac{\gamma^g}{\lambda}(N_t^{g,\lambda})^\top\theta_t(\psi_t)}$ in (7.41) and solve for $N_t^{g,\lambda}$ to obtain the explicit expression

$$N_t^{g,\lambda}=\alpha\left(\mathbf{1}+\frac{y_\lambda}{\gamma^r}\Sigma_t^{-1}\theta_t(\psi_t)\right).$$

Plugging this back into (7.41) we find y_λ by solving the one-dimensional equation

$$y_\lambda = e^{-\frac{\alpha\gamma^g}{\lambda}\{\mathbf{1}^\top\theta_t(\psi_t)+\frac{y_\lambda}{\gamma^r}\theta_t(\psi_t)^\top\Sigma_t^{-1}\theta_t(\psi_t)\}}.$$

Finally the expression for p_0 is obtained from the condition $p_T = D_T$.

First we show existence and uniqueness of the solution y_λ to Eq. (7.20). Writing

$$g(y) := \alpha(\mathbf{1}^\top\theta_t(\psi_t) + \frac{y}{\gamma^r}\theta_t(\psi_t)^\top\Sigma_t^{-1}\theta_t(\psi_t)),$$

Equation (7.20) becomes

$$-\lambda\log y = \gamma^g g(y), \tag{7.42}$$

where it is also worth noticing that $g(1) = \theta_t(\psi_t)^\top N_t^g$. Since g is increasing and linear and the function $y \mapsto -\lambda\log y$ is strictly decreasing, continuous and maps $(0,\infty)$ onto $\mathbb{R}$, there exists one and only one positive solution of (7.20).

Next we show the monotonicity of the map $\lambda \mapsto y_\lambda$ stated in (ii)–(iv). If $g(1) > 0$, the solution belongs to the interval $(0,1)$, and it is easy to see that it is monotonically increasing in λ. If $g(1) = 1$, the solution is constant and equal to 1. If $g(1) < 0$, the solution satisfies $y_\lambda > 1$ and it is easy to see that it is decreasing in λ.

To study the asymptotic behavior of y_λ as $\lambda \to \infty$, we expand the expression on the right hand side of (7.20) using Taylor up to the first order in λ^{-1}. That gives

$$y_\lambda = 1 - \frac{\alpha\gamma^g}{\lambda}\left(\mathbf{1}^\top\theta_t(\psi_t) + \frac{y_\lambda}{\gamma^r}\theta_t(\psi_t)^\top\Sigma_t^{-1}\theta_t(\psi_t)\right) + O(\lambda^{-2}).$$

Then we substitute $y_\lambda = y_0 + y_1\lambda^{-1}$ on both sides of the expression above and, equating terms of the same order in λ^{-1}, we find $y_0 = 1$ and

$$y_1 = -\alpha\gamma^g\left(\mathbf{1}^\top\theta_t(\psi_t) + \frac{1}{\gamma^r}\theta_t(\psi_t)^\top\Sigma_t^{-1}\theta_t(\psi_t)\right).$$

Since $y_\lambda = y_0 + y_1\lambda^{-1}$ it is now immediate to obtain $\lim_{\lambda\to\infty} y_\lambda = 1$ in all cases.

It remains to prove (i). Since ψ_t is fixed, for simplicity we omit it from some of the formulae below. A straightforward computation yields the following expression for $(N_t^{g,\lambda})^\top\Sigma_t N_t^{g,\lambda}$:

$$(N_t^{g,\lambda})^\top\Sigma_t N_t^{g,\lambda} = \|\Sigma_t^{1/2}N_t^{g,\lambda}\|^2 = \alpha^2\|N_t^g\|^2 + \frac{2\alpha^2}{\gamma^r}(y_\lambda - 1)\theta_t^\top\mathbf{1} + \frac{\alpha^2}{(\gamma^r)^2}(y_\lambda^2 - 1)\theta_t^\top\Sigma_t^{-1}\theta_t.$$

Let

$$f(y) := \frac{2}{\gamma^r}(y-1)\theta_t^\top\mathbf{1} + \frac{1}{(\gamma^r)^2}(y^2-1)\theta_t^\top\Sigma_t^{-1}\theta_t$$

and notice that f is a quadratic function with $f'(y) = 2/(\gamma^r\alpha)g(y)$. From the arguments above, if $g(1) > 0$ we have $\lambda \mapsto y_\lambda \in (0,1)$ and increasing so that by Equation (7.42), $g(y^\lambda) > 0$, and therefore $\lambda \mapsto f(y^\lambda)$ is increasing as well. If, on the other hand $g(1) < 0$, then $\lambda \mapsto y_\lambda \in (1,\infty)$ and decreasing so that by Equation (7.42), $g(y^\lambda) < 0$, and therefore $\lambda \mapsto f(y^\lambda)$ is increasing once again. The claim holds trivially in the remaining case $g(1) = 0$.

Proof of Propositions 7 and 8

Proof of Proposition 7. For each $y > 0$ the derivation of Eq. (7.22) follows immediately from (7.21), as in Corollary 1. Next we prove solvability of Eq. (7.23) and some properties of its solution.

Proof of Proposition 8. Writing

$$g^*(y) := \mathbf{1}^\top \theta_t(\psi_t(y)) + \frac{\alpha y}{\gamma^r}\theta_t(\psi_t(y))^\top \Sigma_t^{-1}\theta_t(\psi_t(y)),$$

Equation (7.23) becomes

$$-\lambda \log y = \alpha\gamma^g g^*(y). \tag{7.43}$$

Unlike $g(y)$, used in the proof of Proposition 6, $g^*(y)$ is a nonlinear and possibly non-monotonic function of y, which is more difficult to study. Nevertheless, it is clear that $g^*(0) < \infty$ and $g^*(y) \sim \frac{\alpha\kappa_0^2}{\gamma_r}\mathbf{1}^\top\Sigma_t^{-1}\mathbf{1}\, y$ as $y \to \infty$, which implies that Equation (7.23) admits a solution on $(0, \infty)$. Then (i) holds.

To prove uniqueness (that is (ii)) it is sufficient to show that (7.24) implies that g^* is monotonic increasing. It is immediate to check

$$\frac{d}{dy}\mathbf{1}^\top\theta_t(\psi_t(y)) = \sum_{i=1}^n \left(\frac{d}{d\psi}\theta_t^i(\psi_t^i(y))\right)\left(\frac{d}{dy}\psi_t^i(y)\right),$$

and

$$\begin{aligned}\frac{d}{dy}\left(\frac{\alpha y}{\gamma^r}\,\theta_t(\psi_t(y))^\top\Sigma_t^{-1}\theta_t(\psi_t(y))\right) &= \frac{\alpha}{\gamma^r}\,\theta_t(\psi_t(y))^\top\Sigma_t^{-1}\theta_t(\psi_t(y)) \\ &\quad + \frac{\alpha y}{\gamma^r}\frac{d}{dy}\theta_t(\psi_t(y))^\top\Sigma_t^{-1}\theta_t(\psi_t(y)) \\ &\geq \frac{\alpha y}{\gamma^r}\frac{d}{dy}\theta_t(\psi_t(y))^\top\Sigma_t^{-1}\theta_t(\psi_t(y)).\end{aligned}$$

The latter expression can be expanded as

$$\frac{d}{dy}\theta_t(\psi_t(y))^\top\Sigma_t^{-1}\theta_t(\psi_t(y)) = \sum_{i,j=1}^n \left(\Sigma_t^{-1}\right)_{ij}\theta_t^j(\psi_t^j(y))\left(\frac{d}{d\psi}\theta_t^i(\psi_t^i(y))\right)\left(\frac{d}{dy}\psi_t^i(y)\right).$$

Combining the calculations above we obtain

$$\frac{d}{dy}g^*(y) \geq \sum_{i=1}^n\left(1 + \frac{\alpha}{\gamma^r}y\sum_{j=1}^n\left(\Sigma_t^{-1}\right)_{ij}\theta_t^j(\psi_t^j(y))\right)\left(\frac{d}{d\psi}\theta_t^i(\psi_t^i(y))\right)\left(\frac{d}{dy}\psi_t^i(y)\right).$$

Recalling the expressions for $\psi_t(y)$ and $\theta_t(\psi)$ (see (7.22) and Corollary 1, respectively), we immediately see that

$$\left(\frac{d}{d\psi}\theta_t^i(\psi_t^i(y))\right)\left(\frac{d}{dy}\psi_t^i(y)\right) \geq 0, \quad \text{for all indexes } i.$$

Then the condition (7.24) is sufficient to guarantee that g^* is monotonic increasing and the solution to (7.43) is unique.

All the statements in (iii) are shown using the same arguments of proof as in Proposition 6, using that g^* is increasing and that $\psi_t^*(1) = \psi_t^{*,0}(1)$.

Appendix B: Calibration

We choose the rate of time preference, ρ, equal to 0.01 (Gollier, 2002; Gollier and Weitzman, 2010). We estimate the share of assets managed taking into account climate criteria, α, at 25% (US SIF, 2018).

We estimate κ and κ_0 by using the estimates in Zerbib (2021) of the externality premium of the electrical equipment (-1.11%) and the coal (+0.12%) industries in the U.S. between 2013 and 2019: knowing that this premium is equal to $-\alpha\theta(\psi)$ in the present paper, with $\theta(\psi) = \kappa_0 - \frac{\kappa}{2}\psi^2$, and that the average carbon intensity of the electrical equipment and coal industries are $\psi^{\text{elec}} = 147$ tCO_2e/USDmn and $\psi^{\text{coal}} = 555$ tCO_2e/USDmn, respectively, we get $\kappa = 3 \times 10^{-7}$ and $\kappa_0 = 0.047$. Considering companies' responsiveness to internalize climate externalities following market pressure, we choose the companies' climate sensitivity slightly lower than that of the average investor sensitivity ($\alpha\kappa = 0.25\kappa$): $\kappa_c = 0.2\kappa = 5.5 \times 10^{-8}$.

We calibrate the marginal abatement costs for an electrical equipment company, which has an average carbon intensity of 147 tCO_2/USDmn, and a coal company, which has an average carbon intensity of 555 tCO_2/USDmn. To do so, we use the equilibrium equation of the emission schedule without climate uncertainty (Equation (7.13)) and we assume that the initial emissions of the companies ($\psi_b^{\text{elec}} = 147$ tCO_2e/USDmn and $\psi_b^{\text{coal}}=555$ tCO_2e/USDmn) are adjusted to the optimal level: $\psi_b = \psi_0^*$. Therefore, $c^{\text{elec}} = 8 \times 10^{-6}$ and $c^{\text{coal}} = 3 \times 10^{-5}$.

Finally, to calibrate the parameters needed for the simulations of the model with climate uncertainty, we assume that the correlation between the two assets is 50% and we take regular investors' absolute risk aversion, γ_r, equal to 0.1 (Barberis et al. (2015)). Thus, as $\alpha = \frac{\gamma_r}{(\gamma_r+\gamma_g)} = 0.25$, $\gamma_g = 0.3$. In the simulations, we consider different levels of intensities, λ, and find y_λ^* as the solution of the fixed point Equation (7.23).

Table 7.2 summarizes the calibrated parameters.

TABLE 7.2: **Calibrated parameters.** This table gives the values of the parameters calibrated based on the estimates in this section and used for the simulations presented in Figures 7.2, 7.3, 7.4, and 7.5

Parameter	*Value*
α	0.25
ρ	0.01
κ	3×10^{-7}
κ_0	0.047
κ^c	5.5×10^{-8}
ψ_b^{elec}	147
ψ_b^{coal}	555
c^{elec}	8×10^{-6}
c^{coal}	3×10^{-5}
$Cor(\text{elec}, \text{coal})$	0.5
γ_r	0.1
γ_g	0.3

References

Aggarwal, S., R. Dizon-Ross, and A. D. Zucker, 2020, Incentivizing Behavioral Change: The Role of Time Preferences, *NBER Working Paper*.

Alley, R. B., J. Marotzke, W. D. Nordhaus, J. T. Overpeck, D. M. Peteet, R. A. Pielke Jr., R. T. Pierrehumbert, P. B. Rhines, L. D. Stocker, T. F. Talley, and J. M. Wallace, 2003, Abrupt Climate Change, *Science* 299, 2005–2010.

Amir, R., M. Germain, and V. Van Steenberghe, 2008, On the Impact of Innovation on the Marginal Abatement Cost Curve, *Journal of Public Economic Theory* 10, 985–1010.

Ardia, D., K. Bluteau, K. Boudt, and K. Inghelbrecht, 2021, Climate change concerns and the performance of green versus brown stocks, *SSRN Working Paper*.

Atmaz, A., and S. Basak, 2018, Belief Dispersion in the Stock Market, *Journal of Finance* 73, 1225–1279.

Avramov, D., S. Cheng, A. Lioui, and A. Tarelli, 2021a, Sustainable Investing with ESG rating uncertainty, *Journal of Financial Economics,* Forthcoming.

Avramov, D., A. Lioui, Y. Liu, and A. Tarelli, 2021b, Dynamic ESG Equilibrium, *Working Paper* SSRN.

Baker, M., D. Bergstresser, G. Serafeim, and J. Wurgler, 2018, Financing the Response to Climate Change: The Pricing and Ownership of U.S. Green Bonds , *Working Paper* NBER.

Bank, W., 2020, State and trends of carbon pricing 2020, Technical report, World Bank, Technical Report.

Barber, B. M., A. Morse, and A. Yasuda, 2021, Impact investing, *Journal of Financial Economics* 139, 162–185.

Barberis, N., and A. Shleifer, 2003, Style Investing, *Journal of Financial Economics* 68, 161–199.

Barberis, N., R. Greenwood, L. Jin, and A. Shleifer, 2015, X-CAPM: An extrapolative capital asset pricing model, *Journal of Financial Economics* 115, 1–24.

Barberis, N., R. Greenwood, L. Jin, and A. Shleifer, 2018, Extrapolation and bubbles, *Journal of Financial Economics* 129, 203–227.

Barnett, M., W. Brock, and L. P. Hansen, 2020, Pricing Uncertainty Induced by Climate Change, *Review of Financial Studies* 33, 1024–1066.

Bauer, R., K. Koedijk, and R. Otten, 2005, International evidence on ethical mutual fund performance and investment style, *Journal of Banking and Finance* 29, 1751–1767.

Bauman, Y., M. Lee, and K. Seeley, 2008, Does Technological Innovation Really Reduce Marginal Abatement Costs? Some Theory, Algebraic Evidence, and Policy Implications, *Environmental and Resource Economics* 40, 507–527.

Berk, J. B., and J. H. van Binsbergen, 2021, The Impact of Impact Investing, *Working Paper* SSRN.

Bolton, P., and M. T. Kacperczyk, 2021, Do Investors Care about Carbon Risk?, *Journal of Financial Economics,* 142, 517–549.

Bolton, P., J. Scheinkman, and W. Xiong, 2006, Executive Compensation and Short-Termist Behaviour in Speculative Markets, *Review of Economics Studies* 73, 577–610.

Brammer, S., C. Brooks, and S. Pavelin, 2006, Corporate Social Performance and Stock Returns: UK Evidence from Disaggregate Measures, *Financial Management* 35, 97–116.

Broccardo, E., O. Hart, and L. Zingales, 2020, Exit vs. Voice, *NBER Working Paper*.

Burke, M., W. M. Davis, and N. S. Diffenbaugh, 2018, Large Potential Reduction in Economic Damages Under UN Mitigation Targets, *Nature* 557, 549–553.

Burke, M., S. M. Hsiang, and E. M. Miguel, 2015, Global Non-linear Effect of Temperature on Economic Production, *Nature* 527, 235–239.

Cai, Y., K. L. Judd, T. M. Lenton, T. S. Lontzek, and D. Narita, 2015, Environmental tipping points significantly affect the cost-benefit assessment of climate policies, *Proceedings of the National Academy of Sciences of the United States of America* 112, 4606–4611.

Chava, S., 2014, Environmental externalities and cost of capital, *Management Science* 60, 2223–2247.

Chowdhry, B., S. W. Davies, and B. Waters, 2018, Investing for Impact, *Review of Financial Studies* 32, 864–904.

Commission, E., 2018, Action plan: Financing sustainable growth, Technical report, European Commission, Report.

Derwall, J., N. Guenster, R. Bauer, and K. Koedijk, 2005, The eco-efficency premium puzzle, *Financial Analysts Journal* 61, 51–63.

Dietz, S., 2011, High impact, low probability? An empirical analysis of risk in the economics of climate change, *Climatic Change* 108, 519–541.

Dietz, S., and N. Stern, 2015, Endogenous growth, convexity of damage and climate risk: How Nordhaus' framework supports deep cuts in carbon emissions, *The Economic Journal* 125, 574–620.

Dimson, E., O. Karakaş, and X. Li, 2015, Active ownership, *Review of Financial Studies* 28, 3225–3268.

Eccles, R. G., I. Ioannou, and G. Serafeim, 2014, The impact of corporate sustainability on organizational processes and performance, *Management Science* 60, 2835–2857.

Edmans, A., 2011, Does the stock market fully value intangibles? Employee satisfaction and equity prices, *Journal of Financial Economics* 101, 621–640.

Edmans, A., X. Gabaix, T. Sadzik, and Y. Sannikov, 2012, Dynamic CEO compensation, *Journal of Finance* 67, 1603–1648.

ElGhoul, S., O. Guedhami, C. C. Kowk, and D. R. Mishra, 2011, Does corporate social responsibility affect the cost of capital?, *Journal of Banking and Finance* 35, 2388–2406.

Galema, R., A. Plantinga, and B. Scholtens, 2008, The stocks at stake: Return and risk in socially responsible investment, *Journal of Banking and Finance* 32, 2646–2654.

Global Sustainable Investment Alliance, 2018, 2018 Global Sustainable Investment Review, Technical report, Global Sustainable Investment Alliance.

Gollier, C., 2002, Discounting and uncertain future, *Journal of Public Economics* 85, 149–166.

Gollier, C., and M. Weitzman, 2010, How should the distant future be discounted when discount rates are uncertain?, *Economic Letters* 107, 350–353.

Green, D., and B. Roth, 2020, The allocation of socially responsible capital, *Working Paper* SSRN.

Hartzmark, S., and A. Sussman, 2020, Do Investors Value Sustainability? A Natural Experiment Examining Ranking and Fund Flows, *Journal of Finance* 74, 2789–2837.

Heeb, F., and J. Kölbel, 2020, The investor's guide to impact, Technical report, University of Zurich, Center for Sustainable Finance & Private Wealth.

Heinkel, R., A. Kraus, and J. Zechner, 2001, The effect of green investment on corporate behaviour, *Journal of Financial and Quantitative Analysis* 36, 377–389.

Hong, H., and M. Kacperczyk, 2009, The price of sin: the effects of social norms on markets, *Journal of Financial Economics* 93, 15–36.

Hong, H., and J. C. Stein, 1999, A Unified Theory of Underreaction, Momentum Trading, and Overreaction in Asset Markets, *Journal of Finance* 54, 2143–2184.

Hsu, P.-H., K. Li, and C.-Y. Tsou, 2019, The Pollution Premium, *Working Paper,* SSRN.

Jaffe, A., R. Newell, and R. Stavins, 2002, Environmental Policy and Technological Change, *Environmental and Resource Economics* 22, 41–70.

Jakob, M., and J. Hilaire, 2015, Unburnable fossil-fuel reserves, *Nature* 517, 150–152.

Krüger, P., 2015, Corporate goodness and shareholder wealth, *Journal of Financial Economics* 115, 304–329.

Krüger, P., Z. Sautner, and L. Starks, 2020, The Importance of Climate Risk for Institutional Investors, *Review of Financial Studies,* 33, 1067–1111.

Lambrecht, B. M., and S. C. Myers, 2017, The Dynamics of Investment, Payout and Debt, *Review of Financial Studies* 30, 3759–3800.

Landier, A., and S. Lovo, 2020, ESG Investing: How to Optimize Impact?, *Working Paper,* SSRN.

Larcker, D. F., and B. Tayan, 2019, CEO Compensation: Data Spotlight, *Stanford GSB Corporate Governance Research Initiative*.

Lontzek, T. S., Y. Cai, K. L. Judd, and T. M. Lenton, 2015, Stochastic integrated assessment of climate tipping points indicates the need for strict climate policy, *Nature Climate Change* 5, 441–444.

Marinovic, I., and F. Varas, 2019, CEO Horizon, Optimal Pay Duration, and the Escalation of Short-Termism, *Journal of Finance* 74, 2011–2053.

Mekaroonreung, M., and A. L. Johnson, 2014, A nonparametric method to estimate a technical change effect on marginal abatement costs of U.S. coal power plants, *Energy Economics* 46, 45–55.

Milliman, S. R., and R. Prince, 1989, Firm incentives to promote technological change in pollution control, *Journal of Environmental Economics and Management* 17, 247–265.

Nordhaus, W., 2014, Estimates of the social cost of carbon: Concepts and results from the dice-2013r model and alternative approaches, *Journal of the Association of Environmental and Resource Economists* 1, 273–312.

Oehmke, M., and M. Opp, 2019, A Theory of Socially Responsible Investment, *Working Paper,* SSRN.

on Sustainable Finance, E. U. H. L. E. G., 2018, Final report, Technical report, European Union.

Osambela, E., 2015, Differences of opinion, endogenous liquidity, and asset prices, *Review of Financial Studies* 28, 1914–1959.

Palmer, K., W. E. Oates, and P. R. Portney, 1995, Tightening environmental standards: The benefit-cost or the no-cost paradigm?, *Journal of Economic Perspectives* 9, 119–132.

Pastor, L., R. Stalbaugh, and L. Taylor, 2021, Dissecting Green Returns, *SSRN Working Paper*.

Pastor, L., R. F. Stambaugh, and L. A. Taylor, 2021, Sustainable investing in equilibrium, *Journal of Financial Economics,* 142, 550–571.

Pedersen, L. H., S. Fitzgibbons, and L. Pomorski, 2021, Responsible investing: The ESG-efficient frontier, *Journal of Financial Economics,* 142, 572–597.

Porter, M. E., and C. van der Linde, 1995, Toward a new conception of the environment-competitiveness relationship, *Journal of Economic Perspectives* 9, 97–118.

Raghunandan, A., and S. Rajgopal, 2021, Do ESG Funds Make Stakeholder-Friendly Investments?, *Working Paper* SSRN.

Renneboog, L., J. Ter Horst, and C. Zhang, 2008, The price of ethics and stakeholder governance: The performance of socially responsible mutual funds, *Journal of Corporate Finance* 14, 302–322.

Riedl, A., and P. Smeets, 2017, Why do investors hold socially responsible mutual funds?, *Journal of Finance* 72, 2505–2550.

Scheinkman, J. A., and W. Xiong, 2003, Overconfidence and speculative bubbles, *Journal of Political Economy* 111, 1183–1219.

Sharfman, M., and C. Fernando, 2008, Environmental risk management and the cost of capital, *Strategic Management Journal* 29, 569–592.

Statman, M., and D. Glushkov, 2009, The wages of social responsibility, *Financial Analysts Journal* 65, 33–46.

Statman, M., and D. Glushkov, 2016, Classifying and measuring the performance of socially responsible mutual funds, *Journal of Portfolio Management* 42, 140–151.

Sugathan, A., R. Bhangale, V. Kansal, and U. Hulke, 2018, How can Indian power plants cost-effectively meet the new sulfur emission standards? Policy evaluation using marginal abatement cost-curves, *Energy Policy* 121, 124–137.

Trinks, A., B. Scholtens, M. Mulder, and L. Dam, 2018, Fossil fuel divestment and portfolio performance, *Ecological Economics* 146.

US SIF, 2018, Report on US Sustainable, Responsible and Impact Investing Trends, Technical report, The Forum for Sustainable and Responsible Investment.

van Binsbergen, J. H., and C. C. Opp, 2019, Real anomalies, *Journal of Finance* 74, 1659–1706.

Varas, F., 2018, Managerial short-termism, turnover policy, and the dynamics of incentives, *Review of Financial Studies* 31, 3409–3451.

Veronesi, P., 1999, Stock market overreaction to bad news in good times: A rational expectations equilibrium model, *Review of Financial Studies* 12, 975–1007.

Weitzman, M., 2009, On modeling and interpreting the economics of catastrophic climate change, *Review of Economic Studies* 91, 1–19.

Weitzman, M., 2011, Fat-tailed uncertainty in the economics of catastrophic climate change, *Review of Environmental Economics and Policy* 5, 275–292.

Xian, Y., K. Wang, Y.-M. Wei, and Z. Huang, 2020, Opportunity and marginal abatement cost savings from China's pilot carbon emissions permit trading system: Simulating evidence from the industrial sectors, *Journal of Environmental Management* 271, 110975.

Yang, R., 2021, What Do We Learn From Ratings About Corporate Social Responsibility (CSR)?, *Working Paper* SSRN.

Zerbib, O. D., 2019, The effect of pro-environmental preferences on bond prices: Evidence from green bonds, *Journal of Banking and Finance* 98, 39–60.

Zerbib, O. D., 2021, A Sustainable Capital Asset Pricing Model (S-CAPM): Evidence from green investing and sin stock exclusion, *Working Paper,* SSRN.

Chapter 8

Quantifying the impact of impact investing

Andrew W. Lo
MIT Sloan School of Management, USA

Ruixun Zhang
Peking University, China

We propose a quantitative framework for assessing the financial impact of any form of impact investing, including socially responsible investing (SRI), environmental, social, and governance (ESG) objectives, and other non-financial investment criteria. We derive conditions under which impact investing detracts from, improves on, or is neutral to the performance of traditional mean-variance optimal portfolios, which depends on whether the correlations between the impact factor and unobserved excess returns are negative, positive, or zero, respectively. Using Treynor-Black portfolios to maximize the risk-adjusted returns of impact portfolios, we derive an explicit and easily computable measure of the financial reward or cost of impact investing as compared to passive index benchmarks. We illustrate our approach with applications to biotech venture philanthropy, a semiconductor R&D consortium, divesting from "sin" stocks, ESG investments, and "meme" stock rallies such as GameStop in 2021.

This chapter is adapted from an article originally published in *Management Science* (Lo and Zhang, 2024). Please cite this work as follows:
Andrew W. Lo and Ruixun Zhang, 2024. Quantifying the Impact of Impact Investing. Management Science 70(10), 7161-7186.

8.1 Introduction

Impact investing, broadly defined as investment toward goals that support specific social priorities and agendas in addition to financial objectives, has drawn an increasing amount of attention in recent years. This concept was originally introduced through populist efforts to effect social change by encouraging institutional investors to divest from companies engaged in activities viewed by critics as unethical, immoral, or otherwise objectionable, e.g., the exploitation of child labor; the tacit support of apartheid or religious persecution; or the gambling, pornography, alcohol, tobacco, and firearms businesses, collectively known as "sin stocks". However, its scope has expanded significantly to include investment products that

DOI: 10.1201/9781032636252-8

use environmental, social, and governance (ESG) criteria, "green" bonds, and private equity funds seeking social impact alongside financial returns.[1]

The growth in popularity and assets under management of impact investing has also triggered a backlash. For example, on 4 August 2022, a letter signed by the attorneys general of 19 states was sent to the multinational asset management firm BlackRock expressing concern over the asset manager's ESG policies and how they might influence their holdings of fossil-fuel energy companies:[2] "BlackRock's actions on a variety of governance objectives may violate multiple state laws. Mr. McCombe's letter asserts compliance with our fiduciary laws because BlackRock has a private motivation that differs from its public commitments and statements. This is likely insufficient to satisfy state laws requiring a sole focus on financial return." These concerns are not minor ones, given that the legal penalty for violating one's fiduciary duty involves personally making up any losses suffered by the client and restoring to the client any profits made by the fiduciary's service provision to said client.[3]

How is impact investing to be reconciled with fiduciary duty? The answer lies in developing a framework in which the financial impact of impact investing can be measured and disclosed, which is the subject of this chapter.

Conventional wisdom typically views impact investing as a standard portfolio selection problem with additional constraints related to the degree of social impact of the underlying securities, in turn implying a non-superior risk/reward profile compared to the unconstrained case. Given that the constrained portfolio contains a proper subset of securities of the unconstrained version, mathematical logic suggests that the constrained optimum is, at best, equal to the unconstrained optimum or, more likely, inferior.

However, the non-superiority of constrained optima relies on a key assumption that is almost never explicitly stated: the constraint is assumed to be statistically independent of the returns of the underlying securities. In some cases, such an assumption is warranted. For example, imagine constructing a subset of securities with CUSIP identifiers that contain prime numbers. Clearly, such a constraint has no relation to the returns of any security; hence imposing such a constraint will only reduce the risk-adjusted return of the optimized portfolio.

But what if the constraint is *not* independent of the returns? For example, consider the constraint "invest only in those companies for which their stock prices will appreciate by more than 10% over the next 12 months." Apart from the infeasibility of imposing such a condition, it should be obvious that this constraint would, in fact, increase the risk-adjusted return of the optimized portfolio. Therefore, the answer to the question of what is the impact of impact investing rests entirely on whether and how the impact criteria are related to the performance characteristics of the securities being considered.

In this chapter, we develop a general framework to quantify the financial impact of impact investing. We formalize impact investing as the sorting and selection of an investment universe of N securities based on an *impact factor* X_i, for security i, so that higher values of X_i correspond to greater impact, e.g., lower carbon emissions, greater sustainability, higher ESG score, etc. As a result, other things being equal, impact investors are assumed to prefer securities with higher values of X_i. This impact factor defines a rank ordering for all securities in the universe from which an impact portfolio can be constructed, e.g., the top decile of ESG-ranked securities or the bottom decile of securities ranked by carbon emissions. Therefore, the impact on investment performance is determined by the joint

[1] The term "impact investing" has been narrowly used to refer specifically to these investments (Barber, Morse, and Yasuda, 2021), but we use the term more broadly in this chapter.

[2] `https://www.texasattorneygeneral.gov/sites/default/files/images/executive-management/BlackRock\%20Letter.pdf`, accessed 15 December 2022.

[3] See 9 U.S. Code §1109 - Liability for breach of fiduciary duty.

distribution of the vector $\mathbf{X} \equiv [X_1 \ X_2 \ \cdots \ X_N]^T$ of impact measures with the investment performance of individual securities.

To formalize this idea, we first propose a general linear multi-factor model for asset returns, and define the excess returns or "alpha" as non-zero intercepts that we model as mean-zero random variables. This framework allows for the possibility of superior investment performance for individual securities, but also includes the conventional case of equilibrium or no-arbitrage pricing if we set the variance of the alphas to zero. In fact, the implications of our model are applicable to a broader equilibrium asset-pricing framework where "alpha" is reinterpreted as any omitted factors of which investors are either unaware or unable to acquire information as easily as professional portfolio managers. Such an agnostic approach to investment performance allows us to determine the conditions under which impact investing does or does not change the risk/reward profile of a given investment product.

In particular, we derive the distribution of individual alphas that have been ranked according to their impact factors $\mathbf{X}$, both in finite samples and asymptotically, as the number of securities increases without bound. It is well known that ranked random variables, known as *order statistics*, have different distributions than their unranked versions. However, for our purposes, a more relevant strand of that literature focuses on *induced order statistics*, in which random variables are ranked not by their own values but by the values of other random variables, e.g., ranking the returns of a collection of mutual funds not by the returns of the funds but by their market betas. We use properties of induced order statistics to derive the distribution of an impact portfolio's alphas ranked by an arbitrary impact score $\mathbf{X}$, allowing us to quantify the impact of impact investing.

Using this framework, we show that the expected alpha from the induced ordering is determined by three terms: the correlation between $\mathbf{X}$ and the individual securities' alphas, the standard deviation of individual securities' alphas, and a cross-sectionally standardized impact score that captures whether the impact factor of a security is above or below average. In addition, we provide an alternative characterization of the expected alpha from the induced ordering as a discounted version of the expected alpha from ordering securities based on alpha (i.e., via an all-knowing oracle, which in reality is of course unattainable, since alphas are unobservable). Much like the Sharpe-Lintner Capital Asset-Pricing Model (CAPM) (Sharpe, 1964; Lintner, 1965), which quantifies the expected return of individual securities through market beta,[4] this simple but profound result emphasizes the mechanism through which an impact factor's excess return is earned: it achieves a fraction of the maximum possible alpha with perfect knowledge, where the fraction is simply the correlation between $\mathbf{X}$ and the individual securities' alphas.

Using this insight, we quantify the alphas of portfolios formed using the impact factor, $\mathbf{X}$, and then apply the Treynor and Black (1973) framework to derive the optimal weights when forming both long/short and long-only portfolios to maximize Sharpe ratio, including for the common cases of creating portfolios from the top or bottom impact-factor quantiles. We show that such impact portfolios are associated with "super-efficient frontiers" as long as the impact factor, $\mathbf{X}$, is positively correlated with the unobserved alphas of the individual securities. We also provide an equilibrium/no-arbitrage interpretation of our results in which excess returns arise from omitted factors that investors may not be aware of, but to which impact portfolio managers have access. In this case, the excess returns are simply "excess" with respect to factors that investors observe, and represent risk premia from specific impact factors.

The use of the Treynor-Black portfolio allows us to construct a natural measure of the financial impact of impact investing: an impact factor has positive alpha when it is positively

[4]For standardized returns with unit variances, the market beta is simply the correlation between security returns and market returns.

correlated with the unobserved alphas of individual securities. Likewise, an impact factor can impose a cost—also quantifiable in our framework—when it is negatively correlated with alphas and investors divest of the bottom-ranked securities (which will have positive alphas on average due to their negative correlation with $\mathbf{X}$). This provides a possible explanation for the inconsistent and sometimes contradictory empirical findings on the effects of impacting investing. The correlation between the impact factor and alpha is affected by different measures of impact,[5] different market conditions,[6] and different asset-pricing models for alpha,[7] all of which influence the final estimate of the benefit or cost of impact investing.

To illustrate the practical relevance of our results, we apply our framework to five specific impact-investing contexts. The specific correlation for each form of impact investing, positive or negative, depends on the specific nature of the impact, the risks involved to achieve that impact, and its relationship with the underlying process of alpha generation.

Our first example is biotech venture philanthropy, a particular form of impact investing in biomedicine, in which nonprofit and mission-driven organizations fund initiatives to advance their objectives and potentially achieve returns that can be reinvested toward their mission. We take the case study by Kim and Lo (2019) about the Cystic Fibrosis Foundation to show that a significantly positive alpha can be achieved by advancing drug development for rare diseases, illustrating the feasibility of "doing well by doing good". In this case, the challenges associated with early stage drug development programs from the financial perspective—low probabilities of success, long time horizons, and large capital requirements as highlighted by (Fagnan et al., 2013)—are more than offset by a positive correlation, ρ.

The second application looks at the R&D consortium, another form of impact investing. We consider the example of Sematech, a high-profile R&D consortium formed in 1987 and funded by the U.S. government and 14 U.S. semiconductor firms, whose purpose was to solve common manufacturing problems by leveraging shared R&D resources. Using a simple difference-in-difference approach, we estimate that joining the consortium led to an increased alpha in the stocks of member firms of around 10%, implying significant returns from investments in the consortium. In this case, the R&D consortium reduced R&D duplication and increased profitability for member firms (Irwin and Klenow, 1996a), leading to a positive correlation between impact and returns.

The third application of our framework measures the cost of divesting from "sin stocks," stocks of companies engaged in businesses considered by some to be socially undesirable but that are documented to have positive alphas, which can be explained by Merton's (1987) model of neglected stocks and segmented markets. Calibrating to Hong and Kacperczyk (2009) and Blitz and Fabozzi (2017) as examples, we estimate the cost of divestment ranging from 0.6% to 3.3% in forgone alpha per annum. This example illustrates the dependence of the magnitude of estimated alpha on the specific asset-pricing model, a well-known issue with all performance attribution exercises.

Fourth, we apply our framework to several ESG empirical studies. Correlations between the specific ESG measures in these studies and the unobserved alphas of individual securities determine the final estimate of the benefit (or cost) of ESG investing. They range from −0.05% for bonds (Baker et al., 2022) to 2.65% for equities in certain market conditions (Bansal, Wu, and Yaron, 2022). This result underscores the importance of asset class, impact measurement, and specific market conditions in determining the alpha of impact investing, as well as the need to operationalize the highly dynamic impact of ESG on asset prices beyond equilibrium models of ESG investing.

---

[5]There is a substantial literature documenting the divergence of ESG ratings for the same firms (Dortfleitner, Halbritter, and Nguyen, 2015; Semenova and Hassel, 2015; Berg, Kölbel, and Rigobon, 2022).

[6]See, for example, the "luxury-good effect" of Bansal, Wu, and Yaron (2022).

[7]See, for example, Geczy, Stambaugh, and Levin (2021) and Madhavan, Sobczyk, and Ang (2021).

Finally, we apply our framework to explain the January 2021 price spike in GameStop Corp. and other "meme" stocks, in which a decentralized short squeeze that exploited the short positions of institutional investors caused their prices to increase sharply before crashing. Classifying such phenomena as impact investing may seem strange, but they share similarities with the other examples in how their returns are generated mechanically, and perhaps a separate category called "price-impact investing" would be more appropriate. Our framework helps to quantify the financial impact of price-impact investing. Measuring the magnitude of such investments and understanding their financial implications can better inform regulators and policymakers as to the scope and severity of this phenomenon so they can devote the appropriate sources to addressing it.

Overall, these five examples spotlight the importance of choosing a baseline asset-pricing model and impact variables based on economic, institutional, and market rationales to establish robust relationships between impact and returns. Our model offers a unified framework that can be calibrated to a wide range of settings, and provides greater rigor for how to think about impact investing and quantify its financial consequences. Moreover, our framework provides a systematic and politically neutral approach for portfolio managers to disclose the financial consequences of impact investing, thereby addressing any concerns that their clients are either unaware or unwilling to bear the consequences of their managers' impact criteria.

Related Literature. There is a growing literature theorizing about the impact of socially responsible investing (SRI), ESG, and other non-financial objectives on asset pricing. Fama and French's (2007) taste model shows that, if investors prefer to invest in socially responsible companies, the expected return on such companies will be lower. Pástor, Stambaugh, and Taylor (2021) provide a model for ESG investing where investor tastes for green assets imply lower returns, and assets can be priced in a two-factor model that includes the ESG factor and the market portfolio. Pedersen, Fitzgibbons, and Pomorski (2021) derive an ESG-efficient frontier, and show that ESG investing may either yield benefits to expected returns because it provides information about firm fundamentals (as in our example, in which the constraints contain information about returns), or incur costs because it affects investor preferences and constraints.[8]

While these studies share some of the same inferences as our framework, we add to this literature in several novel ways. The equilibrium frameworks of Fama and French (2007), Pástor, Stambaugh, and Taylor (2021), and Pedersen, Fitzgibbons, and Pomorski (2021) highlight that the expected return of ESG investing depends on the mix of investors and preferences in the market. However, impact investing is still an evolving concept, and its expected returns are dynamic and context-dependent. It is possible that market prices are still adjusting to reach a new equilibrium that reflects these considerations (Cornell and Damodaran, 2020).[9] Our unified econometric framework provides an explicit method to quantify the excess returns of any form of impact investment during different stages of this adaptive process—including, but not limited to, the equilibrium setting. These results are in turn consistent with the equilibrium-based models where the correlation between $\mathbf{X}$ and security returns reflects the particular market condition and shift in preferences over time.

[8]Other theoretical work on sustainable investing includes Heinkel, Kraus, and Zechner (2001), Adler and Kritzman (2008), Friedman and Heinle (2016), Luo and Balvers (2017), Albuquerque, Koskinen, and Zhang (2019), Chen and Mussalli (2020), Berk and van Binsbergen (2021), Goldstein et al. (2021), Idzorek, Kaplan, and Ibbotson (2021), Sorensen, Chen, and Mussalli (2021), and Zerbib (2022).

[9]For example, Bebchuk, Cohen, and Wang (2013) document the disappearance of a return premium associated with highly rated corporate governance during an earlier period, and Bansal, Wu, and Yaron (2022) find a "luxury-good effect," both of which suggest there is a time-dependent performance of stocks as a function of their ESG ratings.

From the adaptive markets (Lo, 2004, 2017) perspective, this correlation could reinforce itself as the amount of assets under management for a given impact factor increases over time, and eventually stabilizing as the size of the new sector reaches a steady state.

Our framework also differs from existing models in that we allow for the possibility of non-zero alphas, or omitted factors in the equilibrium/no-arbitrage interpretation, which is particularly relevant for the highly adaptive and dynamic ESG investment industry. An important insight from Pedersen, Fitzgibbons, and Pomorski (2021) is that ESG information about firm fundamentals can yield benefits to its returns, while screening constraints will incur costs to ESG investing.[10] Our model shows that, when securities have non-zero alphas that are otherwise inaccessible to investors, ESG investing can derive financial benefit from constraints, because of the information about returns implicit in these constraints (see also Lo and Zhang (2025)). This effect is formalized statistically by the correlation between the impact factor, $\mathbf{X}$, and the returns. As a result, we can explicitly construct the optimal super-efficient portfolio from any $\mathbf{X}$ and explicitly quantify its financial impact.

The empirical literature on measuring the returns of SRI and ESG are not always consistent with each other. These inconsistencies raise the question of the real financial impact of impact investing, which is precisely the motivation for our current contribution. Our framework explains not only how to measure the financial impact of impact investing, but also why there is such a wide range of empirical estimates for the expected returns of SRI and ESG investing. It is a consequence of the wide range of date ranges, asset classes, asset-pricing models for alpha, and definitions of impact itself, each of which leads to a different specification that may have a potentially different correlation between the impact factor and asset returns.

More generally, our framework is applicable to portfolios constructed on the basis of any characteristic, including traditional factors such as value, size, momentum, and other variables. As such, our work is also related to several strands of the asset pricing and econometrics literature. This includes a large literature devoted to identifying asset pricing factors and a vast econometrics literature focused on factor models. In particular, we make use of the same statistical results on induced order statistics first applied to financial data by Lo and MacKinlay (1990), albeit in a very different context.

8.2 The framework

We consider a universe of N securities with returns R_{it} that satisfy the following multi-factor model:

$$R_{it} - R_{ft} = \alpha_i + \beta_{i1}\left(\Lambda_{1t} - R_{ft}\right) + \cdots + \beta_{iK}\left(\Lambda_{Kt} - R_{ft}\right) + \epsilon_{it} \tag{8.1}$$

$$\text{such that } \; \mathrm{E}[\epsilon_{it}|\Lambda_{kt}] = 0 \;\; , \;\; k = 1, \ldots, K \tag{8.2}$$

where Λ_{kt} is the k-th factor return, $k = 1, \ldots, K$, R_{ft} is the risk-free rate, α_i and β_{ik} are the excess return and factor betas, respectively, and ϵ_{it} is the idiosyncratic return. Because we consider only a static model in this chapter, we omit the subscript t throughout for notational simplicity.

Under suitable restrictions on the parameters $\{\alpha_i, \beta_{ik}\}$ and the definitions of the factor returns $\{\Lambda_k\}$, the linear multi-factor model (8.1) is consistent with a number of asset-pricing

[10]In their framework, the standard mean-variance tangency portfolio has the highest Sharpe ratio among all portfolios, and restricting portfolios to have any ESG score other than that of the tangency portfolio must yield a lower Sharpe ratio (Pedersen, Fitzgibbons, and Pomorski, 2021, p. 573).

models such as the Sharpe-Lintner Capital Asset-Pricing Model (CAPM) (Sharpe, 1964; Lintner, 1965), Merton's Intertemporal CAPM (Merton, 1973), Ross's Arbitrage Pricing Theory (APT) (Ross, 1976), and the Fama-French multi-factor models (Fama and French, 1993, 2015). In particular, all of these asset-pricing models imply that $\alpha_i = 0$.

However, to measure the impact of impact investing, we take no position as to whether any particular asset-pricing model holds, nor do we make any assumptions about investor belief structures. Instead, we derive the consequences of impact investing on the statistical properties of impact portfolio returns without constraining the excess returns to be zero. These properties lead to a framework that is flexible enough to interpret impact from multiple perspectives.

8.2.1 The no-impact baseline case

We begin by stating the near-trivial result that arbitrary portfolios formed according to criteria unrelated to the parameters of the return-generating processes $\{R_{it}\}$ are necessarily less than or equal to the investment performance of the mean-variance optimal portfolio.[11]

Proposition 9 *If asset returns satisfy (8.1)–(8.2) and* $\alpha_1 = \cdots = \alpha_N = 0$*, then any arbitrary subset* $\mathcal{S} \subseteq \{1, \ldots, N\}$ *formed independently of the joint distribution of returns,* $\{R_{it}\}$*, satisfies:*

$$\max_{\{\omega_1,\ldots,\omega_N \mid \sum_{i=1}^N \omega_i = 1\}} \mathrm{E}[U(W)] \ \geq \max_{\{\omega_1^c,\ldots,\omega_N^c \mid \sum_{i\in\mathcal{S}} \omega_i^c = 1 \text{ and } \omega_i^c = 0 \text{ for } i \notin \mathcal{S}\}} \mathrm{E}[U(W^c)] \quad (8.3)$$

for any non-decreasing concave utility function $U(\cdot)$ *where*

$$W \ \equiv \ \sum_{i=1}^N \omega_i R_i \quad \text{and} \quad W^c \ \equiv \ \sum_{i\in\mathcal{S}} \omega_i^c R_i \ . \quad (8.4)$$

In addition, under certain fairly realistic conditions given in the Appendix, the loss in utility by restricting to the subset $\mathcal{S}$ *is generally small, as long as the number of securities excluded by* $\mathcal{S}$ *is small relative to the total number of securities,* N.

This proposition confirms the common critique that skeptics often level against impact investing. If the constraint $\mathcal{S}$ has nothing to do with the characteristics of the underlying asset returns, $\{R_i\}$, then imposing such constraints can only reduce investment performance, or at best achieve the unconstrained optimum. In particular, the independence of $\mathcal{S}$ and $\{R_i\}$ implies that the excess returns, $\{\alpha_i\}$, are indistinguishable from $\{\epsilon_i\}$, in which case we are essentially assuming zero excess returns, so there is no possibility of generating any excess performance.[12] In addition, although impact investing in this special case cannot improve returns, the underperformance is likely to be small (assuming no transactions costs or fees, of course).

However, suppose we allow for non-zero alphas that are unobserved by investors; in other words, the unconstrained optimization problem in Equation (8.3) does not have the ability to find securities with positive alphas.[13] If we relax the condition that $\mathcal{S}$ is independent of the joint distribution of $\{R_i\}$, then Proposition 9 clearly does not hold. For example, suppose:

$$\mathcal{S} \ = \ \{i : \alpha_i > 0 \ , \ i = 1, \ldots, N\} \ . \quad (8.5)$$

[11] Proofs of all propositions are provided in the appendix of Lo and Zhang (2024).

[12] In this baseline portfolio selection problem, investors do not have information on the impact of individual securities, or **X** defined in the next section.

[13] An alternative interpretation adopted by, among others, Adler and Kritzman (2008) is that certain investors have skills that yield alphas private to themselves.

Clearly in this case, it is possible for the risk-adjusted returns of the $\mathcal{S}$-portfolio to outperform those of the unconstrained portfolio, given that the subset contains all positive-alpha securities and the complement contains the reverse. This conclusion may seem counterintuitive, since the constrained portfolio is, by definition, a feasible solution in the unconstrained case, so how can imposing the constraint ever improve performance? The answer lies in the fact that, in the unconstrained case, information about the $\{\alpha_i\}$ is not available—the constraint contains private information[14] that can dramatically improve performance. Therefore, the constrained solution is actually not feasible in the unconstrained case.

Therefore, the fundamental question of whether an impact investment has positive (or negative) financial impact reduces to the information content in the constraint, i.e., the relation between the constraint and the joint distribution of asset returns. No relation implies no information, and hence no impact. However, the presence of even the slightest amount of dependence between the constraint and returns implies the possibility of some degree of impact. We can quantify this degree by being explicit about the statistical relation between asset returns and the impact factor.

It is true that this counterexample assumes the existence of mispriced or positive-alpha securities in Equation (8.5), but an equally valid equilibrium/no-arbitrage interpretation is that the α_is are omitted factors from the investor's linear-factor benchmark. Either investors are unaware of these factors, or they do not have the ability to access them (e.g., exotic betas from private equity, distressed debt, event-driven opportunities, and so on). Under this interpretation, impact investing can be viewed as providing investors with alternative risk premia.

As described below, our framework accommodates both interpretations, and offers a systematic and quantitative approach to measuring impact in either case.

8.2.2 Impact factors and induced order statistics

We assume that the excess return of the i-the security, α_i, is not observable, whereas the impact factor, X_i, for that security is observable. Contrary to the usual asset-pricing assumptions in which the α_is are assumed to be fixed constants (and, in equilibrium or under no-arbitrage conditions, identically equal to 0), we assume that they are random variables.

Impact investors select a portfolio based on the impact factor, $\mathbf{X}$, and the excess return of their portfolio is determined by the corresponding vector of excess returns of the individual securities in that portfolio, $\boldsymbol{\alpha} \equiv [\alpha_1 \ldots \alpha_N]^T$. Specifically, suppose an investor ranks N securities according to $\mathbf{X}$. Let us reorder the bivariate vector $(X_i, \alpha_i)^T, i = 1, 2, \ldots, N$, according to the magnitudes of their first components:

$$\begin{pmatrix} X_{1:N} \\ \alpha_{[1:N]} \end{pmatrix}, \begin{pmatrix} X_{2:N} \\ \alpha_{[2:N]} \end{pmatrix}, \ldots, \begin{pmatrix} X_{N:N} \\ \alpha_{[N:N]} \end{pmatrix} \tag{8.6}$$

where $X_{1:N} < X_{2:N} < X_{N:N}$ and the notation $X_{i:N}$ denotes the i-th order statistic from a total of N random variables. The notation $\alpha_{[i:N]}$ represents the i-th *induced order statistic*,[15] where the order is *induced* by another variable $\mathbf{X}$.

[14]By private information, we mean that the alphas are assumed to be unobservable by investors. Without the constraint, $\mathcal{S}$, investors have no way to select securities with positive alphas. In this sense, the constraint is, in fact, a mechanism for alpha selection and therefore contains valuable information.

[15]The term was coined by Bhattacharya (1974) to distinguish between random variables ranked by their own realized values versus random variables ranked by the realizations of related random variables. These indirectly ranked statistics are also referred to as *concomitants* of the order statistic, $X_{i:N}$ (David, 1973). Lo and MacKinlay (1990) applied these same statistical tools to quantify data-snooping biases in testing financial asset-pricing models.

Because $\boldsymbol{\alpha}$ is defined with respect to the multi-factor model of Equation (8.1), by studying the interaction between $\mathbf{X}$ and $\boldsymbol{\alpha}$, our framework allows for the existence of any set of predefined asset-pricing factors, and focuses on the incremental role of $\mathbf{X}$ in determining asset returns.

8.2.3 Defining an impact portfolio

Impact investing corresponds to the selection of securities based on the impact factor, $\mathbf{X}$. For example, an investor may choose to invest in the top n_0 securities ranked by $\mathbf{X}$, or form portfolios long the top decile and short the bottom decile. In general, we define an *impact portfolio* to be any portfolio $\mathcal{S}(\mathbf{X})$ formed as a function of the impact factor, $\mathbf{X}$. With portfolio weights $\{\omega_i,\ i \in \mathcal{S}\}$, the return of the impact portfolio is given by:

$$R_{\mathcal{S}} = \sum_{i \in \mathcal{S}} \omega_i R_i. \tag{8.7}$$

To characterize $R_{\mathcal{S}}$, we therefore need to quantify the distribution of the excess returns—or the induced order statistic, $\alpha_{[i:N]}$, given certain assumptions on the joint distribution of $(\mathbf{X}, \boldsymbol{\alpha})$.

Note that $\mathbf{X}$ can represent a variety of characteristics related to metrics for climate change, sustainable farming, tobacco usage, gambling, biomedical R&D, or any other SRI or ESG criteria. Together with the generality of our multi-factor asset-pricing model of Equation (8.1), this corresponds to a wide range of impact investing contexts. Section 8.5 provides five concrete examples, highlighting that the specific economic, institutional, and market variables that matter in each case will depend on the specific context and time period in consideration (see the discussion in Section 8.7).

In fact, our framework applies more generally to any characteristics of a security, including the traditional value, size, and momentum factors, as well as the many denizens of the "factor zoo" described in the recent literature. For the purposes of this study, we will focus on the impact investing interpretation, but will discuss broader interpretations in Section 8.8.

8.3 Characterizing excess returns

To assess the impact of impact portfolios, we require the distribution of $\alpha_{[i:N]}$, which can be derived explicitly under the following assumption:

(A1) $(X_i, \alpha_i)^T, i = 1, 2, \ldots, N$, are independently and identically distributed (IID) bivariate normal random vectors with mean $(\mu_x, \mu_\alpha)^T$, variance $(\sigma_x^2, \sigma_\alpha^2)^T$, and correlation $\rho \in (-1, 1)$.

The assumption that α_i is random is somewhat unconventional, so a few clarifying remarks are in order. This assumption was first used in Lo and MacKinlay (1990) to represent the cross-sectional estimation errors of intercepts from CAPM regressions. However, in our current context, we interpret the randomness in α_i as a measure of uncertainty of the degree of mispricings of securities in our investment universe.[16] This uncertainty can be interpreted from a Bayesian perspective as the degree of conviction that mispricings exist

[16] Pástor and Stambaugh (1999) also model this randomness as a measure of mispricing uncertainty.

in the cross-section. Under this interpretation, we will make the auxiliary assumption (with little loss of generality) that all α_is are mean 0 (μ_α=0). This corresponds to centering the Bayesian prior on zero average deviations from equilibrium or no-arbitrage pricing in our investment universe, a reasonable first approximation that still allows for mispricings which, of course, motivate a significant portion of the asset management industry's products and services.[17] Moreover, we can calibrate the degree of mispricings in our model through σ_α^2, smaller values of which corresponding to greater efficiency, and larger values corresponding to lower efficiency, and thus to more active management opportunities.

However, our framework can also be interpreted from an equilibrium/no-arbitrage perspective, where non-zero α_is are due to the presence of omitted factors that investors are either unaware of or unable to access directly. Under this interpretation, we will see below that the randomness in α_i is due to the cross-sectional variability in security i's omitted-factor betas. In this case, however, it is possible for μ_α to be non-zero to reflect the risk premia of the omitted factors.

We deliberately make the assumption of joint normality for $(X_i, \alpha_i)^T$ so as to capture the interaction between $\mathbf{X}$ and $\boldsymbol{\alpha}$ while still being able to derive explicit and easily interpretable results for induced excess returns. This assumption can be relaxed considerably to allow for cross-sectional dependence (see Section 8.3.3), and general marginal distributions and dependent structures for $(X_i, \alpha_i)^T$ (see Section 8.6 and Lo et al. (2024)), at the expense of simplicity.

Regardless of the interpretation of α_i, the theory of induced order statistics allows us to completely characterize its statistical properties. We first present its finite-sample distribution, followed by asymptotic results when the number of securities, N, increases without bound.

8.3.1 Finite-sample distribution

We first observe that the mean and standard deviation of the impact factor, $\mathbf{X}$, do not actually matter for the distribution of $\alpha_{[i:N]}$'s, because it is only the relative order of X_i's that determines the order of $\alpha_{[i:N]}$'s. Therefore, we assume without loss of generality that $\mu_x = 0$ and $\sigma_x = 1$, so that $\mathbf{X}$ is a standard normal random vector. The following result then characterizes the finite-sample distributions of the induced order statistics $\{\alpha_{[i:N]}\}$:

Proposition 10 *Under Assumption (A1), the expected value of the i-th induced order statistic $\alpha_{[i:N]}, i = 1, 2, \ldots, N$ is given by:*

$$\mu_i \equiv \mathrm{E}\left[\alpha_{[i:N]}\right] = \rho\sigma_\alpha \mathrm{E}\left[X_{i:N}\right]. \tag{8.8}$$

The variance of the i-th induced order statistic $\alpha_{[i:N]}, i = 1, 2, \ldots, N$ is given by:

$$\sigma_i^2 \equiv \mathrm{Var}\left(\alpha_{[i:N]}\right) = \sigma_\alpha^2\left(1 - \rho^2 + \rho^2 \mathrm{Var}\left(X_{i:N}\right)\right). \tag{8.9}$$

The covariance of the i-th and j-th induced order statistic, $\alpha_{[i:N]}$ and $\alpha_{[j:N]}$, for $i \neq j$ is given by:

$$\sigma_{ij} \equiv \mathrm{Cov}\left(\alpha_{[i:N]}, \alpha_{[j:N]}\right) = \sigma_\alpha^2\rho^2 \mathrm{Cov}\left(X_{i:N}, X_{j:N}\right). \tag{8.10}$$

Proposition 10 gives us the first two moments of the induced order statistics, $\alpha_{[i:N]}$'s. In particular, the expected alpha in Equation (8.8) is determined by three terms: the correlation (ρ) between $\mathbf{X}$ and the alphas of individual securities, the standard deviation of individual

[17]In fact, Grossman and Stiglitz (1980) have argued that the presence of occasional mispricings is a prerequisite for achieving informationally efficient markets; otherwise, no one would have an incentive to gather information and incorporate it into market prices.

alphas (σ_α), and a cross-sectionally standardized impact score ($\mathrm{E}[X_{i:N}]$). The correlation ρ here plays a critical role in determining the expected alpha of both individual securities and impact portfolios (see Section 8.4). This resembles the CAPM's market beta—which quantifies security returns attributable to systematic market risk—because the market beta is simply the correlation between security returns and market returns when they are both standardized with unit variances.

If we view the impact factor $\mathbf{X}$ as a signal for predicting asset returns, the expected alpha in Equation (8.8) is closely related to the results by Grinold (1994),[18] who provides a simple decomposition of alpha into the product of three terms: the information coefficient (the correlation ρ in our notation), the volatility of residual returns, and a standardized score that measures the strength of the signal for each asset. In our context, it is the volatility of the unobserved alpha, not the volatility of residual returns, that determines the expected alpha of each asset. In addition, because we only use the rank information in $\mathbf{X}$, the standardized score can be quantified explicitly by $\mathrm{E}[X_{i:N}]$. Finally, our results also provide the variance and covariances of individual alphas, which is crucial for quantifying the uncertainty of these alphas in practice.

We note that all three quantities in Equations (8.8)–(8.10) depend on the distribution of the order statistics of standard normal random variables. In fact, the terms $\mathrm{E}[X_{i:N}]$ in Equation (8.8), $\mathrm{Var}(X_{i:N})$ in Equation (8.9), and $\mathrm{Cov}(X_{i:N}, X_{j:N})$ in Equation (8.10) can be explicitly evaluated by numerical integration over the density function of $X_{i:N}$ (see David and Nagaraja (2004, Section 3.1), for example).

We can also explicitly evaluate the quantities in Equations (8.8)–(8.10) based on the following approximation results:

Proposition 11 *Let $p_i \equiv \frac{i}{N+1}$ denote the relative position of the order i in the population of N securities. The expected value and variance of the i-th order statistic of the standard normal random variable, $X_{i:N}$, can be approximated up to order $(N+2)^{-2}$, when N increases without bound, by:*

$$\mathrm{E}[X_{i:N}] \approx \Phi^{-1}(p_i) + \frac{p_i(1-p_i)}{2(N+2)}Q_i'' + \frac{p_i(1-p_i)}{(N+2)^2}\left[\frac{1}{3}(1-2p_i)Q_i''' + \frac{1}{8}p_i(1-p_i)Q_i''''\right] \tag{8.11}$$

$$\mathrm{Var}(X_{i:N}) \approx \frac{p_i(1-p_i)}{N+2}Q_i'^2 + \frac{p_i(1-p_i)}{(N+2)^2}\left[2(1-2p_i)Q_i'Q_i'' + p_i(1-p_i)\left(Q_i'Q_i''' + \frac{1}{2}Q_i''^2\right)\right] \tag{8.12}$$

for $i = 1, 2, \ldots N$. Their covariances can be approximated up to order $(N+2)^{-2}$, when N increases without bound, by:

$$\begin{aligned}\mathrm{Cov}(X_{i:N}, X_{j:N}) \approx{}& \frac{p_i(1-p_j)}{N+2}Q_i'Q_j' + \frac{p_i(1-p_j)}{(N+2)^2}\Big[(1-2p_i)Q_i''Q_j' + (1-2p_j)Q_i'Q_j'' \\ &+ \frac{1}{2}p_i(1-p_i)Q_i'''Q_j' + \frac{1}{2}p_j(1-p_j)Q_i'Q_j''' + \frac{1}{2}p_i(1-p_j)Q_i''Q_j''\Big]\end{aligned} \tag{8.13}$$

[18] See also Grinold and Kahn (1999).

for $1 \le i < j \le N$. Here Q_i', Q_i'', Q_i''', and Q_i'''' are the first four derivatives of $\Phi^{-1}(p_i)$:

$$Q_i' = \left(\Phi^{-1}(p_i)\right)' = \frac{1}{\phi\left(\Phi^{-1}(p_i)\right)} \tag{8.14}$$

$$Q_i'' = \left(\Phi^{-1}(p_i)\right)'' = \frac{\Phi^{-1}(p_i)}{\phi\left(\Phi^{-1}(p_i)\right)^2} \tag{8.15}$$

$$Q_i''' = \left(\Phi^{-1}(p_i)\right)''' = \frac{1 + 2\left(\Phi^{-1}(p_i)\right)^2}{\phi\left(\Phi^{-1}(p_i)\right)^3} \tag{8.16}$$

$$Q_i'''' = \left(\Phi^{-1}(p_i)\right)'''' = \frac{\Phi^{-1}(p_i)\left(7 + 6\left(\Phi^{-1}(p_i)\right)^2\right)}{\phi\left(\Phi^{-1}(p_i)\right)^4}. \tag{8.17}$$

Φ and ϕ are the cumulative distribution function (CDF) and density function of the standard normal distribution, respectively.

Although the approximations in Proposition 11 may seem daunting, their first-order terms are fairly intuitive. The first term in Equation (8.11) is $\Phi^{-1}(p_i)$, which simply approximates $\mathrm{E}\left[X_{i:N}\right]$ by the inverse CDF applied to the relative rank, $p_i \equiv \frac{i}{N+1}$, of the i-th order statistic, which is a well-known first-order approximation by itself.

Figure 8.1 displays the mean, variance, and covariances of the induced order statistic, $\alpha_{[i:N]}$, for a collection of $N = 50$ securities, as given in Proposition 10 using the approximations in Proposition 11. When the correlation ρ between $\boldsymbol{\alpha}$ and $\mathbf{X}$ is positive, the expected value of the induced order statistic increases as the order i increases (see Figure 8.1a). The dispersion of the mean is larger when the correlation ρ or the dispersion of the unknown $\boldsymbol{\alpha}$, σ_α, is larger.

Figure 8.1b shows that the variances, $\mathrm{Var}\left(X_{i:N}\right)$, stay relatively constant across the ordered securities i, and are primarily determined by ρ and σ_α. Finally, it can be seen that the covariances, $\mathrm{Cov}\left(X_{i:N}, X_{j:N}\right)$, are very close to zero except when i and j are close to the two extremes.

The appendix of Lo and Zhang (2024) provides the asymptotic mean, variance, and covariances of induced order statistics that generalizes Proposition 10.

8.3.2 Comparison with conventional order statistics

To further develop an intuition for the effects of induced ordering, we compare the distributions of induced order statistics with their conventionally ordered counterparts, $\alpha_{i:N}$. Note that this comparison is merely meant to be a thought experiment; $\boldsymbol{\alpha}$ is unobservable in practice, hence such rankings are also not feasible in practice. Nonetheless, this is a useful comparison to what can be achieved by ordering based on the impact factor, $\mathbf{X}$.

Proposition 12 *Under Assumption (A1), the first two moments of the induced order statistic, $\alpha_{[i:N]}$, are related to the order statistic, $\alpha_{i:N}$, by the following identities:*

$$\mu_i \equiv \mathrm{E}\left[\alpha_{[i:N]}\right] = \rho \mathrm{E}[\alpha_{i:N}] \tag{8.18}$$

$$\sigma_i^2 - \sigma_\alpha^2 \equiv \mathrm{Var}\left(\alpha_{[i:N]}\right) - \sigma_\alpha^2 = \rho^2\left[\mathrm{Var}\left(\alpha_{i:N}\right) - \sigma_\alpha^2\right] \tag{8.19}$$

$$\sigma_{ij} \equiv \mathrm{Cov}\left(\alpha_{[i:N]}, \alpha_{[j:N]}\right) = \rho^2 \mathrm{Cov}\left(\alpha_{i:N}, \alpha_{j:N}\right). \tag{8.20}$$

Proposition 12 tells us that the mean, variance, and covariances of the induced order statistics, $\alpha_{[i:N]}$, are essentially a discounted version of the corresponding moments of the conventional order statistics, $\alpha_{i:N}$. The discount factor, ρ, is precisely the correlation between $\mathbf{X}$ and $\boldsymbol{\alpha}$.

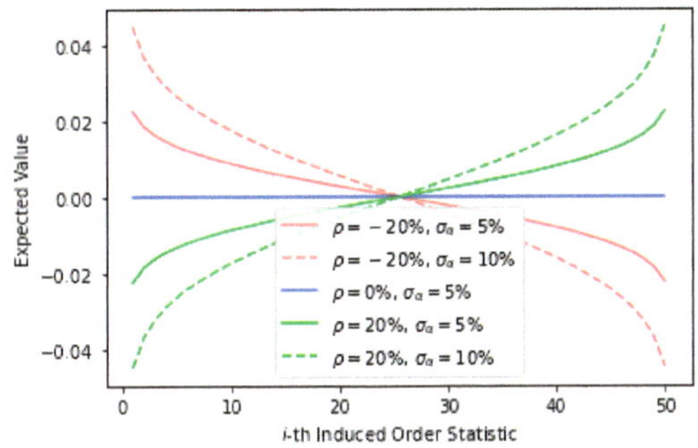

(a) Expected Value

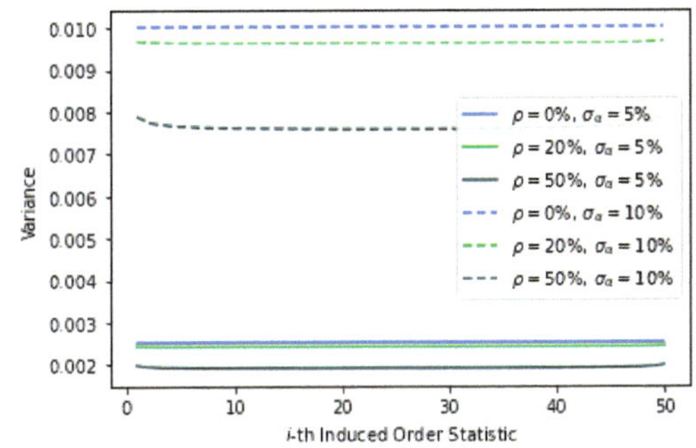

(b) Variance

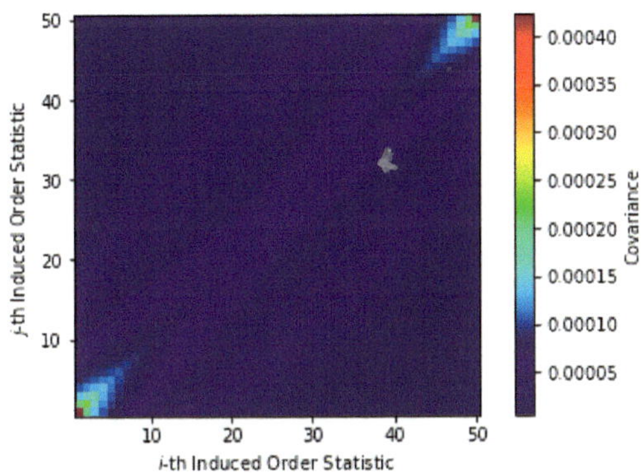

(c) Covariances

FIGURE 8.1: Mean, variance, and covariances of the induced order statistic, $\alpha_{[i:N]}$. The total number of securities is set to 50 for illustrative purposes. In (c) we set $\rho = 20\%$ and $\sigma_\alpha = 5\%$.

To visualize this effect, Figure 8.2 contains a graphed comparison of the expected excess returns of the induced order statistic, $\alpha_{[i:N]}$, and the conventional order statistic, $\alpha_{i:N}$. As the correlation ρ increases to 1, the expected excess return approaches the hypothetical value of sorting based on $\boldsymbol{\alpha}$.

This result emphasizes the role that induced ordering plays in distinguishing securities with positive alpha from those with negative alpha. If alpha could be fully observed by a hypothetical oracle, she could simply pick securities with the highest alphas to construct impact portfolios. In our oracle-free reality, however, the correlation between the sorting variable (the impact factor) and the target variable (the unobserved $\boldsymbol{\alpha}$) determines how much of the mean, variance, and covariances from a hypothetical sorting based on $\boldsymbol{\alpha}$ can actually be achieved via the induced ordering of $\mathbf{X}$.

8.3.3 Interpreting excess return as omitted factors

Having completely characterized the stochastic properties of the excess returns $\boldsymbol{\alpha}$ of securities ranked according to an arbitrary impact factor $\mathbf{X}$, we now provide an explicit derivation of the equilibrium/no-arbitrage interpretation of $\boldsymbol{\alpha}$ as risk premia associated with omitted factors.

Let security returns follow the K-factor asset-pricing model as specified in Equations (8.1)–(8.2), but now assume there are no mispricings. However, suppose that investors only

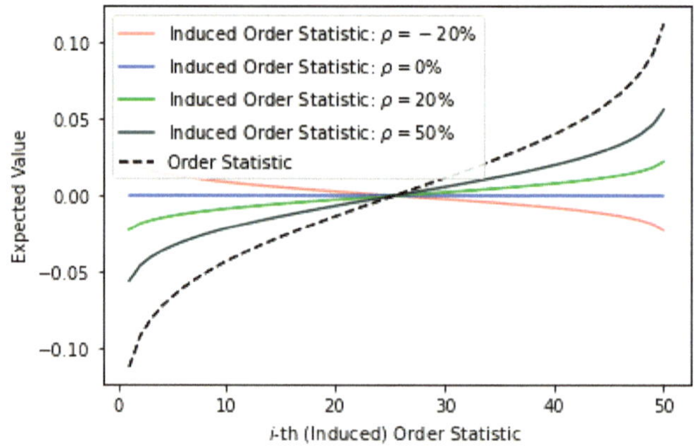

FIGURE 8.2: Comparison between the expected value of the induced order statistic, $\alpha_{[i:N]}$, and the order statistic, $\alpha_{i:N}$, with $N=50$ and $\sigma_\alpha=5\%$.

account for the first factor Λ_1, without loss of generality, and are unaware of the remaining $K-1$ factors $\Lambda_2, \ldots, \Lambda_K$. We define $\lambda_{ik} \equiv \beta_{ik}(\Lambda_k - R_f)$ to be factor k's contribution to security i's return, for $i = 1, \ldots, N$ and $k = 2, \ldots, K$, and $\lambda_i \equiv \sum_{k=2}^{K} \lambda_{ik}$ to be the total net contribution of all the omitted factors to security i's return. Given that investors are unaware of factors $2, \ldots, K$, the total excess expected returns for the securities in our universe appear to be alphas to such investors:

$$\alpha_i \equiv \mathrm{E}[\lambda_i] = \sum_{k=2}^{K} \beta_{ik}(\mathrm{E}[\Lambda_k] - R_f) . \tag{8.21}$$

To characterize the distribution of λ_i after ranking securities based on the impact factor $\mathbf{X}$, we make the following assumption:

(A2) $(X_i, \lambda_i)^T, i = 1, 2, \ldots, N$, are bivariate normal random vectors with their marginal distributions and paired correlations defined by:

$$\mu_x \equiv \mathrm{E}[X_i],\ \mu_\lambda \equiv \mathrm{E}[\lambda_i],\ \sigma_x^2 \equiv \mathrm{Var}(X_i),\ \sigma_\lambda^2 \equiv \mathrm{Var}(\lambda_i),\ \text{and}\ \rho_{x,\lambda} \equiv \mathrm{Corr}(X_i, \lambda_i). \tag{8.22}$$

In addition, for $i \neq j$, the correlations across different securities are defined by:

$$\rho_x \equiv \mathrm{Corr}(X_i, X_j),\ \rho_\lambda \equiv \mathrm{Corr}(\lambda_i, \lambda_j),\ \text{and}\ \tilde{\rho}_{x,\lambda} \equiv \mathrm{Corr}(X_i, \lambda_j). \tag{8.23}$$

Under this assumption, the cross-sectional randomness of λ_i can be interpreted as variations coming from both the factor values and the distribution of factor betas across companies in our universe. $(X_i, \lambda_i)^T$ can be correlated across securities, and their correlation structure is described by the four parameters $\rho_{x,\lambda}$, $\tilde{\rho}_{x,\lambda}$, ρ_x, and ρ_λ.

We can characterize the first two moments of $\lambda_{[i:N]}$. Recall that the notation $\lambda_{[i:N]}$ denotes the i-th induced order statistic where the order is *induced* by the impact factor $\mathbf{X}$. We again assume without loss of generality that $\mu_x = 0$ and $\sigma_x = 1$, so that $\mathbf{X}$ is a standard normal random vector. However, we allow for a non-zero risk premium μ_λ.

Proposition 13 *Under Assumption (A2), define*

$$\rho_{adj} \equiv \frac{\rho_{x,\lambda} - \tilde{\rho}_{x,\lambda}}{1 - \rho_x} \tag{8.24}$$

to be an adjusted correlation. The expected value of the i-th induced order statistic $\lambda_{[i:N]}, i = 1, 2, \ldots, N$ *is given by:*

$$\mathrm{E}\left[\lambda_{[i:N]}\right] - \mu_\lambda = \rho_{adj}\sigma_\lambda \mathrm{E}\left[X_{i:N}\right]. \tag{8.25}$$

The variance of the i-th induced order statistic $\lambda_{[i:N]}, i = 1, 2, \ldots, N$ *is given by:*

$$\mathrm{Var}\left(\lambda_{[i:N]}\right) = \sigma_\lambda^2\left(1 - \rho_{adj}^2 + \rho_{adj}^2 \mathrm{Var}\left(X_{i:N}\right)\right). \tag{8.26}$$

The covariance of the i-th and j-th induced order statistic, $\lambda_{[i:N]}$ *and* $\lambda_{[j:N]}$*, for* $i \neq j$ *is given by:*

$$\mathrm{Cov}\left(\lambda_{[i:N]}, \lambda_{[j:N]}\right) = \sigma_\lambda^2\rho_{adj}^2 \mathrm{Cov}\left(X_{i:N}, X_{j:N}\right) + \left(\rho_\lambda - \rho_x\rho_{adj}^2\right). \tag{8.27}$$

Proposition 13 characterizes the return from omitted factors for the i-th security ranked by $\mathbf{X}$. This result highlights an important implication when estimating the financial impact of impact investing. Given any definition of impact $\mathbf{X}$, if the portfolio selected based on $\mathbf{X}$ produces a non-zero excess return, $\mathbf{X}$ must be correlated with some factors not previously accounted for in the asset-pricing framework. This may imply the existence of a new factor that corresponds to the very definition of $\mathbf{X}$, such as an "ESG factor" or a "carbon factor" (Bolton and Kacperczyk, 2021).

Conversely, Proposition 13 also implies that, when forming a portfolio, if one uses a selection criteria that appears independent of return characteristics such as market betas and factor loadings, it may still be correlated with risk premiums from omitted factors, in which case the selection criteria will produce nonzero excess returns. In other words, what appears to be an "impact factor" (a selection criteria $\mathbf{X}$ based on a particular concept) may simply be correlations with omitted factors that are unrelated to the impact concept one intends to capture. Therefore, one's impact estimates may be inaccurate and misleading without first properly accounting for all known factors.

This observation is supported empirically by both Blitz and Fabozzi (2017) in the case of estimating excess returns for sin stocks, and Madhavan, Sobczyk, and Ang (2021) for ESG scores, both of which we discuss in more detail in Section 8.5.

8.4 Impact portfolio construction

Having quantified the distribution of the induced order statistics, $\alpha_{[i:N]}$, we can now construct portfolios based on the impact factor, $\mathbf{X}$, and characterize the statistical properties of their excess returns. We first quantify the performance of arbitrary impact portfolios. We then use the Treynor and Black (1973) framework to derive the optimal weights for each security, as well as the optimal way to combine an impact portfolio with any existing portfolio such as the passive market index. The latter result follows directly from our ability to completely characterize the statistical properties of individual alphas in our framework.

8.4.1 Properties of arbitrary iimpact portfolios

Consider an arbitrary impact portfolio of n_0 securities with indexes in $\mathcal{S}$:

$$\mathcal{S} \equiv \{i_1, i_2, \ldots, i_{n_0}\} \tag{8.28}$$

which is obtained from a rank-ordering of securities from the investment universe according to the impact factor, $\mathbf{X}$. The excess return of the portfolio is then given by:

$$\tilde{\alpha} \equiv \sum_{i \in \mathcal{S}} \omega_i \alpha_{[i:N]} \tag{8.29}$$

where $\{\omega_i : i \in \mathcal{S}\}$ are arbitrary portfolio weights that sum to 1. Based on the distribution of the induced order statistics in Proposition 10, we have the following result for portfolio excess returns:

Proposition 14 *Under Assumption (A1), the expected excess return of a portfolio $\mathcal{S}$ defined in Equation (8.28) is:*

$$\mathrm{E}\left[\tilde{\alpha}\right] = \sum_{i\in\mathcal{S}} \omega_i \mu_i = \rho\sigma_\alpha \sum_{i\in\mathcal{S}} \omega_i \mathrm{E}\left[X_{i:N}\right], \tag{8.30}$$

and the variance is:

$$\begin{aligned}\mathrm{Var}\left(\tilde{\alpha}\right) =& \sum_{i\in\mathcal{S}} \omega_i^2\sigma_i^2 + 2\sum_{i<j\in\mathcal{S}} \omega_i\omega_j\sigma_{ij}\\ =&\sigma_\alpha^2\left(1-\rho^2+\rho^2\left(\sum_{i\in\mathcal{S}}\omega_i^2\mathrm{Var}\left(X_{i:N}\right)+2\sum_{i<j\in\mathcal{S}}\omega_i\omega_j\mathrm{Cov}\left(X_{i:N},X_{j:N}\right)\right)\right).\end{aligned} \tag{8.31}$$

Proposition 14 quantifies the distribution of excess returns for any portfolio constructed according to the impact factor, $\mathbf{X}$. The appendix of Lo and Zhang (2024) provides several numerical examples where impact portfolios are formed based on top-ranking securities and decile portfolios. This result implies that the full range of tools and results from modern portfolio theory may be applied here, including: the calculation of various performance metrics such as the Sharpe ratio (Sharpe, 1966), Sortino ratio (Sortino and Van Der Meer, 1991; Sortino and Price, 1994), and information ratios (Treynor and Black, 1973); performance attribution (Brinson, Hood, and Beebower, 1986); and active portfolio management and enterprise risk management (Grinold and Kahn, 1999).

8.4.2 Treynor-black portfolios

A key advantage of our framework is the ability to characterize the alphas of arbitrary impact portfolios via induced order statistics. Given this representation, it is clear that equal-weighted portfolios are not optimal in terms of achieving the best risk-adjusted returns.

However, Treynor and Black (1973) provide a methodology that is designed to maximize a portfolio's Sharpe ratio, when investors have access to alpha forecasts for a certain subset of securities.[19] This can be interpreted as a temporary departure from equilibrium in the sense of mispricings that investors can exploit, or differences in information across investors in the sense of omitted risk factors. In either case, Treynor and Black's (1973) methodology allows us to construct a portfolio that maximizes the Sharpe ratio, which can be directly applied in our case to derive optimal weights for securities selected by the impact factor.

To apply the Treynor-Black framework, we rewrite the excess return of the i-th security, α_i, as its mean plus noise:

$$\alpha_i = \mu_i + \zeta_i \tag{8.32}$$

where $\{\zeta_i\}$ are independent random variables with zero means. We can then combine ζ_i with security i's idiosyncratic error, ϵ_i. Because ζ_i and ϵ_i are independent, the combined idiosyncratic variance for security i is simply $\sigma_i^2 + \sigma(\epsilon_i)^2$, where σ_i^2 is given in Equation (8.9).

Given any number of securities selected by $\mathbf{X}$, we can form an optimal portfolio based on the Treynor-Black weights, which we summarize in the following result:

[19] When no alpha or mispricing estimates are available, the optimal portfolio is given by the traditional mean-variance analysis.

Proposition 15 *Under Assumption (A1), the Treynor-Black weight of security i is proportional to its expected alpha divided by its combined idiosyncratic variance:*

$$\omega_i \propto \frac{\mu_i}{\sigma_i^2 + \sigma(\epsilon_i)^2}. \tag{8.33}$$

In addition, if the idiosyncratic volatility, $\sigma(\epsilon_i)$, is constant across securities i, as N increases without bound, the Treynor-Black weight of security i in Equation (8.33) can be further simplified to:

$$\omega_i \propto \frac{\rho\sigma_\alpha \Phi^{-1}(\xi_i)}{\sigma_\alpha^2(1-\rho^2) + \sigma(\epsilon_i)^2} \propto \Phi^{-1}(\xi_i) \cdot \mathit{Constant}. \tag{8.34}$$

Proposition 15 gives an explicit formula for the Treynor-Black weights that optimize the risk-adjusted returns of the impact portfolio, which can easily be implemented in practice. For further intuition behind Equation (8.33), recall that the variance of the i-th induced order statistic, σ_i^2, is approximately a constant when N is large (see Figure 8.1b). The expected excess return, $\mu_i = \rho\sigma_\alpha \mathrm{E}\left[X_{i:N}\right]$, varies with respect to i only through the last term $\mathrm{E}\left[X_{i:N}\right]$. As a result, if each security's idiosyncratic volatility is the same, the Treynor-Black weights of security i in (8.33) depend only on their relative ranking in the universe of N securities, which is specified by the term $\Phi^{-1}(\xi_i)$ in Equation (8.34).

The portfolio selected by ranking $\mathbf{X}$ and applying the Treynor-Black weights in Equation (8.33) is one specific example of an *impact portfolio* we defined in Section 8.2.3. Treynor and Black (1973) call this the "active management" portfolio, and its return characteristics are given by:

$$\alpha_A = \sum_{k=1}^{n} \omega_k \mu_k, \tag{8.35}$$

$$\beta_A = \sum_{k=1}^{n} \omega_k \beta_k, \tag{8.36}$$

$$\sigma(\epsilon_A)^2 = \sum_{k=1}^{n} \omega_k^2 \left(\sigma_k^2 + \sigma(\epsilon_k)^2\right). \tag{8.37}$$

These results, together with the explicit quantification of individual-security alphas in Propositions 10 and 13, and the optimal Treynor-Black weights in Proposition 15, provide a complete characterization of the performance of optimal impact portfolios. In particular, the information ratio of the impact portfolio, defined as $\mathrm{E}[\alpha_A]/\sigma(\epsilon_A)$, is proportional to the correlation ρ between unobserved alpha, $\boldsymbol{\alpha}$, and the impact factor, $\mathbf{X}$. This is closely related to the fundamental law of active management (FLAM) by Grinold (1989),[20] which provides a simple approximation of the information ratio of an active portfolio by the product of information coefficient (ρ in our notation) and the *breadth* of a strategy.[21] The appendix of Lo and Zhang (2024) provides several numerical examples of the performance of Treynor-Black impact portfolios.

---

[20]See also (Grinold and Kahn, 1999, chapter 6) and (Grinold and Kahn, 2019, chapter 4 and 5) for recent developments.

[21]In Grinold's (1989) framework, breadth is defined as the number of independent bets of a strategy in a given year. In our context, breadth is determined by the number of available assets in the universe. Our framework can also be regarded as a generalization of the FLAM because the impact portfolio only uses the rank information in $\mathbf{X}$.

8.4.3 Combining impact and passive portfolios

Once the relative weights of the securities within an impact portfolio are determined, one can combine the portfolio with any other portfolio. For example, we may form an impact portfolio by ranking a company's impact on global warming, which can be combined with other characteristics such as sustainable farming, tobacco usage, and gaming, to form an overall "ESG" portfolio. We can also add the impact portfolio to the suite of portfolios mimicking more traditional asset pricing factors such as value, size, and momentum.

However, perhaps the most natural application is to consider combining the impact portfolio with a passive index fund such as the market portfolio. Let ω_A denote the weight of the impact portfolio, and $1-\omega_A$ the weight of a passive portfolio. To maximize the Sharpe ratio of the combined portfolio, the relative weight is determined by the impact portfolio's excess return and idiosyncratic volatility:

$$\omega_A = \left(\frac{\alpha_A}{\sigma(\epsilon_A)^2}\right) \Big/ \left(\frac{\mathrm{E}[R_m] - R_f}{\sigma_m^2}\right) \tag{8.38}$$

where $\mathrm{E}[R_m]$ and σ_m^2 are the expected return and variance of the passive portfolio, respectively.

We illustrate the impact portfolio's alpha and its corresponding weight, ω_A, using a numerical example. Suppose the passive portfolio has an annualized risk premium of $\mathrm{E}[R_m]-R_f = 6\%$ and volatility of $\sigma_m = 15\%$. The idiosyncratic volatility is a constant $\sigma(\epsilon_i) = 15\%$ for all securities.[22] Consider again a collection of $N = 500$ securities. We divide them into 10 decile portfolios ranked by the impact factor, $\mathbf{X}$. For several different values of ρ and σ_α, Panel A of Table 8.1 reports the expected excess return of the impact portfolio,[23] α_A in Equation (8.35), for the top and bottom two decile portfolios. Panel B of Table 8.1 reports the weight, ω_A, and the corresponding excess return of the impact portfolio combined with the passive portfolio.

We first consider the case in which the cross-sectional standard deviation $\sigma_\alpha = 2\%$. In other words, most securities' excess returns are within $[-2\sigma_\alpha, 2\sigma_\alpha] = [-4\%, 4\%]$. This is a fairly conservative assumption for U.S. equities,[24] but even with such a modest range of $\boldsymbol{\alpha}$, the impact portfolios yield economically significant alpha. For example, α_A for the top decile is 1.1% when the correlation $\rho = 30\%$, and 0.4% when $\rho = 10\%$. Observe that ρ^2 is simply the R^2 of the cross-sectional regression of $\boldsymbol{\alpha}$ on $\mathbf{X}$, so a 30% (10%) correlation implies that only 9% (1%) of the variation in $\boldsymbol{\alpha}$ is explained by $\mathbf{X}$, which is a fairly plausible assumption for a typical impact factor.

When the cross-sectional standard deviation, σ_α, increases to 5%, the alpha of the top decile impact portfolio increases further to 2.8% with a 30% correlation and 0.9% with a 10% correlation.

When combined with the passive index to form long/short portfolios, the optimal portfolio contains significant weight from the impact portfolio, sometimes implying a highly leveraged position (the top half of Panel B). The corresponding gain in expected excess return for the combined portfolio is 8.9% when $\rho = 30\%$ and 1.0% when $\rho = 10\%$. Of course, these calculations are meant only as a proof of concept, because we have not considered specific regulatory and institutional constraints for investors in practice. For example, the leveraged portfolios with short positions in the passive index may be suitable only for certain types of hedge funds.

---

[22] This is an innocuous assumption. and we show later via simulation that cross-sectional heterogeneity in idiosyncratic volatility does not affect our conclusions.

[23] We set the individual security weights of the impact portfolios to be always positive by convention from the perspective of long-only investors.

[24] For example, using several factor models, Pástor and Stambaugh (1999) estimate σ_α to be between 0% to 10% with a Bayesian framework (depending on different priors) for 1,994 stocks.

TABLE 8.1: Expected excess returns for the impact portfolios and their corresponding Treynor-Black weights when combined with a passive portfolio. We set $N = 500$ and assume that the passive portfolio has an annualized expected excess return of $\mathrm{E}[R_m] - R_f = 6\%$ and volatility of $\sigma_m = 15\%$. The idiosyncratic volatility is a constant $\sigma(\epsilon_i) = 15\%$ for all securities. Panel A reports α_A for the impact portfolio and Panel B reports the weight, ω_A, and the corresponding excess return of the impact portfolio combined with the passive portfolio

Panel A: Expected excess return of the impact portfolio (α_A)

Correlation ρ	*Bottom*	2nd	9th	*Top*
		$\sigma_\alpha = 2\%$		
30% ($R^2 = 9\%$)	−1.1%	−0.6%	0.6%	1.1%
10% ($R^2 = 1\%$)	−0.4%	−0.2%	0.2%	0.4%
−10% ($R^2 = 1\%$)	0.4%	0.2%	−0.2%	−0.4%
−30% ($R^2 = 9\%$)	1.1%	0.6%	−0.6%	−1.1%
		$\sigma_\alpha = 5\%$		
30% ($R^2 = 9\%$)	−2.8%	−1.6%	1.6%	2.8%
10% ($R^2 = 1\%$)	−0.9%	−0.5%	0.5%	0.9%
−10% ($R^2 = 1\%$)	0.9%	0.5%	−0.5%	−0.9%
−30% ($R^2 = 9\%$)	2.8%	1.6%	−1.6%	−2.8%

Panel B: Impact portfolios combined with a passive portfolio

Correlation ρ	*Weight* ω_A				*Expected Excess Return* $\omega_A\alpha_A$			
	Bottom	2nd	9th	*Top*	*Bottom*	2nd	9th	*Top*
			Long/Short, $\sigma_\alpha = 2\%$					
30% ($R^2 = 9\%$)	−1.31	−0.78	0.78	1.31	1.4%	0.5%	0.5%	1.4%
10% ($R^2 = 1\%$)	−0.44	−0.26	0.26	0.44	0.2%	0.1%	0.1%	0.2%
−10% ($R^2 = 1\%$)	0.44	0.26	−0.26	−0.44	0.2%	0.1%	0.1%	0.2%
−30% ($R^2 = 9\%$)	1.31	0.78	−0.78	−1.31	1.4%	0.5%	0.5%	1.4%
			Long/Short, $\sigma_\alpha = 5\%$					
30% ($R^2 = 9\%$)	−3.23	−1.93	1.93	3.23	8.9%	3.1%	3.1%	8.9%
10% ($R^2 = 1\%$)	−1.08	−0.64	0.64	1.08	1.0%	0.3%	0.3%	1.0%
−10% ($R^2 = 1\%$)	1.08	0.64	−0.64	−1.08	1.0%	0.3%	0.3%	1.0%
−30% ($R^2 = 9\%$)	3.23	1.93	−1.93	−3.23	8.9%	3.1%	3.1%	8.9%
			Long-Only, $\sigma_\alpha = 2\%$					
30% ($R^2 = 9\%$)	0	0	0.78	1.00	0%	0%	0.5%	1.1%
10% ($R^2 = 1\%$)	0	0	0.26	0.44	0%	0%	0.1%	0.2%
−10% ($R^2 = 1\%$)	0.44	0.26	0	0	0.2%	0.1%	0%	0%
−30% ($R^2 = 9\%$)	1.00	0.78	0	0	1.1%	0.5%	0%	0%
			Long-Only, $\sigma_\alpha = 5\%$					
30% ($R^2 = 9\%$)	0	0	1.00	1.00	0%	0%	1.6%	2.8%
10% ($R^2 = 1\%$)	0	0	0.64	1.00	0%	0%	0.3%	0.9%
−10% ($R^2 = 1\%$)	1.00	0.64	0	0	0.9%	0.3%	0%	0%
−30% ($R^2 = 9\%$)	1.00	1.00	0	0	2.8%	1.6%	0%	0%

The bottom half of Panel B shows the performance of long-only combined portfolios, which corresponds to a constrained weight, ω_A, between 0 and 1. When the correlation, ρ, is positive, only the top two deciles provide positive alpha and, therefore, yield positive weight in the combined portfolio. When $\sigma_\alpha = 5\%$, for example, this leads to a sizable expected excess return of 2.8% when $\rho = 30\%$, and 0.9% when $\rho = 10\%$. Section 8.5 provides two real-world examples with positive impact alpha in the context of venture philanthropy and R&D subsidies.

When the correlation ρ is negative, only the bottom two deciles provide positive excess returns. However, securities in these deciles have the lowest impact factor $\mathbf{X}$. Therefore, impact investors may use these to obtain an estimate of the opportunity cost of not investing in these portfolios. Divesting from sin stocks provides a real-world example of this case, which we discuss further in Section 8.5.

Finally, we show how our impact portfolio improves the efficient frontier and the Capital Market Line to achieve a "super-efficient frontier" under the assumption that $\boldsymbol{\alpha}$ are mispricings. Under the omitted-factor interpretation, the "super-efficiency" of the new frontier may be viewed as the result of additional risk premia not accessible to investors except through impact portfolio managers.

Proposition 16 *Under Assumption (A1), the return of the final portfolio, P, that consists of the impact portfolio with Treynor-Black weights and the passive market portfolio is:*

$$R_P - R_f = \omega_A R_A + (1-\omega_A) R_m - R_f = \omega_A \alpha_A + (R_m - R_f)(\beta_A \omega_A + (1-\omega_A)) + \omega_A \epsilon_A, \quad (8.39)$$

where R_m is the return of the passive portfolio. The expected value and variance of R_P are:

$$\begin{aligned} \mathrm{E}[R_P] - R_f &= \omega_A \alpha_A + (\mathrm{E}[R_m] - R_f)(\beta_A \omega_A + (1 - \omega_A)), & (8.40) \\ \mathrm{Var}[R_P] &= \mathrm{Var}[R_m](\beta_A \omega_A + (1 - \omega_A))^2 + \omega_A^2 \sigma(\epsilon_A)^2. & (8.41) \end{aligned}$$

This forms a super-efficient frontier in comparison to the Capital Market Line associated with the passive portfolio.

Figure 8.3 displays the passive portfolio as well as several combinations with impact portfolios in relation to the efficient frontier. We continue to assume that the passive portfolio has an annualized risk premium of $\mathrm{E}[R_m] - R_f = 6\%$ and volatility of $\sigma_m = 15\%$. In Figures 8.3a and 8.3c, the idiosyncratic volatility is assumed to be a constant $\sigma(\epsilon_i) = 15\%$ for all securities. As the correlation ρ and variance in alpha, σ_α^2, increase, the impact portfolios (defined as the top half of the securities ranked by $\mathbf{X}$) are able to improve the original Capital Market Line for both long/short and long-only portfolios, leading to super-efficient frontiers.

The results in this section have so far relied on the assumption that the idiosyncratic volatility, $\sigma(\epsilon_i)$, is cross-sectionally constant. To check the robustness of our results, we simulate a collection of securities where the i-th security's idiosyncratic volatility follows a lognormal distribution:

$$\log(\sigma(\epsilon_i)) \sim \mathrm{Normal}(\mu_\epsilon, \sigma_\epsilon). \quad (8.42)$$

Calibrating to empirically plausible values in the literature (e.g., Kuntz (2020)), we perform simulations for $\log(\mu_\epsilon) = 15\%$ and $\sigma_\epsilon = 1$. Figures 8.3b and 8.3d confirm that even with such cross-sectional heterogeneity, the Capital Market Line is still improved, leading to super-efficient frontiers.

8.4.4 Qualifications

The Treynor-Black portfolios outlined in Proposition 15–16 provide the optimal impact portfolios when investors face no constraints. In practice, investors may face additional regulatory and institutional constraints, such as limits on short positions and sector concentration. Because our framework quantifies the expected alpha for individual securities, it is easy to derive optimal impact portfolios accounting for these additional constraints as well. Section 8.4.3 provides examples of long-only portfolios. In the meantime, we emphasize

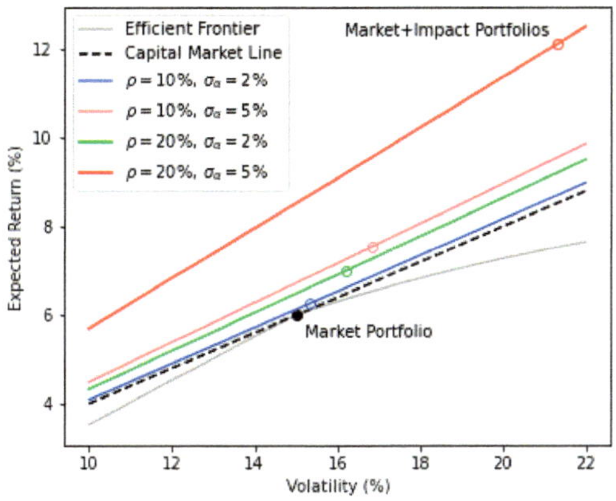

(a) Long/Short, Constant $\sigma(\epsilon_i)$

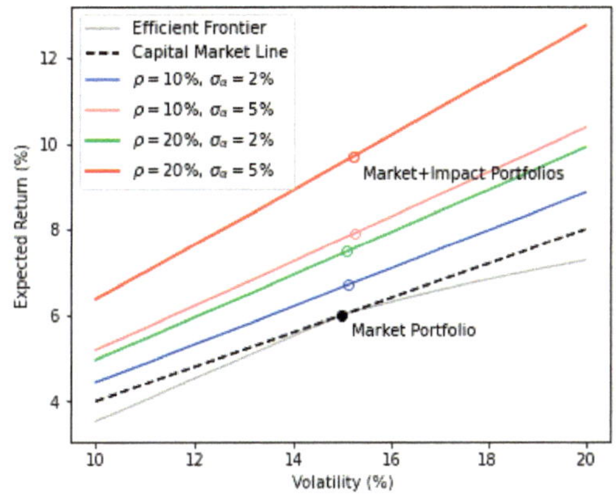

(b) Long/Short, Simulated $\sigma(\epsilon_i)$

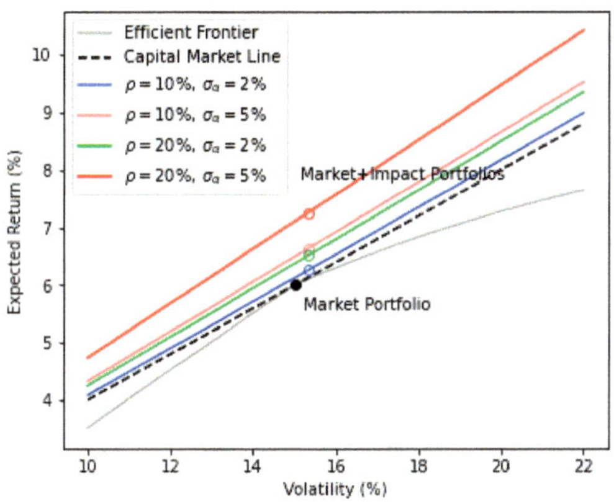

(c) Long-Only, Constant $\sigma(\epsilon_i)$

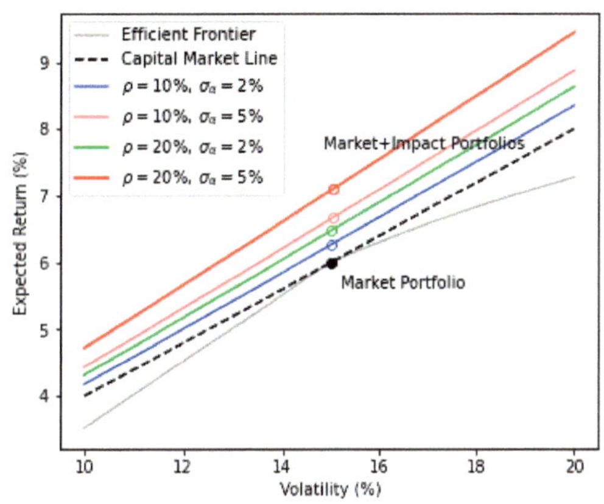

(d) Long-Only, Simulated $\sigma(\epsilon_i)$

FIGURE 8.3: Super-efficient frontiers from the combined portfolio that consists of the impact portfolio and passive market portfolio. (a) and (b) allow short positions in the passive index, while (c) and (d) consider long-only portfolios. In (a) and (c) the idiosyncratic volatility is a constant $\sigma(\epsilon_i) = 15\%$ for all securities. In (b) and (d) we simulated idiosyncratic volatility based on Equation (8.42) and apply a maximum leverage ratio of 3:1.

that the performance metrics associated with these portfolios should be taken as a proof of concept, and real-world investors should account for their specific constraints and trading costs on top of our framework to derive proper benchmarks.

Our framework provides a more robust methodology to construct portfolios in practice compared to traditional approaches. First, investors may follow traditional mean-variance optimization and integrate the level of impact as part of either the objective or constraint.[25] However, estimating the expected return and covariance matrix in traditional Markowitz portfolios is known to be challenging, which often leads to unstable portfolios (Brodie et al., 2009; Tu and Zhou, 2011). Thanks to its ability to quantify individual security alphas

[25] In particular, investors can construct portfolios by maximizing the expected utility that adds a term for portfolio ESG on top of traditional mean-variance of security returns. Alternatively, they can maximize expected utility subject to a constraint that the portfolio ESG is higher than a preset threshold. See Pedersen, Fitzgibbons, and Pomorski (2021) and Pástor, Stambaugh, and Taylor (2021) for example.

based on a small (2) number of parameters, our approach provides a much more regularized approach to constructing optimal impact portfolios.

Second, investors may also build portfolios through a univariate regression, in which security alphas are estimated by regressing returns onto the impact factor $\mathbf{X}$. Compared to this approach, our framework corresponds more closely to actual industry construction of impact portfolios, which usually relies on ordering securities based on $\mathbf{X}$ and assigning weights to each quartile or decile. In fact, our theory requires weaker assumptions than this approach because its results only depend on the rank of $\mathbf{X}$, not its value. As a result, the optimal portfolio relies on quantiles of $\mathbf{X}$, making our approach robust against noise and outliers, as demonstrated empirically by Lo, Zhang, and Zhao (2022). its focus on rank also allows for the generalization of our theory to capture nonlinear dependence between returns and the impact factor, as shown subsequently (Lo et al., 2024).

The key insight of our framework lies in its ability to quantify the distribution of impact-ranked returns when mispricings exist for individual securities. In an omitted factor interpretation, it identifies impact factors $\mathbf{X}$ that explain the cross-section of security returns, and quantifies the excess returns associated with these omitted factors. If alphas are fully observed—either in the mispricing or omitted factor interpretation—then investors have complete information to their tradeoff returns and impact, and optimal impact portfolios can simply be constructed following Treynor and Black (1973). Our framework shows how to quantify the impact of impact investing when alphas are unknown in the real world, and how to construct the optimal portfolio that achieves a "super-efficient frontier" either through mispricings or additional risk premia not accessible to investors except through impact portfolio managers.

When alphas are interpreted as omitted factors whose beta loadings are correlated with the impact factor, $\mathbf{X}$, our model also provides an approach to recovering the stochastic discount factor (SDF).[26] To see this, observe that our results in Proposition 13 quantify the excess returns of impact-ranked securities. In particular, they identify the $\mathbf{X}$ that explains the cross-section of returns. The correlation scaled by the standard deviation, $\rho_{adj}\sigma_\lambda$, can be treated as the risk premium for the omitted factor, while the standardized impact score, $\mathrm{E}[X_{i:N}]$, provides the factor loadings. Because a factor model in expected returns always corresponds to a factor model in SDF space, our framework essentially recovers the SDF with respect to this factor $\mathbf{X}$, though the corresponding SDF is not uniquely determined.[27]

8.5 Applications to five impact investments

In this section, we apply our framework to five particular examples of impact investing: biotech venture philanthropy, semiconductor R&D consortium, divesting from sin stocks, ESG investing, and the GameStop short squeeze during January 2021.

[26] We thank the associate editor for pointing out this interesting connection.

[27] The correspondence between return factor models and SDF factor models is initially due to Ross (1978) and Dybvig and Ingersoll (1982). The example given by Cochrane (2009, Section 6.3) illustrates this connection in its simplest form when asset returns are written in excess of the risk-free rate, which we denote as R^e following Cochrane (2009). Given a return factor model $E[R^e] = \beta'\mathrm{E}[\lambda]$, one can always find a SDF, m, such that m satisfies $m = 1 + b'\lambda$ and $0 = \mathrm{E}[mR^e]$, where b is given by the equation $\mathrm{E}[\lambda] = -\mathrm{Var}[\lambda]b$. Nonetheless, m is not unique because one can always add to m any random variable orthogonal to returns.

8.5.1 Venture philanthropy: the cystic fibrosis foundation

The concept of venture philanthropy (VP) was introduced by Letts, Ryan, and Grossman (1997), who suggested that nonprofit organizations could learn useful practices from venture capitalists. In particular, recent biomedical advances have created significant opportunities for a new generation of therapeutics (Sharp and Hockfield, 2017). However, early stage R&D efforts often face a dearth of funding, given the high risk of failure and significant funding requirements. This has been particularly true for rare disease drug development, where market sizes are often too small to attract much attention and funding (Kim and Lo, 2019).

We consider the example of the Cystic Fibrosis (CF) Foundation—profiled in the case study by Kim and Lo (2019)—and conclude that VP in biomedicine can produce significant positive excess returns. This example illustrates the possibility of an impact investment that is positively correlated with $\boldsymbol{\alpha}$, or an omitted factor that patient advocacy groups can more easily exploit than typical investors.

The CF Foundation is the world's leading philanthropic organization for CF, a rare genetic disease that currently affects more than thirty thousand Americans. Over a period of 12 years, the CF Foundation invested \$150 million to fund CF drug development efforts at Vertex Pharmaceuticals, which led to the identification and development of Kalydeco, the first FDA-approved treatment to address the underlying causes of CF. In 2014, their rights to Vertex royalties were sold to an outside investment firm, New York City-based Royalty Pharma, for \$3.3 billion in cash.

If we assume, for simplicity, that the \$150 million investment was made upfront and the \$3.3 billion sale occurred 12 years later, this implies a compound annual return of 29.4% over this period. To estimate the realized α of this investment, we require an estimate of the cost of capital of Vertex during the 12-year investment period from 2002 to 2013, prior to the 2014 royalty sale. We consider a simple CAPM model:

$$R_i - R_f = \alpha_i + \beta_i(R_M - R_f) + \epsilon_i, \qquad (8.43)$$

where R_i is the return of the i-th security with CF Foundation being one of them, R_f is the risk-free rate, and R_M is the market return. The average beta of Vertex between 2 January 2001 and 31 December 2013 was 1.42. The average 5-year constant-maturity Treasury yield from January 2001 to December 2013 was 2.8%,[28] and the annualized compound return of the CRSP value-weighted returns index with dividends during this period was 5.4%, leading to a simple CAPM estimate of the cost of capital of $1.42 \times (5.4\% - 2.8\%) + 2.82\% = 6.5\%$.

Of course, this crude estimate does not account for the illiquid nature of biomedical assets or the financing risks posed by their multi-year investment horizons. A cost of capital of 20% for privately held biotech investments is a commonly used industry benchmark. Therefore, a plausible range for the α of the CF Foundation's investment in Vertex is 9.4% (using a 20% cost of capital) to 22.9% (using a 6.5% cost of capital).

Using this estimated range for the CF Foundation's α, we can estimate its correlation with a "rare disease impact investing" factor $\mathbf{X}$. Making a few additional assumptions about auxiliary parameters, we can reverse-engineer an implied correlation, ρ, that is consistent with this performance range, which is $[35\%, 86\%]$.[29] Our highly stylized calculations are not

[28] See https://fred.stlouisfed.org/series/GS5/.

[29] Using a Bayesian framework, Pástor and Stambaugh (1999) estimate the posterior σ_α to be between 0% to 10% (depending on different priors) for 1,994 stocks using several factor models. Given that private biopharma companies likely have larger mispricings than the average stock, we assume that the cross-sectional standard deviation of $\boldsymbol{\alpha}$ is $\sigma_\alpha = 10\%$, and the CF Foundation's investment in Vertex ranks at the top 1% of $N = 10,000$ securities based on a rare disease impact factor. If we assume instead that $\sigma_\alpha = 20\%$, the implied correlation range is $[18\%, 43\%]$.

meant to yield a rigorous estimate of the true alpha associated with drug development for rare diseases, and the plausible range of the true alpha is likely larger, potentially including zero. However, more systematic empirical analyses of the biopharma industry show that pharmaceutical companies have become increasingly profitable, with risk-adjusted returns outperforming the aggregate stock market in recent years (Thakor et al., 2017; Lo and Thakor, 2019). The example of the CF Foundation provides additional intuition for how impact and investment performance need not be a zero-sum game in the presence of sufficient correlation between impact and performance.

Nevertheless, there is a deeper message in this striking example, which is that, in certain cases, impact is a *prerequisite* for performance. The CF Foundation's main objective—helping to create a disease-modifying drug for CF—was, in fact, the primary source of Vertex's outsized investment performance. The fact that the Foundation focused on this one long-term goal, to the exclusion of shorter-term financial metrics and milestones, and was willing to continue investing in Vertex over multiple years despite business cycle fluctuations (including the 2008 Financial Crisis) contributed significantly to its success, both in impact and in financial returns. Indeed, many traditional venture capitalists have shied away from investing in projects with such high risks and long-term capital commitments. In other words, in this case, correlation may actually be *causation*; impact can sometimes be responsible for financial success.

8.5.2 An R&D consortium: sematech

The second example of impact investing is Semiconductor Manufacturing Technology ("Sematech"), an R&D consortium established in 1987 by 14 U.S. semiconductor firms and the U.S. government to solve common manufacturing problems by leveraging shared R&D resources and research results. It was initially funded by the U.S. government, and later by its member firms.

To measure the returns to Sematech as a form of impact investing, we follow Irwin and Klenow (1996a) and use a difference-in-difference approach to compare the excess returns to Sematech member firms against those of non-members, controlling for past returns prior to the formation of Sematech. In particular, we collect monthly return data from CRSP for all U.S. firms whose principal business is semiconductors and related devices (SIC 3674) from 1975 to 1999.[30] Our sample consists of 52 firms, including 11 of the original 14 members of Sematech.[31]

For each firm, we divide the sample into pre-Sematech (January 1975 to August 1987) and post-Sematech periods (September 1987 to December 1999) and estimate two CAPM models:

$$\begin{aligned} R_{i,\text{pre}} - R_{f,\text{pre}} &= \alpha_{i,\text{pre}} + \beta_{i,\text{pre}}(R_{M,\text{pre}} - R_{f,\text{pre}}) + \epsilon_i, \\ R_{i,\text{post}} - R_{f,\text{post}} &= \alpha_{i,\text{post}} + \beta_{i,\text{post}}(R_{M,\text{post}} - R_{f,\text{post}}) + \epsilon_i, \end{aligned} \tag{8.44}$$

where $R_{i,\cdot}$ is the return of the i-th security, $R_{f,\cdot}$ is the risk-free rate estimated by the 5-year constant maturity Treasury yield, and $R_{M,\cdot}$ is the market return estimated by the CRSP value-weighted returns index including dividends.

---

[30] Sematech consisted only of U.S. companies during its initial years. In the late 1990s, members from Asia and Europe began to join Sematech in a limited capacity, and Sematech completed its first year of operations as a unified global consortium in 2000. Therefore, we chose our sample period to be from 1975 to 1999, which covers roughly 12 years before and after the formation of Sematech. Returns are winsorized at 15% each side to reduce the impact of outliers.

[31] We require at least six valid monthly returns before and after the formation of Sematech for each firm to be included. The 11 Sematech members that meet these criteria are AT&T Microelectronics, Advanced Micro Devices, International Business Machines, Hewlett-Packard, Intel, LSI Logic, Micron Technology, Motorola, National Semiconductor, Rockwell International, and Texas Instruments.

TABLE 8.2: Estimated annualized CAPM excess returns based on a difference-in-difference (DiD) approach that compares Sematech member firms against non-members, before and after the formation of Sematech. Estimated coefficients are significant at the 1 percent (***), 5 percent (**), or 10 percent (*) levels based on Bootstrap confidence intervals

Firm	*Period*	*Excess return*	*Excess return* (Post minus pre)	*Excess return* (DiD)
Sematech Members	Pre	−2.67%	9.90%**	10.22%*
	Post	7.23%***		
Non-Members	Pre	−8.90%***	−0.33%	
	Post	−9.22%***		

Table 8.2 summarizes the estimated annualized alphas. Sematech member firms have an average excess return of 7.23% after joining the R&D consortium compared to an average of −2.67% prior to Sematech's formation. The average excess returns for non-member firms are around −9% in both periods. This leads to an overall difference-in-difference estimate of 10.22% lift in annualized excess returns for a firm that joined the R&D consortium, with a 90% Bootstrap confidence interval of $[0.56\%, 19.88\%]$. If we consider an impact factor $\mathbf{X}$ with $X_i = 1$ representing Sematech membership and 0 otherwise, this leads to an estimated correlation, ρ, of 25% with a range of $[1.4\%, 49\%]$ between the Sematech impact factor and excess returns.[32]

Our estimated excess returns based on the simple CAPM model should only be interpreted as a crude approximation to the true effect of Sematech, and the true effect may be different once other control variables are accounted for. In particular, firm entry decisions into the consortium may have been endogenously affected by other factors, such as the size and the profitability of the firm. We conduct a simple analysis in the appendix of Lo and Zhang (2024) to mitigate this concern, comparing basic characteristics of Sematech vs. non-Sematech firms, showing that the impact from the Sematech consortium largely remains robust on samples that are constructed to match the distribution of Sematech firms.

Our results are also consistent with prior evidence on Sematech's far-reaching effects on members' R&D spending and profitability (Irwin and Klenow, 1996a,b; Link, Teece, and Finan, 1996). In particular, Irwin and Klenow (1996a) show that joining Sematech improved firm profitability, a potential channel through which excess returns were earned.

This example shows how the impact of R&D consortia can be measured in our framework, and the difference-in-difference approach yields a potentially causal estimate of the positive financial impact of the Sematech impact factor.

8.5.3 Divesting sin stocks

Another particular type of impact investing is avoiding or divesting sin stocks—stocks from companies involved in or associated with activities considered unethical or immoral. Although there may be a degree of subjectivity involved in determining what is considered sinful, common examples include companies involved in producing, distributing, or otherwise supporting alcohol, tobacco, gambling, sex-related industries, and firearms. It has been found that sin stocks are less commonly held by norm-constrained institutions such as pension plans compared to mutual funds or hedge funds, and receive less coverage from

[32] We assume that the cross-sectional standard deviation of $\boldsymbol{\alpha}$ is $\sigma_\alpha = 20\%$, and the universe of firms in the semiconductor industry is approximately 200. If we assume instead that $\sigma_\alpha = 10\%$, the average implied correlation is 51% with a range of $[2.8\%, 99\%]$.

analysts. As a result, sin stocks seem to yield higher expected returns (Fabozzi, Ma, and Oliphant, 2008; Hong and Kacperczyk, 2009; Statman and Glushkov, 2009).

This empirical fact implies a negative correlation between a stock's excess return and an "anti-sin stock" factor. In terms of the super-efficient frontier shown in Figure 8.3, divesting from sin stocks likely yields a negative return and a lower efficient frontier. This leads to a natural definition of the cost to this specific impact factor.

We use Hong and Kacperczyk (2009) to calibrate our model, and focus on tobacco, alcohol, and gambling as proxies for sin stocks. The authors report a monthly excess return of 0.26% for an equal-weighted portfolio long sin stocks and short their comparables, by running a time-series regression controlling for market, size, value, and momentum factors, using equity data in the United States from 1965 to 2006. This corresponds to the following four-factor asset-pricing model:

$$R_i - R_f = \alpha_i + \beta_{i,1}(R_M - R_f) + \beta_{i,2}SMB + \beta_{i,3}HML + \beta_{i,4}MOM + \epsilon_i, \tag{8.45}$$

and an impact factor $\mathbf{X}$ representing whether a stock is considered "sinful." Hong and Kacperczyk's (2009) estimate can be used to calculate the implied correlation between $\boldsymbol{\alpha}$ and $\mathbf{X}$ in our model, using results from Proposition 10.

Panel A of Table 8.3 summarizes these calibration results.[33] The implied correlation is 27% ($R^2 = 7.2\%$), assuming a standard deviation of cross-sectional alpha of $\sigma_\alpha = 5\%$.[34] This leads to a measure of the cost of avoiding sin stocks. If we form an impact portfolio based on the top half of all securities based on an anti-sin factor, it leaves an excess return of 1.7% per annum on the table. If we form a Treynor-Black portfolio based on the omitted sin stocks and the passive market portfolio, we could have achieved a leveraged alpha of 14.4% per annum with a (leveraged) weight of 8.58 for the sin stocks portfolio. Conversely, if we form an impact portfolio by leaving out only the top decile of the most sinful stocks, the opportunity cost is 2.5% , and if we form an impact portfolio by leaving out the top 2%, the opportunity cost is 3.3%.

A few studies have tried to understand why sin stocks appear to show positive excess returns. In particular, Blitz and Fabozzi (2017) show that sin stocks indeed exhibit a significantly positive CAPM alpha, but this alpha disappears completely when controlling not only for classic factors such as size, value, and momentum, but also for exposures to the two new Fama and French (2015) quality factors, profitability and investment. This corresponds to the following six-factor asset-pricing model:

$$R_i - R_f = \alpha_i + \beta_{i,1}(R_M - R_f) + \beta_{i,2}SMB + \beta_{i,3}HML + \beta_{i,4}MOM + \beta_{i,5}RMW + \beta_{i,6}CMA + \epsilon_i. \tag{8.46}$$

We also summarize the implied correlation and cost of divesting sin stocks based on Blitz and Fabozzi (2017) in Panel B of Table 8.3.[35] Both the correlation and the excess returns of sin stocks decrease sharply based on this study.

This example spotlights the fact that the measurement of excess returns of impact investing depends on the specific asset-pricing model used to estimate alpha. Our framework can be applied to any number of factors as specified in Equations (8.1)–(8.2). Indeed, a factor

---

[33] Hong and Kacperczyk (2009) report 193 sin stocks in their selection, and Blitz and Fabozzi (2017) report that sin stocks are about 2.5% of its universe. We calibrate to these parameters when determining the quantiles of the induced order statistics.

[34] We use an intermediate value based on Pástor and Stambaugh's (1999) estimate of σ_α (between 0% to 10% for 1,994 stocks). Results for other values of σ_α follow trivially. Note that different values for σ_α lead to different estimates of ρ, but not of the final estimates of the cost of avoiding sin stocks, since the expected alpha is invariant to the product of the two ($\rho\sigma_\alpha$).

[35] Blitz and Fabozzi (2017) use monthly returns. For U.S. data in 1963–2016, the authors report a non-significant monthly excess return of 0.10%. This number becomes negative when restricted to data after 1990.

TABLE 8.3: Estimated cost in excess return per annum for avoiding sin stocks, calibrated to prior empirical studies. Here we assume that the passive portfolio has an annualized risk premium of $E[R_m] - R_f = 6\%$ and volatility of $\sigma_m = 15\%$, and $\sigma_\alpha = 5\%$

Impact portfolio	*Weight of impact portfolio* ω_A	*Expected excess return* *Impact portfolio* α_A	*Combined with Passive portfolio* $\omega_A\alpha_A$
Panel A: Hong and Kacperczyk (2009)			
Implied correlation ρ=27% (R^2=7.2%).			
Top Half	8.58	1.7%	14.4%
Top Decile	3.78	2.5%	9.35%
Top 2%	1.04	3.3%	3.4%
Panel B: Blitz and Fabozzi (2017)			
Implied correlation ρ=10% (R^2=1.1%).			
Top Half	3.30	0.6%	2.1%
Top Decile	1.45	1.0%	1.4%
Top 2%	0.40	1.3%	0.5%

may yield positive correlation with alpha under one asset-pricing model, implying a positive excess return, that may disappear or change sign after controlling for additional factors.

8.5.4 ESG investing

SRI and ESG-aware investing have both drawn an increasing amount of attention in recent years. Our model provides a systematic framework to measure the financial impact of SRI and ESG—positive or negative—and construct optimal portfolios based on the correlation between the impact characteristic and excess returns.

As with the sin stocks in Section 8.5.3, we also calibrate our model with respect to several studies in Table 8.4. Panel A uses the estimates from Berg et al. (2024), who construct ESG portfolios based on a range of aggregation methods to de-noise ESG metrics from six different data vendors. They report an annualized Fama-French five-factor alpha of 4.5% for the top-bottom decile portfolio in the U.S. using a simple average ESG score, which implies a correlation of 26% ($R^2=6.7\%$) between stock alpha and the ESG impact factor. This is consistent with the opinions of industry advocates of ESG, although the magnitude of excess returns in the literature vary with the specific ESG metric.

In contrast, Baker et al. (2022) study the U.S. bond market, and report a yield difference of 6 basis points at issuance for green bonds below other ordinary bonds.[36] This corresponds to a plausible and economically meaningful 0.6% difference in value on a bond with a 10-year duration. Panel B of Table 8.4 shows the implied correlation of −2%.[37] This result points to a negative to neutral ESG alpha in the bond market, in which case the excess return, $\omega_A\alpha_A$, in the last column should be interpreted as the cost to ESG investing in this particular market.

Several recent studies provide additional insights into which economic state variables may potentially be driving the observed ESG returns. Using stock data from the S&P 500 and Russell 3000 in 1993–2013, Bansal, Wu, and Yaron (2022) document a "luxury-good

[36] 2,083 green U.S. municipal bonds are used in the sample, compared to 643,299 ordinary bonds.

[37] We assume the standard deviation of cross-sectional alpha is $\sigma_\alpha = 1\%$ because of smaller magnitudes for bond returns. This corresponds to a lower value in the range of estimates by Pástor and Stambaugh (1999) (between 0% to 10% in the stock market). like our results for sin stocks, different values for σ_α lead to different estimates of ρ, but not the final estimates of the ESG alpha.

TABLE 8.4: Estimated ESG excess returns per annum, calibrated to prior empirical studies. Here we assume that the passive portfolio has an annualized risk premium of $\mathrm{E}[R_m] - R_f = 6\%$ and volatility of $\sigma_m = 15\%$, and $\sigma_\alpha = 5\%$ for stock markets and $\sigma_\alpha = 1\%$ for bond markets

Impact portfolio	*Weight of impact portfolio* ω_A	*Expected excess return* *Impact portfolio* α_A	*Combined with passive portfolio* $\omega_A\alpha_A$
Panel A: Berg et al. (2024)			
Implied correlation $\rho = 26\%$ ($R^2 = 6.7\%$).			
Top Half	3.20	1.61%	5.15%
Top Decile	1.41	2.38%	3.35%
Top 2%	0.39	3.14%	1.24%
Panel B: Baker et al. (2022)			
Implied correlation $\rho = -2.0\%$ ($R^2 = 0.04\%$).			
Top Half	−1.06	−0.02%	0.03%
Top Decile	−0.47	−0.04%	0.02%
Top 2%	−0.13	−0.05%	0.00%
Panel C: Bansal, Wu, and Yaron (2022) ("good times")			
Implied correlation $\rho = 22\%$ ($R^2 = 4.7\%$).			
Top Half	2.69	1.35%	3.64%
Top Decile	1.19	1.99%	2.37%
Top 2%	0.33	2.65%	0.88%
Panel D: Bansal, Wu, and Yaron (2022) ("bad times")			
Implied correlation $\rho = -0.2\%$ ($R^2 = 0.0\%$).			
Top Half	−0.02	−0.01%	0.00%
Top Decile	−0.01	−0.02%	0.00%
Top 2%	−0.00	−0.02%	0.00%
Panel E: Pástor, Stambaugh, and Taylor (2022)			
Implied correlation $\rho = 5.5\%$ ($R^2 = 0.3\%$).			
Top Half	0.69	0.34%	0.24%
Top Decile	0.30	0.51%	0.15%
Top 2%	0.08	0.68%	0.06%
Panel F: Lo, Zhang, and Zhao (2022)			
Implied correlation $\rho = 3.8\%$ ($R^2 = 0.1\%$).			
Top Half	0.47	0.24%	0.11%
Top Decile	0.21	0.35%	0.07%
Top 2%	0.06	0.46%	0.03%

effect" for an ESG factor that combines analyst ratings, firm announcements, and realized incidents. Stocks with higher ESG ratings significantly outperform lower-ranked ones during good economic times, but not during bad economic times, resembling the demand for luxury goods. Their analysis corresponds to an impact factor **X** representing the customized ESG score, and a four-factor asset-pricing model with an interaction term:

$$\begin{aligned} R_i - R_f = \alpha_i + \beta_{i,0} I + \beta_{i,1}(R_M - R_f) + \beta_{i,2} SMB + \beta_{i,3} HML + \beta_{i,4} MOM \\ + \gamma_{i,1}(R_M - R_f) \cdot I + \gamma_{i,2} SMB \cdot I + \gamma_{i,3} HML \cdot I + \gamma_{i,4} MOM \cdot I + \epsilon_i, \end{aligned} \tag{8.47}$$

where I is a dummy variable representing good economic times defined by the cyclically-adjusted real P/E (CAPE) ratios from Shiller (2005). We report the implied correlation between stock alpha and the ESG factor based on their estimates for good and bad economic times, in Panels C and D of Table 8.4, respectively. During good economic times, Bansal, Wu, and Yaron (2022) report a monthly Fama-French four-factor alpha of 0.315% for the top-bottom ESG portfolio. This implies a 22% correlation between stock alpha and the ESG factor, and sizable positive excess returns for the impact portfolios. However, during bad economic times, the monthly Fama-French four-factor alpha in Bansal, Wu, and Yaron (2022) becomes −0.0026%, rendering all of our estimates of correlation and ESG alpha to be essentially zero.

Pástor, Stambaugh, and Taylor (2022) provide another channel through which ESG returns can be explained. Using the environmental score from MSCI ESG Ratings data, which corresponds to the impact factor **X** in our framework, they document a highly significant monthly Fama-French three-factor alpha of 0.50% for a green-minus-brown portfolio,[38] which reduces to an insignificant level of 0.05% when excess returns are regressed on proxies of unexpected climate shocks:

$$R_i - R_f = \alpha_i + \beta_{i,1}(R_M - R_f) + \beta_{i,2}SMB + \beta_{i,3}HML + \beta_{i,4}DCC + \beta_{i,5}DCC_{\text{Lag1}} + \epsilon_i, \quad (8.48)$$

where DCC stands for a "delta climate concern" variable constructed from Ardia et al.'s (2022) climate change concerns index, and DCC_{Lag1} its one-month lagged values. This implies a much smaller correlation of 5.5% (Panel E) compared to estimates from Berg et al. (2024), for example.

Likewise, Panel F shows the estimates based on Lo, Zhang, and Zhao (2022), who use Trucost Environmental data to find a similar effect to Pástor, Stambaugh, and Taylor's (2022) findings for a wide range of environmental measures, in particular, to carbon emissions. Based on the framework in this chapter, they estimate an annualized alpha of 2.38% for a long/short green portfolio. This alpha reduces to a statistically insignificant level of 0.3% per annum with the same proxy variables for unexpected climate shocks:

$$\begin{aligned} R_i - R_f = \alpha_i &+ \beta_{i,1}(R_M - R_f) + \beta_{i,2}SMB + \beta_{i,3}HML + \beta_{i,4}RMW + \beta_{i,5}CMA \\ &+ \beta_{i,6}DCC + \beta_{i,7}DCC_{\text{Lag1}} + \epsilon_i, \end{aligned} \quad (8.49)$$

which corresponds to a small implied correlation of 3.8% between excess returns and a low-carbon impact factor **X**.[39]

The five studies we spotlight in Table 8.4 underscore the difficulty in measuring consistent excess returns of ESG, which depend on many factors, including the asset class, region, and time period under study. In addition, the specific choice of asset-pricing model also affects the empirical estimates of ESG alpha (see also Madhavan, Sobczyk, and Ang (2021) and Geczy, Stambaugh, and Levin (2021)).

8.5.5 The GameStop phenomenon

In January 2021, the share price of Gamestop Corp. (GME)—a struggling videogame retailer that had recently announced a 30% decline in 2020Q3 net sales, due in part to an 11% reduction in their store base—went from \$17.25 on January 5 to an all-time high of \$347.51 on January 27.[40] Although few investment professionals would consider GME an "impact investment," it is difficult to categorize it as anything else, given the apparent origin of its meteoric price spike.

[38]Stocks ranked by the greenness scores in the top third minus those in the bottom third.

[39]In particular, this estimate corresponds to a portfolio constructed from ranking stocks based on the negative value of the logarithm of scope 1 carbon emissions.

[40]Note that these prices are unadjusted for the subsequent 4-to-1 stock-split that GME underwent on July 22, 2022.

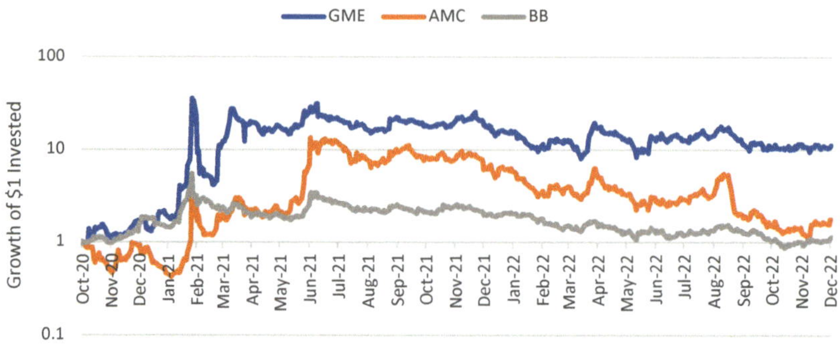

FIGURE 8.4: Growth of one dollar invested in GameStop (GME), AMC Entertainment Holdings (AMC), and Blackberry (BB) on a semi-logarithmic scale.

The key turning point for GME seemed to be growing interest among retail investors affiliated with the Reddit forum "r/WallStreetBets." While it is difficult to determine the exact cause and motivation behind the early initiators, the GME price spike is unlikely to have been driven by changes in the fundamentals of the company,[41] but rather caused by a combination of a grass-roots "David vs. Goliath" conflict between retail investors and hedge-fund short-sellers, and trend followers taking advantage of this dynamic. Other stocks that seemed to be involved in this movement included AMC Entertainment Holdings (AMC) and Blackberry (BB), both of which were facing short-selling pressure from institutional investors. These events attracted substantial media attention due to the populist narrative playing out on social media at the time, as well as the stock's extraordinary price gyrations and wealth transfers. As shown in Figure 8.4, if an investor bought $1 of GME at the beginning of October 2020, she would have gained over $30 at the end of January 2021. Strikingly, GME's stock price stayed at a high level ever since. As of December 2022, the initial $1 is still worth over $10. This sustained price increase is no longer just attributable to temporary supply/demand imbalances—the increased value of GME equity gave the company substantial resources to improve its business, thereby allowing it to make positive changes in its business which permanently altered its future prospects for the better.[42]

In a very real sense, WallStreetBets participants can be viewed as a general form of impact investors. By most accounts, they have been highly successful in achieving the impact they desired, i.e., punishing the short-sellers, pushing up the price of an underdog company bullied by elite institutional investors, and saving the company from bankruptcy. However, to distinguish this type of activity from traditional impact investing, we shall call the GME phenomenon "price-impact investing".

The Gamestop example is likely driven by retail mania and herding, rather than a social welfare-enhancing investment. It has a very different (and perhaps opposite) underlying economic motivation from the previous impact investing examples we gave. However, there are mechanical similarities between the Gamestop mania and, for example, ESG investing. While the former was perhaps driven by investor mania with fast and temporary shifts in

[41] In fact, GME's revenue has been declining since 2017 and its earnings-per-share has been negative since 2018.

[42] For example, on July 6, 2021 Gamestop announced an expansion of its fulfillment center through the leasing of a 530,000 square foot facility in Reno, Nevada; by September 2021 the company hired an additional 500 employees; and in November 2021, the company secured a $500 million global asset-based revolving credit facility, which was oversubscribed, and which replaced its existing $420 million facility due in November 2022 (in addition to delivering enhanced liquidity, the new facility offered reduced borrowing costs, lighter covenants and more flexibility).

preferences toward meme stocks, while the latter is driven by slow but persistent shifts in preferences toward green assets (or concerns toward climate change), as shown by Pástor, Stambaugh, and Taylor (2022) and Lo, Zhang, and Zhao (2022), some behavioral economists might in fact argue that the distinction between manias and preference shifts is largely a matter of semantics.

In the case of GME, it is almost obvious in retrospect that the very act of investing can produce a positive α, at least for a short period of time.[43] However, the same strategy may not work as well for other stocks. In general, all stocks can be affected by such price-impact investors in theory, but the degree to which each of them is susceptible depends on a number of factors, including its market capitalization, liquidity, price dynamics, main shareholders, amount of short interest, sentiment and attention from the general public, and so on. Moreover, manipulating the prices of publicly traded equities clearly violates both securities law and anti-trust regulation,[44] therefore there are significant ethical and legal ramifications of this type of price-impact investing that have yet to be fully explored. Nevertheless, our impact framework provides a means to measure the magnitude of such investments, which have the potential to be an important component of policy debates on whether and how to regulate this type of activity. Because this type of price-impact investing is very different from the other impact investments, in the appendix of Lo and Zhang (2024), we demonstrate how to apply well-known market microstructure models such as Bertsimas and Lo (1998) to quantify the relation between short-term trading programs and market price reactions.

In practice, it is difficult to accurately calibrate the relevant parameters for each stock, and consequently, the expected profit of engaging in GME-like price-impact investing is correspondingly difficult to estimate. However, the fundamental determinant for a price-impact investor's α is the correlation between each stock's trading profit and its susceptibility to price manipulations as a function of its characteristics, e.g., its market capitalization, liquidity, and short interest. In fact, stocks like GME, AMC Entertainment Holdings, and Blackberry were the perfect targets for the short squeezes that occurred at the end of 2020 to early 2021 because of their highly publicized amounts of short interest from hedge funds, and high customer concentration among young people, both features of which are arguably correlated to short-squeeze profits.

8.6 General dependence and marginal distributions

Lo et al. (2024) generalize the results in this chapter to allow for arbitrary marginal distributions of the asset returns and the impact factor, as well as arbitrary dependence structures between the two, characterized by copulas. Our main result is a series of representation theorems which show that, for a *general* bivariate distribution of $(X_i, \alpha_i)^\top$, the distribution of induced order statistics of α_i ranked by X_i can be represented by a mixture of order statistics and uniformly distributed random noise, and the mixture function is determined by the copula of X_i and α_i and the marginal distribution of α_i. Due to page constraints, we only include the main representation theorem under Assumption (A3) below and a smoothness condition to help readers build an intuition for the final result. For the

[43]What constitutes a "short" period of time is clearly subjective and context dependent—as of December 2, 2022, GME's closing price was \$27.52, more than 10 times higher than the \$2.62 closing price on October 30, 2020 (note: these are split-adjusted prices).

[44]See the Securities Exchange Act of 1934, Section 9, and the Sherman Act and the Commodity Exchange Act.

general representation theorem without any smoothness assumptions, please see Lo et al. (2024).

(A3) $(X_i, \alpha_i)^T, i = 1, 2, \ldots, N$, are independently and identically distributed (IID) bivariate random vectors that follow a general distribution F:

$$\begin{pmatrix} X_1 \\ \alpha_1 \end{pmatrix}, \begin{pmatrix} X_2 \\ \alpha_2 \end{pmatrix}, \ldots, \begin{pmatrix} X_N \\ \alpha_N \end{pmatrix} \overset{\text{IID}}{\sim} F(\cdot, \cdot). \tag{8.50}$$

Proposition 17 (Theorem 1 of Lo et al. (2024)) *Under Assumption (A3), if both marginal distributions F_X and F_α are continuous functions, and C is a copula of F with a density, we have:*

$$(\alpha_{[1:N]}, \alpha_{[2:N]}, \ldots, \alpha_{[N:N]}) \overset{d}{=} (g(U_{1:N}, V_1), g(U_{2:N}, V_2), \ldots, g(U_{N:N}, V_N)), \tag{8.51}$$

where "$\overset{d}{=}$" denotes equality in distribution. Function g is defined as:

$$g(u, w) \equiv F_\alpha^{-1} \circ g_u(w), \tag{8.52}$$

where $g_u(w)$ is the inverse function of $v \mapsto \frac{\partial C}{\partial u}(u, v)$,[45] *and "$\circ$" represents function composition. In addition, $U_1, U_2 \ldots, U_N$ and $V_1, V_2 \ldots, V_N$ are mutually independent and IID random variables, each following* Uniform(0, 1)*, and $U_{1:N} \leq U_{2:N} \leq \cdots \leq U_{N:N}$ are the order statistics of $U_1, U_2 \ldots, U_N$.*

Proposition 17 provides an explicit representation for the distribution of induced order statistics by a mixture of order statistics, $U_{i:N}$, and IID uniform noise, V_i, for any smooth bivariate distribution, F. The mixture function given by Equation (8.52), $g(\cdot, \cdot)$, is constructed using F_θ and C, but does not depend on F_X. This result completely characterizes the distribution of induced order statistics, both in finite samples and asymptotically, and allows for the efficient computation of optimal portfolio weights.

8.7 Discussion

Our framework for assessing the financial consequences of impact investing has several limitations but also potential extensions, both of which are reviewed in this section.

8.7.1 Impact measurement error

First, it needs to be emphasized that measurement errors can occur in the impact variable $\mathbf{X}$ itself, as with any other stock-level characteristic. This is especially challenging to mitigate for emerging concepts such as ESG, as documented recently by Berg, Kölbel, and Rigobon (2022). One potential approach to dealing with the noise in ESG measurement is to combine multiple sources of data and improve the signal with instrumented variable regression (Berg et al., 2022) or aggregation methods designed for denoising (Berg et al., 2024). Nonetheless, empirical estimates of ESG returns should be taken with the caveat that they are potentially affected by measurement noise.

[45]The function $g_u(w)$ corresponds to the inverse Rosenblatt transform (Rosenblatt, 1952), which is widely used for sampling random vectors from copula. See, for example, Nelsen (2007, Section 2.9).

8.7.2 Asset-pricing models, impact factor selection, and economic foundations

The wide range of examples in Section 8.5 underscore the fact that any quantitative assessment of impact investing depends critically on an asset-pricing model. In the same way that investment performance attribution requires a benchmark from which to measure deviations, measurements within our framework are always relative to a given benchmark asset-pricing model, such as those in Equations (8.43)–(8.49). Therefore, to determine whether returns are attributable to a given impact variable or any other omitted variable, one can either include the proposed omitted variable in the factor model at the outset, or directly test the omitted variable as $\mathbf{X}$ using our framework. In either case, a concrete alternative hypothesis is required, as in any test of a given asset-pricing model such as the CAPM (Sharpe, 1964; Lintner, 1965), the APT (Ross, 1976), or the Fama-French multifactor models (Fama and French, 1993, 2015).[46] For example, in the case of sin stocks, Blitz and Fabozzi (2017) show that the initial excess returns documented by Hong and Kacperczyk (2009) reduce significantly after accounting for two concrete omitted quality factors—profitability and investment.

More generally, the selection of impact variables $\mathbf{X}$ should be on the basis of *a priori* economic, institutional, scientific, social, and other rationales, which not only helps to establish relationships that are more likely to be robust and causal, but also mitigates the potential for measurement errors discussed above. In the case of biotech venture philanthropy, curing the disease is a prerequisite, and therefore likely a causal impact factor for financial performance. The excess returns are driven by well-documented challenges in early stage drug development programs, including the low probabilities of success, long time horizons, and large capital requirements (Fagnan et al., 2013). In the Sematech example, there exist well-documented economic channels through which superior financial performance can be realized, i.e., the reduced duplication of R&D effort and subsequent increased profitability per R&D dollar for member firms (Irwin and Klenow, 1996a). Our difference-in-difference approach also adds additional confidence to the causal nature of this impact factor. For sin stocks, their excess returns can be explained by Merton's (1987) model of neglected stocks and segmented markets or Fama and French's (2007) taste model, which predicts that non-socially responsible companies that are out of favor by investors will earn higher expected returns. For ESG investing, although theories suggest that green assets should not earn higher returns in equilibrium, they can have prolonged periods of better returns when aggregate tastes are changing (Pástor, Stambaugh, and Taylor, 2021). Our framework is flexible enough to be calibrated to asset-pricing models that account for underlying financial market variables driving ESG returns, such as good versus bad economic times (Bansal, Wu, and Yaron, 2022) or market-wide shifts in attention to climate change (Pástor, Stambaugh, and Taylor, 2022; Lo, Zhang, and Zhao, 2022). Finally, the causal mechanism for the meme stock phenomenon is given by the price-impact model in the appendix of Lo and Zhang (2024) or Pedersen's (2022) model of influencers and thought leaders in social networks.

8.7.3 Nonstationarities and estimating ρ

The specific motivation underlying each impact investment also plays an important role in determining the key parameter ρ, the correlation between the impact factor $\mathbf{X}$ and

[46] It is known that testing an asset-pricing model against an unspecified alternative results in poor power (see, for example, MacKinlay (1987) and Campbell, Lo, and MacKinlay (1997, pp. 261)). From the Bayesian perspective, at least two models are needed to update their posteriors with data. Without a concrete alternative model, the likelihood $P(\text{data}|\text{not } H_0)$ is ill-defined.

returns. In particular, such motivation is often nonstationary, which implies that ρ may be time-varying. The dynamics of such time variation are often related to changes in the most pressing issues facing society at each point in history. For example, the idea that portfolio managers should include company-specific carbon risk exposures in their investment process was greeted with skepticism in a not-so-distant conference in 2010 (Andersson, Bolton, and Samama, 2016), in contrast to today's 5,301 UNPRI signatories.[47] of which 74% are asset managers, 15% are asset owners, and 11% are service providers. In this adaptive process (Lo, 2004, 2017), the correlation between a climate impact factor and returns will rise as the amount of assets under management and the number of products that are attempting to take advantage of a given $\mathbf{X}$ increase over time, and will eventually stabilize as the size of the new sector reaches a steady state. Even in the absence of any direct physical relationship between a company's carbon emissions and its business prospects, if enough investors care about the company's carbon footprint due to general environmental concerns, this factor can have an impact on the company's returns, thereby inducing a risk premium and, consequently, a non-zero ρ.

Because impact investing is so often associated with non-pecuniary preferences of investors, and because behavioral economists and other social scientists have documented the fact that individual preferences change over time and across contexts, nonstationarity is especially relevant for such investments. Therefore, attempting to estimate ρ using decades of historical data is unlikely to be fruitful, given the shift in these correlations over time.[48] A more useful approach is to either estimate ρ using a structural model based on the specific impact measure, or to apply time-series methods that are more robust to nonstationarities such as rolling-window estimators, regime-switching models, or machine-learning techniques which are often more adaptive than standard statistical estimators.

A separate but closely related issue involves the time horizon over which correlation is measured. In our theoretical analysis, ρ is assumed to be a fixed constant, but in practice, ρ is not only time-varying but also horizon-dependent. For some types of impact, the horizon is short, such as the price-impact example of Section 8.5.5, so correlations can be measured using daily or monthly returns. However, for other types of impact, the horizon can be several years, such as the case with the Cystic Fibrosis Foundation of Section 8.5.1). These two extremes reflect the specific mechanisms of impact; in the case of meme stocks, impact occurs trade by trade, whereas in the case of biomedical innovation, impact occurs as drug candidates reach clinical-trial milestones, which may take over a decade. Therefore, when applying the framework of Section 8.2 to a given context, a necessary prerequisite is a clear understanding of the nature of the impact being sought and over the specific time span that it is likely to be observed.

Additionally, when disclosing the financial impact of such an investment, it would be prudent to disclose the estimated correlation ρ over multiple horizons, in the same way that retail investors are now given 1-, 3-, and 5-year historical returns when available to evaluate current and potential future investments. Correlations over different horizons can reveal important nonstationarities as well as other structural and economic features about the impact that investors are hoping to obtain, and will also help to set expectations as to when such impact might be realized.

[47] `https://www.unpri.org/signatories/signatory-resources/signatory-directory` (accessed 2 Dec 2022).

[48] For example, Pástor, Stambaugh, and Taylor (2022) and Lo, Zhang, and Zhao (2022) show that it is important to account for shocks in preference shifts in order to get correlations that are more likely to persist in the future.

8.8 Conclusion

In this chapter, we propose a new framework to quantify the financial value-added (or subtracted) of impact investing. Using the theory of induced order statistics, we show that the correlation between the impact factor, $\mathbf{X}$, and the excess returns of individual securities determines the excess return of the impact portfolio. The impact factor provides a ranking and selection mechanism for portfolio construction, and its correlation with $\boldsymbol{\alpha}$ provides additional information that can be used to achieve better risk-adjusted returns as well as greater chosen impact.

In practice, we require estimates of $\boldsymbol{\alpha}$ to measure the correlation between $\mathbf{X}$ and $\boldsymbol{\alpha}$, which are demonstrated empirically in two projects involving a wide range of ESG metrics (Berg et al., 2024) and environmental metrics such as carbon emission (Lo, Zhang, and Zhao, 2022).

Why do we not merely estimate $\boldsymbol{\alpha}$ and stop there? The reality is that not all investors have access to good estimates of $\boldsymbol{\alpha}$, or to the alphas of more sophisticated impact portfolios. Our framework provides a much simpler one-dimensional quantity (ρ) compared to $\boldsymbol{\alpha}$, to be estimated either based on historical data or economic theory. It also is a simple and unified quantity that asset managers can disclose to investors. In this sense, this correlation is analogous to the CAPM's market beta, which simply enough reduces to the correlation between returns of an individual security and market returns when they are both standardized to have unit variances.

The ability to quantify the distribution of alphas for impact-sorted securities allows us to form Treynor-Black portfolios to exploit the alphas optimally. This is particularly relevant for the investment management industry, as it strives to bridge the gap between traditional investment products and the growing demand for impact investments. Regardless of the nature of the desired impact—whether it is biomedical innovation, promoting ESG, avoiding socially unsavory businesses, or attempting to achieve certain price objectives—our framework can be used to construct the most efficient way of investing in impact portfolios. By comparing the properties of impact portfolios on the Treynor-Black super-efficient-frontier with those of non-impact investments after accounting for investor-specific constraints, we have a concrete metric of the reward (or cost) of impact investing, as demonstrated in our earlier five examples.

In fact, an investment's alpha can itself be influenced by its impact, as demonstrated in our example of venture philanthropy. If the Cystic Fibrosis Foundation were not able to achieve the impact to develop effective drugs for cystic fibrosis, it is unlikely that they could have generated any meaningful return. In this sense, there is an endogenous and likely a *causal* relationship between $\mathbf{X}$ and alpha. Another example of realizing alpha by achieving impact is activist investing, about which it has been empirically documented that activists may help their portfolio companies improve production efficiency (Brav, Jiang, and Kim, 2015), long-term fundamentals (Bebchuk, Brav, and Jiang, 2015), and stock performance (Dimson, Karakaş, and Li, 2015).

More broadly, our framework is relevant not only to impact proxies such as SRI and ESG metrics, but to any characteristics that may be correlated with excess returns. This includes traditional factors such as value, quality, size, and momentum, as well as the hundreds of new factors and anomalies in the "Factor Zoo" discussed in the recent literature (Harvey, Liu, and Zhu, 2016; Feng, Giglio, and Xiu, 2020; Hou, Xue, and Zhang, 2020). From this perspective, our model is able to define a measure for the alpha of any factor.

Our framework may also help inform regulators and policymakers on the most appropriate tools to encourage investments with more socially aware goals. Not all types of impact investing are created equal. When they create positive excess returns, one must understand what drives the initial undervaluation in the first place, and what risks are preventing investors from participating in these opportunities. In the case of venture philanthropy in biomedical research and development, for example, it is crucial to develop new tools to mitigate risks from low probabilities of success, long time horizons, and large capital requirements (Fagnan et al., 2013; Thakor et al., 2017).

On the other hand, when impact investing incurs a cost to investors, at the very least, it suggests the need for more explicit investor disclosures. It may also justify certain incentives and industrial policies, such as tax benefits and R&D grants to encourage the growth of these socially beneficial firms and organizations. One case in point is the area of green energy, where Baker et al. (2022) document a lower yield for green bonds compared to otherwise equivalent bonds. Governments around the world are designing policies to help grow industries such as clean energy and electric vehicles. Even if they incur a cost in the short- to medium-term, as a society we need to invest in them if we value greater sustainability.

Indeed, our analysis underscores the fact that finance need not be a zero-sum game. While impact investing does imply sacrificing excess returns in certain situations, in other situations it is in fact possible to achieve impact and attractive financial returns at the same time. We hope to apply our framework more broadly, not only to allow portfolio managers to satisfy their fiduciary duties, but also to achieve the more ambitious goal of doing well by doing good on behalf of investors.

References

Adler, T., and M. Kritzman, 2008, The cost of socially responsible investing, *The Journal of Portfolio Management* 35, 52–56.

Albuquerque, R., Y. Koskinen, and C. Zhang, 2019, Corporate social responsibility and firm risk: Theory and empirical evidence, *Management Science* 65, 4451–4469.

Andersson, M., P. Bolton, and F. Samama, 2016, Governance and climate change: A success story in mobilizing investor support for corporate responses to climate change, *Journal of Applied Corporate Finance* 28, 29–33.

Ardia, D., K. Bluteau, K. Boudt, and K. Inghelbrecht, 2022, Climate change concerns and the performance of green versus brown stocks, *Management Science* forthcoming.

Baker, M., D. Bergstresser, G. Serafeim, and J. Wurgler, 2022, The pricing and ownership of US green bonds, *Annual Review of Financial Economics* 14, 415–437.

Bansal, R., D. Wu, and A. Yaron, 2022, Socially responsible investing in good and bad times, *The Review of Financial Studies* 35, 2067–2099.

Barber, B. M., A. Morse, and A. Yasuda, 2021, Impact investing, *Journal of Financial Economics* 139, 162–185.

Bebchuk, L. A., A. Brav, and W. Jiang, 2015, The long-term effects of hedge fund activism, *Columbia Law Review* 115, 1085–1156.

Bebchuk, L. A., A. Cohen, and C. C. Wang, 2013, Learning and the disappearing association between governance and returns, *Journal of Financial Economics* 108, 323–348.

Berg, F., J. F. Koelbel, A. Pavlova, and R. Rigobon, 2022, ESG confusion and stock returns: Tackling the problem of noise, (October 12, 2021). Available at SSRN: https://ssrn.com/abstract=3941514.

Berg, F., J. F. Kölbel, and R. Rigobon, 2022, Aggregate confusion: The divergence of ESG ratings, *Review of Finance* 26, 1315–1344.

Berg, F., A. W. Lo, R. Rigobon, M. Singh, and R. Zhang, 2024, Quantifying the returns of ESG investing: An empirical analysis with six ESG metrics, *The Journal of Portfolio Management* 50, 216–238.

Berk, J., and J. H. van Binsbergen, 2021, The impact of impact investing, (August 21, 2021). Available at SSRN: https://ssrn.com/abstract=3909166.

Bertsimas, D., and A. W. Lo, 1998, Optimal control of execution costs, *Journal of Financial Markets* 1, 1–50.

Bhattacharya, P., 1974, Convergence of sample paths of normalized sums of induced order statistics, *The Annals of Statistics* 2, 1034–1039.

Blitz, D., and F. J. Fabozzi, 2017, Sin stocks revisited: Resolving the sin stock anomaly, *The Journal of Portfolio Management* 44, 105–111.

Bolton, P., and M. Kacperczyk, 2021, Do investors care about carbon risk?, *Journal of Financial Economics* 142, 517–549.

Brav, A., W. Jiang, and H. Kim, 2015, The real effects of hedge fund activism: Productivity, asset allocation, and labor outcomes, *The Review of Financial Studies* 28, 2723–2769.

Brinson, G. P., L. R. Hood, and G. L. Beebower, 1986, Determinants of portfolio performance, *Financial Analysts Journal* 42, 39–44.

Brodie, J., I. Daubechies, C. De Mol, D. Giannone, and I. Loris, 2009, Sparse and stable Markowitz portfolios, *Proceedings of the National Academy of Sciences* 106, 12267–12272.

Campbell, J. Y., A. W. Lo, and A. C. MacKinlay, 1997, *The Econometrics of Financial Markets* (Princeton University Press, Princeton, NJ).

Chen, M., and G. Mussalli, 2020, An integrated approach to quantitative ESG investing, *The Journal of Portfolio Management* 46, 65–74.

Cochrane, J., 2009, *Asset Pricing: Revised edition* (Princeton University Press).

Cornell, B., and A. Damodaran, 2020, Valuing ESG: Doing good or sounding good?, *The Journal of Impact and ESG Investing* 1, 76–93.

David, H., 1973, Concomitants of order statistics, *Bulletin of the International Statistical Institute* 45, 295–300.

David, H. A., and H. N. Nagaraja, 2004, *Order Statistics* (John Wiley & Sons).

Dimson, E., O. Karakaş, and X. Li, 2015, Active ownership, *The Review of Financial Studies* 28, 3225–3268.

Dortfleitner, G., G. Halbritter, and M. Nguyen, 2015, Measuring the level and risk of corporate responsibility–an empirical comparison of different ESG ratings approaches, *The Journal of Asset Management* 17, 450–466.

Dybvig, P. H., and J. E. Ingersoll, 1982, Mean-variance theory in complete markets, *Journal of Business* 233–251.

Fabozzi, F. J., K. Ma, and B. J. Oliphant, 2008, Sin stock returns, *The Journal of Portfolio Management* 35, 82–94.

Fagnan, D. E., J. M. Fernandez, A. W. Lo, and R. M. Stein, 2013, Can financial engineering cure cancer?, *American Economic Review* 103, 406–11.

Fama, E. F., and K. R. French, 1993, Common risk factors in the returns on stocks and bonds, *Journal of Financial Economics* 33, 3–56.

Fama, E. F., and K. R. French, 2007, Disagreement, tastes, and asset prices, *Journal of Financial Economics* 83, 667–689.

Fama, E. F., and K. R. French, 2015, A five-factor asset pricing model, *Journal of Financial Economics* 116, 1–22.

Feng, G., S. Giglio, and D. Xiu, 2020, Taming the factor zoo: A test of new factors, *The Journal of Finance* 75, 1327–1370.

Friedman, H. L., and M. S. Heinle, 2016, Taste, information, and asset prices: Implications for the valuation of csr, *Review of Accounting Studies* 21, 740–767.

Geczy, C. C., R. F. Stambaugh, and D. Levin, 2021, Investing in socially responsible mutual funds, *The Review of Asset Pricing Studies* 11, 309–351.

Goldstein, I., A. Kopytov, L. Shen, and H. Xiang, 2021, On ESG investing: Heterogeneous preferences, information, and asset prices, G(April 3, 2021). Available at SSRN: https://ssrn.com/abstract=3823042.

Grinold, R. C., 1989, The fundamental law of active management, *The Journal of Portfolio Management* 15, 30–37.

Grinold, R. C., 1994, Alpha is volatility times IC times score, *Journal of Portfolio Management* 20, 9–16.

Grinold, R. C., and R. N. Kahn, 1999, *Active Portfolio Management: A Quantitative Approach for Producing Superior Returns and Controlling Risk*, second edition (McGraw-Hill Education, New York).

Grinold, R. C., and R. N. Kahn, 2019, *Advances in Active Portfolio Management: New Developments in Quantitative Investing* (McGraw Hill Professional).

Grossman, S. J., and J. E. Stiglitz, 1980, On the impossibility of informationally efficient markets, *American Economic Review* 70, 393–408.

Harvey, C. R., Y. Liu, and H. Zhu, 2016, ... and the cross-section of expected returns, *The Review of Financial Studies* 29, 5–68.

Heinkel, R., A. Kraus, and J. Zechner, 2001, The effect of green investment on corporate behavior, *Journal of Financial and Quantitative Analysis* 36, 431–449.

Hong, H., and M. Kacperczyk, 2009, The price of sin: The effects of social norms on markets, *Journal of Financial Economics* 93, 15–36.

Hou, K., C. Xue, and L. Zhang, 2020, Replicating anomalies, *The Review of Financial Studies* 33, 2019–2133.

Idzorek, T. M., P. D. Kaplan, and R. G. Ibbotson, 2021, The popularity asset pricing model, *(October 25, 2021). Available at SSRN: https://ssrn.com/abstract=3451554*.

Irwin, D. A., and P. J. Klenow, 1996a, High-tech R&D subsidies estimating the effects of sematech, *Journal of International Economics* 40, 323–344.

Irwin, D. A., and P. J. Klenow, 1996b, Sematech: purpose and performance, *Proceedings of the National Academy of Sciences* 93, 12739–12742.

Kim, E., and A. W. Lo, 2019, Venture philanthropy: a case study of the cystic fibrosis foundation, (April 23, 2019). Available at SSRN: https://ssrn.com/abstract=3376673.

Kuntz, L.-C., 2020, Beta dispersion and market timing, *Journal of Empirical Finance* 59, 235–256.

Letts, C. W., W. Ryan, and A. Grossman, 1997, Virtuous capital: What foundations can learn from venture capitalists, *Harvard Business Review* 75, 36–50.

Link, A. N., D. J. Teece, and W. F. Finan, 1996, Estimating the benefits from collaboration: The case of sematech, *Review of Industrial Organization* 11, 737–751.

Lintner, J., 1965, The valuation of risk assets and the selection of risky investments in stock portfolios and capital budgets, *Review of Economics and Statistics* 47, 13–37.

Lo, A. W., 2004, The adaptive markets hypothesis, *Journal of Portfolio Management* 30, 15–29.

Lo, A. W., 2017, *Adaptive Markets: Financial Evolution at the Speed of Thought* (Princeton University Press, Princeton, NJ).

Lo, A. W., and A. C. MacKinlay, 1990, Data-snooping biases in tests of financial asset pricing models, *The Review of Financial Studies* 3, 431–467.

Lo, A. W., and R. T. Thakor, 2019, Risk and reward in the orphan drug industry, *The Journal of Portfolio Management* 45, 30–45.

Lo, A. W., L. Wu, R. Zhang, and C. Zhao, 2024, Optimal impact portfolios with general dependence and marginals, *Operations Research* 72, 1751–2261.

Lo, A. W., and R. Zhang, 2024, Quantifying the impact of impact investing, *Management Science* 70, 7161–7186.

Lo, A. W., and R. Zhang, 2025, Performance attribution for portfolio constraints, *Management Science* forthcoming.

Lo, A. W., R. Zhang, and C. Zhao, 2022, Measuring and optimizing the risk and reward of green portfolios, *The Journal of Impact and ESG Investing* 3, 55–93.

Luo, H. A., and R. J. Balvers, 2017, Social screens and systematic investor boycott risk, *Journal of Financial and Quantitative Analysis* 52, 365–399.

MacKinlay, A. C., 1987, On multivariate tests of the capm, *Journal of Financial Economics* 18, 341–371.

Madhavan, A., A. Sobczyk, and A. Ang, 2021, Toward ESG alpha: Analyzing ESG exposures through a factor lens, *Financial Analysts Journal* 77, 69–88.

Merton, R. C., 1973, An intertemporal capital asset pricing model, *Econometrica* 41, 867–887.

Merton, R. C., 1987, A simple model of capital market equilibrium with incomplete information, *The Journal of Finance* 42, 483–510.

Nelsen, R. B., 2007, *An Introduction to Copulas* (Springer Science & Business Media).

Pástor, L., and R. F. Stambaugh, 1999, Costs of equity capital and model mispricing, *The Journal of Finance* 54, 67–121.

Pástor, L., R. F. Stambaugh, and L. A. Taylor, 2021, Sustainable investing in equilibrium, *Journal of Financial Economics* 142, 550–571.

Pástor, L., R. F. Stambaugh, and L. A. Taylor, 2022, Dissecting green returns, *Journal of Financial Economics* 146, 403–424.

Pedersen, L. H., 2022, Game on: Social networks and markets, *Journal of Financial Economics* 146, 1097–1119.

Pedersen, L. H., S. Fitzgibbons, and L. Pomorski, 2021, Responsible investing: The ESG-efficient frontier, *Journal of Financial Economics* 142, 572–597.

Rosenblatt, M., 1952, Remarks on a multivariate transformation, *The Annals of Mathematical Statistics* 23, 470–472.

Ross, S. A., 1976, The arbitrage theory of capital asset pricing, *Journal of Economic Theory* 13, 341–360.

Ross, S. A., 1978, A simple approach to the valuation of risky streams, *Journal of Business* 51, 453–475.

Semenova, N., and L. G. Hassel, 2015, On the validity of environmental performance metrics, *Journal of Business Ethics* 132, 249–258.

Sharp, P., and S. Hockfield, 2017, Convergence: The future of health, *Science* 355, 589–589.

Sharpe, W. F., 1964, Capital asset prices: A theory of market equilibrium under conditions of risk, *The Journal of Finance* 19, 425–442.

Sharpe, W. F., 1966, Mutual fund performance, *The Journal of Business* 39, 119–138.

Shiller, R. J., 2005, *Irrational Exuberance*, second edition (Princeton University Press, Princeton, New Jersey).

Sorensen, E., M. Chen, and G. Mussalli, 2021, The quantitative approach for sustainable investing, *The Journal of Portfolio Management* 47, 38–49.

Sortino, F. A., and L. N. Price, 1994, Performance measurement in a downside risk framework, *The Journal of Investing* 3, 59–64.

Sortino, F. A., and R. Van Der Meer, 1991, Downside risk, *Journal of Portfolio Management* 17, 27–31.

Statman, M., and D. Glushkov, 2009, The wages of social responsibility, *Financial Analysts Journal* 65, 33–46.

Thakor, R. T., N. Anaya, Y. Zhang, C. Vilanilam, K. W. Siah, C. H. Wong, and A. W. Lo, 2017, Just how good an investment is the biopharmaceutical sector?, *Nature Biotechnology* 35, 1149–1157.

Treynor, J. L., and F. Black, 1973, How to use security analysis to improve portfolio selection, *The Journal of Business* 46, 66–86.

Tu, J., and G. Zhou, 2011, Markowitz meets Talmud: A combination of sophisticated and naive diversification strategies, *Journal of Financial Economics* 99, 204–215.

Zerbib, O. D., 2022, A sustainable capital asset pricing model (S-CAPM): Evidence from environmental integration and sin stock exclusion, *Review of Finance* 26, 1345–1388.

Chapter 9

Performance attribution for portfolio constraints

Andrew W. Lo
MIT Sloan School of Management, USA

Ruixun Zhang
Peking University, China

We propose a new performance attribution framework that decomposes a constrained portfolio's holdings, expected returns, variance, expected utility, and realized returns into components attributable to: (1) the unconstrained mean-variance optimal portfolio; (2) individual static constraints; and (3) information, if any, arising from those constraints. A key contribution of our framework is the recognition that constraints may contain information that is correlated with returns, in which case imposing such constraints can affect performance. We extend our framework to accommodate estimation risk in portfolio construction using Bayesian portfolio analysis, which allows one to select constraints that improve—or are least detrimental to—future performance. We provide simulations and empirical examples involving constraints on ESG portfolios. Under certain scenarios, constraints may improve portfolio performance relative to a passive benchmark that does not account for the information contained in these constraints.

This chapter is adapted from an article originally published in *Management Science* (Lo and Zhang, 2025). Please cite this work as follows:
Andrew W. Lo and Ruixun Zhang, 2025. Performance Attribution for Portfolio Constraints. Management Science.

9.1 Introduction

Constraints are ubiquitous in portfolio management. They are regularly imposed, both directly and indirectly, by portfolio managers, regulators, risk managers, trading desks, and investors. Because these constraints directly affect the portfolio construction process, all stakeholders have become interested in quantifying how constrained portfolios deviate from the unconstrained optimal benchmark, using various metrics and concepts such as unrealized alpha, opportunity cost, and implementation inefficiency. Measuring the impact of constraints on portfolio performance has become particularly important as socially responsible investing (SRI) and environmental, social, and governance (ESG) products have grown in popularity over the last decade, because constraints are usually involved in the construction of these portfolios.

DOI: 10.1201/9781032636252-9

In this chapter, we develop a general framework in which the financial impact of constraints can be measured by attributing the performance of portfolios to contributions from individual constraints. Because a constrained portfolio contains a proper subset of securities of the unconstrained version, mathematical logic suggests that the constrained optimum is at best equal to the unconstrained optimum or, more likely, inferior. However, the nonsuperiority of constrained optima relies on a key assumption that is almost never explicitly stated: the constraint does not provide additional information regarding asset returns. In other words, constraints are assumed to be statistically independent of the returns. In some cases, such an assumption is warranted; for example, one can imagine constructing a subset of securities with CUSIP identifiers that contain prime numbers. Imposing such a constraint will clearly reduce the risk-adjusted return of the optimized portfolio.

But what if the constraint is *not* independent of the returns? For example, consider the constraint, "invest only in stocks whose prices will appreciate by more than 10% over the next 12 months." Apart from the infeasibility of imposing such a condition, it should be obvious that this constraint would, in fact, increase the risk-adjusted return of the optimized portfolio. Therefore, quantifying the impact of constraints rests entirely on whether and how the constraints are related to the performance characteristics of the securities under consideration.

To formalize this idea, we consider investors who construct portfolios by maximizing the standard mean-variance utility. We denote the optimal portfolio without constraints as the mean-variance optimal (MVO) portfolio, which is usually a simple and passive benchmark to which all investors have access. However, the portfolio obtained while imposing constraints will likely differ from the MVO portfolio. Therefore, we develop a methodology to decompose the constrained portfolio's holdings, expected returns, variance, expected utility, and realized returns into different components: those attributable to the MVO portfolio, the individual constraints treated as static, and the information contained in the constraints. This methodology yields a constraint attribution framework for evaluating the performance of a portfolio.

The key to our framework is to model the information content available in portfolio constraints. We assume that each constraint is based on a firm characteristic, $\mathbf{x} \equiv [x_1 \; x_2 \; \cdots \; x_N]'$, where x_i is the characteristic for the i-th asset, such as its ESG score or a label representing its industry. By modeling $\mathbf{x}$ as a random variable and allowing it to be correlated with asset returns, we are able to provide an explicit decomposition of the performance of a portfolio attributable to information in its constraints, which depends critically on the expected value and covariance matrix of returns *conditioned* on $\mathbf{x}$. Furthermore, in the special case of normally and multivariate Student-t distributed returns, we demonstrate that the information contribution from a constraint is determined by the correlation between $\mathbf{x}$ and the individual asset returns. The excess return from information is positive when this correlation is positive and the constraint is binding. The excess variance of a portfolio is negative when the portfolio holdings of a *shrinkage portfolio* (defined in Equation (9.8)) and the holdings attributable to constraints are positively correlated, and the magnitude of the reduction in variance depends on the absolute value of the same correlation.

This simple but profound result highlights the mechanism through which a constraint contributes to the performance of a portfolio. While a constraint treated as static must decrease a portfolio's expected utility, the information in the constraint can contribute either positively or negatively to a portfolio's expected utility and returns, depending on whether the characteristics of the constraint are positively or negatively correlated with asset returns. In this sense, constraints serve as an indirect mechanism for using information that is otherwise unavailable to investors in the passive benchmark.[1]

In addition, we demonstrate that our framework can accommodate estimation risk in the expected value and covariance matrix of asset returns used to construct portfolios. We do this using Bayesian portfolio analysis and modeling the dependence between firm characteristic $\mathbf{x}$ and the posterior predictive returns. This allows for performance attribution of portfolio constraints for out-of-sample returns. Furthermore, we establish an equivalence result between a Bayesian portfolio with constraints and an unconstrained portfolio with a certain prior on asset returns, which implies that our attribution results can also be interpreted as the influence of investors' views on returns (expressed as Bayesian priors) on the portfolio's performance.

We apply our framework to two common classes of portfolio constraints. The first occurs when investors restrict exposure to a certain factor, such as the average ESG score, market capitalization, beta, or book-to-market values of the portfolio. The second is exclusionary investing, in which certain assets are excluded from the portfolio based on criteria such as whether the firm belongs to an industry associated with "sin" stocks. We derive additional analytical results and provide simulated examples to illustrate the attribution in these scenarios, which naturally leads to a method for selecting constraints that may improve portfolio performance.

Finally, we provide an empirical application in the context of SRI and ESG investing. Their growing popularity and assets under management have triggered a backlash recently.[2] Our framework provides a potential solution to properly disclose the financial impact of the constraints imposed by these investments, and to reconcile SRI and ESG investing with fiduciary duty.

In particular, using real-world datasets, we quantify the financial impact of ESG constraints when the average portfolio ESG score is required to be above a certain threshold, as well as exclusionary investing based on sin stocks and stranded assets. While the expected utility contribution of these constraints, treated as static, is indeed negative, the contribution from the information contained in the constraints to portfolio performance is dynamic over time. This contribution is generally negative before 2007, implying that high ESG stocks delivered lower excess returns relative to the Fama–French five-factor model, on average, which is consistent with equilibrium theories of ESG returns (Pástor, Stambaugh, and Taylor, 2021; Pedersen, Fitzgibbons, and Pomorski, 2021). However, after 2008, the information in the constraints starts to contribute positively to portfolio performance in certain years, reflecting the increasing attention toward SRI and ESG-related issues, an effect consistent with Pástor, Stambaugh, and Taylor's (2022) findings of shifted preferences.

We emphasize that our intention in this chapter is not to provide a measure for determining whether SRI and ESG investing deliver positive or negative excess returns. Instead, our primary objective is to illustrate how our framework can be used to attribute performance to any portfolio constraints, and to the information contained in those constraints. We make our software publicly available to researchers and investors to facilitate the application of our performance attribution framework.

[1] This is a key distinction from the literature that shows constraints may improve out-of-sample performance due to the parameter estimation risk in portfolio construction. We provide more details in the related literature.

[2] For example, on 4 August 2022, a letter signed by 19 attorneys general was sent to BlackRock expressing concern over its ESG policies, stating in part: "BlackRock's actions on a variety of governance objectives may violate multiple state laws. ... BlackRock has a private motivation that differs from its public commitments and statements. This is likely insufficient to satisfy state laws requiring a sole focus on financial return." See `https://www.texasattorneygeneral.gov/sites/default/files/images/executive-management/BlackRock\%20Letter.pdf` (accessed 15 December 2022).

Contributions and related literature. Our work contributes to the literature in several respects. First and foremost, we provide an attribution framework that decomposes the performance of constrained portfolios relative to a static benchmark by quantifying the information content in constraints, both in-sample and out-of-sample. The classical literature on style analysis and performance attribution includes Fama (1972), Brinson, Hood, and Beebower (1986), and Sharpe (1992). In addition, the transfer coefficient (Clarke, de Silva, and Thorley, 2002) and other measures based on the shadow cost have been proposed to measure the marginal cost of each constraint (Grinold, 2005; Stubbs and Vandenbussche, 2010; Menchero and Davis, 2011; Goldberg, 2021).

A key contribution of our decomposition framework to this literature is a quantitative measure of the information contained in each constraint. We show that recognizing the information contained in the constraints has important implications for performance analysis and understanding the impact of SRI and ESG investing, for example.

Second, our framework is connected to the extensive literature on robust portfolios that accounts for estimation risk in the parameters of portfolio construction, because imposing constraints is one way to deal with estimation risk. We discuss the relationship between this literature and our method in more detail in Section 9.3. In particular, portfolio constraints *can* mitigate estimation risk as shown by the no short-sale constraints of Jagannathan and Ma (2003), the norm constraints of DeMiguel et al. (2009), the gross exposure constraints of Fan, Zhang, and Yu (2012), and the variance-based constraints of Levy and Levy (2014).

Although our framework can accommodate estimation risk, our goal is *not* to propose a new robust portfolio rule or a better way to deal with estimation risk. Instead, we propose to use our framework to decompose the performance of out-of-sample returns of *existing* robust portfolio rules. In fact, we show that even without estimation error, there exists information in constraints that can affect portfolio performance. While constraints undoubtedly play a role in reducing estimation risk and improving out-of-sample performance as documented by the literature cited above, we demonstrate a different mechanism through which they can affect a portfolio: through the information contained in the constraints, which can contribute either positively or negatively to the performance of a portfolio depending on the statistical correlations of this information with returns. This is a key distinction from the prior literature on estimation risk and robust portfolio optimization. In practice, investors often impose constraints for business or regulation reasons other than reducing estimation risk—such as in SRI and ESG products—and these constraints can serve as an indirect mechanism to incorporate information. Our methodology provides a way to quantify this effect precisely, which, to the best of our knowledge, has not appeared in the existing literature.

Third, our framework and results differ in several respects from those of Brandt, Santa-Clara, and Valkanov (2009) and Hjalmarsson and Manchev (2012), who incorporate firm characteristics into the estimation of portfolio weights to derive robust portfolios. Their frameworks parameterize the portfolio weights of each stock as a (linear) function of the firm characteristic and then estimate the coefficients of this function. They do not model the impact of portfolio constraints, which is our focus. Also, the goal of their frameworks is to derive robust portfolio rules by avoiding the estimation of the return distribution. As discussed above, our framework has a very different goal. While Brandt, Santa-Clara, and Valkanov (2009) and Hjalmarsson and Manchev (2012) have convincingly demonstrated that their frameworks can produce robust portfolios, our results make a different contribution: we provide a quantitative measure of the information contained in portfolio constraints. This is relevant to both the asset pricing literature and the SRI and ESG literature because constraints are routinely used to construct such portfolios.

Fourth, from a technical perspective, our methodology involves solving a quadratic optimization problem subject to stochastic linear constraints. While the literature on stochastic optimization is vast (Powell, 2019), most of it is concerned with solving a sequential problem where the stochasticity in the environment, objective, or constraints affect the optimal policy. The constraint is imposed either on the expected value of the random variables as in Jin, Xu, and Zhou (2008) or on the probability of a certain condition as in Bonami and Lejeune (2009). Our framework is concerned with the static portfolio selection problem which is easy to solve, and the constraint is imposed sample-wise. Our main contribution, instead, is to quantify the information these stochastic constraints contain with respect to the random variables involved in the objective function.

Finally, our results also contribute to the literature on the impact of SRI, ESG, and other non-financial objectives on investment returns. Because ESG measures from different data providers can lead to very different correlations (Berg, Koelbel, and Rigobon, 2022), our findings highlight that the effect of a specific measure of SRI or ESG on investment performance depends on the information contained in the constraints created by these measures. These constraints need not always result in lower risk-adjusted returns.

We develop our main framework in Section 9.2. In Section 9.3, we incorporate estimation risk into our framework using Bayesian portfolio analysis. We consider two common examples of portfolio constraints in Section 9.4 and provide an illustrative application to ESG investing in Section 9.5. We conclude in Section 9.6.

9.2 A framework for constraint attribution

We consider a universe of N assets whose returns are given by the random vector $\mathbf{r}_t = [\, r_{1,t} \ \cdots \ r_{N,t}\,]'$.[3] Because we principally consider the static portfolio selection problem in this chapter, we omit the time subscript t and simply write $\mathbf{r}$ in most cases. We denote by $\boldsymbol{\mu}$ and $\boldsymbol{\Sigma}$ the expected value and covariance matrix of $\mathbf{r}$, respectively. Investors solve the following mean-variance portfolio:

$$\begin{aligned} \max_{\boldsymbol{\omega}} \ \ & \boldsymbol{\omega}'\boldsymbol{\mu} - \frac{\gamma}{2}\boldsymbol{\omega}'\boldsymbol{\Sigma}\boldsymbol{\omega} \\ \text{s.t.} \ \ & \mathbf{A}\boldsymbol{\omega} = \mathbf{b}, \end{aligned} \tag{9.1}$$

where $\boldsymbol{\omega} \equiv [\,\omega_1 \ \cdots \ \omega_N\,]'$ is an N-dimensional vector representing portfolio weights, γ is the relative risk-aversion parameter, $\mathbf{b} \equiv [\,b_1 \ \cdots \ b_J\,]'$ is a J-dimensional vector, and

$$\mathbf{A} \ \equiv \ \begin{pmatrix} \mathbf{A}'_1 \\ \cdots \\ \mathbf{A}'_J \end{pmatrix}$$

is a $J \times N$ full-rank matrix. Together, $\mathbf{b}$ and $\mathbf{A}$ describe J constraints. In particular, $\mathbf{A}'_j$ is the j-th row of $\mathbf{A}$ and b_j is the j-th element of $\mathbf{b}$, which together describe the j-th constraint.

We consider the case of equality constraints in Equation (9.1) for expositional simplicity. It is also easy to derive a parallel set of results under inequality constraints, $\mathbf{A}\boldsymbol{\omega} \le \mathbf{b}$, and we describe ways to generalize our results throughout our exposition. Common examples of portfolio constraints that can be described by Equation (9.1) include $\boldsymbol{\omega}'\mathbf{1} = 1$ representing a full investment constraint, $\omega_i = 0$ representing the exclusion of asset i, and $\boldsymbol{\omega}'\mathbf{A}_1 = b_1$ representing a certain level of factor exposure.

[3] We follow the common convention that all vectors are column vectors and all vectors and matrices are boldface.

A critical implicit assumption in the existing literature on constraint attribution is that the constraints, $\mathbf{A}$, are treated as constants, and are therefore independent of returns, $\mathbf{r}$. Under this setting, the solution to the optimization problem in Equation (9.1) without constraints, which we refer to as the unconstrained MVO portfolio, yields the best portfolio in terms of the objective value, and imposing constraints can only decrease the objective value.[4] The following result summarizes the optimal portfolio weights and the decomposition of portfolio holdings, expected return, and expected utility attributable to each constraint.[5] We provide proofs of all propositions in the appendix to Lo and Zhang (2025).

Proposition 18 (Static Constraints) *The optimal portfolio weight, $\boldsymbol{\omega}^*$, of problem (9.1) is given by:*

$$\boldsymbol{\omega}^* = \frac{1}{\gamma}\boldsymbol{\Sigma}^{-1}(\boldsymbol{\mu} - \mathbf{A}'\boldsymbol{\lambda}^*) \tag{9.2}$$

where the Lagrange multipliers are given by $\boldsymbol{\lambda}^* = \left(\mathbf{A}\boldsymbol{\Sigma}^{-1}\mathbf{A}'\right)^{-1}\left(\mathbf{A}\boldsymbol{\Sigma}^{-1}\boldsymbol{\mu} - \gamma\mathbf{b}\right)$, provided that the feasible region of the constrained optimization problem is nonempty. Here $\boldsymbol{\lambda}^*$ is a measure of the shadow cost of the portfolio's expected utility with respect to each constraint.[6]

Equation (9.2) leads to a series of decompositions.

1. *Portfolio holdings decomposition.*

$$\boldsymbol{\omega}^* = \frac{1}{\gamma}\boldsymbol{\Sigma}^{-1}\boldsymbol{\mu} - \frac{1}{\gamma}\boldsymbol{\Sigma}^{-1}\mathbf{A}'\boldsymbol{\lambda}^*. \tag{9.3}$$

- $\boldsymbol{\omega}_{\text{MVO}} \equiv \frac{1}{\gamma}\boldsymbol{\Sigma}^{-1}\boldsymbol{\mu}$*: holdings of the unconstrained MVO portfolio.*
- $\boldsymbol{\omega}_{\text{CSTR}} \equiv -\frac{1}{\gamma}\boldsymbol{\Sigma}^{-1}\mathbf{A}'\boldsymbol{\lambda}^*$*: components attributable to each constraint.*

2. *Expected return decomposition.*

$$\boldsymbol{\mu}'\boldsymbol{\omega}^* = \frac{1}{\gamma}\boldsymbol{\mu}'\boldsymbol{\Sigma}^{-1}\boldsymbol{\mu} - \frac{1}{\gamma}\boldsymbol{\mu}'\boldsymbol{\Sigma}^{-1}\mathbf{A}'\boldsymbol{\lambda}^*. \tag{9.4}$$

- $\frac{1}{\gamma}\boldsymbol{\mu}'\boldsymbol{\Sigma}^{-1}\boldsymbol{\mu}$*: expected return of the unconstrained MVO portfolio.*
- $-\frac{1}{\gamma}\boldsymbol{\mu}'\boldsymbol{\Sigma}^{-1}\mathbf{A}'\boldsymbol{\lambda}^*$*: components attributable to each constraint.*

3. *Expected utility decomposition.*

$$\boldsymbol{\mu}'\boldsymbol{\omega}^* - \frac{\gamma}{2}\boldsymbol{\omega}^{*'}\boldsymbol{\Sigma}\boldsymbol{\omega}^* = \frac{1}{2\gamma}\boldsymbol{\mu}'\boldsymbol{\Sigma}^{-1}\boldsymbol{\mu} - \frac{1}{2\gamma}\boldsymbol{\lambda}^{*'}\mathbf{A}\boldsymbol{\Sigma}^{-1}\mathbf{A}'\boldsymbol{\lambda}^*. \tag{9.5}$$

- $\frac{1}{2\gamma}\boldsymbol{\mu}'\boldsymbol{\Sigma}^{-1}\boldsymbol{\mu}$*: expected utility of the unconstrained MVO portfolio.*
- $-\frac{1}{2\gamma}\boldsymbol{\lambda}^{*'}\mathbf{A}\boldsymbol{\Sigma}^{-1}\mathbf{A}'\boldsymbol{\lambda}^*$*: components attributable to all constraints combined together. This term can be equivalently written as* $-\frac{\gamma}{2}\boldsymbol{\omega}_{\text{CSTR}}'\boldsymbol{\Sigma}\boldsymbol{\omega}_{\text{CSTR}}$.

A few observations regarding the intuition behind these decompositions are in order. First, constraints may change the portfolio weights in either direction, because the last term in Equation (9.3) can lead to both positive and negative entries.

[4]We are comparing unconstrained and constrained portfolios in the same context, i.e., with or without a risk-free asset for both portfolios.

[5]This decomposition is known in the literature; see, for example, Stubbs and Vandenbussche (2010), Menchero and Davis (2011), and Kan, Wang, and Zhou (2022).

[6]When the optimization problem in Equation (9.1) contains inequality constraints, the Lagrange multipliers $\boldsymbol{\lambda}^* \geq 0$ satisfy the complementary slackness condition: $(b_i - \mathbf{A}_i'\boldsymbol{\omega}^*)\lambda_i^* = 0$, for $i = 1, 2, \ldots, J$.

Second, in the expected return decomposition in Equation (9.4), if the Lagrange multiplier, or the shadow cost, for the i-th constraint $\lambda_i > 0$, the sign of the marginal contribution of that constraint is determined by the sign of $-\boldsymbol{\mu}'\boldsymbol{\Sigma}^{-1}\mathbf{A}_j$, which may be positive. To see that, we observe $-\boldsymbol{\mu}'\boldsymbol{\Sigma}^{-1}\mathbf{A}_j = -|\boldsymbol{\Sigma}^{-1}\boldsymbol{\mu}| \cdot |\mathbf{A}_j| \cdot \cos\theta$, where θ is the angle between $\boldsymbol{\omega}_{\text{MVO}} = \frac{1}{\gamma}\boldsymbol{\Sigma}^{-1}\boldsymbol{\mu}$ (holdings of the MVO portfolio) and $\mathbf{A}_j$ (constraint coefficients). When these two vectors are negatively correlated, $\cos\theta$ is negative, which implies that the i-th constraint *increases* expected returns.

Third, when constraints are static, they always decrease expected utility relative to the unconstrained MVO portfolio, because the last term in Equation (9.5) is always negative. In addition, this last term provides an attribution of expected utility to all constraints combined together. Unlike the decomposition of the portfolio holdings and expected return, both of which provide a simple linear additive attribution to individual constraints, the expected utility cannot be decomposed into a linear combination of each constraint due to the risk term. Nonetheless, there is a clear interpretation of the portfolio holdings decomposition, $\boldsymbol{\omega}_{\text{CSTR}}$, operating on the covariance matrix $\boldsymbol{\Sigma}$.

Finally, the portfolio holdings decomposition in Equation (9.3) provides a way to derive investor views on $\boldsymbol{\mu}$ implied by their holdings. For example, if a portfolio deviates from the benchmark portfolio, $\boldsymbol{\omega}_{\text{MVO}}$, by $\boldsymbol{\omega}_{\text{CSTR}}$, it is equivalent to solving an unconstrained mean-variance problem using $\boldsymbol{\mu} - \mathbf{A}'\boldsymbol{\lambda}^*$ as the expected value of assets. Therefore, the deviation, $-\mathbf{A}'\boldsymbol{\lambda}^*$, can be treated as the investor's view implied by their portfolio holdings. This idea has been explored extensively by, for example, Tu and Zhou (2010) and Ardia and Boudt (2015). We follow Tu and Zhou (2010) and the performance analysis literature mentioned in Section 9.1 to consider a partial equilibrium analysis of the impact of constraints from the investment perspective. The equilibrium analysis that accounts for the impact of investors on asset prices is an important problem, but one beyond the scope of this chapter.

9.2.1 Constraints with information

To extend the framework beyond static constraints, we consider constraints that are potentially correlated with returns. Let the random vector $\mathbf{x}_t \equiv [x_{1,t} \ \cdots \ x_{N,t}]'$ be a characteristic or score associated with each asset at time t. The randomness in $\mathbf{x}$ comes from the fact that these scores may change period by period and, more importantly, may be correlated with $\mathbf{r}_t$. For example, $\mathbf{x}_t$ can represent a firm's ESG score, P/E ratio, momentum measure, or even a proprietary alpha signal at time t. We again omit the subscript t and simply write $\mathbf{x}$ in most cases.

Investors form constraints based on the value of $\mathbf{x}$, and we denote the j-th constraint by $\mathbf{A}_j(\mathbf{x})$. For example, $\mathbf{A}_j(\mathbf{x}) = \mathbf{x}$ corresponds to the factor exposure constraint, $\mathbf{x}'\boldsymbol{\omega} = b_j$. More generally, investors can impose J constraints based on J different characteristics denoted by the vector:

$$\mathbf{X} \equiv [\mathbf{x}_1' \ \mathbf{x}_2' \ \cdots \ \mathbf{x}_J']',$$

where $\mathbf{x}_j$ represents the j-th characteristic that forms the j-th constraint, $\mathbf{A}_j(\mathbf{x}_j)$. We denote:

$$\mathbf{A}(\mathbf{X}) = \begin{pmatrix} \mathbf{A}_1'(\mathbf{x}_1) \\ \vdots \\ \mathbf{A}_J'(\mathbf{x}_J) \end{pmatrix}$$

to be the $J \times N$ coefficient matrix of J constraints that each depend on one characteristic. Investors observe the characteristics $\mathbf{X}$ at the time of portfolio construction, but not the returns $\mathbf{r}$. We assume the following about the distribution of asset characteristics, $\mathbf{X}$:

Assumption 1 *The characteristics* $\mathbf{X}$ *have finite moments up to order 2.*

9.2.2 Attribution with information

We use $\mathbf{r}|\mathbf{X}$ to denote the distribution of returns conditioned on information in $\mathbf{X}$, and $\boldsymbol{\mu}_{\mathbf{X}}$ and $\boldsymbol{\Sigma}_{\mathbf{X}}$ to denote its conditional mean and covariance matrix. We first attribute portfolio performance metrics conditioned on $\mathbf{X}$. This can be interpreted as a per-period attribution, because it is a function of the realizations of asset characteristics $\mathbf{X}$ in each period. After that, the overall attribution is simply the expectation with respect to $\mathbf{X}$, which can be interpreted as the long-run average of the per-period attribution.

Portfolio weights depend on the constraints, and therefore on $\mathbf{X}$. We still use $\boldsymbol{\omega}^*$ and $\boldsymbol{\omega}_{\text{CSTR}}$ to represent the weights of the constrained portfolio and components attributable to constraints, respectively, but it is worth noting that they are functions of $\mathbf{X}$ in the context of random characteristics. The following result summarizes the attribution of expected return and utility due to information in each constraint, conditioned on $\mathbf{X}$.

Proposition 19 (Conditional Attribution with Information) *Under Assumption 1 and conditioned on information in $\mathbf{X}$ that is used to form constraints $\mathbf{A}(\mathbf{X})$, the following decompositions hold for the optimal portfolio $\boldsymbol{\omega}^*$.*

1. *Expected return decomposition.*

$$\mathbb{E}\left[\boldsymbol{\omega}^{*\prime}\mathbf{r}|\mathbf{X}\right] = \boldsymbol{\mu}_{\mathbf{X}}'\boldsymbol{\omega}^* = \boldsymbol{\mu}_{\mathbf{X}}'\boldsymbol{\omega}_{\text{MVO}} + \boldsymbol{\mu}'\boldsymbol{\omega}_{\text{CSTR}} + (\boldsymbol{\mu}_{\mathbf{X}}' - \boldsymbol{\mu}')\boldsymbol{\omega}_{\text{CSTR}}, \tag{9.6}$$

where

- $\boldsymbol{\mu}_{\mathbf{X}}'\boldsymbol{\omega}_{\text{MVO}} = \frac{1}{\gamma}\boldsymbol{\mu}_{\mathbf{X}}'\boldsymbol{\Sigma}^{-1}\boldsymbol{\mu}$: *expected return of the unconstrained MVO portfolio.*
- $\boldsymbol{\mu}'\boldsymbol{\omega}_{\text{CSTR}} = -\frac{1}{\gamma}\boldsymbol{\mu}'\boldsymbol{\Sigma}^{-1}\mathbf{A}(\mathbf{X})'\boldsymbol{\lambda}^*$: *components attributable to each constraint treated as static.*
- $(\boldsymbol{\mu}_{\mathbf{X}}' - \boldsymbol{\mu}')\boldsymbol{\omega}_{\text{CSTR}} = -\frac{1}{\gamma}(\boldsymbol{\mu}_{\mathbf{X}}' - \boldsymbol{\mu}')\boldsymbol{\Sigma}^{-1}\mathbf{A}(\mathbf{X})'\boldsymbol{\lambda}^*$: *components attributable to information in constraints.*

Here the Lagrange multipliers are given by $\boldsymbol{\lambda}^* = \left(\mathbf{A}(\mathbf{X})\boldsymbol{\Sigma}^{-1}\mathbf{A}(\mathbf{X})'\right)^{-1}\left(\mathbf{A}(\mathbf{X})\boldsymbol{\Sigma}^{-1}\boldsymbol{\mu} - \gamma\mathbf{b}\right)$ provided that the feasible region of the constrained optimization problem is nonempty.

2. *Expected utility decomposition.*

$$\begin{aligned}\boldsymbol{\mu}_{\mathbf{X}}'\boldsymbol{\omega}^* - \frac{\gamma}{2}\boldsymbol{\omega}^{*\prime}\boldsymbol{\Sigma}_{\mathbf{X}}\boldsymbol{\omega}^* = {} & \boldsymbol{\mu}_{\mathbf{X}}'\boldsymbol{\omega}_{\text{MVO}} - \frac{\gamma}{2}\boldsymbol{\omega}_{\text{MVO}}'\boldsymbol{\Sigma}_{\mathbf{X}}\boldsymbol{\omega}_{\text{MVO}} \\ & - \frac{\gamma}{2}\boldsymbol{\omega}_{\text{CSTR}}'\boldsymbol{\Sigma}\boldsymbol{\omega}_{\text{CSTR}} \\ & + (\boldsymbol{\mu}_{\mathbf{X}}' - \boldsymbol{\mu}')\boldsymbol{\omega}_{\text{CSTR}} - \gamma\boldsymbol{\omega}_{\text{SHR}}'(\boldsymbol{\Sigma}_{\mathbf{X}} - \boldsymbol{\Sigma})\boldsymbol{\omega}_{\text{CSTR}}.\end{aligned} \tag{9.7}$$

- $\boldsymbol{\mu}_{\mathbf{X}}'\boldsymbol{\omega}_{\text{MVO}} - \frac{\gamma}{2}\boldsymbol{\omega}_{\text{MVO}}'\boldsymbol{\Sigma}_{\mathbf{X}}\boldsymbol{\omega}_{\text{MVO}} = \frac{1}{\gamma}\boldsymbol{\mu}_{\mathbf{X}}'\boldsymbol{\Sigma}^{-1}\boldsymbol{\mu} - \frac{1}{2\gamma}\left(\boldsymbol{\Sigma}^{-1}\boldsymbol{\mu}\right)'\boldsymbol{\Sigma}_{\mathbf{X}}\left(\boldsymbol{\Sigma}^{-1}\boldsymbol{\mu}\right)$: *optimal expected utility of the unconstrained MVO portfolio.*
- $-\frac{\gamma}{2}\boldsymbol{\omega}_{\text{CSTR}}'\boldsymbol{\Sigma}\boldsymbol{\omega}_{\text{CSTR}} = -\frac{1}{2\gamma}\boldsymbol{\lambda}^{*\prime}\mathbf{A}(\mathbf{X})\boldsymbol{\Sigma}^{-1}\mathbf{A}(\mathbf{X})'\boldsymbol{\lambda}^*$: *components attributable to all constraints combined together, treated as static.*
- $(\boldsymbol{\mu}_{\mathbf{X}}' - \boldsymbol{\mu}')\boldsymbol{\omega}_{\text{CSTR}} - \gamma\boldsymbol{\omega}_{\text{SHR}}'(\boldsymbol{\Sigma}_{\mathbf{X}} - \boldsymbol{\Sigma})\boldsymbol{\omega}_{\text{CSTR}} = -\frac{1}{\gamma}(\boldsymbol{\mu}_{\mathbf{X}}' - \boldsymbol{\mu}')\boldsymbol{\Sigma}^{-1}\mathbf{A}(\mathbf{X})'\boldsymbol{\lambda}^* - \frac{1}{\gamma}\left(\boldsymbol{\Sigma}^{-1}\boldsymbol{\mu} + \frac{1}{2}\boldsymbol{\Sigma}^{-1}\mathbf{A}(\mathbf{X})'\boldsymbol{\lambda}^*\right)'(\boldsymbol{\Sigma}_{\mathbf{X}} - \boldsymbol{\Sigma})\boldsymbol{\Sigma}^{-1}\mathbf{A}(\mathbf{X})'\boldsymbol{\lambda}^*$: *component attributable to information in constraints.*

Here, $\boldsymbol{\omega}_{\text{SHR}}$ is a shrinkage portfolio defined as:

$$\boldsymbol{\omega}_{\text{SHR}} \equiv \boldsymbol{\omega}_{\text{MVO}} + \frac{1}{2}\boldsymbol{\omega}_{\text{CSTR}} = \boldsymbol{\omega}^* - \frac{1}{2}\boldsymbol{\omega}_{\text{CSTR}}. \tag{9.8}$$

Proposition 19 provides a decomposition of the expected return and utility into components attributable to the unconstrained MVO portfolio, static constraints, and information.

We provide a few remarks regarding the choice of the benchmark portfolio in this decomposition. First, the unconstrained MVO portfolio here is with respect to Equation (9.1), using the unconditional expected return $\boldsymbol{\mu}$ and covariance matrix $\boldsymbol{\Sigma}$. This is consistent with how performance attribution is typically carried out in the investment management industry; investors pick a widely accepted benchmark portfolio, such as the market portfolio or a passive index, and compare the performance of the actual portfolio, which may contain additional constraints and information, against the chosen benchmark. The decomposition in Proposition 19 provides investors with a quantitative measure of information contained in portfolio constraints relative to any benchmark portfolios they may use.[7]

Second, benchmarking toward the unconstrained MVO portfolio without information in $\mathbf{X}$ provides a way to quantify the value-added performance investors receive by paying for professional portfolio management. Investors delegate the process of incorporating information through constraints, either because of limited cognitive resources or time budgets or because they are not able to fully incorporate such information in $\mathbf{X}$ into return forecasts. The large literature documenting limited attention (Corwin and Coughenour, 2008; Hirshleifer, Lim, and Teoh, 2011) and bounded rationality (Simon, 1955; Hirshleifer, Subrahmanyam, and Titman, 2006) in investor behavior provide support for this view. This is particularly relevant for uninformed and less sophisticated investors who lack the conditioning information available only to informed investors, as in the frameworks of Dybvig and Ross (1985) and Ferson and Siegel (2001).

We emphasize that the decomposition in Proposition 19 is fundamentally different from the traditional constraint attribution given in Proposition 18, in which the coefficients that form constraints are assumed to be constant. Once the constraints depend on asset characteristics $\mathbf{X}$ that are potentially correlated with returns, they provide information. Proposition 19 quantifies this effect explicitly by showing how information contributes to the expected return and utility of a portfolio.

The information component of expected return is $(\boldsymbol{\mu}'_{\mathbf{X}} - \boldsymbol{\mu}')\boldsymbol{\omega}_{\text{CSTR}}$, which implies that the information contributes positively when portfolio holdings attributable to constraints are positively correlated with the excess return vector of all assets, $\boldsymbol{\mu}'_{\mathbf{X}} - \boldsymbol{\mu}'$. This term can be further decomposed into components attributable to each individual constraint under certain distributional assumptions of $\mathbf{X}$. We discuss this in Section 9.2.3.

The information component of expected utility is $(\boldsymbol{\mu}'_{\mathbf{X}} - \boldsymbol{\mu}')\boldsymbol{\omega}_{\text{CSTR}} - \gamma\boldsymbol{\omega}'_{\text{SHR}}(\boldsymbol{\Sigma}_{\mathbf{X}} - \boldsymbol{\Sigma})\boldsymbol{\omega}_{\text{CSTR}}$. The first part is the same as the information component of the expected return, and the second part corresponds to the information component from the variance, which itself consists of two terms:

$$\begin{aligned}
-\gamma\boldsymbol{\omega}_{\text{SHR}}(\boldsymbol{\Sigma}_{\mathbf{X}} - \boldsymbol{\Sigma})\boldsymbol{\omega}_{\text{CSTR}} &= -\gamma\left(\boldsymbol{\omega}'_{\text{MVO}} + \frac{1}{2}\boldsymbol{\omega}'_{\text{CSTR}}\right)(\boldsymbol{\Sigma}_{\mathbf{X}} - \boldsymbol{\Sigma})\boldsymbol{\omega}_{\text{CSTR}} \\
&= -\frac{\gamma}{2}\boldsymbol{\omega}'_{\text{CSTR}}(\boldsymbol{\Sigma}_{\mathbf{X}} - \boldsymbol{\Sigma})\boldsymbol{\omega}_{\text{CSTR}} - \gamma\boldsymbol{\omega}'_{\text{MVO}}(\boldsymbol{\Sigma}_{\mathbf{X}} - \boldsymbol{\Sigma})\boldsymbol{\omega}_{\text{CSTR}}.
\end{aligned} \tag{9.9}$$

The first term corresponds to the excess variance, $\boldsymbol{\Sigma}_{\mathbf{X}} - \boldsymbol{\Sigma}$, from portfolio holdings attributable to constraints, $\boldsymbol{\omega}_{\text{CSTR}}$. The second term corresponds to an interaction effect between the unconstrained MVO portfolio, $\boldsymbol{\omega}_{\text{MVO}}$, and the component attributable to constraints, $\boldsymbol{\omega}_{\text{CSTR}}$.

[7] If one compares the performance of the constrained portfolio with the mean-variance optimal portfolio using the conditional moments, $\boldsymbol{\mu}_{\mathbf{X}}$ and $\boldsymbol{\Sigma}_{\mathbf{X}}$, one recovers the classical result that constraints must decrease the expected utility of the optimal portfolio with respect to $\boldsymbol{\mu}_{\mathbf{X}}$ and $\boldsymbol{\Sigma}_{\mathbf{X}}$. However, this comparison is not very realistic because, in practice, investors often need to understand the impact of a managed portfolio relative to simple and passive benchmarks to which they have easy access, rather than to a hypothetical portfolio that is optimal with respect to the conditional mean and variance with full information in $\mathbf{X}$.

Taken together, Equation (9.9) can be interpreted as the covariance between the shrinkage portfolio, ω_{SHR}, and the portfolio attributable to constraints, ω_{CSTR}. However, this covariance is not with respect to the returns of the original assets, but with respect to a set of hypothetical assets whose covariance is determined by the negative excess covariance matrix, $-(\boldsymbol{\Sigma}_{\mathbf{X}} - \boldsymbol{\Sigma})$.[8] Figure 9.1 shows two examples of the shrinkage portfolio, which shrinks the optimal constrained portfolio, ω^*, toward the unconstrained MVO portfolio, ω_{MVO}.

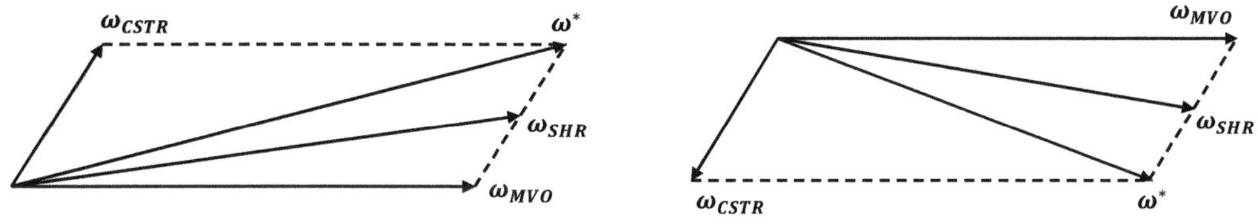

(a) Acute angle between ω_{SHR} and ω_{CSTR} (b) Obtuse angle between ω_{SHR} and ω_{CSTR}

FIGURE 9.1: Illustration of shrinkage portfolios defined in (9.8).

Finally, the Appendix of Lo and Zhang (2025) provides the unconditional version of Proposition 19, which can be interpreted as the average decomposition over multiple time periods.

9.2.3 Decomposing information with specific distributions

The attribution in Proposition 19 depends critically on two terms, $(\boldsymbol{\mu}_{\mathbf{X}} - \boldsymbol{\mu})$ and $(\boldsymbol{\Sigma}_{\mathbf{X}} - \boldsymbol{\Sigma})$, which capture the *excess* return and covariance due to information in $\mathbf{X}$, respectively. They cannot be simplified further for general distributions of $\mathbf{r}$ and $\mathbf{X}$. However, for certain classes of specific distributions, we can decompose these terms further and obtain considerable intuition about their contribution.

We consider multivariate normal and multivariate Student-t (MVT) distributions in this section. The normal distribution is a common choice in the portfolio literature, and the Student-t distribution is widely adopted to model heavy-tail returns. Both appear in the posterior predictive distribution in Bayesian portfolio analysis, which we use to account for estimation risk in Section 9.3.

We make the following two parallel assumptions.

Assumption 2 *The returns and characteristics,* $(\mathbf{r}', \mathbf{x}_1', \ldots, \mathbf{x}_J')$, *are jointly normally distributed.*

Assumption 2' *The returns and characteristics,* $(\mathbf{r}', \mathbf{x}_1', \ldots, \mathbf{x}_J')$, *follow a multivariate Student-t distribution:*

$$\begin{pmatrix} \mathbf{r} \\ \mathbf{X} \end{pmatrix} \sim \text{MVT}\left(\begin{pmatrix} \boldsymbol{\mu} \\ \bar{\mathbf{X}} \end{pmatrix}, \begin{pmatrix} \mathbf{V} & \mathbf{V}_{\mathbf{r},\mathbf{X}} \\ \mathbf{V}_{\mathbf{X},\mathbf{r}} & \mathbf{V}_{\mathbf{X},\mathbf{X}} \end{pmatrix}, \nu \right).$$

Recall that $\mathbf{X}$ represents the $(N \times J)$-dimensional vector $[\mathbf{x}_1' \ \cdots \ \mathbf{x}_J']'$, and we use $\bar{\mathbf{X}} \equiv [\bar{\mathbf{x}}_1' \ \cdots \ \bar{\mathbf{x}}_J']'$ to denote the expected value of $\mathbf{X}$. Assumption 2' implies that the returns $\mathbf{r} \sim \text{MVT}(\boldsymbol{\mu}, \mathbf{V}, \nu)$. The parameters $\boldsymbol{\mu}$ and $\mathbf{V}$ are usually referred to as the location vector

[8]To see this, imagine a set of N hypothetical assets whose returns, $\mathbf{s}$, have a covariance matrix $\boldsymbol{\Sigma} - \boldsymbol{\Sigma}_{\mathbf{X}}$. We have $\text{Cov}\left(\omega_{\text{SHR}}'\mathbf{s}, \omega_{\text{CSTR}}'\mathbf{s}\right) = \omega_{\text{SHR}}'\text{Cov}(\mathbf{s},\mathbf{s})\omega_{\text{CSTR}} = \omega_{\text{SHR}}'\left(\boldsymbol{\Sigma} - \boldsymbol{\Sigma}_{\mathbf{X}}\right)\omega_{\text{CSTR}}$. We show in Section 9.2.3 and the Appendix of Lo and Zhang (2025) that $\boldsymbol{\Sigma} - \boldsymbol{\Sigma}_{\mathbf{X}}$ is always positive semi-definite under certain distributional assumptions of $\mathbf{r}$ and $\mathbf{X}$.

and scale matrix, respectively, and ν is the degree of freedom. The first two moments of an MVT distribution are given by:

$$\mathbb{E}[\mathbf{r}] = \boldsymbol{\mu} \quad \text{and} \quad \boldsymbol{\Sigma} = \mathrm{Cov}(\mathbf{r}) = \frac{\nu}{\nu - 2}\mathbf{V}. \tag{9.10}$$

The expected value is the location vector, but the covariance matrix is not the scale matrix. Hence, we use $\mathbf{V}$ to denote the scale matrix to emphasize its difference from the covariance matrix, $\boldsymbol{\Sigma}$.

The next assumption describes the dependence between characteristics and returns.

Assumption 3 *The joint distribution of the return vector,* $\mathbf{r}$*, and characteristics* $\mathbf{X} = [\mathbf{x}_1' \ \mathbf{x}_2' \ \cdots \ \mathbf{x}_J']'$ *satisfies the following conditions.*

1. *The characteristic values are independent both across different assets and between the* J *different constraints.*

2. *For the* j*-th constraint, the correlation between the return and characteristic value of each asset is* ρ_j*, and there is no cross-correlation between the return and characteristic value of different assets. In other words, the covariance between returns* $\mathbf{r}$ *and characteristics* $\mathbf{x}_j$ *is given by* $\mathrm{Cov}(\mathbf{r}, \mathbf{x}_j) = \rho_j \sigma_{\mathbf{r}} \sigma_{\mathbf{x}_j} \mathbf{I}$*, with* $\sigma_{\mathbf{r}}$ *the cross-sectional standard deviation of returns,* $\sigma_{\mathbf{x}_j}$ *the cross-sectional standard deviation of the* j*-th characteristic, and* $\mathbf{I}$ *the identity matrix.*

Assumption 3.1 asserts that multiple asset characteristics are independent of each other, which makes it possible to decompose the contribution from each constraint in mathematically intuitive forms (see Propositions 20–21). More generally, when characteristics are dependent, a similar decomposition is also possible with additional interaction terms between different constraints, and we provide this generalization in the appendix to Lo and Zhang (2025).

Assumption 3.2 simplifies the dependence between returns and characteristics, while still allowing the returns, $\mathbf{r}$, to be arbitrarily dependent cross-sectionally. This assumption was first used in Lo and MacKinlay (1990) to describe cross-sectional estimation errors of intercepts in CAPM regressions, and later in Lo and Zhang (2024) to describe the dependence structure between returns and an impact factor such as ESG. The Appendix of Lo and Zhang (2025) provides a generalization that relaxes this assumption as well.

Given Assumption 3, the covariance matrix of $[\mathbf{r}' \ \mathbf{x}_1' \ \cdots \ \mathbf{x}_J']$ can be written as:

$$\begin{pmatrix} \boldsymbol{\Sigma} & \rho_1 \sigma_{\mathbf{r}} \sigma_{\mathbf{x}_1} \mathbf{I} \ \cdots \ \rho_J \sigma_{\mathbf{r}} \sigma_{\mathbf{x}_J} \mathbf{I} \\ \begin{matrix} \rho_1 \sigma_{\mathbf{r}} \sigma_{\mathbf{x}_1} \mathbf{I} \\ \vdots \\ \rho_J \sigma_{\mathbf{r}} \sigma_{\mathbf{x}_J} \mathbf{I} \end{matrix} & \begin{pmatrix} \sigma_{\mathbf{x}_1}^2 \mathbf{I} & 0 & 0 \\ 0 & \ddots & 0 \\ 0 & 0 & \sigma_{\mathbf{x}_J}^2 \mathbf{I} \end{pmatrix} \end{pmatrix}. \tag{9.11}$$

The following result characterizes the *excess* return and covariance due to the information in $\mathbf{X}$.

Proposition 20 (Information Decomposition) *Under Assumptions 1 and 3,*

- *if Assumption 2 holds,* $\mathbf{r}|\mathbf{X}$ *is normally distributed, with an expected value given by:*

$$\boldsymbol{\mu}_{\mathbf{X}} = \mathbb{E}[\mathbf{r}|\mathbf{X}] = \boldsymbol{\mu} + \sum_{j=1}^{J} \rho_j \sigma_{\mathbf{r}} \frac{(\mathbf{x}_j - \bar{\mathbf{x}}_j)}{\sigma_{\mathbf{x}_j}}, \tag{9.12}$$

and a covariance matrix given by:

$$\boldsymbol{\Sigma}_{\mathbf{X}} = \mathrm{Cov}(\mathbf{r}|\mathbf{X}) = \boldsymbol{\Sigma} - \sum_{j=1}^{J} \rho_j^2 \sigma_{\mathbf{r}}^2 \mathbf{I}; \tag{9.13}$$

- *if Assumption 2' holds,* $\mathbf{r}|\mathbf{X}$ *follows a multivariate Student-t distribution:*

$$\mathrm{MVT}\left(\boldsymbol{\mu} + \mathbf{V}_{\mathbf{r},\mathbf{X}}\mathbf{V}_{\mathbf{X},\mathbf{X}}^{-1}(\mathbf{X} - \bar{\mathbf{X}}),\ s(\mathbf{V} - \mathbf{V}_{\mathbf{r},\mathbf{X}}\mathbf{V}_{\mathbf{X},\mathbf{X}}^{-1}\mathbf{V}_{\mathbf{X},\mathbf{r}}),\ \nu + NJ\right), \tag{9.14}$$

where $s = \frac{\nu + (\mathbf{X}-\bar{\mathbf{X}})'\mathbf{V}_{\mathbf{X},\mathbf{X}}^{-1}(\mathbf{X}-\bar{\mathbf{X}})}{\nu + NJ}$ *is a scaling parameter that approaches* 1 *as the number of assets,* N*, increases without bound. In particular, its expected value is given by:*

$$\boldsymbol{\mu}_{\mathbf{X}} = \mathbb{E}[\mathbf{r}|\mathbf{X}] = \boldsymbol{\mu} + \sum_{j=1}^{J} \rho_j \sigma_{\mathbf{r}} \frac{(\mathbf{x}_j - \bar{\mathbf{x}}_j)}{\sigma_{\mathbf{x}_j}}, \tag{9.15}$$

for all N*, and its covariance matrix is given by:*

$$\boldsymbol{\Sigma}_{\mathbf{X}} = \mathrm{Cov}(\mathbf{r}|\mathbf{X}) \overset{p}{=} (1 - 2/\nu)\left(\boldsymbol{\Sigma} - \sum_{j=1}^{J} \rho_j^2 \sigma_{\mathbf{r}}^2 \mathbf{I}\right), \tag{9.16}$$

where $\overset{p}{=}$ *denotes equality in probability as* N *increases without bound.*

Both the normal distribution and the multivariate Student-t distribution are special cases of the elliptical distribution family (Fang, Kotz, and Ng, 1990, p. 42–47). Therefore, it is not surprising that their conditional moments conform to similar analytical forms.

Proposition 20 allows for more explicit decompositions of the expected return and utility of the constrained portfolio by substituting Equations (9.12)–(9.16) into Proposition 19, which we summarize in Proposition 21.

Proposition 21 (Attribution with Normal and MVT Returns) *Under Assumptions 1, 2 (or 2'), and 3 and conditioned on information in* $\mathbf{X}$ *that is used to form constraints* $\mathbf{A}(\mathbf{X})$*, the following decompositions hold for the optimal portfolio,* $\boldsymbol{\omega}^*$*.*

1. *Expected return decomposition.*

$$\mathbb{E}\left[\boldsymbol{\omega}^{*\prime}\mathbf{r}|\mathbf{X}\right] = \boldsymbol{\mu}_{\mathbf{X}}'\boldsymbol{\omega}^* = \boldsymbol{\mu}_{\mathbf{X}}'\boldsymbol{\omega}_{\mathrm{MVO}} + \boldsymbol{\mu}'\boldsymbol{\omega}_{\mathrm{CSTR}} + \sum_{j=1}^{J} \rho_j \sigma_{\mathbf{r}} \frac{(\mathbf{x}_j' - \bar{\mathbf{x}}_j')\boldsymbol{\omega}_{\mathrm{CSTR}}}{\sigma_{\mathbf{x}_j}}, \tag{9.17}$$

where the Lagrange multipliers are the same as in Proposition 19 provided that the feasible region of the constrained optimization problem is nonempty.

- $\boldsymbol{\mu}_{\mathbf{X}}'\boldsymbol{\omega}_{\mathrm{MVO}}$: *expected return of the unconstrained MVO portfolio.*
- $\boldsymbol{\mu}'\boldsymbol{\omega}_{\mathrm{CSTR}}$: *components attributable to each constraint treated as static.*
- $\rho_j \sigma_{\mathbf{r}} \frac{(\mathbf{x}_j' - \bar{\mathbf{x}}_j')\boldsymbol{\omega}_{\mathrm{CSTR}}}{\sigma_{\mathbf{x}_j}}$: *component attributable to information in the j-th constraint.*

2. *Expected utility decomposition.*

$$\begin{aligned}\boldsymbol{\mu}_{\mathbf{X}}'\boldsymbol{\omega}^* - \frac{\gamma}{2}\boldsymbol{\omega}^{*\prime}\boldsymbol{\Sigma}_{\mathbf{X}}\boldsymbol{\omega}^* &= \boldsymbol{\mu}_{\mathbf{X}}'\boldsymbol{\omega}_{\mathrm{MVO}} - \frac{\gamma}{2}\boldsymbol{\omega}_{\mathrm{MVO}}'\boldsymbol{\Sigma}_{\mathbf{X}}\boldsymbol{\omega}_{\mathrm{MVO}} \\ &\quad - \frac{\gamma}{2}\boldsymbol{\omega}_{\mathrm{CSTR}}'\boldsymbol{\Sigma}\boldsymbol{\omega}_{\mathrm{CSTR}} \\ &\quad + \sum_{j=1}^{J}\left(\rho_j \sigma_{\mathbf{r}} \frac{(\mathbf{x}_j' - \bar{\mathbf{x}}_j')\boldsymbol{\omega}_{\mathrm{CSTR}}}{\sigma_{\mathbf{x}_j}} + \gamma \rho_j^2 \sigma_{\mathbf{r}}^2 \boldsymbol{\omega}_{\mathrm{SHR}}'\boldsymbol{\omega}_{\mathrm{CSTR}}\right).\end{aligned} \tag{9.18}$$

- $\boldsymbol{\mu}'_{\mathbf{X}}\boldsymbol{\omega}_{\mathrm{MVO}} - \frac{\gamma}{2}\boldsymbol{\omega}'_{\mathrm{MVO}}\boldsymbol{\Sigma}_{\mathbf{X}}\boldsymbol{\omega}_{\mathrm{MVO}}$: *optimal expected utility of the unconstrained MVO portfolio.*
- $-\frac{\gamma}{2}\boldsymbol{\omega}'_{\mathrm{CSTR}}\boldsymbol{\Sigma}\boldsymbol{\omega}_{\mathrm{CSTR}}$: *components attributable to all constraints combined together, treated as static.*
- $\rho_j\sigma_{\mathbf{r}}\frac{(\mathbf{x}'_j-\bar{\mathbf{x}}'_j)\boldsymbol{\omega}_{\mathrm{CSTR}}}{\sigma_{\mathbf{x}_j}} + \gamma\rho_j^2\sigma_{\mathbf{r}}^2\boldsymbol{\omega}'_{\mathrm{SHR}}\boldsymbol{\omega}_{\mathrm{CSTR}}$: *component attributable to information in the j-th constraint.*

When Assumption 2' is true, Equation (9.18) holds when N and ν increase without bound.[9]

The last term of Equation (9.17) (see also Equation (9.12)) shows that the excess expected return, $\boldsymbol{\mu}_{\mathbf{X}} - \boldsymbol{\mu}$, is linear in $\mathbf{x}$. More importantly, it is determined by three terms: The first term, ρ_j, determines the correlation between asset characteristics and returns; the second term, $\sigma_{\mathbf{r}}$, measures the standard deviation of returns; and the third term, $(\mathbf{x}_j - \bar{\mathbf{x}}_j)/\sigma_{\mathbf{x}_j}$, determines whether each asset's characteristic value is above or below the average characteristic value, much like a z-score. When the asset characteristics are positively correlated with returns, those assets with above-average characteristic values have positive excess returns. When the asset characteristics are negatively correlated with returns, those assets with below-average characteristic values have positive excess returns.

Proposition 21 also highlights a connection between the information component of our decomposition and traditional regression-based performance attribution (Fama, 1972; Sharpe, 1992). The last term of Equation (9.17) can be written equivalently as the product of two terms. The first term, $\rho_j\sigma_{\mathbf{r}}/\sigma_{\mathbf{x}_j}$, is the beta of the asset returns regressed on each of the j-th characteristics, which is a standard procedure in classical style analysis and performance attribution. The second term, $(\mathbf{x}'_j - \bar{\mathbf{x}}'_j)\boldsymbol{\omega}_{\mathrm{CSTR}}$, is the average loading on the j-th characteristic of the portfolio holdings directly attributable to constraints. Our results show that an interaction effect between the two determines the information contribution from constraints.

The last term of Equation (9.18) (see also Equation (9.13)) shows that the excess covariance, $(\boldsymbol{\Sigma}_{\mathbf{X}} - \boldsymbol{\Sigma})$, is always negative. In other words, when $\boldsymbol{\omega}'_{\mathrm{SHR}}\boldsymbol{\omega}_{\mathrm{CSTR}} > 0$, i.e., the shrinkage portfolio holdings and the holdings attributable to constraints are positively correlated, incorporating information in $\mathbf{X}$ always reduces the variance of a portfolio. In addition, the magnitude of reduction in variance due to the j-th constraint depends on ρ_j^2, the squared correlation between the j-th asset characteristics and returns. The constraints with a higher magnitude of correlation lead to larger reductions in variance. Conversely, when $\boldsymbol{\omega}'_{\mathrm{SHR}}\boldsymbol{\omega}_{\mathrm{CSTR}} < 0$, incorporating information in $\mathbf{X}$ increases the variance and reduces the expected utility of a portfolio. These two scenarios correspond to the two cases in Figure 9.1, respectively.

9.3 Accounting for estimation risk via bayesian portfolio analysis

Our framework so far relies on knowing the *population* value of $\boldsymbol{\mu}$ and $\boldsymbol{\Sigma}$, but does not account for the estimation risk when these parameters are estimated using a *sample*. In this section, we apply Bayesian portfolio analysis to accommodate estimation risk and analyze out-of-sample returns. We emphasize that we present our main framework in the previous section in the population model because it highlights the fundamental role of information in constraints in determining the performance of a portfolio. This holds true even if investors

[9]This is true for Bayesian portfolios with common choices of priors, as shown in Section 9.3.2.

know the true $\boldsymbol{\mu}$ and $\boldsymbol{\Sigma}$ without any estimation risk. The goal of this section is *not* to propose a new robust portfolio rule or a better way to deal with estimation risk, but rather to generalize our framework to out-of-sample returns of *existing* robust portfolio rules.

9.3.1 Bayesian portfolio analysis

Bayesian portfolio analysis is a classical and natural framework to account for estimation risk (Avramov and Zhou, 2010; Fabozzi, Huang, and Zhou, 2010; Jacquier and Polson, 2011). We follow the classical literature to assume that returns $\mathbf{r}_t$, $t = 1, 2, \ldots$ are independently and identically distributed (IID) over time with mean $\boldsymbol{\mu}$ and covariance matrix $\boldsymbol{\Sigma}$. Investors at time T have observed returns $\Phi_T = \{\mathbf{r}_1, \mathbf{r}_2, \ldots, \mathbf{r}_T\}$, and the predictive density of the next return is:

$$\mathbb{P}(\mathbf{r}_{T+1}|\Phi_T) = \int_{\boldsymbol{\mu}}\int_{\boldsymbol{\Sigma}} \mathbb{P}(\mathbf{r}_{T+1}, \boldsymbol{\mu}, \boldsymbol{\Sigma}|\Phi_T)\, d\boldsymbol{\mu}\, d\boldsymbol{\Sigma} = \int_{\boldsymbol{\mu}}\int_{\boldsymbol{\Sigma}} \mathbb{P}(\mathbf{r}_{T+1}|\boldsymbol{\mu}, \boldsymbol{\Sigma})\mathbb{P}(\boldsymbol{\mu}, \boldsymbol{\Sigma}|\Phi_T)\, d\boldsymbol{\mu}\, d\boldsymbol{\Sigma}, \tag{9.19}$$

where $\mathbb{P}(\mathbf{r}_{T+1}|\boldsymbol{\mu}, \boldsymbol{\Sigma})$ describes the data generating process, i.e., the distribution of returns conditioned on known parameters, and $\mathbb{P}(\boldsymbol{\mu}, \boldsymbol{\Sigma}|\Phi_T)$ is the posterior density of $\boldsymbol{\mu}$ and $\boldsymbol{\Sigma}$. Given an appropriate prior, $\mathbb{P}(\boldsymbol{\mu}, \boldsymbol{\Sigma})$, the posterior density is $\mathbb{P}(\boldsymbol{\mu}, \boldsymbol{\Sigma}|\Phi_T) \propto \mathbb{P}(\Phi_T|\boldsymbol{\mu}, \boldsymbol{\Sigma}) \times \mathbb{P}(\boldsymbol{\mu}, \boldsymbol{\Sigma})$.

Unlike the conditional distribution, the Bayesian predictive distribution of Equation (9.19) accounts for estimation error by integrating over the unknown parameter space. The optimal portfolio is then constructed by maximizing the expected utility with respect to the predictive distribution. For notational convenience, we denote the $(N \times 1)$-vector, $\tilde{\mathbf{r}} \equiv \mathbf{r}_{T+1}|\Phi_T$, as the predictive return vector that follows Equation (9.19), and $\tilde{\boldsymbol{\mu}} \equiv \mathbb{E}[\tilde{\mathbf{r}}]$ and $\tilde{\boldsymbol{\Sigma}} \equiv \mathrm{Cov}(\tilde{\mathbf{r}})$ are the expected value and covariance matrix of $\tilde{\mathbf{r}}$, respectively. By convention, we always consider predictive returns conditioned on the previous T returns in the Bayesian framework, and we omit the time subscript $T+1$ in $\tilde{\mathbf{r}}$ and its moments for notational convenience without causing any confusion.

In the case of mean-variance utility, the optimal portfolio with respect to the predictive distribution solves the optimization problem in Equation (9.1) with $\boldsymbol{\mu}$ and $\boldsymbol{\Sigma}$ replaced by $\tilde{\boldsymbol{\mu}}$ and $\tilde{\boldsymbol{\Sigma}}$:

$$\begin{aligned} \max_{\boldsymbol{\omega}} \;& \boldsymbol{\omega}'\tilde{\boldsymbol{\mu}} - \frac{\gamma}{2}\boldsymbol{\omega}'\tilde{\boldsymbol{\Sigma}}\boldsymbol{\omega} \\ \text{s.t.} \;& \mathbf{A}(\mathbf{X}_T)\boldsymbol{\omega} = \mathbf{b}, \end{aligned} \tag{9.20}$$

where $\mathbf{X}_T$ is the asset characteristics at time T. We use $\tilde{\boldsymbol{\omega}}^*$ to denote the weights for the optimal Bayesian portfolio by solving the optimization problem of Equation (9.20).

9.3.2 Attribution with information using predictive distribution

We assume that the asset characteristics, $\mathbf{X}_t$, $t = 1, 2, \ldots$, are identically distributed over time, and $\mathbf{X}_t$ may be correlated with $\mathbf{r}_{t+1}$ in general. Therefore, at time T, $\mathbf{X}_T$ is *not* independent of the predictive returns, $\tilde{\mathbf{r}}$, that follow Equation (9.19). Following our notations so far and again omitting the time subscript T in asset characteristics without causing any confusion, we use $\tilde{\mathbf{r}}|\mathbf{X}$ to denote the predictive returns conditioned on additional information in $\mathbf{X}_T$, and $\tilde{\boldsymbol{\mu}}_{\mathbf{X}} \equiv \mathbb{E}\left[\tilde{\mathbf{r}}|\mathbf{X}\right]$ and $\tilde{\boldsymbol{\Sigma}}_{\mathbf{X}} \equiv \mathrm{Cov}\left[\tilde{\mathbf{r}}|\mathbf{X}\right]$ to denote the conditional mean and covariance matrix.

With this notation, the performance attribution in Proposition 19 can be generalized to accommodate the Bayesian predictive distribution by replacing the population moments ($\boldsymbol{\mu}$, $\boldsymbol{\Sigma}$, $\boldsymbol{\mu}_{\mathbf{X}}$, $\boldsymbol{\Sigma}_{\mathbf{X}}$) with the corresponding moments from the predictive distribution ($\tilde{\boldsymbol{\mu}}$, $\tilde{\boldsymbol{\Sigma}}$,

$\tilde{\boldsymbol{\mu}}_{\mathbf{X}}$, $\tilde{\boldsymbol{\Sigma}}_{\mathbf{X}}$). We provide the formal statement of this result in the appendix to Lo and Zhang (2025).

We emphasize that so far we have not made any assumptions about the specific form of the prior $\mathbb{P}(\boldsymbol{\mu}, \boldsymbol{\Sigma})$, nor the distribution of returns conditioned on known parameters $\mathbb{P}(\mathbf{r}_{T+1}|\boldsymbol{\mu}, \boldsymbol{\Sigma})$. Therefore, the decomposition holds for any Bayesian portfolios constructed using the predictive distribution. In general, the predictive distribution in Equation (9.19) and its moments, $\tilde{\boldsymbol{\mu}}$, $\tilde{\boldsymbol{\Sigma}}$, $\tilde{\boldsymbol{\mu}}_{\mathbf{X}}$, and $\tilde{\boldsymbol{\Sigma}}_{\mathbf{X}}$, can be obtained via Monte Carlo simulation in the Bayesian statistics literature.[10] Therefore, as long as a portfolio rule can be expressed as a Bayesian portfolio, our framework can be applied to derive a decomposition of the performance into the different components outlined in Proposition 19, at least numerically.

For certain Bayesian portfolios in the literature with specific distributions of the prior, we can use our results in Section 9.2.3 to derive more explicit decompositions, which we discuss next. We follow the majority of this literature to assume that the return distribution conditioned on $\boldsymbol{\mu}$ and $\boldsymbol{\Sigma}$ is normal to allow for tractability of the posterior predictive distribution.

Normal predictive returns with unknown $\boldsymbol{\mu}$ and known $\boldsymbol{\Sigma}$. It is known that the expected return vector $\boldsymbol{\mu}$ is much more challenging to estimate than the covariance matrix $\boldsymbol{\Sigma}$, partially because using higher sampling frequencies improves the accuracy of the covariance estimator but not the mean estimator (Merton, 1980). Several studies have documented that the estimation risk is mainly due to errors in the estimated expected return (Broadie, 1993; Chopra and Ziemba, 1993; Ceria and Stubbs, 2006), which suggests that uncertainty in the mean is the first-order effect to address in portfolio optimization (Jacquier and Polson, 2011).

Therefore, we first consider uncertainties in $\boldsymbol{\mu}$ and treat $\boldsymbol{\Sigma}$ as known. In this case, the posterior predictive distribution is usually normal. The appendix to Lo and Zhang (2025) derives an explicit expression of the information component for predictive returns, which is similar to Proposition 20 except that the population moments ($\boldsymbol{\mu}$, $\boldsymbol{\Sigma}$, $\boldsymbol{\mu}_{\mathbf{X}}$, $\boldsymbol{\Sigma}_{\mathbf{X}}$) are replaced by the predictive moments ($\tilde{\boldsymbol{\mu}}$, $\tilde{\boldsymbol{\Sigma}}$, $\tilde{\boldsymbol{\mu}}_{\mathbf{X}}$, $\tilde{\boldsymbol{\Sigma}}_{\mathbf{X}}$).

This result can be applied to several classical models such as Klein and Bawa (1976), Jorion (1986), and Black and Litterman (1992). In each model, they adopt different priors that lead to normal predictive distributions with different parameters. In our framework, these portfolios without constraints correspond to different benchmark portfolios $\tilde{\boldsymbol{\omega}}_{\mathrm{MVO}}$.

Student-t predictive returns with unknown $\boldsymbol{\mu}$ and $\boldsymbol{\Sigma}$.

When $\boldsymbol{\mu}$ and $\boldsymbol{\Sigma}$ are both unknown and modeled as random parameters in the Bayesian framework, the predictive density of returns is typically a multivariate Student-t distribution, $\tilde{\mathbf{r}} \sim \mathrm{MVT}\left(\tilde{\boldsymbol{\mu}}, \tilde{\mathbf{V}}, \nu\right)$.

Similar to the case of normal predictive returns, our results can be applied to a broader class of Bayesian portfolios and shrinkage portfolios with certain equivalence to Bayesian portfolios, including the non-informative diffuse prior $\mathbb{P}(\boldsymbol{\mu}, \boldsymbol{\Sigma}) \propto |\boldsymbol{\Sigma}|^{-(N+1)/2}$ as in Klein and Bawa (1976) and Brown (1978), and the normal–inverse–Wishart conjugate prior for $(\boldsymbol{\mu}, \boldsymbol{\Sigma})$ as in Frost and Savarino (1986), Stambaugh (1997), Pástor (2000), Pástor and Stambaugh (2000), Zhou (2009), and Lai, Xing, and Chen (2011). In these examples, the predictive distribution is MVT with different parameters.

[10] Pástor (2000) and Harvey et al. (2010) provide examples of using Monte Carlo simulation to obtain Bayesian portfolio weights. Tu and Zhou (2010) suggest using Monte Carlo simulation to derive the relationship between expected return and optimal Bayesian portfolio weights for non-mean-variance utility functions.

In addition, Tu and Zhou (2010) establish a certain equivalence between priors on portfolio weights, $\boldsymbol{\omega}$, and priors on $\boldsymbol{\mu}$ and $\boldsymbol{\Sigma}$. DeMiguel et al. (2009) show that certain norm-constrained portfolios can be equivalently interpreted as Bayesian portfolios with prior beliefs on portfolio weights. The insights of these two seminal articles provide an important link between the literature on robust portfolios with priors on portfolio weights[11] and the literature on Bayesian portfolio analysis. This implies that our attribution framework can, in principle, be applied to robust portfolio rules that directly impose priors or shrinkage on portfolio weights through their Bayesian equivalents. However, deriving the specific Bayesian formulation for each of them is not the focus and beyond the scope of this chapter. The appendix to Lo and Zhang (2025) discusses this literature in more detail.

9.3.3 Interpreting constraints as bayesian priors

In this section, we formally establish that, in the context of Bayesian portfolios, solving a portfolio with constraints can be equivalently interpreted as solving an unconstrained portfolio with a certain prior on $\boldsymbol{\mu}$ and $\boldsymbol{\Sigma}$.

In particular, following Section 9.3.2, we consider normally distributed returns conditioned on $\boldsymbol{\mu}$ and $\boldsymbol{\Sigma}$: $\mathbf{r} \mid (\boldsymbol{\mu}, \boldsymbol{\Sigma}) \sim N(\boldsymbol{\mu}, \boldsymbol{\Sigma})$, and two priors, π_1 and π_2, on the unknown parameters, which lead to posterior predictive returns with moments $(\tilde{\boldsymbol{\mu}}_1, \tilde{\boldsymbol{\Sigma}}_1)$ and $(\tilde{\boldsymbol{\mu}}_2, \tilde{\boldsymbol{\Sigma}}_2)$, respectively. We consider a constrained Bayesian portfolio (I) with prior π_1 that solves the problem in Equation (9.20):

$$\tilde{\boldsymbol{\omega}}_1^* = \frac{1}{\gamma}\tilde{\boldsymbol{\Sigma}}_1^{-1}(\tilde{\boldsymbol{\mu}}_1 - \mathbf{A}'\tilde{\boldsymbol{\lambda}}) \quad \text{where} \quad \tilde{\boldsymbol{\lambda}} = \left(\mathbf{A}\tilde{\boldsymbol{\Sigma}}_1^{-1}\mathbf{A}'\right)^{-1}\left(\mathbf{A}\tilde{\boldsymbol{\Sigma}}_1^{-1}\tilde{\boldsymbol{\mu}}_1 - \gamma\mathbf{b}\right), \tag{9.21}$$

and an unconstrained Bayesian portfolio (II) with prior π_2 that maximizes the same predictive mean-variance utility in Equation (9.20) but without any constraint:

$$\tilde{\boldsymbol{\omega}}_2^* = \frac{1}{\gamma}\tilde{\boldsymbol{\Sigma}}_2^{-1}\tilde{\boldsymbol{\mu}}_2. \tag{9.22}$$

The following result establishes the equivalence between the two portfolios.

Proposition 22 (Constraints as Priors) *With portfolios I and II in Equations (9.21) and (9.22), respectively,*

1. *When $\boldsymbol{\mu}$ is unknown and $\boldsymbol{\Sigma}$ is known, consider two priors on $\boldsymbol{\mu}$:*

$$\pi_1 : \boldsymbol{\mu} \sim N\left(\boldsymbol{\mu}_1, \frac{\boldsymbol{\Sigma}}{\tau}\right),$$
$$\pi_2 : \boldsymbol{\mu} \sim N\left(\boldsymbol{\mu}_2, \frac{\boldsymbol{\Sigma}}{\tau}\right),$$

where $\boldsymbol{\mu}_1$ and $\boldsymbol{\mu}_2$ are prior means, and τ is a precision hyperparameter. Portfolios I and II are identical ($\tilde{\boldsymbol{\omega}}_1^* = \tilde{\boldsymbol{\omega}}_2^*$) if the following condition holds:

$$\boldsymbol{\mu}_2 = \boldsymbol{\mu}_1 - \mathbf{A}'\tilde{\boldsymbol{\lambda}}_{1,\mathrm{PR}} - \frac{T}{\tau}\mathbf{A}'\tilde{\boldsymbol{\lambda}}_{1,\mathrm{DT}} \tag{9.23}$$

where

$$\tilde{\boldsymbol{\lambda}}_{1,\mathrm{PR}} \equiv (\mathbf{A}\boldsymbol{\Sigma}^{-1}\mathbf{A}')^{-1}\left(\mathbf{A}\boldsymbol{\Sigma}^{-1}\boldsymbol{\mu}_1 - (1+1/\tau)\gamma\mathbf{b}\right),$$
$$\tilde{\boldsymbol{\lambda}}_{1,\mathrm{DT}} \equiv (\mathbf{A}\boldsymbol{\Sigma}^{-1}\mathbf{A}')^{-1}(\mathbf{A}\boldsymbol{\Sigma}^{-1}\hat{\boldsymbol{\mu}} - \gamma\mathbf{b})$$

[11] In addition to the studies mentioned above, see, for example, Brodie et al. (2009), Fan, Li, and Yu (2012), and Fan, Zhang, and Yu (2012), as well as combinations of different portfolio rules such as Kan and Zhou (2007) Tu and Zhou (2011), Kan, Wang, and Zhou (2022), and Kan and Wang (2023).

are Lagrange multipliers of two hypothetical constrained portfolios (9.20) with predictive distributions

$$\tilde{\mathbf{r}} \sim N\left(\boldsymbol{\mu}_1, \left(1+\frac{1}{\tau}\right)\boldsymbol{\Sigma}\right) \quad \text{and} \quad \tilde{\mathbf{r}} \sim N(\hat{\boldsymbol{\mu}}, \boldsymbol{\Sigma}),$$

respectively. $\hat{\boldsymbol{\mu}}$ *is the sample mean of the returns.*

2. *When* $\boldsymbol{\mu}$ *and* $\boldsymbol{\Sigma}$ *are both unknown, consider two priors on* $\boldsymbol{\mu}$ *and* $\boldsymbol{\Sigma}$*:*

$$\pi_1 : \boldsymbol{\mu} \mid \boldsymbol{\Sigma} \sim N\left(\boldsymbol{\mu}_1, \frac{\boldsymbol{\Sigma}}{\tau}\right), \quad \boldsymbol{\Sigma} \sim IW(\boldsymbol{\Sigma}_1, v_0),$$

$$\pi_2 : \boldsymbol{\mu} \mid \boldsymbol{\Sigma} \sim N\left(\boldsymbol{\mu}_2, \frac{\boldsymbol{\Sigma}}{\tau}\right), \quad \boldsymbol{\Sigma} \sim IW(\boldsymbol{\Sigma}_2, v_0),$$

where $\boldsymbol{\mu}_1$ *and* $\boldsymbol{\mu}_2$ *are prior means,* $\boldsymbol{\Sigma}_1$ *and* $\boldsymbol{\Sigma}_2$ *are prior covariance matrices,* τ *and* v_0 *are hyperparameters, and IW stands for the Inversed-Wishart distribution. Portfolios I and II are identical* ($\tilde{\boldsymbol{\omega}}_1^* = \tilde{\boldsymbol{\omega}}_2^*$) *if the following conditions hold:*

$$\boldsymbol{\mu}_2 = \boldsymbol{\mu}_1 - \mathbf{A}'\tilde{\boldsymbol{\lambda}}_{2,\text{PR}} - \frac{T}{\tau}\mathbf{A}'\tilde{\boldsymbol{\lambda}}_{2,\text{DT}}, \tag{9.24}$$

$$\boldsymbol{\Sigma}_2 = \boldsymbol{\Sigma}_1 + \frac{T\tau}{T+\tau}(\boldsymbol{\mu}_1 - \boldsymbol{\mu}_2)(\boldsymbol{\mu}_1 + \boldsymbol{\mu}_2 - 2\hat{\boldsymbol{\mu}})', \tag{9.25}$$

where

$$\tilde{\boldsymbol{\lambda}}_{2,\text{PR}} \equiv \left(\mathbf{A}\tilde{\boldsymbol{\Sigma}}_1^{-1}\mathbf{A}'\right)^{-1}\left(\mathbf{A}\tilde{\boldsymbol{\Sigma}}_1^{-1}\boldsymbol{\mu}_1 - \gamma\mathbf{b}\right),$$

$$\tilde{\boldsymbol{\lambda}}_{2,\text{DT}} \equiv \left(\mathbf{A}\tilde{\boldsymbol{\Sigma}}_1^{-1}\mathbf{A}'\right)^{-1}\left(\mathbf{A}\tilde{\boldsymbol{\Sigma}}_1^{-1}\hat{\boldsymbol{\mu}} - \gamma\mathbf{b}\right)$$

are Lagrange multipliers of two hypothetical constrained portfolios as in Equation (9.20) with predictive distributions

$$\tilde{\mathbf{r}} \sim MVT\left(\boldsymbol{\mu}_1,\ \frac{v_0+T-N-1}{v_0+T-N+1}\tilde{\boldsymbol{\Sigma}}_1,\ v_0+T-N+1\right),$$

$$\tilde{\mathbf{r}} \sim MVT\left(\hat{\boldsymbol{\mu}},\ \frac{v_0+T-N-1}{v_0+T-N+1}\tilde{\boldsymbol{\Sigma}}_1,\ v_0+T-N+1\right),$$

respectively, and

$$\tilde{\boldsymbol{\Sigma}}_1 \equiv \left(1+\frac{1}{T+\tau}\right)\left(\frac{1}{v_0+T-N-1}\right)\left(\boldsymbol{\Sigma}_1 + T\hat{\boldsymbol{\Sigma}} + \frac{T\tau}{T+\tau}(\boldsymbol{\mu}_1-\hat{\boldsymbol{\mu}})(\boldsymbol{\mu}_1-\hat{\boldsymbol{\mu}})'\right)$$

is the predictive covariance matrix of returns under prior π_1.

Proposition 22 shows that Bayesian portfolio I with constraints is equivalent to Bayesian portfolio II with a different prior and without constraints. The relationships between their priors are characterized by Equation (9.23) when $\boldsymbol{\Sigma}$ is known, and Equations (9.24)–(9.25) when $\boldsymbol{\Sigma}$ is unknown. In both cases, imposing a constraint is equivalent to moving the prior mean on $\boldsymbol{\mu}$ by two terms, both of which yield clear financial intuitions. The first term reflects the effect of prior π_1 (see $\tilde{\boldsymbol{\lambda}}_{1,\text{PR}}$ and $\tilde{\boldsymbol{\lambda}}_{2,\text{PR}}$), and the second term reflects the effect of the observed data (see $\tilde{\boldsymbol{\lambda}}_{1,\text{DT}}$ and $\tilde{\boldsymbol{\lambda}}_{2,\text{DT}}$).[12]

[12] For example, as a sanity check, $\tilde{\boldsymbol{\lambda}}_{1,\text{PR}}$ vanishes if the unconstrained optimal portfolio under $\tilde{\mathbf{r}} \sim N\left(\boldsymbol{\mu}_1, \left(1+\frac{1}{\tau}\right)\boldsymbol{\Sigma}\right)$ already satisfies the constraint: $\mathbf{A}\left(\frac{1}{\gamma}(1+\frac{1}{\tau})\boldsymbol{\Sigma}\right)^{-1}\boldsymbol{\mu}_1 = \mathbf{b}$. Similarly, $\tilde{\boldsymbol{\lambda}}_{1,\text{DT}}$ vanishes if the optimal portfolio under the plug-in mean, $\tilde{\mathbf{r}} \sim N(\hat{\boldsymbol{\mu}}, \boldsymbol{\Sigma})$, already satisfies the constraint: $\mathbf{A}\left(\frac{1}{\gamma}\boldsymbol{\Sigma}^{-1}\hat{\boldsymbol{\mu}}\right) = \mathbf{b}$.

The connection between portfolio constraints and investor views is closely related to the seminal work of Tu and Zhou (2010), DeMiguel et al. (2009), and Ardia and Boudt (2015). Tu and Zhou (2010) establish the equivalence between (objective-based) priors on $\boldsymbol{\omega}$ and priors on $\boldsymbol{\mu}$ and $\boldsymbol{\Sigma}$, but do not explicitly consider the effect of portfolio constraints. DeMiguel et al. (2009) consider minimum variance portfolios and show that certain norm constraints on $\boldsymbol{\omega}$ are equivalent to Bayesian portfolios with a prior on $\boldsymbol{\omega}$. We consider mean-variance portfolios and priors on $\boldsymbol{\mu}$ and $\boldsymbol{\Sigma}$, and our constraints are different from theirs. Ardia and Boudt (2015) point out the connection between portfolio constraints and investor views but do not explore the equivalence in the context of Bayesian portfolios.

9.3.4 Summary

We summarize the procedure of applying our framework to Bayesian portfolios that account for estimation risk. This allows for performance attribution and constraint selection based on out-of-sample returns.

- Step 1. Pick a Bayesian method to derive the predictive distribution of asset returns that accounts for estimation risk. Different methods correspond to different benchmark portfolios $\tilde{\boldsymbol{\omega}}_{\text{MVO}}$.
- Step 2. Performance attribution either directly with Monte Carlo methods or using the analytical expressions given in the Appendix of Lo and Zhang (2025). For the latter, one estimates the correlation, ρ_j, between the characteristics of the j-th constraint, $\mathbf{x}_j$, and the predictive returns, $\tilde{\mathbf{r}}$, using a rolling window of panel data in the past T periods. This is a low-dimensional problem because there is only one parameter to estimate for each constraint, ρ_j, which can be achieved with a fairly high degree of accuracy (Lo and Zhang, 2024; Lo et al., 2024).
- Step 3. The results can be used either for purposes of performance attribution and disclosure, or to choose the constraint that provides the highest contribution to performance. Because portfolio constraints can be interpreted as Bayesian priors, as shown in Section 9.3.3, the attribution result also reflects the influence of investors' views on returns (expressed as Bayesian priors) on the portfolio's performance.

Different Bayesian portfolios correspond to different predictive distributions of returns and different benchmark portfolios. While our framework can be applied to Bayesian portfolios, we emphasize that it should not be treated *only* as a way to deal with estimation risk. In fact, even without estimation error, the same decomposition holds. Constraints can serve as a mechanism for incorporating information in both population and out-of-sample returns. Our methodology provides a way to quantify this effect.

We consider two common types of portfolio constraints and demonstrate how to select constraints that improve portfolio performance using simulations in Section 9.4. Our empirical analysis in Section 9.5 applies the framework to portfolios that account for estimation risk using two robust portfolio rules—Jorion's (1986) rule and the $1/N$ rule which is shown to be equivalent to Bayesian portfolios with a particular prior (DeMiguel, Garlappi, and Uppal, 2009; DeMiguel et al., 2009).

9.4 Common examples of portfolio constraints

In this section, we consider two common examples of portfolio constraints: factor exposures and exclusions. We derive additional analytical results for these constraints and conduct simulations to illustrate the attribution of expected returns and utility. All results in this section can be derived from either the unconditional $\boldsymbol{\mu}$ and $\boldsymbol{\Sigma}$ or the posterior predictive moments, $\tilde{\boldsymbol{\mu}}$ and $\tilde{\boldsymbol{\Sigma}}$. The latter can be used to select constraints *ex-ante*.

9.4.1 Factor exposure

A common constraint in portfolio construction arises when investors wish to control the average value of a characteristic or the exposure to a certain factor such as the average ESG score, market capitalization, beta, or book-to-market values of the portfolio.

We consider the case of a single constraint, $\mathbf{A}(\mathbf{x}) = \mathbf{x}'$, for simplicity. In this case, the Lagrange multiplier is a scalar, which leads to the following result.

Proposition 23 (Factor Exposure) *Under Assumptions 1, 2 (or 2'), and 3, and assuming without loss of generality that the cross-sectional average factor value $\bar{\mathbf{X}} = 0$, the expected return of the optimal portfolio with a factor exposure constraint can be decomposed into:*

$$\mathbb{E}\left[\boldsymbol{\omega}^{*\prime}\mathbf{r}|\mathbf{x}\right] = \frac{1}{\gamma}\boldsymbol{\mu}_{\mathbf{X}}'\boldsymbol{\Sigma}^{-1}\boldsymbol{\mu} + \frac{\mathbf{x}'\boldsymbol{\Sigma}^{-1}\boldsymbol{\mu}}{\mathbf{x}'\boldsymbol{\Sigma}^{-1}\mathbf{x}}\left(b - \mathbf{x}'\boldsymbol{\Sigma}^{-1}\boldsymbol{\mu}/\gamma\right) + \frac{\rho\sigma_{\mathbf{r}}}{\sigma_{\mathbf{x}}}\left(b - \mathbf{x}'\boldsymbol{\Sigma}^{-1}\boldsymbol{\mu}/\gamma\right), \tag{9.26}$$

if the constraint is binding. In the case of a non-binding inequality constraint, the Lagrange multiplier $\lambda^* = 0$ and Equation (9.26) reduces to: $\mathbb{E}\left[\boldsymbol{\omega}^{*\prime}\mathbf{r}|\mathbf{x}\right] = \boldsymbol{\mu}_{\mathbf{X}}'\boldsymbol{\Sigma}^{-1}\boldsymbol{\mu}/\gamma$.

If the moments of returns in Proposition 23 are replaced by their counterparts from the predictive distribution, we derive the same decomposition for the expected return of the predictive distribution.

There is no information contribution when the constraint is not binding in the case of an inequality constraint. When the constraint is binding, the information component depends on the correlation, ρ, between returns and asset characteristics. In addition, $\mathbf{x}'\boldsymbol{\Sigma}^{-1}\boldsymbol{\mu}$ is the characteristic value of the unconstrained MVO portfolio, so $b - \mathbf{x}'\boldsymbol{\Sigma}^{-1}\boldsymbol{\mu}$ measures the constrained characteristic value b *relative to* the unconstrained MVO portfolio. If a particular asset characteristic, e.g., ESG, is positively correlated with returns ($\rho > 0$), a positive desired ESG level relative to the unconstrained MVO portfolio ($b - \mathbf{x}'\boldsymbol{\Sigma}^{-1}\boldsymbol{\mu} > 0$) adds value to expected returns. However, if ESG is negatively correlated with returns ($\rho < 0$), the same constraint will hurt expected returns. This simple intuition informs the selection of constraints using our framework—pick constraints with a positive correlation, ρ, with the predictive returns. We demonstrate this further with the following simulation.

Simulation. We consider a world with 10 assets in which investors adopt a certain Bayesian method to account for estimation risk and derive the predictive density of returns. The expected predictive returns are generated from a normal distribution $\tilde{\mu}_i \sim N(0.05, 0.05^2)$, for $i = 1, 2, \ldots, 10$. The covariance matrix, $\tilde{\boldsymbol{\Sigma}} = (\sigma_{i,j})_{10\times10}$, is generated by using the algorithm of Davies and Higham (2000)—implemented by `scipy.stats.random_correlation` in Python—which generates a random correlation matrix given a set of eigenvalues. We set the 10 eigenvalues to range from 0.9 to 1.1 and then multiply the correlation matrix by 0.1 to get $\tilde{\boldsymbol{\Sigma}}$.[13]

[13]Results are not sensitive to the specific parameterization of $\tilde{\boldsymbol{\Sigma}}$ as long as it is positive definite.

Investors solve the following optimization problem with two constraints on factor exposures:

$$\begin{aligned} \max_{\boldsymbol{\omega}} \quad & \boldsymbol{\omega}'\tilde{\boldsymbol{\mu}} - \frac{\gamma}{2}\boldsymbol{\omega}'\tilde{\boldsymbol{\Sigma}}\boldsymbol{\omega} \\ \text{s.t.} \quad & \boldsymbol{\omega}'\mathbf{x} \geq 0.5 \quad \text{and} \quad \boldsymbol{\omega}'\mathbf{y} \geq 0.5. \end{aligned} \tag{9.27}$$

Here, $\mathbf{x}$ and $\mathbf{y}$ represent two characteristics such as an ESG score and return momentum. They are both 10-dimensional $N(0,1)$ random vectors that are IID over time. We denote the correlations between asset returns and these two characteristics by $\rho_1 \equiv \text{Corr}(\mathbf{x}, \tilde{\mathbf{r}})$ and $\rho_2 \equiv \text{Corr}(\mathbf{y}, \tilde{\mathbf{r}})$.

Figure 9.2 shows the attribution of expected returns following Proposition 23. Figure 9.2a shows the expected return of the constrained portfolio, which ranges between 1.6 and 3.0 as ρ_1 and ρ_2 vary between -0.8 and 0.8. Figure 9.2b shows the expected return of the unconstrained MVO portfolio, which is a constant value around 2.5 regardless of values of ρ_1 and ρ_2.

The source of the difference in expected returns between the unconstrained MVO portfolio and the constrained portfolio becomes clear in Figures 9.2c–9.2f. Figures 9.2c and 9.2d show the expected returns attributable to the two constraints, respectively, as if they are static. They each contribute to the expected returns with a negative constant value of around -0.1. Figure 9.2e shows the expected returns attributable to information in the first constraint, which increase as ρ_1 increases, but remain constant as ρ_2 varies. Figure 9.2f shows the expected returns attributable to information in the second constraint, which increase as ρ_2 increases, but remain constant as ρ_1 varies. Similar patterns are seen in decomposing expected utility, as shown in the appendix to Lo and Zhang (2025).

Overall, these results demonstrate how to understand the expected return and utility of a constrained portfolio by decomposing them into an unconstrained MVO portfolio, static constraints, and information in each constraint. In particular, when the information in constraints is sufficiently positively correlated with returns, they can lead to higher expected returns and utilities for the constrained portfolio.

9.4.2 Exclusionary investing

Another common form of constraint in portfolio construction is the exclusionary constraint, where assets are excluded from the portfolio based on criteria such as a minimum ESG score, or the particular industry in which a firm operates.

For convenience, we introduce some new notation. Suppose that assets are excluded based on the ranking of a characteristic $\mathbf{x}$. We use the notation $x_{1:N} < x_{2:N} < x_{N:N}$ to denote the ranked values of $\mathbf{x}$, or their order statistics. We then denote by $r_{[i:N]}$ the return associated with the i-th order statistic, $x_{i:N}$. This return is called the i-th *induced order statistic* in the literature, to emphasize the fact that its ranking is determined not by its own value, but by the value of $\mathbf{x}$ (David, 1973). For simplicity, we also use the subscript "$[N]$" to denote a vector or a matrix that is reordered based on values of $\mathbf{x}$. For example, $\boldsymbol{\omega}_{[N]}$ represents the vector of weights for assets that are reordered based on values of $\mathbf{x}$.

Investors solve the following optimization problem in which the top N_0 assets ranked by $\mathbf{x}$ are allowed to enter the portfolio, while the bottom $N - N_0$ assets are excluded:

$$\begin{aligned} \min_{\boldsymbol{\omega}_{[N]}} \quad & \boldsymbol{\omega}'_{[N]}\boldsymbol{\mu}_{[N]} - \frac{\gamma}{2}\boldsymbol{\omega}'_{[N]}\boldsymbol{\Sigma}_{[N]}\boldsymbol{\omega}_{[N]} \\ \text{s.t.} \quad & \omega_{[i:N]} = 0 \quad \text{for } i \leq N - N_0. \end{aligned} \tag{9.28}$$

The optimal portfolio for Equation (9.28) is simply the optimal portfolio restricted to N_0 assets. Therefore, the optimal portfolio weights are given by $\boldsymbol{\omega}^*_{N_0} = \boldsymbol{\Sigma}^{-1}_{N_0}\boldsymbol{\mu}_{N_0}$.

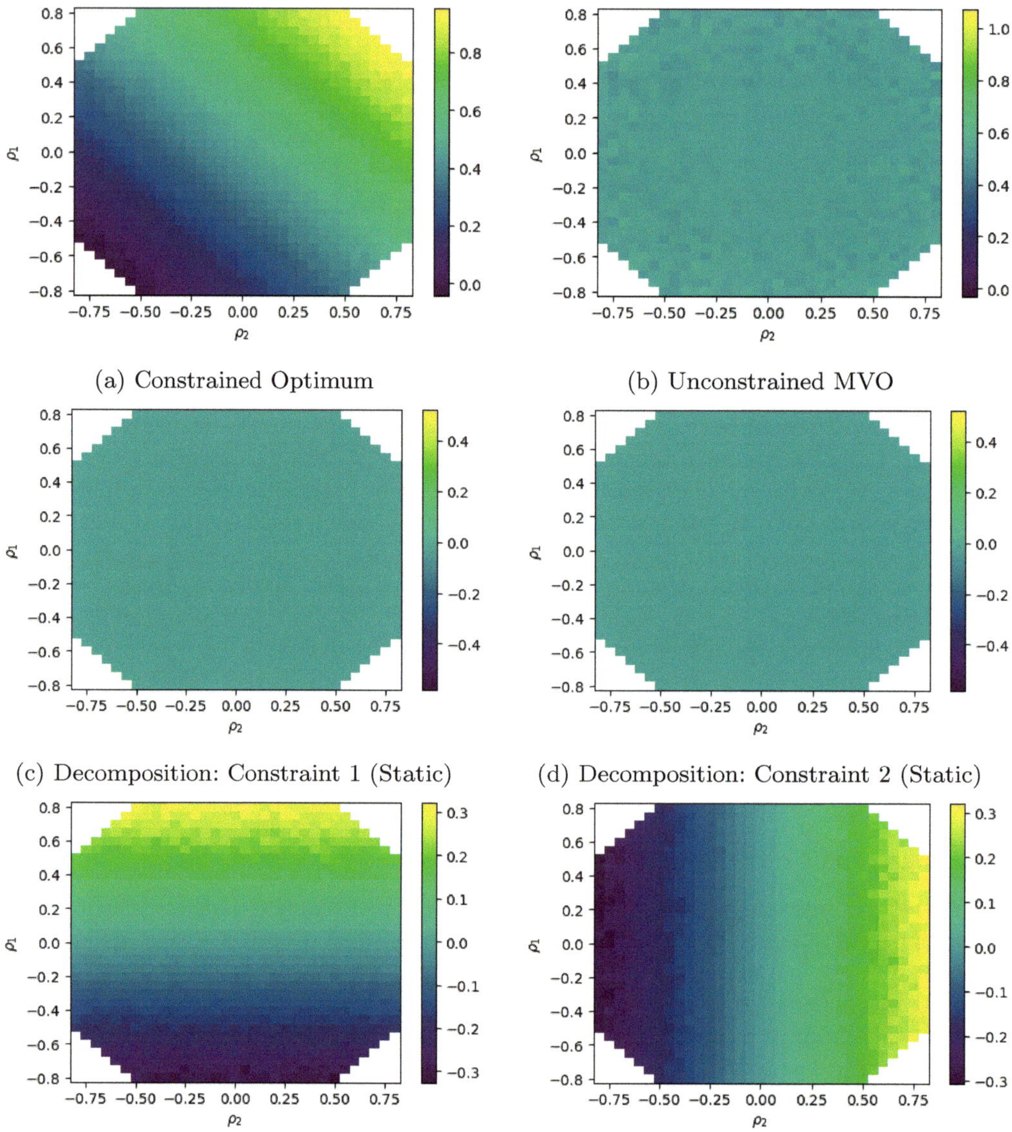

(a) Constrained Optimum (b) Unconstrained MVO

(c) Decomposition: Constraint 1 (Static) (d) Decomposition: Constraint 2 (Static)

(e) Decomposition: Constraint 1 (Information) (f) Decomposition: Constraint 2 (Information)

FIGURE 9.2: Decomposition of expected return for the optimization problem in (9.27) with two constraints that depend on random characteristics, as correlations (ρ_1 and ρ_2) between random characteristics and asset returns vary. The expected return of the constrained portfolio (a) is decomposed into components corresponding to the unconstrained MVO portfolio (b), static constraints (c–d), and information in the constraints (e–f).

If $\mathbf{x}$ is a vector of continuous random variables, the expected return of the optimal portfolio with an exclusionary constraint can be decomposed by Equation (9.17) in Proposition 21. If $\mathbf{x}$ is a vector of binary random variables, as is the case for exclusion based on industry or "sin stock" labels, the following result provides a more intuitive form of attribution for the expected return of this portfolio.

Proposition 24 (Exclusion based on Binary Characteristic) *Under Assumptions 1, 2 (or 2'), and 3, if* $\mathbf{x}$ *is a vector of binary random variables and* x_i *follows a Bernoulli distribution, the expected return of the optimal portfolio with an exclusionary constraint in (9.28) can be decomposed into:*

$$\mathbb{E}\left[\boldsymbol{\omega}^{*\prime}\mathbf{r}|\mathbf{x}\right] = \boldsymbol{\mu}'_{\mathbf{x},N_0}\boldsymbol{\omega}^*_{N_0} = \boldsymbol{\mu}'_{\mathbf{x}}\boldsymbol{\omega}_{\text{MVO}} + \boldsymbol{\mu}'\boldsymbol{\omega}_{\text{CSTR}} + \rho\sigma_{\mathbf{r}}\left(\mathbf{x}\odot\mathbf{u} - (1-\mathbf{x})\odot\mathbf{v}\right)'\boldsymbol{\omega}_{\text{CSTR}} \quad (9.29)$$

where

$$\mathbf{u} = \left(\sqrt{\frac{\psi_{x_1=0}}{\psi_{x_1=1}}},\ldots,\sqrt{\frac{\psi_{x_N=0}}{\psi_{x_N=1}}}\right)' \quad \textit{and} \quad \mathbf{v} = \left(\sqrt{\frac{\psi_{x_1=1}}{\psi_{x_1=0}}},\ldots,\sqrt{\frac{\psi_{x_N=1}}{\psi_{x_N=0}}}\right)'$$

are two vectors of the odds ratio (that is, the relative chance) of each asset being excluded from the portfolio, $\psi_{x_i=0} \equiv \mathbb{P}(x_i=0)$, $\psi_{x_i=1} \equiv \mathbb{P}(x_i=1)$, *and* $\odot$ *represents element-wise multiplication of two vectors.*

If the return moments in Proposition 24 are replaced by their counterparts from the predictive distribution, we derive the same decomposition for the expected return of the predictive distribution. Proposition 24 shows that the component attributable to information depends on the correlation, ρ, between returns and characteristics $\mathbf{x}$, and the chance of being excluded from the portfolio, $\frac{\psi_{x_i=0}}{\psi_{x_i=1}}$.

Simulation.

We consider a world with 10 assets, and, as before, investors adopt a certain Bayesian method to account for estimation risk and derive the predictive density of returns. The expected predictive return, $\tilde{\boldsymbol{\mu}}$, and covariance matrix, $\tilde{\boldsymbol{\Sigma}}$, are generated in the same way as in Section 9.4.1.

Investors solve the following problem:

$$\begin{aligned} &\max_{\boldsymbol{\omega}} \ \boldsymbol{\omega}'\tilde{\boldsymbol{\mu}} - \frac{\gamma}{2}\boldsymbol{\omega}'\tilde{\boldsymbol{\Sigma}}\boldsymbol{\omega} \\ &\text{s.t.} \ \ \omega_{[i:10]} = 0, \ \text{for } i \leq N - N_0 \text{ assets ordered by } \mathbf{x}. \end{aligned} \quad (9.30)$$

Here, $\mathbf{x}$ represents the asset characteristic that is used to exclude assets (e.g., minimum ESG score), a 10-dimensional $N(0,1)$ random vector that is IID over time. We denote the correlations of the asset characteristics with returns by $\rho \equiv \text{Corr}(\mathbf{x}, \tilde{\mathbf{r}})$.

Figure 9.3 shows the attribution of expected returns following Proposition 24, with ρ varying between -0.8 and 0.8 and the number of excluded assets varying between 1 and 9. Figure 9.3a shows the expected return of the constrained problem, which varies from a low of -0.2 to a high of 3.0 as ρ and the number of excluded assets vary. Figure 9.3b shows the expected return of the unconstrained MVO portfolio, which has a constant value of around 2.5.

The source of the difference in expected returns between the unconstrained MVO and the constrained portfolio becomes clear in Figures 9.3c–9.3d. Figure 9.3c shows the expected returns attributable to the constraints as if they are static. As more assets are excluded, the contributions from static constraints also increase. However, this component remains unchanged as the correlation, ρ, varies. Figure 9.3d shows the expected returns attributable to the information in the constraint, which increase as ρ increases. This result highlights a trade-off when a greater number of assets are excluded: When the correlation, ρ, is nonzero, excluding more assets implies that only assets with positive or negative returns are included in the portfolio, but excluding too many assets allows the portfolio too little choice in the universe of available assets. As a result, the highest returns are achieved when an intermediate number of assets are excluded, given a positive correlation ρ. Similarly, the lowest returns are achieved with an intermediate number of assets given a negative correlation. These results are consistent with Proposition 24.

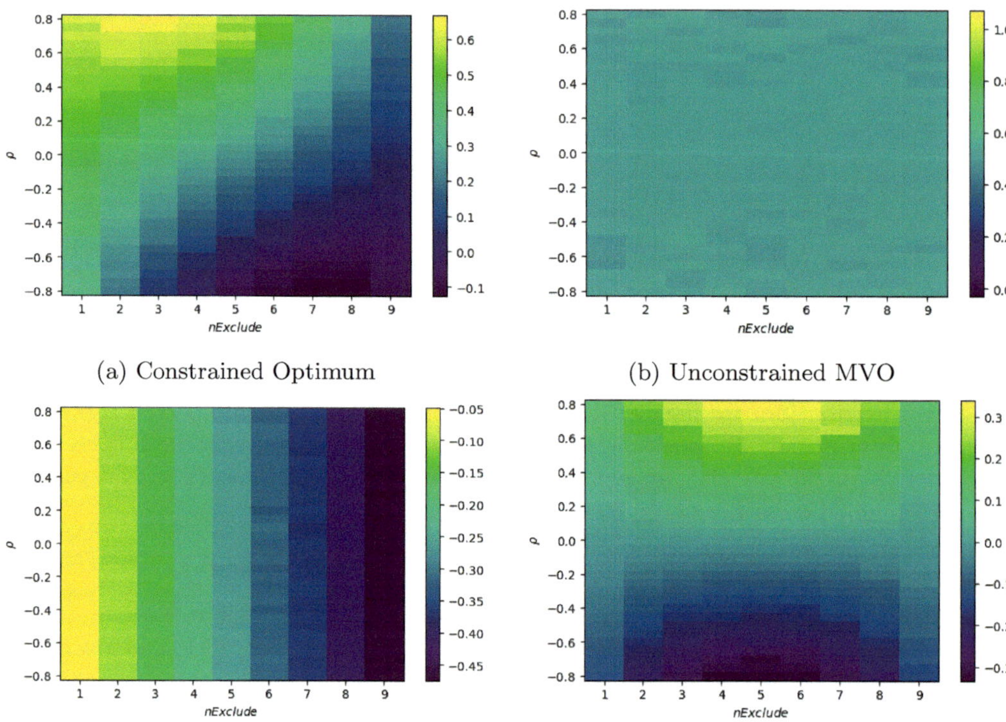

(a) Constrained Optimum

(b) Unconstrained MVO

(c) Decomposition: Constraint (Static)

(d) Decomposition: Constraint (Information)

FIGURE 9.3: Decomposition of expected return for the problem in (9.30) with one exclusionary constraint that depends on random characteristics, as the number of excluded assets (*nExclude*) and the correlation (ρ) between the random characteristic and asset returns vary. The expected return of the constrained portfolio (a) is decomposed into components corresponding to the unconstrained MVO portfolio (b), static constraints (c), and information in the constraints (d).

In the same way, the appendix to Lo and Zhang (2025) includes similar demonstrations for the attribution of expected utility. The patterns are similar to those in Figure 9.3.

Overall, these results demonstrate how to understand the expected return and utility of an exclusionary portfolio by decomposing them into an unconstrained MVO portfolio, static constraints, and information in each constraint.

9.4.3 Selection of constraints

These examples highlight how investors can use our framework to select constraints that help improve performance *ex-ante*. If investors can select one or several constraints from a collection of factor exposure constraints, they should select the *binding* constraint(s) with information that is sufficiently positively correlated with returns, because they can lead to higher expected returns and utilities for the constrained portfolio, as demonstrated by the simulation in Section 9.4.1.

The analysis in Section 9.4.2 demonstrates that with exclusionary investing, to decide which characteristic to use and how many assets to include, investors should select the characteristic with the highest correlation with returns and the optimal number of assets to exclude. The latter depends on the correlation of the specific characteristic and may lie between 0 and N, reflecting a trade-off between information and diversification. As a result, the best performance is achieved when an intermediate number of assets are excluded.

Our empirical analysis in Section 9.5.4 gives an example with real world data in which different forms of ESG constraints lead to different impacts on performance.

9.5 Empirical analysis

In this section, we apply our framework to real-world datasets, and consider an example of ESG investing in which the average portfolio ESG score is required to be above a certain threshold. In the appendix to Lo and Zhang (2025), we also consider exclusionary investing, excluding sin stocks and stranded assets.

9.5.1 Data

Returns. We obtain daily return data for all U.S. stocks from 2001 to 2020 from the Center for Research in Security Prices (CRSP) available through Wharton Research Data Services (WRDS). The CRSP dataset also contains basic firm characteristics such as market capitalization. We obtain the daily Fama–French factor data from Kenneth French's website.[14]

Because the ESG data is updated annually, we require that a stock has at least ten years of valid return data to be included in our analysis. For each stock, we estimate a Fama–French five-factor model (Fama and French, 2015) based on daily returns:

$$R_{i,t} = \alpha_i + \beta_{i,1}(R_{M,t} - R_{f,t}) + \beta_{i,2}SMB_t + \beta_{i,3}HML_t + \beta_{i,5}RMW_t + \beta_{i,6}CMA_t + \epsilon_{i,t}. \quad (9.31)$$

We follow the common practice in the empirical asset pricing literature of winsorizing the raw returns used in the regression at 2.5% on both sides for each stock in each year in order to mitigate the impact of extreme outliers on the regression results. We then use the residual returns $r_{i,t} \equiv \alpha_i + \epsilon_{i,t}$ as the main target of interest in our analysis. We summarize the residual returns annually to match the frequency of the ESG data.

ESG. We merge the CRSP dataset with the MSCI KLD ESG dataset which contains yearly environmental, social, and governance ratings of roughly 3,000 large U.S. publicly

[14]See `https://mba.tuck.dartmouth.edu/pages/faculty/ken.french/data_library.html`. Accessed 20 July 2022.

traded companies from 2003 to 2018. It has been used in numerous studies examining the effect of ESG ratings on firm performance.[15] We follow Lins, Servaes, and Tamayo (2017) to aggregate the raw data into an ESG score. Details are provided in the appendix to Lo and Zhang (2025).

9.5.2 Descriptive statistics

Table 9.1 shows for each year the number of firms in our dataset, the summary statistics of its annualized residual returns, and the summary statistics of the aggregate ESG score we construct lagged by one year.

TABLE 9.1: Summary statistics of the annualized residual returns (in percentage) from the Fama–French five-factor model and the aggregate ESG score lagged by one year

Year	*#Firms*	*Annualized residual return (%)*							*ESG score (lag one year)*						
		mean	*std*	*min*	25%	50%	75%	*max*	*mean*	*std*	*min*	25%	50%	75%	*max*
2004	1,246	4.9	30.0	−69.1	−12.6	1.9	16.9	330.3	−0.1	0.5	−3.4	−0.3	0.0	0.1	2.9
2005	1,344	4.0	29.0	−64.0	−14.1	−0.3	17.6	156.5	−0.1	0.7	−3.0	−0.6	−0.2	0.2	2.2
2006	1,305	3.3	27.2	−74.4	−13.1	−0.1	15.5	234.1	−0.3	0.6	−3.3	−0.5	−0.2	0.0	2.5
2007	1,383	8.1	43.1	−81.3	−17.0	1.4	23.0	511.0	−0.3	0.6	−3.7	−0.7	−0.2	0.0	3.0
2008	1,487	−0.7	51.3	−95.3	−32.7	−7.5	24.2	647.1	−0.3	0.6	−3.5	−0.7	−0.3	0.0	3.4
2009	1,614	13.1	75.4	−98.0	−20.5	1.5	27.9	1,768.7	−0.3	0.6	−3.6	−0.7	−0.3	0.0	2.8
2010	1,611	5.5	35.6	−86.0	−13.9	0.8	18.3	549.0	−0.3	0.6	−3.5	−0.7	−0.3	0.0	2.8
2011	1,710	2.3	30.4	−96.5	−15.5	1.7	18.4	202.9	−0.4	0.7	−2.8	−0.7	−0.6	0.0	3.9
2012	1,643	2.3	31.0	−81.6	−13.0	−1.0	13.0	407.6	−0.4	0.8	−2.7	−0.9	−0.6	−0.2	4.2
2013	1,573	3.7	31.0	−92.1	−13.5	0.5	16.0	456.9	0.1	0.7	−2.3	−0.3	0.0	0.5	3.8
2014	1,621	1.1	23.9	−93.4	−12.4	1.0	14.2	142.6	0.1	0.8	−2.4	−0.3	0.0	0.3	3.2
2015	1,330	2.7	28.3	−86.4	−14.2	3.9	19.8	194.9	0.1	0.5	−3.7	0.0	0.0	0.3	3.2
2016	1,374	2.7	27.6	−82.2	−12.3	0.4	14.7	245.4	0.2	0.6	−2.6	−0.1	0.1	0.5	3.3
2017	1,320	4.0	27.3	−83.0	−10.7	2.0	16.0	216.5	0.2	0.7	−2.4	0.0	0.1	0.5	3.1
2018	1,440	1.0	26.0	−71.9	−15.6	−0.5	15.2	130.0	0.2	0.7	−2.9	0.0	0.2	0.6	4.1
2019	1,476	8.0	26.5	73.8	−6.8	7.8	20.9	259.8	0.7	0.8	−2.3	0.1	0.5	1.0	4.7

Figure 9.4 shows the year-over-year cross-sectional correlations between the residual returns and lag-1 ESG scores.[16] The correlations are generally negative before 2007, implying that high ESG stocks, on average, delivered lower excess returns, consistent with equilibrium theories of ESG returns (Pástor, Stambaugh, and Taylor, 2021; Pedersen, Fitzgibbons, and Pomorski, 2021). After 2008, the correlations fluctuate around zero, and are positive in certain years, reflecting increasing attention toward ESG and climate-related issues, an effect consistent with Pástor, Stambaugh, and Taylor's (2022) and Lo, Zhang, and Zhao's (2022) findings.

Although the magnitude of the correlations—between 2% and 10%—may not seem large, they can still have a non-negligible impact on asset prices. In fact, as shown by Lo and MacKinlay (1990) and Lo and Zhang (2024), the correlation between stock alphas and firm characteristics such as beta, size, or the ESG score of a firm also ranges from 2% to 20%, depending on the year. Nevertheless, correlations can have significant implications for tests of asset pricing models or yield portfolios with significant alphas. In fact, our empirical analysis that follows will show that even a seemingly low level of correlation can still contribute to the performance of a portfolio by a significant amount.

[15] See, for example, Hong and Kostovetsky (2012), Lins, Servaes, and Tamayo (2017), and Berg, Koelbel, and Rigobon (2022). Makridis and Simaan (2023) use Refinitiv ESG data to compare the performance of different shrinkage estimators of the covariance matrix and different ESG-based rules.

[16] The ESG scores are available from 2003 to 2018, so the correlations are computed from 2004 to 2019.

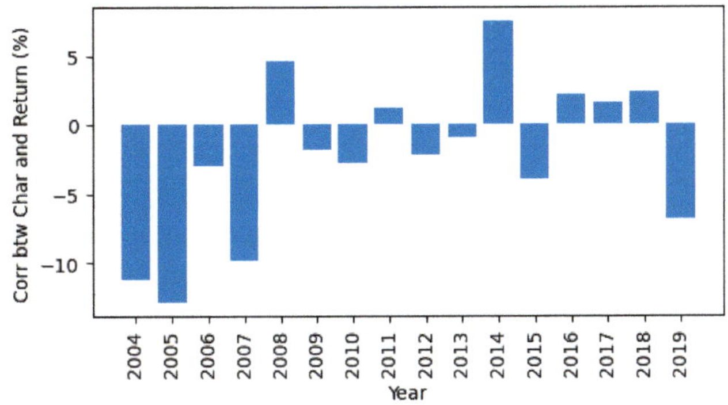

FIGURE 9.4: Cross-sectional correlations between asset returns and lag-1 ESG scores each year.

We also emphasize that our intention is not to find the best ESG score or to provide a comprehensive study of whether ESG delivers positive or negative excess returns. While Berg, Koelbel, and Rigobon (2022) show that there exists substantial noise in ESG measures and ESG scores from different data providers may lead to very different correlations, Berg et al. (2024) further show that aggregating individual ESG ratings improves portfolio performance, and there exists a significant signal in ESG rating scores that can be used for portfolio construction despite their noisy nature. What we hope to demonstrate in this chapter is that, given any ESG score, it is possible to attribute portfolio performance metrics to different constraints and to the information in those constraints, and we use the MSCI KLD ESG dataset as an illustrative example.

9.5.3 Performance attribution

Portfolio construction. We consider investors who construct portfolios each year by solving the following problem:

$$\begin{aligned} &\max_{\boldsymbol{\omega}} \ \boldsymbol{\omega}'\boldsymbol{\mu} - \frac{\gamma}{2}\boldsymbol{\omega}'\boldsymbol{\Sigma}\boldsymbol{\omega} \\ &\text{s.t.} \ \ \boldsymbol{\omega}'\mathbf{1} = 1 \quad \text{and} \quad \boldsymbol{\omega}'\mathbf{x}_{\text{ESG}} \geq b. \end{aligned} \tag{9.32}$$

The first constraint implies that the portfolio is fully invested while allowing for both long and short positions in individual assets. The second constraint imposes a minimum level of portfolio ESG score. In our example, we set $b = 1$ and $\gamma = 5$.

To construct these portfolios, investors need to estimate the expected residual return, $\boldsymbol{\mu}$, and the covariance matrix of the residual returns, $\boldsymbol{\Sigma}$, each year. We consider the two sets of estimators discussed in Section 9.3, both of which account for estimation risk. The goal of our analysis is *not* to exhaustively compare different portfolio rules and find the best one in terms of their ability to account for estimation risk, but rather to demonstrate how to attribute performance to constraints, given *any* such portfolio rules.

In our empirical analysis, we implement two portfolio rules. The first is Jorion's (1986) portfolio rule which is also summarized in Avramov and Zhou (2010, p. 32). In each period t, the predictive mean and covariance matrix of returns are given by:

$$\hat{\boldsymbol{\mu}}_t^{\text{Jorion}} = (1-\xi_1)\hat{\boldsymbol{\mu}}_s + \xi_1\hat{\boldsymbol{\mu}}_g\mathbf{1} \quad \text{and} \quad \hat{\boldsymbol{\Sigma}}_t^{\text{Jorion}} = \left(1 + \frac{1}{T+\xi_2}\right)\bar{\boldsymbol{\Sigma}} + \frac{\xi_2}{T(T+1+\xi_2)\mathbf{1}'\bar{\boldsymbol{\Sigma}}^{-1}\mathbf{1}}\mathbf{1}\mathbf{1}' \tag{9.33}$$

where
$\xi_1 = \frac{N+2}{(N+2)+T(\hat{\boldsymbol{\mu}}_s-\hat{\mu}_g\mathbf{1})'\bar{\boldsymbol{\Sigma}}^{-1}(\hat{\boldsymbol{\mu}}_s-\hat{\mu}_g\mathbf{1})}$, $\xi_2 = \frac{N+2}{(\hat{\boldsymbol{\mu}}_s-\hat{\mu}_g\mathbf{1})'\bar{\boldsymbol{\Sigma}}^{-1}(\hat{\boldsymbol{\mu}}_s-\hat{\mu}_g\mathbf{1})}$, $\bar{\boldsymbol{\Sigma}} = \frac{T}{T-N-2}\hat{\boldsymbol{\Sigma}}_s$,

$\hat{\boldsymbol{\mu}}_s$ and $\hat{\boldsymbol{\Sigma}}_s$ are the sample mean and sample covariance matrix of returns over the last T periods, respectively, and $\hat{\mu}_g = \mathbf{1}'\bar{\boldsymbol{\Sigma}}^{-1}\hat{\boldsymbol{\mu}}_s/\mathbf{1}'\bar{\boldsymbol{\Sigma}}^{-1}\mathbf{1}$ is the estimated return of the global minimum variance portfolio.[17]

The second is the $1/N$ rule which is extensively studied by DeMiguel, Garlappi, and Uppal (2009). As discussed in Section 9.3, the $1/N$ rule can be interpreted as a Bayesian approach with an appropriate prior. We specify the predictive mean and covariance matrix of returns so that they are consistent with an unconstrained optimal portfolio $\boldsymbol{\omega}_{\text{MVO}} = \mathbf{1}/N$:

$$\hat{\boldsymbol{\mu}}_t^{\text{Equal}} = \frac{\gamma}{N}\mathbf{1} \quad \text{and} \quad \hat{\boldsymbol{\Sigma}}_t^{\text{Equal}} = \mathbf{I}. \tag{9.34}$$

Combining Equation (9.32) and the estimators in Equations (9.33) or (9.34), we can solve for the optimal constrained portfolios. The appendix to Lo and Zhang (2025) shows the decomposition of portfolio holdings based on Proposition 18.

Expected return and utility decomposition. Figure 9.5 demonstrates the decomposition of the expected return and utility of the portfolio into different components for Jorion's (1986) rule as an example. The upper panel of Figure 9.5a shows that the expected utility of the optimal portfolio is generally negative in the first half of our sample period, and starts to turn positive toward the second half. In the lower panel, this expected utility is decomposed into three components using Equation (9.18) in Proposition 21 and its predictive return version in the appendix to Lo and Zhang (2025).[18] The expected utility of the unconstrained MVO portfolio (blue) is positive over the 16 years in our sample. As the conventional wisdom of constrained optimization would suggest, the expected utility contribution of the two constraints (orange), treated as static, is indeed negative.

However, the expected utility contribution from the information contained in the constraints (green) varies over time. During the first four years in our period, the expected utility contribution from information is negative. After 2008, it alternates in sign, with 2008, 2011, 2014, 2016, 2017, and 2018 being positive years in information. This pattern coincides with the signs of correlations between asset returns and ESG scores in Figure 9.4. The magnitude of the information contribution, however, differs from the patterns of correlations, because the former is jointly determined by several terms as shown in Propositions 20 and 21.

This example vividly demonstrates that while constraints must decrease the overall expected utility of a portfolio when treated as static, they can sometimes increase the expected utility relative to a passive benchmark, depending on the information contained in the constraints.

Figure 9.5b shows the expected return of the optimal portfolio and its decomposition based on Equation (9.17) in Proposition 21 and its predictive return version in the appendix to Lo and Zhang (2025). While the two constraints (orange and green) can contribute either positively or negatively to the expected returns,[19] the main source of contribution in terms

[17]We encounter the high-dimensional problem in our analysis whereby the total number of assets, N, in the portfolio—typically between 1,000 to 4,000—is larger than the number of daily , T, in a year. Therefore, we use the shrinkage covariance estimator of Chen et al. (2010), implemented by `sklearn.covariance.OAS` in Python, multiplied by a constant factor (100) to replace $\bar{\Sigma}$ in order to ensure numerical stability.

[18]We estimate the correlations, ρ, between lagged ESG values and residual returns, rather than predictive returns, because the $1/N$ rule implies constant predictive returns for all assets which lead to a trivial decomposition.

[19]The sign is determined by the correlation between the holdings of the MVO portfolio and the coefficients of the constraint. See the remarks after Proposition 18.

of expected return is from the information in the constraints (red). Like the expected utility decomposition, the expected return contribution from information is strongly negative between 2004 and 2007 and is strongly positive in 2008 and 2014.

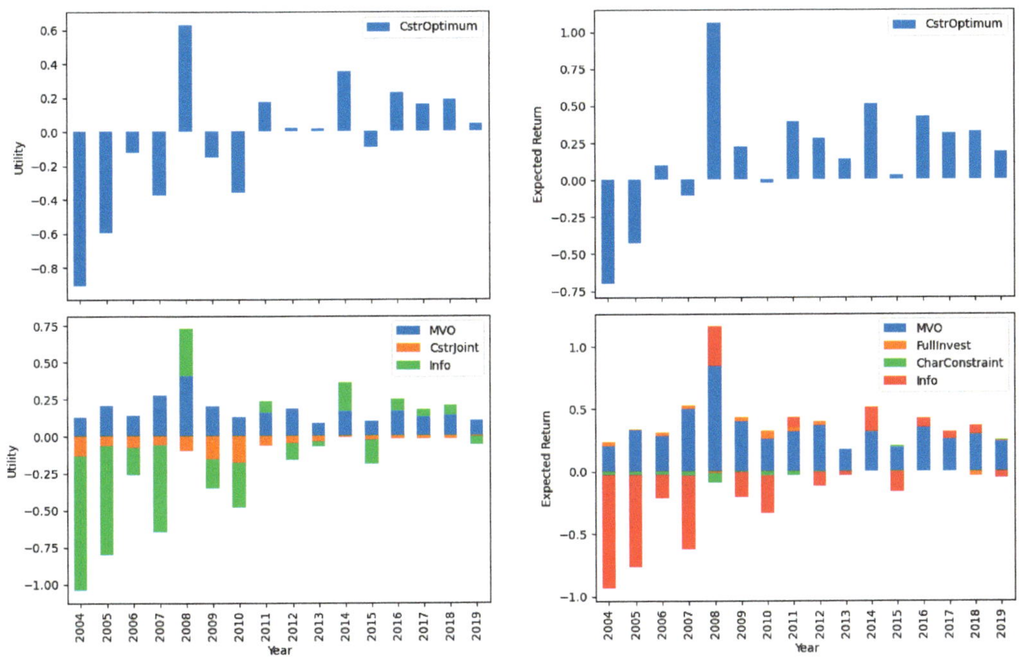

(a) Expected Utility

(b) Expected Return

FIGURE 9.5: Expected return and utility and their decomposition, for the portfolio in (9.32) with a constraint on the average portfolio characteristic value ($\boldsymbol{\omega}'\mathbf{x}_{\text{ESG}} \geq 1.0$) and Jorion's (1986) estimates of predictive moments. In (a), the top panel shows the expected utility of the constrained portfolio, and the bottom panel shows its decomposition into components corresponding to the unconstrained MVO portfolio (blue), all constraints treated as static (orange), and the information from the ESG constraint (green). In (b), the top panel shows the expected return in excess of the Fama–French five-factor model of the constrained portfolio, and the bottom panel shows its decomposition into components corresponding to the unconstrained MVO portfolio (blue), the full investment constraint (orange), the ESG constraint ($\boldsymbol{\omega}'\mathbf{x}_{\text{ESG}} \geq 1.0$) treated as static (green), and the information from the ESG constraint (red).

Realized return decomposition. Figure 9.6 shows the realized returns of the optimal portfolio for both Jorion's (1986) rule and the $1/N$ rule as the unconstrained MVO portfolio.

The upper panel of Figure 9.6a shows that, for Jorion's (1986) rule, the realized residual returns in excess of the Fama–French five-factor model of the constrained portfolio is generally positive over the 16 years in our sample except 2008.

The lower panel decomposes the realized return of the constrained portfolio. The full investment constraint generally contributes positively to the returns. This is consistent with the fact that the average residual returns during these years are positive, as shown in Table 9.1. The contribution of the ESG constraint, treated as static, also varies over time. In contrast, the information component contributes negatively to realized returns before 2007 and positively in certain years after 2008. Notably, in 2008, the information component from

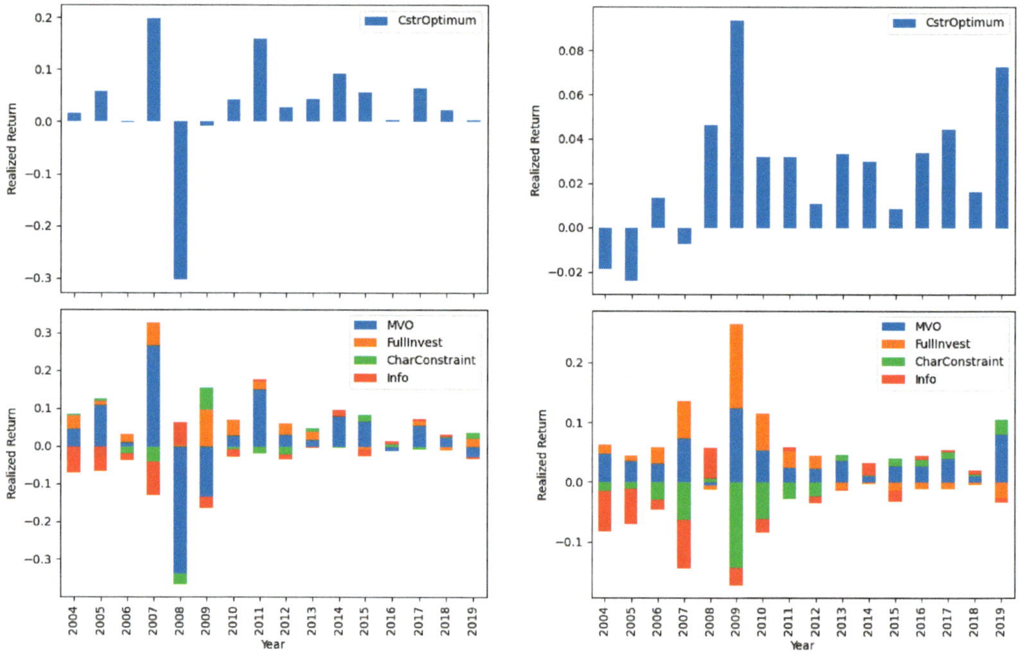

(a) Jorion's (1986) rule (b) $1/N$ rule

FIGURE 9.6: Realized return for the portfolio defined in (9.32) with a constraint on the average portfolio characteristic value ($\boldsymbol{\omega}'\mathbf{x}_{\text{ESG}} \geq 1.0$). (a) corresponds to Jorion's (1986) estimates of predictive moments in (9.33), and (b) corresponds to predictive moments consistent with the $1/N$ rule in (9.34). In each subfigure, the top panel shows the realized return in excess of the Fama–French five-factor model of the constrained portfolio, and the bottom panel shows its decomposition into components corresponding to the unconstrained MVO portfolio (blue), the full investment constraint (orange), the ESG constraint ($\boldsymbol{\omega}'\mathbf{x}_{\text{ESG}} \geq 1.0$) treated as static (green), and the information from the ESG constraint (red).

the ESG constraint contributed positively to the large negative returns from the unconstrained portfolio. These patterns are consistent with the correlations between asset returns and ESG scores shown in Figure 9.4. Overall, these components explain the difference in residual returns between the unconstrained MVO and the constrained portfolio.

Figure 9.6b shows parallel results, this time with the $1/N$ rule as the unconstrained MVO portfolio. The realized residual returns for both the unconstrained and the constrained portfolio are much more stable over time. The decomposition is also similar to the case of Jorion's (1986) rule, except that the contributions from the information in the constraints, in relative terms, are bigger. For example, the negative residual returns before 2007 were primarily driven by the negative contribution from the ESG constraints.

9.5.4 Other portfolios and selection of constraints

The portfolio given by Equation (9.32) is one way to construct ESG portfolios. In practice, investors need to determine the threshold, b. In addition, they may face long-only constraints, and we demonstrate how to decompose the performance of these portfolios using our framework in the appendix to Lo and Zhang (2025).

While the choice of the specific form of constraints depends on many factors such as the regulatory requirements and other business considerations, here we show how to use our framework to compare the impact of different forms of ESG constraints purely from the perspective of financial performance.

We consider 32 portfolios with different constraints, including the factor exposure constraint in Equation (9.32) with $b = 0, 0.1, \ldots, 2$, and the factor exposure constraint and long-only constraint in the appendix to Lo and Zhang (2025) with $b = 0, 0.1, \ldots, 1$. We use a rolling window of three years to estimate the financial performance of these portfolios and select constraints for the next year based on their expected utilities with and without the information component.

Table 9.2 summarizes the key out-of-sample performance metrics and their decomposition, averaged over 13 years in our sample. For portfolios that allow short positions (the first two columns of Table 9.2) and long-only portfolios (the last two columns of Table 9.2), we compare the performance of portfolios whose constraints are selected based on their estimated expected utilities with and without the information component. For both Jorion's (1986) rule and the $1/N$ rule, if the information component is ignored, investors achieve a lower average ESG score, lower expected utility, and lower realized return. In other words, if the information in constraints is not properly accounted for, investors may select constraints that are more detrimental to performance, leading to suboptimal portfolios.

TABLE 9.2: Out-of-sample performance metrics of ESG portfolios with different criteria to select constraints, for Jorion's (1986) rule and the $1/N$ rule. $\tilde{\omega}^*$ = the optimal constrained portfolio; $\tilde{\omega}_{\text{MVO}}$ = the unconstrained MVO portfolio; $\tilde{\omega}_{\text{CSTR}}$ (static) = the component attributable to constraints treated as static; and $\tilde{\omega}_{\text{CSTR}}$ (info) = the information component in constraints. The portfolio with a better performance metric in each comparison is in bold

Selection criteria		*Allowing short positions* *Exp util*	*Allowing short positions* *Exp util w/o info*	*Long-only* *Exp util*	*Long-only* *Exp util w/o info*
Panel A: Jorion's (1986) rule					
ESG	$\tilde{\omega}^*$	**0.33**	0.13	**0.15**	0.11
Expected utility	$\tilde{\omega}_{\text{MVO}}$	17.16	17.16	17.16	17.16
	$\tilde{\omega}_{\text{CSTR}}$ (static)	−1.45	**−1.15**	−9.60	**−9.51**
	$\tilde{\omega}_{\text{CSTR}}$ (info)	**0.72**	−1.39	**−1.04**	−1.33
	$\tilde{\omega}^*$	**16.43**	14.62	**6.49**	6.29
Realized return	$\tilde{\omega}_{\text{MVO}}$	1.59	1.59	1.59	1.59
	$\tilde{\omega}_{\text{CSTR}}$ (static)	1.52	**1.68**	1.67	1.67
	$\tilde{\omega}_{\text{CSTR}}$ (info)	**0.07**	−0.14	**−0.11**	−0.14
	$\tilde{\omega}^*$	**3.17**	3.13	**3.16**	3.12
Panel B: $1/N$ rule					
ESG	$\tilde{\omega}^*$	**0.52**	0.13	**0.33**	0.13
Expected utility	$\tilde{\omega}_{\text{MVO}}$	0.17	0.17	0.17	0.17
	$\tilde{\omega}_{\text{CSTR}}$ (static)	−0.23	**−0.02**	−0.09	**−0.02**
	$\tilde{\omega}_{\text{CSTR}}$ (info)	**1.64**	−0.36	**0.63**	−0.36
	$\tilde{\omega}^*$	**1.58**	−0.21	**0.71**	−0.21
Realized return	$\tilde{\omega}_{\text{MVO}}$	4.03	4.03	4.03	4.03
	$\tilde{\omega}_{\text{CSTR}}$ (static)	0.00	0.00	−0.07	**−0.03**
	$\tilde{\omega}_{\text{CSTR}}$ (info)	**0.36**	−0.14	**0.11**	−0.14
	$\tilde{\omega}^*$	**4.39**	3.89	**4.07**	3.86

Overall, these comparisons demonstrate how to use our framework to select constraints, and the importance of accounting for information in the constraints. In this example, the ESG constraints are detrimental to the financial performance of the portfolio because the ESG scores are negatively correlated with residual returns in 7 out of the 13 years in our out-of-sample period, and the ESG constraints are beneficial to its performance because the ESG scores are positively correlated with residual returns in the remaining 6 years (see Figure 9.4). Because there is substantial noise in the ESG measures (Berg, Koelbel, and Rigobon, 2022), the specific numbers of our empirical example should be taken with a grain of salt. Nonetheless, this example demonstrates that there exists a significant signal in ESG scores that may be used for portfolio construction despite their noisy nature, which is consistent with Berg et al.'s (2024) findings, and that adequately accounting for information in the constraints is important for the performance attribution of a portfolio.

More generally, in other applications where constraints contain information that is positively correlated with returns, investors will miss out on gains from information if they are not properly accounted for. Examples include other company sustainability measures that may provide *material* information to the operations and risks of the company (Khan, Serafeim, and Yoon, 2016), and proprietary constraints that may contain private information about future returns.

9.6 Conclusion

Constraints are an integral part of the portfolio construction process, and they have become particularly relevant as investors and regulators debate whether investing with ESG constraints or excluding stranded assets is to the benefit or detriment of investors. We propose a framework for constraint attribution that decomposes portfolio holdings, expected returns, variance, expected utilities, and realized returns into components attributable to each constraint and the information contained in each constraint.

Our framework provides a quantifiable and easily implementable measure of the information content from constraints when they are stochastic and potentially correlated with asset returns. The correlation between characteristics that form constraints and the individual asset returns plays a key role in determining the sign and magnitude of the excess return and variance from information. Constraints serve as an indirect mechanism for incorporating this information into the portfolio that is otherwise unavailable to investors.

We demonstrate that our methodology can accommodate estimation risk in parameter values of the portfolio construction process using Bayesian portfolio analysis, which provides the same decomposition of performance out-of-sample and a method to select constraints that improve—or are least detrimental to—portfolio performance *ex-ante*.

Our framework can be applied to common examples of constraints including the level of a characteristic, such as ESG scores, and exclusion constraints, such as divesting from sin stocks and energy stocks. Our results show that these constraints can contribute positively (or negatively) to portfolio performance compared to a passive benchmark, when the information contained in the constraints is not accounted for in the passive benchmark but is sufficiently positively (or negatively) correlated with asset returns.

References

Ardia, D., and K. Boudt, 2015, Implied expected returns and the choice of a mean–variance efficient portfolio proxy, *The Journal of Portfolio Management* 41, 68–81.

Avramov, D., and G. Zhou, 2010, Bayesian portfolio analysis, *Annual Review of Financial Economics* 2, 25–47.

Berg, F., J. F. Koelbel, and R. Rigobon, 2022, Aggregate confusion: The divergence of ESG ratings, *Review of Finance* 26, 1315–1344.

Berg, F., A. W. Lo, R. Rigobon, M. Singh, and R. Zhang, 2024, Quantifying the returns of ESG investing: An empirical analysis with six ESG metrics, *The Journal of Portfolio Management* 50, 216–238.

Black, F., and R. Litterman, 1992, Global portfolio optimization, *Financial Analysts Journal* 48, 28–43.

Bonami, P., and M. A. Lejeune, 2009, An exact solution approach for portfolio optimization problems under stochastic and integer constraints, *Operations Research* 57, 650–670.

Brandt, M. W., P. Santa-Clara, and R. Valkanov, 2009, Parametric portfolio policies: Exploiting characteristics in the cross-section of equity returns, *The Review of Financial Studies* 22, 3411–3447.

Brinson, G. P., L. R. Hood, and G. L. Beebower, 1986, Determinants of portfolio performance, *Financial Analysts Journal* 42, 39–44.

Broadie, M., 1993, Computing efficient frontiers using estimated parameters, *Annals of Operations Research* 45, 21–58.

Brodie, J., I. Daubechies, C. De Mol, D. Giannone, and I. Loris, 2009, Sparse and stable markowitz portfolios, *Proceedings of the National Academy of Sciences* 106, 12267–12272.

Brown, S. J., 1978, The portfolio choice problem: Comparison of certainty equivalence and optimal bayes portfolios, *Communications in Statistics-Simulation and Computation* 7, 321–334.

Ceria, S., and R. A. Stubbs, 2006, Incorporating estimation errors into portfolio selection: Robust portfolio construction, *Journal of Asset Management* 7, 109–127.

Chen, Y., A. Wiesel, Y. C. Eldar, and A. O. Hero, 2010, Shrinkage algorithms for mmse covariance estimation, *IEEE Transactions on Signal Processing* 58, 5016–5029.

Chopra, V. K., and W. T. Ziemba, 1993, The effect of errors in means, variances, and covariances on optimal portfolio choice, *The Journal of Portfolio Management* 19, 6–11.

Clarke, R., H. de Silva, and S. Thorley, 2002, Portfolio constraints and the fundamental law of active management, *Financial Analysts Journal* 58, 48–66.

Corwin, S. A., and J. F. Coughenour, 2008, Limited attention and the allocation of effort in securities trading, *The Journal of Finance* 63, 3031–3067.

David, H., 1973, Concomitants of order statistics, *Bulletin of the International Statistical Institute* 45, 295–300.

Davies, P. I., and N. J. Higham, 2000, Numerically stable generation of correlation matrices and their factors, *BIT Numerical Mathematics* 40, 640–651.

DeMiguel, V., L. Garlappi, F. J. Nogales, and R. Uppal, 2009, A generalized approach to portfolio optimization: Improving performance by constraining portfolio norms, *Management Science* 55, 798–812.

DeMiguel, V., L. Garlappi, and R. Uppal, 2009, Optimal versus naive diversification: How inefficient is the 1/N portfolio strategy?, *The Review of Financial Studies* 22, 1915–1953.

Dybvig, P. H., and S. A. Ross, 1985, Differential information and performance measurement using a security market line, *The Journal of Finance* 40, 383–399.

Fabozzi, F. J., D. Huang, and G. Zhou, 2010, Robust portfolios: contributions from operations research and finance, *Annals of Operations Research* 176, 191–220.

Fama, E. F., 1972, Components of investment performance, *The Journal of Finance* 27, 551–567.

Fama, E. F., and K. R. French, 2015, A five-factor asset pricing model, *Journal of Financial Economics* 116, 1–22.

Fan, J., Y. Li, and K. Yu, 2012, Vast volatility matrix estimation using high-frequency data for portfolio selection, *Journal of the American Statistical Association* 107, 412–428.

Fan, J., J. Zhang, and K. Yu, 2012, Vast portfolio selection with gross-exposure constraints, *Journal of the American Statistical Association* 107, 592–606.

Fang, K.-T., S. Kotz, and K. W. Ng, 1990, *Symmetric Multivariate and Related Distributions* (Chapman and Hall, New York).

Ferson, W. E., and A. F. Siegel, 2001, The efficient use of conditioning information in portfolios, *The Journal of Finance* 56, 967–982.

Frost, P. A., and J. E. Savarino, 1986, An empirical bayes approach to efficient portfolio selection, *Journal of Financial and Quantitative Analysis* 21, 293–305.

Goldberg, L. R., 2021, Constraint implied returns and stress in mean-variance optimized portfolios, Working paper.

Grinold, R., 2005, Implementation efficiency, *Financial Analysts Journal* 61, 52–64.

Harvey, C. R., J. C. Liechty, M. W. Liechty, and P. Müller, 2010, Portfolio selection with higher moments, *Quantitative Finance* 10, 469–485.

Hirshleifer, D., S. S. Lim, and S. H. Teoh, 2011, Limited investor attention and stock market misreactions to accounting information, *The Review of Asset Pricing Studies* 1, 35–73.

Hirshleifer, D., A. Subrahmanyam, and S. Titman, 2006, Feedback and the success of irrational investors, *Journal of Financial Economics* 81, 311–338.

Hjalmarsson, E., and P. Manchev, 2012, Characteristic-based mean-variance portfolio choice, *Journal of Banking & Finance* 36, 1392–1401.

Hong, H., and L. Kostovetsky, 2012, Red and blue investing: Values and finance, *Journal of Financial Economics* 103, 1–19.

Jacquier, E., and N. Polson, 2011, Bayesian methods in finance, in *The Oxford Handbook of Bayesian Econometrics*, chapter 9, 439–512 (Oxford University Press).

Jagannathan, R., and T. Ma, 2003, Risk reduction in large portfolios: Why imposing the wrong constraints helps, *The Journal of Finance* 58, 1651–1683.

Jin, H., Z.-Q. Xu, and X. Y. Zhou, 2008, A convex stochastic optimization problem arising from portfolio selection, *Mathematical Finance* 18, 171–183.

Jorion, P., 1986, Bayes-stein estimation for portfolio analysis, *Journal of Financial and Quantitative Analysis* 21, 279–292.

Kan, R., and X. Wang, 2023, Optimal portfolio choice with unknown benchmark efficiency, *Management Science* 70, 6117–6138.

Kan, R., X. Wang, and G. Zhou, 2022, Optimal portfolio choice with estimation risk: No risk-free asset case, *Management Science* 68, 2047–2068.

Kan, R., and G. Zhou, 2007, Optimal portfolio choice with parameter uncertainty, *Journal of Financial and Quantitative Analysis* 42, 621–656.

Khan, M., G. Serafeim, and A. Yoon, 2016, Corporate sustainability: First evidence on materiality, *The Accounting Review* 91, 1697–1724.

Klein, R. W., and V. S. Bawa, 1976, The effect of estimation risk on optimal portfolio choice, *Journal of Financial Economics* 3, 215–231.

Lai, T. L., H. Xing, and Z. Chen, 2011, Mean–variance portfolio optimization when means and covariances are unknown, *The Annals of Applied Statistics* 5, 798 – 823.

Levy, H., and M. Levy, 2014, The benefits of differential variance-based constraints in portfolio optimization, *European Journal of Operational Research* 234, 372–381.

Lins, K. V., H. Servaes, and A. Tamayo, 2017, Social capital, trust, and firm performance: The value of corporate social responsibility during the financial crisis, *The Journal of Finance* 72, 1785–1824.

Lo, A. W., and A. C. MacKinlay, 1990, Data-snooping biases in tests of financial asset pricing models, *The Review of Financial Studies* 3, 431–467.

Lo, A. W., L. Wu, R. Zhang, and C. Zhao, 2024, Optimal impact portfolios with general dependence and marginals, *Operations Research* forthcoming.

Lo, A. W., and R. Zhang, 2024, Quantifying the impact of impact investing, *Management Science* 70, 7161–7186.

Lo, A. W., and R. Zhang, 2025, Performance attribution for portfolio constraints, *Management Science* forthcoming.

Lo, A. W., R. Zhang, and C. Zhao, 2022, Measuring and optimizing the risk and reward of green portfolios, *The Journal of Impact and ESG Investing*.

Makridis, C., and M. Simaan, 2023, Balancing returns and responsibility: Evidence from shrinkage-based portfolios, Available at SSRN 4597152.

Menchero, J., and B. Davis, 2011, Risk contribution is exposure times volatility times correlation: decomposing risk using the x-sigma-rho formula, *The Journal of Portfolio Management* 37, 97–106.

Merton, R. C., 1980, On estimating the expected return on the market: An exploratory investigation, *Journal of Financial Economics* 8, 323–361.

Pástor, L., 2000, Portfolio selection and asset pricing models, *The Journal of Finance* 55, 179–223.

Pástor, L., and R. F. Stambaugh, 2000, Comparing asset pricing models: an investment perspective, *Journal of Financial Economics* 56, 335–381.

Pástor, L., R. F. Stambaugh, and L. A. Taylor, 2021, Sustainable investing in equilibrium, *Journal of Financial Economics* 142, 550–571.

Pástor, L., R. F. Stambaugh, and L. A. Taylor, 2022, Dissecting green returns, *Journal of Financial Economics* 146, 403–424.

Pedersen, L. H., S. Fitzgibbons, and L. Pomorski, 2021, Responsible investing: The ESG-efficient frontier, *Journal of Financial Economics* 142, 572–597.

Powell, W. B., 2019, A unified framework for stochastic optimization, *European Journal of Operational Research* 275, 795–821.

Sharpe, W. F., 1992, Asset allocation: Management style and performance measurement, *Journal of Portfolio Management* 18, 7–19.

Simon, H. A., 1955, A behavioral model of rational choice, *The Quarterly Journal of Economics* 69, 99–118.

Stambaugh, R. F., 1997, Analyzing investments whose histories differ in length, *Journal of Financial Economics* 45, 285–331.

Stubbs, R. A., and D. Vandenbussche, 2010, Constraint attribution, *The Journal of Portfolio Management* 36, 48–59.

Tu, J., and G. Zhou, 2010, Incorporating economic objectives into Bayesian priors: Portfolio choice under parameter uncertainty, *Journal of Financial and Quantitative Analysis* 45, 959–986.

Tu, J., and G. Zhou, 2011, Markowitz meets Talmud: A combination of sophisticated and naive diversification strategies, *Journal of Financial Economics* 99, 204–215.

Zhou, G., 2009, Beyond black–litterman: Letting the data speak, *The Journal of Portfolio Management* 36, 36–45.

Chapter 10

Portfolio alignment and net zero investing

Thierry Roncalli

Amundi Asset Management & University of Evry-Paris-Saclay, France

In this chapter, we examine portfolio alignment in net-zero investing, distinguishing it from portfolio decarbonization. We explore carbon reduction strategies, self-decarbonization, and the balance between reducing carbon intensity and increasing green intensity. The study outlines portfolio optimization methodologies for equities, corporate bonds, and government bonds, highlighting key constraints and challenges. The core-satellite approach is introduced as a way to separate decarbonization from transition financing. Findings show that net-zero implementation is complex, especially for equities, and highlight the role of investor engagement in promoting a low-carbon economy.

10.1 Introduction

The concept of portfolio alignment emerged at the end of the 2010s with the European Commission's willingness to create climate benchmarks linked to the goals of the Paris Agreement. The development of the CTB and PAB benchmarks[1] is then an important step that has popularized the practice of portfolio alignment (EU Technical Expert Group on Sustainable Finance, September 2019). The two benchmarks use a pre-defined decarbonization pathway based on the IPCC recommendation to limit the temperature anomaly to below 2°C. They also incorporate other constraints aimed at financing the transition to a low-carbon economy. In a sense, they combine two well-established climate investment strategies: portfolio decarbonization and climate solutions. Portfolio decarbonization is the process of reducing the carbon footprint of an investment portfolio relative to a benchmark portfolio (Andersson, Bolton, and Samama, 2016). It involves shifting investments from higher carbon emitting assets to lower carbon emitting assets. Climate solutions are thematic investments and strategies that directly contribute to mitigating climate change and adapting to its impacts. These solutions focus on financing technologies, projects and practices that promote the green economy, such as investments in solar, wind and hydro power, carbon capture solutions, sustainable infrastructure or green real estate. To include climate solutions, the TEG proposed two constraints: a minimum exposure to sectors highly exposed to climate change and a minimum share of green revenues (or capex) in line with the EU taxonomy. Unfortunately, in the final version of the CTB and PAB benchmarks, the reference to green footprint (*e.g.*, green revenues or capex) disappear (EU Technical Expert Group on Sustainable Finance, December 2019). In addition, the final version of the climate

[1] Although they are not labels, we also use the abusive terms CTB and PAB labels.

DOI: 10.1201/9781032636252-10

impact sectors includes a large number of industries, making this criterion irrelevant for the inclusion of climate solutions (Le Guenedal and Roncalli, 2022).

The second factor in the development of portfolio alignment is the publication of the net-zero scenario by the International Energy Agency in May 2021. Until the IEA report, most of the decarbonization pathways are exogenous. With the publication of the IEA report, the decarbonization pathway of the net-zero scenario is the result of in-depth sector analysis. For each sector, the IEA creates a decarbonization scenario that takes into account the development of energy efficiency, the implementation of decarbonization tools and techniques, behavioral changes, and other factors, including the deployment of carbon capture and storage and massive investments to finance the transition to a low-carbon economy. The release of the IEA report gives hope that it is possible to achieve net zero by 2050 because it is a credible roadmap to follow and not just a hypothetical scenario. Investors and the financial community will study this net-zero scenario closely and adopt it as their own. So it's no coincidence that 2021 will see a proliferation of net zero initiatives in the financial sector, culminating in the launch of the Glasgow Financial Alliance for Net Zero (GFANZ). During COP26 in Glasgow in November 2021, the GFANZ Chair announced that more than 450 financial companies (banks, asset managers, asset owners, insurance companies, etc.) from 45 countries, representing more than \$130 trillion in assets under management, are committed to tackling climate change. While the PAB/CTB decarbonization pathways are global, the IEA scenario emphasizes that decarbonization efforts must be differentiated. Not all sectors are the same. Indeed, electricity must be decarbonized as a priority, as shown in Figure 10.1. By 2035, this sector must achieve a decarbonization rate of 85% if we are to reach net zero by 2050. Any delay in the decarbonization of the electricity sector will jeopardize the net zero goal, because the decarbonization of the other sectors depends on this first intermediate step. Figure 10.1 also shows that there is a sequence of decarbonization and a sequence of transition. While electricity must be the first sector to decarbonize, industrials is the last sector to decarbonize because it needs green electricity, green materials, green buildings, and green transportation. The IEA report was not written specifically for financial institutions, but rather for policymakers and carbon-intensive industries. However, the report has had a resounding impact on the financial industry by showing what a net zero economy would look like. We note that the report talks more about investment and financing than about decarbonization, with the following frequency: 220 occurrences for investment, 42 references for financing, and only 35 times for decarbonization[2].

Strictly speaking, portfolio alignment means that the carbon emissions trajectory of the portfolio is aligned or follows a decarbonization path of a given climate scenario. However, the IEA's net-zero scenario shows that decarbonization is not exogenous. So what does it mean for investors to align their portfolios with a net-zero scenario? Over the past decade, we have seen a shift among investors to view portfolio alignment not only as a portfolio construction exercise with a carbon footprint constraint, but also as an investment exercise that contributes to achieving net zero by 2050. In this approach, portfolio alignment is not a rebalancing process to reduce the carbon footprint, but rather a selection of investments to increase the greenness of the economy. The differences between carbon footprint and green footprint are at the heart of the concept of portfolio alignment and explain why it differs from portfolio decarbonization.

[2]We count the word and its derivatives. For example, the 35 occurrences of decarbonization are broken down as follows: decarbonization (13 occurrences), decarbonized (8 occurrences), decarbonize (7 occurrences), and decarbonizing (7 occurrences).

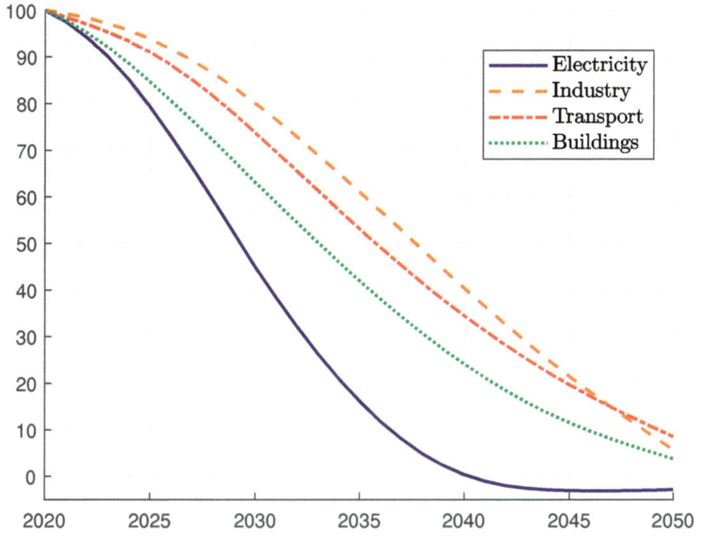

Source: (International Energy Agency, May 2021, Figure 3.1, page 100) & Author's calculations.

FIGURE 10.1: Sectoral decarbonization pathways

10.2 Portfolio decarbonization vs. portfolio alignment

As explained by Barahhou et al. (2022) and Ben Slimane et al. (2023), portfolio alignment is not portfolio decarbonization. However, many investors continue to confuse the two concepts (Roncalli, 2024). There are several reasons for this confusion (Raynaud et al., 2024, 2020). First, portfolio alignment involves decarbonizing the portfolio, which implies that alignment is related to decarbonization. Second, while portfolio decarbonization is well defined, this is not the case for portfolio alignment, which is a fuzzy term. It is a common term in the financial industry, but there is no accepted definition among investors. We know that portfolio alignment goes beyond portfolio decarbonization, but like many terms in sustainable finance, the concept can vary across countries and investor types. For example, portfolio alignment may or may not include a social pillar, leading to the concept of just transition. Third, portfolio alignment is closely related to net zero. When investors talk about portfolio alignment today, they implicitly assume that their investments are aligned with a net-zero scenario. The difficulty is that a climate scenario has a low probability of being realized and is, by definition, the path of a stochastic system. This means that there are many net-zero scenarios that change every time we have new information about the climate system.

CTB and PAB decarbonization pathways

CTB (Climate Transition Benchmark) and PAB (Paris-Aligned Benchmark) are two types of climate benchmarks introduced by the European Union (EU) to assess the alignment of investment portfolios with climate goals. The CTB and PAB decrabonization pathways are defined as:

$$\mathcal{CI}(t) = (1 - \Delta\mathcal{R})^{t-t_0} \left(1 - \mathcal{R}^{-}\right) \mathcal{CI}(t_0)$$

where t_0 is the baseline date, $\mathcal{CI}(t)$ is the carbon equivalent intensity (preferably measured with Scope 1+2+3 GHG emissions, $\Delta\mathcal{R} = 7\%$ and $\mathcal{R}^-$ takes the values 30% (CTB) and 50% (PAB), respectively. Below we report the reduction rate in % of CTB and PAB labels when the base year is 2020:

Year	2020	2025	2030	2035	2040	2045	2050
CTB	30.0	51.3	66.1	76.4	83.6	88.6	92.1
PAB	50.0	65.2	75.8	83.2	88.3	91.9	94.3

For example, in 2025, this implies a reduction of 51.3% and 65.2% from the 2020 carbon intensity of the benchmark for the CTB and PAB labels, respectively. The CTB decarbonization pathway is less aggressive. In fact, the CTB pathway follows the PAB pathway with a lag of 4.6 years (Le Guenedal and Roncalli, 2022).

10.2.1 The various dimensions of net zero

Defining a net-zero investment portfolio is not easy because it has several dimensions. First, a net-zero portfolio is not a low-carbon portfolio, because when we build a low-carbon portfolio, the goal is very simple. We want to reduce the carbon footprint of a given portfolio. So a low-carbon portfolio is a static approach to decarbonization. A net-zero portfolio is a dynamic approach to decarbonization to get to net zero by 2050. So the concept of self-decarbonization is important. And we can't look at a net-zero portfolio as an investment process where we just apply a sequence of decarbonization rates and a sequence of portfolio rebalancing. In fact, some of the decarbonization has to be endogenous. Another important difference is that a net-zero portfolio has to take into account the financing of the transition. Thus, the greenness or green intensity of the portfolio is also an important component of a net zero policy. This dimension is sometimes called the contribution part and is related to the concept of climate solutions. A net zero investment policy therefore has at least two main dimensions: portfolio decarbonization and financing the transition. Net zero investing is not just a carbon footprint issue, it is also a green footprint issue, and these two concepts are different.

10.2.2 The issue of portfolio rebalancing and the concept of self-decarbonization

To illustrate the concept of self-decarbonization, we assume that the decarbonization rate at the beginning of the year is 30%. At the end of the year, we want to have a decarbonization rate of 35% relative to the benchmark. We look at two extreme cases. At the end of the year, the decarbonization rate of the portfolio is 25%, which means that the carbon footprint of our portfolio has increased during the year. In this case, we have to rebalance the portfolio to get to the 35% level. This is the bad case because the self-decarbonization of the portfolio is zero. In the second case, we assume that at the end of the year, the decarbonization rate of the portfolio is greater than 35%, which means that we do not need to rebalance the portfolio. This is the good case because we do not have to rebalance the portfolio. So we can always follow a decarbonization path by rebalancing a portfolio of liquid assets, but it does not mean that the investment process is a net zero investment policy. In particular, there are some financial businesses where it is difficult to rebalance the portfolio because the assets are not liquid, for example banking and credit or insurance.

How to implement self-decarbonization in portfolio alignment? We recall that the carbon intensity at time $t+1$ is related to the carbon intensity at time t by the following equation[3]:

$$\mathcal{CI}(t+1) = (1 - \mathcal{R}(t,t+1))\,\mathcal{CI}(t)$$

where $\mathcal{R}(t,t+1)$ is the carbon reduction between t and $t+1$. This equation is valid at time $t+1$ once we know the value of carbon intensity at time $t+1$. We can decompose the carbon reduction into a predictable component and an unpredictable component:

$$\mathcal{R}(t,t+1) = \mathbb{E}\left[\mathcal{R}(t,t+1) \mid \mathcal{F}_t\right] + \varepsilon(t+1)$$

where $\mathcal{F}_t$ is the filtration at time t and $\varepsilon(t+1)$ is a stochastic noise process. In this case, $\mathcal{R}(t,t+1)$ is random and we have[4]:

$$\mathbb{E}\left[\mathcal{CI}(t+1) \mid \mathcal{F}_t\right] = \left(1 - \mathbb{E}\left[\mathcal{R}(t,t+1) \mid \mathcal{F}_t\right]\right)\mathcal{CI}(t)$$

Implementing self-decarbonization requires defining a process to control the expected carbon reduction $\mathbb{E}\left[\mathcal{R}(t,t+1) \mid \mathcal{F}_t\right]$. Two main approaches are used by professionals. The first is to predict the carbon reduction $\mathcal{R}(t,t+1)$ using time series modeling of discrete processes. Since the number of observations is relatively small, we can only use simple econometric models such as the linear (or log-linear) trend model. The second is to incorporate forward-looking measures of the future carbon path. This is typically done using the targets announced by issuers. In this case, investors use implied temperature ratings and map these ratings to expected carbon reductions.

Carbon momentum

Le Guenedal et al. (2022) define the carbon trend by considering the linear trend model:

$$\mathcal{CE}(t) = \beta_0 + \beta_1(t - t_0) + u(t) \tag{10.1}$$

where $\mathcal{CE}(t)$ is the carbon emissions and $u(t) \sim \mathcal{N}\left(0, \sigma_u^2\right)$. An alternative model is the log-linear trend model proposed by Roncalli (2024):

$$\ln \mathcal{CE}(t) = \gamma_0 + \gamma_1(t - t_0) + v(t) \tag{10.2}$$

where $v(t) \sim \mathcal{N}\left(0, \sigma_v^2\right)$. The parameters β_0, β_1, γ_0 and γ_1 are estimated with the least squares method and a sample of observations. Le Guenedal et al. (2022) define the carbon momentum as the historical growth rate of carbon emissions. In the case of the linear trend model, we have $\mathcal{CM}(t) = \dfrac{\hat{\beta}_1}{\mathcal{CE}(t)}$ while it is directly equal to $\hat{\gamma}_1$ in the case of the log-linear trend model. For example, $\mathcal{CM}(t) = -5\%$ means that the issuer has reduced his carbon emissions by 5% per year. The carbon momentum can also be estimated using the carbon intensity instead of the carbon emissions.

10.2.3 Carbon intensity and green intensity

Reducing the carbon footprint of a portfolio is not equivalent to increasing the green footprint of the portfolio or improving the greenness of the economy. The two goals are only equivalent if the correlation between the carbon intensity and the green intensity of

[3]The rebalancing dates are assumed to be synchronized at times t, $t+1$, and so on.
[4]We assume that $\mathbb{E}\left[\varepsilon(t+1)\right] = 0$.

issuers is -100%. This is currently not the case. For example, Table 10.1 shows the carbon and green[5] intensities of sectors in the MSCI EMU Index at the end of June 2024. We find that high carbon sectors also have high green intensities. On average, utilities stocks in the MSCI EMU stocks have a Scope 1 of 433 tons of CO_2 per million dollars of revenue, while the green turnover, capex and opex are 31.6%, 75.4% and 55.7%, respectively. Table 10.2 gives the sector breakdown in % of the carbon and green intensities. The utilities sector represents 5.4% of the weight in the MSCI EMU index, but it contributes 35.5% of the scope 1 emissions and 44% of the green capex of the MSCI EMU index. Three sectors account for about 80% of the greenness of the MSCI EMU index: consumer discretionary, industrials and utilities. However, two of these sectors are among the largest contributors to the carbon footprint (industrials and utilities). So when we reduce the carbon footprint of a portfolio, we run the risk of reducing the green footprint of the portfolio and the economy.

Green intensity

Like carbon intensity, which is a relative measure of carbon footprint, green intensity is a relative measure of green footprint (or greenness). However, the two measures are conceptually different. In fact, carbon intensity is the ratio of an issuer's carbon emissions to its financial size, such as EVIC or revenue. Carbon intensity is then expressed in $tCO_2e/\$$ mn. In contrast, green intensity measures the proportion of the issuer's operations that are considered green. Green intensity is therefore expressed as a percentage. For example, the most common measure is based on revenue (or turnover) and is also called green revenue share:

$$\mathcal{GI} = \frac{\mathcal{GR}}{\mathcal{TR}}$$

where $\mathcal{GR}$ is the green revenue and $\mathcal{TR}$ is the total revenue of the issuer. Another approach to calculating green intensity is to use green capex. While the green revenue ratio measures the current green footprint of a company, the green capex ratio measures the future green footprint of a company because the future greenness of a company is related to the current investment and R&D in green activities.

10.3 The portfolio optimization approach

Let w be a portfolio. The general optimization problem is to minimize a financial risk measure $\mathcal{R}(w)$ subject to a set of constraints Ω that take into account climate risk, portfolio alignment and portfolio construction:

$$w^{\star}(t) = \arg\min \mathcal{R}(w) \quad \text{s.t.} \quad w \in \Omega \tag{10.3}$$

A typical portfolio alignment constraint is the decarbonization pathway:

$$\mathcal{CI}(t, w) \leq (1 - \mathcal{R}(t_0, t))\, \mathcal{CI}(t_0, b(t_0)) \tag{10.4}$$

[5]We use the three measures of green intensity defined in the EU taxonomy: green turnover, green capex (capital expenditure) and green opex (operating expenditure). The data correspond to the figures reported by the companies.

TABLE 10.1: Carbon intensity and green intensity by sector (MSCI EMU, June 2024)

Sector	*Weight* (*in* %)	$\mathcal{SC}_1$	$\mathcal{SC}_{1-2}$	$\mathcal{SC}_{1-3}^{\text{up}}$	$\mathcal{SC}_{1-3}$	*Turnover*	*Capex*	*Opex*
		Carbon intensity (*in* tCO_2e/\$ mn)				*Green intensity* (*in* %)		
Communication Services	4.3	2.6	32.2	73.6	133.4	1.1	0.5	0.2
Consumer Discretionary	14.7	6.7	19.5	170.9	617.7	2.0	7.9	7.4
Consumer Staples	7.1	19.4	34.3	296.0	471.2	0.0	7.9	4.2
Energy	4.1	220.7	237.8	462.5	2 120.9	3.9	19.1	8.7
Financials	19.1	1.2	4.7	25.8	684.5	0.1	0.1	0.1
Health Care	6.9	11.9	28.1	106.5	128.9	0.0	1.1	0.0
Industrials	17.1	20.7	30.7	147.3	3 383.6	9.1	10.7	12.1
Information Technology	15.1	3.6	14.8	106.5	653.2	0.3	0.4	0.3
Materials	5.3	477.4	780.9	1 026.4	1 418.3	1.7	5.5	3.1
Real Estate	1.0	48.1	90.7	119.3	515.0	17.5	35.0	17.2
Utilities	5.4	433.3	497.7	601.0	911.8	31.6	75.4	55.7
MSCI EMU	100.0	65.6	95.5	209.3	1 162.5	4.1	9.2	7.2

Source: MSCI, Trucost & Author's calculations.

TABLE 10.2: Sector breakdown in % of the carbon intensity and taxonomy-aligned green intensity (MSCI EMU, June 2024)

Sector	*Weight*	*Carbon intensity* $\mathcal{SC}_1$	$\mathcal{SC}_{1-2}$	$\mathcal{SC}_{1-3}^{\text{up}}$	$\mathcal{SC}_{1-3}$	*Green intensity* *Turnover*	*Capex*	*Opex*
Communication Services	4.3	0.2	1.4	1.5	0.5	1.1	0.2	0.1
Consumer Discretionary	14.7	1.5	3.0	12.0	7.8	7.1	12.6	15.1
Consumer Staples	7.1	2.1	2.5	10.0	2.9	0.0	6.1	4.1
Energy	4.1	13.8	10.2	9.1	7.5	3.9	8.5	5.0
Financials	19.1	0.3	0.9	2.4	11.3	0.4	0.3	0.2
Health Care	6.9	1.3	2.0	3.5	0.8	0.0	0.9	0.0
Industrials	17.1	5.4	5.5	12.0	49.7	38.2	19.9	28.7
Information Technology	15.1	0.8	2.3	7.7	8.5	1.1	0.7	0.7
Materials	5.3	38.4	43.1	25.9	6.4	2.2	3.2	2.3
Real Estate	1.0	0.7	0.9	0.6	0.4	4.2	3.7	2.3
Utilities	5.4	35.5	28.0	15.4	4.2	41.8	44.0	41.7

Source: MSCI, Trucost & Author's calculations.

where t_0 is the baseline date, $b(t_0)$ is the benchmark at time t_0 and $\mathcal{R}(t_0, t)$ is the targeted rate of carbon intensity reduction between t_0 and t. Of course, the set Ω can include many other constraints that manage the exclusions or the green footprint. In fact, there is not one specification, but many, depending on the sustainable preferences of the investor. In contrast, the measure of financial risk is relatively standard. In most cases, it is the tracking error variance for equity portfolios and active risk for bond portfolios. This means that the portfolio is optimized relative to a benchmark, which may be a passive benchmark or an active portfolio.

10.3.1 Equity portfolios

For example, Barahhou et al. (2022) use the following optimization problem:

$$\begin{aligned} w^{\star}(t) &= \arg\min \frac{1}{2}(w - b(t))^{\top} \Sigma(t)(w - b(t)) && (10.5) \\ \text{s.t.} & \begin{cases} \mathcal{CI}(t, w) \leq (1 - \mathcal{R}(t_0, t))\,\mathcal{CI}(t_0, b(t_0)) & \leftarrow \text{Alignment} \\ \mathcal{CM}(t, w) \leq \mathcal{CM}^{\star} & \leftarrow \text{Self-decarbonization} \\ \mathcal{GI}(t, w) \geq (1 + \mathcal{G})\,\mathcal{GI}(t, b(t)) & \leftarrow \text{Greenness} \\ 0 \leq w_i \leq \mathbb{1}\left\{\mathcal{CM}_i(t) \leq \mathcal{CM}^{+}\right\} & \leftarrow \text{Exclusion} \\ w \in \Omega & \leftarrow \text{Other constraints} \end{cases} \end{aligned}$$

where Σ is the covariance matrix of asset returns. The alignment constraint is defined using the traditional decarbonization pathway formulation based on carbon intensity. For the self-decarbonization dimension, the authors impose an absolute upper bound on carbon momentum. In general, it is common to impose that at least half of the reduction rate comes from self-decarbonization. For example, if we assume that $\mathcal{R}(t-1,t)$ is equal to 7%, then $\mathcal{CM}^{\star}(t)$ is set to -3.5%. Green footprint management consists of improving the green footprint of the current benchmark. Indeed, while the alignment refers to the carbon intensity $\mathcal{CI}(t_0, b(t_0))$ of the benchmark at the baseline time t_0, the greenness refers to the green footprint $\mathcal{GI}(t, b(t))$ of the benchmark at the rebalancing date t. The coefficient $\mathcal{G}$ measures the rate of increase of the green footprint. A typical value is 100%, which means that investors aim to double the green footprint of their portfolios relative to the benchmark. In addition to alignment, self-decarbonization and greenness, investors see exclusionary constraints as key. They are of two types. A first set of exclusionary constraints is often implemented and concerns ESG exclusions. In this case, assets may be excluded because the issuer has a poor ESG rating or is on a blacklist of companies. The second set of exclusionary constraints concerns assets that do not meet a net zero investment policy. This second set of constraints defines the "*net zero enemies*". Here, the authors decide to impose a maximum carbon momentum. Indeed, if $\mathcal{CM}_i(t) > \mathcal{CM}^{+}$, then $w_i = 0$. For example, if $\mathcal{CM}^{+} = 10\%$, this means that all issuers that have increased their carbon intensity beyond 10% are excluded.

As we have said, there is not just one, but many approaches to implementing portfolio alignment. For example, we can consider several variants of the optimization problem (10.5). First, we can change the objective function. Instead of using a benchmark portfolio such as a stock index, we can use an initial portfolio w_0 that we want to adjust to satisfy the net zero constraints. w_0 can be an active portfolio defined by the fund manager or a systematic portfolio derived from a quantitative process. For example, this type of optimization process is used to transform an equity multi-factor portfolio into a net-zero equity multi-factor portfolio. Instead of using the tracking error variance, the risk measure can be the standard mean-variance utility function, which may or may not include sustainability preferences (Blitz et al., 2024; Bouyé et al., 2024; Pedersen, Fitzgibbons, and Pomorski, 2021). Second, we can use an objective function that depends on the active share:

$$\mathcal{AS}(w \mid b(t)) = \frac{1}{2}\sum_{i=1}^{n} |w_i - b_i(t)|$$

We can combine the stock-based active share, the sector-based active share, and the country-based active share to construct the objective function. For example, this is the approach taken by the S&P PAB indices, which use the following objective function[6]:

$$\frac{1}{n}\sum_{i=1}^{n}\frac{(w_i-b_i(t))^2}{b_i(t)}+\frac{1}{n_{Sector}}\sum_{s=1}^{n_{Sector}}\frac{(w_s-b_s(t))^2}{b_s(t)}+\frac{1}{n_{Country}}\sum_{c=1}^{Country}\frac{(w_c-b_c(t))^2}{b_c(t)}$$

where $w_s=\sum_{i\in s}w_i$ and $b_s(t)$ are the sector weights of the portfolio and benchmark, and $w_c=\sum_{i\in c}w_i$ and $b_c(t)$ are the country weights of the portfolio and benchmark. The third approach is to use the multi-period portfolio optimization method proposed by Le Guenedal and Roncalli (2022) and solved by Lezmi, Roncalli, and Xu (2022).

Regarding the alignment constraint, two different approaches can be used. First, we can replace the carbon intensity with the carbon budget (Le Guenedal et al., 2022), which is the sum of the carbon emissions of the portfolio between two dates t_1 and t_2:

$$\mathcal{CB}(t_1,t_2,w)\leq\int_{t_1}^{t_2}\mathcal{CE}(t,w(t))\,\mathrm{d}t$$

In this case, we want the carbon emissions of the portfolio between the base date t_0 and the target date $t^\star$ not to exceed a given carbon budget:

$$\mathcal{CB}(t_0,t^\star,w)\leq\mathcal{CB}^+(t_0,t^\star)$$

In most cases, the maximum carbon budget is calculated using a global net-zero scenario[7]:

$$\mathcal{CB}^+(t_0,t^\star)=\mathcal{CE}(t_0,b(t_0))\int_{t_0}^{t^\star}(1-\mathcal{R}(t_0,t))\,\mathrm{d}t$$

The alignment constraint between two rebalancing dates t and $t+1$ becomes:

$$\mathcal{CB}(t_0,t+1,w)-\mathcal{CB}(t_0,t,w)\leq\mathcal{CB}^+(t,t+1):=\mathcal{CB}^+(t_0,t+1)-\mathcal{CB}^+(t_0,t)\qquad(10.6)$$

The difficulty is modeling the future carbon emissions between the current date t and the next rebalancing date t. For instance, we can use the carbon trend model. Another solution is to assume that carbon emissions are constant between t and $t+1$, which is acceptable if we consider annual periods. In this case, the constraint (10.6) reduces to:

$$\mathcal{CE}(t,w)\leq\mathcal{CB}^+(t,t+1):=(1-\mathcal{R}(t_0,t))\,\mathcal{CE}(t_0,b(t_0))$$

The carbon budget approach is then equivalent to a carbon emissions approach and has been proposed by Bolton, Kacperczyk, and Samama (2022). The second approach is to define sectoral decarbonization pathways rather than a global one. Indeed, as we have said, we need to differentiate some sectors because they are key to achieving a low-carbon economy. This is the case of utilities, for example. Ben Slimane et al. (2023) then propose to use a specific alignment constraint for the utilities sector and a global alignment constraint for the other sectors.

The self-decarbonization constraint can also be measured using the implied temperature rating. In this case, a temperature score is calculated for each issuer using multiple metrics: historical carbon emissions, carbon trends, issuer reduction targets, market-based climate

[6]The choice of the $\mathcal{L}_2$ norm instead of the $\mathcal{L}_1$ norm to define the active share is due to the tractability of the objective function and, in particular, to its quadratic property. In fact, as explained by Perrin and Roncalli (2020), from an industrial point of view, it is important to cast portfolio optimization problems into quadratic programming problems in order to obtain numerical solutions and avoid convergence and computational issues.

[7]For example, this could be the IPCC or IEA scenario.

scenarios, and so on. The idea is to compare the historical trajectory and its trend, the issuer's announced carbon targets, and the net-zero decarbonization scenario. The assessment of the company's future trajectory then has three dimensions: (historical) participation, ambition and credibility. The temperature score is then converted into an implied temperature score $\mathcal{T}$. For example, if $\mathcal{T} \leq 2°C$, this means that the company's decarbonization trajectory is compatible with a climate scenario of 2°C or less. Conversely, if $\mathcal{T} \geq 4°C$, this implies that the company is making no effort and is participating in a world with a temperature anomaly of 4°C. By assuming that the portfolio's temperature rating is the weighted average of the issuers' temperature ratings, self-decarbonization can be controlled by imposing an upper bound on the portfolio's temperature rating.

The greenness dimension of the net zero pathway is certainly the most difficult component to model and incorporate into the portfolio. While the decarbonization dimension is an exclusion process to avoid high-carbon issuers, the green dimension is a selection process to finance the issuers necessary to achieve a low-carbon economy. The decarbonization dimension is then implemented using a backward-looking framework, as it is based on past carbon intensity or emissions. In contrast, the green dimension should be implemented using a forward-looking framework, as it should be based on the future contribution to the net zero transformation. Barahhou et al. (2022) chose to model the green dimension using the green revenue share. However, this is a current measure of green intensity. A better measure would be green capex, but data on this metric are not really available or noisy today[8]. Of course, this situation will improve with the new CSRD reporting and the implementation of the EU green taxonomy, but these regulations are European, which means that we will not have the data for many American, Japanese and emerging market companies. This is a real problem when implementing a net zero policy for large global indices such as the MSCI ACWI IMI Index. Sometimes the green dimension is captured by the green-to-brown ratio, but this measure has many biases because it depends on both brown and green activities[9]. It is a relative measure and does not indicate the absolute green intensity of the portfolio.

The next set of constraints is present in all alignment portfolio construction and defines the exclusion policy. It is a source of significant tracking error variance. The idea is to define a list of issuers, sectors, or activities that are inconsistent with a net zero investment policy. For example, the PAB label implies the exclusion of fossil fuel companies that derive 1%, 10%, and 50% or more of their revenues from coal, oil, and natural gas exploration or processing, respectively. Another example is the exclusion of power generation companies that generate electricity above a certain threshold, typically above $100\,\mathrm{gCO_2e/kWh}$. In addition to these activity exclusions, we also find ESG exclusions for some sectors, such as controversial weapons and tobacco. The last type of exclusion concerns issuers for many reasons. They may be excluded because they are involved in ESG controversies, they do not respect social norms (UN Declaration of Human Rights, ILO Declaration and UN Global Compact), they have a low ESG score, etc. They may also be excluded because they have a high carbon momentum or temperature score. Typical values are $\mathcal{CM} \geq 10\%$ and $\mathcal{T} \geq 4°C$.

The last set of constraints is necessary to control the active bets of the optimized portfolio. In fact, the optimization typically concentrates the allocation in a few stocks, a few sectors, and a few countries. Without these constraints, the portfolio is not diversified and has too many sector and country biases. Thus, MSCI (2023) uses six diversification constraints to construct its PAB indexes. For example, the active country and sector weights cannot deviate by more than ±5% from the country and sector weights in the parent index.

[8]It is also the case of many climate risk measures such as the scope 3 upstream and downstream emissions, whose quality is very poor, both those reported by companies and those estimated by ESG rating agencies.

[9]This ratio compares the amount of green investment or financing to the amount of brown investment or financing.

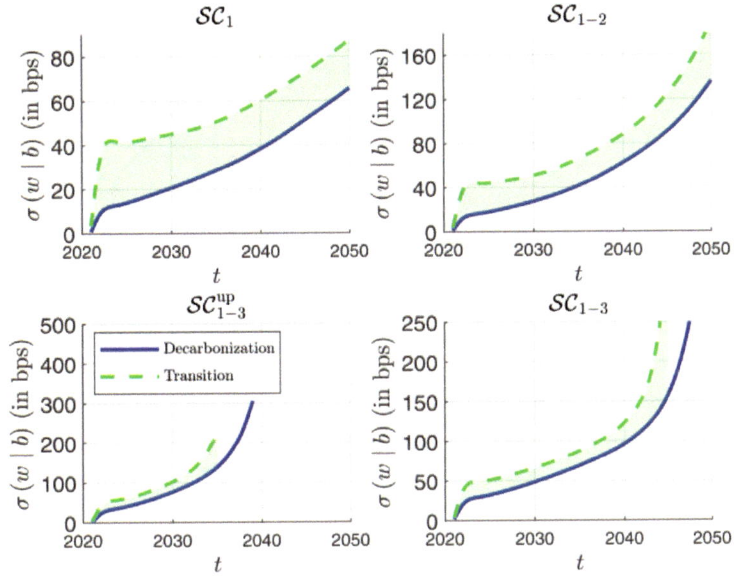

Source: MSCI, Trucost & Barahhou et al. (2022).

FIGURE 10.2: Tracking error volatility of net-zero portfolios (MSCI World, June 2022, PAB, $\mathcal{G} = 100\%$, $\mathcal{CM}^\star = -5\%$).

The active stock weight is limited to $\pm 2\%$ with an upper limit that it cannot exceed 20 times the weight in the parent index. Thus, for a stock with weights of 0.01%, 1%, and 5% in the parent index, the upper limits are 0.2%, 3%, and 7%, respectively. Barahhou et al. (2022) use the following standard constraints, which are classic in the ETF market: $w_i \leq 10 b_i$ and $b_s/2 \leq w_s \leq 2b_s$. The first constraint imposes that the stock weighting w cannot exceed 10 times the weighting b_i in the benchmark portfolio, while the second constraint imposes that the sector allocation w_s cannot be less than half and greater than twice the sector allocation b_s in the benchmark portfolio.

Figures 10.2 and 10.3 are taken from Barahhou et al. (2022) when implementing the PAB decarbonization pathway, the self-decarbonization constraint with $\mathcal{CM}^\star = -5\%$, the greenness constraint with $\mathcal{G} = 100\%$, and the diversification constraint on the MSCI World index. The first figure shows the relationship between time t and the tracking error volatility $\sigma(w(t) \mid b(t))$, measured in bps, considering several scopes of emissions (Scope 1, Scope 1+2, Scope 1+2+3 upstream, Scope 1+2+3). The figure also shows the decomposition between the decarbonization and transition dimensions. The results of these simulations clearly show that the transition dimension entails significant and additional costs. Moreover, there may be no solution to the optimization problem by 2050, especially if the carbon footprint is based on upstream/downstream Scope 3 emissions. Of course, all these results are very sensitive to the choice of the green multiplier $\mathcal{G}$ and the carbon threshold $\mathcal{CM}^\star$. In practice, many constraints are used to construct net-zero portfolios, especially exclusion constraints, which imply that the portfolio is more concentrated than the benchmark. Therefore, we may face not only diversification risk, but also liquidity risk. These risks will be reduced if the economy decarbonizes in the coming years. However, we are not immune to the possibility that carbon emissions will continue to rise in the short term. In this case, solutions will be very sensitive to the gap between the carbon target of net-zero portfolios and the carbon footprint of the economy. Figure 10.3 illustrates the shrinkage risk of the investment universe in the first year of PAB implementation when excluding or including issuers with a positive carbon

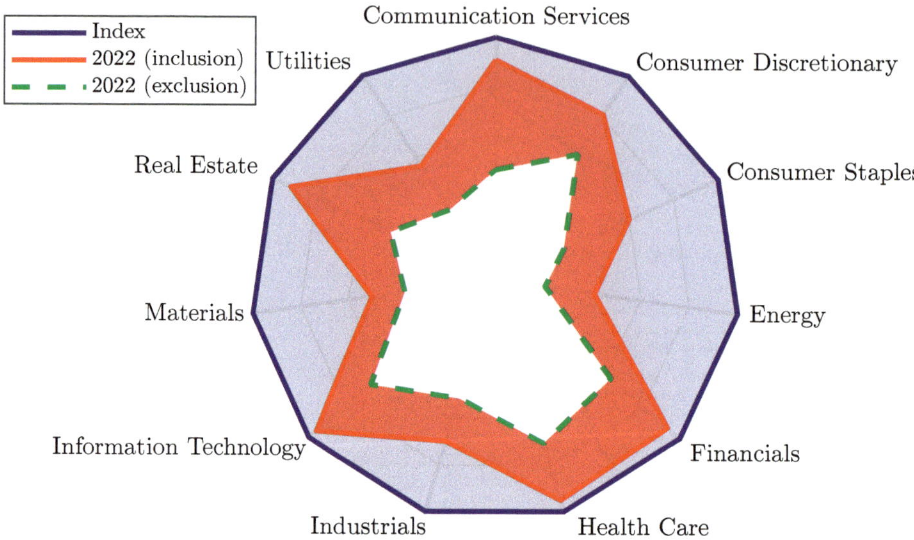

Source: MSCI, Trucost & Barahhou et al. (2022).

FIGURE 10.3: Universe shrinkage of net-zero portfolios (MSCI World, June 2022, PAB, $\mathcal{G} = 100\%$, $\mathcal{CM}^{\star} = -5\%$, Scope $\mathcal{SC}_{1-3}$, $\mathcal{CM}^{+} = 0\%$). This radar chart shows the percentage of the benchmark's allocation remaining in each sector after excluding net-zero portfolios. The baseline of 100% represents the cap-weighted index (blue curve).

momentum ($\mathcal{CM}^{+} = 0\%$). The blue area shows the reduction of the investment universe when the optimization process (10.5) is implemented without the exclusion policy. The red area shows the additional reduction of the investment universe when the optimization process (10.5) is implemented with the exclusion policy. The effect of exclusions is very important.

What lessons can we learn from these simulations? First, reducing a portfolio's carbon footprint is an investment process that is long financials and short energy, industrials, materials and utilities, even after accounting for Scope 3 emissions. In contrast, increasing a portfolio's green footprint is an investment process that is long industrials and utilities. This is a dilemma, especially for the utilities sector. This apparent contradiction explains the increase in active risk of net-zero portfolios. Managing decarbonization and transition at the same time is not an easy task. Second, exclusion policy can have a big impact on portfolio construction, especially if we stack multiple layers of exclusion. Being too conservative can lead to having no solution for portfolio optimization. There is a real risk that existing net zero investment processes may not survive in the future, especially if the decarbonization rate of the economy is not sufficient. Third, while portfolio decarbonization is primarily an exclusion process, portfolio alignment is also a selection process. This third point is critical because investors take a lot of time to define exclusion policies, but the alignment to a net-zero scenario also requires selecting and investing in the issuers that will transform the high-carbon economy into a low-carbon economy.

10.3.2 Corporate bond portfolios

Managing a portfolio of corporate bonds is very similar to managing a portfolio of stocks, except that the measure of risk is completely different. For example, Barahhou et al. (2022)

use the following optimization problem:

$$w^{\star}(t) = \arg\min \frac{1}{2} \sum_{i \in b(t)} |w_i - b_i(t)| + \varphi \sum_{s=1}^{n_{Sector}} \left| \sum_{i \in s} (w_i - b_i(t)) \, \mathrm{DTS}_i \right| \tag{10.7}$$

$$\text{s.t.} \quad \begin{cases} \mathcal{CI}(t,w) \leq (1 - \mathcal{R}(t_0,t)) \, \mathcal{CI}(t_0, b(t_0)) & \leftarrow \text{Alignment} \\ \mathcal{CM}(t,w) \leq \mathcal{CM}^{\star}(t) & \leftarrow \text{Self-decarbonization} \\ \mathcal{GI}(t,w) \geq (1+\mathcal{G}) \, \mathcal{GI}(t, b(t)) & \leftarrow \text{Greenness} \\ 0 \leq w_i \leq \mathbb{1}\{\mathcal{CM}_i(t) \leq \mathcal{CM}^{+}\} & \leftarrow \text{Exclusion} \\ \sum_{i=1}^{n} (w_i - b_i(t)) \, \mathrm{MD}_i = 0 & \leftarrow \text{MD constraint} \\ w \in \Omega & \leftarrow \text{Other constraints} \end{cases}$$

where DTS_i and MD_i are the duration-times-spread and modified duration factors. Compared to the equity optimization problem (10.5), the objective function is the sum of the active share and the duration-time-spread active risk. In addition, the bond optimization problem requires that the optimized portfolio and the benchmark have the same modified duration.

As with equity portfolios, there are many variations of the optimization problem (10.7). For example, we can include a modified duration risk measure directly in the objective function, and S&P combines the stock-based active share and the sector-based active share to construct the objective function of the iBoxx EUR Corporates Net Zero 2050 Paris-Aligned ESG Index. The above discussion of different constraints also applies to corporate bonds.

The conclusions we draw for equity portfolios remain valid for corporate bond portfolios. However, we find that it is easier to construct a net-zero bond portfolio than a net-zero equity portfolio. This is because the investment universe is larger for corporate bonds and the active risk of MD and DTS can be better controlled than the variance of the tracking error. Another reason is the impact of the primary market, which is small in equity markets but significant in bond markets. As bonds mature, they are replaced by new bonds that are greener. So the primary market helps to achieve net zero in corporate bonds.

10.4 The core-satellite approach

We have seen that the comprehensive integrated approach can sometimes be difficult to implement because today, on average, carbon intensities are positively correlated with green intensities. This means that the greenness of the economy is not necessarily found in companies with low carbon footprints. Therefore, a second approach has emerged that is easier to implement. Ben Slimane et al. (2023) propose a core-satellite strategy, where decarbonization is applied to the core portfolio, while the objective of the satellite portfolio is to finance the transition to a low-carbon economy. In the financial literature, the core portfolio is called the net-zero decarbonization portfolio, while the satellite portfolio is called the net-zero contribution portfolio.

This is equivalent to splitting the problem into two sub-problems. The goal of the first sub-problem is to decarbonize and manage the carbon footprint of the investment. The core portfolio is more of a top-down allocation process and exclusion strategy, where the central climate risk metric is carbon intensity. The goal of the second sub-problem is to contribute to increasing the green footprint of the economy. The satellite portfolio is more of a bottom-up allocation process and asset selection strategy, where the central climate risk metric is green intensity. This approach also has the advantage of making the allocation between the two net-zero strategies clear. Of course, the allocation to the satellite can be dynamic and change over time as the world and economy progresses toward net zero.

10.4.1 The core portfolio

The core-satellite portfolio can be applied to equity, fixed-income or multi-asset classes. For equity and fixed-income asset classes, we can use the optimization problems (10.5) and (10.7) without including the greenness constraint (and also the self-decarbonization constraint). For example, the equity or fixed-income core portfolio can be a CTB index portfolio. In the case of multi-asset strategies, the solution is to allocate between an equity core portfolio and a fixed-income core portfolio. The allocation process is generally standard in the multi-asset class and can be a mean-variance optimized, constant-mix weighting or risk parity process (Roncalli, 2013).

10.4.2 The satellite portfolio

The construction of satellite portfolios is more complex. Indeed, while the core portfolio aims to implement decarbonization policies, the satellite portfolio aims to finance the transition to a low-carbon economy and monitor green intensity. By definition, the core portfolio can be seen as an exclusion process applied to a traditional investment portfolio, while the satellite portfolio can be seen as a selection or asset-picking process. As the core-satellite approach is mainly implemented in strategic asset allocation or multi-asset portfolios, the investment universe is diversified and typically consists of the following asset classes: (1) green, sustainability and sustainability-linked (GSS) bonds, (2) green stocks, (3) green infrastructure and (4) sustainable real estate.

In order to determine more precisely the asset universe of the satellite portfolio, we need to understand the funding requirements of the net zero transition. According to McKinsey (January 2022), the world will need about \$275 trillion of investment in physical assets between 2021 and 2050, or \$9.2 trillion per year, to finance the transition to a low-carbon economy. This represents an increase of about \$3.5 trillion per year over today's allocation. More than 85% of this \$275 trillion will go to the buildings, power, transportation sectors. By region, the most important sectors are transportation in developed markets and energy in emerging markets, including China and India. These figures are roughly in line with those calculated by the Energy Transitions Commission (Energy Transitions Commission, March 2023). Figure 10.4 shows the distribution of net-zero investments. At the global level, the power sector must represent 70% of the investments with the following breakdown: 38% in power generation (green electricity), 26% in power networks (electricity infrastructure and grids), and 6% in power storage (electricity efficiency). If we include buildings (14%) and transport (8%), the figure is 92%. The remaining 8% concerns removals (waste management, recycling), hydrogen and finally industry. This confirms that all sectors are not equal in terms of net-zero investments.

Because the satellite portfolio is invested in the few sectors necessary to achieve net zero, it is far from a traditional investment portfolio. Therefore, the satellite portfolio has a higher active risk relative to a classical benchmark than the core portfolio. Ben Slimane et al. (2023) estimated that the tracking error volatility of the satellite portfolio is about 10 times the tracking error volatility of the core portfolio. They also found that the tracking error volatility of a net-zero core-satellite portfolio is currently about 3% for a 60/40 constant-mix strategy.

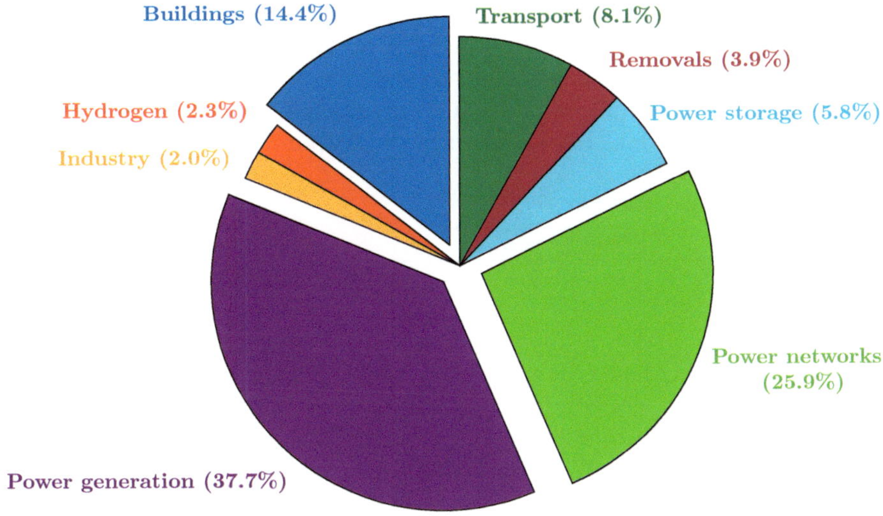

Source: Energy Transitions Commission (March 2023) & Authors' calculations.

FIGURE 10.4: Net zero capital investments.

10.5 The case of sovereign bond portfolios

The case of government bonds is studied by Barahhou, Ferreira, and Maalej (2023). As for corporate bonds, the integrated approach[10] for sovereign bonds consists of several steps: (1) we need to define the decarbonization scenario at the country level; (2) we can assess the self-decarbonization of a country by considering the government's credible commitments and decarbonization plans toward a low-carbon economy; (3) a specific green intensity measure needs to measure the country's contribution to the climate transition and its greenness.

In addition to the IEA net-zero scenario, we can also choose the NGFS net-zero scenario. An important point is the distinction between advanced and developing economies. Another difference with corporate bonds is the definition of the scope of emissions in particular the choice between production-based and consumption-based emissions. If we prefer a carbon intensity measure, the normalization variable can be population, GDP or public debt. Using data on carbon emissions and commitments, Barahhou, Ferreira, and Maalej (2023) also derived four types of forward-looking metrics: (1) carbon trend, (2) nationally determined contribution, (3) NDC ambition, and (4) NDC fulfillment. These metrics can be combined to measure country's self-decarbonization. For the green intensity, the most common approach is to measure the share of renewable electricity in the country or the share of government spending on renewable electricity.

Simulations show that it is more difficult to build a net-zero government bond portfolio than a net-zero corporate bond portfolio. There are several reasons for this. The first reason is the limited number of countries compared to the number of companies. Therefore, we can quickly and dramatically shrink the investment universe. The second reason is the impact of the decarbonization dimension, because there are not many countries that have decarbonized their economies. In fact, we find that companies can change their carbon footprint faster than countries. The last reason is the liquidity issue that net-zero portfolios raise. In

[10]In the core-satellite approach, the satellite consists of green government bonds.

fact, some countries, such as the United States, have a significant share of the government bond market, and their bonds are very liquid. So a net-zero portfolio generally implies a reduction in liquidity. Managing this issue, the high active bets and the idiosyncratic risks of net-zero strategies is then a challenge and this explains why these strategies are less popular in the government debt market than in the corporate debt market.

10.6 Conclusion

The previous empirical results for equities, corporate bonds, and sovereign bonds suggest the following lessons. First, the solution is parameter and data sensitive. In particular, we need to be careful in choosing the carbon scope metric to assess the decarbonization rate. A net-zero investment policy only makes sense for a closed system. Therefore, Scope 3 and consumption-based emissions need to be taken into account to align a portfolio with a net-zero scenario. The problem is that we see a lack of data reliability on these indirect emissions today. Similarly, the solution is highly dependent on the green intensity target and the level of self-decarbonization we want to achieve. Then we have to be careful because there may be no solution to the optimization problem in the medium term. The question of no solution depends on the relative speed of the portfolio's decarbonization path relative to the economy's decarbonization path and the initial starting point.

The second key finding is that portfolio decarbonization and net-zero construction lead to different solutions. In particular, decarbonizing a portfolio is easier than constructing a net-zero portfolio. We find that decarbonizing along CTB or PAB pathways never leads to exploding tracking errors by 2030. In fact, the real problem with decarbonization is the diversification and liquidity risk that an investor may face. These results are amplified when we add the transition dimension to the optimization program. In addition to higher tracking risk, there is no guarantee that there will always be a solution. Moreover, the introduction of the transition pillar highlights the difficulty of choosing an appropriate set of constraints for net-zero portfolios, as some metrics may be negatively correlated with others. Portfolio decarbonization is systematically a strategy that is long financial issuers and short energy, materials and utilities issuers. Therefore, we have a situation where the transition dimension of a decarbonized portfolio is weaker than that of the benchmark portfolio, as green solutions are also located in carbon-intensive sectors. It is therefore crucial to distinguish between issuers with a high carbon footprint that will not participate in the transition and those that will reduce their carbon emissions and find low-carbon solutions.

The third key finding is that portfolio decarbonization and alignment are two processes of exclusion. This means that it is quite impossible to achieve net zero alignment without allowing the algorithm to exclude companies (or countries) from the benchmark. For example, the optimization program will generally not find a solution if it imposes non-zero lower bounds. As a result, some key players in the transition, such as energy and utility companies, unfortunately disappear. Furthermore, imposing sector neutrality can lead to similar problems in finding a solution.

The final lesson is that it is easier to implement net zero in bonds than in equities. At first glance, this result may seem surprising, since there is no reason why net zero should affect the equity and bond markets differently. In fact, there are two possible explanations. First, the structure of equity and bond indexes is different, with the latter having a more balanced allocation across sectors and a high exposure to financial issuers. Second, bond indexes are strongly influenced by new fresh capital, while equity indexes are sticky to the stock of existing capital. This is because the primary bond market is very active, which implies a significant impact on the secondary market. Indeed, bonds mature and are replaced by new greener bonds. The primary market then helps to achieve net zero in bonds. This is not the case in the stock market, where IPOs and capital increases are only a small part of the secondary market. This means that portfolio holdings change faster for bond indexes than for equity indexes. Therefore, the greenness of bond indexes increases faster than the greenness of equity indexes. All of these factors suggests that the cost of implementing net zero investments relative to traditional investments will be higher for equity portfolios than for bond portfolios, and that the bond market will benefit more quickly from the transition to a low-carbon economy.

Beyond these empirical results, there is another important point that is missing from our analysis. This is the issue of engagement. Indeed, it may be surprising to talk about net zero investing without integrating this dimension into portfolio construction. The reason is that engagement is difficult to model quantitatively. Nevertheless, Cenedese, Han, and Kacperczyk (2023) has recently developed a quantitative framework and an objective metric called distance-to-exit that can help identify the most critical companies that investors need to engage with when building a net-zero portfolio. Whatever the approach, the results of engagement must of course be integrated into the net zero investment policy. This can be done, for example, by adjusting carbon and/or green footprints. Engagement is also important because dialogue with issuers also helps investors understand the industrials issues surrounding net zero. Too often, investors assume that net zero investing is a matter of overweighting or underweighting a benchmark. This is an oversimplified view of net zero. For example, real asset investors know the importance of understanding the industrial processes involved in a net zero transition. Again, this ties into the top-down versus bottom-up debate about net-zero portfolio construction. Even if we take a top-down approach (which this chapter largely does), it is important to ensure that the investment of each stock in the portfolio has a bottom-up rationale.

References

Andersson, M., P. Bolton, and F. Samama, 2016, Hedging Climate Risk, *Financial Analysts Journal* 72, 13–32.

Barahhou, I., M. Ben Slimane, N. Oulid Azouz, and T. Roncalli, 2022, Net Zero Investment Portfolios — Part 1. The Comprehensive Integrated Approach, *SSRN* 4283998, 103 pages.

Barahhou, I., P. Ferreira, and Y. Maalej, 2023, A Framework to Align Sovereign Bond Portfolios with Net Zero Trajectories, *SSRN* 4515462, 55 pages.

Ben Slimane, M., D. Lucius, T. Roncalli, and J. Xu, 2023, Net Zero Investment Portfolios — Part 2. The Core-Satellite Approach, *SSRN* 4611418, 69 pages.

Blitz, D., M. Chen, C. Howard, and H. Lohre, 2024, 3D Investing: Jointly Optimizing Return, Risk, and Sustainability, *Financial Analysts Journal forthcoming*, 17 pages.

Bolton, P., M. Kacperczyk, and F. Samama, 2022, Net-zero Carbon Portfolio Alignment, *Financial Analysts Journal* 78, 19–33.

Bouyé, E., R. Deguest, E. Jurczenko, and J. Teiletche, 2024, Biodiversity and Climate: Friends or Foes?, *SSRN* 4778320, 66 pages.

Cenedese, G., S. Han, and M. T. Kacperczyk, 2023, Carbon-transition Risk and Net-zero Portfolios, *SSRN* 4565220, 64 pages.

Energy Transitions Commission, March 2023, Financing the Transition: How to Make the Money Flow for a Net-zero Economy, *Report* 101 pages.

EU Technical Expert Group on Sustainable Finance, December 2019, *Handbook of Climate Transition Benchmarks, Paris-Aligned Benchmarks and Benchmarks' ESG Disclosures* (Report, 171 pages).

EU Technical Expert Group on Sustainable Finance, September 2019, *TEG Final Report on Climate Benchmarks and Benchmarks' ESG Disclosures* (Report, 75 pages).

International Energy Agency, May 2021, *Net Zero by 2050: A Roadmap for the Global Energy Sector* (Report, 223 pages).

Le Guenedal, T., F. Lombard, T. Roncalli, and T. Sekine, 2022, Net Zero Carbon Metrics, *SSRN* 4033686, 74 pages.

Le Guenedal, T., and T. Roncalli, 2022, Portfolio Construction with Climate Risk Measures, in E. Jurczenko, ed., *Climate Investing: New Strategies and Implementation Challenges*, chapter 3, 49–86 (Wiley).

Lezmi, E., T. Roncalli, and J. Xu, 2022, Multi-period Portfolio Optimization, *SSRN* 4078043, 64 pages.

McKinsey, January 2022, The Net-zero Transition — What it Would Cost, What it Could Bring, *Report* 224 pages.

MSCI, 2023, MSCI Climate Paris Aligned Indexes Methodology, *Index Methodology* 26 pages.

Pedersen, L. H., S. Fitzgibbons, and L. Pomorski, 2021, Responsible Investing: The ESG-Efficient Frontier, *Journal of Financial Economics* 142, 572–597.

Perrin, S., and T. Roncalli, 2020, Machine Learning Optimization Algorithms & Portfolio Allocation, in E. Jurczenko, ed., *Machine Learning for Asset Management: New Developments and Financial Applications*, chapter 8, 261–328 (Wiley).

Raynaud, J., Voisin, S., Tankov, P., Bouchet. V., Ferrand, A., and Serta, J.P., 2024, The Alignment Cookbook 2 — A Technical Panorama of the Alignment Methodologies and Metrics used by and applied to the Financial Sector, with a View to Inform Consolidated Alignment Assessments, *Institut Louis Bachelier Report* 233 pages.

Raynaud, J., S. Voisin, P. Tankov, A. Hilke, and A. Pauthier, 2020, The Alignment Cookbook — A Technical Review of Methodologies Assessing a Portfolio's Alignment with Low-carbon Trajectories or Temperature Goal., *Institut Louis Bachelier Report* 233 pages.

Roncalli, T., 2013, *Introduction to Risk Parity and Budgeting* (Chapman and Hall/CRC Financial Mathematics Series).

Roncalli, T., 2024, *Handbook of Sustainable Finance* (SSRN, 4277875, 930 pages).

Part III

Data, Measurement, and AI

Chapter 11

Data for climate and sustainable finance

Alexis Tantet
Ecole Polytechnique, France

Peter Tankov
CREST – ENSAE, Institut Polytechnique de Paris, France

Sustainable finance uses a multitude of data sources with varying degrees of reliability, from company sustainability reports to physical measurements and climate model projections. The goal of this chapter is to present a high-level overview of these data sources, describe existing classification attempts and important open source initiatives. We shall then make a deep dive into physical risk assessment, increasingly based on forward looking climate scenarios, and review the main sources of climate data, to provide an access point to this field for academics and finance professionals. Other types of data used in sustainable finance, such as greenhouse gas emission estimates and ESG ratings are discussed in depth in subsequent chapters.

11.1 Introduction

Unlike financial data, which usually corresponds to prices of financial transactions, recorded without error, and disseminated by major financial data providers on a subscription basis, sustainable finance uses a multitude of data sources, for a multitude of different applications. Among the different types of extra-financial data used in sustainable finance, one can mention

- Numeric and textual data on environmental, social and other impacts reported by companies and compiled by data providers with some quality checks;
- Textual data from news;
- Data observed by physical measurements: satellite images, climate / weather etc.
- Asset-level data compiled by NGOs, data providers etc., based on a variety of sources;
- Scenarios and projections produced by various organisms using a variety of models (NGFS, IPCC, IEA, World Bank);
- Indices, ratings and scores computed by data providers (such as ESG scores).

Data format and access options depend on the specific type of data: for example, climate data is usually available free of charge but the users must be familiar with data formats and software which have been developed in the climate science community over many years and

DOI: 10.1201/9781032636252-11

are not easy to manipulate for laymen. On the other hand, impact data, such as company-level carbon emissions, are available from data providers in convenient formats but at a cost which may be relatively high, especially for an academic institution.

These extra-financial data sources are employed in a wide variety of use cases, from analyzing social and environmental impact of companies and their scenario alignment, to regulatory compliance analysis. In addition, new uses cases and data initiatives are constantly appearing, so that it is impossible to give a precise definition of what constitutes data for sustainable finance. Given this "Wild West" character of the domain, some classification initiatives have appeared, aiming to put some order in the field and direct the users to reliable data sources depending on their specific needs. These initiatives typically start with use cases for extra-financial data, and categorize datasets and data providers depending on the type of application they can be used for. We shall discuss two such initiatives, the ESG Data Cartography by the Louis Bachelier Institute (`https://institutlouisbachelier.org/en/esg-data-cartography`), and the NGFS Data Directory (`https://ngfs.dev.masdkp.io`).

The ILB ESG Data Cartography is a listing of extra-financial data providers. It identifies around 100 use cases (see Figure 11.1), and proposes a 4-level classification. The highest level (with broadest categories) includes alignment analysis, risk, impact, governance and sustainability tools. For each use case, the cartography presents a list of providers catering for this use case, along with a detailed description of each provider. For example, flood risk may be found under `risk->Physical risk->Acute risk->floods` and corresponds to a list of 22 data providers including both commercial ones and open data initiatives such as Climada and EM-DAT. On the other hand, Scope 3 carbon emissions are found under `Sustainability Tools->Footprints->Carbon footprint->Scope 3` (34 providers).

The NGFS data directory focuses on data items for the assessment of climate-related risks and opportunities in the financial sector, and was developed as part of the workstream of NGFS on bridging the data gaps (NGFS, 2021, 2022). The goal of this directory is to identify metrics and data items required for the main use cases,[1] and associate each data item to a data source. When no source can be found for a given data item, this constitutes a potential data gap. At the time of publication of the NGFS final report on data gaps (NGFS, 2022), the directory identified 1262 raw data items, 748 among which were linked to data sources.

11.2 Deep dive: climate data for physical risk assessment

Historically, risk assessment has relied mainly on some measure of damages, rather than on the climatic origin of the damage itself. As our exposure to various kinds of risks increases and since climate change is already changing the pattern of the physical climate risk in the world (Stocker et al., 2013), physical climate risk assessments will increasingly need to rely on climate data.

In the rest of this chapter we shall give a brief overview of the climate datasets useful for physical climate risk assessment studies. Economic, sociological and anthropological data on the assets and populations of interest, as well as the risk assessment methodology is not discussed here, but some of these issues are discussed elsewhere in this book.

[1] The top-level use case categories are: Financial stability monitoring, Climate-related financial disclosure, Stress test, Scenario analysis, Exposure quantification, Macroeconomic modeling, Investment and lending decisions, Economic growth analysis.

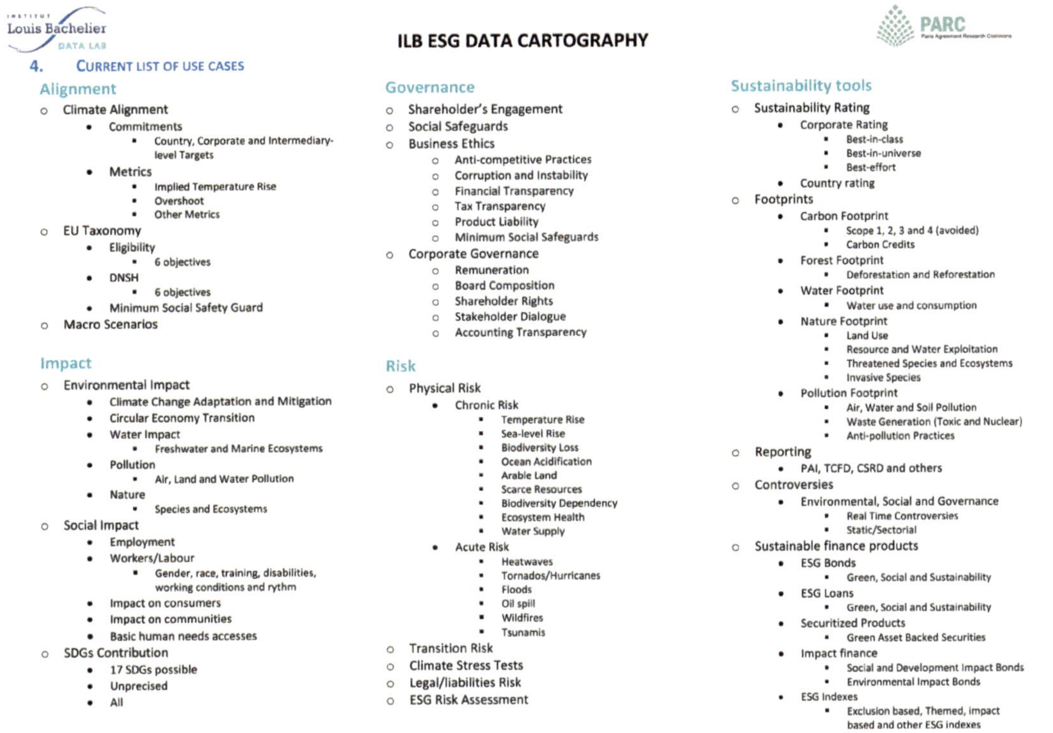

Institut Louis Bachelier DATA LAB — ILB ESG DATA CARTOGRAPHY — PARC Paris Agreement Research Commons

4. CURRENT LIST OF USE CASES

Alignment
- Climate Alignment
  - Commitments
    - Country, Corporate and Intermediary-level Targets
  - Metrics
    - Implied Temperature Rise
    - Overshoot
    - Other Metrics
- EU Taxonomy
  - Eligibility
    - 6 objectives
  - DNSH
    - 6 objectives
  - Minimum Social Safety Guard
- Macro Scenarios

Impact
- Environmental Impact
  - Climate Change Adaptation and Mitigation
  - Circular Economy Transition
  - Water Impact
    - Freshwater and Marine Ecosystems
  - Pollution
    - Air, Land and Water Pollution
  - Nature
    - Species and Ecosystems
- Social Impact
  - Employment
  - Workers/Labour
    - Gender, race, training, disabilities, working conditions and rythm
  - Impact on consumers
  - Impact on communities
  - Basic human needs accesses
- SDGs Contribution
  - 17 SDGs possible
  - Unprecised
  - All

Governance
- Shareholder's Engagement
- Social Safeguards
- Business Ethics
  - Anti-competitive Practices
  - Corruption and Instability
  - Financial Transparency
  - Tax Transparency
  - Product Liability
  - Minimum Social Safeguards
- Corporate Governance
  - Remuneration
  - Board Composition
  - Shareholder Rights
  - Stakeholder Dialogue
  - Accounting Transparency

Risk
- Physical Risk
  - Chronic Risk
    - Temperature Rise
    - Sea-level Rise
    - Biodiversity Loss
    - Ocean Acidification
    - Arable Land
    - Scarce Resources
    - Biodiversity Dependency
    - Ecosystem Health
    - Water Supply
  - Acute Risk
    - Heatwaves
    - Tornados/Hurricanes
    - Floods
    - Oil spill
    - Wildfires
    - Tsunamis
- Transition Risk
- Climate Stress Tests
- Legal/liabilities Risk
- ESG Risk Assessment

Sustainability tools
- Sustainability Rating
  - Corporate Rating
    - Best-in-class
    - Best-in-universe
    - Best-effort
  - Country rating
- Footprints
  - Carbon Footprint
    - Scope 1, 2, 3 and 4 (avoided)
    - Carbon Credits
  - Forest Footprint
    - Deforestation and Reforestation
  - Water Footprint
    - Water use and consumption
  - Nature Footprint
    - Land Use
    - Resource and Water Exploitation
    - Threatened Species and Ecosystems
    - Invasive Species
  - Pollution Footprint
    - Air, Water and Soil Pollution
    - Waste Generation (Toxic and Nuclear)
    - Anti-pollution Practices
- Reporting
  - PAI, TCFD, CSRD and others
- Controversies
  - Environmental, Social and Governance
    - Real Time Controversies
    - Static/Sectorial
- Sustainable finance products
  - ESG Bonds
    - Green, Social and Sustainability
  - ESG Loans
    - Green, Social and Sustainability
  - Securitized Products
    - Green Asset Backed Securities
  - Impact finance
    - Social and Development Impact Bonds
    - Environmental Impact Bonds
  - ESG Indexes
    - Exclusion based, Themed, impact based and other ESG indexes

FIGURE 11.1: Sustainable finance use cases of the ILB ESG Data Cartography (as of August 6, 2024).

After a quick primer on physical risk analysis, we introduce general guidelines to help select a dataset in Section 11.3. In Section 11.4, the main types of climate data sources are presented. Then, Section 11.5 introduces downscaling, an often unavoidable, yet sensitive, step to translate climate data into physical risk. Finally, Section 11.6 warns about model biases and uncertainty. A fully worked-out example related to heat waves in the Paris area is provided in Section 11.7. Section 11.7.2 gives a few links to web pages that may prove useful in finding an appropriate dataset. A glossary of main climate-related terms is provided in the Appendix for reader's convenience. Additional references and data sources are given throughout the text.

A primer on physical risk analysis Physical climate risks are the risks of destruction of assets and / or disruption of operations, trade routes, supply chains and markets by weather events related to climate change and climate variability. One may distinguish chronic climate risks related to slowly evolving phenomena, like sea-level rise, temperature rise, changing precipitation patterns etc., and the acute climate risks related to extreme weather phenomena (floods, extreme drought, heatwaves, wildfires, hurricanes), whose frequency and severity may increase with the onset of climate change.

The physical climate risk is a combination of three factors:

- The climate hazard: specific weather pattern or event whose frequency or severity may change;
- The exposure: presence of an asset or a system which may be affected by the climate hazard;

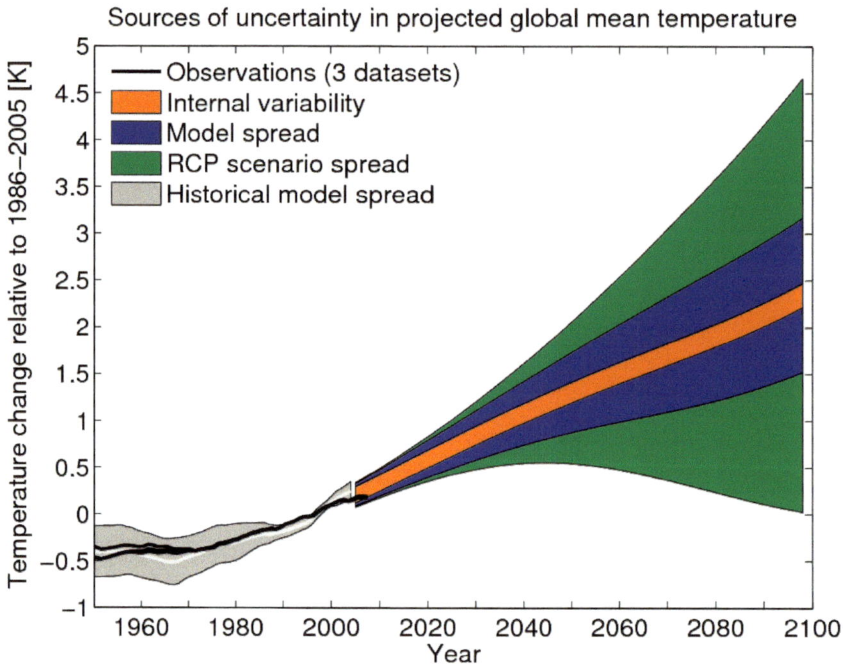

FIGURE 11.2: Uncertainty in global mean temperature projections. Source: IPCC AR5.

- The vulnerability: extent to which the asset or system may suffer from exposure to the climate hazard.

Climate data is necessary to evaluate the first factor, that is, the type, frequency and severity of expected climate hazards. Evaluating the two other factors requires access to asset level data: the geographical position and the specific characteristic of exposed assets or systems.

The analysis of physical risks typically starts with the evaluation of climate hazards expected in the geographical area of interest. The following step is to identify the potential impact of these hazards on the economic objects of interest: assets (factories, infrastructure, agriculture), transport links, supply links, markets, economic environment etc. These are then translated into estimates of balance sheets of specific companies and then possibly into the value of financial assets such as stocks and bonds.

The characterization of climate exposures requires different scales of climate information, from global to local level. Climate models provide a resolution of about 100km; they can be downscaled to lower resolutions either dynamically or statistically. Regional information on non-climatic components (relief etc.) should also be used.

Finally, one needs to keep in mind that there are large uncertainties both about the local response of the earth system to the radiative forcing associated with increased concentrations of greenhouse gases (GHGs) and about the capacity of the economic system to mitigate and adapt to changing climate. Internal variability of the climate and inter-model uncertainty are dominant in the short-term, while factors such as the level of GHG emissions are dominant at longer time scales, see Figure 11.2.

11.3 Choosing a dataset

Recognizing that no dataset is perfect and that risk estimates should always be taken with care, several characteristics can guide the choice of a dataset: the period covered, whether historic or future; the spatial coverage; the presence of biases; the spatio-temporal scale of interest; the climate variable; the type of hazard (chronic or acute).

Past data vs. forward looking data Assessing a risk based on past data relies on the assumption that the system may be considered statistically stationary on time scales relevant for the study and that the available data allows one to sample the underlying physical phenomena sufficiently well (in terms of frequency, duration and quality).

Considering that the state of the climate system varies on a continuous range of time scales, the statistical stationarity assumption may be violated due to (i) low-frequency variability — variability on time scales longer than the length of the time series from which risk is assessed, and sufficiently large (e.g., in terms of variance) to impact the risk — (ii) climate change — associated with changes in the forcing of the climate system, in particular, anthropogenic climate forcing through GHG emissions, land-use change, etc.

The climate change effects become material for horizons starting from 20–30 years ahead, but climate variability may manifest itself at shorter horizons (5–10 years).

Climate models vs. integrated assessment Climate models produce projections of future climate corresponding to specific scenarios of future economic activity (CO_2 emissions, pollutants, land use etc.). These scenarios are created with economic reasoning, however the feedbacks of the climate system onto the economic system (such as the increase in mitigation actions after the widespread adverse consequences of climate change start to be felt) are not modeled. By contrast, integrated assessment models aim to describe jointly the evolution of the climate system and the economic system, sometimes with two-way feedbacks. Unfortunately, most integrated assessment models to date only contain a very basic climate module which may not even be based on the equations of physics, and therefore are not realistic enough to be used for physical risk analysis. These models are therefore not considered in this chapter, which focuses on climate models/data sets.

The Climate Data Guide[2] provides a starting point to get a quick overview of available climate datasets and their characteristics. It is not, however, exhaustive.

11.4 Selected sources of climate data

In this section, we describe the main types of climate data that may be used for risk assessment. We have ordered them by increasing dependence on modeling. We also focus on the following non-exhaustive list of key climate variables related to hazards: temperature, drought, precipitation, sea level and winds.

[2]`https://climatedataguide.ucar.edu`

11.4.1 Observations

Observations may be used to assess past climate conditions and relate them to historical damages. Moreover, if the environmental sub-system of interest may be considered statistically stationary, observations may also be used to assess future risks.

Observational data being upstream of the modeling chain, they offer the advantage over reanalyses and projections to be less dependent on model assumptions and errors. In situ observations, such as from weather stations, tend to be as close as possible to the object studied, although observational devices need to be calibrated and never give a perfect representation of the measured variables. Teledetection from the ground or from space via satellites tend to offer a greater spatial (vertical and/or horizontal) coverage. However, they also tend to require more data processing and calibration than in situ observations.

Both in situ and teledetected data may be further processed to facilitate their use. For instance, NOAA's MLOST, NASA's GISTEMP, and the UK's HadCRUT datasets[3] provide a compilation of in situ observations on a regular grid. One should keep in mind that the strategy used to compile and interpolate the data impacts its quality.

In any case, the *representativity* of the data should always be questioned. For instance, wind observations from a weather station on grass land may not be representative of the wind conditions over a nearby forest. On the other hand, the wind averaged over a 100km large grid box may not be representative of the wind conditions at a particular location within the box if the wind conditions vary greatly within this box.

11.4.2 Reanalyses

Motivation

Purely observational datasets are limited by their spatial (horizontal and vertical) and temporal coverage, and by the number of observed variables. This is particularly true in sparsely populated, economically little developed regions. On the other hand, while numerical weather predictions give a complete description of the atmosphere, they diverge from the future state of the climate system within a few days, due to combined observational errors and irreducible chaotic dynamics, and to model errors.

To alleviate this issue, one approach could be to interpolate variables of interest. Whatever the interpolation method used, there is no reason for the interpolated data to represent well the state of the climate system. In particular, the interpolated state may not be consistent with the physical laws governing the climate system. For instance, the important role of topography or of land-ocean contrasts may be taken into account.

Another approach that has proved particularly efficient for numerical weather prediction and for climate studies is to produce analyses by assimilating observational data into simulations. In brief, given present observations, data assimilation seeks to modify the past state of a climate model simulation in order for the present climate state predicted by model integration to best reproduce observations.

Data assimilation can thus be seen as a way to seamlessly interpolate observational data that is consistent with physical equations as encoded in a climate model. Yet, due to the sparsity of observations and to the strong dependence of analyses on model assumptions, data assimilation is better seen as a methodology to constrain model simulations with observations.

---

[3]National Center for Atmospheric Research Staff (Eds). "The Climate Data Guide: Global Temperature Data Sets: Overview & Comparison Table." Retrieved from `https://climatedataguide.ucar.edu/climate-data/global-temperature-data-sets-overview-comparison-table`. Last retrieved on August 6, 2024.

As a climate model, or a data assimilation method, is improved and as more observations become available, past analyses become deprecated. For that reason, new analyses are produced over the past record with the new version of the climate model and with the integration of new observational datasets. The resulting dataset is thus called a *reanalysis*. Another advantage of reanalysis datasets is that they provide a number of variables represented in, or diagnosed from, the model that may be difficult to observe.

The quality of a reanalysis strongly depends on the density and quality of observations injected in the data assimilation system. The sparser the observations at a given location, the less constrained is the model used to produce the reanalysis. This means that the data is more prone to model errors and more chaotic. This last point deserves attention. Indeed, risk assessment may require climate data that is synchronized with observed damages. However, integrating sparse observations in a reanalysis tends to constrain large scale dynamics only. This means that, while the statistical properties of small-scale processes may be well resolved by a reanalysis, chaotic dynamics may prevent considering the state of the reanalysis at a particular moment to be representative of the state of the climate system at that moment and at these scales.

Available reanalyses

The reanalyses.org (`https://reanalyses.org`) project provides an overview of most available reanalyses. Atmospheric reanalyses may be divided into lower-resolution global reanalyses and higher-resolution regional reanalyses. State-of-the-art global reanalyses include ECMWF's ERA5 (Copernicus Climate Change Service (C3S), 2017), NCEP's CFSR (Saha et al., 2010), NASA's MERRA-2 (Gelaro et al., 2017) and JMA's JRA-55 (Kobayashi et al., 2015). A detailed comparison of the design and performance of these reanalyses is given by Fujiwara et al. (2017) and Long et al. (2017). See also Buizza et al. (2005) and Wang et al. (2011).

Let us also mention reanalyses for the entire 20th century, such as ECMWF's ERA-20C (Poli et al., 2013), although the latter should be used with great care, since the observational constraints weaken as one goes further back in time (this is particularly true before the early 80s, the beginning of the satellite observation era). As an example of regional reanalysis, let us mention the COSMO reanalyses over Europe (Bollmeyer et al., 2015; Wahl et al., 2017). Such regional reanalyses may improve the representation of small scales — which may in turn impact large scales — relevant for impact studies.

11.4.3 21st century simulations

Motivation

Observations and reanalyses only provide information on past climate conditions. Due to the chaotic nature of the climate system, the exact future states of the climate may not be predicted. Much can be said, however, about the future evolution of statistics of the climate system, and, in particular, about physical climate risk. Due to the impact of low-frequency variability and changes in the forcing due to anthropogenic climate change, statistics of some key climate variables have changed and are expected to change further (Taylor, Stouffer, and Meehl, 2012). This is particularly true regarding surface temperature and sea level.

While low-frequency variability and anthropogenic climate change are both sources of non-stationarity, the two do not have the same implications, as far as risk assessment is concerned. The long term — say over a period of several decades — response of the statistics of the climate system, is mainly constrained by changes in the radiative forcing associated with increasing greenhouse gases. Simulating this response thus depends first on the socio-economic scenario followed by current and future emitting countries, rather than on the initial state of the climate system. Such simulations are referred to as *projections*. On the other hand, decadal predictions is an emerging field that attempts to predict the slow evolution of the climate system associated with low-frequency variability. In this case, information about the initial state of the climate system is used, although most of it is lost. Whether this information is useful to impact studies is still being investigated and is not further discussed in this chapter.

Projections and scenarios

Global 21st century projections are structured by the WCRP Coupled[4] Model Intercomparison Project (CMIP). CMIP's 5th phase (Taylor, Stouffer, and Meehl, 2012) contributed to IPCC's 5th Assessment Report (AR5), while CMIP's 6th phase (Eyring et al., 2016) contributed to IPCC's 6th Assessment Report (AR6). As the resolution of models increases, together with the number of processes represented, projections are expected to improve from one phase to the next. As more data becomes available, processing this data for impact studies may, however, prove challenging.

The main source of uncertainty in projections is the scenario used to force the models. Since the 5th phase of CMIP, Radiative Concentration Pathways (RCPs) are used to impose changing greenhouse gases concentrations in climate models. It is these changing concentrations that are responsible for changes in the radiative forcing eventually resulting in global warming. In parallel, different socio-economic scenarios leading to a particular RCP, so-called Shared Socio-Economic Pathways (O'Neill et al., 2014) were developed by economists and social scientists.

In addition to scenario-related uncertainties, model errors as well as spread due to intrinsic variability also impact the projections, see Figure 11.2. Since all models have errors and since projections are not constrained by observations, it is important to (i) be aware of model biases and (ii) use multi-model ensembles. Regarding this last point, it is indeed known that aggregating information from several models allows one to improve projection skills compared to that of a particular model. Moreover, multi-model ensembles allow one to partially estimate model errors and the spread due to intrinsic variability, see Figure 11.3.

CMIP data may be found on the ESGF data portal[5].

Regional projections

Computational resources limit the resolution of global climate projections. For these reasons, global projections may not be suited for risk assessment. This is why Regional Climate Models (RCMs) are sometimes used. The latter resolve a limited area, allowing to increase the resolution of the model and the number of resolved processes, for the same computational resource. To produce regional projections, RCMs are forced at their boundaries[6] by projections from global models.

Just as the CMIP project for global projections, the CORDEX project organizes regional climate projection efforts. It defines standard domains over multiple regions as well as standard variables and entry points. Evaluation of models participating in the CORDEX project is provided by the RCMES.[7] For more information on evaluation and decision

[4]While weather models tend to represent dynamically the atmosphere only, the long-term evolution of the climate system strongly depends on the interaction of the atmosphere with other components of the Earth system. Thus, climate projections rely on general circulation models coupling the atmosphere with oceans, sea ice, land surfaces, bio-geo-chemistry, etc.

[5]`https://aims2.llnl.gov/search/cmip6/`

[6]Note that the boundary of the domain of RCMs may be source of spurious instabilities. Modelers attempt to design RCMs so as to dump these instabilities, but some may persist and lead to spurious results in impact studies.

[7]`https://rcmes.jpl.nasa.gov/content/cordex-evaluation`

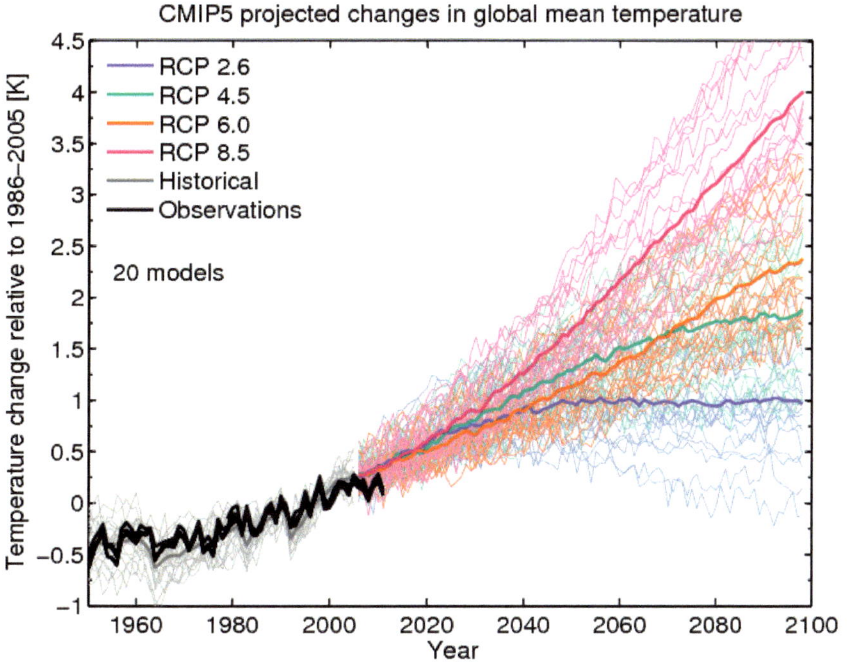

FIGURE 11.3: Changes in global temperature projected with 20 CMIP5 models for different RCP scenarios. Source: IPCC AR5. Note the bias between model averages and observations over the period where the two intersect, and the variability of projections by different models within the same RCP scenario.

support, see Whitehall et al. (2012). CORDEX data may also be found on the ESGF data portal.

11.4.4 Indices

Some hazards may not be directly characterized by a particular climate variable. For instance the severity of a drought depends, in particular, on precipitation, temperature and on soil parameters and variables. One option is to define an index translating and/or aggregating information into a hazard intensity.

In the case of droughts, the *Standardized Precipitation Index* (SPI) quantifies the number of standard deviations by which the observed anomaly in precipitation deviates from the long-term mean. The *Standardized Precipitation Evapotranspiration Index* (SPEI) extends the SPI to take into account potential evapotranspiration in order to capture the main impact of increased temperatures on water demand[8]. Yet another index, the *self-calibrating Palmer Drought Severity Index* (sc-PDSI) is also sometimes used. The relevance of a particular index depends on the hazard, geographical location, local environmental conditions and index design.

[8]Vicente-Serrano, Sergio M. & National Center for Atmospheric Research Staff (Eds). Last modified 18 Jul 2015. "The Climate Data Guide: Standardized Precipitation Evapotranspiration Index (SPEI)." Retrieved from `https://climatedataguide.ucar.edu/climate-data/standardized-precipitation-evapotranspiration-index-spei`.

Last, `climdex.org` collects a number of indices of extreme events computed from different global datasets. While such tools can facilitate physical climate risk assessment, care should however be taken to evaluate the local representativity of global indices.

11.5 Dynamical and statistical downscaling

The objects of risk assessment, and of impact studies in general, tend to depend on climatic conditions at relatively small scale, that is, the scale of a field, a city, a plant, a building, a river, and so on. On the other hand, global projections and reanalyzes, with a resolution of about a degree or half a degree, tend to properly resolve climatic variations around a few degrees (i.e., about 110 km at the equator, 60 km at a latitude of 45°). To pass from the scale resolved by such global models to the scale of interest for impact studies several methodologies, known as *downscaling* are often used. These methodologies all rely on the addition of small-scale information in some way or another. Moreover, some approaches attempt to downscale climatic information into a small scale climate variable or index, while other directly downscale climatic information into some measure of impact or damage.

Downscaling may be divided in two broad categories: *dynamical* and *statistical* downscaling (for an introduction to the topic, see Benestad (2016)). Dynamical downscaling relies on the resolution of earth science equations at small scales, via the integration of a regional, or limited-area, climate model. This the case of the COSMO reanalysis and of the CORDEX simulations discussed in Sections 11.4.2 and 11.4.3, respectively, whereby scales of a few kilometers or tens of kilometers are reached. Even smaller scales, about a 100 m, can be reached with the use of computational fluid dynamics models. At such scales, dynamics are turbulent, so that only statistical information is relevant.

Dynamical downscaling presents the advantage of relying on fundamental equations and on integrating information at small scales about topography, land use, etc. Because of this, one expects dynamical downscaling approaches to be relatively robust to parameter or forcing changes, although some parameters of the model are still calibrated on the historical period. It is known to add value to global climate data. However, dynamical downscaling is also a computationally intensive exercise. In addition, the information downscaled by RCMs is still climatic, so that integrated approaches are needed to translate this small-scale information into an impact.

Statistical downscaling provides an alternative that does not rely on earth science laws but rather on local information and statistical learning (Hastie, Tibshirani, and Friedman, 2009). Take as an example the assessment of heat wave occurrence (a statistic) in the 21st century in the city center of Grenoble, France. We know that historical temperatures in the city center of Grenoble are not well reproduced by global reanalyses. Regional reanalyses tend to improve these results by better resolving the impact of the Alps surrounding Grenoble, but are still not satisfactory: the urban heat island is not well resolved. Thus, not only reanalyses cannot be used here to project the increase in temperatures in Grenoble associated with global warming, but global climate projections may not be expected to resolve well local temperature distributions. Assume, however, that a number of temperature sensors are scattered in Grenoble so that time series of temperatures there for the last few years are available. This information on the local climatic conditions could be combined with information about global warming provided by projections to estimate future changes in the temperature distribution in Grenoble, and in particular the occurrence of heat waves. For instance, perhaps the most basic approach would be to estimate a historical temperature distribution from the local measurements, compute the difference between the mean of this distribution and the temperature predicted by the global model during the same period, and predict future temperature distributions by shifting the local temperature distribution to follow the changes in the mean temperature projected by the global model. This particular approach is known as bias correction.

The design and choice of a particular downscaling method is an art in itself. The particular impact considered, its location, the data and resources available, need to be considered. Thus, Estrada et al. (2013) advise against applying downscaling automatically. The VALUE project Maraun et al. (2015) provides a framework to assess the skills of different downscaling approaches which has been applied other Europe by Gutiérrez et al. (2018). In particular, downscaling studies often rely on reanalyses, the skills of which have a strong impact on the result of downscaling (Brands et al., 2011).

11.6 Biases and uncertainty in climate data

The goal of this section is not to give an overview of all known biases in climate data, as their relevance depends on the object of the risk assessment study. Rather, we would like to warn that the availability of a large amount of climate data should not lead to overconfidence in their use for risk assessment. While global mean changes in temperature are relatively well resolved by reanalyses and projections, this is not the case of all variables and statistics:

- Some variables are more sensitive to measurement errors and modeling assumptions. For instance, precipitation directly depends on cloud microphysics which still constitutes one of the main weaknesses of earth system models (Stocker et al., 2013).
- The finer the spatial scale, the more biased the models. This is true for both reanalyses and projections.
- Statistics giving more weight to tails of distributions, such as extreme values, tend to be more biased than, for instance, averages.

It is thus essential, in risk assessments relying on climate data, to estimate errors originating from the climate data itself. Indeed, it is often found in impact studies that a dominant source of errors comes from the climate data. To compute a first, yet incomplete, estimate of errors coming from the climate data, multiple data sources may be used. In the case of projections, multiple CMIP and CORDEX models may be used. The number of reanalyses is not as large, but several of them are available, and errors are reduced thanks to observational constraints.

When dealing with 21st century projections, the main source of uncertainty stems from the socio-economic scenario that the economic actors with the largest footprint will follow. For that reason, several typical scenarios need to be considered. As discussed in Section 11.4.3, the CMIP and CORDEX databases collect simulations from different models that are forced by different RCP scenarios. The latter may be used to prepare assessments or indicators corresponding to different socio-economic scenarios.

11.7 Example: heatwaves in the Paris area

To illustrate the key messages of this report, we give the following example. Our objective is to compute the evolution of the number of days of heatwave over the end of the 20th century and over the 21st century at a given location and from various climate data-sources. This example is described in Section 11.7.1, while the methodology is further detailed in Section 11.7.2.

11.7.1 Description and interpretation

A heatwave is a type of extreme event increasingly impacting societies. The risk associated with a heatwave depends on the vulnerability of the population, such as its age, its access to water, etc. Here, we instead focus on the intensity of a heatwave as a climate event and follow the definition given in Ouzeau et al. (2015), Section 5.3.1:

Definition 1 *A heatwave is defined as an abnormally hot period lasting more than five consecutive days. Hot days are defined as days for which the maximum daily temperature is* 5 °C *larger than a climatological reference value, counting only days belonging to a series of more than five consecutive hot days. To obtain this reference value for each day of the year, one computes the mean annual-cycle of the daily maximum temperature simulated for the reference period (1976–2005) [1980–2010, here], applying a running average over five days to this annual cycle.*

The SIRTA observational platform provides hourly temperature observations from 2003 to 2019 (Haeffelin et al., 2005). We choose to study heatwaves at the location of the platform, 48.718°N, 2.202°E (20 km south of Paris). This data is compared to a global reanalysis and to Regional Climate Model (RCM) runs. The MERRA-2 reanalysis Gelaro et al. (2017) covers the 1980–2019 period at a resolution of about half a degree. The RCM runs from the Med-CORDEX exercise Ruti et al. (2016) are concatenated to cover both the historical period (1950–2006) and the 21st century (2006–2100), using projections for the RCP 8.5 scenario (the warmest, business-as-usual) at a little less than half a degree of resolution. Two runs from the GUF and the CMCC institutes are selected. Records from the nearest grid-point of these gridded datasets to the SIRTA platform are analyzed. MERRA-2's nearest grid point is at 48.500°N, 2.500°E (about 33 km from SIRTA) , while GUF's and CMCC's is at 48.524°N, 2.094°E (about 23 km from SIRTA).

Four records of daily-maximum temperature are obtained from these datasets. In addition, as a simple example of statistical downscaling, we remove from the reanalysis and the two RCM runs the bias with respect to the SIRTA data. These two sets of time series are represented in Figure 11.4a and 11.4b. We then compute heatwave indices for each run and represent them in Figure 11.4d and 11.4c. Thick lines correspond to smoothed versions of the yearly time-series (thin lines) using a rolling average over 20 years. Note that the rolling average uses fewer points when less than 20 years are available. For instance, at the middle of the 12-years-long SIRTA time-series, only 12 points are available and used.

We first observe that a strong bias (half a degree to several degrees) exists between the SIRTA observations (black) and both the reanalysis (gray) and the RCM runs (blue and orange). This bias may be explained by modeling errors, a poor representativity of the selected grid-box of the gridded datasets with respect to the SIRTA station, or by (less likely) observational errors in the reanalysis and the in situ observations. The RCM runs are also significantly different from each other. This shows the importance of comparing

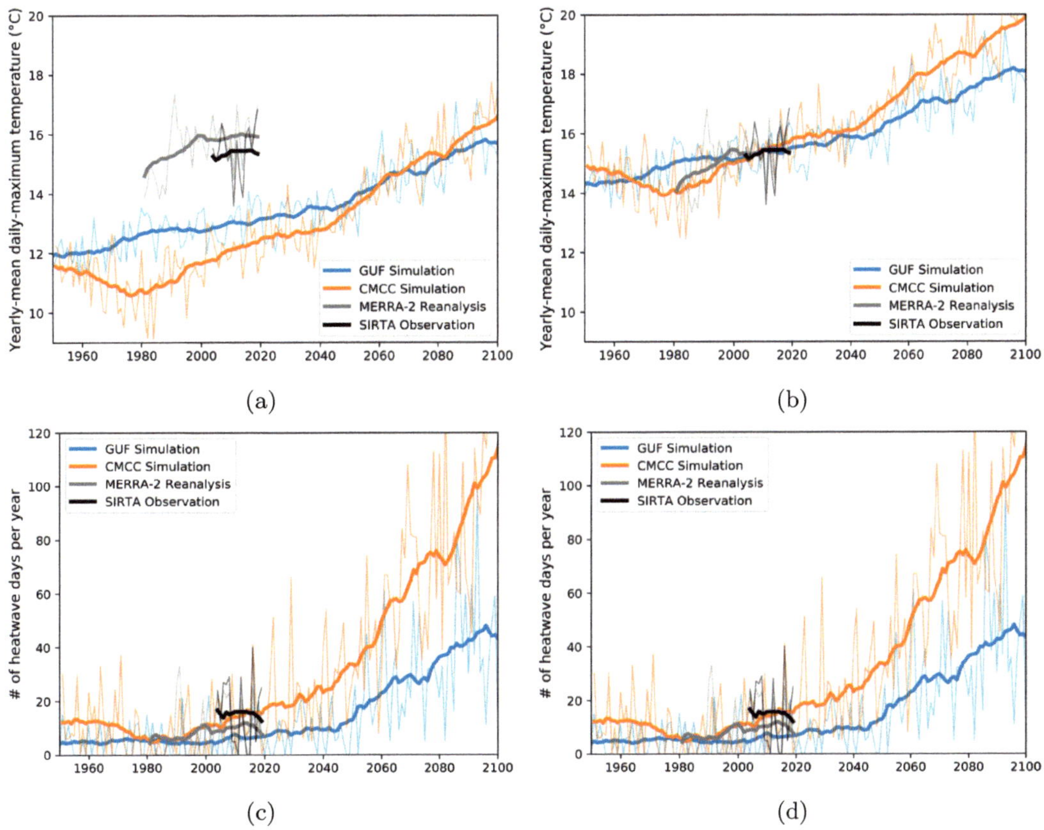

FIGURE 11.4: Yearly-mean daily-maximum temperature (top) and number of heatwave days (bottom) without (left) and with (right) bias-correction. Thick lines correspond to smoothed versions of the thin lines using a rolling average over 20 years. Black lines are obtained from the SIRTA observational data from 2003 to 2019. Gray lines are obtained from the MERRA-2 reanalysis data from 1980 to 2019. Blue lines are obtained from the concatenation of the historical and RCP 8.5 CORDEX runs by the GUF institute. Orange lines are obtained from the concatenation of the historical and RCP 8.5 CORDEX runs by the CMCC institute. The number of heatwave days are computed following the methodology described in Ouzeau et al. (2015), Section 5.3.1, with 1980–2010 as reference period. Since the SIRTA record is too short, we use the bias-corrected MERRA-2 data as reference to compute the number of heatwave days in the SIRTA record. The bias correction is simply performed by removing from all non-SIRTA records their mean over the SIRTA period and adding instead the mean of the SIRTA record.

multiple data sources and validating estimations against observation over some available period.

We also note from the erratic behavior of the thin curves compared to the thick, low-pass filtered, curves that there is a strong interannual variability in the maximum temperature. Zooming on the period covered by the SIRTA record, Figure 11.5, we can see that the correlation between MERRA-2 temperatures and the SIRTA temperatures is stronger than that between the SIRTA temperatures and the RCM runs. Considering the chaotic nature of the climate system, this may be understood from the fact that, contrary to the RCM runs, the reanalysis is constrained by observations, and thus more likely to follow the observed

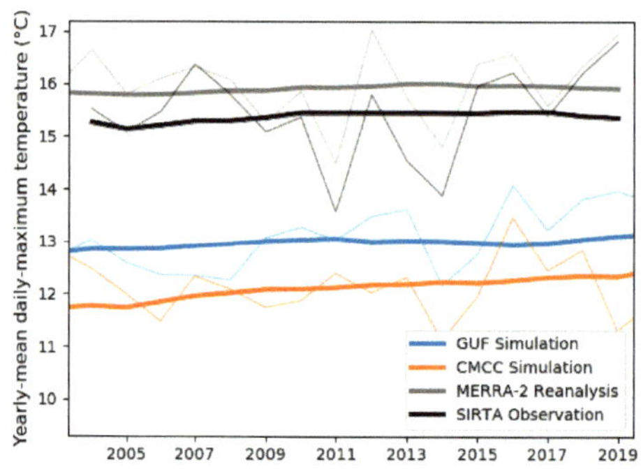

FIGURE 11.5: Zoom of Figure 11.4a.

interannual variability. On the other hand, by design, the reanalysis does not provide projections for the 21st century.

The bias correction, Figure 11.4b, allows us to reduce the gap between the observations and the other records. However, we are not guaranteed that this bias is stationary over time. For instance, during the 21st century, a warmer climate may induce changes in physical processes which could result in weaker or stronger biases.

Let us now focus on the heatwave index, Figure 11.4d and 11.4c. We can first note that the observational record (black) appears to be too short (2002–2019) to observe any trend in the number of heatwave days with this methodology. From 1980 to 2019, a small trend is visible in the reanalysis (gray) and the RCM runs (blue and orange), although its significance should be properly tested. A very clear trend is, however, visible over the 21st century for both RCM runs, although the magnitude of this trend differs significantly (by about 70 days).

The heatwave index is less sensitive to biases, as it is defined as a difference with respect to a reference annual cycle computed from the same dataset (except for the SIRTA heatwave index, which is computed using the bias-corrected MERRA-2 annual cycle).

Finally, note that we have focused here on the estimation of the number of heatwave days from different climate data to illustrate basic methodologies and their pitfalls. It remains then to estimate the actual damages associated with these climate events.

11.7.2 Methodology

We now explain in more details the methodology used to obtain the results of Section 11.7.1. The in-situ observations, the reanalysis and the projections are obtained from the SIRTA observatory, NASA's MERRA-2 and the Med-CORDEX initiative.

SIRTA

The SIRTA database is accessible free of cost for public research and teaching applications[9]. We download the SIRTA (Haeffelin et al., 2005) temperature record via File Transfer Protocol (FTP) from SIRTA's data-download page (`https://sirta.ipsl.fr/data-overview/`), selecting *atmospheric state/surface meteorology/Air temperature/Meteorological variables with LMD station in zone 2* for the 2003/01/03–2019/01/01 period. The data is sampled every minute.

[9] See SIRTA's Data Policy (`https://sirta.ipsl.fr/data_policy.html`).

MERRA-2

NASA promotes the full and open sharing of all data with the research and applications communities, private industry, academia, and the general public. The MERRA-2 data is accessible from NASA's GES DISC (`https://disc.gsfc.nasa.gov/`), after registration. To get the daily maximum temperature, the single-level daily-statistics dataset should be selected.

Med-CORDEX

Data from the Med-CORDEX project available on this server is provided without charge and may be used for research and education only. Commercial use of the data is not permitted. The Med-CORDEX data is accessible through the Med-CORDEX FTP (`ftp://www.medcordex.eu/`). For instance, the daily-maximum temperatures from the GUF historical run and RCP8.5 scenario were respectively downloaded from `ftp://www.medcordex.eu/MED-44/GUF/MPI-ESM-LR/historical/r1i1p1/GUF-CCLM4-8-18/v1/day/tasmax/` and `ftp://www.medcordex.eu/MED-44/GUF/MPI-ESM-LR/rcp85/r1i1p1/GUF-CCLM4-8-18/v1/day/tasmax/`.

Bias correction

The bias correction adjusts for differences in the mean between the SIRTA maximum-temperature record and that of the reanalysis and RCM records. To do so, we first select the period over which both records overlap. The mean over this period of the record to be corrected is removed from it and we add instead the mean of the SIRTA record over the same period. As a result, the mean of the corrected record is the same as the mean of the SIRTA record over the overlapping period.

Further reading

- The Climate Data Guide (`https://climatedataguide.ucar.edu/`) describes different sources of climate data.
- Reanalyses.org project (`https://reanalyses.org`) focuses on reanalyses.
- The SPARC Reanalysis Intercomparison Project (S-RIP) `https://s-rip.ees.hokudai.ac.jp/pubs/index.html` provides data from different reanalyses on common grids.
- Data used for downscaling from the VALUE project `http://www.value-cost.eu/data`.
- The S-RIP report Fujiwara et al. (2017) gives an overview of the different reanalysis systems.
- Chapter 12 of the Working Group I on *The Physical Science Basis*[10] contribution to the 6th Assessment Report of the IPCC and Chapter 16 of the Working Group II on *Impacts, Adaptation and Vulnerability*[11] cover climate risk assessment.

[10] `https://www.ipcc.ch/report/sixth-assessment-report-working-group-i/`

[11] `https://www.ipcc.ch/report/sixth-assessment-report-working-group-ii/`

Acknowledgment

The authors would like to acknowledge SIRTA for providing the temperature data used in this study.

Glossary

Climate model	A physical model for the evolution of the Earth climate in the long term (100+ years). Climate models are used to make climate projections.
Climate variability	Natural variability of the Earth climate at the scales longer than one year.
CMIP	Coupled Model Intercomparison Project, an exercise led by IPCC wherein different climate models are run under the same conditions to produce multiple climate scenarios which may be compared to evaluate the uncertainty of future climate.
Downscaling	A procedure to refine the scale of climate models down to that of local phenomena of interest.
ECMWF	European Centre for Medium-Range Weather Forecasts
Integrated assessment model	A model which aims to describe the joint evolution of the economy and the climate system.
Physical climate risk	Risk of destruction of assets / disruption of systems by weather events related to climate change and climate variability.
Reanalysis	In a reanalysis exercise, a climatological model is run over a long period of time in the past, assimilating all available observations as model constraints. This allows to interpolate available observations over regular time and space domain. The standard practice in climatology, which we follow in this paper, is to replace historical data with reanalysis data.
RCP	Representative Concentration Pathway, a scenario of greenhouse gas emissions used to parameterize climate models

References

Benestad, R., 2016, Downscaling Climate Information, *Oxford Research Encyclopedia of Climate Science*.

Bollmeyer, C., J. D. Keller, C. Ohlwein, S. Wahl, S. Crewell, P. Friederichs, A. Hense, J. Keune, S. Kneifel, I. Pscheidt, S. Redl, and S. Steinke, 2015, Towards a high-resolution regional reanalysis for the European CORDEX domain, *Quarterly Journal of the Royal Meteorological Society* 141, 1–15.

Brands, S., J. M. Gutiérrez, S. Herrera, and A. S. Cofiño, 2011, On the Use of Reanalysis Data for Downscaling, *Journal of Climate* 25, 2517–2526.

Buizza, R., P. L. Houtekamer, G. Pellerin, Z. Toth, Y. Zhu, and M. Wei, 2005, A Comparison of the ECMWF, MSC, and NCEP Global Ensemble Prediction Systems, *Monthly Weather Review* 133, 1076–1097.

Copernicus Climate Change Service (C3S), 2017, ERA5: Fifth generation of ECMWF atmospheric reanalyses of the global climate, *Copernicus Climate Change Service Climate Data Store (CDS)*.

Estrada, F., V. M. Guerrero, C. Gay-García, and B. Martínez-López, 2013, A cautionary note on automated statistical downscaling methods for climate change, *Climatic Change* 120, 263–276.

Eyring, V., S. Bony, G. A. Meehl, C. A. Senior, B. Stevens, R. J. Stouffer, and K. E. Taylor, 2016, Overview of the Coupled Model Intercomparison Project Phase 6 (CMIP6) experimental design and organization, *Geoscientific Model Development* 9, 1937–1958.

Fujiwara, M., J. S. Wright, G. L. Manney, L. J. Gray, J. Anstey, T. Birner, S. Davis, E. P. Gerber, V. L. Harvey, M. I. Hegglin, C. R. Homeyer, J. A. Knox, K. Krüger, A. Lambert, C. S. Long, P. Martineau, A. Molod, B. M. Monge-Sanz, M. L. Santee, S. Tegtmeier, S. Chabrillat, D. G. H. Tan, D. R. Jackson, S. Polavarapu, G. P. Compo, R. Dragani, W. Ebisuzaki, Y. Harada, C. Kobayashi, W. McCarty, K. Onogi, S. Pawson, A. Simmons, K. Wargan, J. S. Whitaker, and C.-Z. Zou, 2017, Introduction to the SPARC Reanalysis Intercomparison Project (S-RIP) and overview of the reanalysis systems, *Atmospheric Chemistry and Physics* 17, 1417–1452.

Gelaro, R., W. McCarty, M. J. Suárez, R. Todling, A. Molod, L. Takacs, C. A. Randles, A. Darmenov, M. G. Bosilovich, R. Reichle, K. Wargan, L. Coy, R. Cullather, C. Draper, S. Akella, V. Buchard, A. Conaty, A. M. da Silva, W. Gu, G.-K. Kim, R. Koster, R. Lucchesi, D. Merkova, J. E. Nielsen, G. Partyka, S. Pawson, W. Putman, M. Rienecker, S. D. Schubert, M. Sienkiewicz, and B. Zhao, 2017, The Modern-Era Retrospective Analysis for Research and Applications, Version 2 (MERRA-2), *Journal of Climate* 30, 5419–5454.

Gutiérrez, J. M., D. Maraun, M. Widmann, R. Huth, E. Hertig, R. Benestad, O. Roessler, J. Wibig, R. Wilcke, S. Kotlarski, D. S. Martín, S. Herrera, J. Bedia, A. Casanueva, R. Manzanas, M. Iturbide, M. Vrac, M. Dubrovsky, J. Ribalaygua, J. Pórtoles, O. Räty, J. Räisänen, B. Hingray, D. Raynaud, M. J. Casado, P. Ramos, T. Zerenner, M. Turco, T. Bosshard, P. Štěpánek, J. Bartholy, R. Pongracz, D. E. Keller, A. M. Fischer, R. M. Cardoso, P. M. M. Soares, B. Czernecki, and C. Pagé, 2018, An intercomparison of a large ensemble of statistical downscaling methods over Europe: Results from the VALUE perfect predictor cross-validation experiment, *International Journal of Climatology* 1–36.

Haeffelin, M., L. Barthès, O. Bock, C. Boitel, S. Bony, D. Bouniol, H. Chepfer, M. Chiriaco, J. Cuesta, J. Delanoë, P. Drobinski, J.-L. Dufresne, C. Flamant, M. Grall, A. Hodzic, F. Hourdin, F. Lapouge, Y. Lemaître, A. Mathieu, Y. Morille, C. Naud, V. Noël, W. O'Hirok, J. Pelon, C. Pietras, A. Protat, B. Romand, G. Scialom, and R. Vautard, 2005, SIRTA, a ground-based atmospheric observatory for cloud and aerosol research, *Annales Geophysicae* 23, 253–275.

Hastie, T., R. Tibshirani, and J. Friedman, 2009, *The Elements of Statistical Learning*, second edition (Springer, New York).

Kobayashi, S., Y. Ota, Y. Harada, A. Ebita, M. Moriya, H. Onoda, K. Onogi, H. Kamahori, C. Kobayashi, H. Endo, K. Miyaoka, and K. Takahashi, 2015, The JRA-55 Reanalysis: General Specifications and Basic Characteristics, *Journal of the Meteorological Society of Japan. Ser. II* 93, 5–48.

Long, C. S., M. Fujiwara, S. Davis, D. M. Mitchell, and C. J. Wright, 2017, Climatology and interannual variability of dynamic variables in multiple reanalyses evaluated by the SPARC Reanalysis Intercomparison Project (S-RIP), *Atmospheric Chemistry and Physics* 17, 14593–14629.

Maraun, D., M. Widmann, J. M. Gutiérrez, S. Kotlarski, R. E. Chandler, E. Hertig, J. Wibig, R. Huth, and R. A. I. Wilcke, 2015, VALUE: A framework to validate downscaling approaches for climate change studies, *Earth's Future* 3, 1–14.

NGFS, 2021, Progress report on bridging data gaps, Technical report, Network for Greening the Financial System.

NGFS, 2022, Final report on bridging data gaps, Technical report, Network for Greening the Financial System.

Ouzeau, G., M. Déqué, M. Jouini, S. Planton, R. Vautard, and J. Jouzel, 2015, *Le Climat de La France Au XXIe Siècle: Scénarios Régionalisés : Édition 2014 Pour La Métropole et Les Régions d'outre-Mer*, volume 4 (Ministère de l'Écologie, du Développement Durable et de l'Énergie).

O'Neill, B. C., E. Kriegler, K. Riahi, K. L. Ebi, S. Hallegatte, T. R. Carter, R. Mathur, and D. P. Van Vuuren, 2014, A new scenario framework for climate change research: the concept of shared socioeconomic pathways, *Climatic Change* 122, 387–400.

Poli, P., H. Hersbach, D. Tan, D. Dee, J.-N. Thépaut, A. Simmons, C. Peubey, P. Laloyaux, T. Komori, P. Berrisford, R. Dragani, Y. Trémolet, E. Hólm, M. Bonavita, L. Isaksen, and M. Fisher, 2013, The data assimilation system and initial performance evaluation of the ECMWF pilot reanalysis of the 20th-century assimilating surface observations only (ERA-20C), Technical report, ECMWF, Shinfield Park, Reading.

Ruti, P. M., S. Somot, F. Giorgi, C. Dubois, E. Flaounas, A. Obermann, A. Dell'Aquila, G. Pisacane, A. Harzallah, E. Lombardi, B. Ahrens, N. Akhtar, A. Alias, T. Arsouze, R. Aznar, S. Bastin, J. Bartholy, K. Béranger, J. Beuvier, S. Bouffies-Cloché, J. Brauch, W. Cabos, S. Calmanti, J. C. Calvet, A. Carillo, D. Conte, E. Coppola, V. Djurdjevic, P. Drobinski, A. Elizalde-Arellano, M. Gaertner, P. Galàn, C. Gallardo, S. Gualdi, M. Goncalves, O. Jorba, G. Jordà, B. L'Heveder, C. Lebeaupin-Brossier, L. Li, G. Liguori, P. Lionello, D. Maciàs, P. Nabat, B. Önol, B. Raikovic, K. Ramage, F. Sevault, G. Sannino, M. V. Struglia, A. Sanna, C. Torma, and V. Vervatis, 2016, Med-CORDEX initiative for Mediterranean climate studies, *Bulletin of the American Meteorological Society* 97, 1187–1208.

Saha, S., Moorthi, S., Pan, H.-L., Wu, X., Wang, J., Nadiga, S., Tripp, P., Kistler, R., Woollen, J., Behringer, D., Liu, H., Stokes, D., Grumbine, R., Gayno, G., Wang, J., Hou, Y.-T., Chuang, H.-y., Juang, H.-M. H., Sela, J., Iredell, M., Treadon, R., Kleist, D., Van Delst, P., Keyser, D., Derber, J., Ek, M., Meng, J., Wei, H., Yang, R., Lord, S., and van den Dool, H., Kumar, A., Wang, W., Long, C., Chelliah, M., Xue, Y., Huang, B., Schemm, J.-K., Ebisuzaki, W., Lin, R., Xie, P., Chen, M., Zhou, S., Higgins, W., Zou, C.-Z., Liu, Q., Chen, Y., Han, Y., Cucurull, L., Reynolds, R.W., Rutledge, G., and Goldberg, M., 2010, The NCEP Climate Forecast System Reanalysis, *Bulletin of the American Meteorological Society* 91, 1015–1058.

Stocker, T., D. Qin, G.-K. Plattner, M. Tignor, S. Allen, J. Boschung, A. Nauels, Y. Xia, V. Bex, and P. Midgley, eds., 2013, *IPCC, 2013: Climate Change 2013: The Physical Science Basis. Contribution of Working Group I to the Fifth Assessment Report of the Intergovernmental Panel on Climate Change* (Cambridge University Press, Cambridge, United Kingdom and New York, NY, USA).

Taylor, K. E. K., R. R. J. Stouffer, and G. G. A. Meehl, 2012, An overview of CMIP5 and the experiment design, *Bulletin of the American Meteorological Society* 93, 485–498.

Wahl, S., C. Bollmeyer, S. Crewell, C. Figura, P. Friederichs, A. Hense, J. D. Keller, and C. Ohlwein, 2017, A novel convective-scale regional reanalysis COSMO-REA2: Improving the representation of precipitation, *Meteorologische Zeitschrift* 345–361.

Wang, W., P. Xie, S.-H. Yoo, Y. Xue, A. Kumar, and X. Wu, 2011, An assessment of the surface climate in the NCEP climate forecast system reanalysis, *Climate Dynamics* 37, 1601–1620.

Whitehall, K., C. Mattmann, D. Waliser, J. Kim, C. Goodale, A. Hart, P. Ramirez, P. Zimdars, D. Crichton, G. Jenkins, C. Jones, G. Asrar, and B. Hewitson, 2012, Building Model Evaluation and Decision Support Capacity for CORDEX, *WMO* 61.

Chapter 12

Estimating corporate greenhouse gas emissions

Thibaud Barreau
Institut Louis Bachelier, France

Mohamed Fahmaoui
Institut Louis Bachelier, France

Peter Tankov
CREST – ENSAE, Institut Polytechnique de Paris, France

This chapter reviews existing machine learning models for estimating scope 1, 2, and 3 greenhouse gas emissions of companies using publicly available data, and develops a new interpretable approach, focusing on out-of-sample evaluation as well as on the data pre-processing steps such as outlier filtering, missing value imputation, and predictor selection. Two frameworks are presented: one without restrictions and another one with fewer inputs, better suited for small and mid-caps and for private companies.

12.1 Introduction

Given the urgency of the climate crisis, policymakers and investors are increasingly interested in assessing companies' carbon footprint to measure their direct and indirect impact on the environment and mitigate climate risk. The concept of carbon footprint emerged in the ecological literature by the end of the 20th century (Ress and Wackernagel, 1996; Wiedmann and Minx, 2008). According to Wiedmann and Minx (2008), the carbon footprint is defined as the amount of carbon dioxide and other greenhouse gases emitted and expressed in tons or kilograms of CO_2 equivalent. Carbon footprint can be measured at multiple scales, e.g., product, company, national, and time horizons. The most commonly used framework standardizing these calculation methodologies is the Greenhouse Gas (GHG) Protocol, first published in 2001 (Ranganathan et al., 2004). This framework divides the carbon inventory into three scopes corresponding to direct and indirect emissions:

- Scope 1 is the sum of direct GHG emissions from sources that are owned or controlled by the company, such as burning oil, gas, coal, and other fuels in boilers or furnaces, fuel-burning cars, vans or trucks owned or controlled by the company;
- Scope 2 is the sum of indirect GHG emissions associated with the generation of purchased electricity, steam, heat, or cooling consumed by the company;

DOI: 10.1201/9781032636252-12

- Scope 3 is the sum of all other indirect emissions that occur in the company's value chain, including production of raw materials (upstream scope 3), utilization of company's products (downstream scope 3) and even financed emissions via investments. It is composed of 15 sub-scopes (see Greenhouse Gas Protocol Team (2011), Table 5.4).

Measuring and reporting corporate GHG emissions is crucial for research in green and sustainable finance as outlined, e.g., in Abdul Gafoor et al. (2024). This issue is also crucial for financial practitioners, as emissions are the most widely used proxy for determining a given portfolio's exposure to transition risks.[1] Besides, carbon emissions are increasingly used in portfolio construction and reallocation, sometimes resulting in significant outflows toward greener investment options when available. However, reporting and auditing of GHG emissions at the corporate level is not yet mandatory worldwide, and the methodologies used are not standardized. This has led to a reliance on sparse and questionable reported data or non-transparent estimation methodologies from a wide variety of data providers, limiting the reliability of assessments of companies' environmental impact.

This study is mainly based on data from the CDP (ex Carbon Disclosure Project), an independent, not-for-profit organization that helps companies, cities, states, regions, and public authorities disclose their environmental impacts. CDP data is recognized as one of the most trusted sources and is widely used in academic publications. Yet, even CDP data cannot be considered entirely reliable according to Stanny (2018), who suggests systematic caution when working with such reported data, since "most firms do not have their reported amounts verified by third parties and the data that some of the largest emitters disclose to the CDP are unreliable and incomparable".

The first attempts at standardization of emission data are very recent compared to financial reporting, and in practice, the GHG data used by practitioners can take many different forms. Academic publications usually use reported data to reduce bias, but this has strong coverage limitations since, e.g., less than 10,000 companies reported to CDP in 2022. On the other hand, financial practitioners need to collect data on all the companies in their investment universe and, therefore, usually rely on estimated values.

Even if the emission data is not estimated using a statistical model, it can vary significantly depending on the reporting option chosen: the operational boundary (organization vs. financial control), the calculation standards (GHG protocol or others), or the scope 2 calculation (market-based vs. location-based). These potential differences make this data especially difficult to model.

To address the shortcomings and biases in the existing GHG emission estimation methodologies, we propose a machine learning model for the estimation of GHG emissions of scopes 1, 2, 3, and 123 (i.e., the sum of the three scopes above) of companies based on publicly available data. The proposed methodology is open source and easily replicable, providing a transparent and reliable approach to estimating corporate GHG emissions. Additionally, we propose a new method for assessing out-of-sample performance of emission estimates, both globally and at the level of individual sectors, countries, or revenue levels. We compare our model to state-of-the-art published methodologies, demonstrating superior accuracy and reliability. We present two alternative frameworks, one using all available data and another limited to the most commonly available predictors, to provide flexibility for different data availability scenarios.

The rest of this chapter is structured as follows. In Section 12.2, we review the currently available methodologies for GHG emission estimation and their shortcomings. In Section

[1] Transition risks refer to the financial risks that result from the process of transitioning toward a lower-carbon and more sustainable global economy. These risks are associated with shifts in policy and regulation, changes in market preferences and technologies, and evolving legal landscapes.

12.3, we describe the datasets used in our approach. The methodology is then described in Section 12.4. In Section 12.5, we evaluate the performance of our methodology and compare it with other methods. Section 12.6 concludes the paper.

12.2 Literature

12.2.1 Estimating GHG emissions using statistical methods

Due to the lack of reliable reported data, major data providers such as Bloomberg (Quants, 2022), MSCI (Shakdwipee and Lee, 2016; Andersson, Bolton, and Samama, 2016; De Jong and Nguyen, 2016), Refinitiv (Refinitiv, 2017), S&P Global Trucost and CDP (Carbon Disclosure Project, 2020) have developed models to estimate unreported GHG emissions. In the academic literature, linear regression models were first used to calculate proxies, as in Goldhammer, Busse, and Busch (2017) and in Griffin, Lont, and Sun (2017). Such estimations are based on a standard regression framework, i.e., each target variable $S_t^{c,i}$ (e.g., emission scope) $i \in \{1, \dots, T\}$ for a given company c and year t is approximated by

$$\widehat{S}_t^{c,i} = f(\theta, (X_t^{c,j})_{j=1,\dots,F}), \tag{12.1}$$

where $(X_t^{c,j})_{j=1,\dots,F}$ is the vector of explanatory variables and the approximating function f and the parameter vector θ is chosen to minimize a given loss function (e.g., Mean Squared Error).

It was then natural to deploy state-of-the-art data science algorithms such as Extreme Gradient Boosting (Friedman, 2001) to maximize the accuracy of these estimates. These algorithms typically perform better on tabular data than more computationally intensive deep learning models (Grinsztajn, Oyallon, and Varoquaux, 2022; Shwartz-Ziv and Armon, 2022).

The paper of Nguyen, Díaz-Rainey, and Kuruppuarachchi (2021) was among the first to introduce such models and to combine them by introducing a meta-algorithm based on several classical algorithms. This approach allowed for better out-of-sample performance, with an accuracy gain between 25% and 30% compared to the existing approaches. However, these models still produce high prediction errors, especially for some specific sectors such as insurance with high negative R^2 values (-378% and -151% for Scope 1 and 2, respectively). These results were improved in more recent publications (Heurtebize et al., 2022; Assael et al., 2023). We now present a detailed overview of the most widely cited references on the subject (as of 2023), published in peer-reviewed journals, namely Nguyen, Díaz-Rainey, and Kuruppuarachchi (2021) and Assael et al. (2023). In these and other papers, several common methodological steps may be identified:

1. Choice of data providers and datasets, preparation of data, treatment of missing values;
2. Choice of the target variable and the predictors;
3. Choice of the machine learning models, model selection, possibly combination of several models into a meta-model to achieve optimal performance;
4. Evaluation of the resulting model on a series of benchmarks.

In Nguyen, Díaz-Rainey, and Kuruppuarachchi (2021) the dataset consists of 13 435 annual observations spanning 2 289 global firms that disclosed their carbon emissions between 2005 and 2017, obtained from the Thomson Reuters Eikon database. Data preparation involved managing missing values and outliers. For missing values in the target variable, records were entirely excluded from all estimation models. However, to preserve sample sizes, missing predictor values were imputed using historical data or peer-group averages. Outliers, identified as emissions values falling below the 1st percentile or above the 99th percentile, were eliminated from the dataset.

The variables used in the study include four target variables and 22 main predictors. A structured approach was employed to determine the best predictors for modeling; this involved selecting the target variable, an industry classification method, firm characteristics, environmental factors, time variables, and energy data. The optimal predictors were identified through exhaustive OLS regressions with 1920 variable combinations. The study's best-performing model, a two-step framework, integrates predictions from six base models using a meta-model (Elastic Net).

This approach outperformed previous models by generating more accurate predictions, even in out-of-sample scenarios, where the model predicted emissions not used during its construction. However, our analysis revealed that the meta-model did not exhibit performance enhancement compared to the Extreme Gradient Boosting methods. Additionally, while the model exhibited improvement over existing methodologies, the authors acknowledged the presence of relatively high prediction errors, even in their best model. Notably, the model demonstrated strong predictive efficacy in forecasting aggregate direct and indirect emission scopes rather than predicting each scope individually.

In Assael et al. (2023), the dataset covers the years 2010 to 2020 and incorporates information from a diverse array of databases, including Refinitiv, Bloomberg, CDP, World Bank, and IEA. During this period, data was collected from 65,673 companies operating in over 115 countries. To evaluate Scope 1 and Scope 2 emissions, the data is sourced primarily from two origins: CDP data, which provides non-modeled and audited emissions, and Bloomberg's greenhouse gas emissions compilation. In instances where both CDP and Bloomberg report emissions for the same company, the approach prioritizes CDP-reported emissions. A LightGBM regressor model, an Extreme Gradient Boosting type of statistical regressor, was used to estimate the emissions. This method requires minimal preprocessing; in particular, missing values are left intact as the LightGBM model can manage them without compromising performance.

The methodology uses 21 predictors to train two different LightGBM models. Similarly to the previous approach, variables fall into several categories, namely Financial, Industry classification, Energy, and Regional. Authors also introduce another categorization called "New Energy Exposure Rating", which corresponds to an estimation of the percentage of an organization's value that is attributable to its activities in renewable energy, smart energy technologies, Carbon Capture and Storage (CCS) and carbon markets. A randomized search on hyper-parameters is performed, fitting and evaluating 180 models. The selection of the best models is based on Mean Squared Error performance on validation sets. In addition, the authors employ data refinement techniques based on SHAP (Shapley Additive exPlanations) values to enhance the model performance, a method referred to as "data polishing". This involved applying a clustering algorithm to identify clusters based on their SHAP values' distance, subsequently removing smaller clusters. They applied this to level 4 industrial sectors (i.e., the most granular ones), and it resulted in the removal of approximately 11.5% of training data for scope 1 and 5% of training data for scope 2, leading to an improvement in overall model performance.

This model yields a high and stable accuracy, notably reaching an average RMSE of 0.5 for scope 1 and 0.46 for scope 2 (in log scale). Moreover, the authors perform detailed robustness checks across countries, sectors and revenue variations. Yet, contrary to what is suggested in this paper, our tests have revealed a significant impact of preprocessing techniques on the model performance. Similarly to their findings, we also observe that the most important predictor across all scopes is the Energy Consumption. The industry classification predictors also demonstrate substantial importance for both Scope 1 and Scope 2 estimations. As expected due to the definition of Scope 2, predictors like Employees are more critical for Scope 2 estimation than for Scope 1.

12.2.2 Limitations of the current methodologies and proposed improvements

Based on the above literature review and our own analysis, the following biases / limitations in existing methodologies can be identified:

1. No difference is usually made between overestimating and underestimating emissions in the loss function. Yet, a methodology which underestimates true emissions is more dangerous because it creates an incentive for companies not to report their emissions, since reported emissions are likely to be above the estimated ones, and thus make the company appear less "green".
2. Reported emissions are subject to several sources of error, and numerous filters and verification steps are required to obtain a dataset, which can be used in a statistical study.
3. Most accurate and efficient models rely on a large number of predictors that are usually not available for companies which are small and / or not publicly listed, in other words the emitters for which estimates are most useful.
4. The distribution of countries and sub-industries in the datasets is highly unbalanced. As a result, when a statistical model is tested, most of the samples come from countries and sub-industries that are already well covered, while the model is usually applied to countries and sub-industries that are poorly covered. Therefore, the future reliability of a model is often overestimated.

Beyond raw performance improvement, our approach includes important differentiating methodological elements to account for the above issues. To mitigate extreme prediction errors and restrict the output to a reasonable window, we introduce multiple outlier removal steps, which are more advanced than those found in the literature. These help limiting the range that our estimates can take, and as a result, our models will usually not underestimate the true emissions by too much due to sectoral lower bounds. The presence of such bounds sometimes leads to overestimation for the lowest emitters. However, this behavior acts as an incentive for companies to report emissions and helps us limit the risk of underestimating true emissions mitigating the limitations 1 and 2. In addition, we explored models with restricted predictors to account for bias 3. We also exclude companies with extreme emissions claims, such as zero Scope 2 or 3 emissions, to avoid lowering our estimates. Further, we develop a novel test set selection method to overcome the limitation 4. Our final contribution to the literature is the modeling of scope 3 emissions, which are more difficult to fit (Nguyen et al., 2022), as well as of the sum of the three scopes (which we denote scope 123).

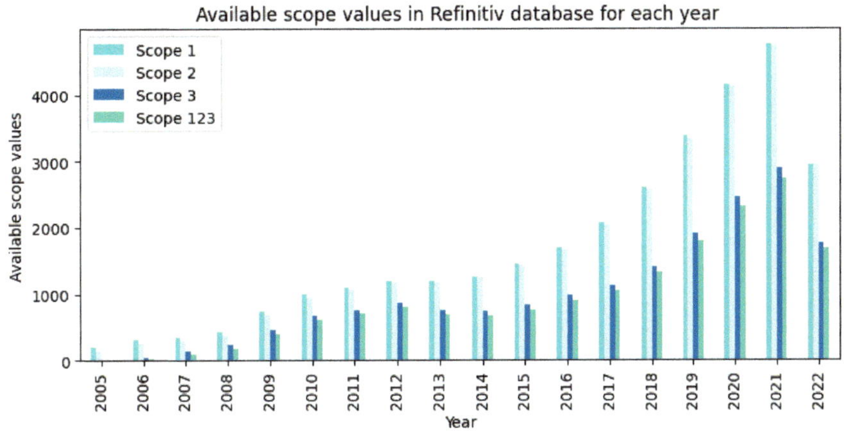

FIGURE 12.1: Available scope values in Refinitiv database for each year.

12.3 Datasets

12.3.1 Proprietary datasets

To develop our models, we rely on proprietary data for relevant predictors (financial data and sector classification) and reported emissions. We explore open-source alternatives to obtain similar data on which to train alternative models, but so far, it appears that proprietary datasets are required to produce sufficiently robust GHG emissions estimates.

Refinitiv The first dataset is based on the Refinitiv Worldscope database, covering the years 2000 to 2022 and comprising 1,146,497 data samples. This dataset contains information on 49,588 companies across 136 countries and 163 sub-industries based on the GICS classification. However, the Refinitiv data contains a significant number of missing values. As Figure 12.1 shows, the coverage of reported emissions has increased over time. The exception is 2022, which can be explained by the delay in reporting emissions, which can be up to two years for some companies. Restricting the dataset to observations with at least one reported scope, we end up with 35 012 data samples for 6 033 companies.

CDP The second dataset is the CDP raw emissions database. It covers the period from 2014 to 2022 and is built from companies' aggregated raw responses to CDP's annual questionnaires, which correspond to quality level 7 (the highest level possible) in the consolidated dataset.[2] It contains 31,174 samples representing 13,345 individual companies located in 130 countries. The number of common companies between Refinitiv and CDP data is 3,107, located in 68 countries. The coverage of reported emissions also increases over the years, as shown in Figure 12.2.

[2]For more details on the CDP, please refer to the Carbon Disclosure Project (2020) documentation.

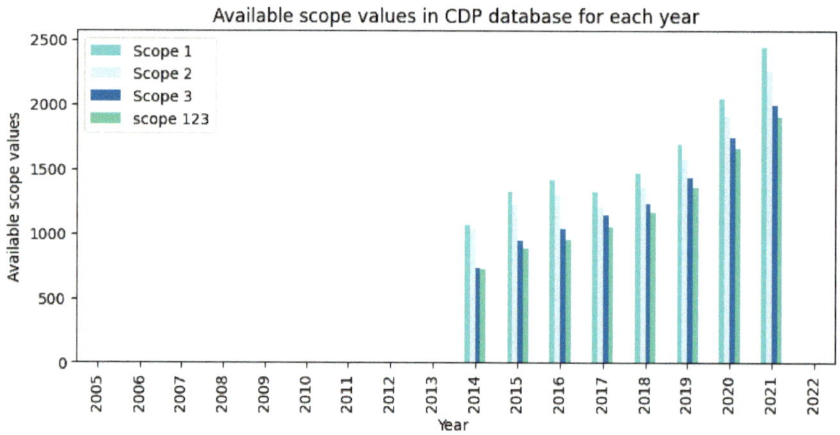

FIGURE 12.2: Available scope values in CDP database for each year.

12.3.2 Open-source datasets

We augment the above datasets with open-source data, allowing us to extend the dataset with key information on regional indicators that significantly impact companies and their operations. These regional indicators include critical factors such as national electricity generation carbon intensities, national carbon tariffs, national carbon prices, and country income group classifications. This approach allows for a more holistic understanding of companies' sustainability efforts and their potential impact on financial performance.

Electricity carbon-intensity – Ember We use the Ember data on electricity generation, which is publicly available and includes the carbon intensity of electricity generation in gCO_2/kWh for 211 countries from 2000 to 2022.

The carbon intensity of an energy mix refers to the amount of carbon dioxide equivalent (TCO_2eq) emissions produced per kilowatt hour (kWh) of energy generated. It is a key measure of how environmentally friendly a country's energy production is. A lower carbon intensity indicates cleaner and more sustainable energy sources, while a higher carbon intensity indicates greater reliance on fossil fuels, which release more CO_2 into the atmosphere.

A quick analysis shows that there is a decreasing trend in carbon intensity for the most represented countries, e.g., the United States, Japan, the United Kingdom, and China, see Figure 12.4. Among the world's continents (Figure 12.3), Oceania has the highest carbon intensity due to its reliance on coal, while South America has the cleanest energy mix, largely based on renewables.

Carbon pricing Dashboard – World Bank The Carbon Pricing Dashboard, developed by the World Bank, provides information on carbon pricing, including the type of carbon pricing – carbon tax or Emissions Trading System (ETS) – the coverage – national or regional – the implementation status and the carbon price, for 251 countries and regions from 1990 to 2022.

As of 2023, out of the 251 countries and regions in the database, 200 have no carbon pricing legislation, 39 have such legislation, and 12 are in the process of implementing it. Specifically, of the 39 jurisdictions with carbon pricing laws, 21 have introduced an emissions trading system, 16 have introduced a carbon tax, and 8 have both an ETS and a carbon

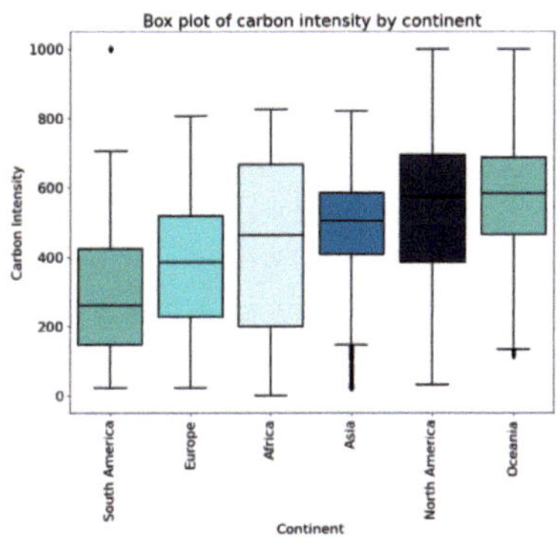

FIGURE 12.3: Box plot of carbon intensity by continent.

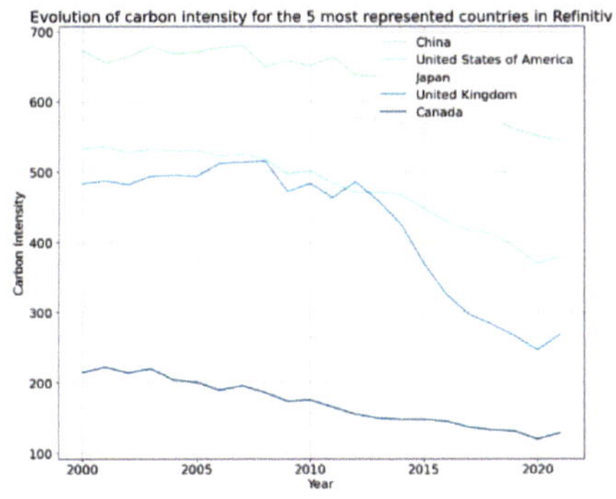

FIGURE 12.4: Evolution of carbon intensity for the 5 most represented countries in Refinitiv.

tax – the two are not mutually exclusive as they can apply to different economic activities. Notably, around 23% of companies listed on Refinitiv are based in jurisdictions that have introduced carbon pricing. The Europe and Central Asia region has some of the highest carbon pricing rates, with the highest being 168 euros per ton of carbon dioxide equivalent (TCO_2e) for Sweden in 2014.

Income group – World Bank Finally, we use the income groups provided by the World Bank. The World Bank classifies countries into four income groups: Low, Low-Middle, Upper-Middle, and High Income. The data cover the period from 1987 to 2021, and these categories are based on gross national income levels.

12.4 Methodology

We follow the general regression framework (12.1) to train our machine learning models for emission estimation. As the emissions of all three scopes tend to follow a log-normal distribution, we work on predicting the base 10 logarithm instead of the raw values (see appendix A in the online version of this chapter (Barreau, Fahmaoui, and Tankov, 2024)). Thus, target variables $S_t^{c,1}$, $S_t^{c,2}$, $S_t^{c,3}$ correspond, respectively, to logarithms of scope 1, 2, and 3 emissions of company c in year t, and the target variable $S_t^{c,123}$ corresponds to the logarithm of the sum of the three scopes for company c in year t. Following the literature, we use the standard mean-square error (MSE) as loss function for fitting our models. In the following paragraphs, the main steps of model construction, outlined in Section 12.2.1, are described in detail.

12.4.1 Choice of targets and predictors

We first describe the full set of explanatory variables used in our modeling approach. Specific predictor selection procedures for each individual model and emission scope will be addressed later on. A detailed list of all variables is available in Tables 4 and 5 in Appendix C of the online version of this chapter (Barreau, Fahmaoui, and Tankov, 2024).

Financial data Financial predictors are crucial to our model as they provide information about the size of a company and how it manages its business. They are generally available for large companies. We extract the Revenue, the EBIT, the EBITDA, the Capital Expenditure (CapEx), the Gross Property Plant & Equipment (GPPE), the Net Property Plant & Equipment (NPPE), the Gross Margin (GMAR), the Long Term Debt LTDebt, the Cost Of Goods Sold (COGS) and the Accumulated Depreciation (Accu Dep), and the Intangible Assets (Intan) annually from 2000 to 2022, for each available company in Refinitv. From the extracted financial predictors, three additional predictors are derived: the Life Expectancy of Assets, the CapEx intensity (CapInten), and the Leverage. They are obtained according to Griffin, Lont, and Sun (2017) and Nguyen, Díaz-Rainey, and Kuruppuarachchi (2021) using the following formulas (when the denominator is not zero):

$$LifeExpectancy = \frac{GPPE}{DepreciationExpense} = \frac{GPPE}{EBITDA - EBIT}$$

$$CapInten = \frac{NPPE}{Revenue}$$

$$Leverage = \frac{LongTermDebt}{Asset}$$

ESG data In addition to financial characteristics, some ESG predictors are significantly correlated with GHG emissions. In this modeling exercise we use the number of employees, which is a good proxy for company size for a given sector as well as the energy produced and consumed by the company, which is often directly related to GHG emissions, especially for Scope 1 and 2. The "Energy produced" attribute has a low coverage but turns out to be key for Scope 1 estimation for specific sectors, especially for "Energy" and "Utilities". Other ESG information may be correlated with emissions, but it usually suffers from too low coverage. Since we aim to use the model for companies that do not disclose their emissions – often smaller ones – we assume that using scarce ESG data is not relevant.

Regional data The geographical region in which a company operates also has a strong influence on its emissions. We use two quantitative predictors: the carbon intensity of electricity generation (FuelMix) from Ember and the carbon price, when available, from the Carbon Pricing Dashboard. We use the headquarters country to match a country to each company, although this has limitations. A more complex approach based on operating countries could be explored, but this would lead to additional assumptions and complexity. In addition, we use two categorical predictors: the existence of a CO_2 pricing scheme and the income group classification. The first is binary and indicates whether the country of the headquarters has a national, sub-national or regional CO_2 pricing scheme. The second is based on the World Bank's classification of countries by income (Fantom and Serajuddin, 2016).

Industry classification data

Finally, industry classification is key to emissions projections, as emission reduction costs and opportunities vary significantly from one sector to another. For example, for heavily emitting industries such as steel, cement, or power generation, emissions will vary greatly depending on technology, production, or location. By using the most accurate classification scheme, we can assess the nature of the facilities and the sources of emissions. We use the GICS classification as it is one of the most widely used in the industry and allows for granularity. It has four hierarchical levels, listed in Table 5 in Appendix C of the online version of this chapter (Barreau, Fahmaoui, and Tankov, 2024).

Target variables Recall that we use two sources for companies' yearly GHG emissions: the CDP and Refinitiv datasets. We use the Refinitiv dataset to complete the CDP dataset, restricting ourselves to data sourced from corporate sustainability reports (as opposed to modeled or estimated data). In the CDP dataset, both market-based and location-based reporting is available for scope 2 emissions: following the recommendation of Brander, Gillenwater, and Ascui (2018) we use the location-based method. All emissions are expressed in TCO_2e. To associate each scope to a given year, we considered that any value disclosed within the first six months (January to June) is associated with the previous year, and any value disclosed within the last six months (July to December) is associated with the current year. Indeed, most corporations report on December 31st, but some use other dates. Zero values for scope 2 and 3 emissions are considered suspicious and treated as missing values. Manual analysis of suspicious values also led us to replace some values, for example, using company sustainability reports.

Once both target datasets are cleaned, we derive our last target, scope 123 for all companies where all three scopes are available by summing them. Then, we merge the two datasets using available keys (ISIN, Tickers, and Name). We consider CDP data as more reliable than emissions from Refinitiv since it is supposed to undergo more rigorous review procedures (Assael et al., 2023). Thus, we use CDP scope values by default and only use Refinitiv ones when no CDP emissions are available.

12.4.2 Construction of the test set

We follow the mainstream machine learning approach of splitting the data into a training set and a test set to evaluate the out-of-sample performances of our models. As highlighted in Section 3, sub-industry and country distribution across the data is highly unbalanced. This bias increases when restricting ourselves to companies reporting emissions, as they are based in the most developed countries, with the United States, China, and the United Kingdom (the three most represented countries) accounting for around 40% of available observations. However, trained regression models will most likely be applied not for companies in these countries/sectors but in low-represented ones where regulation is less binding on environmental issues. To address this bias, our test set is constructed using the following procedure:

- We restrict the set of "candidates" to companies having at least one year after 2011 for which they report on all three scopes. This ensures a balanced representation of all scopes in the test set, without companies reporting only some specific ones. We used the year 2011 as a threshold because Scope 3 standard was published during this year (Greenhouse Gas Protocol Team, 2011). This reduces the number of "eligible" companies from 6 473 to 3 963.

- We then select companies for the test set based on their sub-industry and country. In our restricted dataset of eligible companies, there are 1 765 unique (sub-industry-country) couples. We randomly select one representative for each unique couple. Using this methodology, we end up with 45% of "eligible" companies in the test set.

We performed several statistical checks to ensure that the scopes and predictor distributions are similar for both the training and the test set. This selection procedure produces a test set with every available sub-industry and country. However, due to this selection, we will test our models with some companies in sectors and/or countries that were not present in the training set, which will make the estimations more difficult. Thus, our metrics may not be directly comparable to the current state-of-the-art, but we consider our test set to be much closer to real-life applications and our performance measures to be more realistic.

12.4.3 Data preprocessing

This phase is decomposed into five steps:

1. Outliers detection and treatment;
2. Encoding of categorical variables;
3. Missing values imputation;
4. Logarithmic transformation;
5. Choice of the starting date.

This crucial phase results in four different preprocessed datasets corresponding to our four target variables. These are then used to train machine learning regression models on the train set and test them on the test set.

1. Detection and treatment of outliers We filter the outliers in our dataset based on two criteria. First, we remark that for a given company, changes in the reporting methodology, corporate actions such as mergers acquisitions, and other similar events can strongly impact the reported emissions. To remove the outliers related to these discontinuities, we use the following methodology, adapted from Heurtebize et al. (2022). For each original emission scope $i \in \{1, 2, 3\}$, we plot a histogram of the ratio

$$\frac{\max_t S_t^{c,i}}{\min_t S_t^{c,i}}, \quad c = 1, \ldots, N,$$

where the maximum/minimum is computed over all available years and N is the number of companies for which the given emission scope is available. Then we estimate an appropriate outlier threshold and filter out all suspicious companies, *i.e.*, companies spotted as outliers for at least one of the three scopes. We estimate the appropriate outlier threshold by visually analyzing the histograms of each scope. Each threshold corresponds to the last "non negligible" bin in the histogram. This approach allows us to identify abnormal scopes variations. The histograms and threshold values can be found in Appendix B of the online version of this chapter (Barreau, Fahmaoui, and Tankov, 2024).

Second, we filter outliers based on the Log-Carbon Intensity (LCI), defined for a given company c, year t, and scope i as:

$$LCI_t^{c,i} = \log \frac{S_t^{c,i}}{R_t^{c,i}},$$

where R denotes the company revenue.

For a given observation, we consider it an outlier and remove it if both requirements are met:

- its LCI is bellow/above a specific threshold;
- its sub-industry is sufficiently represented, (i.e., there are more than 10 distinct companies in the sub-industry).

The removal thresholds are sub-industry and scope-specific and are defined, for a given sub-industry SI and given scope i, as follows

$$min\_threshold^{SI,i} = median(LCI_t^{c,i}, c \in SI, t = 1, \ldots, T) - 1.5IQR$$
$$max\_threshold^{SI,i} = median(LCI_t^{c,i}, c \in SI, t = 1, \ldots, T) + 2.5IQR$$
$$IQR = Q3(LCI_t^{c,i}, c \in SI, t = 1, \ldots, T) - Q1(LCI_t^{c,i}, c \in SI, t = 1, \ldots, T),$$

where $Q1$ and $Q3$ are the first and third empirical quartiles, respectively. According Tukey (1977), the most natural factor to use would be 1.5 for both thresholds. However, we decided to increase the upper one to 2.5 since reported emissions that are above the true emissions are far less likely than under-reported ones (Stanny, 2013). This selection method leads us to remove around 5% of under-reported and only 0.5% of over-reported emission scope values. This asymmetry allows us to limit the risk of under-estimation in our model, which could be considered as greenwashing.

2. Encoding We use two different encodings for categorical predictors:

- **Ordinal Encoding**: here, labels are translated to numbers based on their ordinal relationship to one another. It requires categories to have an ordering.
  - Year: being sequential, years have a natural ordering.
  - Income Group: the four income groups have the following order: $L < LM < UM < H$, and are thus encoded using this mapping: $\{L : 0, LM : 1, UM : 2, H : 3\}$.
- **One Hot Encoding**: here, we create k new predictors for each of the k possible values of a given categorical variable.
  - CO_2 pricing scheme: this variable is binary and is thus encoded in only one predictor $has\_co_2\_law$ taking values zero or one.
  - Sectorial Classification: for all used sectorial classification levels (namely GICSSector, GICSGroup, GICSInd and GICSSubInd), for each existing label sector_X in the dataset, we create a new category is_sector_X. When all four levels are used, this adds up to $11 + 25 + 74 + 163 = 273$ columns, as a result, most of our models' predictors are generated from this one hot encoding.

3. Missing value imputation Even though some of the best models can be used with data containing missing values, our empirical tests show that imputing them with appropriate values leads to better performance. We have thus implemented an imputation methodology based on precomputed "sub-industrial means per Revenue deciles" on the training set. The full algorithm is detailed in Appendix D of the online version of this chapter (Barreau, Fahmaoui, and Tankov, 2024).

4. Logarithmic transformation We use a log 10 transformation on the following predictors, since their distribution is close to log-normal (as shown in Appendix A of the online version of this chapter (Barreau, Fahmaoui, and Tankov, 2024)): Revenue, EBIT, EBITDA, CapEx, GPPE, NPPE, Accu Dep, Intan, COGS, Asset, LTDebt, Employees, Energy Consumed, Energy Produced, scope 1, scope 2, scope 3, scope 123.

5. Choice of the starting date Although it is possible to find reported emissions before 2000, standard methodologies have only been stabilized recently. We base ourselves on the

GHG protocol updates to define the starting year for the time series used for training our models. The latest major updates took place in 2005 for scope 1 and 2 and in 2011 for scope 3 (see Ranganathan et al. (2004); Greenhouse Gas Protocol Team (2011)). We thus drop the data for years before 2005 for scope 1 and 2 and before 2011 for scope 3. Our experiments show that considering the chosen cut-off year improves performance compared to using a more recent one.

12.4.4 Model choice and training methodology

Among several classical machine learning models tested (from linear regressions with penalization such as Ridge, Lasso and Elastic Net to random forest regressors), gradient boosting model achieves the best performance for our task, both in terms of bias and variance.

In machine learning, boosting is an ensemble meta-algorithm that consists of the iterative training of weak algorithms, such as shallow decision trees, on re-weighted data to obtain a stronger one when combining all the algorithms (Barreau, 2020). Let us say that we want to predict values of the form $\hat{y} = F(x)$ by minimizing the mean squared error:

$$\frac{1}{n}\sum_{i=1}^{n}(\hat{y}_i - y_i)^2$$

where i indexes a given training set of size n of real values of the output variable y_i. Here, we also have explanatory predictors x_i associated with each y_i. Now, let us consider a gradient boosting algorithm with M stages. Each stage m satisfying $1 \leq m \leq M$ of gradient boosting corresponds to an imperfect model F_m (for $m = 1$, this model may simply return $\hat{y}_i = \bar{y}$, where $\bar{y}$ is the mean of the y_i). To improve F_m, the algorithm adds a new estimator $h_m(x)$ to form $F_{m+1}(x) = F_m(x) + h_m(x)$. At each stage, F_{m+1} attempts to correct the errors of its predecessor. A more detailed description of boosting can be find in Li (2016).

We explored the following three models, for each of our targets:

- XGBoostRegressor: the regressor from the most commonly used library for gradient boosting (XGBoost) that stands for Extreme Gradient Boosting and won many machine learning competitions since its release in 2014. It is a highly efficient implementation of gradient boosting algorithm known for its speed and performance. It builds trees sequentially, where each new tree tries to correct the errors made by the previous one. XGBoost provides various regularization parameters to prevent overfitting and improve generalization.

- CatBoostRegressor: short for Categorical Boosting, it is another popular library regressor. Released in 2017, it aims at handling categorical predictors efficiently. It uses an innovative method called Ordered Boosting, which treats categorical predictors in a unique way during the tree building process. CatBoost automatically handles missing data and reduces the need for extensive preprocessing.

- LGBMRegressor: the regressor from LGBM, a performance/scalabilty focused gradient boosting framework released by Microsoft in 2016 (Microsoft, 2024). It uses a technique called Gradient-based One-Side Sampling (GOSS) to select only the most informative instances for building trees, which speeds up training. It also supports parallel and distributed learning, making it suitable for large datasets.

All three models have a large number of hyperparameters. We optimize these using Bayesian optimization. It operates by constructing a probabilistic model (a Gaussian process) to represent the objective function (such as a model's performance metric). The process iteratively explores the hyperparameter space, effectively balancing exploration (trying new hyperparameters) and exploitation (focusing on promising regions). This iterative approach helps in gradually refining the search space, ultimately leading to the identification of the most promising hyperparameters configuration, thereby enhancing the overall model's performance. However, we note a relatively minor influence of this optimization on the results. Such low influence may be due to several factors, including the differences between the training and the testing sets and the impossibility to reduce bias and variance beyond some threshold.

Predictor selection In the established methodologies like Assael et al. (2023), models usually rely on the same set of predictors, independently of the target (scope 1 or 2). In our study, we used SelectKBest, a feature selection technique in machine learning used for selecting the top k features (variables) that are most important or contribute the most to the predictive model's performance, to identify a specific set of predictors for each target, leading to the following findings:

(a) The presence or absence of the majority of predictors had minimal influence on the overall performance of the models. This suggests that the predictive power of these variables may not be as strong as initially presumed.

(b) Additionally, we discovered that the optimal set of predictors varies depending on the chosen target variable. This underscores the importance of tailoring predictor selection to the specific objectives and characteristics of each target.

As a result, we have identified the final set of predictors for each scope studied:

- For scope 1: [Revenue, Employees, Asset, NPPE, GPPE, CapEx, Cap intern, GMAR, LTDebt, CO_2 pricing scheme, Carbon Intensity, Year, Energy consumed, Energy produced, Intan, Accu Dep, COGS, EBIT, EBITDA, CO_2 pricing scheme, Income group, GICSsubind, GICSind, GICSgroup, GICSsector]
- For scope 2: [Revenue, Employees, Asset, GPPE, CapEx, Age, Cap intern, GMAR, LTDebt, CO_2 pricing scheme, Carbon Intensity, Year, Energy consumed, Energy produced, Intan, Accu Dep, COGS, EBIT, EBITDA, LTDebtt, CO_2 pricing scheme, Income group, GICSsubind, GICSind, GICSgroup, GICSsector]
- For scope 3: [Revenue, Employees, Asset, NPPE, GPPE, CapEx, Age, Cap intern, GMAR, LTDebt, CO_2 pricing scheme, Carbon Intensity, Year, Energy consumed, Intan, Accu Dep, COGS, EBIT, EBITDA, LTDebt, CO_2 pricing scheme, Income group, GICSsubind, GICSind, GICSsector]
- For scope 123: [Revenue, Employees, Asset, GPPE, CapEx, Age, Carbon Intensity, Year, Energy consumed, Energy produced, EBIT, LTDebt, Income group, GICSsubind, GICSind, GICSgroup]

Performance evaluation metrics

We employ several metrics and visualization tools to assess the quality of our models. To ensure comparability with existing studies on GHG emissions models, which often use the Root Mean Squared Error (RMSE), we employ the RMSE as the main performance metric. For each emission scope i, the RMSE is defined as:

$$\text{RMSE}^i = \sqrt{\frac{1}{NT}\sum_{c=1}^{N}\sum_{t=1}^{T}\left|S_t^{c,i} - \widehat{S}_t^{c,i}\right|^2}. \tag{12.2}$$

Here, $S_t^{c,i}$ represents the ground truth (log-transformed reported emissions), and $\widehat{S}_t^{c,i}$ is the GHG estimation (log-transformed) from the model for company c, year t and scope i. Since our models use the MSE as the risk function, they are optimized to minimize the RMSE. Moreover, as a quadratic function, RMSE gives more importance to large errors, which is a desirable property in our case, since far-off estimations compromise the credibility of our models. To assess the variability of the performance of our models, we also compute the variance of the RMSE with respect to the choice of the test set. We do so by splitting our test set in several batches, and calculating the RMSEs on each one. We then use the classical unbiased variance estimator:

$$variance = \frac{\sum_{i=1}^{n}(x_i - \bar{x})^2}{n-1}, \tag{12.3}$$

where n is the number of batches, x_i is the RMSE for batch i, and $\bar{x}$ is the empirical mean. Other sources of variance have been explored, such as the seed of the machine learning models, but were found to be negligible compared to this one.

Other commonly used performance measures are the Mean Absolute Error (MAE), defined as:

$$\text{MAE}^i = \frac{1}{NT}\sum_{c=1}^{N}\sum_{t=1}^{T}\left|S_t^{c,i} - \widehat{S}_t^{c,i}\right|, \tag{12.4}$$

with the same notation as above, and the coefficient of determination R^2, defined as

$$R_i^2 = 1 - \frac{\sum_{c=1}^{N}\sum_{t=1}^{T}(S_t^{c,i} - \widehat{S}_t^{c,i})^2}{\sum_{c=1}^{N}\sum_{t=1}^{T}(S_t^{c,i} - \overline{S}^i)^2}, \tag{12.5}$$

where $\overline{S}^i$ denotes the empirical mean of observations for scope i.

The Mean Absolute Percentage Error (MAPE) was also used to evaluate the accuracy predictions, it can be defined as follows:

$$\text{MAPE}^i = \frac{1}{NT}\sum_{c=1}^{N}\sum_{t=1}^{T}\left|\frac{S_t^{c,i} - \widehat{S}_t^{c,i}}{S_t^{c,i}}\right| \times 100, \tag{12.6}$$

where $\widehat{S}_t^{c,i}$ represents the predicted value and $S_t^{c,i}$ represents the actual value for scope i at time t and category c.

In addition to the numerical performance metrics, to explore the influence of some explanatory variables on model performance, we also use graphical tools, such as RMSE boxplots.

12.4.5 Uncertainty estimation

To provide results beyond a single point estimation, we also estimate prediction intervals. Given a probability p and a multi-index $\{c, t, i\}$ representing a given company, year, and scope, a prediction interval $[Q_1, Q_2]$ is an interval where $S_t^{c,i}$ belongs to $[Q_1, Q_2]$ with probability p. Following Geisser (2017), we consider them to be a better method than confidence intervals to account for prediction uncertainty. Indeed, a confidence interval for a deterministic but unknown parameter and is designed so that, for a given level of confidence $1 - \alpha$, a proportion $1 - \alpha$ of the intervals obtained by repeating the estimation procedure with independent data samples would contain the true parameter. This differs from our predictive intervals, sometimes called "predictive confidence intervals" (Geisser, 2017).

A common method to derive a predictive interval is to compute $[Q_\alpha, Q_{1-\alpha}]$, where $\alpha = \frac{p}{2}$ and Q_α is an α-quantile of the distribution of the variable one aims to predict. To do so using boosting algorithms, one needs to perform quantile regression by changing the objective function to quantile loss function. This can be done for any boosting algorithm with a tunable loss function.

Another approach is to consider that the residuals follow a known (e.g., Gaussian) distribution, and to derive the quantiles based on point-specific variances. Such variances can be computed based on approach proposed by Malinin, Prokhorenkova, and Ustimenko (2020). For example, CatBoost already has an implemented loss function, "RMSEWithUncertainty", that allows to get a set a variances in addition to a set of predictions.

To compare the quality of the produced prediction intervals, two main types of metrics must be analyzed:

- A "sharpness" metric, referring to the width of the prediction interval. Indeed, we want the smallest intervals possible. To account for this, we will compare the average lengths of the intervals produced by several algorithms, on both train and test data.
- A "calibration" metric, referring to the requirement that the interval effectively contains the true value with probability p, even on unseen data. To account for this, we will compute the effective coverage of the true values by the estimated intervals, on both train and test data.

12.5 Results

As mentioned above, three models were trained, namely LightGBM, CatBoost, and XGBoost. Among these, CatBoost emerged as the top performer across all four target variables. The main explanation of CatBoost superior performance is its capacity of handling categorical variables, which is very important since they represent the vast majority of our predictors. Additionally, CatBoost is more recent than the other two and its built-in regularization methodology and other specificities can also explain its performances. LGBM yields very comparable yet slightly inferior metrics in most tests, sometimes with a lower variance, which makes it a viable option.

12.5.1 Global performances

Table 12.1 presents the global results for scopes 1, 2 3 and 123 obtained using the three models, and evaluated using three key metrics: RMSE, MAE and R^2. These metrics are computed on the decimal logarithm of greenhouse gas emissions. This transformation is employed for the sake of comparability across the literature. Metrics can also be computed on raw values (by taking the power of ten), but then most of the errors will be associated with large values of the target. A RMSE of 1 in log scale corresponds to an error of one order of magnitude (factor 10) in terms of raw greenhouse gas emissions.

In addition to these metrics, the plot of residuals in Figure 12.5 and the plot of predicted against actual values, in Figure 12.6, reveal a distinctive trend within Scope 1. The models consistently exhibit a pattern of overestimation for smaller values, which is related to our outlier filtering that involves a lower bound for the scopes.

TABLE 12.1: Summary of model performance

Target	*Model*	*RMSE +/− STD*	*MAE*	*MAPE*	*R2*
scope 1	xgboost	0.5493 +/− 0.0960	0.3909	0.09882	0.8442
scope 1	catboost	**0.5224 +/− 0.1011**	**0.3751**	**0.0953**	**0.8591**
scope 1	lgbm	0.5265 +/− 0.0906	0.3757	0.0952	0.8569
scope 2	xgboost	0.4499 +/− 0.0702	0.3230	0.072	0.7852
scope 2	catboost	**0.4184 +/− 0.0744**	**0.2969**	**0.067**	**0.8142**
scope 2	lgbm	0.4337 +/− 0.0738	0.3094	0.070	0.8004
scope 3	xgboost	0.9609 +/− 0.0782	0.7375	0.162	0.5797
scope 3	catboost	**0.9029 +/− 0.0820**	**0.6973**	**0.154**	**0.6288**
scope 3	lgbm	0.9246 +/− 0.0748	0.7179	0.157	0.6108
scope 123	xgboost	0.5755 +/− 0.0520	0.4350	0.079	0.7754
scope 123	catboost	**0.5517 +/− 0.0550**	**0.4139**	**0.075**	**0.7935**
scope 123	lgbm	0.5588 +/− 0.0509	0.4239	0.076	0.7882

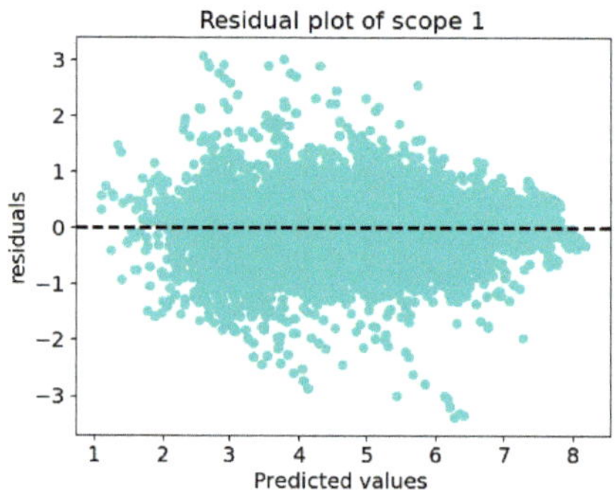

FIGURE 12.5: Residual plot for scope 1 (unrestricted predictors).

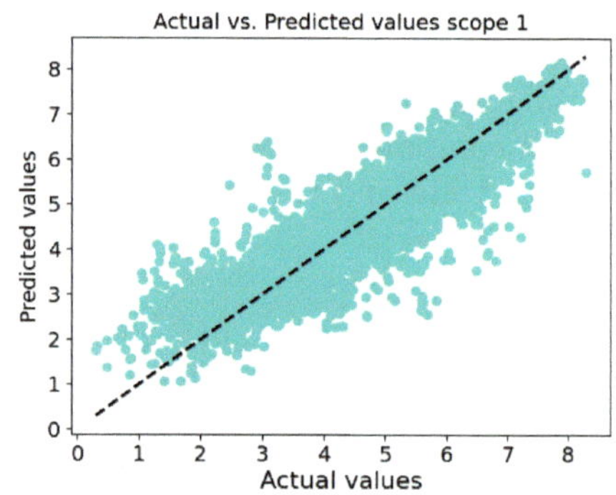

FIGURE 12.6: Predicted vs. actual plot for scope 1 (unrestricted predictors).

12.5.2 Performance breakdown by sector, country and revenue

In addition to the evaluation of the overall performance of our models, we conduct a thorough analysis of their robustness across various dimensions, including sectors, countries, years and revenue. We have summarized these findings in Figures 12.7, 12.8, 12.9 for Scope 1; detailed analyses for other scopes can be found in Appendix F of the online version of this chapter (Barreau, Fahmaoui, and Tankov, 2024).

Figure 12.7 shows the top 5 countries where the best-performing model provides the most accurate estimations, as well as the 5 countries where its estimations are the least accurate. This breakdown allows us to pinpoint areas that require improvement. One can note that performances can seem unsatisfactory for some countries, due to lack of data and standards and general uncertainty regarding some disclosed reported emissions.

Figure 12.8 expands our evaluation by assessing the performance of the model across eleven GICS Sectors. For Scope 1, we observe that the sector "Material" has the highest level of accuracy, which is due to direct links between the company size and its greenhouse gas emissions. Utilities, on the other hand, is the most difficult to predict and this is probably due to large variation between methodologies of emission reporting and their respective boundaries. The accuracy of the model estimations across sectors varies strongly depending on the specific scope being evaluated.

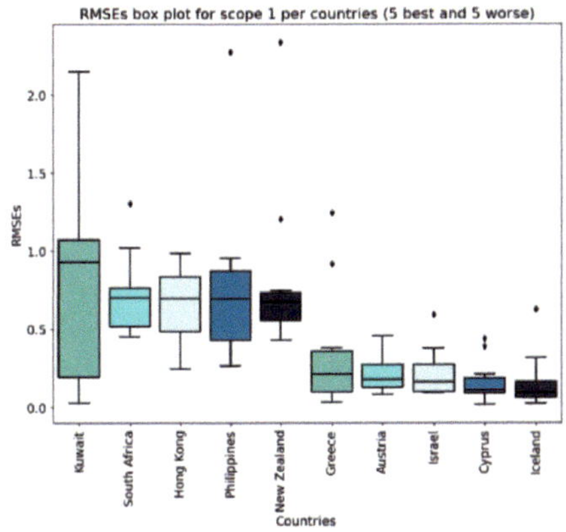

FIGURE 12.7: Box Plot Comparing RMSE (log) Estimates: Top 5 Accurate vs. Bottom 5 Less Accurate Countries for scope 1 (unrestricted predictors).

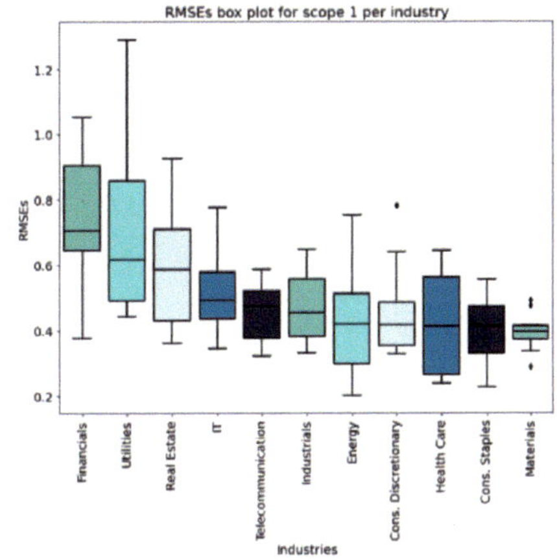

FIGURE 12.8: Box Plot Comparing Accuracy of Scope 1 (unrestricted predictors) Estimations by Industries (RMSE log).

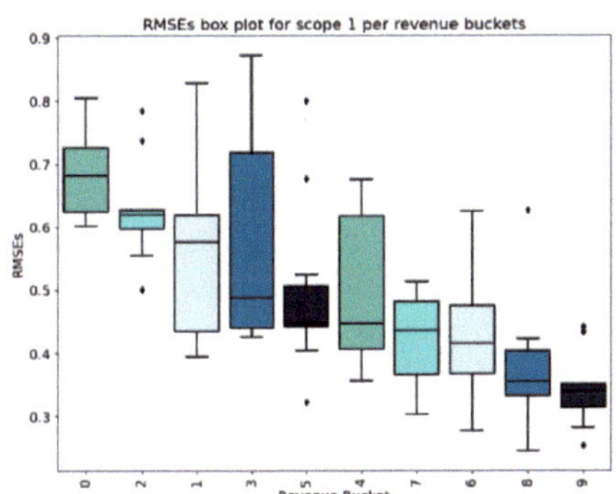

FIGURE 12.9: Box Plot Comparing Accuracy of Scope 1 (unrestricted predictors) Estimations by revenue buckets (RMSE log).

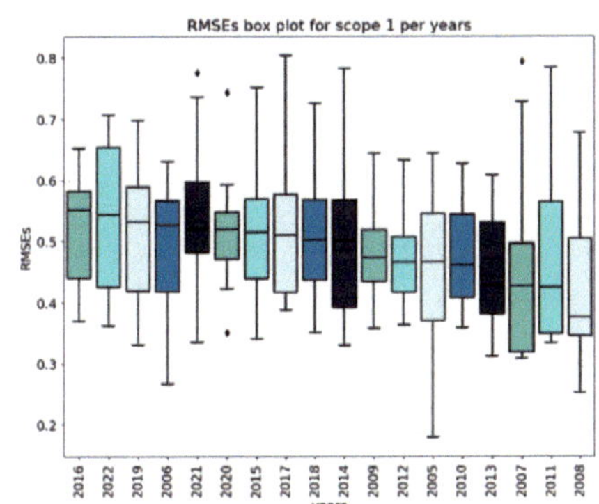

FIGURE 12.10: Box plot comparing accuracy of scope 1 (unrestricted predictors) estimations years (RMSE log).

Moving on to Figure 12.9, we analyze the performances of our top-performing model across different revenue categories. We observe that the model tends to deliver more accurate estimates for high-revenue companies, perhaps due to the fact that high-revenue companies have stronger incentives to maintain a positive environmental reputation and provide accurate emissions data.

In Figure 12.10, there is a considerable spread in the RMSE values for different years, as indicated by the height of the boxes, suggesting variability in model performance. There are outliers present in the years 2006, 2007, and 2020 (indicated by the dots), which could suggest occasional extreme deviations from typical error rates. Overall, while there is a trend of improving accuracy, the presence of outliers and the year-on-year variability indicate areas where the model could potentially be improved for consistency.

12.5.3 SHAP values

In recent years, the data science community has seen an increasing use of SHAP (SHapley Additive explanations) values associated with predictors for Machine Learning interpretation. Introduced by Shapley (1997), these values allow to understand predictor utilization by a given model (Molnar, 2020). Below is a brief explanation of how SHAP values are computed after the model prediction:

- Predictor Permutation: SHAP values are based on the idea of predictor permutation. The values of each predictor for the instance being explained are permuted (exchanged), and predictions are made for each permutation.
- Difference Calculation: For each permutation, the difference between the model's prediction with the original predictors and the prediction with the permuted predictors is calculated. This captures the change in prediction due to the alteration of a specific predictor.
- Shapley Value Estimation: The SHAP values are then calculated using Shapley values from cooperative game theory, which assign a value to each predictor indicating its contribution to the prediction. These values consider all possible combinations of predictors and their interactions.
- Final Interpretation: The SHAP values provide insight into how each predictor contributes to the model's output for a particular instance. Positive SHAP values indicate a predictor that pushes the prediction higher, while negative values indicate a predictor that pushes it lower.

We use "beeswarm" plots to visualize the predictors-target relations, i.e., a summary of how the top predictors in the dataset (training) impact the model's output.

We present an illustrative example in Figure 12.11, for the scope 1 emissions as the target. The "Energy Consumption" predictor emerges as the most influential factor in our model's predictions for all scopes but scope 123. This finding aligns with our expectations and previous literature. The energy produced also impacts significantly scopes 1 and 2, as it is also tightly linked to GHG emissions. Additionally, the financial metrics and some sector predictors also hold considerable importance in our model's decision-making process depending on the target.

Due to the One Hot Encoding, we have a large number of predictors, which limits the possibilities of interpreting all predictors. A stricter predictor selection can easily reduce this number without losing too much performance. In the next section, we specifically explore models using far fewer predictors, which are thus easier to interpret, as well as probably better suited for unreported emission estimations of small and mid-caps companies since they rely on less financial and extra financial information.

12.5.4 Alternative models based on smaller predictor sets

To make our methodology more widely applicable, we employ the same learning procedure on a limited set of predictors which are almost always available for all companies, including private and smaller ones. These predictors are revenue, profits, assets and the sector of the company. When employing the restricted predictors approach, CatBoost still consistently outperforms the other models as shown in Table 12.2. As expected, these models demonstrate lower accuracy levels.

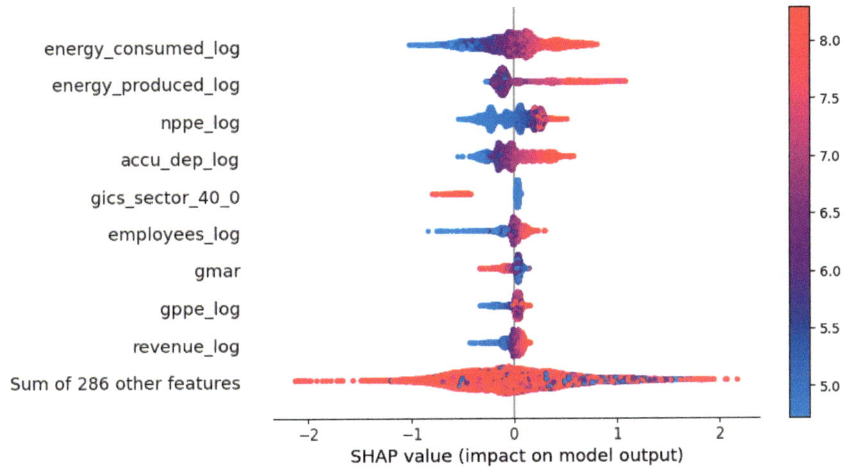

FIGURE 12.11: SHAP value for the estimation of scope 1 (unrestricted predictors).

TABLE 12.2: Summary of model performance for restricted predictors

Target	*Model*	*RMSE* ± STD	*MAE*	*MAPE*	R^2
scope 1	catboost	0.6203 ± 0.1143	0.4550	0.0856	0.8013
scope 2	catboost	0.4861 ± 0.0812	0.3620	0.0709	0.7492
scope 3	catboost	0.9143 ± 0.1578	0.7127	0.1578	0.6195
scope 123	catboost	0.5799 ± 0.0815	0.4389	0.0815	0.7719

Figures 12.12 to 12.15 show the performance breakdown by country, sector, revenue bucket and year. In Figure 12.12, we see that while with restricted data we have more outliers, overall results are coherent with the previous ones. Figure 12.13 focuses on GICS sub-industries, the only sector categorization used within this methodology. Within Scope 1, a discernible trend emerges: certain sub-industries within the financial and utilities sectors exhibit notably poor predictions. Conversely, companies operating within the Consumer Discretionary and Staples sectors demonstrate the most accurate predictions.

Turning our attention to Figure 12.14, we observe that conclusions for revenue categories mirror the observations made in the first methodology: the model consistently produces more precise estimates for high-revenue companies.

Figure 12.15 presents the RMSE for scope 1 estimations across different years for a limited set of predictors. There seems to be variability in the RMSE from year to year, with some years showing a tighter range of values (e.g., 2015) and others displaying a broader range. Several years show outliers, suggesting occasional estimations with significantly higher errors. Overall, the box plot suggests that while the model performance varies from year to year, there is no clear trend of improvement or degradation over the period.

Similarly to the previous section, Figure 12.16 plots the SHAP values for the restricted set of predictors. It highlights that revenues and assets are key predictors, together with sub-sector predictors. This is consistent with our expectations, as these two financial metrics and greenhouse gas emissions depend on the size of the company. Compared to financial and sector-specific predictors, it seems that geographical information plays a less important role. This could be different if, instead of using the country of headquarters, we used all countries in which the company operates, but this extension is left for further research.

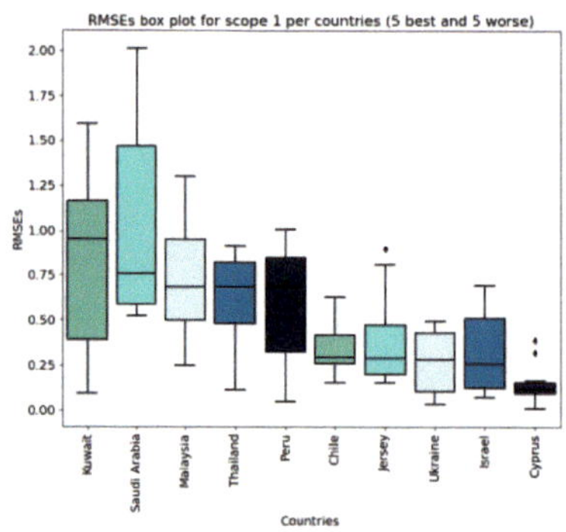

FIGURE 12.12: Box Plot Comparing RMSE (log) Estimates: Top 5 Accurate vs. Bottom 5 Less Accurate Countries for scope 1 (restricted predictors).

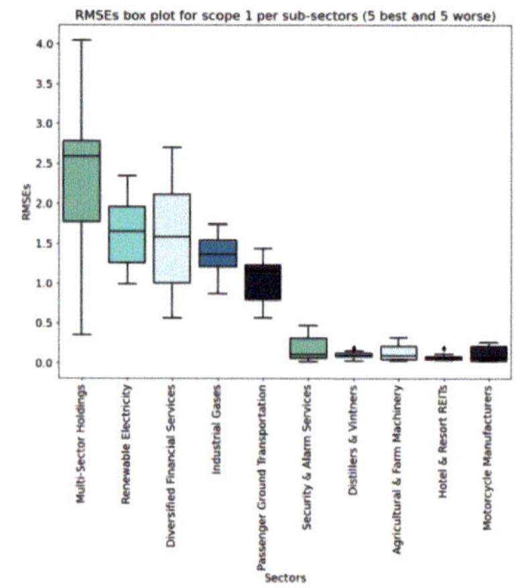

FIGURE 12.13: Box Plot Comparing Accuracy of Scope Estimations by Industries (RMSE log) for scope 1 (restricted predictors).

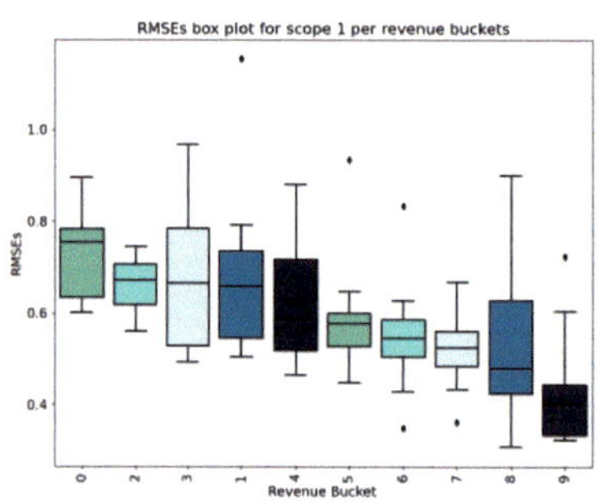

FIGURE 12.14: Box Plot Comparing Accuracy of Scope 1 (restricted predictors) Estimations by revenue buckets (RMSE log).

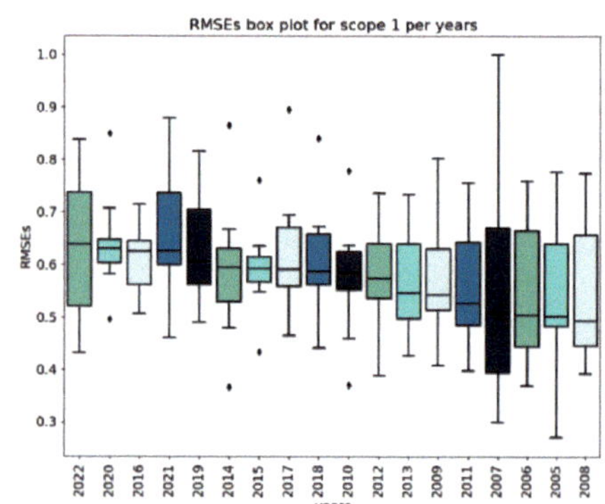

FIGURE 12.15: Box plot comparing accuracy of scope 1 (restricted predictors) estimations years (RMSE log).

12.5.5 Uncertainty estimation performances

In this section, we discuss prediction intervals produced with Catboost. The results are presented in table 12.3, for prediction intervals with probability 90%. All displayed results are on test data, which explains why the coverage values are below the expected value of 0.9.

The CatBoost implementation uses the "RMSEWithUncertainty" loss function. A model trained using such parameter predicts two outputs instead of one: for each sample, it returns both the mean prediction and the associated variance. We then use the hypothesis that the distributions of the residuals is Gaussian. In order to get a 90% prediction interval, we derive, for each sample, the 5% and the 95% quantiles using the following formulas:

$$Q_{5\%} = \hat{y} - z_{90\%} * \sqrt{\hat{V}}, \qquad Q_{95\%} = \hat{y} + z_{90\%} * \sqrt{\hat{V}},$$

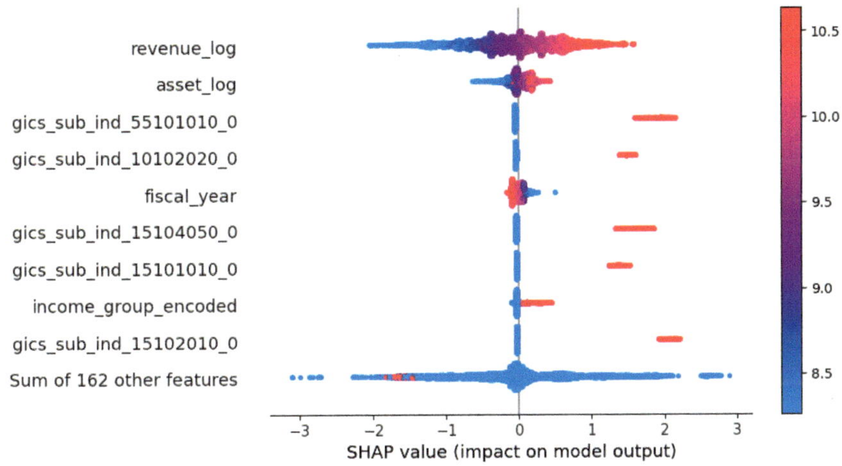

FIGURE 12.16: SHAP value for the estimation of scope 1 (restricted predictors) with restricted predictors.

where $\hat{y}$ is the mean prediction, $\hat{V}$ the variance prediction and $z_{90\%}$ the z value for a 90% interval. Under the Gaussian distribution hypothesis, we have $z_{90\%} = 1.645$.

TABLE 12.3: Summary of uncertainty estimation performances

Target	*Coverage*	*Mean length*
scope 1	0.8273	1.602
scope 2	0.8271	1.200
scope 3	0.8080	2.257
scope 123	0.7936	1.416

Table 12.3 shows that the produced intervals are quite large, from over one order of magnitude for Scope 2 to over two orders of magnitude for Scope 3. The coverage (probability that an element of the test set belongs to the prediction interval) is lower than 90%, perhaps due to not quite Gaussian shape of the distribution. More work on this specific issue is probably required to obtain prediction intervals that would be genuinely useful in practice.

12.6 Conclusion

Interest to corporate GHG emission estimation has grown in the recent years, especially since the COP 21 in 2015. This data is becoming a must use in most risk analysis studies including an environmental component, i.e., a significant amount of the published literature in the past decade. For both academic and industrial applications, proxies are crucial and the lack of confidence in existing estimations, especially regarding the scope 3, is one of the main obstacles for the development of sustainable finance. With our in-depth study and our efforts to identify and overcome biases that were not considered previously, we hope to help the field and facilitate future use of this kind of data.

Our goal was to propose accurate models for estimating GHG emissions of the three scopes while accounting for multiple biases. The proposed methodology is more complex than the ones available in the literature, but the additional steps are required for the gains in performance we observe. However, some of our experiments did not yield entirely satisfactory results and there remain avenues for improvement, which include using learning weights, improving the uncertainty measures and exploring additional variables (sectoral information, environmental engagements, operating countries, third validation of scopes, etc.).

Data Availability

Input data Restrictions apply to the availability of some data. Data obtained from Refinitiv and CDP cannot be shared by the authors without the permission of respectively Refinitiv and CDP. Data from World Bank and Ember are publicly available and can be found here:

- `https://datahelpdesk.worldbank.org/knowledgebase/articles/906519-world-bank-country-and-lending-groups`;
- `https://carbonpricingdashboard.worldbank.org`;
- `https://ember-climate.org/data-catalogue/yearly-electricity-data/`.

Output data Our methodology is open-source and has been developed as part of the Pladifes project (ANR-21-ESRE-0036). The goal of this project is to develop an innovative methodology for assessing corporate GHG emissions and to make it easily available to the academic community.

The computer implementation of the models can be downloaded from `https://github.com/Pladifes/corporate-ghg-emissions-estimations` and the Pladifes project distributes off-the-shelf databases of estimated emissions at `https://pladifes.institutlouisbachelier.org/data/`. We invite any potential user to contact the authors of this study for more details.

Acknowledgment

This work originates is developed and funded in the context of the Pladifes project (Equipex ANR-21-ESRE-0036). The authors would like to thank Louis Boulanger, Marine Neyret and Capucine Nobletz for their helpful comments that greatly improved the article.

References

Abdul Gafoor, C., S. Perumbalath, P. Daimari, and K. Naheem, 2024, Trends and patterns in green finance research: A bibliometric study, *Innovation and Green Development* 3, no. 2 (2024): 100119.

Andersson, M., P. Bolton, and F. Samama, 2016, Hedging climate risk, *Financial Analysts Journal* 72, no. 3 13–32.

Assael, J., T. Heurtebize, L. Carlier, and F. Soupé, 2023, Greenhouse gases emissions: Estimating corporate non-reported emissions using interpretable machine learning, *Sustainability* 15, no. 4 3391.

Barreau, T., 2020, Strategic optimization of a global bank capital management using statistical methods on open data. (Dissertation). Retrieved from https://urn.kb.se/resolve?urn=urn:nbn:se:kth:diva-273413

Barreau, T., M. Fahmaoui, and P. Tankov, 2024, Estimating corporate greenhouse gas emissions. Available at SSRN 4922542. 2024 Aug 11.

Brander, M., M. Gillenwater, and F. Ascui, 2018, Creative accounting: A critical perspective on the market-based method for reporting purchased electricity (scope 2) emissions, *Energy Policy* 112, 29–33.

Carbon Disclosure Project, 2020, Cdp full ghg emissions dataset – technical annex iii: Statistical framework. Available from: https://www.cdp.net/en/data

De Jong, M., and A. Nguyen, 2016, Weathered for climate risk: A bond investment proposition, *Financial Analysts Journal* 72, 34–39.

Fantom, N. J., and U. Serajuddin, 2016, The world bank's classification of countries by income. World bank policy research working paper. 2016(7528).

Friedman, J. H., 2001, Greedy function approximation: A gradient boosting machine, *Annals of Statistics* 1189–1232.

Geisser, S., 2017, *Predictive Inference* (Chapman and Hall/CRC).

Goldhammer, B., C. Busse, and T. Busch, 2017, Estimating corporate carbon footprints with externally available data, *Journal of Industrial Ecology* 21, 1165–1179.

Greenhouse Gas Protocol Team, 2011, Corporate value chain (scope 3) accounting and reporting standard. Available from: https://ghgprotocol.org/sites/default/files/standards/Corporate-Value-Chain-Accounting-Reporing-Standard_041613_2.pdf

Griffin, P. A., D. H. Lont, and E. Y. Sun, 2017, The relevance to investors of greenhouse gas emission disclosures, *Contemporary Accounting Research* 34, 1265–1297.

Grinsztajn, L., E. Oyallon, and G. Varoquaux, 2022, Why do tree-based models still outperform deep learning on typical tabular data?, in *Advances in Neural Information Processing Systems*, volume 35, 507–520.

Heurtebize, T., Chen, F., Soupé, F., and Carvalho, R. L. D., 2022, Corporate carbon footprint: A machine learning predictive model for unreported data. The Journal of Impact and ESG Investing, Winter. 2022;3(2):36–54.

Li, C., 2016, A gentle introduction to gradient boosting. Available from: https://www.chengli.io/tutorials/gradient_boosting.pdf

Malinin, A., L. Prokhorenkova, and A. Ustimenko, 2020, Uncertainty in gradient boosting via ensembles, *arXiv preprint arXiv:2006.10562*.

Microsoft, 2024, Lightgbm documentation. Available from: https://lightgbm.readthedocs.io/en/latest/index.html

Molnar, C., 2020, *Interpretable Machine Learning* (Lulu.com).

Nguyen, Q., I. Díaz-Rainey, A. Kitto, B. McNeil, N. Pittman, and R. Zhang, 2022, Scope 3 emissions: Data quality and machine learning prediction accuracy. PLOS climate, 2(11), e0000208.

Nguyen, Q., I. Díaz-Rainey, and D. Kuruppuarachchi, 2021, Predicting corporate carbon footprints for climate finance risk analyses: A machine learning approach, *Energy Economics* 95, 105129.

Quants, B. E., 2022, Distributional greenhouse gas emissions estimates: Data challenges and modeling solutions. Available from: https://assets.bbhub.io/professional/sites/10/1025089_CDS_REF_Greenhouse_Emissions_WP.pdf

Ranganathan, J., L. Corbier, P. Bhatia, S. Schmitz, P. Gage, and K. Oren, 2004, *The Greenhouse Gas Protocol: A Corporate Accounting and Reporting Standard (Revised Edition)* (World Resources Institute and World Business Council for Sustainable Development, Washington, DC).

Refinitiv, 2017, Refinitiv esg carbon data and estimate models.

Ress, W. E., and M. Wackernagel, 1996, Ecological footprints and appropriated carrying capacity: Measuring the natural capital requirements of the human economy, *Focus* 6, 45–60.

Shakdwipee, M., and L.-E. Lee, 2016, Filling the blanks: Comparing carbon estimates against disclosures. MSCI ESG Research Issue Brief.

Shapley, L. S., 1997, A value for n-person games. contributions to the theory of games ii (1953), in H. W. Kuhn, ed., *Classics in Game Theory*, 69–79 (Princeton University Press, Princeton).

Shwartz-Ziv, R., and A. Armon, 2022, Tabular data: Deep learning is not all you need, *Information Fusion* 81, 84–90.

Stanny, E., 2013, Voluntary disclosures of emissions by us firms, *Business Strategy and the Environment* 22, 145–158.

Stanny, E., 2018, Reliability and comparability of ghg disclosures to the cdp by us electric utilities, *Social and Environmental Accountability Journal* 38, 111–130.

Tukey, J. W., 1977, *Exploratory Data Analysis*, volume 2 (Reading, MA).

Wiedmann, T., and J. Minx, 2008, A definition of 'carbon footprint', *Ecological Economics Research Trends* 1, 1–11.

Chapter 13

Measurement of firm climate risks

Meng Rui
China Europe International Business School (CEIBS), China

Hongyu Shan
China Europe International Business School (CEIBS), China

This chapter presents firm-level measures of climate risk exposure, distinguishing between physical and transition risks while capturing both time-series and cross-sectional variation. These measures also help identify firms that respond proactively to climate challenges. We demonstrate that these measures have significant implications for firm performance in capital markets and for predicting real economic outcomes. In particular, firms with high transition risk - especially those that fail to respond proactively - have been increasingly discounted by investors as aggregate attention to climate issues has grown. We further show that these measures help explain the systematic differences in firm activities, including variations in investment, green innovation, and employment strategies.

13.1 Introduction

Climate change presents significant challenges for both businesses and society. Scientists anticipate that climate change will result in more frequent and severe chronic and acute climate events, creating unprecedented risks and disruptions that could impact corporations, the financial system, and the overall economy (Litterman et al., 2020). Building on the foundational work of Nordhaus (1977), numerous economists have examined the interplay between climate change and economic factors (e.g., Golosov et al., 2014; Nordhaus, 2019); however, explorations in climate finance - such as assessing, mitigating, and hedging climate risk across different firms and asset classes - have only recently begun to receive significant attention. A key obstacle in furthering this research is the absence of reliable measures for climate risk exposure across various asset classes, particularly in relation to equity assets (Hong, Li, and Xu, 2019; Engle et al., 2020; Giglio, Kelly, and Stroebel, 2021).

Several issues contribute to the current gap in measures for assessing firm-level climate risk exposure. First, despite enhanced regulatory requirements and increased pressure from investors, firms are often reluctant to reveal their climate risk exposure. For example, available carbon emissions data typically cover only a few traditional sectors (e.g., manufacturing and utilities), and firms frequently overlook the indirect carbon costs associated with their supply chains (Shapiro, 2021). Second, the dynamic nature of climate change creates considerable uncertainty about how future climate shifts will impact firms, which complicates decision-making for both governments and corporations (Barnett, Brock, and Hansen, 2020). Third, while historical emissions data are useful for analyzing past business

DOI: 10.1201/9781032636252-13

practices, forward-looking data are more critical for evaluating a firm's climate risk exposure and its ability to adapt in the transition to a sustainable economy—a key focus for climate finance research (Giglio, Kelly, and Stroebel, 2021).

In this chapter, we discuss several methodologies that address the gap by quantifying climate risk exposure at the individual firm level. We focus on the methodology developed by Li et al. (2024), which employs Natural Language Processing (NLP, thereafter) techniques on the earnings call transcript data from U.S. public companies. They analyze earnings call textual information for several reasons. First, these calls are held regularly by most U.S. public firms with analysts and investors to discuss performance and related factors. Crucially, these conversations involve detailed and insightful discussions about a firm's climate risks that go beyond public sources. Second, earnings call content is timelier and more variable compared to other firm disclosures like regulatory filings, which are often scripted and less informative (e.g., Brown and Tucker, 2011). Third, the focused nature of earnings calls, where only material factors are discussed, helps in assessing the significance of climate risks. Finally, these earnings calls also reveal how firms are responding to climate risks from the management's perspective, allowing us to evaluate their proactive measures.

Li et al. (2024) assess the climate risk faced by a firm at a specific point in time by analyzing the proportion of earnings call conversations focused on physical climate risk and transition risk, respectively. This method aligns with approaches used in previous research (e.g., Baker, Bloom, and Davis, 2016; Hassan et al., 2019, 2023, 2020). Additionally, the authors evaluate whether a company's responses to climate risk are proactive by examining the verbs used in discussions about climate risk. To achieve this, they address several challenges associated with standard NLP methods. First, any such analysis must consider the various categories of climate risk (e.g., Giglio, Kelly, and Stroebel, 2021; Stroebel and Wurgler, 2021), which include: (i) physical climate risks, associated with the impacts of acute climate events (e.g., hurricanes and wildfires) or chronic conditions (e.g., unusual winter weather), and (ii) transition risks. Due to the complex nature of climate risk, creating a single measure to capture all aspects of a firm's climate risk exposure is challenging. Therefore, distinct climate risks are separately measured using a dictionary-based approach.

The second challenge in measuring climate risk is the lack of a well-constructed dictionary of climate-related keywords in existing literature, which can lead to a high number of false positives and negatives when commonly known weather or climate keywords are applied to a large set of transcripts. The authors choose a dictionary-based approach over machine learning (ML) methods, emphasizing careful human oversight to minimize these inaccuracies. This method allows researchers to make informed and deliberate judgments when classifying text based on the complex nature of climate risks, while also ensuring transparency and replicability.[1] Through meticulous selection over multiple iterations, Li et al. (2024) develop three comprehensive dictionaries comprising more than 1,600 climate keywords that do not pertain to energy costs or general environmental risks.

To develop climate risk measures, the authors require that specific physical climate risk keywords be mentioned within one sentence of at least one risk synonym, ensuring that firms are genuinely exposed to uncertainties related to climate events (as in Hassan et al., 2019). In contrast, transition risk may not manifest in the short term and is assessed based solely on discussions of keywords from our transition risk dictionary, without needing proximity to a risk synonym. This approach yields three climate risk measures for each firm on a quarterly

[1] Humans excel at understanding the nuances of language in specific contexts, such as earnings calls. Our approach rests on the belief that no algorithm comprehends the subtleties of human conversation better than people do. For example, studies involving advanced conversational AI algorithms like Google Meena (Adiwardana et al., 2020) and Facebook BlenderBot (Roller et al., 2020; Xu, Szlam, and Weston, 2021) illustrate this point. Additional details regarding the benefits of using human-constructed dictionaries over ML methods can be found in Section 13.2.1.

basis. Additionally, by utilizing a list of verbs that reflect firms' proactive attitudes toward discussing transition risk, the authors decompose their transition risk measure into the proactive and nonproactive components.

We also discuss the alternative measures of firm climate risk, in addition to the one constructed by Li et al. (2024). These include (i) a transition climate risk measure derived from textual data in firms' 10-K/10-Q filings using the same dictionary, (ii) a measure based on climate-related news from Dow Jones Newswires, (iii) MSCI CCI or ESG ratings, (iv) the plant level emissions data, and (v) measures from Sautner et al. (2023) utilizing different dictionaries and methodologies. The measures developed by Li et al. (2024) maintain a high correlation with the alternative ones while offering enhanced coverage and quantification of firm-level exposure to climate risk. For example, unlike ESG ratings, which are often subject to selection bias, the measures by Li et al. (2024) are available quarterly for 4,719 public firms over an extended period. In contrast to the EPA's plant-level CO_2 emissions data, which is limited to specific sectors like manufacturing, mining, and trade, the measures by Li et al. (2024) encompass all sectors represented in earnings call data, providing a comprehensive evaluation of climate risk across the economy. Among the public firms with accessible earnings call data, approximately 61.8% (2,918 firms) report at least one positive value in the transition risk measure, accounting for 34.7% of firm-years with positive transition risk values. Even when accounting for the availability of MSCI CCI data, the measures by Li et al. (2024) expand coverage to an additional 952 firms with non-missing values and 480 firms with positive values.

The rest of the chapter continues to summarize the time-series and cross-sectional properties of the measures by Li et al. (2024) at the industry and firm levels. First, we present the most commonly discussed keywords in each measure. Next, we show the time-series trends and variations at both the industry and firm levels. While the relative rankings among industries differ for various climate risks, all show significant differences that correspond with the expectations of the industry-specific exposure to these risks. Additionally, variance decomposition analysis reveals that most variations in the three climate risk measures occur at the firm level, effectively capturing both cross-firm and within-firm variations in climate risk exposure.

The rest of the chapter examines the real-world implications of the measures. We first discuss a key issue in climate finance: how climate risk, particularly transition risk, is reflected in capital markets (e.g., Giglio, Kelly, and Stroebel, 2021). Li et al. (2024) examine the relationship between their firm-level transition risk measure and a firm's market valuation, as indicated by Tobin's Q. Their findings reveal a negative correlation, implying that transition risk exposure is indeed priced in equity markets. Notably, this relationship has gained significance primarily since 2010, likely due to increased investor attention to climate risks (e.g., Choi, Gao, and Jiang, 2020; Engle et al., 2020) and the introduction of climate-related regulations around that time.[2] Furthermore, when analyzing the proactive and nonproactive components of transition risk, the authors discover that only the nonproactive component shows a significant negative correlation with Tobin's Q. This indicates that equity markets tend to penalize firms that do not actively manage their transition risks, while those that take proactive measures are not similarly disadvantaged. Importantly, these results hold steady even after accounting for firm fixed effects, reinforcing the notion that changes in climate risk discussions are linked to fluctuations in Tobin's Q. They also include the alternative measures in a horse-race. The coefficients for the transition risk measure and its nonproactive component consistently remain negative and significant at the 1% level, highlighting the distinctive contribution of the earnings call data and the

[2]For example, the SEC issued its first guidance in January 2010 on how existing disclosure requirements pertain to climate-related issues for public firms.

supervised dictionary construction approach. Overall, the transition risk measure by Li et al. (2024) not only delivers new, valuable information missing from existing public sources but also ensures broad coverage across a substantial sample of public firms from 2002 onward.

The final section of this chapter examines how this measure can be applied to explain corporate real outcomes. Li et al. (2024) investigate how firms respond to transition risk concerning investments, innovation, and employment. Their findings reveal that a firm's proactiveness on climate issues significantly influences its responses in these areas. Specifically, nonproactive firms show no significant link between transition risk and total capital expenditures (CapEx), whereas proactive firms increase their investments in response to climate risk. Moreover, a negative correlation is found between transition risk and R&D expenditures, driven exclusively by nonproactive firms. Conversely, proactive firms demonstrate enhanced innovation, as evidenced by an increase in green patents in subsequent years. Further analysis indicates that equity markets place a higher value on firms proactively engaging in green patenting compared to those focused on non-green initiatives. Lastly, the employment analysis reveals that nonproactive firms tend to reduce their workforce following an increase in transition risk, while proactive firms maintain employment levels. Overall, the corporate climate risk measures not only enhance the understanding of how transition risk is priced in capital markets but also help predict real-world corporate outcomes as firms adapt proactively to climate risk.

13.2 The supervised approach

13.2.1 Constructing climate dictionaries

In this section, we focus on the supervised approach introduced by Li et al. (2024), and detail their dictionary construction methodology. The authors build on recent literature that leverages textual information from earnings call data to identify risks (e.g., Hassan et al., 2019, 2020, 2023) in order to develop their firm-level climate risk measures. However, employing textual analysis techniques to construct these climate risk measures entails several challenges.

The authors emphasize, as noted by Giglio, Kelly, and Stroebel (2021), the importance of recognizing the different categories of climate risks and their timelines when assessing their effects on underlying assets. Climate-related risks are generally divided into two main categories: (1) physical risks, which encompass acute events (such as droughts, floods, extreme precipitation, and wildfires) and chronic conditions (like rising temperatures and biodiversity loss), and (2) transition risks, which stem from the inability to adapt to climate change and shift to sustainable practices (e.g., low-carbon manufacturing). Transition risks are largely influenced by policies, regulations, societal expectations, and market pressures. Given the complex and multifaceted nature of climate risks, developing a single measure to encompass all aspects of a firm's climate risk exposure is difficult. Therefore, the authors adopt a dictionary-based approach to separately quantify three types of climate-related risks: (1) acute physical risk, (2) chronic physical risk, and (3) transition risk. They argue that the dictionary approach is preferable to machine learning (ML) methods for several reasons. First, ML methods often function as black-box models, lacking the transparency inherent in the dictionary approach. Second, ML methods are highly sensitive to initial inputs and parameter settings. Third, the effectiveness of ML predictions depends on having a large, representative training dataset, which is not readily available for the intricate and multifaceted nature of climate issues.

Second, unlike established training libraries (as in, e.g., political or accounting textbooks), creating a dictionary of climate-related keywords involves substantial human effort. When applying a set of commonly known weather or climate keywords to a large volume of transcripts, the authors encounter two significant issues. First, there are numerous false positives, where keywords refer to unrelated topics (e.g., "business climate," "public cloud," "economic storm"). Second, weather and climate anomalies are often described with combinations of contrasting terms (e.g., "warm winter," "unseasonably cold," "cool summer"). Relying solely on a dictionary of unigrams (single words) would likely miss critical terms, such as "winter" or "warm," leading to numerous false negatives. To overcome this, the authors manually construct a hybrid dictionary that includes both unigrams and bigrams (adjacent two-word combinations). This approach helps reduce both false positives and false negatives.

Specifically, the authors' method is founded on the belief that no algorithm can grasp the context of human conversation as effectively as humans themselves can.[3] They begin constructing their dictionaries with a compilation of unigrams sourced from: (a) disaster "incident-type" indicators in the Disaster Declarations Summary from the Federal Emergency Management Agency (FEMA), and (b) Wikipedia's list of severe weather phenomena,[4] and (c) additional seed words that they added manually, namely, "temperature," "cold," "unseasonable," and so on. The authors utilize this list to extract all bigrams that include at least one unigram from the full set of earnings call transcripts. They then manually review the top 500 associated bigrams for each unigram. If these bigrams are clearly used in the context of climate-related discussions, the corresponding unigrams are added to the unigram dictionary. If not, the bigrams are placed in a pending bigram library for further examination. To minimize false negatives, the bigram library is enhanced with climate-related bigrams gathered from additional sources, such as (a) white papers and reports on climate issues referenced by Engle et al. (2020), (b) news articles from The Weather Channel, and (c) an undergraduate textbook on meteorology (Ahrens, 2008). Finally, the authors iteratively screen the library to eliminate false positives and include any previously overlooked terms.

In constructing the dictionaries, the authors emphasize the importance of differentiating climate risk from other types of risk. While companies may address climate-related subjects that pertain to energy prices, these discussions do not inherently indicate climate risk. To ensure that the climate risk measures remain unaffected by fluctuations in energy costs, the dictionaries specifically exclude keywords associated with energy prices.[5] Instead, the authors develop a firm-specific, time-varying energy-price exposure index to serve as a control variable in the main analysis. Furthermore, while there may be a connection between companies' environmental responsibility initiatives and greenhouse gas emissions, these concepts are not identical. As a result, the authors exclude keywords related to general environmental risks (such as air pollution, environmental issues, EPA, and sulfur dioxide) from their climate dictionaries.

The final dictionaries comprise 37 unigrams and 1,649 bigrams. Specifically, the acute physical risk dictionary includes 21 unigrams and 350 bigrams, while the chronic physical risk dictionary contains 16 unigrams and 977 bigrams. The transition risk dictionary consists of 322 bigrams. The predominance of bigrams in these dictionaries reflects a deliberate focus

[3]For instance, this is highlighted in studies of advanced conversational AI algorithms like Google Meena (Adiwardana et al., 2020) and Facebook BlenderBot (Roller et al., 2020; Xu, Szlam, and Weston, 2021).

[4]See `https://en.wikipedia.org/wiki/List_of_severe_weather_phenomena`.

[5]Examples of excluded keywords include "energy cost," "energy costs," "fuel bill," "fuel cost," "fuel costs," "fuel expense," "fuel expenses," "gas cost," "gas costs," "wind cost," and "wind costs."

on enhancing text identification and quantification accuracy. Previous research suggests that using bigrams improves text classification accuracy compared to relying solely on unigrams (e.g., Tan, Wang, and Lee, 2002; Bekkerman and Allan, 2004).

13.2.2 Measuring climate risk

Next, these dictionaries are used to construct the firm-level climate risk measures. Specifically, Li et al. (2024) begin by breaking down each earnings call transcript into a list of unigrams and bigrams. Since acute or chronic physical risks are frequently mentioned alongside short-term climate or weather events in news headlines (e.g., *hurricane*, *wildfire*, and *warm winter*), they require that their respective keywords appear within one sentence (± 1 sentence) of at least one synonym for risk. This ensures that the discussion is genuinely related to the firm's exposure to climate risks (similar to Hassan et al., 2019). Merely referencing a well-known weather or climate event without explicitly linking it to a firm's risk profile may indicate an attempt to attract attention or deflect blame, which does not contribute to their physical climate risk measures. The authors calculate the frequency of these relevant occurrences relative to the transcript length and then multiply the result by 10^4 to simplify the decimal representation. Essentially, these measures reflect the proportion of the conversation in which acute or chronic weather or climate events are discussed in conjunction with the firm's risk exposure.

Transition risk differs from physical climate risk in that it relates to policies and regulations, technological improvements, and evolving climate patterns. Unlike physical risks, transition risk may not materialize in the short run and thus does not pose immediate threats or introduce any uncertainty to a firm's business operations. As a result, the authors measure transition risk exposure based on discussions of the keywords in the transition risk dictionary only, without requiring these discussions to appear near a risk keyword. Moreover, firms exhibit varying perceptions of and attitudes toward climate risk, with some discussing and addressing transition risk more proactively than others. With this in mind, they develop an additional measure that captures a firm's proactiveness when discussing transition risk. To achieve this, they analyze verbs that appear near (within ± 1 sentences of) discussions of transition risk keywords in earnings calls, and manually identify a list of 30 verbs that suggest more proactive attitudes when discussing climate issues.[6] Using proactive verbs, they separately identify their transition risk measures with and without proactiveness.

By implementing the aforementioned procedures, Li et al. (2024) create three distinct firm-level climate risk measures: (1) acute physical climate risk, (2) chronic physical climate risk, and (3) transition risk. The transition risk measure is further divided into proactive and nonproactive components. All measures are available at the firm-quarter level.

13.3 The alternative methods

In this section, we present the alternative measures of firm climate risk, in addition to the one constructed by Li et al. (2024). These include (i) a transition climate risk measure derived from textual data in firms' 10-K/10-Q filings using the same dictionary, (ii) a

[6]The complete list of the proactive verbs includes *achieve, acquire, add, announce, build, change, create, develop, enhance, evaluate, expand, generate, grow, hedge, help, improve, increase, initiate, integrate, invest, make, prepare, produce, purchase, rebuild, reduce, replace, respond, restructure* and *spend*.

measure based on climate-related news from Dow Jones Newswires, (iii) MSCI CCI or ESG ratings, (iv) the plant level emissions data, and (v) measures from Sautner et al. (2023) utilizing different dictionaries and methodologies.

The first set of alternative measures can be constructed using the firms' regulatory filings, in particular 10-K and 10-Q filings. Generally speaking, people focus on the two most relevant sections in 10-K/10-Q filings: (1) management discussion and analysis (MD&A) and (2) Item 1A "Risk Factors." MD&A section contains management discussions of firms' performance, risks, and future plans. The risk factors (RF) section provides information about the risk factors a firm identifies that might influence the company or its equity return. We should note that, compared with the earnings call data, three major drawbacks of using the Risk Factors section is that it contains only information about the risk factors themselves, with no discussion of how a company addresses or responds to those risks. Furthermore, the SEC filings are more likely to be scripted by lawyers and thus lack the informative perspective from the management of the firm. Lastly, RF section is available only from 2006 onward following the implementation of Regulation S-K Item 105.

The second alternative measure is constructed using publicly available company news as another source of textual data. We can obtain such data from RavenPack, which provides a comprehensive sample of firm-specific news stories from Dow Jones Newswires.[7] To identify news stories about specific firms, we can use relevance scores from RavenPack; these scores range from 0 to 100, capturing how closely the underlying news is related to a particular company. We may identify relevant news stories for a given firm by requiring the relevance score to be 75 or above, as recommended by RavenPack. Then we can calculate the alternative transition risk measure as the ratio between the number of news articles related to a firm's transition climate risk exposure and the number of all news articles related to the company.

The third set of alternative measures comprise several external ESG index or ratings. These scores measure how well a company manages ESG risks and opportunities based on information published in news coverage and/or corporate disclosures, such as sustainability reports and corporate websites, surveys, and information provided by other stakeholders, such as regulatory agencies and industry associations (e.g., Berg, Koelbel, and Rigobon, 2022; Christensen, Serafeim, and Sikochi, 2022). Leading vendors include MSCI, RepRisk, and Refinitiv, among others. These ratings include overall scores as well as three individual scores (environmental, social, and governance) at the monthly or annual level. In particular, MSCI presents the CCI score — a climate change theme score that is directly comparable to the climate risk exposure measures developed by Li et al. (2024). We should also note that the environmental components of ESG ratings provided by rating agencies focus on environmental risk that is entangled with, but different from, climate risk.

We can also use the CO_2 emissions data from the EPA's Greenhouse Gas Reporting Program (GHGRP) as an additional proxy for the transition risk exposure. Since October 2009, the GHGRP program has mandated that sources that emit 25,000 metric tons or more of CO_2 greenhouse gases per year must report their emissions, and the data are made publicly available on an annual basis starting in 2010 at the plant level; and these data include plant identity, geographic location, parent company, industry (NAICS), and greenhouse gas emissions. Following Bartram, Hou, and Kim (2022), we can obtain plant-level emissions data from the EPA and match them with firm-level data from Compustat based on the names of parent companies.

The study by Sautner et al. (2023) introduces a methodology to measure firm-level exposure to climate change by analyzing the language used in earnings conference calls.

[7] News include *The Wall Street Journal*, *Barron's*, *MarketWatch*, all major PR newswires and regulatory feeds.

Unlike Li et al. (2024), which takes a different approach to dictionary construction, Sautner et al. (2023) utilizes a machine learning keyword discovery algorithm to identify bigrams that indicate climate-related topics, categorizing them into four groups: general climate change, opportunities, physical risks, and regulatory shocks. We argue that both the machine-learning-based methodology and the supervised dictionary approach complement each other in effectively uncovering firm-level climate change risks.

13.4 Properties of the firm-level climate risk measures

This section offers preliminary validation of the measures developed by Li et al. (2024) by analyzing the underlying keywords. We explore the measures' time-series and cross-sectional characteristics, highlighting how they capture variations in climate risk exposure across firms and over time.

13.4.1 Top keywords

We analyze the most frequently used keywords - both unigrams and bigrams - essential for constructing their climate risk measures. We rank these keywords by their mention frequency and frequency weight at the transcript level, with results presented in Table 13.1.[8] The findings in columns 1–3 indicate that *hurricanes* and *hurricane* are the most frequently mentioned acute climate unigrams near risk synonyms, with other keywords like *storms*, *drought*, *flooding*, and *wildfire(s)* also increasingly discussed in recent years. Columns 4–6 reveal that *weather* is the most commonly referenced chronic climate keyword in this context, followed by terms related to specific weather conditions, such as *temperatures* and *snow*. These results strongly support the effectiveness of their measures in capturing acute and chronic climate risks.

In contrast to physical climate keywords, transition risk indicators are more widely distributed across various terms. Frequently mentioned keywords include *energy efficiency*, *renewable energy*, *solar*, *clean energy*, and *alternative energy*. Other notable terms are *superior energy*, *higher energy*, *new energy*, *the renewable*, and *the ecosystem*, all of which effectively represent discussions around transition climate risk. The calculation of *fweight* for transition climate risk follows a similar method; however, the authors do not require these keywords to be near risk synonyms, resulting in higher average frequencies and *fweights*. Table 13.2 contrasts the frequency of climate-related bigrams and unigrams with political-risk-related bigrams from a prior study (Hassan et al., 2019) and leading climate keywords from another research effort (Sautner et al., 2023). This table includes the number of earnings calls and firms that referenced each climate-related term, along with their frequency and *fweight*. The findings indicate that the frequency of top climate-related bigrams identified by Li et al. (2024) is significantly higher — about 1,600 times — than that of the leading political-risk-related bigrams (e.g., *the constitution*) noted in Hassan et al. (2019), and comparable to the top climate keywords identified in Sautner et al. (2023).

[8]The frequency weight, denoted as *fweight*, is calculated by dividing the occurrence frequency by the transcript length, multiplying by 10^4 to manage decimal precision, and summing across all transcripts. The average length of the earnings call transcripts is around 4,200 words before cleaning and 2,440 words afterward, aligning with existing literature (e.g., Chen, Nagar, and Schoenfeld, 2018).

TABLE 13.1: Top climate-related keywords. This table lists the top-30 unigrams or bigrams in each category of $ClimateRisk_{i,t}$ measures, ranked by *fweight*. To calculate the *fweight* for acute and chronic climate risk measures, Li et al. (2024) first identify the frequency of mentions of individual unigrams and bigram b in proximity to risk synonyms ($Freq_{b,P}$). They then divide this frequency by the length of the transcript P (B_P), multiply the quotient by 10^4, and sum the resultant values across all transcripts in our sample. The calculation of *fweight* in the case of transition climate risk is the same except that the mention of the unigrams and bigrams are not required to be in the proximity of risk synonyms, which leads to higher *Freq* and *fweight* for that specific category

Physical climate risk						*Transition climate risk*		
Acute risk			*Chronic risk*					
Bigram/ Unigram	*Freq*	*fweight* $= \frac{Freq_{b,P}}{B_P} \times 10^4$	*Bigram/ Unigram*	*Freq*	*fweight* $= \frac{Freq_{b,P}}{B_P} \times 10^4$	*Bigram/ Unigram*	*Freq*	*fweight* $= \frac{Freq_{b,P}}{B_P} \times 10^4$
(1)	(2)	(3)	(4)	(5)	(6)	(7)	(8)	(9)
hurricane	1560	6371.9	weather	6154	26342.7	energy efficiency	7738	32512.0
hurricanes	552	2243.5	temperatures	122	596.0	renewable energy	6663	29104.3
storms	409	1622.7	the snow	75	299.4	the solar	6623	28819.0
drought	294	1177.2	high water	72	266.2	clean energy	5117	21372.2
flooding	185	728.7	heating season	49	260.4	alternative energy	4160	18367.0
the flood	108	440.6	precipitation	46	252.1	superior energy	3354	12482.7
wildfire	110	356.4	wind season	60	237.1	higher energy	2806	11273.8
windstorm	75	333.8	the ice	57	216.7	new energy	2503	10878.1
wildfires	54	201.6	mild winter	48	188.8	the renewable	2389	10564.8
storm losses	30	155.4	snowfall	42	186.8	the ecosystem	2590	10036.0
severe winter	33	134.0	rainfall	42	175.4	energy management	2156	8861.2
storm related	31	132.5	degree days	34	173.9	energy efficient	2171	8459.6
wind storm	28	125.0	normal winter	36	170.7	the carbon	2243	8414.0
the floods	24	102.0	winter conditions	43	170.5	green energy	2224	8303.4
storm activity	25	100.8	warm winter	36	161.0	wind energy	1893	7817.5
storm costs	21	86.8	rains	34	138.0	the climate	1926	7300.8
water flood	22	82.4	cold winter	33	126.4	fuel efficiency	1874	6730.5
polar vortex	22	76.8	hot summer	30	124.9	shale gas	1655	6350.9
storm season	14	69.7	unseasonably warm	24	110.1	lower energy	1553	6290.3
storm damage	10	64.4	the fog	28	107.4	fuel efficient	1592	5925.9
droughts	14	57.4	harsh winter	27	103.5	energy technologies	1643	5883.5
tropical storm	13	55.3	unseasonably cold	19	99.6	solar power	1344	5836.2
snowstorms	13	52.6	the clouds	23	96.7	alternative fuel	1301	5776.1
snowstorm	12	50.1	the warmest	13	74.5	wind farm	1283	5696.7
winter storm	14	50.1	early winter	13	74.1	fuel economy	1586	5487.9
hailstorm	11	49.6	cool summer	13	72.3	the co_2	1479	5476.3
extreme cold	11	48.1	cold season	17	70.9	solar cell	1170	5457.9
extremely cold	10	40.0	the rain	16	64.7	gas drilling	1286	4947.8
storm cost	11	39.0	wind hail	11	63.2	energy future	1214	4715.9
the volcano	11	38.3	the winds	17	62.8	solar projects	1076	4667.6

13.4.2 Time-series patterns

We now turn our attention to analyzing the properties of the constructed measures to validate their effectiveness through time-series and cross-sectional variations. Figure 13.1 illustrates the average values of the climate risk measures over time. In panel A, the acute risk series shows six significant spikes over the past 17 years, with corresponding topics discussed in the conference calls identified and labeled for each spike. For instance, the spike in 2005 is attributed to the devastating and lasting impact of *Hurricane Katrina*, which inundated the New Orleans area. Conversely, the chronic risk series has remained relatively stable over the past two decades, with notable spikes occurring only between 2012 and 2014, during which the term *abnormal weather* was frequently mentioned.

Panel B plots the time series for the transition climate risk measure, revealing a consistent rise from the beginning of the sample period until 2008Q3, followed by a gradual decline back to 2005 levels. This recent downtrend aligns closely with the trajectory of U.S. greenhouse gas emissions. Notably, spikes in 2006, 2008, 2011, and 2015 correspond to increased discussions around *energy efficiency* and *renewable energy*. Panel C compares the average transition risk measures with and without proactive keywords, normalized by their

TABLE 13.2: Comparison of Frequency of Keywords. The table presents the statistics of the top-10 bigram/unigram in the transition risk measure developed by Li et al. (2024), and compares them to the reported statistics of the two measures constructed using earnings call transcripts data by Hassan et al. (2019) and Sautner et al. (2023), respectively. In Panel A, we show the fweight, the total frequency of mentions, the number of earnings calls that mentioned the bigram/unigram, and the number of unique firms discussing the bigram/unigram, respectively. fweight is calculated in the same way as described in Table 13.1. In Panel B, we present the statistics of the other two measures as reported in Table 2 of Hassan et al. (2019) and Sautner et al. (2023). Note that the fweight of the political risk measures in Hassan et al. (2019) is scaled up by 10^5 - here we divide their reported number by 10 to make their fweight comparable to that of Li et al. (2024). Also, note that the frequency reported by Sautner et al. (2023) is summarized across the global sample from 2002 to 2020 - here we multiply it by the percentage of U.S. firm transcripts from 2002 to 2018 to make their frequency comparable to that of Li et al. (2024)

Panel A: The transition risk measure by Li et al. (2024)

Transition risk measure				
U.S. Sample				
Bigram	*fweight* $= \frac{Freq_{b,P}}{B_P} \times 10^4$	*Freq*	*No of Earnings calls*	*No of Firms*
energy efficiency	32,512.0	7,738	3,086	568
renewable energy	29,104.3	6,663	2,591	456
the solar	28,819.0	6,623	2,475	418
clean energy	21,372.2	5,117	1,171	281
alternative energy	18,367.0	4,160	1,638	483
superior energy	12,482.7	3,354	120	34
higher energy	11,273.8	2,806	2,026	694
new energy	10,878.1	2,503	1,193	609
the renewable	10,564.8	2,389	1,467	355
the ecosystem	10,036.0	2,590	1,866	651

Panel B: Other risk measures

Political risk (Hassan et al., 2019, Table 2)			*Climate change exposure* (Sautner et al., 2023, Table 2)	
U.S. sample			*U.S. sample*	
Bigram	*fweight* $= \frac{Freq_{b,P}}{B_P} \times 10^4$	*Freq*	*Bigram*	*Freq*
the constitution	20.1	9	renewable energy	6,205
the states	13.4	203	electric vehicle	3,780
public opinion	11.9	4	clean energy	2,557
interest groups	11.8	8	new energy	1,807
of government	11.6	316	climate change	1,739
the GOP	10.2	1	wind power	1,691
in Congress	7.8	107	wind energy	1,604
national government	6.8	7	energy efficient	1,550
social policy	6.2	1	greenhouse gas	1,358
the civil	6.1	64	solar energy	998

(A) Physical risk measures

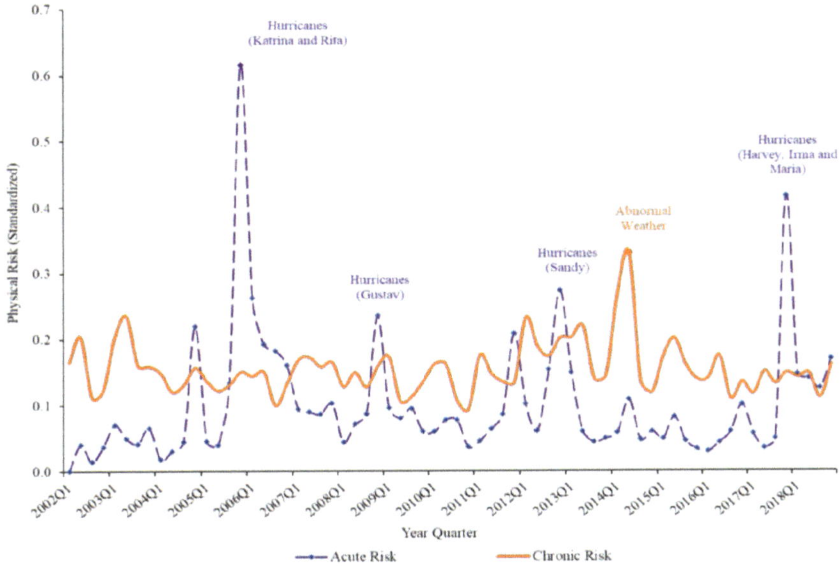

(B) Transition risk measure

respective standard deviations. Since 2008, the two series have diverged significantly, with firms that adopt proactive strategies showing considerably lower transition risk compared to their levels in 2008.

13.4.3 Industry variations

Industries have distinct exposures to climate risk, prompting us to analyze variations in the climate risk measures developed by Li et al. (2024) across different sectors. The authors conduct regressions of the various climate risk measures against industry dummies,

(c) Proactive and nonproactive components of transition risk measure

FIGURE 13.1: **Firm-level** $ClimateRisk_{i,t}$. These panels report the average of firm-level $ClimateRisk_{i,t}$ over time. Panels A and B show the time-series average of firm-level *acute risk*, *chronic risk*, and *transition risk* (divided by its SD in the time series), respectively. Each spike is labeled with the corresponding topics discussed in the conference calls which contribute to the increase in each type of climate risk. Panel C plots the time-series average of proactive and nonproactive components of transition risk, divided by their corresponding SDs, based on a subsample of firms with positive transition risk.

while controlling for time and state fixed effects. Figure 13.2 displays the coefficients for the NAICS two-digit industry dummies, using other services (NAICS 81) as the reference category.

Panel A indicates that utilities encounter the highest acute physical climate risk of all industries, trailed by agriculture, mining, transportation, and construction. Many activities in these sectors occur outdoors, making them vulnerable to disruptions from natural disasters. Panel B reveals similar trends, with some notable exceptions: utilities still have a high chronic physical climate risk (second-highest overall), but arts and recreation show the highest chronic risk, followed by agriculture in third place. The observed variations across industries largely align with their respective exposures to both acute and chronic climate risks.

Panel C reveals broader variations in transition risk compared to physical climate risk, with utilities and transportation experiencing significantly higher transition risk than other sectors. Conversely, service industries exhibit notably lower transition risk. Panel D highlights variations in proactive transition risk measures, showing that utility firms are more inclined to employ proactive keywords during discussions of transition risk. In contrast, firms in mining, information, and real estate tend to use these proactive terms less frequently. These patterns align well with the overall industry-level exposure to climate regulatory risk.

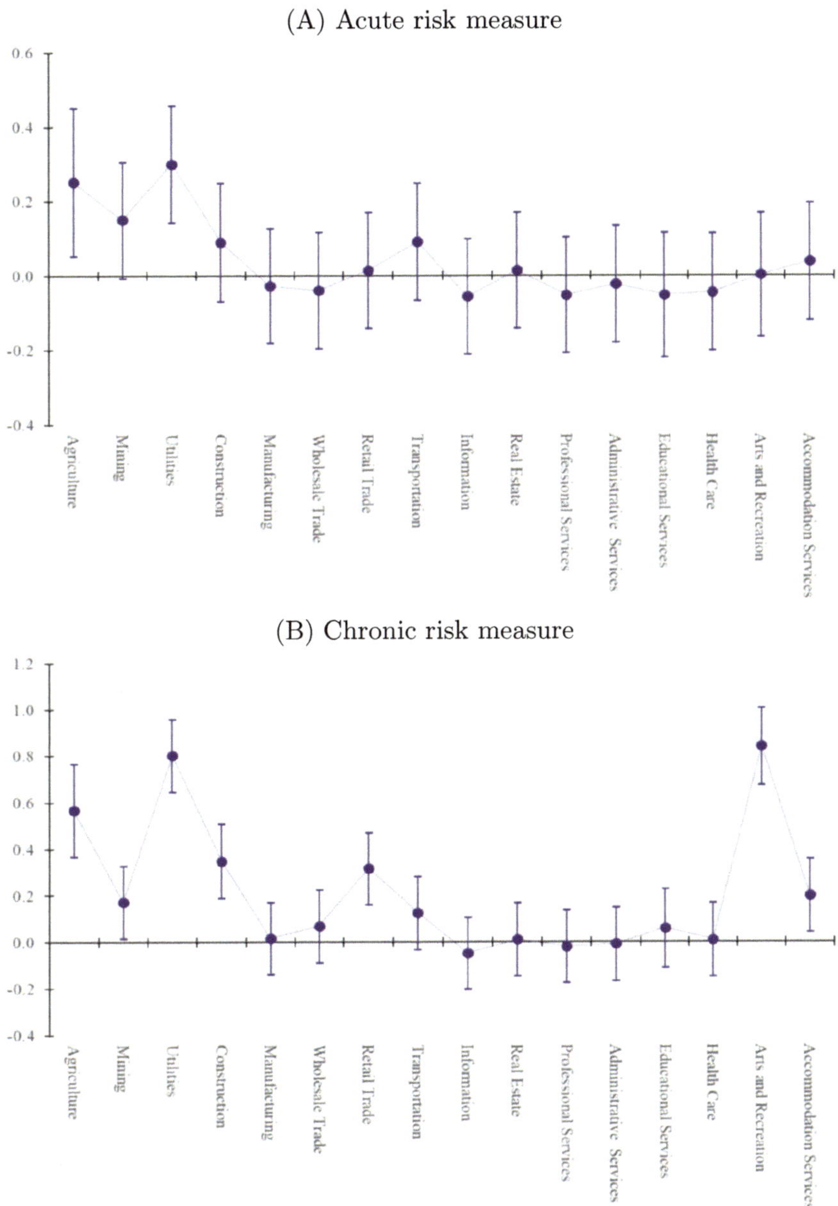

13.4.4 Firm-level variations

Table 13.3 presents excerpts from transcripts with the highest $ClimateRisk_{i,t}$. The transcripts reflecting the highest acute climate risk are from Edison International and PG&E Corporation, California's largest utility companies, both associated with some of the state's deadliest wildfires. In terms of chronic risk, discussions of abnormal weather and weather variability are highlighted, with the highest chronic climate risk transcript sourced from Suburban Propane Partners, a utility company focused on propane heating.

The transcript with the highest transition climate risk comes from CDTI Advanced Materials, a company focused on automotive emissions control solutions in the U.S. During a discussion on August 11, 2011, the company highlighted how "states such as California

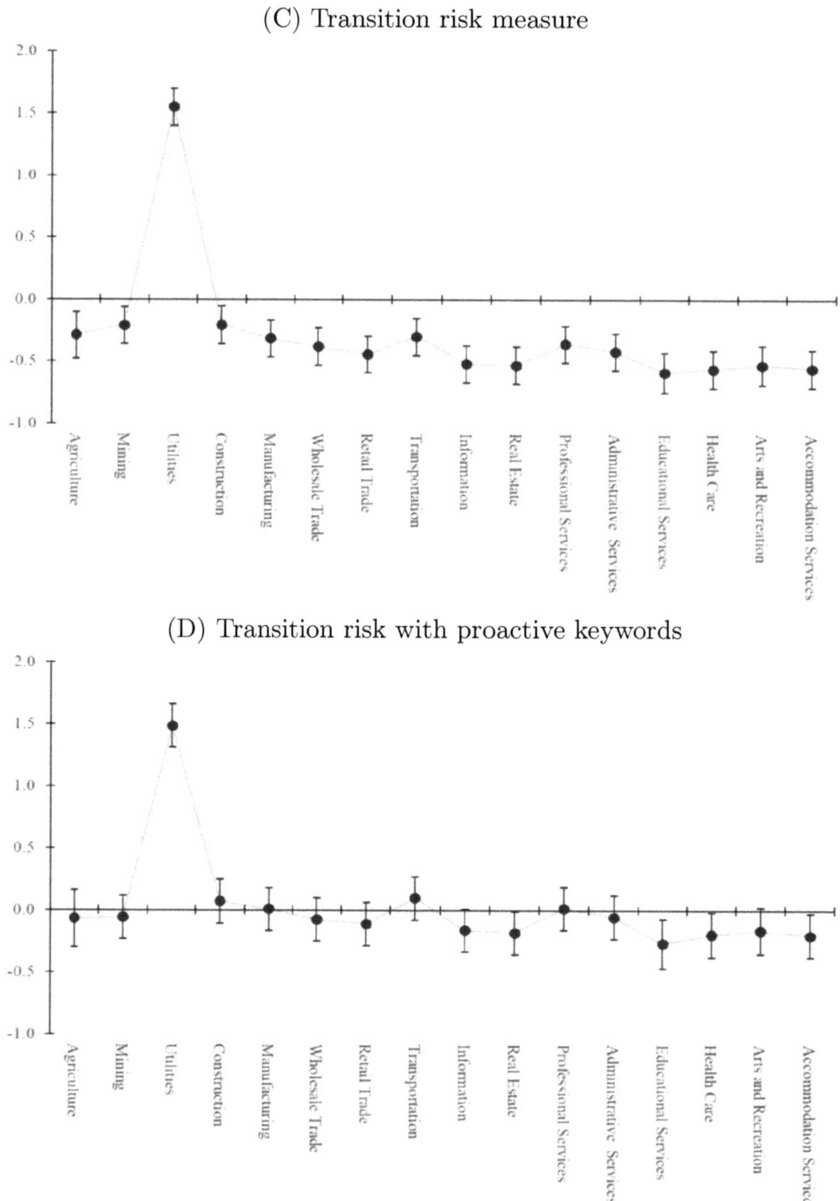

FIGURE 13.2: Industry variations in $ClimateRisk_{i,t}$. These panels plot the coefficients for industry (NAICS two-digit) fixed effects and their corresponding 95% interval from regressions of acute climate risk (panel A), chronic climate risk (panel D), transition risk (panel C), and the proactive transition risk (panel D). Time and state fixed effects are controlled in each regression. The reference industry is other services (NAICS 81).

continue to demonstrate their commitment to on-road diesel emission reductions through innovative programs to drive early adoption by truck operators." Other notable transcripts with high transition climate risk are from New Jersey Resources Corp, Magnetek Inc., and Lime Energy Co, all of which provide clean or renewable energy services.

TABLE 13.3: Excerpts in transcripts with highest climate risks. The table presents the excerpts in the earnings call transcripts with the highest acute, chronic and transition climate risks, respectively. The values of climate risk measures are ranked before winsorization. For acute and chronic climate risks, the corresponding climate-related keywords and risk synonyms are reported. For transition climate risk, only the climate-related keywords are reported

Firm	*Date*	*Climate risk*	*Value*	*Keywords*	*Text surrounding the keywords*
Edison International	Oct. 30, 2018	Acute risk	40.00	Wildfire; Uncertainty	We also have the flexibility at these entities to obtain both short and long-term debt while we continue to evaluate options as we work through uncertainty around the wildfire liability and cost recovery.
PG&E Corp	Nov. 5, 2018	Acute risk	39.85	Wildfire; Risks	Our expanded Community Wildfire Safety Program was established after the 2017 wildfires to implement additional precautionary measures intended to reduce or further reduce wildfire risks.
Patriot Transportation	Nov. 30, 2017	Acute risk	35.63	Hurricane; Unpredictable	Hurricane Irma more directly impacted our operations as the state of Florida shut down for 2 or 3 days. This type of business is generally less productive with long lines, unpredictable traffic patterns and other negative occurrences leading to inefficient utilization of our equipment.
Sotherly Hotels Inc	Nov. 8, 2016	Acute risk	32.40	Hurricane; Unsure	Heading into that markets' high winter season we are unsure what the effects may be. The impact of hurricane Matthew on our portfolio in early October was significant.
Talos Petroleum LLC	Nov. 5, 2008	Acute risk	29.00	Storm; Risk	We're also actively engaged in a program of accelerated idle well abandonment to mitigate the ongoing risk of future storms.
Suburban Propane Partners	Nov. 15, 2018	Chronic risk	77.72	Weather; Variability	While the heating season presented some extreme weather variability, average temperatures across our service territories were 8% cooler than the prior year.
Sport Chalet Inc	Feb. 6, 2013	Chronic risk	63.22	Unseasonably warm; Uncertainty	Unseasonably warm and dry weather coming on top of a bad winter sports season last year, combined with our customers' general economic uncertainty along with our desire to be less promotional, all contributed to the slight decrease in comparable store sales.
Idacorp Inc	Feb. 18, 2016	Chronic risk	61.79	Precipitation; Chance	According to the National Oceanic Atmospheric Administration, in March through May, we are looking at about a 33% to 40% chance of above-normal precipitation in the southern portion of our service area and normal precipitation levels in the northern portion.
CH Energy Group inc	Apr. 24, 2002	Chronic risk	52.52	Weather; Risk	A certain amount of variation from normal, either above or below normal degree days was a variation or risk that we retained. Then there was a wider range where we would be compensated if weather were warmer than normal.
Southern Company Gas	Oct. 30, 2013	Chronic risk	51.63	Weather; Unpredictable	Given where you see the rates today, when you're coming up for the 2014 expirations, do you expect – doesn't seem to have been much movement in the market. Is there anything out there that you think might have a significant impact, other than unpredictable weather?
CDTI Advanced Materials Inc	Aug. 11, 2011	Transition risk	464.9	Emission Reductions	Looking at the domestic growth opportunities, we think that the economic recovery, although a little bumpy, is spurring growth in our business and with our distributor network. Additionally, states such as California continue to demonstrate their commitment for on-road diesel emission reductions through innovative programs to drive early adoption by truck operators.
New Jersey Resources Corp	May. 4, 2018	Transition risk	298.2	Clean Energy	I talked about our strategy to provide our customers with reliable, affordable and clean energy services. To execute that strategy, we remain focused on natural gas, energy efficiency, and clean energy investments.
Magnetek Inc.	May. 9, 2012	Transition risk	267.5	Renewable Energy	Some of the growth we experienced in our served industrial markets was offset by lower sales in renewable energy, namely, wind inverters, which declined by more than $3 million year over year to about $2.4 million in the quarter.
Lime Energy Co	Aug. 12, 2009	Transition risk	267.2	Energy Efficiency	This counterbalance truly reflects the underlying strength of our business model and supports our efforts to date in the rapid deployment of tailored energy efficiency solutions to the public and utility marketplaces.
Enel X North America Inc	Aug. 7, 2008	Transition risk	256.7	Clean Energy	Various factors, ranging from unprecedented regulatory support for clean energy solutions, to rising fuel and construction costs, have made the value proposition of our scalable solutions stronger and more important than ever.

13.5 The determinants of the firm-level climate risk measures

Firm climate risks are endogenous, and correlated with firm fundamentals. In this section, we show how aggregate, sectoral, and firm-level variations, along with specific firm characteristics, contribute to the climate risk measures developed by Li et al. (2024). This analysis helps understand the distinct factors influencing climate risk at different levels.

13.5.1 Variance decomposition

We begin with a variance decomposition analysis to determine the extent to which firm-level characteristics and various fixed effects account for the variations in the three climate risk measures developed by Li et al. (2024). Panel A of Table 13.4 presents the R^2 values from multiple specifications that aim to explain these measures. The results show that the combination of time, state, and industry fixed effects accounts for only 2%, 3.4%, and

TABLE 13.4: Characteristics of climate risk measures. Panel A reports the results on the adjusted R^2 from a projection of $ClimateRisk_{i,t}$ on various sets of fixed effects. Column 1 reports the adjusted R^2 of the regressions with acute climate risk as the dependent variable and different sets of fixed effects as the independent variables. Column 2 reports the change/improvement in adjusted R^2 relative to a benchmark. The benchmark for regressions in the first block is zero (no fixed effects). The benchmark for regressions in the second and third blocks is the fourth row in the first block (Time + NAICS2 fixed effects). The analysis in columns 3 and 4 use chronic climate risk as the dependent variable, and columns 5 and 6 use transition climate risk as the dependent variable. Panel B presents regressions of acute risk, chronic risk, all transition risk, and proactive transition risk on a variety of lagged deterministic variables. Industry by time fixed effects are included in all regressions in panel B. Standard errors are clustered at the firm level. t-statistics are shown in parentheses. *p <.1; **p <.05; ***p <.01

A. Variance decomposition

	Dep var					
	Acute $risk_{i,t}$		*Chronic* $risk_{i,t}$		*Transition climate* $risk_{i,t}$	
Model specification	*Adj.* R^2	Δ	*Adj.* R^2	Δ	*Adj.* R^2	Δ
Time	.009		.001		.005	
Time + State	.015	.015	.008	.008	.018	.018
Time + County	.025	.025	.040	.040	.073	.073
Time + NAICS2	.016	.016	.030	.030	.118	.118
Time + NAICS3	.026	.026	.043	.043	.161	.161
Time + NAICS4	.028	.028	.075	.075	.199	.199
Time + State + NAICS2	.020	.020	.034	.034	.124	.124
State + NAICS2 × Time	.028	.012	.042	.012	.136	.018
State × Time + NAICS2	.037	.021	.037	.007	.118	.000
State × Time + NAICS2 × Time	.042	.026	.045	.015	.130	.012
County × Time + NAICS2 × Time	.063	.047	.064	.034	.121	.003
Firm + Time	.080	.064	.200	.170	.655	.537
Firm + Time + Firm Attributes	.080	.064	.200	.170	.655	.537
Firm + Time + Firm Attributes + NAICS2 × Time	.088	.072	.209	.179	.673	.555
Firm + Time + Firm Attributes + State × Time	.097	.081	.209	.179	.657	.539
Residual	.903		.791		.343	

B. Firm characteristics of climate risk measures

Dep var	*Physical* $risk_{i,t}$		*Transition* $risk_{i,t}$	
	Acute	*Chronic*	*All*	*Proactive*
	(1)	(2)	(3)	(4)
$\log(\text{Asset})_{i,t-1}$	0.0074**	0.0055	0.0138**	0.0104***
	(2.147)	(0.992)	(1.982)	(2.989)
$\text{CapEx}_{i,t-1}$	−0.0011	−0.0025	−0.0008	0.0007
	(−0.845)	(−1.184)	(−0.314)	(0.480)
$\text{PPE}_{i,t-1}$	0.1204***	0.1410***	0.2768***	0.0943**
	(4.687)	(2.907)	(2.773)	(1.976)
$\text{Book Leverage}_{i,t-1}$	−0.0095	0.0194	−0.1163***	−0.0328*
	(−0.463)	(0.578)	(−3.318)	(−1.819)
$\log(\text{No\_Analysts})_{i,t-1}$	−0.0094	−0.0455***	−0.0135	−0.0218***
	(−1.463)	(−3.414)	(−0.854)	(−3.190)
$\text{Institution\%}_{i,t-1}$	0.0304*	−0.0028	−0.0767	−0.0122
	(1.680)	(−0.067)	(−1.132)	(−0.498)
$\text{Institution HHI}_{i,t-1}$	0.0133	−0.0724	0.0413	0.0239
	(0.444)	(−1.240)	(0.430)	(0.553)
$\text{Transition Risk}_{i,t}$				0.5858***
				(11.711)
Industry × Time FE	Yes	Yes	Yes	Yes
N	124,682	124,682	124,682	124,682
Adj. R^2	.0243	.0419	.129	.386

12.4% of the variations in the acute, chronic, and transition risk measures, respectively. While including interactions among state, industry, and time does enhance the model's explanatory power, the improvements are minimal. Even in the most rigorous specification controlling for county-by-time and industry-by-time fixed effects, the model still explains less than 12.5% of the variations in any of the climate risk measures. This leaves over 87% of the variations attributed to firm-level or other unique factors. This finding highlights that, unlike data on natural disasters or market-wide news regarding long-term climate risks used by Engle et al. (2020), the bulk of the variations in their climate risk measures stem from firm-specific dynamics.

When we include firm and time fixed effects, the model explains 9.7%, 20.9%, and 65.7% of the variations in the acute, chronic, and transition climate risk measures, respectively. Adding firm-level attributes and interactions between industry and time or state and time improves the predictive power for the two physical risk measures but not for the transition risk measure. This indicates that the climate risk measures capture both cross-firm differences and variations within firms regarding climate risk exposure. For instance, the transition risk measure for Sempra Texas Holdings rose significantly from 11.10 in Q1 2006 to 184.97 in Q3 2013.

13.5.2 Correlation with firm characteristics

Panel B of Table 13.4 presents the results of regressions relating climate risk measures to a list of important firm-level attributes, all lagged by one quarter, to better understand the first-stage question - what types of firms tend to have higher values in the climate risk measures that Li et al. (2024) constructed. Li et al. (2024) control for industry-by-time fixed effects to account for time-varying heterogeneity across industries. Among all the variables, the first set is related to a firm's physical exposure to climate risk. They find an overall positive relationship between the firm's physical assets and the climate risk measures: the coefficients for PPE and total assets are positive and significant in most regressions. The results suggest that firms that hold more physical assets tend to face higher climate risk exposure.

A second set measures the firm's financial leverage. The authors find it to be negatively correlated with the transition risk, but not with the two physical risk measures, suggesting that highly leveraged firms tend to be associated with lower transition risk exposure. This evidence is consistent with the evidence documented by Ginglinger and Moreau (2023), who find that firms with greater climate risk have lower leverage even after controlling for firm characteristics known to determine leverage.

The final set of measures included in their regressions capture external characteristics of firms, such as the number of analysts covering the firm and institutional ownership. These measures could be correlated with how climate issues are discussed in earnings calls. They find a negative relationship between the number of analysts and our climate risk measures, with one measure being statistically significant. This suggests that firms are less likely to discuss climate-related topics during earnings conference calls when a large number of analysts cover the firm. This could be because with higher analyst coverage, ample information already may be available regarding the firm's climate exposure, leading to less need for discussion during earnings calls. They do not find a significant correlation between institutional ownership and our climate risk measures.

Lastly, they analyze the correlations between the proactive component of the transition risk measure and firm-level attributes, controlling for transition risk itself. Their results show that firms that carry low leverage, hold more physical assets, and are followed by fewer analysts tend to respond more proactively to rising climate risk.

13.6 Applications

13.6.1 Asset prices

The pricing of climate risks in financial markets represents a critical topic in the climate finance literature, as underscored by recent studies (Giglio, Kelly, and Stroebel, 2021; Stroebel and Wurgler, 2021). In particular, regulatory risks related to transition risk are perceived as a significant climate risk over the next 5 to 30 years. In this section, we discuss whether transition risk is reflected in the stock market valuations based on the findings in Li et al. (2024). To assess a firm's valuation, the authors utilize Tobin's q, defined as the ratio of a firm's market value to the replacement value of its physical assets. Tobin's q is a well-established measure in the literature, as it accounts for the value of intangible assets in addition to physical capital. A high (or low) Tobin's q indicates that a firm possesses more (or fewer) valuable intangible assets, making it particularly suitable for analyzing the predictable effects of a firm's transition risk on its overall value. The authors proceed to estimate the following regression specification:

$$Tobin's\ q_{i,t+k} = \beta \cdot TransitionRisk_{i,t} + \gamma \cdot X_{i,t-1} + \zeta_{j,t} + \epsilon_{i,t}, \tag{13.1}$$

where the dependent variable is Tobin's q in quarter $t + k$ ($k = 1, 3, 5$); The primary explanatory variable is $TransitionRisk_{i,t}$; $X_{i,t-1}$ encompasses the firm's assets, CapEx, PPE, book leverage, ROA, and energy price exposure, all of which were constructed using earnings call data. Additionally, industry-by-quarter fixed effects are included to control for both observable and unobservable time-varying heterogeneity across different industries.

In panel A of Table 13.5, the authors present the baseline results based on the entire sample, reporting the regression outcomes of Tobin's q over various lead times k (1, 3, and 5). In columns 1–3, $TransitionRisk_{i,t}$ serves as the primary explanatory variable. The coefficients for $TransitionRisk_{i,t}$ are consistently negative and significant at the 1% level. For instance, the findings in column 1 indicate that a one-standard-deviation increase in the transition risk measure is associated with approximately a 0.0389 decrease in Tobin's q in the following quarter, which equates to about 1.9% of the mean value.[9] Furthermore, the coefficient's magnitude increases slightly when Tobin's q is used as the dependent variable over longer horizons ($k = 3, 5$), suggesting that the estimated pricing effect does not reverse. Thus, the results presented in this table indicate that transition risk is indeed priced in equity markets.

In columns 4–6, the authors incorporate both proactive and nonproactive components of their transition risk measures as the main explanatory variables. They also include the firm-level *Action Index* as an additional control, which captures the overall proactiveness of firms not facing high transition risk. This index is calculated based on the total frequency of mentions of proactive verbs throughout the entire transcript (excluding those within ± 1 sentences of climate-related discussions), normalized by the transcript length. Interestingly, the findings reveal that while the coefficient for nonproactive transition risk is negative and

[9]This estimate aligns with findings in several studies examining the pricing effect of carbon emissions. For example, Matsumura, Prakash, and Vera-Munoz (2014) find that an increase in carbon emissions from the 25th to the 75th percentile correlates with a 4.2% decrease in market value of equity (calculated as the product of the number of shares outstanding and year-end stock price). Additionally, both Bolton and Kacperczyk (2021) and Chava (2014) estimate a significant carbon premium, with impacts of 2.85% on stock returns per one-standard-deviation change in total emissions levels across countries and 1.04% on the expected cost of equity for U.S. firms with greater net environmental concerns, respectively.

TABLE 13.5: Pricing of climate risk. This table presents results from firm level regressions testing the relation between the transition climate risk measures (standardized) and Tobin's q. It reports the results from regression analysis of firm's Tobin's q in different lead time periods (*t+1*, *t+3*, and *t+5*) on the lagged transition climate risk (in quarter *t*). In columns 1–3, the key explanatory variable is the overall transition risk measure. In columns 4–6, the transition risk measure is decomposed into the proactive and nonproactive components and add *Action Index* as an additional control variable. All specifications include time-varying firm-level control variables, including lagged (i.e., *t*-1) *log(Asset)*, *CapEx*, *PPE*, *Book Leverage*, and *ROA (%)*. Industry (NAICS three-digit) by quarter fixed effects are also included in all tests. Standard errors are double clustered at the firm and quarter levels. *t*-statistics are shown in parentheses. *$p < .1$; **$p < .05$; ***$p < .01$

Dep Var	*Tobin's* $q_{i,t+h}$					
	$h = 1$	$h = 3$	$h = 5$	$h = 1$	$h = 3$	$h = 5$
	(1)	(2)	(3)	(4)	(5)	(6)
Transition risk$_{i,t}$	−0.0389***	−0.0404***	−0.0418***			
	(−3.828)	(−3.978)	(−4.179)			
Transition risk/Nonproactive$_{i,t}$				−0.0416***	−0.0407***	−0.0405***
				(−4.764)	(−4.466)	(−4.719)
Transition risk/Proactive$_{i,t}$				0.0047	0.0005	−0.0024
				(0.618)	(0.081)	(−0.326)
Energy Price Exposure$_{i,t}$	−0.0634***	−0.0577***	−0.0547***	−0.0601***	−0.0545***	−0.0517***
	(−5.814)	(−5.382)	(−5.059)	(−5.503)	(−5.077)	(−4.784)
Action Index$_{i,t}$				−0.0583***	−0.0520***	−0.0462***
				(−4.458)	(−3.941)	(−3.455)
Firm attributes$_{i,t-1}$	Yes	Yes	Yes	Yes	Yes	Yes
Industry × Time FE	Yes	Yes	Yes	Yes	Yes	Yes
N	111,691	104,442	97,470	111,691	104,442	97,470
Adj. R^2	.182	.210	.171	.218	.211	.210
F-test				−0.0463***	−0.0412***	−0.0381***

statistically significant, the coefficient for the proactive measure is not significant. Furthermore, the difference between these two coefficients is statistically significant at the 1% level. This suggests that equity markets tend to discount firms that do not actively manage their transition risk, whereas those that are proactive in addressing the risk do not face the same penalty.

13.6.2 Corporate real outcomes

13.6.2.1 Investment

The theoretical literature presents varied predictions concerning investment in uncertain conditions. Authors such as Bernanke (1983), Pindyck (1991), Pindyck and Solimano (1993), and Dixit and Pindyck (1994) suggest that investment declines during periods of high uncertainty. In contrast, other studies, including Oi (1961), Hartman (1972, 1976), Abel (1983), Roberts and Weitzman (1981), and Bar-Ilan and Strange (1996), indicate a positive correlation. Ultimately, the empirical investigation of how firm-level investment responds to climate risk exposure remains an open question.

Table 13.6 showcases the results from an analysis where CapEx, scaled by total assets, serves as the dependent variable. The findings in columns 1–3 reveal a positive but statistically insignificant coefficient for $ClimateRisk_{i,t}$, indicating no significant difference in future investment between firms experiencing high transition risk and those facing low transition

TABLE 13.6: Predicting the firm's investment. This table reports estimates of the regressions of capital expenditures (in different lead time periods) on transition risk. Columns 1–3 shows the results using *Transition risk* as the key explanatory variable. In columns 4–6, the transition risk measure is replaced with its two components: nonproactive and proactive transition risk. *Action index* is added as an additional control variable. Lagged *log(Asset)* and industry by quarter fixed effects are included in all tests. Standard errors are double clustered at the firm and quarter levels. *t*-statistics are shown in parentheses. *p <.1; **p <.05; ***p <.01

Dep Var	$CapEx_{i,t+h}$					
	$h=1$	$h=3$	$h=5$	$h=1$	$h=3$	$h=5$
	(1)	(2)	(3)	(4)	(5)	(6)
Transition $\text{Risk}_{i,t}$	0.0480	0.0498	0.0421			
	(1.460)	(1.499)	(1.256)			
Transition Risk/$\text{Nonproactive}_{i,t}$				0.0236	0.0158	0.0074
				(0.783)	(0.547)	(0.252)
Transition Risk/$\text{Proactive}_{i,t}$				0.0460***	0.0623***	0.0641***
				(2.951)	(3.455)	(3.287)
Energy Price $\text{Exposure}_{i,t}$	0.0896*	0.1120**	0.1186**	0.0867*	0.1030**	0.1136**
	(1.901)	(2.435)	(2.557)	(1.836)	(2.236)	(2.452)
Action $\text{Index}_{i,t}$				0.0058	0.0782***	0.0201
				(0.274)	(3.442)	(0.909)
Firm $\text{Attributes}_{i,t-1}$	Yes	Yes	Yes	Yes	Yes	Yes
Industry × Time FE	Yes	Yes	Yes	Yes	Yes	Yes
N	126,099	118,043	110,313	126,099	118,043	110,313
Adj. R^2	.439	.437	.435	.439	.438	.435
F-test				−0.0224	−0.0465	−0.0567*

risk. In columns 4–6, Li et al. (2024) explore the differences in responses between firms that proactively address climate risk and those that do not. This is accomplished by regressing the same set of firm-level outcomes against transition risk, both with and without proactive keywords. The coefficients for two transition risk measures are positive; however, only the coefficient for proactive transition risk reaches statistical significance at the 1% level. This suggests that firms taking proactive measures tend to increase their CapEx in response to heightened transition risk. Specifically, a one-standard-deviation increase in transition risk with proactive keywords in t correlates with a 0.046-percentage-point increase in CapEx in $t+1$ and approximately a 0.06-percentage-point increase in $t+3$ and $t+5$.[10] The estimates are economically significant, accounting for approximately 1.6%–2.3% of the average investment level. The final row presents the differences between the two coefficients and their significance levels based on F-tests, indicating that the disparity in CapEx between proactive and nonproactive firms, both facing high climate risk, is significant at the 10% level in $t+5$.

13.6.2.2 Innovation

To achieve net-zero emissions or decarbonization, firms must inevitably innovate or alter their business practices. Consequently, innovation emerges as a crucial response for those

[10] Although not fully detailed in this table, the analysis indicates that the coefficients for the firm-level action index (i.e., *Action index*) remain positive over five consecutive quarters, with varying magnitudes. The coefficient is 0.0058 in $t+1$, increasing to 0.0782 in $t+3$ before declining toward zero. While not significant in $t+1$, it becomes statistically significant at the 1% level in $t+2$ and $t+3$, then loses significance thereafter. These findings imply that a higher action index generally correlates with increased CapEx investments with a two-quarter lag, even for firms not experiencing high climate risk.

encountering high transition risk. Li et al. (2024) consider two indicators of innovation: R&D expenditure scaled by assets and the proportion of green patents. In panel A of Table 13.7, the results for R&D expenditures are presented. The analysis reveals negative and statistically significant coefficients for $ClimateRisk_{i,t}$ in columns 1-3, indicating that high transition risk correlates with lower R&D expenditures. Specifically, a one-standard-deviation increase in transition risk corresponds to a decrease of 0.0529 to 0.0565 percentage points in future R&D expenditures. Notably, these coefficients remain relatively stable across different horizons of R&D expenditures. The findings in columns 4-6 further indicate that the negative association between transition risk and future R&D expenditures is significant only for firms that do not respond proactively, whereas proactive firms do not exhibit the same trend.

In panel B of Table 13.7 , Li et al. (2024) present the results of regressions that utilize measures of green patents as the dependent variable. The findings in columns 1–4 encompass all firms and employ an indicator for having at least one green patent. The analysis reveals a positive yet statistically insignificant coefficient for $ClimateRisk_{i,t}$ in columns 1 and 2, indicating no statistically significant difference in future green patent filings between firms with high and low transition risk. In columns 3 and 4, the authors further examine the distinctions in responses between firms that proactively address climate risk and those that do not. The coefficients for two transition risk measures are both positive, but only the coefficient for proactive transition risk attains statistical significance at the 5% level. This suggests that firms responding proactively are more inclined to innovate through green patenting when faced with high transition risk. Specifically, a one-standard-deviation increase in transition risk associated with proactive keywords in t correlates with a 0.01-percentage-point increase in the likelihood of filing a green patent in $t+1$ and an additional 0.01-percentage-point increase in $t+2$. The estimates are economically meaningful, representing approximately 12.5% of the average probability of filing a green patent.

The findings in columns 5–8 focus exclusively on patenting firms, employing the ratio of green patents to the total number of patents filed by a firm as the dependent variable. The authors find positive and statistically significant coefficients for $ClimateRisk_{i,t}$ at the 1% level in columns 5 and 6, indicating that firms facing high transition risk are associated with a higher ratio of green patents. Specifically, a one-standard-deviation increase in transition risk linked to proactive keywords in t corresponds to a 0.0321-percentage-point increase in the ratio of green patents in $t+1$ and a 0.0331-percentage-point increase in $t+2$. In columns 7 and 8, the results indicate that the coefficients for two transition risk measures are both positive and significant; however, the coefficient for proactive transition risk is slightly higher and more significant (at the 5% level or lower). A one-standard-deviation increase in transition risk with proactive keywords in t is associated with a 0.0251-percentage-point increase in the ratio of green patents in $t+2$.

13.6.2.3 Employment

Another strategy available to firms in response to increasing climate risk involves adjusting their employment levels, such as through plant closures, layoffs, or hiring freezes. Executives at public companies often implement layoffs and plant closings to enhance productivity, manage ongoing risks, and satisfy capital markets. The findings by Li et al. (2024) presented in panel C of Table 13.7 reveal a negative and statistically significant relationship at the 5% level between transition risk and the logarithm of employment levels over the subsequent two years. Specifically, a one-standard-deviation increase in transition risk correlates with an approximate 0.02-percentage-point decrease in a firm's employment stock. This negative relationship is predominantly driven by firms that do not respond proactively, as it does not reach statistical significance for those that do take proactive measures.

TABLE 13.7: Predicting the firm's other responses. In panel A, *R&D Investment* (in *t+1*, *t+3*, *t+5*) is regressed on overall transition risk measure (in columns 1-3) and decomposed transition risk measures (in columns 4–6), respectively. In columns 1–4 of panel B, the dependent variable is *I(Green patents)*, a dummy variable equals one if a firm has at least one green patent, and zero otherwise. The sample includes all firms. In columns 5–8 of panel B, the dependent variable is *Green patents ratio*, the number of green patents scaled by the total number of patents in the year. The sample is restricted to the firms with patents. In panel C, the dependent variable is the natural logarithm of the firm's employment level. All specifications include lagged (i.e., *t*-1) *log(Asset)* as the control variable. Industry by quarter fixed effects are included in all tests. Standard errors are double clustered at the firm and quarter levels. *t*-statistics are shown in parentheses. $^{*}p<.1$; $^{**}p<.05$; $^{***}p<.01$

A. R&D expenditures

Dep Var	*R&D investment$_{i,t+h}$*					
	$h=1$	$h=3$	$h=5$	$h=1$	$h=3$	$h=5$
	(1)	(2)	(3)	(4)	(5)	(6)
Transition Risk$_{i,t}$	−0.0556**	−0.0529**	−0.0565**			
	(−2.393)	(−2.269)	(−2.391)			
Transition Risk/Nonproactive$_{i,t}$				−0.0548***	−0.0550**	−0.0557***
				(−2.697)	(−2.611)	(−2.675)
Transition Risk/Proactive$_{i,t}$				−0.0033	0.0026	−0.0025
				(−0.252)	(0.179)	(−0.167)
Energy Price Exposure$_{i,t}$	−0.1797***	−0.1782***	−0.1741***	−0.1603***	−0.1605***	−0.1561***
	(−7.917)	(−7.756)	(−7.701)	(−7.174)	(−7.107)	(−7.015)
Action Index$_{i,t}$				−0.2684***	−0.2463***	−0.2485***
				(−11.879)	(−10.685)	(−10.758)
Firm Attributes$_{i,t-1}$	Yes	Yes	Yes	Yes	Yes	Yes
Industry × Time FE	Yes	Yes	Yes	Yes	Yes	Yes
N	128,503	119,997	111,971	128,503	119,997	111,971
Adj. R^2	.388	.381	.373	.398	.389	.382
F-test				−0.0515**	−0.0576**	−0.0532**

B. Green patents (annual)

Dep Var	*I(Green patents)$_{i,t+h}$*				*Green patents ratio$_{i,t+h}$*			
	$h=1$	$h=2$	$h=1$	$h=2$	$h=1$	$h=2$	$h=1$	$h=2$
Sample	*All Firms*				*Firms with Patents Only*			
	(1)	(2)	(3)	(4)	(5)	(6)	(7)	(8)
Transition Risk$_{i,t}$	0.0115	0.0080			0.0321***	0.0331***		
	(1.598)	(1.276)			(3.914)	(3.883)		
Transition Risk/Nonproactive$_{i,t}$			0.0057	0.0014			0.0189**	0.0165*
			(0.804)	(0.200)			(2.252)	(1.836)
Transition Risk/Proactive$_{i,t}$			0.0090**	0.0103**			0.0193**	0.0251***
			(2.396)	(2.168)			(2.959)	(3.395)
Energy Price Exposure$_{i,t}$	0.0373***	0.0352***	0.0359***	0.0340***	0.0207**	0.0231**	0.0188**	0.0208**
	(4.930)	(4.683)	(4.903)	(4.708)	(2.525)	(2.367)	(2.339)	(2.307)
Action Index$_{i,t}$			0.0063	0.0026			−0.0019	−0.0018
			(1.580)	(0.745)			(−0.820)	(−0.908)
Firm Attributes$_{i,t-1}$	Yes	Yes	Yes	Yes	Yes	Yes	Yes	Yes
Industry × Time FE	Yes	Yes	Yes	Yes	Yes	Yes	Yes	Yes
N	32,713	32,713	32,713	32,713	9,372	8,186	9,372	8,186
Adj. R^2	.199	.193	.192	.193	.103	.110	.109	.122
F-test			−0.0033	−0.0089			−0.0004	−0.0086

C. Employment (annual)

Dep Var	*log(Employment)$_{i,t+h}$*			
	$h = 1$	$h = 2$	$h = 1$	$h = 2$
	(1)	(2)	(3)	(4)
Transition risk$_{i,t}$	−0.0195**	−0.0202**		
	(−2.050)	(−2.047)		
Transition risk/nonproactive$_{i,t}$			−0.0188*	−0.0197*
			(−1.731)	(−1.692)
Transition risk/proactive$_{i,t}$			−0.0000	0.0002
			(−0.003)	(0.022)
Energy price exposure$_{i,t}$	0.0032	0.0007	−0.0041	−0.0067
	(0.249)	(0.050)	(−0.325)	(−0.508)
Action index$_{i,t}$			0.0634***	0.0624***
			(6.647)	(6.267)
Firm attributes$_{i,t-1}$	Yes	Yes	Yes	Yes
Industry × Time FE	Yes	Yes	Yes	Yes
N	32,165	30,533	32,165	30,533
Adj. R^2	.776	.771	.778	.773
F-test			−0.0188	−0.0199

13.7 Glossary

Acute physical climate risk: The risks that stem from extreme weather events linked to climate change, such as hurricanes, floods, wildfires, and heatwaves. These risks are sudden and severe, often leading to significant immediate damage to infrastructure, supply chains, and production capabilities.

Chronic physical climate risk: The risks refer to the long-term, gradual effects of climate change. These include persistent changes like rising average temperatures, sea-level rise, changing precipitation patterns, and prolonged droughts. These risks can disrupt operations, impact productivity, and require long-term adaptation measures.

Transition climate risk: The risks associated with the transition to a low-carbon economy. These risks arise from changes in policy, technology, market conditions, or consumer preferences aimed at mitigating climate change. Examples include stricter environmental regulations, carbon pricing, shifts in investor sentiment, and increased competition from renewable technologies.

CO_2 intensity: Sum of CO_2 emissions of all plants operated by the firm, scaled by the total assets.

MSCI Climate Change Index (CCI): From MSCI ESG rating database, the climate change materiality weight × the climate change risk rating. The materiality weight measures the importance of climate change to a firm's financial performance. The climate change risk rating is calculated as (10 - climate change theme score). Climate change theme score is a continuous variable ranging from 0 to 10, with higher value indicating better performance (i.e., lower risk).

References

Abel, A. B., 1983, Optimal investment under uncertainty, *American Economic Review* 73, 228–233.

Adiwardana, D., M.-T. Luong, D. R. So, J. Hall, N. Fiedel, R. Thoppilan, Z. Yang, A. Kulshreshtha, G. Nemade, Y. Lu, et al., 2020, Towards a human-like open-domain chatbot, *arXiv, preprint, https://arxiv.org/abs/2001.09977.*

Ahrens, C. D., 2008, *Meteorology Today: An Introduction to Weather, Climate, and the Environment*, 9th edition (Cengage Learning).

Baker, S. R., N. Bloom, and S. J. Davis, 2016, Measuring economic policy uncertainty, *Quarterly Journal of Economics* 131, 1593–1636.

Bar-Ilan, A., and W. C. Strange, 1996, Investment lags, *American Economic Review* 86, 610–622.

Barnett, M., W. Brock, and L. P. Hansen, 2020, Pricing uncertainty induced by climate change, *Review of Financial Studies* 33, 1024–1066.

Bartram, S. M., K. Hou, and S. Kim, 2022, Real effects of climate policy: Financial constraints and spillovers, *Journal of Financial Economics* 143, 668–696.

Bekkerman, R., and J. Allan, 2004, Using bigrams in text categorization, Technical Report IR-408, Center of Intelligent Information Retrieval, University of Massachusetts.

Berg, F., J. F. Koelbel, and R. Rigobon, 2022, Aggregate confusion: The divergence of esg ratings, *Review of Finance* 26, 1315–1344.

Bernanke, B. S., 1983, Irreversibility, uncertainty, and cyclical investment, *Quarterly Journal of Economics* 98, 85–106.

Bolton, P., and M. Kacperczyk, 2021, Global pricing of carbon-transition risk, Technical report, Columbia Business School.

Brown, S. V., and J. W. Tucker, 2011, Large-sample evidence on firms' year-over-year md&a modifications, *Journal of Accounting Research* 49, 309–346.

Chava, S., 2014, Environmental externalities and cost of capital, *Management Science* 60, 2223–2247.

Chen, J. V., V. Nagar, and J. Schoenfeld, 2018, Manager-analyst conversations in earnings conference calls, *Review of Accounting Studies* 23, 1315–1354.

Choi, D., Z. Gao, and W. Jiang, 2020, Attention to global warming, *Review of Financial Studies* 33, 1112–1145.

Christensen, D. M., G. Serafeim, and A. Sikochi, 2022, Why is corporate virtue in the eye of the beholder? the case of esg ratings, *The Accounting Review* 97, 147–175.

Dixit, A. K., and R. S. Pindyck, 1994, *Investment Under Uncertainty* (Princeton University Press).

Engle, R. F., S. Giglio, B. Kelly, H. Lee, and J. Stroebel, 2020, Hedging climate change news, *Review of Financial Studies* 33, 1184–1216.

Giglio, S., B. Kelly, and J. Stroebel, 2021, Climate finance, *Annual Review of Financial Economics* 13(1), 15–36.

Ginglinger, E., and Q. Moreau, 2023, Climate risk and capital structure, *Management Science* 69, 7492–7516.

Golosov, M., J. Hassler, P. Krusell, and A. Tsyvinski, 2014, Optimal taxes on fossil fuel in general equilibrium, *Econometrica* 82, 41–88.

Hartman, R., 1972, The effects of price and cost uncertainty on investment, *Journal of Economic Theory* 5, 258–266.

Hartman, R., 1976, Factor demand with output price uncertainty, *American Economic Review* 66, 675–681.

Hassan, T. A., S. Hollander, L. van Lent, and A. Tahoun, 2019, Firm-level political risk: Measurement and effects, *Quarterly Journal of Economics* 134, 2135–2202.

Hassan, T. A., S. Hollander, L. van Lent, and A. Tahoun, 2024, The global impact of brexit uncertainty, The Journal of Finance, 79(1), pp. 413–458.

Hassan, T. A., S. Hollander, L. Van Lent, M. Schwedeler, and A. Tahoun, 2023, Firm-level exposure to epidemic diseases: Covid-19, sars, and h1n1, *Review of Financial Studies* Advance Access published May 17, 2023, 10.1093/rfs/hhad044.

Hong, H., F. W. Li, and J. Xu, 2019, Climate risks and market efficiency, *Journal of Econometrics* 208, 265–281.

Li, Q., H. Shan, Y. Tang, and V. Yao, 2024, Corporate climate risk: Measurements and responses, *The Review of Financial Studies* 37, 1778–1830.

Litterman, R., C. E. Anderson, N. Bullard, B. Caldecott, et al., 2020, Managing climate risk in the us financial system, *Report of the Climate-Related Market Risk Subcommittee, U.S. Commodity Futures Trading Commission.*

Matsumura, E. M., R. Prakash, and S. C. Vera-Munoz, 2014, Firm-value effects of carbon emissions and carbon disclosures, *The Accounting Review* 89, 695–724.

Nordhaus, W., 2019, Climate change: The ultimate challenge for economics, *American Economic Review* 109, 1991–2014.

Nordhaus, W. D., 1977, Economic growth and climate: the carbon dioxide problem, *American Economic Review* 67, 341–346.

Oi, W. Y., 1961, The desirability of price instability under perfect competition, *Econometrica* 29, 58–64.

Pindyck, R. S., 1991, Irreversibility, uncertainty, and investment, *Journal of Economic Literature* 29, 1110–1148.

Pindyck, R. S., and A. Solimano, 1993, Economic instability and aggregate investment, *NBER Macroeconomics Annual* 8, 259–303.

Roberts, K., and M. L. Weitzman, 1981, Funding criteria for research, development, and exploration projects, *Econometrica* 49, 1261–88.

Roller, S., E. Dinan, N. Goyal, D. Ju, M. Williamson, Y. Liu, J. Xu, M. Ott, K. Shuster, E. M. Smith, et al., 2020, Recipes for building an open-domain chatbot, *arXiv, preprint, https://arxiv.org/abs/2004.13637*.

Sautner, Z., L. Van Lent, G. Vilkov, and R. Zhang, 2023, Firm-level climate change exposure, *Journal of Finance* 78, 1449–1498.

Shapiro, J. S., 2021, The environmental bias of trade policy, *Quarterly Journal of Economics* 136, 831–886.

Stroebel, J., and J. Wurgler, 2021, What do you think about climate finance?, *Journal of Financial Economics* 142, 487–498.

Tan, C.-M., Y.-F. Wang, and C.-D. Lee, 2002, The use of bigrams to enhance text categorization, *Information Processing & Management* 38, 529–546.

Xu, J., A. Szlam, and J. Weston, 2021, Beyond goldfish memory: Long-term open-domain conversation, *arXiv, preprint, https://arxiv.org/abs/2107.07567*.

Chapter 14

ESG scores to the rescue: objectives, procedures, and uses

Roberto Rigobon
MIT Sloan School of Management, USA

In this chapter I would like to address three questions regarding the Environmental, Social and Governance rating scores. First, what is the objective or purpose of the scores? In other words, what is the problem they are supposed to solve? Second, how the scores are produced, and what are the weaknesses and challenges of those procedures? Finally, why they are needed? And in my personal view, why they are a fundamental piece for a plausible solution to the problems we are facing.

14.1 What is the objective of the ESG scores?

It's easy to feel overwhelmed by the immense challenges our world is facing today. We are navigating a landscape fraught with unsustainability on multiple fronts. Broadly, these challenges can be categorized into three critical dimensions: how we treat the environment, how we treat each other, and how we treat ourselves.

First, the way we interact with the environment is increasingly unsustainable. Deforestation, pollution, and the overconsumption of natural resources are pushing our planet to its limits. Climate change is accelerating, leading to more frequent and severe natural disasters. Our reliance on fossil fuels continues to deplete the earth's reserves, contributing to global warming. Second, on a societal level, the way we treat each other is often marked by unkindness, discrimination, and conflict. Across the globe, social injustices are rampant—whether it's racial inequality, gender discrimination, or economic disparity. The lack of empathy and understanding between different groups only exacerbates these problems, leading to further alienation and social breakdown. Finally, at the individual level, many of us struggle with how we treat ourselves. The pressures of modern lifework, social expectations, and the constant influx of information—can lead to mental health issues such as anxiety and depression. We often neglect our well-being, prioritizing productivity over rest, and external validation over self-care.

When the firms contribute to these problems it can be argued that it is a form of externality. Firms affecting the environment, treating workers incorrectly, or creating circumstances in which mental health problems are exacerbated, are all externalities. If we take this point of view the first question is how economics can help address those problems.

DOI: 10.1201/9781032636252-14

14.1.1 The externalities

Externalities occur because there is a gap between social benefits and private benefits. The person or institution making the individual decision faces different returns than what the society does.

The economic solutions to externalities are all in principle similar—create an intervention that aligns private benefits with social benefits. There are three possible solutions: First, the creation of property rights. Clearly defining and enforcing property rights can lead to negotiations between parties to address externalities efficiently, provided transaction costs are low (Coase, 1960). The property rights include the use of marketable permits—such as the case of cap-and-trade systems used frequently to deal with greenhouse emissions. After all, a permit is indeed a property right.[1]

Second, there are institutional arrangements that can collectively solve the problem. This is the main contribution from Ostrom (1990) where she argues that institutions that create collective action can internalize their externalities. For example, if one firm creates an externality on another, by merging the two firms, the problem of externalities is resolved.

Third, government intervention in the form of taxation and subsidies offer another possibility. The taxation is known as Pigouvian Taxes (Pigou, 1920), but the regulation can take other forms, such as a cap on emissions. All those solutions imply that a cost is imposed unto the firm causing a negative externality (and a reward is given to the one producing a positive externality).

ESG rating agencies deal with many forms of externality. In our research (Berg, Kölbel, and Rigobon, 2022) we categorized the raw data from the rating agencies and one striking fact is that about 60 % of the indicators and categories are indeed associated with criminal activity. So, that is an externality which almost every country is already regulating and punishing—still they are occurring. But the ESG rating agencies go beyond measuring outcomes. They measure (and estimate) actions and intentions. The next section addresses how the scores are produced, and what are the challenges.

14.2 How ESG scores are computed?

In this section I follow closely the decomposition from Berg, Kölbel, and Rigobon (2022). The procedure of most rating agencies can be approximated as follows:

1. The rating agency decides which environmental, social, and governance issues it considers important. In some circles this is called materiality or double materiality. In the end this step answers the question: What matters?

2. The rating agency decides which indicator (or indicators) to use to measure the desired issue. Here the decision is not easy. What to measure? Outcomes, Procedures, Intentions? Most will be tempted to jump to outcomes. That is an incorrect approach—almost for certain.

   A simple example might help: the outcome of a person running over (and killing) a pedestrian can be an accident or a first-degree murder. The difference between the two is NOT the outcome, but the context and intention.

   Many people in ESG are obsessed with measuring outcomes just because they believe intentions are impossible to measure. This is a mistake and one that is very likely to produce unfairness and inefficiencies.

[1] See Tietenberg (2006) and the references therein.

3. Once the indicators have been computed, the rating agency puts them together and aggregates them into a firm-level score: the ESG scores. This step is also quite complicated. The aggregation of issues into a single score implies trade-offs across ESG issues that most individuals might not agree with.

In our paper we reclassify all the indicator level data from several rating agencies (as of this moment we are updating our paper to see what has happened the last 10 years). We created a more detailed set of categories—significantly more granular than SASB for instance—to be able to help us compare across rating agencies.

Figure 14.1 lists a keyword from each of the categories to have a sense of what is included. The definitions are significantly longer, and almost all of them are defined with negative connotation. For instance, "human rights" refers to the likelihood of violating human rights. "packaging" is about harmful, wasteful, and environmentally damaging packaging.

- access to basic services,
- access to healthcare,
- animal welfare,
- anticompetitive practices,
- audit,
- biodiversity,
- board,
- board cultural
- diversity, board
- gender diversity,
- business ethics,
- chairman ceo separation,
- children and forced labor,
- climate risk mgmt,
- collective bargaining,
- community and society,
- corporate governance,
- corruption,
- customer relationship,
- diversity,
- electromagnetic fields,
- employee development,
- employee turnover,
- energy,
- environmental fines,
- environmental mgmt. system,
- environmental policy,
- environmental reporting,
- equator principles,
- esg incentives,
- financial inclusion,
- fleet emissions,
- forests,
- ghg emissions,
- ghg policies,
- global compact membership,
- gmos,
- green buildings,
- green products,
- hazardous waste,
- health and safety,
- hiv programes,
- human rights,
- labor practices,
- lobbying,
- media ethics,
- non ghg air emissions,
- organic products,
- ozone depleting gases,
- packaging,
- philanthropy,
- pri signatory,
- privacy and it,
- product quality and safety,
- public health,
- qms certification,
- remuneration,
- reporting quality,
- resource efficiency,
- responsible marketing,
- risk management,
- shareholders,
- site closure,
- supply chain,
- sustainable finance,
- systemic risk,
- taxes, toxic spills,
- waste,
- water

FIGURE 14.1: Categories.

14.2.1 Measurement problem

As I hope it is clear, this is not an easy task. First, there are challenges on how to measure: Does anyone know how to measure the probability that the supply chain uses children and force labor? Or the intention to use? Or the negligence not to check? Furthermore, how to operationalize the meaning of absence of discrimination through statistics? Is it the share of women, or how they are treated? Would you prefer a firm that has 47 percent women in management that treats every women horribly, or one that has 32 percent that treats women equally, and it has 32 just because too few women applied for the job?

I'm obsessed with measurement. I know what you're thinking: "Measurement? Are you kidding me?" Trust me, I get it—I hear that reaction daily from my wife. But measurement fascinates me because, quite frankly, I think we're terrible at it. We don't measure the quality of relationships; we mostly document breakups. We don't measure sadness; we diagnose depression. We have no clue when a society is simmering with frustration; we notice only when riots erupt. We don't measure human rights abuses until there's a mass exodus or the discovery of mass graves. We don't track workplace discrimination; we wait for a scandal.

Across the board, we measure poorly, and the fact that we've made good decisions in the past feels more like a miracle than evidence of wisdom.

What Are We Doing Wrong? We measure infrequently, focusing on the extremes of the distribution, relying on perception, and using insufficient statistics.

First, infrequent measurement: Take health, for example. If you're a responsible person, you might visit the doctor once a year. I remember my annual checkups vividly. For 20 years, my doctor would say the same three things: I'm older, balder, and half a pound heavier than the year before. One day, I asked her, "How does telling people they're older, balder, and fatter encourage them to eat more broccoli?" The truth is, expecting people to change their behavior after a 30-minute annual lecture—filled with regret and humiliation—is naive at best.

Think about the economy. We measure income distribution, discrimination, and population growth once every ten years. GDP gets measured quarterly. Inflation is tracked monthly. But the stock market and exchange rates? Those are measured every second. So, which of these grabs our attention? Which ones have the power to influence our behavior? Naturally, it's the stock market and exchange rates—not because they're more important than GDP or income distribution, but simply because they're measured continuously.

Second, focusing on extremes: We tend to measure extremes rather than averages. For example, we don't measure drug use in a city; we count overdose cases in emergency rooms. This means we're focusing only on the heavy users while ignoring casual users, even though the latter represent the majority. Yet we still make decisions about the average based on the extremes. Drug cartels? They're smarter than us—they measure averages and the entire distribution.

In the state of Massachusetts, during the worst time of the opium epidemic, the state decided distribution of Naloxone to fire and police departments. The objective was to have the saving drug close to the individual that was suffering the overdose. The state measured the consumption of drugs by counting how many kids were dying or being admitted to the emergency room. The distribution of Naloxone reduced deaths dramatically. Is the reduction of death due to a reduction in the consumption of opioids? In other words, is the epidemic resolved? If anything, it is possible that the consumption increases given that the probability of dying of an overdose is smaller.

Third, relying on perception: Too often, our measurements are driven by perception, particularly media coverage. In 2013, stories of refugees dying in rafts while fleeing Syria dominated the news. But by 2023, even though the European Commission was tracking the same crisis, the media was silent. We only react to scandals, and our short bursts of outrage are insufficient for tackling complex societal problems that demand sustained effort.

Finally, using insufficient statistics: Sometimes, we measure things that, while meaningful, fail to capture the root issue. Take workplace gender equality. Instead of measuring hostility toward women, we count how many women there are. When only 5% of management positions are held by women, it's clear the environment is hostile. As that share increases to 10% or 20%, we can infer hostility is decreasing. But when the number reaches 35% or more, does moving from 35% to 40% really reflect a significant reduction in discrimination? While achieving 50% is a worthy goal, assuming that discrimination disappears at 50% is misguided. At lower levels, the share of women is a strong indicator of bias, but as the share grows, it becomes a weaker proxy.

The bottom line is that we measure infrequently, focus on extremes, rely on insufficient statistics, and act only when outrage compels us. If we want to solve problems, we need to measure thoughtfully, consistently, and comprehensively—long before they escalate into crises.

The ESG rating agencies are by far the best institution we have right now to address this problem.

14.2.2 Aggregation and scope problem

The second challenge is how to put these different categories into a single score. In our research we find that the aggregation rules—even though they are not exactly linear—are very close to being linear. In our original paper, we obtained R-squared close to 90 percent and we were not even controlling by industry (it was all cross sectional and we did not have that many firms—less than a thousand). Nevertheless, if the aggregation rules are linear or quasi-linear that implies that there are simple tradeoffs across ESG issues that might be inappropriate.

An example might be useful. Imagine that there are 4 firms that have scores on women treatment and greenhouse emissions as the table below. Assume that a higher number means better performance—meaning less emissions and better treatment. Assume those are the only two ESG issues measured

	Firm 1	Firm 2	Firm 3	Firm 4
GHG Emission	30	90	30	90
Women Treatment	30	30	90	90

I have chosen the numbers to make the point quite trivial. Clearly Firm 4 is the best. It has the highest score on all the ESG issues. Firm 1 is the worst. But how would you rank Firm 2 and Firm 3? One is destroying the environment, while the other one is hurting women. Of course, one answer is that every person can do the ranking they want. But this is not ice cream versus chocolate. As a society, which one would we chose as the better one? In my classes the students never agreed.

Furthermore, imagine that Firm 2 realizes they are treating women very badly and they decide to do an effort to change it. The score increases from 30 to 32, but in that process their GHG emissions come down to 85. The reason is that management has a limited capacity to pay attention to the externalities the firm creates, and has limited resources, and therefore by devoting attention and resources to one, they let the other slip. Would you find that outcome acceptable?

I know that if GHG emissions remains at 90, and women treatment improves from 30 to 32 we would see that as desirable. But what if that is not what really happens? If the decline in GHG emissions score is large would anyone recommend Firm 2 to stop improving women treatment issues?

When I ask anyone in class what they prefer, apples, bananas, or coconuts, everybody is able to rank the products within seconds. When I ask them to rank priorities on environment, poverty, and civil rights it takes them a little bit more time, but they can rank them. The next question is the complicated one.

When I ask people their willingness to exchange apples and bananas—i.e., how many bananas will you be willing to exchange for one apple so you feel equally happy—almost everybody can tell me their exchange. Why can we do that? Because we have made that decision many times in the supermarket where a clear price is the medium of comparison. So, we have practiced the tradeoff and understand it.

When I ask, however, what is your willingness to exchange an improvement in civil rights while environment will be deteriorated, people have a very complicated time making a choice. There are no prices to compare, we have very little practice asking the question—because we truly find the conversation complicated and most like easy conversations—and therefore we tend not to have an answer.

ESG rating agencies are fundamental in this process. It is, to my knowledge, the only entities that actually ask this question repeatedly. They have to answer on our behalf and conceivably their answer is not good enough to encompass what everybody feels.

Because of the complexity of the ESG scores procedures there are big discrepancies across rating agencies. In our paper we decided to construct a decomposition using the raw indicators from the rating agencies. Figure 14.2 depicts our procedure. We split the reasons of the discrepancy between measurement, scope, and aggregation.

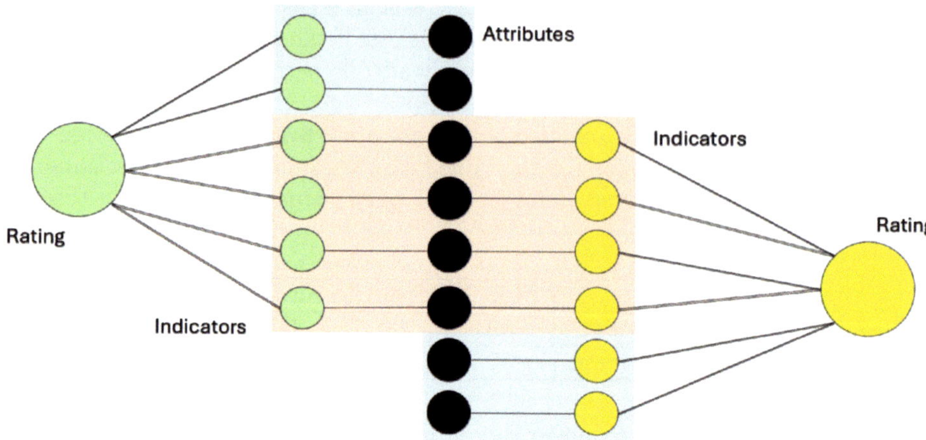

FIGURE 14.2: Decomposition.

The measurement problem (pink square) occurs at the category level, when indicators in the same categories differ across rating agencies—the difference is not on the actual number but on the relative rankings that those indicators imply. We find that about 40 to 50 percent of the discrepancy can take place due to measurement errors.

The scope problem is when categories are excluded from a rating agency—meaning that such rating agency does not even have a score for that category. In fact, in our paper it is the case that not every rating agency agrees with what needs to be measured. From our list of 60+ categories, most rating agencies cover about 30 to 40 of them. So, "what matters" is a question that has been implemented quite differently across rating agencies. And a very simple reason why this is the case, is that what matters is different depending on the audience.

For an investor, how the different categories are related to the risk of the firm matters. In this case, a small restaurant in Miami is affected by the water level rise. The investment will be not very valuable when it is under the water—except as tourist attraction. However, for an environmentalist, the restaurant has a very small impact on emissions and on the water level rise. Rating agencies construct different indexes with the objective to tailor to different informational requirements.

We find that the scope contributes to a sizeable proportion of the discrepancy—between 20 to 40 percent. Although what the ESG rating agencies do is limited and probably incorrect, most people in the world do not know what to put into an index and what to leave out, what are the tradeoffs, and more importantly, they do not want to engage in these extremely difficult conversations. This is understandable in the sense that most of us would not believe that women treatment and CO_2 emissions are substitutable. So, treating them as such is indeed uncomfortable. However, regardless of how we feel, the ESG rating agencies have to deal with the fact that, in reality, they tend to be. The reason that occurs is because there is a common resource used to deal with the different ESG issues, and because that resource—our time—is bounded. For example, when we pay attention to the Russian-Ukranian conflict, the Israel-Hamas conflict, and every single trade restriction the Trump administration issues and takes away within the day, it is hard to pay attention to sexual harassment regulation. Like it or not, we are paying less attention to some issues that are quite important.

14.2.3 The verification problem

There is another problem with the ESG rating agencies—which is the problem of verification.

A credit rating agency has similar difficulties. They need to measure the creditworthiness of a client, and that includes two categories (using our language): willingness to pay, and ability to pay. The ability to pay can be confirmed by the assets the individual owns; its debts and any liquidity needs. These measures are not perfect because unobservable contingent debts will be difficult to assess.

The willingness to pay is much harder to measure. Especially for someone who has never received a credit. The way this is assessed is by looking at the behavior of the individual. In practice there are many credit transactions individuals are involved with—paying the rent, the phone company, all the other utilities, the credit card, the mortgage, and the car payment, etc. In the end, the idea is to assume that prior behavior is an estimate of future behavior.

Both aspects of the credit score are indeed substitutable. A person with a higher willingness to pay can compensate a smaller ability to pay. So, the fact that we treat them as substitutes in the calculation of the score is not something that will make us feel uncomfortable.

The credit rating agencies have an additional advantage. Even when the measurement is imperfect, there are many credit events. That means that their models are checked and calibrated. That is not the case of ESG.

We do not have that many events to which we can compare and calibrate the ESG scoring. Some occur with certain regularity. For example, in the US we only know of 1.2 thousand cases a year of women being fired due to pregnancy related issues, from a universe of 75.54 million working women. So, in principle there are many "events". But these are only the cases when the women were discriminated against badly enough so that they decided to go to court. There is a selection bias and therefore this is a low bound of the actual treatment. However, some statistics could be performed. There are other issues that occur very irregularly: water level rise, global warming, etc. And when you include specificity—such as a hurricane hitting New Orleans—the number of events is virtually zero. Even more so when we think about the attribution problem: estimating how much of the emissions caused by the Miami restaurant led to the severity of the typhoon hitting Bangkok three years later is a daunting task.

ESG rating agencies cannot verify how good their models are. There are no clear events and attributions that could help them tailor their models to capture reality better.

14.3 Why ESG ratings are needed?

14.3.1 The CO_2 example

The problems of environment, labor treatment, and many that ESG rating agencies measure are externalities that are much harder to deal with than the textbook case taught in economics. Let us concentrate on the case of CO_2 emissions as an example.

There are several dimensions to evaluate the policies. (i) a solution should be implementable—meaning that if something needs to be taxed then it should be taxed. (ii) the ESG issues are global in nature—so, solutions have to be able to be adopted and implemented worldwide. (iii) the ESG issues are multidimensional and solutions should be able to incorporate them.

Let's start with government intervention. If firms (directly or indirectly through consumers) are producing too many CO_2 emissions, in economics we know that a solution is to tax the activity that is producing it, and to subsidize any activity that reduces or have negative emissions (such as forestry). So, in practice any action that emits is taxed, while any action that reduces CO_2 in the atmosphere is subsidized and encouraged.

The quota works is similar way—it limits the activity that can be performed instead of taxing it. A market trading for such CO_2 emissions permits works as a tax because those that benefit the most from the emission pay the highest price, making it onerous to the rest.

CO_2 emissions are a global problem so therefore this taxation should be global. Furthermore, if two communities experience the same cost of emitting CO_2, then they should experience the same taxation. So, the taxes paid for every ton of CO_2 emissions in a poor country should be the same as the taxes paid by a firm in a rich country when both have the same cost. In practice if the poorer suffers a higher marginal damage it should actually pay more, not less. This is one of those merciless outcomes of optimality.

From this perspective it should be clear that it is unlikely to solve the problem of CO_2 emissions based on pure regulation. The likelihood that countries such as Russia, China, and India accept a CO_2 tax that has the exact same dollars-per-ton price than Germany, Sweden and Denmark is small. It is also difficult to imagine a world in which Indonesia and Bangladesh pay a much higher tax than Switzerland because the water level rise is likely to affect Indonesia and Bangladesh more than Switzerland.

Although carbon taxes are the best possible solution, I think they are mostly feasible in rich countries.

The second solution to the externalities is the creation of a communal revenue pool. This works relatively well in local problems. For example, assume there is a small community of fishermen, and their problem is that they need to reduce overfishing to allow the fish pool to regenerate. The fishermen can combine their revenues in such a way that if anyone of them overfish the additional profits are shared among everybody.[2] That means that the fisherman that is deviating from the desired behavior (fishing too much) does not benefit from the deviation. Of course, this is very similar to the communist ideals that fail so dramatically in the world. It is the case that this might work in small villages, but it is unlikely to work in the economy. Especially when differences in effort imply different amounts of fish being caught.

Some international accords are trying to create such institutions for CO_2 emissions. This is clearly a good idea. The effect of these institutions, unfortunately, will be (and has been) small—if any. The reason is that it is difficult to share international revenues and cost among firms and countries that have little incentive to cooperate. This is the reason why the community-based solutions tend to work very well in exactly that—small communities.

This solution is extremely effective in small settings: For example, Cabo Pulmo in Baja California Sur, Mexico. This is an extraordinary example how the community with a clever design (by the University of San Diego) was able to reduce the biodiversity impact due to tourism. It did require the prohibition of fishing in the area, but it was truly the community and NGO coalitions that made the transformation from fishing to eco-tourism what made

[2]This is in essence the brilliant proposal from Elinor Ostrom.

everything possible. It is a very clever implementation of Elinor Ostrom's design using random assignments, which increased biodiversity in 20 years by more than 4 times.

One possibility of creating an international institution is the creation of a "environmental club". Where membership is hard to obtain, and membership is conditional on behavior and outcomes. These clubs, however, are only useful if consumers believe it is a useful sign of commitment to "protect the environment". And the problem of the clubs is that you need a club for each one of the ESG issues. However, the advantage of clubs is that are supra-national. A proper design has not occurred yet.

The limitations of global taxation and subsidies, as well as the ineffectiveness of international institutions, leaves us with the third option only. One that relies on the private sector solution to be able to provide the right incentives to reduce CO_2 emissions.

Property rights are usually associated with the ownership of a home, or a stock, or any claim on the equity of a firm or asset. However, property rights can be more general and abstract. For example, in a cap-and-trade system, the property right is the right to pollute. The simple property right (ownership of the asset) is extremely hard to implement in ESG issues. For the CO_2 problem, it is hard to define a one cubic meter of air; for the labor mistreatment problem it is immoral and impossible to define the ownership of 10 percent of a human.

Some cap-and-trade systems have worked, but they have problems been implemented globally. For example, the ETS in Europe covers only Europe, some key sectors (Power generation, large plants, aviation, and shipping), and has many exceptions. The plan is to include buildings in 2027. No plan for consumers' CO_2 emissions yet. The scheme covers about 40 percent of the EU's greenhouse gas emissions.

There are some concerns about the ETS that I believe have not been materialized. Some criticize the system because it can lead to carbon leakages—that the dirty plant is moved away from EU and into another country. Although some anecdotal evidence exists, there is little evidence of this behavior in the overall economy. But this possibility creates a problem in the initial allocation of permits. Many industries have received free permits because the EU is concerned about the possibility of leakages in sectors that are highly competitive. From the economic point of view this is a terrible idea. The EU gives the sectors something that should have a positive price—a right to pollute—at a zero price defeating the whole point of the taxation to avoid the externality. And this is done because they are concerned the sector cannot handle the competitive pressure which in a globalized world have zero chance of disappearing. More research needs to be done, but if sectors that are not competing internationally as strongly as consumer products can barely manage the taxation implied by the cap-and-trade, it is clear to me that a better design is needed.[3]

One alternative has been the use of the carbon border adjustment mechanism—where the leakage and the free allowances can be eliminated. The carbon border adjustment mechanism is equivalent to an import substitution protection policy. These policies have historically failed because this is also equivalent to an export tax (Lerner Symmetry Theorem). The exporting firms are going to request exceptions—as they have been requesting already, and the issuance of free allowances shows.

In conclusion, the externality caused by CO_2 emissions cannot be solved entirely with a property right—because the property right is badly defined, nor it can be solved with regulation—given that regulation is local and nor global—and it cannot be solved by communal ownership or international institutions because they are quite weak in their enforcement. Not surprisingly we have not been able to limit emissions in a significant manner.

[3]About 40 percent of the general allowances are free. In the aviation industry more than 80 percent of the allowances are free. The plan is to reduce them to zero.

14.3.2 ESG and product labels

Therefore, for most of the ESG issues we are forced to explore other possibilities. The case of product labels is particularly interesting. Let us concentrate on the case of "fair trade" label. Let me describe a well-functioning market:

International competition in the supply of food products creates pressures to reduce the margins of poor rural producers. These pressures can lead—and have led—to very bad economic conditions for millions of individuals in poor countries. This can be thought as an unintended consequence of competition—an externality. Especially when the distribution of the agricultural products can be oligopolistic.

Consumers care about the unfairness caused by the excessive competition. Some firms also care and want to signal that they are different—that they would be willing to pay more for the products—and trade in a relatively more benign condition.

So, an independent institution measures the degree of "fair trade" and for the firms that are doing something to ameliorate the situation, it places the tag of "fair trade". That label cannot be purchased (in principle) and needs the firm to change their competitive behavior. The label can be displayed publicly, and consumers can confidently know that their expenses are being directed to companies that are ultimately helping poor farmers in emerging markets.

Furthermore, to make the market even better, there is an auditor that checks the behavior of the firms every year and verifies (and assures) that the firm placing the label is indeed independent.

14.3.3 Why are ESG ratings needed?

Why ESG ratings are needed? ESG ratings can become a solution to this problem. They work as product labels—such as organic, fair trade, etc.—therefore they are global in nature, the labeling is the enforcement, and the claim is the property right.

Assume that consumers, workers and investors care about the un-ethical behavior of firms. I believe there is enough evidence that this is true by looking at the emphasis of investment and consumers pursuing brands that are "green". Additionally, it should be clear that this is also the case when we observe firms desperate to make "green" statement, even when unfounded.

In a world where consumer and investors care about the greenness of firms, the ESG score can become a label. The firm uses the score as a claim, consumers and investors worldwide care about such claim, and the enforcement comes from the almost continuous scoring of the firms by the rating agencies. An example, assume that an ESG score from MSCI that gives a firm AAA. Being AAA means that the firm is green. This implies that consumers might pay a premium for that label, workers feel higher levels of fulfillment by working in a firm that is doing things correctly, and investors are willing to take a smaller rate of return because the firm is less risky in the future—lower retaliation, lower fines, etc. The enforcement mechanism is that the firm violates the conditions that make it a AAA. Finally, the AAA is not only about CO_2, but about a set of important ESG issues.

In summary, the firm benefits from the AAA label, the workers, investors, and consumers appreciate the label, and the rating agency—together with the auditing—become the enforcement mechanism, and the label deals with the multidimensional problem embodied in ESG. The label is global, the enforcement is global, and the label is enforceable and multidimensional.

As famously said by Denzel Washington (Robert McCall) in The Equalizer, "Progress, not Perfection". ESG is not perfect, but it has the chance to offer some progress.

References

Berg, F., J. F. Kölbel, and R. Rigobon, 2022, Aggregate confusion: The divergence of ESG ratings, *Review of Finance* 26, 1315–1344.

Coase, R., 1960, The problem of social cost, *Journal of Law and Economics* 3, 1–44.

Ostrom, E., 1990, *Governing the commons: The Evolution of Institutions for Collective Action* (Cambridge University Press).

Pigou, A. C., 1920, *The Economics of Welfare* (Macmillan and Co., London).

Tietenberg, T., 2006, *Environmental and Natural Resource Economics* (Pearson Education).

Part IV

Product Design and Specific Markets

Chapter 15

The value of green energy: optimal investment in mutually exclusive projects and operating leverage

Jerome Detemple
Questrom School of Business, USA

Yerkin Kitapbayev
Khalifa University of Science and Technology, UAE

This chapter studies investments in exclusive projects with different cost structures. The analysis incorporates the possibility of producing a stochastic revenue stream from two alternative technologies with a stochastic variable cost and a fixed cost, respectively, and accounts for endogenous operating decisions. The optimal investment decision is characterized by two possibly non-monotone boundaries. The chapter examines the effects of operating leverage on managerial policies, investment decisions and values, and carries out an application to power generation projects. The impact of knowledge acquisition, i.e., preliminary investments in knowledge assets, is assessed.

15.1 Introduction

This chapter studies investments in mutually exclusive projects with different cost structures, specifically a variable cost project (VCP) and a fixed cost project (FCP). It derives the optimal investment policy accounting for the endogenous operating decisions for each underlying project. Optimal investment is characterized by two nonlinear and possibly non-monotone boundaries satisfying a system of coupled integral equations. The option to invest in the best of the two projects increases the value of investment and postpones the optimal investment in the VCP relative to the stand-alone VCP. The optimal delay to investment in either project can be substantial even if individual projects are infinitely valuable. Operating leverage emerges as a key determinant of the investment policy. An extension of the model examines the behavior of preliminary investments in knowledge assets for the choice between alternative projects.

Choices between exclusive alternatives are prevalent in corporate decision-making. Common examples include choices between different production technologies (manufacturing), applications of a given technology (product choice), target firms (mergers and acquisitions), research programs (research and development), or advertising strategies (marketing). They are also quite common in individual decision-making, for example, education choices,

DOI: 10.1201/9781032636252-15

employment decisions, or entertainment selections. Alternatives may entail different revenue streams and cost profiles.

This chapter focuses on situations in which the underlying projects exhibit fundamental differences in cost structures, specifically in the mix between fixed and variable costs. An illustrative example is that of utilities, nowadays faced with nontrivial choices between competing technologies when they invest in new power production units. On the one hand they can build plants using fossil fuels to generate power (e.g., coal- or gas-fired plants). On the other hand, they can use Green Energy (hereafter GE; e.g., wind or solar plants). The distinguishing feature between these alternatives is their respective cost structures. The former case entails fixed and variable costs. Fixed costs consist of wages and maintenance costs, among others. Variable costs relate to the fuel used as an input in the production process. The price of fuel varies stochastically and is typically positively correlated with the electricity price and, hence increases with it. Such a technology has lower operating leverage, depends less on demand and is less risky as a stand-alone project. In the latter case, the GE input is free, and costs are independent of the price of electricity. Such a technology entails fixed costs, is more dependent on demand and has higher operating leverage. Hence, in essence, the choice between the two technologies is a choice between projects with lower and higher operating leverage.[1]

Motivated by this example, we study the problematic of investing in mutually exclusive alternatives consisting of a VCP and a FCP.[2] The timing of the decision to invest is endogenous. Moreover, each of the possible selections involves subsequent timing choices regarding operations. Hence, the firm holds an American-style real option on the maximum of two asset values, which are themselves American contingent claims. Some of the questions arising in this context are as follows. What is the optimal time to invest? Under what conditions is it optimal to invest in the FCP versus the VCP? What is the impact of operating leverage on these decisions? Applied to power generation projects, our model can be used to examine a host of issues. In particular, it can shed light on the relevance of subsidies and on the value of GE. Such issues are important not only for power operators but also for consumers and policy makers.

The model developed considers two underlying sources of uncertainty, that is, two state variables, one driving revenue and the other costs. The total contribution margin, that is, the difference between revenue and variable cost, determines the value of the VCP. Revenue is the main factor underlying the value of the FCP. Our first contribution is to solve the operating problem of the VCP manager in this setting with *two state variables*. In that regard, the value functions associated with the decisions to operate or idle the project are solutions to a system of coupled optimal stopping time problems. We provide an early switching premium (ESP) representation of the solution which decomposes each value function into the value achieved if the state of the project (operating or idling) is maintained forever and a premium, the ESP, which tallies the present value of the gains realized by optimally switching back and forth between operations states. Optimal decisions are then described by a pair of boundaries that depend on the cost factor. The ESP representation of the value functions enables us to show that these boundaries satisfy a system of coupled

[1]A second example is a decision to outsource. Outsourcing has become a common alternative for various functions traditionally located within a firm, for example, research, product development, manufacturing, sales, information technology, and administration. A typical decision involves a choice between exclusive alternatives entailing fixed costs (in-house) and variable cost (outsourcing). A third example is a decision, common for multinationals, to produce domestically or abroad. Because of exchange rate uncertainty, any FCP designed to meet a given domestic demand becomes a VCP in domestic currency if production is located abroad.

[2]A more general problem is the choice between projects with different combinations of fixed and variable costs. The methods developed in this chapter apply to such cases as well, although the details of those cases can be more challenging.

integral equations of Fredholm type. Although these equations are nonrecursive in nature, we are able to design an iterative algorithm to compute the boundaries. This scheme is easy to implement and quickly converges in our numerical experiments.

The operating decisions for the FCP are the same as for the VCP and involve a similar coupled optimal stopping time problem. But the problem is simpler because cash flows depend on revenue alone. The resultant switching boundaries are constant, and the problem reduces to a free-boundary problem. Optimal boundaries are solutions to a system of algebraic equations.

Our second and main contribution pertains to the investment decision of the firm, namely, the decision to undertake the VCP or the FCP. We show that the decision to invest is characterized by two boundaries. When the variable cost factor exceeds the first boundary, then immediate investment in the FCP is optimal. When it falls below the second boundary, it is optimal to invest in the VCP. Each of these boundaries depends on the other source of uncertainty, namely, revenue. Moreover, both boundaries can fail to be monotone in revenue. In the infinite horizon case, we show that they satisfy a system of coupled integral equations of Fredholm type. In the finite horizon case, the system of coupled integral equations is recursive in nature and a dynamic programming procedure can then be applied.

A project to invest in the best of two production technologies is an American-style dual strike compound max-option. It is a compound option, because the value of each underlying project embeds options to shut down or restart production when the relevant state variables reach certain thresholds; hence, it is inherently nonlinear. It is a dual strike option, because the cost of building a VCP differs from the cost of building a FCP. Finally, it is a max-option because of the availability of competing technologies for production. We show that the value of a perpetual project has an early investment premium (EIP) representation. That is, it can be written as the present value of the cumulative local gains realized by optimally investing in the best technology. This premium splits as a premium for investing when the VCP is optimal and one for investing when the FCP is optimal. Each of these components is parametrized by both investment boundaries, reflecting the inherent interdependence between the two decisions. The boundaries can be computed, so the value of the project can be calculated as well.

The optimal investment decision has notable features. First, it is always suboptimal to invest when the values of the two underlying projects coincide and this is true even if both values are extremely large. In this instance, it pays to delay investment even if it is optimal to invest when a given technology is considered in isolation. Moreover, the premium for postponing investment can be substantial. Second, the optimal behavior when the two project values are close to each other is unusual. There are two cases. In the first case, delaying investment even if one of the projects has significantly more value than the other one can be optimal. In fact, we identify a cone within which immediate investment is suboptimal. The edges of this cone diverge as the underlying project values increase. This means that the value of waiting can remain positive even when the difference between the underlying projects increases to infinity, that is, when one project becomes infinitely more valuable than the other one. In the second case, the no-investment region shrinks as the revenue factor becomes large. In this instance the optimal waiting time when both projects have the same net present value (NPV) becomes asymptotically negligible, and the premium for waiting vanishes. We identify and discuss the factors leading to either configuration.

Operating leverage affects decisions and value functions. When the exposure (sensitivity) to the stochastic cost factor increases, that is, when the VCP has lower operating leverage, the project manager will speed up the decision to idle and postpone the decision to restart operations. Lower operating leverage also decreases the instantaneous benefits from investing in the VCP. This leads to a decrease in the value of the project to invest in the best technology and a substitution of investment in the FCP at the expense of the VCP.

When the correlation between revenue and cost increases, shocks to revenue have a more systematic effect on variable cost leading to a decrease in the riskiness of the operating profit of the VCP and its value. This also reduces the value of the project to invest in the best technology and tends to increase the likelihood of investing in the FCP, as opposed to the VCP.

The model is then applied to power generation projects and we conduct a numerical study to gauge the relevance of GE projects for power generation. We focus on a gas-fired plant (GFP) as the technology based on traditional fuels. This reflects the fact that natural gas has become the most efficient source of power production from fossil fuels. A wind plant (WP) is the GE technology considered.[3] Wind power was the second fastest growing segment in that space (behind SPV) and had the second highest share of global electricity production of all renewable energies (behind hydroelectric power) in 2016. A GFP (resp. WP) is an example of a VCP (resp. FCP). We thus consider an operator with a project to build a power plant and who seeks to choose between building a GFP or a WP.

Numerical results underscore the importance of GE for power generation projects. We find that considering an alternative GE-based technology, such as a WP, as a possible choice can lead to a substantial increase in the value of a power project over typical ranges of electricity and gas prices. The value added, that is, the wind premium, can be large even if the current value of a GFP is significantly larger than that of a WP. It decreases slowly as the gas price decreases and increases when the cost of building a WP decreases. Subsidies play an important role. An increase in subsidies provides incentives to invest earlier in wind power and boosts the wind premium. It also delays or eliminates investments in gas plants. The type of subsidy determines the limits of regulation: a constant (resp. proportional) premium on top of the electricity price is unable (resp. able) to shrink the waiting region beyond an asymptotic cone.

For an operator of fossil fuel plants, investing in GE power may require a preliminary acquisition of technological expertise and information.[4] Our final contribution is to examine this issue in a two-stage extension of the basic model. In the first stage the firm has three alternatives. It can invest in the familiar technology, the VCP (i.e., GFP); invest in the unfamiliar technology, the FCP (i.e., the WP); or acquire technological expertise about the FCP and postpone investments. If it chooses the second alternative, it bears an efficiency cost due to imperfect knowledge and hence suboptimal design. If it chooses the last one, it can in a second stage still choose the familiar technology or invest in the FCP instead. The project then becomes a two-stage compound max-option. We show that the first-stage decision, which involves the alternative to acquire technological information, is driven by a triplet of boundaries. Each of these endogenous barriers, depends on the subsequent decisions and their determinants. We examine the properties of these boundaries and, in the context of power generation, study the impact of exogenous parameters of the model, such as operating leverage, on the information acquisition decision and the value of the option to invest in knowledge assets.

The chapter relates to several branches of the literature. First, it contributes to the broad literature on growth options, operating leverage and firm value. Lev (1974) examines the relation between operating leverage and risk. Real options models drawing the link between

[3]The differences between technologies available for power production are significant. Valuation of specific technologies and choices between technologies should be carried out on a case-by-case basis.

[4]Energy firms often invest to build real options positions. An example is Statoil's (now Equinor) 2007 acquisition of equity stakes in two wind power technology firms, Sway and ChapDrive. Statoil's objective at the time was to strengthen its position as a potential future producer of wind power. It has since used the technology developed to build the first floating wind farm in the world, the Hywind farm in Scotland, which became operational in 2017. Another example is the global energy provider ENGIE, who founded ENGIE New Ventures in 2014, a fund dedicated to investments in technology startups, to foster internal innovation.

operating leverage and value premium can be found in Carlson, Fisher, and Giammarino (2004), Zhang (2005), and Novy-Marx (2011). In contrast to this literature, our analysis studies the choice between exclusive projects with different leverage characteristics, specifically cost structures (fixed versus variable cost). It highlights, in particular, the role of the variable cost structure for the optimal timing and selection decisions, and the value of the project.

Second, it also relates to the more specialized literature on power generation. Most of the literature in that area focuses on projects involving a given technology.[5] Two exceptions are Décamps, Mariotti, and Villeneuve (2006) and Siddiqui and Fleten (2010), who consider a choice between two exclusive alternatives. The first study develops a general method to deal with investments in the best of two technologies with values driven by a common underlying state variable. They show that the optimal investment region can consist of two intervals separated by a continuation interval. The second study adds the possibility of learning about the costs of one of the technologies after investing in it, but before deciding to deploy it. Assuming that the cost of deployment is incurred at the time of investment, they show that the model reduces to a single state variable problem, like in Décamps, Mariotti, and Villeneuve (2006). They proceed to examine the impact of learning on the optimal investment decision and the project valuation.[6]

The model considered in this chapter cannot be reduced to a single state variable. It involves two distinct underlying processes, for revenues and costs, as well as embedded (two-dimensional) optimal stopping time problems for the operating decisions of the projects and the timing and selection decisions of the manager. We solve this two-state variables compound optimal stopping problem in generality, assuming that the underlying processes follow geometric Brownian motions (GBM). In the application to power projects, the first state variable is the price of electricity and the second one the price of gas. The GBM specification allows the spark spread to take negative values, when the price of gas adjusted by the heat rate exceeds the price of electricity and ensures that the price of gas cannot become negative. GBM, as a model of the long-run behavior of the electricity price, is standard in the literature on power generation.[7]

Third, the chapter is also related to broader literatures on real options and American options. Seminal contributions in the real options area highlight the values of flexibility and of waiting to invest (e.g., Brennan and Schwartz (1985); McDonald and Siegel (1985); McDonald and Siegel (1986); see also Dixit and Pindyck (1994) and the references therein). This literature relies on option pricing methods and typically uses partial differential equations to price projects. In contrast, we exploit a characterization of the value function based on the EIP representation, which is rooted in probabilistic/martingale methods (see, e.g., Kim (1990) and Carr, Jarrow, and Myneni (1992) for American options). We also extend the literature on the valuation of multiasset American claims (e.g., Broadie and Detemple (1997)), by considering dual strike compound max-options and designing an implementable algorithm that solves some systems of coupled integral equations of Fredholm type for optimal exercise boundaries. The compound structure of our problem and the presence of uncertain operating leverage modifies some properties characterizing optimal exercise decisions for max-options. An advantage of our algorithm is the ability to tackle certain types of

[5]See Deng, Johnson, and Sogomonian (1999), Maribu, Galli, and Armstrong (2007), and Fleten and Näsäkkälä (2010) for GFPs and Fleten, Maribu, and Wangensteen (2007) and Boomsma, Meade, and Fleten (2012) for WPs.

[6]The choice between exclusive alternatives has been studied in other settings as well. See, for instance, Geltner, Riddiough, and Stojanovic (1996) for an application to land use and Bakke et al. (2016) for the selection between transmission asset locations. The structure of our problem is more complex, because it involves underlying projects with operating decisions, resulting in a dual strike compound max-option.

[7]GBM is also common as a model for the long-run behavior of commodity and energy prices. See, for example, Pindyck (1999) and Schwartz and Smith (2000).

perpetual or long horizon problems, in addition to short horizon problems. This approach could help to address other challenging multidimensional stopping problems in real options and American option pricing applications.

15.2 Investments in exclusive projects

15.2.1 Variable cost project

1. In this paragraph, we describe the variable cost project and comment on the special case with constant cost factor. We consider a project generating maximal instantaneous revenue X at cost κY, where Y is a stochastic cost factor (state variable) and $\kappa > 0$ a constant exposure coefficient. Let us assume that the revenue and cost factor processes X and Y follow correlated geometric Brownian motion processes under the risk-neutral measure Q

$$
\begin{aligned}
dX_t &= (r - \delta_X)X_t dt + \sigma_X X_t dW_t \\
dY_t &= (r - \delta_Y)Y_t dt + \sigma_Y Y_t dB_t,
\end{aligned}
$$

where $r > 0$ is the interest rate, δ_Z and σ_Z are the (implicit) yield[8] and volatility parameters, respectively, for $Z = \{X, Y\}$. We denote the correlation between W and B by $\rho \in [-1, 1]$.

We suppose that the project can be operated in two states: 0, or idle, and 1, or full operation. There is a fixed running cost $k_{i,v} > 0$ for state $i = 0, 1$ with $k_{0,v} < k_{1,v}$, and a fixed switching cost $c_{i,v} \geq 0$ when the manager shifts from state i to state $j \neq i$. We also define the scale factor $\gamma_v \in (0, 1)$, representing the effective operational capacity. The operating profit at time t is $\gamma_v(X_t - \kappa Y_t) - k_{1,v}$. For notational convenience, we absorb the exposure coefficient κ into the initial value of Y, which can be done due to the multiplicative structure of Y. Then the value of the VCP once it is started is given by the system of coupled optimal timing problems

$$
V_{0,v}(x, y) = \sup_{\tau \geq 0} \mathsf{E}_{x,y}\left[-\int_0^\tau e^{-rt} k_{0,v} dt + e^{-r\tau}(V_{1,v}(X_\tau, Y_\tau) - c_{0,v})\right] \tag{15.1}
$$

$$
V_{1,v}(x, y) = \sup_{\tau \geq 0} \mathsf{E}_{x,y}\left[\int_0^\tau e^{-rt}(\gamma_v(X_t - Y_t) - k_{1,v}) dt + e^{-r\tau}(V_{0,v}(X_\tau, Y_\tau) - c_{1,v})\right] \tag{15.2}
$$

for $x > 0$ and $y > 0$, where the supremum is taken over stopping times of (X, Y) that represent the times of switching from one state to another. The goal now is to find the value functions $(V_{0,v}, V_{1,v})$ and the optimal operating policy. The problem is two-dimensional, so, to the best of our knowledge, it cannot be solved in a closed form.

First, consider the optimal operating policy when the cost factor $Y \equiv 0$ permanently. The problem becomes one-dimensional, and the method of solution is well known (see, e.g., Dixit and Pindyck (1994)). Starting from state 0, the manager should switch and generate revenue when X becomes sufficiently high, that is, $X \geq b_{0,v}$ for some constant $b_{0,v} > 0$ to be determined. Similarly, when in state 1, the project should be switched to the idle state when X falls below some threshold $b_{1,v} > 0$. It is clear that the threshold $b_{0,v}$ is above $b_{1,v}$ and that $b_{1,v} < (k_{1,v} - k_{0,v})/\gamma_v < b_{0,v}$. Appendix A offers details about the solution.

[8] The price of a nonstorable commodity, such as electricity, can diverge from risk-neutral valuation, leading to the emergence of an implicit yield.

2. Now we turn to the general case with stochastic cost factor, when the initial cost is $Y_0 = y > 0$. Like in the previous case, there are two boundaries $b_{0,v}(y)$ and $b_{1,v}(y)$ such that the project is optimally switched from state i to state $j \neq i$ when X hits $b_{i,v}(Y)$. However, now these boundaries are not constants but functions of Y. It is clear that the limits $b_{i,v}(0+)$ were already found in the previous paragraph as the problem becomes one-dimensional due to the absorbing nature of the point $y = 0$. Also, one can show that both curves are increasing in $y > 0$. Using hedging arguments, we find that the value functions $V_{0,v}$ and $V_{1,v}$ solve the free boundary PDE problems

$$\mathbb{L}_{X,Y}V_{0,v}(x,y) - rV_{0,v}(x,y) = k_{0,v} \quad \text{for } x < b_{0,v}(y) \tag{15.3}$$
$$V_{0,v}(x,y) = V_{1,v}(x,y) - c_{0,v} \quad \text{for } x \geq b_{0,v}(y) \tag{15.4}$$
$$\mathbb{L}_{X,Y}V_{1,v}(x,y) - rV_{1,v}(x,y) = -\gamma_v(x-y) + k_{1,v} \quad \text{for } x > b_{1,v}(y) \tag{15.5}$$
$$V_{1,v}(x,y) = V_{0,v}(x,y) - c_{1,v} \quad \text{for } x \leq b_{1,v}(y) \tag{15.6}$$

where $\mathbb{L}_{X,Y}$ is the joint infinitesimal operator

$$\begin{aligned}\mathbb{L}_{X,Y}f =& \tfrac{1}{2}\sigma_X^2 x^2 f_{xx} + \tfrac{1}{2}\sigma_Y^2 y^2 f_{yy} + \rho\sigma_X\sigma_Y xy f_{xy} \\ &+ (r-\delta_X)xf_x + (r-\delta_Y)yf_y\end{aligned}$$

for $f \in C^{2,2}$ on $(0,\infty)^2$. We also have smooth-pasting conditions at $b_{0,v}(y)$ and $b_{1,v}(y)$.

In contrast to the previous case with $Y = 0$, PDEs (15.3) and (15.5) cannot be solved in closed form.[9] Hence, we will have to rely on numerical methods. To the best of our knowledge, the representation formulas below are new and offer an approach to tackle this type of problem numerically. In most of the literature on switching problems or coupled timing problems, the case of a single stochastic state variable was considered (e.g., Brennan and Schwartz (1985)). In multidimensional settings, finite difference methods were mainly employed. However, these methods are not very convenient for our purposes as we will need to use the value functions $V_{0,v}$ and $V_{1,v}$ further in the following sections. The representation formulas presented next also have economic meaning as will be discussed below.

Theorem 1 *The value functions satisfy the ESP formulas:*

$$V_{0,v}(x,y) = -k_{0,v}/r + \pi_0(x,y;b_{0,v}) \tag{15.7}$$
$$V_{1,v}(x,y) = \gamma_v x/\delta_X - \gamma_v y/\delta_Y - k_{1,v}/r + \pi_1(x,y;b_{1,v}) \tag{15.8}$$

for $x, y > 0$, where $\pi_i(x,y;b_{0,v},b_{1,v})$, $i = 0,1$, represent the ESPs defined as,

$$\pi_0(x,y;b_{0,v}) = \mathsf{E}_{x,y}\left[\int_0^\infty e^{-rt}(\gamma_v(X_t - Y_t) - k_{1,v} + k_{0,v} - rc_{0,v})I(X_t \geq b_{0,v}(Y_t))dt\right]$$
$$\pi_1(x,y;b_{1,v}) = \mathsf{E}_{x,y}\left[\int_0^\infty e^{-rt}(-\gamma_v(X_t - Y_t) + k_{1,v} - k_{0,v} - rc_{1,v})I(X_t \leq b_{1,v}(Y_t))dt\right].$$

The pair of switching boundaries $(b_{0,v}, b_{1,v})$ satisfy the system of coupled integral equations:

$$\begin{aligned}&\gamma_v b_{0,v}(y)/\delta_X - \gamma_v y/\delta_Y - k_{1,v}/r + \pi_1(b_{0,v}(y),y;b_{1,v}) - c_{0,v} \\ &\quad = -k_{0,v}/r + \pi_0(b_{0,v}(y),y;b_{0,v})\end{aligned} \tag{15.9}$$
$$\begin{aligned}&\gamma_v b_{1,v}(y)/\delta_X - \gamma_v y/\delta_Y - k_{1,v}/r + \pi_1(b_{1,v}(y),y;b_{1,v}) \\ &\quad = -k_{0,v}/r + \pi_0(b_{1,v}(y),y;b_{0,v}) - c_{1,v}\end{aligned} \tag{15.10}$$

for $y > 0$.

[9] Operating costs and switching costs preclude the existence of a closed-form solution. In the absence of fixed costs, the option to switch becomes an exchange option (see Margrabe (1978)).

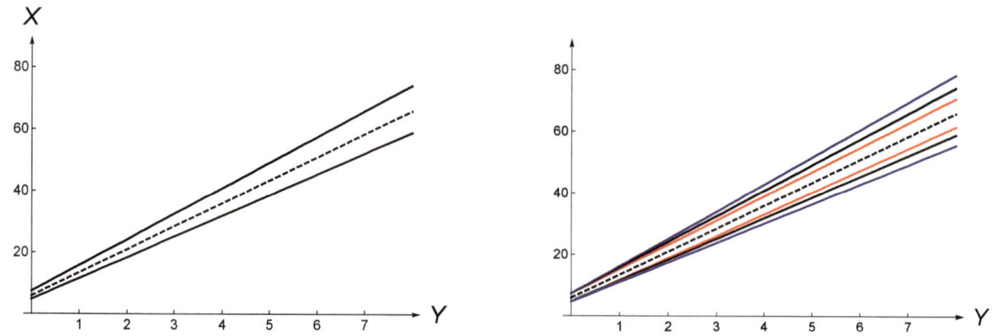

FIGURE 15.1: Optimal operating boundaries for VCP. The left panel shows the optimal operating boundaries $b_{0,v}$ (upper solid) and $b_{1,v}$ (lower solid) for the VCP when $\rho = 0.3$. The right panel displays the boundaries for different correlation coefficients: $\rho = 0$ (blue), $\rho = 0.3$ (black), $\rho = 0.6$ (red). The black dashed curve corresponds to $x = y+(k_{1,v}-k_{0,v})/\gamma_v$, where the cash flows for each state have the same value. In state 0, switching to state 1 when $X \geq b_{0,v}(Y)$ is optimal. In state 1, switching to state 0 when $X \leq b_{1,v}(Y)$ is optimal. Section 15.3.2 describes other parameters.

The ESP formulas (15.7) and (15.8) in Theorem 1 have a clear interpretation. The value $V_{0,v}(x,y)$ of the project, in state 0, can be decomposed as the present value of cash flows under indefinite operation in the current state, that is, $-k_{0,v}/r$, plus the present value of the local gains from switching to state 1. When the manager switches, the instantaneous benefit is the cash flow $(\gamma_v X_t - \gamma_v Y_t - k_{1,v})dt$ and she stops paying the cost $k_{0,v}\,dt$ but at the same time she forgoes the interest, $rc_{0,v}\,dt$ on the switching cost $c_{0,v}$. One can explain formula (15.8) for the value function $V_{1,v}(x,y)$ in a similar manner. Obviously, (15.7) and (15.8) do not provide a full characterization of the problem as they depend on unknown optimal switching boundaries. To proceed, one must first solve the system of coupled integral equations (of Fredholm type) (15.9) and (15.10) for the boundaries $b_{0,v}(y)$ and $b_{1,v}(y)$.

3. We complete the presentation of the variable cost case by commenting on the numerical implementation procedure and providing insights about the behavior of optimal policies. Appendix C develops a numerical procedure to solve the system (15.9) and (15.10). This procedure is iterative and produces the pair of boundaries $b_{0,v}(y)$ and $b_{1,v}(y)$ efficiently (typically 2–4 iterations are enough for acceptable accuracy). The boundaries then can be plugged back into (15.7) and (15.8) to find the value of the project in both states. Figures 15.1 and 15.2 illustrate the results.

Remark 1 *The value of the project under perpetual operation is an affine function of the state variables*

$$\begin{aligned} V_{cont,v}(x,y) =& \mathsf{E}_{x,y}\left[\int_0^\infty e^{-rt}(\gamma_v(X_t - Y_t) - k_{1,v})dt\right] \\ =& \gamma_v\left(\frac{x}{\delta_X} - \frac{y}{\delta_Y}\right) - \frac{k_{1,v}}{r} \end{aligned}$$

for $x, y > 0$. The value of operational flexibility, that is, of the option to switch to the idle state, is the difference $V_{1,v}(x,y) - V_{cont,v}(x,y) = \pi_1(x,y;b_{1,v})$. See Figure 15.2 for an illustration.

Figure 15.1 illustrates the effect of the correlation coefficient ρ on the optimal operating policy. As ρ increases, the instantaneous variance of the operating profit decreases. This

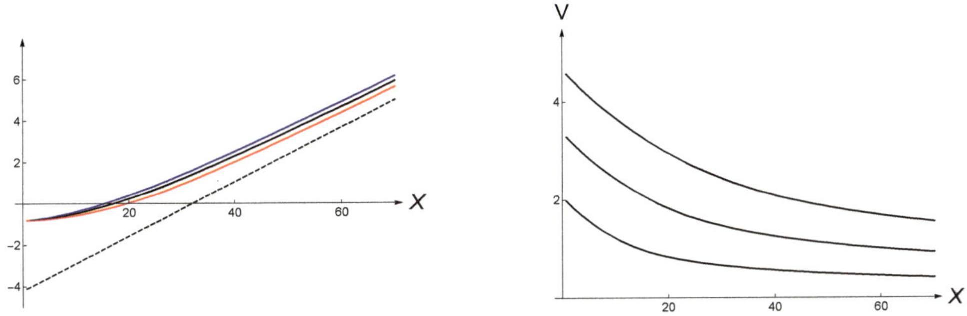

FIGURE 15.2: Value function for VCP. The left panel plots the value function $V_{1,v}(x, y)$ of the VCP versus the value of perpetual operation $V_{cont,v}(x, y)$ (black dashed) as functions of x for given $y > 0$ and different levels of the correlation parameter: $\rho = 0$ (blue), $\rho = 0.3$ (black), $\rho = 0.6$ (red). The right panel plots the value of operational flexibility $V_{1,v}(x, y) - V_{cont,v}(x, y)$ as a function of x for three different levels of y: $y = 1.3$ \$/MBtu (upper), $y = 2.6$ \$/MBtu (middle), and $y = 3.9$ \$/MBtu (lower). Section 15.3.2 describes other parameters.

has two effects on the options to switch. The direct effect is the standard impact of risk on the value of a derivative with convex payoff. A decrease in riskiness reduces value and hence incentivizes early action. The indirect effect is the associated decrease in the project values $V_{0,v}$ and $V_{1,v}$. This reduction implies that the payoff from switching also decreases, which raises the value of waiting and mitigates the risk effect. The ESP representation in Theorem 1, which expresses the switching premium entirely in terms of the fundamentals of the project, shows that the risk effect is the driving force. Hence, the upper boundary $b_{0,v}(Y)$ decreases, the lower boundary $b_{1,v}(Y)$ increases, and the area between the boundaries shrinks.

15.2.2 Fixed cost project

We now turn to the alternative project that also produces revenues driven by the process X. However, the cost factor is a constant in this case and the factor process Y is not relevant. The profit generated by this FCP is an affine function of the revenue factor X, namely,

$$\gamma_f(X_t + s) - k_{1,f}. \tag{15.11}$$

The term s stands for an incremental constant revenue stream, for example, a government subsidy, like in the case of green energy projects. The subsidy is a premium on top of the random revenue factor and is assumed to be constant $s > 0$. The running cost of operations, which is also constant $k_{1,f} > 0$, consists of wages and fixed maintenance costs.

The FCP can operate in two states: 0, or idle, and 1, or operating at capacity level $\gamma_f \in (0, 1]$. Perpetual operation is optimal if $\gamma_f s - k_{1,f}$ exceeds the cost of maintenance $k_{0,f}$ when the project sits idle. We consider the general case and solve the optimal switching problem for the FCP. That problem is one-dimensional with single state variable X, thus can be easily solved. In fact, we already provided the solution in our treatment of the VCP when $Y = 0$. Drawing on that result, the value of the FCP (in states 0 and 1, respectively) is

$$V_{0,f}(x) = -k_{0,f}/r + A_{0,f}x^p \quad \text{for } x < b_{0,f} \tag{15.12}$$

$$V_{1,f}(x) = \gamma_f(x/\delta_X + s/r) - k_{1,f}/r + A_{1,f}x^q \quad \text{for} \quad x > b_{1,f}, \tag{15.13}$$

where $p > 1$ and $q < 0$ are roots to the associated quadratic equation, and constants $A_{0,f}$, $A_{1,f}$, $b_{0,f}$, and $b_{1,f}$ can be numerically computed like in Appendix A for the VCP case ($Y = 0$). If we start from state 0, we switch to state 1 when $x \geq b_{0,f}$, pay the cost $c_{0,f}$ and obtain the value $V_{1,f}(x)$ of the project in state 1. Thus, for $x \geq b_{0,f}$, we define $V_{0,f}(x) := V_{1,f}(x) - c_{0,f}$ where $V_{1,f}(x)$ is as in (15.13). Similarly, we switch from state 1 to state 0 when $x \leq b_{1,f}$. For $x \leq b_{1,f}$, we define $V_{1,f}(x) := V_{0,f}(x) - c_{1,f}$, where $V_{0,f}(x)$ is as in (15.12). The first two terms in (15.13) give the value of the project under perpetual operation

$$\mathsf{E}_x\left[\int_0^\infty e^{-rt}(\gamma_f(X_t + s) - k_{1,f})dt\right] = \gamma_f(x/\delta_X + s/r) - k_{1,f}/r \tag{15.14}$$

for $x > 0$. The last term on the right-hand side of (15.13) is the value of the option to idle.

15.2.3 Investment problem

We consider a firm with the option to invest in two exclusive alternatives, one project with variable cost, the other with fixed cost. For now we assume that the option is infinitely lived. We suppose that the projects are built in an idle state and can be immediately run according to the optimal operating policies derived in Sections 15.2.1 and 15.2.2. For example, if the operator invests in the VCP when $x \geq b_{0,v}(y)$, then it is optimal to pay the cost $c_{0,v}$ and immediately switch to the operating state with value $V_{1,v}(x, y)$ given in (15.8). On the other hand, one should not invest in a project when it is currently optimal to leave it idle, because the immediate local profit is negative. Hence, we need only consider investing in the VCP when $x \geq b_0(y)$ and the project is optimally switched to state 1 upon paying the fixed cost $c_{0,v}$. Similar arguments can be applied for the FCP. The relevant value of the VCP is $V_v\,(x, y) = V_{0,v}(x, y)$ given in (15.7) and for the FCP it is $V_{0,f}(x)$ as defined in the paragraph following (15.12).

The option to choose between the two types of projects has payoff,

$$\max\{V_v(x, y) - K_v, \alpha(V_f(x) - K_f)\},$$

where $K_v > 0$ and $K_f > 0$ represent the sunk costs for each project, including startup costs, and $\alpha > 0$ is a scale factor. The latter can be used to capture various situations of interest. If the firm is capital constrained, for instance, with constraint K_v, the scale factor can be used to equalize the costs of investments in the two alternatives so that $K_v = \alpha K_f$. If the firm is demand constrained, that is, there is a fixed market size for the output produced, for instance, γ_v, then $\gamma_v = \alpha\gamma_f$ ensures that each of the two projects can service the demand.

The firm can also choose the timing τ of the investment. The value of the project to engage in one of the two alternative projects is therefore given by

$$V(x, y) = \sup_{0 \leq \tau \leq \infty} \mathsf{E}_{x,y}\left[e^{-r\tau}\max(V_v(X_\tau, Y_\tau) - K_v, \alpha(V_f(X_\tau) - K_f))\right] \tag{15.15}$$

for $x > 0$ and $y > 0$, where the supremum is taken over stopping times $\tau \geq 0$. This is the value of an American option on the maximum of two assets with different strikes, that is, a dual strike max-option. One of the assets (VCP) depends on two state variables.

The option payoff can be written as

$$G(x, y) = \max(V_v(x, y) - K_v, \alpha(V_f(x) - K_f))$$

for $x > 0$ and $y > 0$. The firm clearly should not enter into any project when the payoff is negative, that is, $G(x, y) \leq 0$.

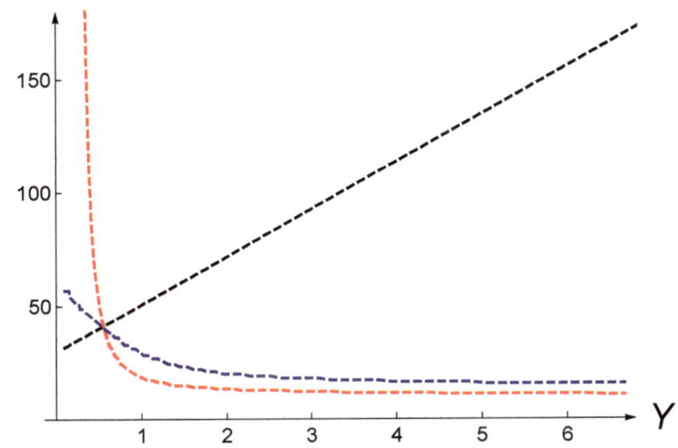

FIGURE 15.3: Indifference curve. This figure shows different configurations of the indifference curve depending on α: (a) $\widehat{y}$ is upward sloping (black dashed, for low α); (b) $\widehat{y}$ converges to the vertical line (red dashed, for medium α); and (c) $\widehat{y}$ is downward sloping and becomes zero for large x (blue dashed, for high α). Section 15.3.2 describes the set of parameters.

The two projects have the same NPVs when $V_v(x,y) - K_v = \alpha(V_f(x) - K_f))$. Solving for y gives the indifference curve $\widehat{y}(x)$ for $x > 0$. If, for a given x_0, the payoff $\alpha(V_f(x_0) - K_f)$ dominates (is dominated by) $V_v(x_0, y) - K_v$ for all $y > 0$, we define $\widehat{y}(x_0) = 0$ ($\widehat{y}(x_0) = +\infty$). Along this curve, none of the two projects dominates. As $V_v(x,y)$ is decreasing in y, it is clear that for fixed $x > 0$

$$G(x,y) = \begin{cases} G_v(x,y) := V_v(x,y) - K_v & \text{for } y < \widehat{y}(x) \\ G_f(x) := \alpha\left(V_f(x) - K_f\right) & \text{for } y > \widehat{y}(x). \end{cases}$$

The curve $\widehat{y}(x)$ can take different forms depending on the scale factor α. Three configurations are possible: $\widehat{y}(x)$ (a) is upward sloping when x goes to $+\infty$, for example, in the case of low values of α; (b) converges asymptotically to the vertical line in (y,x)-space, for example, when $\alpha = \gamma_v/\gamma_g$; and (c) is decreasing for large x and eventually becomes zero (note that Y cannot be negative), for example, when α is high. As the value function $V_v(x,y)$ is not obtained in closed form and has a nonlinear structure, one has to recover the indifference curve numerically by solving an algebraic equation for every $x > 0$. See Figure 15.3 for an illustration. Simplifications occur if the projects are run (state 1) in perpetuity. In this case the value functions have explicit affine structures and the indifference curve is obtained in closed form.

15.2.4 Optimal investment strategy

1. *Investment region.* We now characterize the optimal investment strategy. First, note that immediate investment is optimal when $V(x,y) = G(x,y)$ and waiting is optimal when $V(x,y) > G(x,y)$. The immediate investment region, furthermore, can be split in two parts, one where it is optimal to invest in the VCP, $V(x,y) = G_v(x,y)$, and another where it is optimal to invest in the FCP, $V(x,y) = G_f(x,y)$. As the payoff G is a continuous function, the optimal investment rule is the first time at which either of these two regions is reached.

Some properties of the regions are obvious. First, when $V_v(x,y) < K_v$ and $V_f(x) < K_f$, it is clear that immediate investment is suboptimal. Second, when $x < b_{0,v}(y)$, one should not enter into the VCP because it is in the idling state and negative cash flows are immediately collected. Similarly, when $x < b_{0,f}$ immediate investment in the FCP is suboptimal.

When y goes to 0, the project valuation reduces to a single state variable problem with payoff $\max(V_v(x,0) - K_v, \alpha(V_f(x) - K_f))$, which can be solved in closed form by standard methods, that is, a free-boundary problem with a smooth-pasting condition. On other hand, when y goes to $+\infty$, the FCP dominates the VCP and the manager faces a one-dimensional problem with payoff $V_f(x) - K_f$ and optimal investment boundary $\bar{X}_{f,\infty}$ that can be easily found.

To gain further insights into the structure of the investment region, let us consider the following simple suboptimal strategy: wait until some deterministic time $\tau > 0$ to invest in the project with largest NPV. Exploiting the local time-space formula (see Peskir (2007)) and the optional sampling theorem, we can write the value of this waiting strategy as

$$\begin{aligned}
\mathsf{E}_{x,y}\left[e^{-r\tau}G(X_\tau, Y_\tau)\right] =& G(x,y) + \mathsf{E}_{x,y}\left[\int_0^\tau e^{-rt}H_v(X_t,Y_t)I(Y_t < \widehat{y}(X_t))dt\right] \\
&+ \mathsf{E}_{x,y}\left[\int_0^\tau e^{-rt}H_f(X_t)I(Y_t > \widehat{y}(X_t))dt\right] \\
&- \frac{1}{2}\mathsf{E}_{x,y}\left[\int_0^\tau e^{-rt}\frac{\partial G_v}{\partial y}(X_t,Y_t)I(Y_t = \widehat{y}(X_t))dL_t^{\widehat{y}}\right]
\end{aligned} \tag{15.16}$$

for $(x,y) \in (0,\infty)^2$, where $L^{\widehat{y}}$ is the local time process that Y spends at the curve $\widehat{y}(X)$ (this term arises because the payoff G is not smooth along $\widehat{y}$ and one finds its definition, e.g., in Peskir (2007)); H_v and H_f represent the local gains of waiting to invest in the VCP and FCP, respectively, which are defined as

$$\begin{aligned}
H_v(x,y) &= \mathbb{L}_{X,Y}G_v(x,y) - rG_v(x,y) = -\gamma_v(x-y) + k_{1,v} + r(K_v + c_{0,v}) \\
H_f(x) &= \mathbb{L}_X G_f(x) - rG_f(x) = -\alpha(\gamma_f(x+s) - k_{1,f} - r(K_f + c_{0,f}))
\end{aligned}$$

for $(x,y) \in (0,\infty)^2$. Hence, the value of waiting until τ can be decomposed as the immediate investment payoff augmented by the present values of the local gains realized by delaying when (i) the VCP dominates, (ii) the FCP dominate and (iii) the two are equal (on the indifference curve). The local gains in regions (i) and (ii) consist of the instantaneous profits foregone augmented by the investment opportunity cost saved. In region (iii), that is, on the indifference curve, the local time term positively affects the value of waiting as G_v is decreasing in y and $dL^{\widehat{y}} \geq 0$.

There are three immediate implications of (15.16). First, immediate investment in the VCP is suboptimal when $\gamma_v(x-y) < k_{1,v} + r(K_v + c_{0,v})$ as the local gains H_v of waiting are positive. Second, immediate investment in the FCP is suboptimal when $H_f > 0$, that is, $\gamma_f(x+s) < k_{1,f} + r(K_f + c_{0,f})$. Third, and most strikingly, investment along the indifference curve is suboptimal even if the current NPVs of the individual projects are arbitrarily large. This follows from the fact that the local time term in (15.16) behaves as $dt/\sqrt{t}$ and dominates dt terms corresponding to H_v and H_f for t sufficiently small. For example, one can choose τ to be deterministic and small enough for any given (x,y) on the indifference curve. This property is robust and holds for any configuration of the curve $\widehat{y}$. We will discuss the intuition behind this result below. The next theorem describes the structure of the immediate investment set.

Theorem 2 *There exist two boundaries b_v: $(0,\infty) \to (0,\infty)$ and $b_f : (0,\infty) \to (0,\infty)$ such that immediate investment in the VCP is optimal for $(x,y) \in (0,\infty)^2 : y \leq b_v(x)$ and in the FCP for $(x,y) \in (0,\infty)^2 : y \geq b_f(x)$. Also, $b_v(x) \leq b_f(x)$ for $x \in (0,\infty)$.*

The optimal investment problem can be easily solved in the extreme cases: when Y approaches 0 and ∞. The associated problems can be written as follows

$$\sup_{0\leq\tau\leq\infty} \mathsf{E}_x \left[e^{-r\tau} \max\left(V_v(X_\tau, 0) - K_v, V_f(X_\tau) - K_f\right)\right] \tag{15.17}$$

$$\sup_{0\leq\tau\leq\infty} \mathsf{E}_x \left[e^{-r\tau} (V_f(X_\tau) - K_f)\right] \tag{15.18}$$

for $x > 0$. Then $b_f(x) = +\infty$ for $x < \bar{X}_{f,\infty}$, where $\bar{X}_{f,\infty}$ is the optimal threshold for the stand-alone FCP investment project problem (15.18).

The problem (15.17) *can have three types of solutions:*
(i) it is optimal to invest only in the VCP, and this occurs when X exceeds a threshold $\bar{X}_v$: then $b_v(x) \to 0$ as $x \to \bar{X}_v+$ (see the upper panels of Figure 15.4);
(ii) it is optimal to invest in the VCP when $X \geq \bar{X}_v$ and in the FCP when $\bar{X}_{f,1} \leq X \leq \bar{X}_{f,2}$ where $\bar{X}_v$, $\bar{X}_{f,2}$ are optimal thresholds and $\bar{X}_{f,1} = \bar{X}_{f,\infty}$: then $b_v(x) \to 0$ as $x \to \bar{X}_v+$ and $b_f(x) \to 0$ as $x \to \bar{X}_{f,i}$, $i = 1, 2$ (see lower-left panel in Figure 15.4);
(iii) it is optimal to invest in the VCP when $\bar{X}_{v,1} \leq X \leq \bar{X}_{v,2}$ and in the FCP when $X \geq \bar{X}_f$ where $\bar{X}_{v,1}$, $\bar{X}_{v,2}$ and $\bar{X}_f$ are optimal thresholds: $b_v(x) \to 0$ as $x \to \bar{X}_{v,i}$, $i = 1, 2$ and $b_f(x) \to 0$ as $x \to \bar{X}_f$ (see lower-right panel in Figure 15.4).

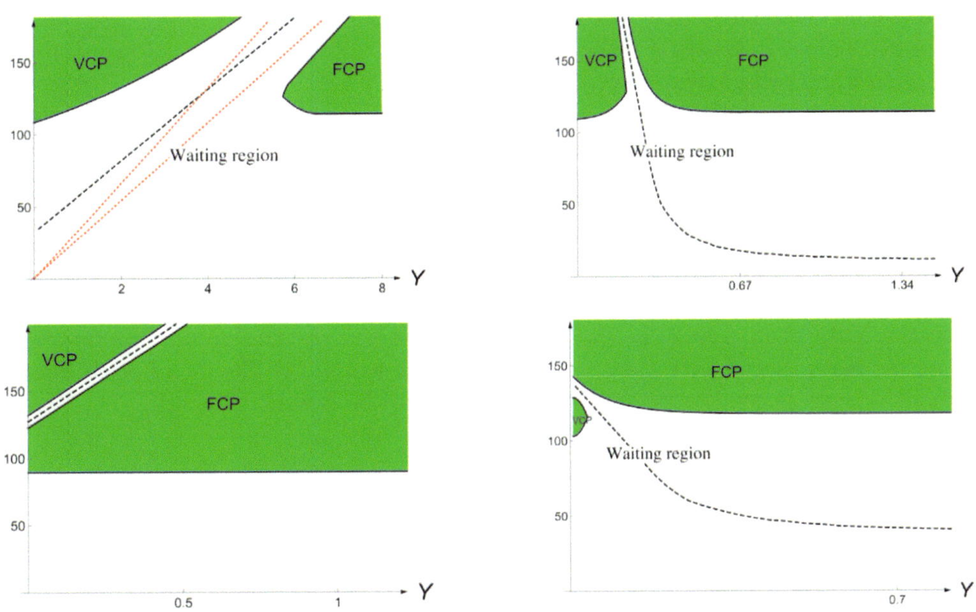

FIGURE 15.4: Immediate investment region. This figure shows the immediate investment region (green) for a (1) capital constraint (upper-left panel); (2) demand constraint (upper-right panel); (3) low K_f and a capital constraint (lower-left panel); and (4) high α (lower-right panel). Investment in the VCP (FCP) is optimal in the upper-left (upper-right) region. The dashed curve is the indifference curve $\widehat{y}(x)$. The red-dotted lines in the upper-left panel correspond to the asymptotic inaction cone. Section 15.3.2 describes the central set parameters. Note the different scale on the horizontal axis.

The first statement of Theorem 2 says that the VCP (FCP) investment region is left-(right-) connected in (Y, X)-space. For instance, if immediate investment in the FCP is optimal at (X, Y), then it remains optimal at $(X, Y + \Delta)$ for any $\Delta > 0$ (i.e., the region is right-connected). This is intuitively clear, as an increase in the cost factor reduces the payoff from investing in the VCP. We also note that, for convenience, we defined the investment boundaries as functions of X in contrast to Section 15.2.1, where switching boundaries were functions of Y.

The second statement describes the investment strategy in extreme cases. When Y approaches 0, the investment problem becomes one-dimensional (see (15.17)). The theorem describes three possible regimes. For example, in the first case, one obtains the optimal threshold $\bar{X}_v$, and investment in the VCP must be made only when $X \geq \bar{X}_v$, that is, $b_v(X) \to 0$ as $X \to \bar{X}_v+$. The other two cases involve disconnected investment regions when $Y = 0$. When Y goes to ∞, the FCP dominates the VCP and the problem reduces to a single dimension as well (see (15.18)). It also then can be solved with a critical threshold $\bar{X}_f$. One therefore should not invest when Y is large and $X < \bar{X}_f$, that is, $b_f(x) = +\infty$ for $x < \bar{X}_f$.

The important question is the following: what is the optimal investment policy along the indifference curve $\widehat{y}$ for large X? In the case of a single project, it is well known that immediate investment is optimal when the project has sufficiently high NPV. However, when there are several mutually exclusive alternatives, it may well be optimal to wait when two or more projects have the same NPV. This is the case, in particular, if project values follow geometric Brownian motions (see, e.g., Broadie and Detemple (1997)). It is then optimal to delay investment along the indifference curve even if all projects have infinitely large NPVs. The reason is that the likelihood of an increase in the payoff from waiting is quite substantial (it increases if any of the two project values increases) and this offsets any benefit from immediate investment.

In our setting in which the projects' revenues are driven by the same stochastic factor X and the VCP has a stochastic cost component Y, we face a similar situation; that is, it is always optimal to wait along the indifference curve. The formal explanation was already provided above using (15.16) and a property of the local time process. There are two intuitive explanations for this phenomenon. First, recall that the cost factor for the VCP is stochastic and that the decision-maker has the option to choose between the two projects. Consider a point (X, Y) on the indifference curve. If Y decreases, the value of the VCP increases and so does the investment payoff. In contrast, if Y increases, the decision-maker does not lose much as she can proceed with the FCP. Hence, if she waits, the likelihood of an increase in the payoff is high. Second, note that the payoff G can be rewritten as

$$G(x, y) = \alpha(V_f(x) - K_f) + \max(V_v(x, y) - \alpha V_f(x) + \alpha K_f - K_v, 0)$$

for $x, y > 0$. Hence, the option to invest can be decomposed in two parts: an option to invest in the FCP and an exchange option (to give up the FCP and receive the VCP). On the indifference curve $y = \widehat{y}(x)$, the immediate payoff of the second part is 0. Now consider the strategy of waiting until some deterministic time $t > 0$ to invest. The value of this strategy is

$$\mathsf{E}_{x,y}\left[e^{-rt}(\alpha(V_f(X_t) - K_f) + \max(V_v(X_t, Y_t) - \alpha V_f(X_t) + \alpha K_f - K_v, 0))\right].$$

As t goes to 0, this value converges to the immediate payoff $G(x, y)$. The first term depreciates at a finite rate as t increases but the second term appreciates at an infinite rate for small t due to the exchange option nature. Hence, for any given $x > 0$, we can select $t > 0$ sufficiently small such that the value of this suboptimal strategy dominates immediate investment.

There are, however, major differences in the shapes taken by the waiting region around $\widehat{y}$ (see Figure 15.4). There are two cases. The first is when the indifference curve converges to a vertical line as the cost factor y becomes small (right panels). This case arises when the FCP is more sensitive to the revenue factor x than the VCP, so the cost factor y

has to decrease to preserve the equality of NPVs. In this situation, the value of waiting along $\widehat{y}$ becomes very small due to the substantial benefits of early investment for large x. In fact, the optimal waiting time $t > 0$ is close to 0. Hence, both optimal investment boundaries converge asymptotically to the indifference curve. From an economic point of view, this waiting region does not add significant economic value as investment will occur quickly. The second case is when the indifference curve is upward sloping (left panels). In this instance, the VCP is more sensitive to x than the FCP, so that y increases to preserve indifference. Here, variations in y are also important (as x increases, y goes up too), and, in fact, there is a cone in which it is optimal to wait.

The behavior of the indifference curve also influences the shapes of the immediate investment regions. In the downward-sloping case, the VCP region can be up-connected for some values of y (upper-right panel), or bounded (lower-right panel), depending on the relative sensitivities of values to x. If the FCP value is very sensitive to x, for instance, when α is large, the FCP eventually dominates the VCP for any value of y. In the upward-sloping case, the FCP region always extends indefinitely to the right: investing remains optimal when the cost factor y increases because the NPV of the VCP decreases. If the cost K_f of investing in the FCP is sufficiently low, it can also extend all the way to the vertical axis (lower left panel). In this instance, investing in the FCP can dominate for all y. This occurs when the VCP value (i.e., x) is sufficiently low so that the benefits of investing in the VCP are never strong enough.

2. *Integral equations and EIP.* Theorem 3 is the main result of this section. The theorem characterizes the optimal investment strategy and the value of the option to invest in one of the projects.

Theorem 3 *Given the optimal investment boundaries (b_v, b_f), the value of the option to invest has the early investment premium (EIP) representation*

$$V(x, y) = \pi(x, y; b_v, b_f) \tag{15.19}$$

for $x, y > 0$, where $\pi(x, y; b_v, b_f)$ represents the early investment premium (EIP) defined as

$$\pi(x, y; b_v, b_f) = \pi_v(x, y; b_v) + \pi_f(x, y; b_f) \tag{15.20}$$

$$\pi_v(x, y; b_v) = -\mathsf{E}_{x,y}\left[\int_0^\infty e^{-rt} H_v(X_t, Y_t) I(Y_t \leq b_v(X_t))dt\right] \tag{15.21}$$

$$\pi_f(x, y; b_f) = -\mathsf{E}_{x,y}\left[\int_0^\infty e^{-rt} H_f(X_t) I(Y_t \geq b_f(X_t))dt\right] \tag{15.22}$$

for $x, y > 0$. The optimal investment boundaries solve the pair of coupled integral equations

$$V_v(x, b_v(x)) - K_v = \pi(x, b_v(x); b_v, b_f) \tag{15.23}$$

$$\alpha(V_f(x) - K_f) = \pi(x, b_f(x); b_v, b_f) \tag{15.24}$$

for $x > 0$.

Theorem 3 shows that the value of the option to invest in either of the two projects, hence the best of the two, entirely stems from the EIP (15.19). This premium is the present value of the cumulative instantaneous gains from investing in the VCP when such a choice is optimal and those from investing in the FCP when the latter is optimal. The expression (15.19) is similar to the early exercise premium formula for pricing American-style finite maturity options. However, because of the perpetual nature of the option to invest, there is no European-style component.

The theorem also states that the boundaries (b_v, b_f) solve a system (15.23)-(15.24) of coupled integral equations. These equations are derived from the EIP representation of the value of the option held by the manager. With two subregions, the premium has two components corresponding to the value of investing when the VCP is optimal $(\pi_v(x, y; b_v))$ and when the FCP is optimal $(\pi_f(x, y; b_f))$, respectively. Because each premium component depends on its corresponding boundary, the equations are coupled and cannot be solved separately. Hence, the boundary b_v depends on b_f, and conversely. Also, these equations do not have a recursive structure in time or space. The absence of recursivity in time comes from the infinite horizon nature of the problem. The absence of recursivity in space is because boundaries are integrated over their respective domains in the premium. As a result, the integral equations are not of the typical Volterra type as in standard American option pricing problems.

The effects of model parameters, such as the operating leverage parameters κ and ρ, on the value of waiting and the optimal investment decision are examined in Section 15.3.6, in the application to power generation technologies.

15.2.5 Model extensions

1. *Finite planning horizon.* In practice, decision-makers do not have an infinite planning horizon (e.g., firms have short-lived shareholders). Here, we describe the case of a project subject to a deadline $T > 0$. Straightforward modifications of the previous results enable us to handle this situation. Theorem 2 continues to apply, except that the boundaries are now functions of time. Thus, there exist two boundaries $b_v : (0, \infty) \times [0, T] \to (0, \infty)$ and $b_f : (0, \infty) \times [0, T] \to (0, \infty)$ such that, at time $t \in [0, T)$, it is optimal to invest in the VCP when $y \le b_v(x, t)$ or in the FCP when $y \ge b_f(x, t)$. Boundaries are characterized as in Theorem 3 where the value of the option to invest, in (15.19), has an additional component $V^e(x, y, t)$ corresponding to the value of a European option to invest in the best of the VCP and FCP at T and the EIP tallies the gains from early investment up to T. These modifications also appear in the integral equations (15.23) and (15.24) for the boundaries. As $t \to T$, the boundaries converge to

$$b_v(x, T-) = \min\left(\widehat{y}(x), \underline{y}_v(x), x - (k_{1,v} + r(K_v + c_{0,v}))/\gamma_v\right) \quad \text{for } x \ge 0$$

$$b_f(x, T-) = \begin{cases} \widehat{y}(x) & \text{for } x \ge \max(\underline{x}_f, -s + (k_{1,f} + r(K_f + c_{0,f}))/\gamma_f) \\ +\infty & \text{for } x < \max(\underline{x}_f, -s + (k_{1,f} + r(K_f + c_{0,f}))/\gamma_f) \end{cases}$$

where $\underline{y}_v(x)$ is the curve defined as $V_v(x, \underline{y}_v(x)) = K_v$ for $x > 0$ and $\underline{x}_f$ is the break-even threshold for the FCP $V_f(\underline{x}_f) = K_f$.

The structure of the investment region and the cases that can arise are the same as those in Figure 15.4. Relative to the infinite horizon case, finite horizon reduces the value of the project to invest in the best project. This means that the boundaries are closer to the indifference curve. For a given planning horizon T, the investment region expands as the horizon approaches (i.e., as $t \to T$). At the limit, the boundary of the VCP region approaches the indifference curve from above (upward-sloping case) or from the left (downward-sloping case) and the boundary of the FCP approaches it from below (upward-sloping case) or from the right (downward-sloping case). Figure 15.5 illustrates these effects for the capital-constrained case.

2. *The value of innovation.* Suppose that the FCP is based on a new technology, whereas the VCP employs established techniques. The value of the option to invest in the stand-alone VCP is $V(x, y; \alpha = 0)$, obtained by taking the scaling factor $\alpha = 0$ in (15.19). The value of the option to invest in the best of the VCP and the FCP is $V(x, y; \alpha)$ as given by

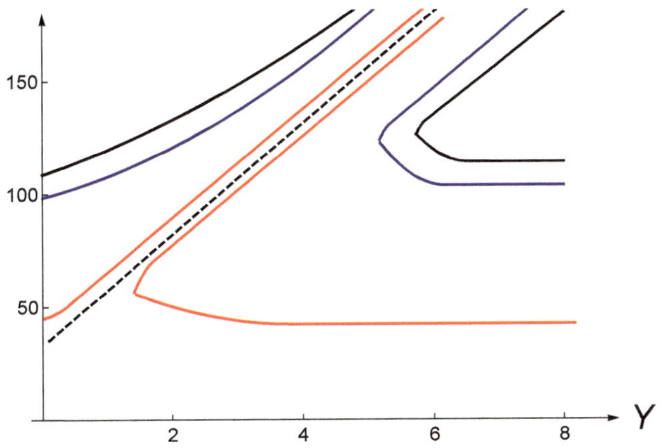

FIGURE 15.5: Optimal investment boundaries for finite horizon. This figure plots the optimal investment boundaries for finite horizon and capital constraint: $T = 1$ year (red line), $T = 10$ years (black line), and $T = \infty$ (black solid line). The black dashed line represents the indifference curve. Section 15.3.2 describes the set of parameters.

(15.19). The value of innovation is the difference

$$\Pi(x, y; \alpha) = V(x, y; \alpha) - V(x, y; \alpha = 0). \tag{15.25}$$

It represents the incremental value generated by the possibility of investing in the new technology as opposed to being limited to the familiar one.

The value of innovation is generally nonnegative (in fact strictly positive when the optimal investment region for the FCP is nonempty) because a possible strategy for the option to invest in the best project is to simply ignore the FCP and invest in the VCP. This strategy is suboptimal therefore reduces the value of the option to invest in the best project. The choice of the best time to invest in the stand-alone VCP clearly reduces the value of innovation. Indeed, investing in the stand-alone VCP at the optimal stopping time for the project where choices are considered simultaneously is suboptimal for the stand-alone VCP, therefore magnifies the value of innovation. Hence, the difference in optimal waiting times between the two options has a mitigating effect: it reduces, but does not offset, the gain associated with the availability of the new technology. Finally, it is worth noting that an increase in the subsidy s raises the profit from the FCP and hence the value of innovation.

3. *Obsolescence.* Projects, in particular those based on new technologies, can become obsolete. The emergence of alternative technologies, the impact of new regulations or the occurrence of natural phenomena, such as those tied to climate change, can contribute to such a scenario. Events of this nature affect the option to invest and the value of innovation.

Suppose that the FCP is subject to such risk. Let λ be the arrival rate of an adverse event, which if it occurs reduces capacity by a fraction δ; that is, effective capacity becomes $\gamma_f(1-\delta)$. The case $\delta = 1$ corresponds to complete obsolescence; $\delta \in (0, 1)$ is partial obsolescence.

The solution is modified as follows. The first adjustment pertains to the value of the FCP and hence the associated EIP. Suppose that the adverse event has not yet occurred at t. The future has two possible states. If the adverse event occurs, the local benefit of investing at v is $\gamma_f(1-\delta)(x+s) - k_{1,f} - r(K_f + c_{0,f})$. Otherwise, it is $\gamma_f(x+s) - k_{1,f} - r(K_f + c_{0,f})$. These events occur with respective probabilities $1 - e^{-\lambda(v-t)}$ and $e^{-\lambda(v-t)}$. The second adjustment pertains to the boundaries which now depend on the occurrence or not of the adverse event. Boundaries are characterized as in Theorem 3, but based on the modified EIP.

Obsolescence risk reduces the operating profit of the FCP by the expected cost $(1 - e^{-\lambda(s-t)})\gamma_f \delta(x+s)$. The likelihood of occurrence λ and the capacity cost δ both raise the expected cost and hence weaken the incentives for investing in the FCP. The investment boundaries for the FCP are therefore expected to move out, whereas those for the VCP to move toward the indifference curve. Investments in the VCP substitute for those in the FCP. The value of the option to invest in the best project also decreases, because to payoff from investing in the FCP decreases. As the stand-alone VCP is not affected, the value of innovation decreases.

15.3 Application: investments in power projects

Across industries, firms often face exclusive choices between fixed and variable cost projects. This section focuses on an application to the electric power industry.

15.3.1 Technology choice for power projects

Nowadays, electric power can be produced using a variety of technologies. Fossil fuel technologies, such as coal- and gas-fired plants, traditionally have been used to supply base loads in many countries worldwide. New technologies, based on renewable energy sources, such as wind and solar photovoltaic, have progressively become more competitive alternatives. We examine the optimal strategy of a power generator who seeks to develop a new plant and can choose between the two types of technologies.

More specifically, we focus on the choice between a gas-fired plant (GFP) and a wind plant (WP). The latest generation of GFP, with carbon-capture technology, is the most efficient among fossil-fuel technologies. With a low heat rate near 7.5 MBtu/MWh it has an efficiency of about 60%. Among technologies based on renewable energy sources, WPs have become very competitive. Taller and more efficient turbines have led to a steady increase in capacity, which reached an average of 37% for 2016 U.S. onshore installations. WPs have the ability to operate 24/7, which is attractive for utilities. In 2017, the majority of utility-scale renewable capacity addition came from wind (see EIA, January 10, 2018).

A GFP, a particular case of a technology with variable costs, sells electricity at the price of X and buys gas at the price of Y. The contribution margin is the spark spread $X - \kappa Y$ where $\kappa > 0$ is the heat rate. The operating profit is $\gamma_g(X - \kappa Y) - k_{1,g}$ where γ_g is the capacity factor and $k_{1,g}$ the operating cost. The operator of the plant has the option to idle or operate the plant.[10] In the idling mode, the costs are $k_{0,g} < k_{1,g}$. Section 15.2.1 describes the optimal operating strategy. Theorem 1 gives the value of a GFP, $V_g(x, y)$.

A WP is a special case of a fixed cost technology. It sells electricity at the price X and has operating profit $\gamma_w(X + s) - k_w$ where γ_w is the capacity factor, s the government subsidy and k_w the operating cost. We assume that government subsidies are discontinued after $T_s = 10$ years of operations, which corresponds to current U.S. regulation. Consistent with data, we also assume that there is no difference between the running costs of operation

[10]The model with two states captures the basic operational regimes of a GFP. A more elaborate analysis framework would allow for additional states, for instance, to capture differences in restart costs associated with the duration of the idling regime.

and idleness. Perpetual operation is then optimal, and the value of the WP is

$$V_w(x) = \gamma_w \left(\frac{x}{\delta_X} + s\frac{1 - e^{-rT_s}}{r} \right) - \frac{k_w}{r} \quad \text{for } x > 0. \tag{15.26}$$

The option to choose between the two types of plants has a payoff of

$$\max \{V_g(x, y) - K_g, \alpha(V_w(x) - K_w)\}$$

where $K_g > 0$ and $K_w > 0$ are the costs per MW of capacity for the GFP and the WP, respectively, and $\alpha > 0$ is the scale factor. The case $\alpha = 0$ represents a project to invest solely in the GFP. The power producer can also choose the timing τ of the investment. The value of the investment project is therefore given by

$$V(x, y; \alpha) = \sup_{0 \le \tau \le \infty} \mathsf{E}_{x,y} \left[e^{-r\tau} \max(V_g(X_\tau, Y_\tau) - K_g, \alpha(V_w(X_\tau) - K_w)) \right]$$

for $x > 0$ and $y > 0$, where the supremum is over all stopping times $\tau \ge 0$. This is the value of an American option on the maximum of two assets with generally different strikes (unless $K_g = \alpha K_w$), that is, a dual strike max-option. One of the assets, the gas-fired plant, depends on two state variables. Theorem 3 shows that $V(x, y; \alpha)$ is the present value of the cumulative instantaneous gains from investing in the WP when such a choice is optimal and those from investing in the GFP when the latter is optimal.[11,12]

15.3.2 Estimation and calibration

Parameter values under the risk-neutral measure are described in Table 15.1, where the drift $\mu_Z := r - \delta_Z$ for $Z \in \{X, Y\}$. The interest rate is $r = 0.04$. To estimate the volatility parameters of the electricity and gas prices, we collect daily data on Henry Hub Natural Gas Spot Prices from January 2001 to December 2016 and PJM WH Electricity prices for the same time period. As we are interested in the long-term trend and would like to remove short-term fluctuations, we consider moving averages for the prices with window of 1 year. Finally, for the resultant time series, we estimate $\sigma_X = 0.3$ and $\sigma_Y = 0.25$.[13] The correlation between the two processes is $\rho = 0.3$. Next, to calibrate the drift parameters under the risk-neutral measure, we collect the data for the corresponding forward contracts. We estimate $\mu_X = 0$ and $\mu_Y = 0$. For reference, typical values of wholesale electricity prices were in the range $10 - 120$ \$/MWh during the period 2016-2017. Typical values of the gas price fluctuated between $2.3 - 4.3$ \$/MBtu. Parameter estimates in Table 15.1 are comparable to estimates reported in Lucia and Schwartz (2002), Maribu, Galli, and Armstrong (2007), and Boomsma, Meade, and Fleten (2012).

We assume that the capacity factors for the WP and GFP are, respectively, $\gamma_w = 0.37$ and $\gamma_g = 0.6$,[14] corresponding to average capacity factors in the US. The costs of operations

[11] A power producer seeking to build a new plant has the choice between multiple individual technologies and multiple hybrid technologies. A project with N alternatives is an American multistrike compound max-options with N underlying assets, each corresponding to a particular technology. It can be analyzed using the same methods that we employ in the chapter, that is, EIP representation for the project value and integral equations for the relevant boundaries, albeit at the cost of a significant increase in computational difficulty.

[12] Instead of adding capacity, a power producer can also progressively phase out an existing technology and replace it by a new one. A replacement project of this type can be viewed as an American compound spread option. Valuation can be performed using the sequential approach in Section 15.4. In practice, replacement and expansion projects may be considered simultaneously.

[13] Drift estimates under the physical measure are 0.03 for electricity and 0.06 for gas.

[14] See tables 6.7.A and 6.7.B in U.S. Energy Information Administration, Electricity Power Monthly, July 2018.

TABLE 15.1: Parameter values

	Electricity price	*Gas price*
	$Z = X$	$Z = Y$
Drift μ_Z	0	0
Volatility σ_Z	0.3	0.25

for the GFP are $k_{0,g} = 3.79$ and $k_{1,g} = 7.33$ \$/MWh.[15] Switching costs are set at $c_{0,g} = 0.004$ M and $c_{1,g} = 0.001$ M (see Fleten et al. (2016)). These costs capture switches between full capacity operation and idleness for extended periods of time. Costs of restarts increase significantly after a few days due to equipment failure. The running cost for the wind plant is $k_w = 5.33$ \$/MWh,[16] and we assume that the subsidy rate is $s = 23$ \$/MWh (see Boomsma, Meade, and Fleten (2012)).

The average cost per MW of a wind turbine was about 1.68 M in the United States. in 2016.[17] For natural gas power plants, typical construction costs vary between 0.7 M and 2.2 M per MW depending on the type of technology deployed. For combined cycle plants with carbon sequestration, which have a heat rate of about 7.5 MBtu/MWh, the 2016 cost was about 2.10 M.[18] We use these cost estimates, namely, 1.68 M and 2.10 M, with a correction for future replacement costs. This gives adjusted investment costs of $K_w = 2.66$ and $K_g = 3.32$.[19]

15.3.3 Capital-constrained investments

We first consider the case of a firm facing a capital constraint and choosing between two projects with identical investment costs. Letting $K := K_g = \alpha K_w$ ensures that this is the case.

The left panel of Figure 15.4 displays the structure of the immediate investment region along with the indifference curve $\widehat{y}(x)$ and the rays that are asymptotically tangent to the investment boundaries. The most significant aspect is the emergence of an asymptotic inaction cone (between the two rays) within which it is suboptimal to invest, even if the underlying values become extremely large. The existence of this cone stems from the projects' different exposures to the electricity price, which ensures that both the electricity price X and the gas price Y matter asymptotically. As constant components of revenues and costs become asymptotically irrelevant, the limit continuation region is determined by the relative sizes of the two factors. The left panel of Figure 15.6 shows the value of waiting along $\widehat{y}(x)$. This value is positive throughout. It increases with $\widehat{y}(x)$ past a certain point, and it can be shown that it converges to $+\infty$. The right panel of Figure 15.6 plots the probability of investment along $\widehat{y}(x)$. In comparison to the stand-alone GFP project, the joint project leads to a substantial decrease in the probability of investing in the GFP, while maintaining the overall probability of investment at roughly the same level.

[15] Cost figures are calculated based on estimates reported in U.S. Energy Information Administration, Annual Energy Outlook 2017. We focus on Advanced Natural Gas plants with Combined Cycle and Carbon Sequestration, which have a heat rate of 7.5 M Btu/MWh. For this technology, fixed costs are estimated at 33,210 \$/MWy for 2016, which corresponds to 3.79 \$/MWh. Variable costs are 7.08 \$/MWh. The cost of operations is therefore $k_{1,g} = 3.79 + 50\% \times 7.08 = 7.33$ \$/MWh. The cost if the plant is idle is $k_{0,g} = 3.79$ \$/MWh.

[16] The 2017 report of the U.S. Energy Information Administration estimated the average fixed cost of operations for a wind turbine to be 46,710 \$/MWy, that is, 5.33 \$/MWh, in 2016.

[17] U.S. Energy Information Administration, Annual Energy Outlook 2017.

[18] U.S. Energy Information Administration, Annual Energy Outlook 2017.

[19] We assume that costs depreciate by 1% per year and that equipment needs to be replaced every 20 years.

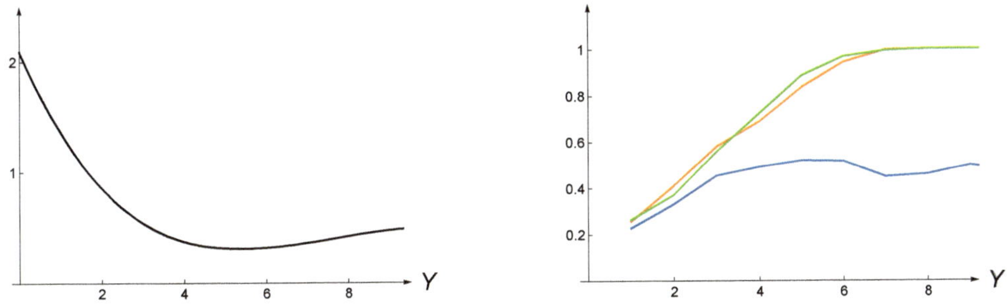

FIGURE 15.6: Value of waiting, and probability of investment. The left panel plots the value of waiting along the indifference curve, as a function of the gas price Y. The right panel displays the probability of investment in (1) the GFP for the project to invest in the best technology (blue curve); (2) the GFP for the stand-alone GFP project (orange curve); and (3) either the GFP or the WP for the project to invest in the best (green curve), as a function of the gas price Y and when (X, Y) moves along the indifference curve. Both panels correspond to the capital-constrained case. Section 15.3.2 describes the set of parameters.

15.3.4 Demand-driven investments

We now consider the case in which the firm seeks to meet a given demand for power by choosing between projects that deliver the same effective capacity. Setting the scale factor α so that $\gamma_g = \alpha\gamma_w$ ensures that both technologies deliver the same output.

The upper-right panel of Figure 15.4 displays the structure of the immediate investment region and the indifference curve $\widehat{y}(x)$ along which the two projects have the same value. Consistent with the results in Section 15.2, the boundaries eventually converge to the indifference curve and the continuation region vanishes as the price of electricity becomes very large. Even for moderate electricity prices, the continuation region shrinks to a narrow band. In this region immediate investment remains suboptimal, but the impact on the project value is small. The likelihood of hitting the boundaries in a small time interval is indeed large, implying limited gains for waiting. The reason the continuation region vanishes asymptotically is that the volatility of the gas price becomes negligible relative to the volatility of the electricity price as we move up the indifference curve. At the limit, the two projects are driven by a common state variable, the electricity price, and the value of waiting converges to zero, because, in the case of single state variable, threshold-type strategies are typically optimal.

15.3.5 Green energy valuation and welfare

The value of green energy is the value of innovations based on GE technologies. In the present context, it is the incremental value generated by the possibility of investing in the alternative wind technology instead of being limited to the gas-fired technology. This wind premium is given by $\Pi(x, y; \alpha)$ in (15.25), where $V(x, y; \alpha)$ is the value of the project to invest in the best of the two technologies and $V(x, y; \alpha = 0)$ is the value of the project to build a GFP.

The wind premium measures the welfare gained by considering the option to invest in the best technology instead of being limited to the GFP. To see this, recall that the Marshalian surplus is the sum of consumer surplus and producer surplus. The consumer surplus for an extra unit of capacity, in our model with exogenous prices, is null. This is true for each of the two projects, because the additional quantity of electricity produced has no price impact,

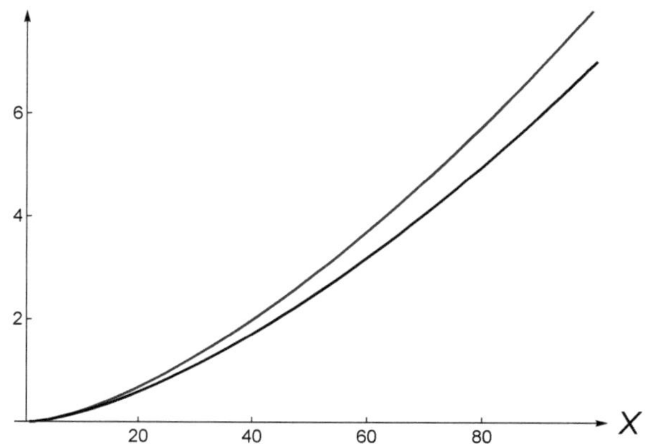

FIGURE 15.7: Project value. This figure shows the value of the project $V(X, Y)$ in the case of a (1) capital constraint (black line) and (2) demand constraint (blue line), as a function of the electricity price X when the gas price $Y = 2.67$ \$/MBtu. Section 15.3.2 describes other parameters.

and hence the incremental area under the demand curve is null.[20] The producer surplus for an extra unit of capacity is given by the profit, that is, marginal revenue minus marginal cost, for each project under consideration.[21] Under these circumstances, the present value of the producer surplus is the present value of the Marshalian surplus, and it corresponds to the value of the project under consideration. The welfare gain is the incremental present value of the Marshalian surplus realized by considering the project to invest in the best technology as opposed to the project to invest in the GFP. This gain is then the difference in project values, that is, the wind premium.

The wind premium, or welfare gain, depends on all the parameters of the model. It inherits the properties of the value of innovation described at the end of Section 15.2. In particular, it is increasing in the subsidy rate s. Subsidizing wind power is therefore welfare enhancing.

Figure 15.7 displays $V(x, y; \alpha)$ for the capital- and demand-constrained cases. Figure 15.8 illustrates the incremental welfare benefits associated with a potential investment in the wind farm. The figure plots the wind premium $\Pi(x, y; \alpha)$ for constant values of the electricity price. The left (right) panel shows the capital-constrained (demand-constrained) case. This premium can be substantial. It increases with the gas price and converges to an upper bound when the latter becomes large. The convex-concave shape is similar in the two panels, but the convex (concave) segment is compressed (magnified) in the demand-constrained case due to the greater efficiency of the WP. The premium is larger in that case for the same reason. The convex-concave shape is because an increase in the gas price, at low levels, decreases the values of both projects, whereas at high levels decreases only the value of the GFP project.

[20]For a small price change $\varepsilon \to 0$, the incremental consumer surplus at time t for an additional unit of installed GFP capacity is $\int_{X-\varepsilon}^{X} \gamma_g dv \to 0$. The same argument applies for an additional unit of WP capacity.

[21]For a small price change $\varepsilon \to 0$, the incremental producer surplus at time t for an additional unit of installed GFP capacity is $\gamma_g(X - \varepsilon) - \gamma_g \kappa Y - k_{1,g} - \int_{X-\varepsilon}^{X} S(v)dv \to \gamma_g(X - \kappa Y) - k_{1,g}$, where $S(v)$ is the original supply curve. A similar argument applies for an additional unit of WP capacity.

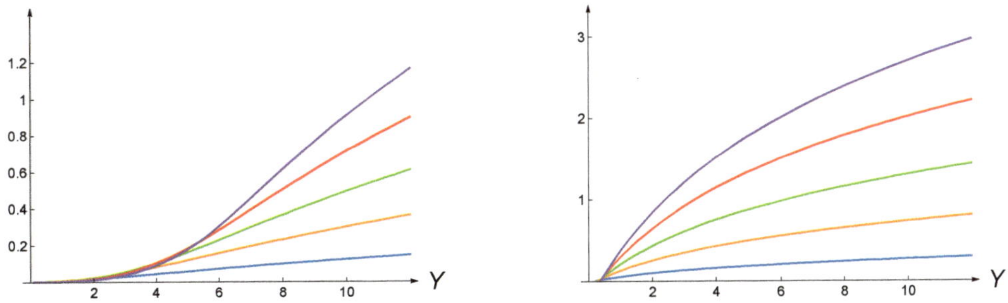

FIGURE 15.8: Wind premium. This figure shows the wind premium $\Pi(x, y; \alpha) = V(x, y; \alpha) - V(x, y; \alpha = 0)$ in the case of a (1) capital constraint (left panel) and (2) demand constraint (right panel), as a function of the gas price Y for different electricity prices X: 20 \$/MWh (blue), 40 \$/MWh (orange), 60 \$/MWh (green), 80 \$/MWh (red), and 100 \$/MWh (purple). Section 15.3.2 describes other parameters. Note the different scale on the vertical axis.

15.3.6 Operating income and leverage

The GFP is equivalent to a long position in the revenue stream (partially) hedged by a short position in the operating cost. The variable part of the operating cost depends on the heat rate κ and the correlation coefficient ρ between the electricity price and the gas price. Variations in these coefficients modify the effectiveness of the hedge and hence affect the performance of the GFP and the value of the project to invest in the best technology.

An increase in the heat rate amounts to a decrease in the efficiency of the GFP as more gas is required to produce a MWh of electricity. The variable cost thus increases and both the total contribution margin and the operating leverage decrease. The increase in κ therefore reduces the payoff from an investment in the GFP, causing the gas boundary to shift left (decrease). It also shifts the wind boundary to the left because of the coupling of optimal decisions. Investments in the gas-fired technology are postponed, whereas those in the wind technology are sped up. The left panel of Figure 15.9 illustrates these effects.

The effects of an increase in the correlation between the electricity and gas prices are similar to those above, as displayed in the right panel of Figure 15.9, albeit less transparent. When the correlation coefficient increases, the value of the GFP decreases (see Section 15.2.1 and Figure 15.2). This leads to a direct reduction in the payoff from investing in that technology. The increase in correlation has a second effect because it reduces the riskiness of the GFP. By convexity, this reduces the value of the option to invest in the best of the two technologies. The combined effect is a decrease in the value of the investment project along with an unambiguous leftward shift in the wind boundary. The direct effect in the GFP value induces a leftward shift in the gas boundary, which is partly mitigated by the reduction in the overall value of the project.

15.3.7 Subsidies and investment costs

The left panel of Figure 15.10 shows the impact of the subsidy s in the capital-constrained case. As s increases the payoff from investing in the WP increases and so does the incentive to invest in that technology. The wind boundary therefore shifts left and the immediate wind-investment region expands. At the same time, because optimal decisions are coupled, the gas boundary also shifts left and the immediate gas-investment region contracts. Investments in

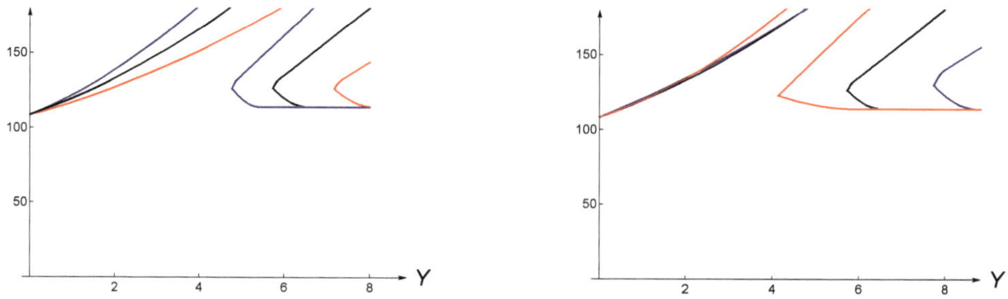

FIGURE 15.9: Investment boundaries: effects of κ and ρ. This figure shows the effects of the heat rate κ and correlation parameter ρ on optimal boundaries for the capital constraint case. Left panel: $\kappa = 6$ MBTu/MWh (red line), $\kappa = 7.5$ MBTu/MWh (black line), and $\kappa = 9$ MBTu/MWh (blue line). Right panel: $\rho = 0$ (blue line), $\rho = 0.3$ (black line), and $\rho = 0.6$ (red line). Section 15.3.2 describes other parameters.

the WP therefore substitute for investments in the GFP, both directly and indirectly. The indirect effect is because it is optimal to wait longer before committing to build a GFP.

This plot also shows that the impact on the wind boundary is substantial. Moreover, it illustrates the limits of regulations based on subsidies in the capital-constrained case. As subsidies go to infinity, the upper part of the wind boundary converges to the lower edge of the no-investment cone. The cone itself is not affected. Hence, a regulation based on a feed-in tariff (FIT) in the form of a constant premium s over the electricity price is unable to reduce the size of the waiting region beyond the asymptotic cone.[22]

As a result of technological advances, the cost of wind plants has substantially decreased during the last decade. For a given cost constraint K, such a reduction amounts to an increase in the scale of an investment in wind power. The right panel of Figure 15.10 shows the impact of a decrease in K_w. As cost decreases, the payoff from an investment in the WP increases, thereby increasing the value of the timing option. The wind boundary therefore shifts left (decreases) and the immediate wind-investment region expands. At the same time, because optimal decisions are coupled, the gas boundary also shifts in the leftward direction (decreases) and the immediate gas-investment region contracts. Investments in the WP therefore substitute for investments in GFP, both directly and indirectly. The indirect effect is because it is optimal to wait longer before committing to build a GFP.

15.4 Investing in growth options: knowledge acquisition

This section examines the decision to invest in knowledge-building assets. Appendix A offers a motivating example. Sections 15.4.1 and 15.4.2 present general results. Section 15.4.3 provides an application to power projects.

[22]Likewise, a regulation based on a constant total FIT s does not affect the asymptotic cone. In contrast, a proportional premium sX affects the cone when s changes.

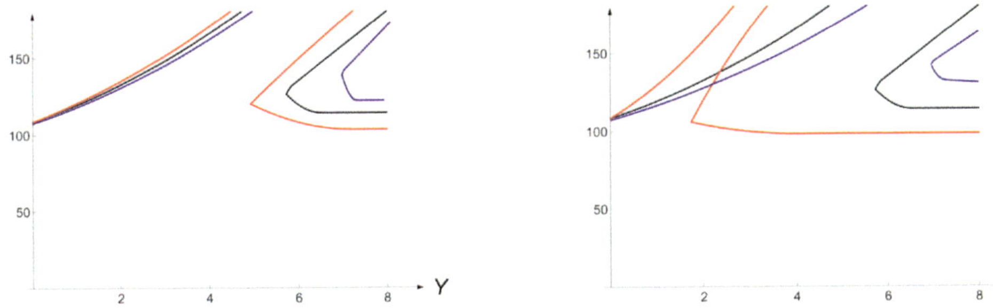

FIGURE 15.10: Investment boundaries: effects of s and K_w. This figure shows the effects of the subsidy s and investment cost K_w on the optimal boundaries for the capital constraint case. Left panel: $s = 15$ \$/MWh (blue line), $s = 23$ \$/MWh (black line), and $s = 30$ \$/MWh (red line). Right panel: $K_w = 1.5$ (red line), $K_w = 1.68$ (black line), and $K_w = 1.75$ (blue line). Section 15.3.2 describes other parameters.

15.4.1 Two-stage model

As the example of Xcel Energy in Appendix A illustrates, a firm specializing in a certain type of projects with variable costs, may not have the expertise to deploy an alternative technology with different cost structure and operate it in the most efficient way. It also may be unable to precisely assess the value of such an alternative using its human capital resources. To improve its decision process, the firm can then invest in knowledge-building assets prior to the actual selection of the best project. This section develops a stylized model with an information acquisition (IA), that is, knowledge-building, stage.

The decision process has now two stages. In the first stage, the firm has three alternatives. It can invest in a VCP using a technology it is familiar with, invest in the FCP with unfamiliar technology bearing an efficiency cost or invest in the acquisition of information about the alternative technology. If it decides to acquire information, it can in a second stage either invest in the VCP or in the FCP. In this setting, both the decision to acquire information and the timing of the acquisition are endogenous. Decisions regarding the choice between the three types of projects and the timing of that choice are also endogenous.

The model for the VCP is described in Section 15.2. The NPV of investing in it is $V_v(x, y) - K_v$. The model for the FCP is a generalization of the previous one, allowing for an endogenous efficiency choice tied to information acquisition. First, we assume that the theoretical capacity factor γ_f is unknown ex ante, given by a random variable with prior distribution function $F(\gamma)$ and support $[0, 1]$. Second, if the project is not optimally designed for the theoretical capacity factor γ_f, there is an efficiency loss ϕ resulting in an effective capacity

$$\gamma_e = \gamma_f(1 - \phi) \tag{15.27}$$

where ϕ is assumed to be constant. Under full information, $\phi = 0$ so that effective capacity equals theoretical capacity. In this instance the FCP model corresponds to the model in Section 15.2. Under incomplete information, suboptimal design results in an efficiency loss $\phi \in (0, 1]$.

Technological information permitting the optimization of capacity utilization can be acquired at a cost K_I. After the information is acquired, the theoretical capacity is revealed (with a short delay) and equal to γ_f. Optimal project design implies that the effective capacity is then also equal to γ_f. The NPV of the FCP conditional on complete information about γ_f is $V_f(x; \gamma_f) - K_f$, with $V_f(x; \gamma_f)$ as in Section 15.2. If the firm chooses to forgo IA, it operates under incomplete information. It can still invest in the FCP, but bears the efficiency cost due to suboptimal design. The NPV of the FCP conditional on imperfect information is $V_f^i(x;\phi) - K_f$, where

$$V_f^i(x;\phi) = \int_0^1 V_f(x;\gamma(1-\phi))dF(\gamma) \tag{15.28}$$

for $x > 0$ and $\phi \in (0,1]$.

The stage two decision of the firm was solved in Section 15.2 with the option value denoted by $V(x,y;\gamma_f)$ for a known value of γ_f. In stage one, the firm has three alternatives and solves

$$V_I(x,y) = \sup_{0\le\tau\le\infty} \mathsf{E}\big[e^{-r\tau}\max(V_v(X_\tau,Y_\tau) - K_v, \alpha(V_f^i(X_\tau;\phi) - K_f), \\ V_{II}(X_\tau,Y_\tau) - K_I)\big] \tag{15.29}$$

where $V_f^i(x;\phi)$ is defined above in (15.28) and the value of the project in stage II is given by

$$V_{II}(x,y) = \mathsf{E}_F[V(x,y;\gamma_f(\omega))] = \int_0^1 V(x,y;\gamma)dF(\gamma) \tag{15.30}$$

for $x, y > 0$. This section focuses on the first-stage problem (15.29), that is, whether or when to acquire information.

Theorem 4 *There exist three boundaries $b_{v,I} : (0,\infty) \to (0,\infty)$, $b_{f,I} : (0,\infty) \to (0,\infty)$ and $b_{f,I}^i : (0,\infty) \to (0,\infty)$ such that it is optimal to invest in the VCP when $y \le b_{v,I}(x)$, to acquire information when $y \ge b_{f,I}(x)$ and to invest in the FCP by forgoing information acquisition when $y \ge b_{f,I}^i(x)$.*

When Y goes to infinity, the optimal timing decision can be inferred from the one-dimensional (state variable) problem

$$\sup_{0\le\tau\le\infty} \mathsf{E}_x\left[e^{-r\tau}\max(\alpha(V_f^i(X_\tau;\phi) - K_f), V_{II}(X_\tau,\infty) - K_I)\right] \tag{15.31}$$

for $x > 0$.

Theorem 4 shows that the immediate investment region splits into three subregions (see the left panel of Figure 15.11 for an illustration). In the upper subregion, where $y \le b_{v,I}(x)$, investment in the VCP is optimal. In the intermediate subregion, where $y \ge b_{f,I}^i(x)$, direct investment in the FCP is optimal. In the lower subregion, where $y \ge b_{f,I}(x)$, investment in IA is optimal. The subregions are separated by two indifference curves along which waiting dominates. The first indifference curve is the set of points along which the NPVs of investments in the VCP and in the inefficient FCP are the same. The second one is the set of points where the NPVs of investments in the inefficient FCP and IA are the same. Along these curves, waiting is optimal even if the values of the underlying stand-alone projects become extremely large. The intuition for this property is the same as in Section 15.2. When the efficiency cost of direct investment increases, the intermediate subregion shrinks. It eventually vanishes when ϕ becomes sufficiently large (see the right panel of Figure 15.11).

15.4.2 Valuation of knowledge assets

Theorem 5 is the main result of this section. The theorem characterizes the optimal investment boundaries for the first stage.

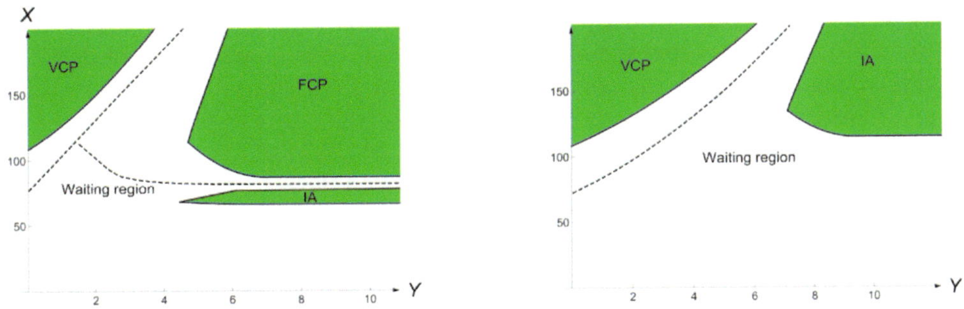

FIGURE 15.11: Two-stage problem: immediate investment region for stage 1. This figure shows the immediate investment region (green) for (1) $\phi = 0$ and low K_I (left panel) and (2) $\phi = 0.2$ and $K_I = 0.6$ (right panel). Dashed lines represent indifference curves. Section 15.3.2 describes the central set of parameters.

Theorem 5 *The optimal investment boundaries* $(b_{v,I}, b^i_{f,I}, b_{f,I})$ *solve the system of three integral equations:*

$$V_v(x, b_{v,I}(x)) - K_v = \pi_I(x, b_{v,I}(x); b_{v,I}, b^i_{f,I}, b_{f,I}) \tag{15.32}$$

$$\alpha(V^i_f(x;\phi) - K_f) = \pi_I(x, b^i_{f,I}(x); b_{v,I}, b^i_{f,I}, b_{f,I}) \tag{15.33}$$

$$V_{II}(x, b_{f,I}(x)) - K_I = \pi_I(x, b_{f,I}(x); b_{v,I}, b^i_{f,I}, b_{f,I}) \tag{15.34}$$

for $x > 0$ *and where* $\pi_I(x, y; b_{v,I}, b^i_{f,I}, b_{f,I})$ *is the early investment premium (EIP) defined as*

$$\begin{aligned}
\pi_I(x, y; b_{v,I}, b^i_{f,I}, b_{f,I}) &= \pi_{v,I}(x, y; b_{v,I}) + \pi^i_{f,I}(x, y; b^i_{f,I}) + \pi_{f,I}(x, y; b_{f,I})\\
\pi_{v,I}(x, y; b_{v,I}) &= -\mathsf{E}_{x,y}\left[\int_0^\infty e^{-rt} H_v(X_t, Y_t) I(Y_t \le b_{v,I}(X_t))dt\right]\\
\pi^i_{f,I}(x, y; b^i_{f,I}) &= -\mathsf{E}_{x,y}\left[\int_0^\infty e^{-rt} H^i_{f,I}(X_t, Y_t) I(Y_t \ge b^i_{f,I}(X_t))dt\right]\\
\pi_{f,I}(x, y; b_{f,I}) &= -\mathsf{E}_{x,y}\left[\int_0^\infty e^{-rt} H_{f,I}(X_t, Y_t) I(Y_t \ge b_{f,I}(X_t))dt\right]
\end{aligned}$$

where

$$\begin{aligned}
-\, H_v(x, y) &= \gamma_v\,(x - \kappa y) - k_{1,v} - r(K_v + c_{0,v})\\
-\, H^i_{f,I}(x, y) &= -\int_0^1 H_f\,(x; \gamma(1-\phi))\, dF\,(\gamma)\\
-\, H_{f,I}(x, y) &= -\int_0^1 H_f\,(x;\gamma)\, I(y \ge b_f(x;\gamma))dF\,(\gamma) - rK_I\\
-\, H_f\,(x;\gamma) &= \alpha\,(\gamma\,(x + s) - k_{1,f} - r(K_f + c_{0,f}))
\end{aligned}$$

for $x, y > 0$. *The value of the project is* $V_I(x, y) = \pi_I(x, y; b_{v,I}, b^i_{f,I}, b_{f,I})$ *for* $x, y > 0$.

Theorem 5 shows that the boundaries for the first stage also satisfy a system of integral equations. These equations follow from the EIP representation of the value of the two-stage project, $V_I(x, y)$. There are now three premium components associated with the local net gains from investing in the VCP, those from investing in technological information acquisition, and those from directly investing in the FCP. For the VCP investment, the structure of the local net gains is similar to that in the benchmark model of Section 15.2. The trigger boundary $b_{v,I}(X)$ is nevertheless different from the VCP boundary $b_v(X;\gamma_f)$ for stage II, which also corresponds to the VCP boundary in the benchmark model. For the information acquisition investment, the structure of the local net gains is different. It is interesting to note that benefits are only collected when information acquisition is optimal, and it is optimal to immediately invest in the FCP for some possible values of γ_f, that is, when $Y > \max(b_{f,I}(X), b_f(X;\gamma_f))$ for some γ_f. Hence, there is no value creation if waiting to invest is ex post optimal for all possible γ_f, that is, after information is acquired. This surprising result follows because the second-stage project does not produce benefits until it is implemented: prior to investment its value grows at the risk-free rate and therefore has a null dividend yield. The opportunity cost rK_I, in contrast, is always incurred starting immediately after information is acquired. Finally, for the direct investment in the FCP, the local net gain is the expected value of the gain under full information reduced by the efficiency loss associated with suboptimal design.

The respective positions of the boundaries for stages 1 and 2 are also interesting to note. Figure 15.12 displays a situation with three possible values of the FCP capacity factor, $\gamma_f = 0.34, 0.37$ and 0.4, each with equal probability 1/3. The case displayed corresponds to parameter values such that direct investment in the FCP is suboptimal in stage 1. Observe that the information boundary (solid curve) is above the second-stage FCP boundary corresponding to the highest capacity factor $\gamma_f = 0.4$ (lowest dashed curve). This is a structural result that will always hold. It reflects the fact that the local benefits of investing in the FCP are null (i.e., waiting to invest is optimal) if the cost factor is below the lowest possible ex post FCP boundary. For similar reasons, the first-stage VCP boundary is always above the lowest possible second-stage VCP boundary. Indeed, if it were lower, it would be optimal to invest in the VCP in the first stage regardless of the FCP capacity factor. Information would then be irrelevant. As long as information provides ex post benefits for some value of γ_f, this cannot happen. As for the FCP boundary for direct investment, it is also greater than the second-stage FCP boundary for the highest capacity factor $\gamma_f = 0.4$. This is because of the lower expected capacity and the efficiency loss for suboptimal design.

Figure 15.12 shows that the first-stage information boundary can exceed the second-stage FCP boundary corresponding to the lowest capacity factor $\gamma_f = 0.34$. This seemingly surprising situation can occur when the marginal contribution of the VCP is sufficiently low. In such an event the value of the VCP tends to zero and the project reduces to a choice between two alternatives, an efficient FCP with an information acquisition stage and an investment stage, or an inefficient FCP under incomplete information (in the plot, the second alternative is suboptimal). Because information is costly, it can then pay to wait before acquiring information even if the cost factor exceeds the highest possible second-stage FCP boundary.

15.4.3 Application: power growth options

In the context of power projects, the effects of government subsidies are especially relevant. As the subsidy size increases, the local gains from information acquisition and wind power investment increase, leading to an increase in the overall value of the project. This raises the value of information—and hence the first-stage payoff from information acquisition—and shifts the information boundary to the left. Because of the interactions between the three boundaries, it also shifts the first-stage gas boundary and the first-stage wind boundary for direct investment to the left and hence optimally delays potential first-stage investments in gas power and speeds up potential first-stage direct investments in

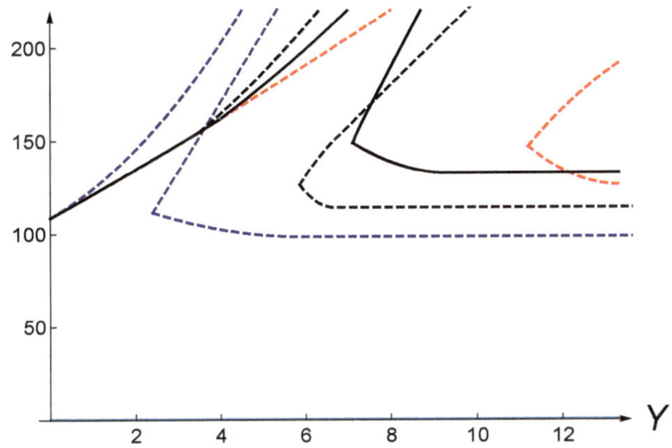

FIGURE 15.12: Two-stage problem: optimal investment boundaries. This figure plots the optimal investment boundaries for stage 1 (solid black line) and stage 2: $\gamma_f = 0.34$ (dashed red), $\gamma_f = 0.37$ (dashed black), and $\gamma_f = 0.4$ (dashed blue). We use $\phi = 0.2$ and $K_I = 0.6$, and Section 15.3.2 describes the other parameters.

wind power. The right panel of Figure 15.13 illustrates these effects for parameter values leading to suboptimal direct investment in the first stage.

Operating leverage effects are displayed in the left panel of Figure 15.13, for the same parameters. As shown in Section 15.3, an increase in the heat rate κ, which amounts to a reduction in the efficiency of the GFP, reduces the GFP value. This decreases the second-stage value V_{II} and hence the payoff from IA, as well as the first-stage payoff from investing in the GFP. The first (second) effect tends to increase (decrease) the value of waiting to invest in information. In the figure, the second effect dominates and reduces the value of waiting, thus moving the IA boundary to the left. Similar effects would combine and tend to move the wind boundary for direct investment to the left, if parameter values led to the optimality of such investment in the first stage.

15.5 Conclusion

In this chapter, we examined the optimal investment in the best of two projects with different types of cost structures, variable and fixed. We showed that the immediate investment policy is determined by a pair of boundaries satisfying a system of coupled integral equations of the Fredholm type. The value of the investment project has an early investment premium representation consisting of two components, each tallying the local gains achieved when investment in a given project is optimal. In this context, we showed that it is always optimal to postpone investment when the NPVs of the two projects are equal. This is true even if these values are arbitrarily large. In fact, for some configurations of parameter values, waiting is optimal as long as the underlying revenue and cost factors lie in a cone. Thus, waiting can be optimal even if the discrepancy between the individual project values becomes infinitely large. Operating leverage emerges as an important factor. When

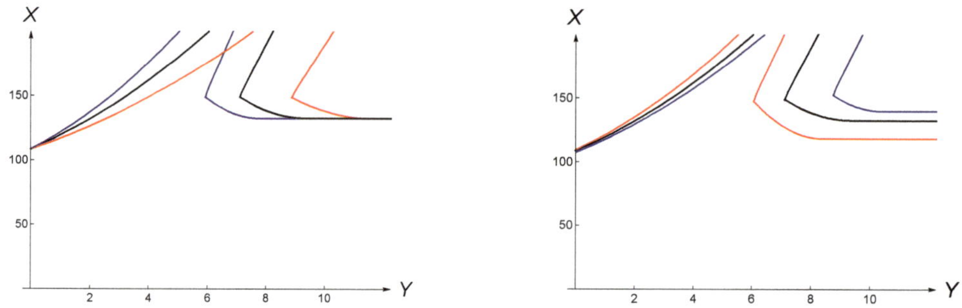

FIGURE 15.13: Two-stage problem: effects of κ and s on stage 1 boundaries. This figure shows the effects of the heat rate κ and the subsidy rate s on the optimal investment boundaries of stage 1. Left panel: $\kappa = 6$ MBTu/MWh (red line), $\kappa = 7.5$ MBTu/MWh (black line), and $\kappa = 9$ MBTu/MWh (blue line). Right panel: $s = 15$ (blue line), $s = 23$ (black line), and $s = 30$ (red line). We use $\phi = 0.2$ and $K_I = 0.6$, and Section 15.3.2 describes other parameters.

it decreases, the welfare gain from investing in the FCP increases, speeding up investments in the FCP at the expense of those in the VCP.

The model was applied to examine optimal investment decisions in the power generation sector. Power can be produced using various technologies entailing different types of costs. We focused on the choice between a GFP and a WP. The possibility of using wind to produce electricity significantly affects the value of a power generation project. The welfare gain (premium) associated with the renewable source is an increasing function of the price of gas. It is large for typical values of the underlying prices and significantly affects the value of a power project. It can remain positive even when the natural gas price decreases and the spark spread increases. This provides an economic rationale as to why renewable power generation, for example, wind, remains competitive even in the face of sharp declines in the cost certain fossil fuels.

Our analysis provides additional perspective regarding the importance of renewable energy for modern societies. Renewables clearly have a major role to play in alleviating the negative consequences, such as global warming and other climate change phenomena, associated with human reliance on fossil fuels. This study shows that renewable energies provide immediate welfare benefits for agents even if the economics of the problem would suggest that they are dominated by fossil fuels for power generation. The mere availability of power production based on renewable sources implies that it might be optimal to postpone investments and consider green power as a viable alternative to fossil fuel power generation.

The chapter also provides new insights about investments in growth options, that is, knowledge-building investments designed to increase information about components of growth options. For power projects, information about proven technologies is more widely available and the cost of acquiring technical expertise to build and optimize a plant has decreased. This is no longer true for advanced technologies that are either experimental or proprietary. Implementation, in these cases, involves a transfer of technology likely to be very costly. Alternatively, it may involve the formation of a partnership with or the outright acquisition of the firm owning the technology. Exploration of these issues might be a fruitful avenue for future research.

Finally, it might be worth commenting on additional applications of the model. One possible area is human capital investments and pertains to decisions involving alternative career paths. For instance, a choice to invest in an engineering degree or a medical degree involves an initial decision about the timing of the investment and about the path to follow, each of which entails several subsequent decisions about whether to stop with the initial degree or pursue subsequent more advanced degrees at later endogenous times. Another application is in the area of conglomerate formation, where a firm grows by acquiring several entities, each engaged in activities that usually differ from that of the acquiring firm and with its own growth options. In such intertwined structures, the risks associated with any given firm's cash flows have an effect on the timing of the acquisition and the selection among the candidate targets. Decisions pertaining to such alternatives are more complex because revenues and/or costs can be driven by different factors.[23] The analysis developed in the present chapter nevertheless can be applied. It shows that the value functions for such multistage timing and selection problems can be written as the present values of cash flows collected over specific future events and provides tools for the computation of the relevant decision boundaries.

Appendix A. Xcel Energy: Investing in Knowledge Acquisition

Xcel Energy is a utility holding company based in Minnesota, resulting from the merger of three predecessors, Public Service Company of Colorado, Southwestern Public Service Company and Northern States Prower. It focuses on the production, transmission and distribution of power and the production, transport and distribution of natural gas. The company operates in Western and Midwestern states and services over 3.5 million customers with electricity and over 2 million with natural gas (see Xcel Energy 2017-Corporate Responsibility Report).

In 2007, most of Xcel's power generation capacity came from about 60 coal and natural gas plants, 20 hydroelectric plants, 2 nuclear plants and 1 wind plant. Total dependable capacity reached 15,699 MW. The wind plant, located in Ponnequin-Weld County, Colorade, was Colorado's first wind farm. It consisted of 44 wind turbines built over 1999–2001 with total capacity of 25 MW. Xcel owned 37 of the 44 turbines. In addition to the power generated by its own plants, Xcel also purchased from other operators under long-term contracts. At the end of 2007, it was the nation's leading utility for wind power purchases, with about 2,500 MW under contract.

December 2007 marked a turning point in Xcel's green energy strategy. On Christmas day, an extreme weather event hit Northeastern Colorado, bringing a record snowfall along with high winds that delivered three quarters of Xcel's generating capacity in an hour, nearly overloading the grid. The event underscored the variability of wind and the unreliability of the day-ahead forecasts based on public information that were used at the time to predict load and available capacity. In the aftermath, Xcel partnered with the National Center for Atmospheric Research (NCAR) to develop reliable wind forecasts and with the National Renewable Energy Laboratory (NREL) to develop a reliable model for the relation between turbine output and wind speed. As part of the project, it installed a network of weather recording stations located at turbine height to provide accurate measurements of wind power. The project led to the development of high-resolution models that update every 15 minutes and produce wind forecasts up to 168 hours ahead. The density of the data collection grid permits the observation of terrain effects within a single wind farm and individual

[23]Even when the same factor drives the costs of the two alternatives, investment subregions could, in principle, fail to be connected. In the special case of perpetual operation, value functions are affine, and subregions have the general structure derived in this chapter. But, regardless of the cost structure, waiting is optimal when alternatives have the same NPV.

predictions for each turbine (see Mahoney et al. (2012) for an overview of the system).[24] In conjunction with the wind forecasting model, the firm developed an automatic dispatch model, that is, an information technology infrastructure that automatically manages load and generation. This system optimizes resources necessary for rapid ramp-up in case of an adverse wind event. Xcel estimates that the project led to a 17%–38% improvement in operations and savings of 22 M from 2009 to 2012 (Montgomery, 2013). Savings through end of 2016 were estimated at 60 M for an investment around 3.8 M (Baskin, 2016).

This acquisition of knowledge assets paved the way for subsequent investments in wind plants. By 2013 Xcel owned and operated 3 WP. By 2015, this number grew to 5, for a total capacity of 850 MW. It added further wind generation capacity by entering long-term supply contracts. Since that time, Xcel has continued an ambitious expansion in wind power using a multipronged approach. In 2017, it introduced a program to retire aging coal pants and replace them with renewable power. It also announced the development of 12 new wind farms across 7 states and a goal of exceeding 10,000 MW of wind capacity in its system by 2022. At that time, it anticipates producing 48% of its power supply from renewable sources.

Appendix B. Optimal Switching Problem for VCP when $Y = 0$

This Appendix briefly explains how to solve the switching problem for VCP when $Y = 0$. Using standard hedging arguments, one can show that the value functions $V_{0,v}$ and $V_{1,v}$ satisfy the second-order linear ODEs in the continuation region

$$
\begin{aligned}
(r-\delta_X)xV_{0,v}'(x) + \frac{\sigma_X^2}{2}x^2V_{0,v}''(x) - rV_{0,v}(x) &= k_{0,v} \quad \text{for } x < b_{0,v} \\
(r-\delta_X)xV_{1,v}'(x) + \frac{\sigma_X^2}{2}x^2V_{1,v}''(x) - rV_{1,v}(x) &= -\gamma_v x + k_{1,v} \quad \text{for } x > b_{1,v}.
\end{aligned}
$$

These equations can be solved, respectively, as

$$
\begin{aligned}
V_{0,v}(x) &= A_0x^p + B_0x^q - k_{0,v}/r \quad \text{for } x < b_{0,v} \\
V_{1,v}(x) &= A_1x^p + B_1x^q + \gamma_v x/\delta_X - k_{1,v}/r \quad \text{for} \quad x > b_{1,v},
\end{aligned}
$$

where A_0, B_0, A_1, B_1 are unknown constants and p, q are roots of the associated quadratic equations with $p > 1$ and $q < 0$. It is clear that $V_{0,v}(0+) = -k_{0,v}/r$ and $V_{1,v}(x)$ is growing at a slower rate than linear. Hence, $B_0 = 0$ and $A_1 = 0$. Next, we impose smooth-pasting and value-matching conditions at $b_{i,v}$, $i = 0, 1$, leading to the system of equations

$$
\begin{aligned}
A_0b_{0,v}^p - k_{0,v}/r &= B_1b_{0,v}^q + \gamma_v b_{0,v}/\delta_X - k_{1,v}/r - c_{0,v} \\
pA_0b_{0,v}^{p-1} &= qB_1b_{0,v}^{q-1} + \gamma_v/\delta_X \\
A_0b_{1,v}^p - k_{0,v}/r - c_{1,v} &= B_1b_{1,v}^q + \gamma_v b_{1,v}/\delta_X - k_{1,v}/r \\
pA_0b_{1,v}^{p-1} &= qB_1b_{1,v}^{q-1} + \gamma_v/\delta_X
\end{aligned}
$$

with four unknowns $(b_{0,v}, b_{1,v}, A_0, B_1)$. This system of algebraic equations has a unique solution that is easy to find by numerical methods.

[24]See also https://ral.ucar.edu/solutions/products/empirical-wind-energy-conversion-algorithm

Appendix C. Numerical Method

This appendix develops a new numerical approach to solve the system (15.9)-(15.10). First, we rewrite the premium π_1 in terms of $1 - I(X_t \geq b_1(Y_t))$ instead of $I(X_t \leq b_1(Y_t))$. Then, by integrating over dt on $(0, \infty)$ we obtain

$$\pi_1(x, y; b_1) = -\gamma_v x/\delta_X + \gamma_v y/\delta_Y + k_{1,v}/r - k_{0,v}/r - c_{1,v} + \tilde{\pi}_1(x, y; b_1)$$

and the alternative characterization $V_1(x, y) = -k_{0,v}/r - c_{1,v} + \tilde{\pi}_1(x, y; b_1)$, where we defined

$$\tilde{\pi}_1(x, y; b_1) = \mathsf{E}_{x,y}\left[\int_0^\infty e^{-rt}(\gamma_v(X_t - Y_t) - k_{1,v} + k_{0,v} + rc_{1,v})I(X_t \geq b_1(Y_t))dt\right]$$

for $x > 0$ and $y > 0$. Now if we insert the new expression of π_1 into (15.9)-(15.10) some of the terms will disappear and the system becomes

$$\tilde{\pi}_1(b_0(y), y; b_1) - \pi_0(b_0(y), y; b_0) = c_{0,v} + c_{1,v} \tag{C35}$$
$$\tilde{\pi}_1(b_1(y), y; b_1) - \pi_0(b_1(y), y; b_0) = 0 \tag{C36}$$

for $y > 0$. Taking expectations inside the integrals, we can then derive more explicit representations and rewrite the first equation (C35) (we proceed analogously with the second one)

$$\begin{aligned} c_{0,v} + c_{1,v} = {} & \gamma_v b_0(y) \int_0^\infty e^{-\delta_X t}(\mathsf{P}^X_{b_0(y),y}(X_t \geq b_1(Y_t)) - \mathsf{P}^X_{b_0(y),y}(X_t \geq b_0(Y_t)))dt \\ & - \gamma_v y \int_0^\infty e^{-\delta_Y t}(\mathsf{P}^Y_{b_0(y),y}(X_t \geq b_1(Y_t)) - \mathsf{P}^Y_{b_0(y),y}(X_t \geq b_0(Y_t)))dt \\ & - (k_{1,v} - k_{0,v} - rc_{1,v}) \int_0^\infty e^{-rt}\mathsf{P}_{b_0(y),y}(X_t \geq b_1(Y_t))dt \\ & + (k_{1,v} - k_{0,v} + rc_{0,v}) \int_0^\infty e^{-rt}\mathsf{P}_{b_0(y),y}(X_t \geq b_0(Y_t))dt \end{aligned}$$

for $y > 0$ where we defined new probability measures $d\mathsf{P}^X = e^{-(r-\delta_X)t}(X_t/X_0)d\mathsf{P}$ and $d\mathsf{P}^Y = e^{-(r-\delta_Y)t}(Y_t/Y_0)d\mathsf{P}$, respectively. The system of equations (C35) and (C36) becomes

$$b_0(y) = \frac{y \cdot I^Y(b_0(y), y; b_0, b_1) + I_1(b_0(y), y; b_1) - I_0(b_0(y), y; b_0) + c_0 + c_1}{I^X(b_0(y), y; b_0, b_1)} \tag{C37}$$
$$b_1(y) = \frac{y \cdot I^Y(b_1(y), y; b_0, b_1) + I_1(b_1(y), y; b_1) - I_0(b_1(y), y; b_0)}{I^X(b_1(y), y; b_0, b_1)} \tag{C38}$$

for $y > 0$, where

$$\begin{aligned} I^X(x, y; b_0, b_1) &= \gamma_v \int_0^\infty e^{-\delta_X t}(\mathsf{P}^X_{x,y}(X_t \geq b_1(Y_t)) - \mathsf{P}^X_{x,y}(X_t \geq b_0(Y_t)))dt \\ I^Y(x, y; b_0, b_1) &= \gamma_v \int_0^\infty e^{-\delta_Y t}(\mathsf{P}^Y_{x,y}(X_t \geq b_1(Y_t)) - \mathsf{P}^Y_{x,y}(X_t \geq b_0(Y_t)))dt \\ I_0(x, y; b_0) &= (k_{1,v} - k_{0,v} + rc_{0,v}) \int_0^\infty e^{-rt}\mathsf{P}_{x,y}(X_t \geq b_0(Y_t))dt \\ I_1(x, y; b_1) &= (k_{1,v} - k_{0,v} - rc_{1,v}) \int_0^\infty e^{-rt}\mathsf{P}_{x,y}(X_t \geq b_1(Y_t))dt \end{aligned}$$

for $x, y > 0$. The boundaries are the fixed point of this system of integral equations. They can be computed using the induction procedure:

$$b_0^{(n)}(y) = \frac{y \cdot I^Y(b_0^{(n-1)}(y), y; b_0^{(n-1)}, b_1^{(n-1)}) + I_1(b_0^{(n-1)}(y), y; b_1^{(n-1)}) - I_0(b_0^{(n-1)}(y), y; b_0^{(n-1)}) + c_0 + c_1}{I^X(b_0^{(n-1)}(y), y; b_0^{(n-1)}, b_1^{(n-1)})} \tag{C39}$$

$$b_1^{(n)}(y) = \frac{y \cdot I^Y(b_1^{(n-1)}(y), y; b_0^{(n-1)}, b_1^{(n-1)}) + I_1(b_1^{(n-1)}(y), y; b_1^{(n-1)}) - I_0(b_1^{(n-1)}(y), y; b_0^{(n-1)})}{I^X(b_1^{(n-1)}(y), y; b_0^{(n-1)}, b_1^{(n-1)})} \tag{C40}$$

for $y > 0$, $n \geq 1$ with, for example, $b_i^{(0)}(y) \equiv b_i(0+)$, $i = 0, 1$, as initial curves. The integrals in I^X, I^Y, I_0, I_1 can be computed using any efficient quadrature scheme.

References

Bakke, I., S.-E. Fleten, L. I. Hagfors, V. Hagspiel, and B. Norheim, 2016, Investment in mutually exclusive transmission projects under policy uncertainty, *Journal of Commodity Markets* 3, 54–69.

Baskin, J. S., 2016, Xcel tames variability in wind power, *Forbes*.

Boomsma, T. K., N. Meade, and S.-E. Fleten, 2012, Renewable energy investments under different support schemes: A real options approach, *European Journal of Operational Research* 220, 225–237.

Brennan, M. J., and E. S. Schwartz, 1985, Evaluating natural resource investments, *Journal of Business* 135–157.

Broadie, M., and J. Detemple, 1997, The valuation of American options on multiple assets, *Mathematical Finance* 7, 241–286.

Carlson, M., A. Fisher, and R. Giammarino, 2004, Corporate investment and asset price dynamics: Implications for the cross-section of returns, *The Journal of Finance* 59, 2577–2603.

Carr, P., R. Jarrow, and R. Myneni, 1992, Alternative characterizations of American put options, *Mathematical Finance* 2, 87–106.

Décamps, J.-P., T. Mariotti, and S. Villeneuve, 2006, Irreversible investment in alternative projects, *Economic Theory* 28, 425–448.

Deng, S., B. Johnson, and A. Sogomonian, 1999, Spark spread options and the valuation of electricity generation assets, in *Proceedings of the 32nd Annual Hawaii International Conference on Systems Sciences. 1999. HICSS-32. Abstracts and CD-ROM of Full Papers*, 7–pp, IEEE.

Dixit, A. K., and R. S. Pindyck, 1994, *Investment Under Uncertainty* (Princeton University Press).

Fleten, S.-E., M. Johansen, A. Pichler, and C. J. Ullrich, 2016, A structural estimation analysis of the strategic switching of peak generators, Working Paper, Norwegian University of Science and Technology.

Fleten, S.-E., K. M. Maribu, and I. Wangensteen, 2007, Optimal investment strategies in decentralized renewable power generation under uncertainty, *Energy* 32, 803–815.

Fleten, S.-E., and E. Näsäkkälä, 2010, Gas-fired power plants: Investment timing, operating flexibility and co2 capture, *Energy Economics* 32, 805–816.

Geltner, D., T. Riddiough, and S. Stojanovic, 1996, Insights on the effect of land use choice: The perpetual option on the best of two underlying assets, *Journal of Urban Economics* 39, 20–50.

Kim, I. J., 1990, The analytic valuation of American options, *The Review of Financial Studies* 3, 547–572.

Lev, B., 1974, On the association between operating leverage and risk, *Journal of Financial and Quantitative Analysis* 9, 627–641.

Lucia, J. J., and E. S. Schwartz, 2002, Electricity prices and power derivatives: Evidence from the nordic power exchange, *Review of Derivatives Research* 5, 5–50.

Mahoney, W. P., K. Parks, G. Wiener, Y. Liu, W. L. Myers, J. Sun, L. Delle Monache, T. Hopson, D. Johnson, and S. E. Haupt, 2012, A wind power forecasting system to optimize grid integration, *IEEE Transactions on Sustainable Energy* 3, 670–682.

Margrabe, W., 1978, The value of an option to exchange one asset for another, *The Journal of Finance* 33, 177–186.

Maribu, K. M., A. Galli, and M. Armstrong, 2007, Valuation of spark-spread options with mean reversion and stochastic volatility., *International Journal of Electronic Business Management.* 5, 173–181.

McDonald, R., and D. Siegel, 1986, The value of waiting to invest, *The Quarterly Journal of Economics* 101, 707–727.

McDonald, R. L., and D. R. Siegel, 1985, Investment and the valuation of firms when there is an option to shut down, *International Economic Review* 331–349.

Montgomery, J., 2013, How xcel saved 22 m with weather and wind forecasting, *Renewable Energy World.*

Novy-Marx, R., 2011, Operating leverage, *Review of Finance* 15, 103–134.

Peskir, G., 2007, A change-of-variable formula with local time on surfaces, *Séminaire de probabilités XL* 70–96.

Pindyck, R. S., 1999, The long-run evolution of energy prices, *The Energy Journal* 20, 1–27.

Schwartz, E., and J. E. Smith, 2000, Short-term variations and long-term dynamics in commodity prices, *Management Science* 46, 893–911.

Siddiqui, A., and S.-E. Fleten, 2010, How to proceed with competing alternative energy technologies: A real options analysis, *Energy Economics* 32, 817–830.

Zhang, L., 2005, The value premium, *The Journal of Finance* 60, 67–103.

Chapter 16

The effect of pro-environmental preferences on bond prices: Evidence from green bonds

Olivier David Zerbib

CREST – ENSAE, Institut Polytechnique de Paris, France

We use green bonds as an instrument to identify the effect of non-pecuniary motives, specifically pro-environmental preferences, on bond market prices. We perform a matching method, followed by a two-step regression procedure, to estimate the yield differential between a green bond and a counterfactual conventional bond from July 2013 to December 2017. The results suggest a small negative premium: the yield of a green bond is lower than that of a conventional bond. On average, the premium is -2 basis points for the entire sample and for euro and USD bonds separately. We show that this negative premium is more pronounced for financial and low-rated bonds. The results emphasize the low impact of investors' pro-environmental preferences on bond prices, which does not represent, at this stage, a disincentive for investors to support the expansion of the green bond market.

16.1 Introduction

In response to environmental crises, financial investors have recently taken up the challenge and become key actors in the energy and environmental transition. This pivotal role is notably due to their ability to mobilize a considerable amount of funds: the global stock of manageable assets,[1] which amounted to USD 160 trillion in 2016 (Financial Stability Board, 2018), can be compared to the infrastructure investment needs of 6.9 trillion over the next 15 years to be consistent with the 2 degrees Celsius threshold (OECD, 2017a). Several initiatives have been launched to redirect assets toward green investments. For example, by signing the Montreal Carbon Pledge,[2] more than 120 investors with assets under management worth more than USD 10 trillion have committed to supporting the development of the green bond market and to measuring and publishing the carbon footprint of their investments.

The drivers to invest in assets with a low environmental impact (*green assets* hereafter) can be related to financial motives, such as the expectation of better financial performance (Nilsson, 2008, Bauer and Smeets, 2015, Hartzmark and Sussman, 2018) or a lower risk (Krüger, 2015). These drivers can also be attributable to non-pecuniary motives.

[1] This amount corresponds to the *Monitoring Universe of Non-bank Financial Intermediation*, including all non-bank financial intermediation: insurance corporations, pension funds, other financial intermediaries and financial auxiliaries.

[2] http://montrealpledge.org/

DOI: 10.1201/9781032636252-16

Preferences linked to pro-social and pro-environmental[3] norms and attitudes lead investors to increase their investments in the assets of companies behaving more ethically (Hong and Kacperczyk, 2009; Riedl and Smeets, 2017; Hartzmark and Sussman, 2018). The incentive is not necessarily a proprietary choice of the asset manager: it can be delegated by the asset owner through the *delegated philanthropy* mechanism described by Benabou and Tirole (2010).

The price impact of investors' preferences for green assets has been broadly documented in the literature. Although there is no unanimity on the subject, most of the works focusing on the bond market suggest that companies with high environmental performance benefit from a lower cost of capital (see Section 16.2 for an extensive literature review). Authors mainly attribute this negative yield differential to a financial reality: intangible asset creation (Porter and van der Linde, 1995, Hart, 1995, Jones, 1995, Ambec and Lanoie, 2008, Flammer, 2015) as well as better risk management and mitigation (Ambec and Lanoie, 2008, Bauer and Hann, 2014), both being imperfectly captured by rating agencies' models (Ge and Liu, 2015, Oikonomou, Brooks, and Pavelin, 2014). However, the existing literature does not identify whether, and by how much, this yield differential is driven by non-pecunary motives.

By integrating into the utility function of a group of investors an appetite for certain types of assets in addition to their expectations regarding return and risk, Fama and French (2007) show that investors' *tastes* modify equilibrium prices. Nevertheless, few studies have empirically isolated the impact of non-pecuniary motives on market prices. Focusing on sin stocks and controlling for a battery of financial indicators, Hong and Kacperczyk (2009) show that social norms lead to a 2.5% higher return for sin stocks than non-sin stocks. However, the analysis of non-pecuniary motives on the stock market implies comparing the financial securities of different companies and thus makes it very difficult to identify the effect.

In this paper, we exploit the bond market to clearly identify the impact of pro-environmental preferences on prices. To do so, we use green bonds as an instrument: we compare each green bond with an otherwise identical counterfactual conventional bond. Unlike two bonds issued by companies with different environmental performances, green and conventional bonds of the same company are subject to the same financial risk once all their differences have been controlled. Comparing the yield of a green bond and that of a conventional counterfactual thus makes it possible to isolate the impact of pro-environmental preferences on bond prices.

Therefore, this paper aims to provide answers to the following two questions:
Research question 1: Do pro-environmental preferences translate into bond market prices?
Research question 2: If so, do they apply uniformly across the entire bond market?

The study of the green bond market is made possible by the recent accelerated development of this asset class, which has been supported by the definition of the Green Bond Principles[4] providing issuers with guidance and investors with reliable information about environmental impacts. The labeled green bond market reached USD 301 billion outstanding in December 2017. Green bond issuances rose to USD 163 billion in 2017, up 68% from the previous year. Government-related bonds, including government, national and supranational agencies, account for 30% of the total, while 32% are financial bonds and 21% are bonds issued by energy companies. Among the 44% of the bonds rated by S&P, Moody's or Fitch in the entire database, 90% are investment-grade bonds, and 10% are high-yield bonds.

[3]Pro-social and pro-environmental motives refer to investors' interest in social and environmental issues per se in their investment decisions.

[4]The 2017 voluntary process guidelines for issuing green bonds are summarized in https://www.icmagroup.org/assets/documents/Regulatory/Green-Bonds/GreenBondsBrochure-JUNE2017.pdf.

The main currencies involved are the USD and the euro (EUR), each of which accounts for one-third of the total outstanding green bond debt.

We identify the effect of pro-environmental preferences through a *green bond premium*, which is defined as the yield differential between a green bond and an otherwise identical conventional bond. We perform an analysis on 110 green bonds on the secondary market between July 2013 and December 2017. This sample accounts for 10% of the number and 17% of the amount of green bonds issued worldwide at the end of 2017. We first use a matching method to estimate the yield of an equivalent synthetic conventional bond for each live green bond issued in the global universe on December 31, 2017. To do so, we build a counterfactual conventional bond from the same issuer, having the same maturity, currency, rating, bond structure, seniority, collateral and coupon type, as well as a limited difference in issue date and size. In the second stage, we control for the residual difference in liquidity between each green bond and its counterfactual to extract a green premium by performing a fixed-effects panel regression: the green premium is the unobserved specific effect of the regression of the yield differential on the bonds' liquidity differential. By performing a panel regression on matched pairs of bonds for which the characteristics are identical except for the green feature of one of the two, we circumvent two biases inherent in a cross-sectional regression of yields on bonds' characteristics: an omitted variables bias and a bias related to overweighting assets with the longest price history. Finally, to identify the factors affecting the costliness of a green bond, we explain these green premia according to the specific characteristics of the bonds through a cross-sectional regression.

We show that there exists a small, albeit significant, negative green bond premium of -2 basis points (bps) in our sample. The sector and the rating are significant drivers of the premium: the negative premium is greater[5] for financial bonds and low-rated bonds. Through several robustness tests, we verify that the premium is neither a risk premium nor a market premium, that the matching method is sufficiently stringent, and that the average and median premia remained negative on a monthly basis from May 2016 onward. We also show that the estimated premium in our sample has a reasonable chance of reflecting a similar phenomenon across the total sample of green bonds.

Our contribution to the literature is threefold. First, we contribute to the literature on non-pecuniary motives in ethical investing. We use green bonds as an instrument to cleanly identify the effect of pro-environmental preferences on the bond market. Although social and environmental preferences can have a substantial positive impact on investment inflows in ethical funds and assets, the 2-bps negative yield premium on green bonds shows that the impact of pro-environmental motives on bond prices is still limited. We also contribute to the literature linking the cost of debt and the company's environmental performance. The low negative green bond premium, which is related to the price impact of pro-environmental preferences, suggests that the lower cost of debt for companies with good environmental performances should be more related to a lower level of risk than to non-pecuniary motives. Third, this study on the valuation of green bonds complying with the Green Bonds Principles is the most extensive in terms of geographical scope, number of bonds studied as well as price history. The methodology developed–which, more generally, can be used to estimate the valuation of other types of bonds of which the proceeds are directed to a specific use, such as *social bonds*[6]–includes strict liquidity control and is supplemented by numerous robustness tests.

[5]When the premium is negative, we use the terms *greater negative premium* and *lower premium* interchangeably to mean that the negative premium has a higher absolute value.

[6]The ICMA recently published voluntary guidelines for issuing *social bonds*, which are a nascent asset class: https://www.icmagroup.org/assets/documents/Regulatory/Green-Bonds/June-2018/Social-Bond-Principles—June-2018-140618-WEB.pdf

The negative yield premium of 2 bps has distinct implications for the various market players; it does not represent a notable disincentive for investors who should not substitute their purchase of green bonds with conventional bonds. Moreover, although low, this premium demonstrates investors' appetite for green bond issues and supports the hypothesis that this instrument offers issuers the opportunity to broaden their debtholder base. Finally, from the supervisory authority perspective, while this negative premium underlines a certain buying pressure on green bonds, it does not yet reveal any substantial valuation discrepancy between green and conventional bonds.

This paper is organized as follows. In the second section, the literature on the topic of interest is reviewed. The method used to build the data on which this study is based is described in the third section. Our empirical approach is described in the fourth section, and the results obtained using the empirical model are presented in section five. The robustness checks run are described in the sixth section, and the results are discussed in section seven. The conclusions of our findings are summarized in section eight.

16.2 Literature review

Numerous authors have addressed the effects of corporate social performance (CSP)[7], especially the effects related to good environmental performance[8], on companies' stock returns (Konar and Cohen, 2001, Derwall et al., 2005, Kempf and Osthoff, 2007, Semenova and Hassel, 2008, Statman and Glushkov, 2009, Dixon, 2010). Although no consensus has been reached, most of the articles published have suggested that CSP has a positive impact on companies' financial performance. Moreover, CSP has been found to have similar effects on the cost of equity capital: firms with better CSP (ElGhoul et al., 2011, Dhaliwal et al., 2011[9]) or a low environmental impact (Heinkel, Kraus, and Zechner, 2001, Sharfman and Fernando, 2008, Chava, 2014) benefit from a lower cost of equity capital. However, these findings are not necessarily transferable to the debt market for at least two reasons. First, the payoff profile of a debtholder differs from that of a stockholder (Oikonomou, Brooks, and Pavelin, 2014 and Ge and Liu, 2015): Merton (1973) specifies that a bond payoff can be replicated by the purchase of a stock and the sale of a call option on the same asset. Since bondholders have little upside available, it is crucial for them to analyze and assess all the downside risks, including environmental hazards. This need for insurance against a market downturn is all the more relevant for socially responsible investing, as CSP leads to better credit ratings (Jiraporn et al., 2014) and has a strong effect on a company's default risk reduction (Sun and Cui, 2014). Second, as previously suggested by Oikonomou, Brooks, and Pavelin (2014), firms are more sensitive to the pressure exerted by bond market investors because firms refinance via the debt market more frequently than they increase their capital.

[7]Luo and Bhattacharya (2009) clarify the difference between corporate social responsibility (CSR) and CSP: CSP (i) refers to stakeholders' assessment of the quality of CSR investments, (ii) can be a proxy for a firm's cumulative involvement in CSR and (iii) is a notion relative to the competition in the industry.

[8]According to the Forum for Sustainable and Responsible Investment, "Sustainable, responsible and impact investing (SRI) is an investment discipline that considers environmental, social and corporate governance (ESG) criteria to generate long-term competitive financial returns and positive societal impact." (https://www.ussif.org/sribasics, answer to the question "What is sustainable, responsible and impact investing?"). Investments with a positive environmental impact (or *good environmental performance*) are therefore a form of sustainable investment for which the expected benefits specifically concern the environment.

[9]Dhaliwal et al., 2011 focus on the initiation of a voluntary disclosure of CSR activities and show that it leads to a reduction in a firm's cost of capital.

This pressure can be all the more easily exerted because debt instruments are frequently held by institutional investors with advanced risk analysis capacities.

Although several studies have focused on the effects of CSP on corporate bond yields, no unequivocal conclusions have yet been reached on this topic. Magnanelli and Izzo (2017), using a database of 332 companies worldwide with 1641 observations from 2005 to 2009, are among the few authors showing that CSP increases the cost of debt. In line with the shareholder theory, their results support the assertion that CSR is considered "a waste of resources that can negatively affect the performance of the firm." Conversely, Menz (2010) focuses on the European corporate bond market and observes that socially responsible firms suffer more from a greater credit spread than do non-socially responsible companies, although this finding is weakly significant. Likewise, Stellner, Klein, and Zwergel (2015) obtain relatively weak evidence that good CSP systematically reduces credit risks. Other authors, however, report a significant negative relationship between CSP and the cost of debt. Oikonomou, Brooks, and Pavelin (2014) show that for U.S. corporate debt, good CSR performance is rewarded by lower bond yields and CSR irresponsibility is positively correlated with financial risk. Based on information provided by a cross-industrial sample of U.S. public corporations, Bauer and Hann (2014) establish that environmental strengths are associated with lower bond yields. Other authors, such as Klock, Mansi, and Maxwell (2005), using U.S. data, and Ghouma, Ben-Nasr, and Yan (2018), using Canadian data, report that bond spreads decline with the quality of corporate governance. Klock, Mansi, and Maxwell (2005) notably show that compared to firms with the strongest shareholder rights (proxied by weak antitakeover provisions), firms with the strongest management rights (strongest antitakeover provisions) benefited from a 34 bps reduction in the cost of debt for the period 1990-2000. Ge and Liu (2015) focus on the effects of CSP disclosure on the spreads of new corporate bonds issued in the U.S. primary market and establish that firms reporting favorable CSPs enjoy lower bond spreads. Hasan, Hoi, and Zhang (2017) also examine the primary market of U.S. firms from 1990 to 2012 and find that firms headquartered in U.S. counties with higher levels of social capital benefit from lower at-issue bond spreads. Finally, although the financing of private loans and public bonds must be analyzed differently mainly because banks have access to more information than bondholders, Goss and Roberts (2011) reach similar conclusions after examining the impact of the CSR scores of 3996 U.S. companies on the cost of the companies' bank loans. They also establish that firms with the lowest CSR scores pay between 7 and 18 bps more than the most responsible firms.

However, few articles have been published on the specific cost of green bonds. Table 16.1 summarizes the results of and differences between these studies.

In contrast to the analyses in the papers presented above, the analysis of the green bond yield is not based on the CSP of the issuing company because the green bond label is associated with the funded projects and not with the issuer type. Thus, we can compare a green bond yield with the yield of a similar conventional bond from the same issuer.

HSBC (2016) and Climate Bonds Initiative (2017) study the difference in yield at issuance between a green bond and a conventional bond by calculating the difference between the two yields for samples of 30 and 14 bonds, respectively. These two works do not find any significant differences on the primary market, which confirms the analyses conducted in OECD (2017b) and I4CE (2016) showing that investors are not willing to pay a premium to acquire a green bond at issuance ("flat pricing"). Barclays (2015) and Bloomberg (2017) focus on the yield differential on the secondary market. Through an OLS regression of the credit spread on several market risk factors, Barclays (2015) points to a negative premium of 17 bps between March 2014 and August 2015. By analyzing 12 bonds between March 2014 and December 2016, Bloomberg (2017) highlights a negative 25 bps premium

TABLE 16.1: Research methods and findings on green bond pricing. This table summarizes the research methods and empirical findings of studies on the relative pricing of green bonds in relation to conventional bonds

	Barclays (2015)	HSBC (2016)	Bloomberg (2017)	Climate Bonds Initiative (2017)
Green bonds (Alignment with the GBP)	Yes	Yes	Yes	Yes
Scope	Global	Euro and US	European Investment Bank, Nordic Investment Bank and International Bank for Reconstruction and Development	EUR- and USD-denominated Govt.-related and corporate bonds
Primary / Secondary market	Secondary	Primary / Secondary	Secondary	Primary
Number of bonds	N.A.	30 / 4	12	14
Time period	Mar. 2014 - Aug. 2015	Nov. 2015 - Sep. 2016	Mar. 2014 - Dec. 2016	Jan. 2016 - Mar. 2017
Method	OLS regression	Comparison	Comparison	Comparison
Liquidity control	Date of issuance	No	No	No
Strict maturity control	No	No	No	No
Yield premium	−17bps	No	EUR-denominated Govt.-related bonds: -25ps USD-denominated and corporate bonds: No	No

(a) Literature prior to this paper.

	Ehlers and Packer (2017)	Karpf and Mandel (2018)	Baker et al. (2018)	Hachenberg and Schiereck (2018)
Green bonds (Alignment with the GBP)	Yes	No	No	Yes
Scope	Euro and US	US Municipal bonds with a Bloomberg green flag	US Corporate and Municipal bonds with a Bloomberg green flag	Global
Primary / Secondary market	Primary	Secondary	Primary	Secondary
Number of bonds	21	1880	2083	63
Time period	2014-2017	2010-2016	2010-2016	Oct. 2015 - March. 2016
Method	Comparison	Oaxaca-Blinder decomposition	OLS regression	Matching + panel regression based on our paper's method
Liquidity control	No	Number of transactions within the past 30 days	Issue amount	Issue amount
Strict maturity control	Yes	Yes	Yes	Yes
Yield premium	−18 bps	+7.8 bps	−7 bps	−1 bp

(b) Literature subsequent to and building on this paper.

on EUR-denominated government-related bonds but does not identify any premium on USD-denominated and corporate bonds.

Subsequent works have built on the first version of this paper (Ehlers and Packer, 2017, Karpf and Mandel, 2018, Hachenberg and Schiereck, 2018 and Baker et al., 2018). Ehlers and Packer (2017) and Hachenberg and Schiereck (2018) study samples of 21 and 63 green bonds aligned with the Green Bond Principles, respectively. Ehlers and Packer (2017) focus on the primary market between 2014 and 2017, whereas Hachenberg and Schiereck (2018) analyze the secondary market over 6 months between 2015 and 2016 using a matching procedure and a panel regression based on the methodology of our paper. Both papers find a negative premium but of very different magnitudes: -18 bps for the former and -1 bp for the latter. Karpf and Mandel (2018) and Baker et al. (2018) study a less restrictive framework than that of green bonds aligned with the Green Bond Principles: U.S. bonds with a Bloomberg green flag. Karpf and Mandel (2018) focus on municipal bonds on the secondary market, and Baker et al. (2018) analyze municipal and corporate bonds on the primary market. By controlling bonds' liquidity through the number of transactions within the past 30 days, Karpf and Mandel (2018) find a positive premium of 7.8 bps. In contrast, using the issue amount as a proxy of the liquidity, Baker et al. (2018) find evidence of a 7 bps negative premium.

Existing works on the relative valuation of green bonds aligned with the Green Bond Principles therefore suffer from both a limited scope of analysis as well as imperfect control of the liquidity premium, leading to mixed results. This paper aims to estimate the fair yield of green bonds compared to that of conventional bonds over an extensive scope, ensuring that all the discrepancies between the two types of bonds are duly controlled.

16.3 Data description and matching method

The empirical method primarily used in the CSR literature to analyze bond spreads consists in performing an appropriate regression on a suitable specification. This step requires determining the financial and extra-financial independent variables likely to explain the intrinsic value of the bond spread as exhaustively as possible while ensuring the robustness of the specification. Analyzing the yield of a green bond allows us to forgo this method because we can match two similar bonds from the same issuer, for which most of the factors explaining the yield are identical. We therefore use a matching method, also known as a model-free approach or a direct approach, which is a useful technique for analyzing the intrinsic value of a specialized financial instrument. This method consists of matching a pair of securities with the same properties except for the one property whose effects we are interested in. This method has been used to assess the additional return of ethical funds in comparison with identical conventional funds or indices (Kreander et al., 2005, Renneboog, Ter Horst, and Zhang, 2008 and Bauer, Koedijk, and Otten, 2005) as well as the cost of liquidity by matching and comparing pairs of bonds issued by the same firm (Helwege, Huang, and Wang, 2014).

We set up this database to evaluate the yield spread between a green bond and an equivalent synthetic conventional bond. For this purpose, we take matched pairs consisting of a green and a conventional bond with identical characteristics except for their liquidity. The variable construction procedure used here is closely related to that used by Helwege, Huang, and Wang (2014) to assess the effects of liquidity on corporate bond spreads. However, while building on the latter study, we add a new parameter–the greenness of a bond: determining the impact of this parameter on the bond yield is the goal of our assessment. The difference between the green bond yield and the equivalent synthetic conventional bond yield is therefore precisely the cumulative effect of the liquidity differential and the green bond premium.

We examine the entire sample of 1065 green bonds complying with the Green Bond Principles indexed by Bloomberg on December 31, 2017. This set includes bonds of various kinds: supranational, sub-sovereign and agency (SSA), municipal, corporate, financial and covered bonds. To build this synthetic conventional bond, for each green bond, we first search for the two conventional bonds with the closest maturity from the same issuer and having exactly the same characteristics: they all have the same currency, rating,[10] bond structure, seniority, collateral and coupon type. Since the maturities cannot be equal, we collect conventional bonds with a maturity that is neither two years shorter nor two years

[10]Since an institution can issue various bonds of different kinds or seniority levels and, thus, with different ratings, we make sure that the rating is the same.

TABLE 16.2: Average issue amount broken down per type of bond and currency. This table gives the average amount of green bonds, CB1 and CB2 issued in each currency

	Average issue amount (bn USD)		
	Green bonds	*Conventional bonds 1*	*Conventional bonds 2*
AUD	0.45	0.63	0.64
CAD	0.40	1.11	0.95
CHF	0.33	0.29	0.35
CNY	0.23	0.15	0.08
EUR	1.05	1.95	1.98
GBP	1.89	6.84	2.47
INR	0.08	0.30	0.22
JPY	0.09	0.15	0.17
RUB	0.01	0.10	0.01
SEK	0.11	0.13	0.13
TRY	0.07	0.03	0.10
USD	0.65	1.50	1.38
Average	0.65	1.34	1.24
Median	0.28	0.29	0.28

longer than the green bond's maturity. The difference in maturity is limited in this way to estimate more accurately the equivalent synthetic conventional bond yield in the next stage. The other difference between the two categories of bonds is their liquidity, which can be assessed from either their issue amount or their issue date (see Bao, Pan, and Wang, 2011 and Houweling, Mentink, and Vorst, 2005). A substantial difference in liquidity can have a notable effect on the yield level and must therefore be limited.[11] Here again, to ensure a fair approximation in this first stage, we combine a double constraint on the difference in liquidity: we restrict the eligible conventional bonds to those (i) with an issue amount of less than four times the green bond's issue amount and greater than one-quarter of this amount (Table 16.2) and (ii) with an issue date that is, at most, six years earlier or six years later than the green bond's issue date[12] (see Figure 16.1). This double restriction in the matching method allows us to better control for any residual liquidity bias in the estimation step of the green bond premium (see Section 16.4.1). Any green bonds for which fewer than two of the corresponding conventional bonds comply with these requirements is excluded from the database.

In a second stage, the maturity bias is eliminated by building a panel composed of pairs of bonds: an equivalent synthetic conventional bond with the same maturity is assigned to each green bond. The ask yields of each triplet of bonds (the green bond and the two corresponding conventional bonds) are retrieved from the issue date of the green bond up to

[11]It is widely agreed that bond credit spreads incorporate a positive illiquidity premium (see for example Chen, Lesmond, and Wei, 2007, Beber, Brandt, and Kavajecz, 2009, Bao, Pan, and Wang, 2011, Dick-Nielsen, Feldhütter, and Lando, 2012, de Jong and Driessen, 2012).

[12]Authors controlling for the difference in liquidity solely through the date of issuance suggest different levels, from 1 year (Elton et al., 2004) to 2 years (Alexander, Edwards, and Ferri, 2000, Houweling, Mentink, and Vorst, 2005). In this paper, we combine three different liquidity controls (two in the matching method and one in the estimation process), with less stringent restrictions for the first two controls, to enable a closer maturity matching and a wider sample. We verify in the robustness checks (Section 16.6) that these liquidity controls are acceptable. Furthermore, Wulandari et al., 2018 find that the impact of illiquidity on green bonds' yield spread has become negligible in most recent years.

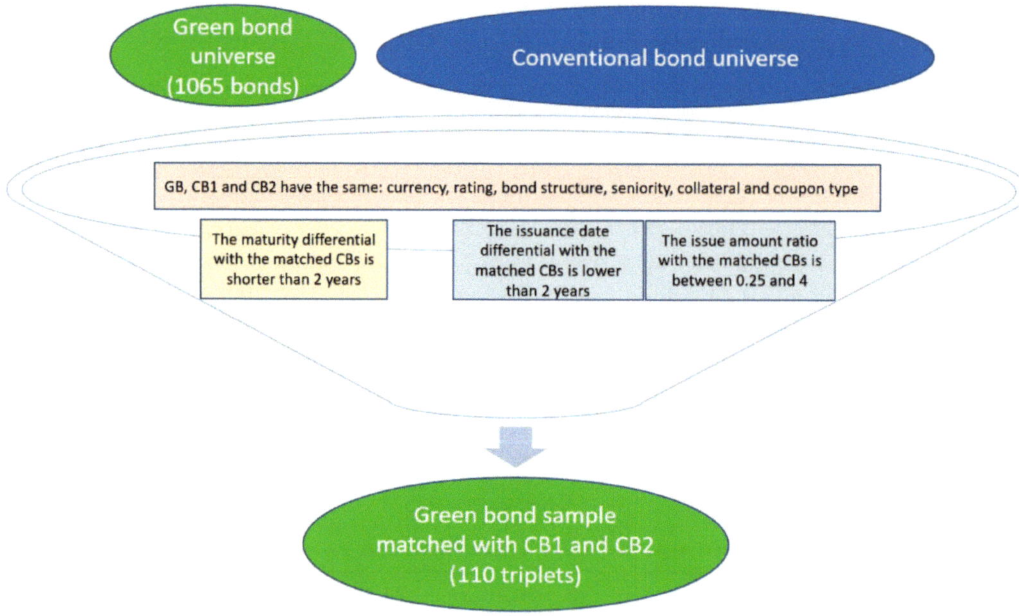

FIGURE 16.1: Matching process. This figure illustrates the matching process. We match each green bond (GB) of the universe on December 31, 2017, with two conventional bonds (CB1 and CB2). Green and conventional bonds are required to have the same currency, rating, bond structure, seniority, collateral and coupon type. Moreover, the maturity of the conventional bond is neither two years shorter nor two years longer than that of the green bond. Also, we select the conventional bonds (i) with an issue amount of less than four times the green bond's issue amount and greater than one-quarter of this amount. We therefore collect 110 triplets of (GB, CB1, CB2).

December 31, 2017. The source used for this purpose is Bloomberg BGN,[13] which provides end-of-day market prices and yields based on multiple contributors' market prices as well as all the characteristics of the bonds. As green bonds are not all listed in TRACE, we cannot take advantage here of the richness of this source, especially with respect to the volumes traded. Since this study focuses on the investors' demand and the issuers' supply of green bonds, we focus on the ask yields of each triplet for a more precise analysis. If, on a specific day, at least one of the three ask yields is not available, we remove the line from our panel. We then interpolate (or extrapolate) the two conventional bonds' yields linearly at the green bond maturity date to obtain a synthetic conventional bond yield, which thus shows the same properties as the green bond except for the difference in liquidity. Practically, for each triplet, with a^* the slope and b^* the intercept of the affine function passing through (Maturity_{CB1}, y^{CB1}) and (Maturity_{CB2}, y^{CB2}), the yield of the synthetic conventional bond is $\tilde{y}^{CB} = a^*\text{Maturity}_{GB} + b^*$ (see Figure 16.2). Because of the linear interpolation (or extrapolation), this method differs slightly from that used in Helwege, Huang, and Wang (2014), in which the closest bond is selected, which gives rise to a tiny maturity bias. The constitution of the database is finalized by defining the yield spread between the green bond and the equivalent synthetic conventional bond. Let $y_{i,t}^{GB}$ and $\tilde{y}_{i,t}^{CB}$ be the green bond and the conventional bond i's ask yields, respectively, on day t. We take $\Delta\tilde{y}_{i,t} = y_{i,t}^{GB} - \tilde{y}_{i,t}^{CB}$.

[13]We voluntarily exclude Bloomberg BVAL prices that combine market data with model pricing.

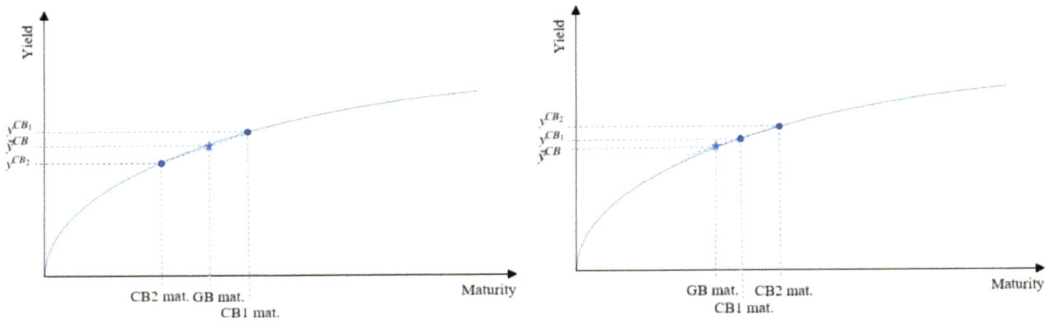

(a) Interpolation (b) Extrapolation

FIGURE 16.2: Interpolation and extrapolation of the synthetic conventional bond yield. This figure shows how we calculate the yield of the synthetic conventional bond through (a) a linear interpolation or (b) a linear extrapolation of the yields of CB1 and CB2 at the maturity date of the green bond.

This approach enables us to remove all the unobservable factors common to both bonds in the matched pairs and to significantly reduce the liquidity bias. The process leaves us with 110 matched green bonds accounting for 10% of the global green bond universe and 17% of the total outstanding green bond debt. All of the bonds in our sample are senior, bullet, fixed-coupon bonds. Except for one BB and 12 non-rated, all of them are investment-grade bonds. Significant variations are observed in the yield levels, notably between the various issue currencies, i.e., across the corresponding rate and credit curves (see Table 16.3). For example, while the average AAA government-related green bond yield in Turkish lira is 10.28%, it only amounts to 0.26% in the same market segment for the bond labeled in EUR.

The sample comprises a 37,503-line unbalanced bond-day panel in which the earliest information dates back to July 18, 2013, and the latest is dated December 29, 2017. For the sample, the statistics of the green and conventional bonds' yields, maturities and issue amounts are presented in Table 16.4.

Upon focusing on the time average difference in yield ($\Delta\tilde{y}_i$), the distribution across bonds is found to be skewed to the left: There are 63% negative values, giving an average of −2 bps[14] and a median value of −1 bp. In the next section, we will therefore study $\Delta\tilde{y}_{i,t}$ to determine whether there is a premium attributable to the greenness of a bond.

16.4 Empirical methodology

16.4.1 Step 1: Estimation of the green bond premium

The first step of the empirical methodology aims at controlling for the residual difference in liquidity between both bonds of each pair and estimating the green bond premium. We

[14]Note that one cannot infer the −2-bps average yield difference with y^{GB} and $\tilde{y}^{CB}$ because the average in i of the average in t of the yield differences is not equal to the yield difference on the average in i of the average in t of the green bonds' yields and the conventional bonds' yields. The same applies to the medians and quartiles.

TABLE 16.3: Description of the sample of 110 green bonds. This table shows the average yield and maturity of the sample of 110 green bonds, broken down by sector, rating, and currency

		AUD	*CAD*	*CHF*	*CNY*	*EUR*	*GBP*	*INR*	*JPY*	*RUB*	*SEK*	*TRY*	*USD*	*Total*
Basic Materials														
NR	Average yield (%)										0.96			0.96
	Average maturity (years)										4.74			4.74
	Nb. of GB										1			1
Consumer, Non-cyclical														
BBB	Average yield (%)					0.78								0.78
	Average maturity (years)					5.51								5.51
	Nb. of GB					1								1
Financial														
AAA	Average yield (%)	2.43				0.07	0.79				0.10		1.98	0.83
	Average maturity (years)	2.50				4.94	2.43				2.96		3.52	4.13
	Nb. of GB	1				10	1				1		6	19
AA	Average yield (%)	3.00				0.28							2.10	1.11
	Average maturity (years)	3.37				5.68							2.70	4.86
	Nb. of GB	3				8							1	12
A	Average yield (%)				3.70	0.36					0.77		2.17	1.34
	Average maturity (years)				0.53	4.25					4.13		1.98	3.09
	Nb. of GB				1	8					2		8	19
BBB	Average yield (%)					0.61							3.65	2.13
	Average maturity (years)					4.49							2.92	3.70
	Nb. of GB					1							1	2
BB	Average yield (%)												5.23	5.23
	Average maturity (years)												3.38	3.38
	Nb. of GB												1	1
NR	Average yield (%)										0.66			0.66
	Average maturity (years)										2.77			2.77
	Nb. of GB										11			11
Government														
AAA	Average yield (%)	2.41	1.57	0.03		0.26	0.59	5.70		6.65	0.49	10.28	1.73	1.92
	Average maturity (years)	1.33	2.85	7.10		5.54	2.18	3.15		1.57	4.75	1.24	3.15	3.50
	Nb. of GB	1	2	1		3	1	1		1	4	1	15	30
AA	Average yield (%)			0.31									2.16	1.23
	Average maturity (years)			11.92									1.64	6.78
	Nb. of GB			2									2	4
A	Average yield (%)								0.39					0.39
	Average maturity (years)								14.79					14.79
	Nb. of GB								3					3
BBB	Average yield (%)												2.68	2.68
	Average maturity (years)												2.25	2.25
	Nb. of GB												1	1
Industrial														
BBB	Average yield (%)					0.83								0.83
	Average maturity (years)					5.94								5.94
	Nb. of GB					1								1
Utilities														
A	Average yield (%)					0.49								0.49
	Average maturity (years)					2.85								2.85
	Nb. of GB					2								2
BBB	Average yield (%)					0.94								0.94
	Average maturity (years)					6.41								6.41
	Nb. of GB					3								3
	Average yield	2.77	1.57	0.22	3.70	0.34	0.69	5.70	0.39	6.65	0.62	10.28	2.09	1.31

therefore design a variable, $\Delta\text{Liquidity}_{i,t}$, capturing the difference in liquidity and defined as the difference between a green bond and a conventional bond's liquidity indicator:

$$\Delta\text{Liquidity}_{i,t} = \text{Liquidity}_{i,t}^{GB} - \text{Liquidity}_{i,t}^{CB} \tag{16.1}$$

The green bond premium p_i is therefore defined as the unobserved effect in the fixed-effects panel regression of $\Delta\tilde{y}_{i,t}$ on $\Delta\text{Liquidity}_{i,t}$:

$$\Delta\tilde{y}_{i,t} = p_i + \beta\Delta\text{Liquidity}_{i,t} + \epsilon_{i,t}, \text{ with } \epsilon_{i,t} \text{ being the error term} \tag{16.2}$$

Given the data sources and the type of regression, the liquidity proxies that can be used here are subject to three constraints. Firstly, since we cannot use intraday data to calculate intraday liquidity indicators, such as the Amihud measure (Amihud, 2002), Range measure (Han and Zhou, 2016) or intraday Roll and Gamma measure (Roll, 1984; Bao, Pan, and

TABLE 16.4: Descriptive statistics of the bonds in the sample. This table gives the distribution of several variables of interest in all 110 triplets of bonds in our sample. The number of days per bond is the length of the time series per pair of bonds since their inception. The distribution of the ask yield is presented for green bonds (y^{GB}), the two closest conventional bonds (y^{CB_1} and y^{CB_2}) and the interpolated (or extrapolated) conventional bonds ($\tilde{y}^{CB}$). The difference in yield ($\Delta\tilde{y}_{i,t}$) is the difference between the green bonds' ask yield and the interpolated (or extrapolated) conventional bonds' ask yield. To compare the accuracy of the interpolations (or extrapolations), this table also shows the distribution of maturities and the issue amounts of the green bonds and the two closest conventional bonds

	Sample					
	Min.	*1st Quart.*	*Median*	*Mean*	*3rd Quart.*	*Max*
Number of days per bond	12	99	306	341	518	1 150
Ask yield of the GB (y^{GB})	−0.35	0.26	0.92	1.31	1.90	10.28
Ask yield of the interp. CB ($\tilde{y}^{CB}$)	−0.43	0.27	0.94	1.33	1.92	10.19
Ask yield of the CB1 (y^{CB1})	−0.34	0.22	0.88	1.29	1.98	10.17
Ask yield of the CB2 (y^{CB2})	−0.33	0.24	0.81	1.25	1.95	10.28
Yield difference % ($\Delta\tilde{y}_{i,t}$)	−0.46	−0.03	−0.01	−0.02	0.01	0.10
Green bond maturity on Dec. 30, 2017 (years)	0.14	2.20	3.45	4.15	4.87	29.74
Conventional bond 1 maturity	0.07	1.86	3.29	4.03	4.72	28.99
Conventional bond 2 maturity	0.26	1.82	3.11	3.79	4.93	28.23
Green bond issue amount (USD bn)	0.01	0.30	0.50	0.65	0.80	3.60
Conventional bond 1 issue amount	0.01	0.32	1.00	1.34	1.48	7.20
Conventional bond 2 issue amount	0.01	0.28	0.90	1.24	1.24	7.48

Wang, 2011), for example, we focus on low-frequency data. Secondly, in contrast to what can be done with the TRACE database, we do not have any information about the daily trading volumes that might have been used as liquidity proxies (Beber, Brandt, and Kavajecz, 2009; Dick-Nielsen, Feldhütter, and Lando, 2012). Thirdly, to ensure the full rank condition of a *within* regression, any variable that does not change over time with a given bond is not suitable. Proxies such as the issue amount, the issue date or off-the-run versus on-the-run indicators (Bao, Pan, and Wang, 2011 and Houweling, Mentink, and Vorst, 2005) therefore cannot be used.

We take the closing percent quoted bid-ask spread as a proxy of the liquidity, consistent with Fong, Holden, and Trzcinka (2017), who show, through an extensive analysis of the quality of high- and low-frequency liquidity proxies, that it is the best low-frequency liquidity proxy. Indeed, bid-ask spread has been widely used as a major measure of the degree of illiquidity of a bond (Chen, Lesmond, and Wei, 2007; Beber, Brandt, and Kavajecz, 2009; Dick-Nielsen, Feldhütter, and Lando, 2012).

Since the synthetic conventional bonds are based on the two closest conventional bonds, the conventional bond's bid-ask spread is defined as the distance-weighted average of CB1's and CB2's bid-ask spreads. In practical terms, let $d_1 = |\text{Green Bond maturity - CB1 maturity}|$ and $d_2 = |\text{Green Bond maturity - CB2 maturity}|$. The synthetic conventional bond's bid-ask spread is therefore as follows:

$$\text{BA}_{i,t}^{CB} = \frac{d_2}{d_1 + d_2}\text{BA}_{i,t}^{CB1} + \frac{d_1}{d_1 + d_2}\text{BA}_{i,t}^{CB2} \tag{16.3}$$

$\Delta\text{BA}_{i,t} = \text{BA}_{i,t}^{GB} - \text{BA}_{i,t}^{CB}$ is consequently the independent variable used in Equation (16.2) to estimate the fixed-effects linear panel.

Table 16.5 shows that ΔBA is concentrated around zero and has a low standard deviation. This condition indicates that the first liquidity controls on the issue amount and the date of issuance in the matching method yielded acceptable results.

TABLE 16.5: Descriptive statistics of the liquidity proxy ΔBA. This table summarizes the distribution of the liquidity control: ΔBA is the difference between the green bonds' bid-ask spread and the conventional bonds' distance-weighted average bid-ask spread, in a specific pair of bonds, during the period under consideration

	Min.	*1st Quart.*	*Median*	*Mean*	*3rd Quart.*	*Max*	*Std. Dev.*
ΔBA	−0.436%	−0.021%	0.000%	0.006%	0.032%	0.758%	0.11%

We use a *within* regression to estimate the fixed effects p_i in Equation (16.2) for various reasons. Firstly, we want to bring out the bond-specific time-invariant unobserved effect without imposing any distribution or using any information about the other bonds. Secondly, these data do not hold for a broader category but, rather, give the characteristics of a specific bond. From the technical point of view, strict exogeneity holds and ensures unbiasedness and consistency of the estimator. Finally, the fact that we do not require the difference in liquidity proxy to be uncorrelated with the unobserved specific effects provides for a wide range of potential control parameters.

Several individual effect tests and a Hausman test are performed to check the efficiency of the fixed-effects estimator. Moreover, controlling the difference in yield by the difference in liquidity prevents the occurrence of any simultaneity effects: the difference between two yields does not have any retroactive effect on the liquidity of the bonds. Lastly, various robustness tests are performed and, to address the loss of efficiency due to heteroscedasticity and serial correlation, we use the Newey-West and Beck-Katz robust estimations of the standard errors.[15]

16.4.2 Step 2: The determinants of the green premium

In the first step, we isolated the yield premium of a green bond linked to the specific nature of the debt security. The second step highlights the determinants of the green bond premium since it may not be stable across bonds. We therefore consider the characteristics through which bonds differ to determine where, and to what extent, the premium applies. The variables considered are the rating, the sector, the currency, the maturity and the issue amount of the green bond. Table 16.6 provides details on the variables and their construction.

After performing robustness tests, we estimate several cross-sectional specifications, including the main specification described in the following equation, through an OLS regression with robust estimation of the standard errors. Taking η_i to denote the error term, we set the following:

$$\begin{aligned}\widehat{p_i} =& \alpha_0 + \sum_{j=1}^{N_{rating}-1} \alpha_{1,rating_j} 1_{rating_j} + \sum_{j=1}^{N_{sector}-1} \alpha_{2,sector_j} 1_{sector_j} \\ &+ \sum_{j=1}^{N_{currency}-1} \alpha_{3,currency_j} 1_{currency_j} + \alpha_4 \text{Maturity} + \alpha_5 log(\text{Issue Amount}) + \eta_i\end{aligned} \tag{16.4}$$

[15]The results are robust to performing a Fixed Effects Generalized Least Squares regression: the estimated premia are equal to those estimated with a Fixed Effects Ordinary Least Squares (*within*) regression by a factor of 0.1 bps. Since the number of bonds studied is lower than the average number of days and for the sake of simplicity, we present here the results of the Fixed Effect OLS regression with robust estimation of the standard errors.

TABLE 16.6: Description of the independent variables of the step 2 regression. This table provides details about the independent variables used in the step 2 regression, namely, their type, unit and description

Variable	*Type*	*Unit*	*Description*
Rating	Qualitative		The rating of the bond can be AAA, AA, A, BBB, BB or Non-rated (NR) in our sample. The reference value is AAA. To attribute a single rating to the bond, the following procedure is used. The issuer ratings of the three agencies S&P, Moody's and Fitch are rounded off by removing the potential + or −. We then take the majority rating among those available. If there are only two different ratings available, we take the highest one.
Sector	Qualitative		We use the level 1 Bloomberg classification (BICS level 1) for the issuer-type breakdown procedure, which leaves us, in the case of the present sample, with six categories: (i) *Basic Materials*; (ii) *Consumer, Non-cyclical*; (iii) *Financials*, which encompasses non-public banks and financial services; (iv) *Government*, also referred to as *Government-related*, which includes public institutions, municipalities, regional and sovereign agencies, and national, supranational and development banks; (v) *Industrial*; and (vi) *Utilities*. The reference value is *Government*.
Currency	Qualitative		The currency of the bond issuance. In our sample, the referenced currencies are as follows: AUD, CHF, EUR, JPY, SEK, USD, CAD, RUB, GBP, TRY, CNY and INR. The reference level is the USD. See the online appendix for the meaning of each acronym.
Maturity	Quantitative	Years	The maturity of the bond on December 31, 2017.
Issue Amount	Quantitative	bn USD	The issue amount of the green bond considered December 31, 2017.

We take the logarithm of the issue amount to linearize the values of the variable that can be interpolated by an exponential function. Moreover, as an alternative to including the variables representing rating and sector alone, we also consider the dummy variables that capture rating × sector cross effects because descriptive statistics indicate that this segmentation may promote the variation of the premium.

16.5 The green bond premium

16.5.1 A small, albeit significant, negative green bond premium

The first step in the analysis aims to estimate the green bond premium, including its significance, sign and magnitude. We confirm the presence of an unobserved heterogeneous effect via an F-test, a Wooldridge test, a Breusch-Pagan test and a Honda test.[16] We also conduct a Hausman test that indicates that the fixed-effects within estimator is more robust than the random-effect estimator. The within estimator is unbiased and consistent: although it is intuitive that the idiosyncratic error term may not be correlated with either the previous or future differences in liquidity (neither feedback effect nor financial periodicity), we confirm the strict exogeneity hypothesis through Su, Zhang, and Wei's (2016) test.[17] This estimation is all the more satisfactory as the average number of days is higher than the number of bonds (Goldstein, 2003) and $\Delta \mathrm{BA}_{i,t}$ varies substantially with time.

Moreover, we run Breusch-Godfrey, Durbin Watson, and Wooldridge tests, all of which indicate the existence of serial correlation. In addition, a Breusch-Pagan test shows the presence of heteroscedasticity. To account for heteroscedasticity and serial correlation, we complement the regression with Newey-West and Beck-Katz[18] robust estimations of the standard errors.

Although the regression evidences a weak R^2 equal to 1%, the bid-ask spread differential used to control for the difference in liquidity proves to be highly significant for the three different estimators of the standard errors (Table 16.7). Although small in the present case, the residual liquidity differential has significant explanatory power and the step used for its control should not be discarded, *a fortiori* in situations in which the matching constraints are less stringent and because it is useful for developing a general method. Thus, a 1-bp increase in the percentage price bid-ask spread differential induces a 9.88-bps decrease in $\Delta \tilde{y}_{i,t}$.

The value of the 110 fixed-effects p_i constituting each of the green bonds' premia is more important for the present purposes. The distribution ranges from -38 bps to +10 bps with a mean and a median value of -1.76 bps and -1.04 bps, respectively (Table 16.8). A total of 63% of the premia are negative, and the amplitudes are greater on the downside than on the upside (Figure 16.3). It is worth noting that the extreme values of $\widehat{p_i}$ appear for currencies presenting a high yield (such as INR, RUB or TRY).

Lastly, we break down the sample in several subsamples by the main characteristics of the bond: its rating, sector and currency. We calculate the average premium by subsample and test whether it is significantly different from zero for subsamples with at least ten bonds. Through a Shapiro-Wilk normality test, we reject the normality hypothesis for all subsamples except AA bonds and SEK-denominated bonds. We therefore use the

[16]See the online appendix for the details of the tests performed.
[17]We test strict exogeneity for a two-day lag and lead period. The P-value is equal to 73.1%.
[18]Beck and Katz (1995) prove that their robust estimator performs well in small panels.

TABLE 16.7: Results of the step 1 regression. This table gives the results of the step 1 regression: $\Delta\tilde{y}_{i,t} = p_i + \beta\Delta\text{BA}_{i,t} + \epsilon_{i,t}$. In addition to a classical within regression, Newey-West and Beck-Katz robust standard error tests are performed

	Dependent variable: $\Delta\tilde{y}_{i,t}$		
	Within	*Newey-West robust std. err.*	*Beck-Katz robust std. err.*
ΔBA	−9.881***	−9.881***	−9.881***
	(0.440)	(2.774)	(3.334)
Observations		37,504	
R^2		0.013	
Adjusted R^2		0.010	
F Statistic		504.125***	
		(df = 1; 37393)	
Note:			*p<0.1; **p<0.05; ***p<0.01

TABLE 16.8: Distribution of the estimated green bond premia. This table summarizes the distribution of the estimated green bond premia in our full green bond sample, i.e., the fixed effect of the following regression: $\Delta\tilde{y}_{i,t} = p_i + \beta\Delta\text{Liquidity}_{i,t} + \epsilon_{i,t}$

$\hat{p}_i$ (%)					
Min.	*1st Quart.*	*Median*	*Mean*	*3rd Quart.*	*Max*
−0.381	−0.029	−0.01	−0.018	0.008	0.100

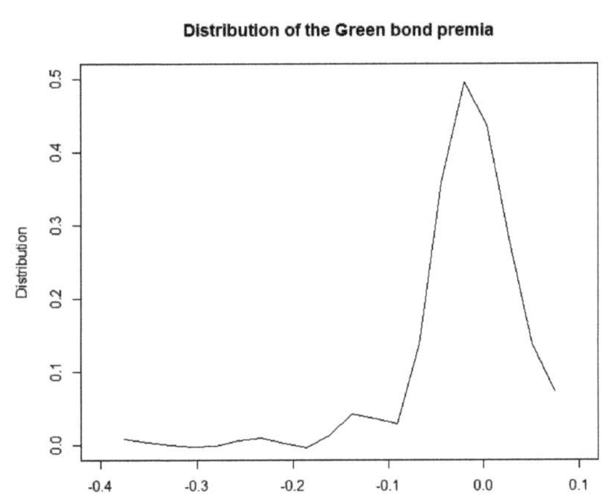

FIGURE 16.3: Green bond premia distribution. This figure gives the distribution of the green bond premia $\widehat{p_i}$ across all bonds included in this study.

TABLE 16.9: Green bond premia in several market segments. This table shows the mean and median green bond premia in several market segments, the level of significance at which we rejected $H_0 : \widehat{p_i} = 0$, and the number of green bonds in each of the subsamples. We use a Wilcoxon signed-rank test with continuity correction

		$Mean(\widehat{p_i})$	$Median(\widehat{p_i})$	$\widehat{p_i} \neq 0$	# GB
	Total	−0.018	−0.010	***	110
Sector	Basic Materials	−0.016	−0.016		1
	Consumer, NC	−0.011	−0.011		1
	Financial	−0.025	−0.013	***	64
	Government	−0.009	0.000		38
	Industrial	0.005	0.005		1
	Utilities	0.002	−0.003		5
Currency	AUD	−0.031	−0.019		5
	CAD	−0.010	−0.010		2
	CHF	0.000	0.001		3
	CNY	0.024	0.024		1
	EUR	−0.017	−0.011	**	37
	GBP	−0.001	−0.001		2
	INR	0.055	0.055		1
	JPY	0.033	0.051		3
	RUB	−0.381	−0.381		1
	SEK	−0.009	−0.007		19
	TRY	0.079	0.079		1
	USD	−0.023	−0.019	***	35
Rating	AAA	−0.010	−0.003		49
	AA	−0.029	−0.024	***	16
	A	−0.018	−0.011		24
	BBB	−0.021	−0.009		8
	BB	−0.206	−0.206		1
	NR	−0.012	−0.007		12
Note:			*p<0.1; **p<0.05; ***p<0.01		

non-parametric Wilcoxon signed-rank test, which is applied to our specific framework,[19] to assess the significance of the premia per subsample. The results are robust to a test under the hypothesis of normality for A and SEK-denominated bonds.

Table 16.9 shows the average and median premia per subsample. The −1.8-bp average premium on the entire sample is significantly different from zero at a 99% level of confidence. Financial green bonds carry a −2.3-bps average premium with the same degree of significance. EUR-denominated and USD-denominated green bonds also have a significant

[19]For a subsample of n premia, the test is built as follows. We rank the n premia in ascending order of their absolute value and assign them a rank, R, from 1 to n. Let sgn represent the sign of the premium; we consider the following statistic:

$$W = \sum_{i=1}^{n} sgn(\widehat{p_i}) R_i$$

Under the null hypothesis H_0: $\widehat{p} = 0$, with $\sigma_w = \sqrt{\frac{n(n+1)(2n+1)}{6}}$, $\frac{W \pm 0.5}{\sigma_w}$ converges to a normal distribution. We add (subtract, resp.) 0.5 if $W < 0$ ($W > 0$, resp.) as a *continuity correction* since we compare discrete data to a continuous probability function.

negative premium of −1.7 bp and −2.3 bps, respectively. Lastly, AA green bonds show a −2.9-bps premium. Although the average and median premia of the other categories are not significantly different from zero, most of them are negative.

The literature analyzing the liquidity of off-the-run vs. on-the-run bonds highlights a significant liquidity premium of approximately 1.5 bp on U.S. Treasury bonds with the same characteristics except for their issue date. The comparison can be of interest because this premium affects bonds from the same issuer that have the same characteristics except for their issue date and, therefore, their degree of liquidity. By matching 55 pairs of bonds between 1994 and 2000, Goldreich, Hanke, and Nath (2005) show a yield differential of 1.5 bp between off-the-run and on-the-run US Treasury bonds. Pasquariello and Vega (2009) also find a yield difference of 1.6 bp on 5-year U.S. Treasury bonds by matching 86 bonds over the period 1992-2000.

We therefore provide evidence that investors in the secondary market pay a small negative yield premium inherent to green bonds, which is of a magnitude comparable to that of the on-the-run liquidity premium on U.S. Treasury bonds.

16.5.2 The determinants of the green bond premium

To determine and evaluate the determinants of a green bond premium, a linear regression of $\widehat{p_i}$ is performed on the characteristics of the green bonds. Table 16.10 shows the four regression specifications considered: (a) represents the most general specification, based on equation 16.4; (b) excludes the variables *Maturity* and *log(Issue Amount)*; (c) further excludes the currency dummies and the independent variables; and (d) represents solely the Rating × Sector cross effects. To avoid artificially high R^2s, the four regressions are performed on samples in which each of the dummy variables captures more than three observations. The R^2s therefore range from 11.3% (d) to 14.1% (a). The regression on the entire sample, of which the results are in line with that on restricted samples, is shown in Appendix (Table 16.11) and has an R^2 equal to 60.6%. Since the results of the Breusch-Pagan test evidence the presence of heteroscedasticity for the first three specifications,[20] we estimate White robust standard errors. Besides, the VIF calculation does not lead to a suspicion of multicollinearity.

Specifications (a) and (b) show that neither the maturity, the issue amount, nor the currency has a significant impact on the level of the premia in the considered subsample. The first two conclusions hold for the regression on the entire sample (Table 16.11); however, although the number of observations is limited, we suspect that the currency involved may have an impact in less mature financial markets. Specifications (a), (b), and (c) show that the rating significantly affects the premium: the lower the rating of the green bond is, the lower the green premium. The effect is particularly significant for AA and A bonds, with both values -2.3 bps with respect to AAA bonds (specification (b)). The study of Rating × Sector cross effects (specification (d)) shows that the level of premia varies between government-related bonds and financial bonds: while the negative impact of a lower rating is maintained in both sectors, the premia on financial bonds (-2.7 bps and -2.5 bps for AA and A, respectively) are lower than those on government-related bonds (-1.7 bps for AA).

These findings can be linked with the literature on the liquidity premium. Similar to the liquidity premium, the green bond premium fades with the increase of the credit quality (Longstaff, Mithal, and Neis, 2005, Chen, Lesmond, and Wei, 2007, Bao, Pan, and Wang, 2011, Dick-Nielsen, Feldhütter, and Lando, 2012, Huang and Huang, 2012, Abudy and A., 2016). In addition, the absolute value of the negative green bond premium is greater for financial bonds, similar to the situation regarding the liquidity premium (Longstaff,

[20]See the online appendix for the details of the tests performed.

TABLE 16.10: Results of step 2 regressions. This table gives the results of step 2 regressions in which the green bond premium is explained by the characteristics of the bonds through specifications (a),(b), (c), and (d). The premium is expressed as a percentage. The rating is a qualitative variable, the four modalities of which are AAA (reference modality), AA, A, and BBB. Maturity is the maturity of the bond expressed in years on December 31, 2017. The issue amount is the amount of green bonds issued expressed in USD billions. Sector is a qualitative variable, of which the three modalities are Government (reference modality), Financials and Utilities. We also consider Rating × Sector cross effects. Currency is a qualitative variable, of which the four modalities are USD (reference modality), AUD, EUR, and SEK

	Dependent variable: $\widehat{p}_i$			
	Cross-sectional regressions with White robust standard errors			
	(a)	(b)	(c)	(d)
Constant	−0.003 (0.015)	−0.004 (0.010)	−0.002 (0.009)	−0.007 (0.009)
Rating AA	−0.025** (0.010)	−0.023** (0.010)	−0.024** (0.010)	
Rating A	−0.026* (0.014)	−0.023* (0.013)	−0.022* (0.013)	
Rating BBB	−0.043 (0.043)	−0.040 (0.041)	−0.041 (0.040)	
Non-rated	−0.018 (0.020)	−0.009 (0.018)	−0.001 (0.014)	
Sector Financial	−0.008 (0.012)	−0.009 (0.012)	−0.008 (0.012)	
Sector Utilities	0.039 (0.034)	0.035 (0.032)	0.037 (0.031)	
AA × Government				−0.017* (0.009)
AAA × Financial				0.004 (0.011)
AA × Financial				−0.027** (0.013)
A × Financial				−0.025* (0.013)
NR × Financial				−0.005 (0.015)
Currency AUD	−0.009 (0.014)	−0.006 (0.013)		
Currency EUR	0.009 (0.010)	0.004 (0.010)		
Currency SEK	0.004 (0.016)	0.010 (0.011)		
Maturity	−0.001 (0.002)			
log(Issue Amount) (bn USD)	−0.006 (0.009)			
Observations	92	92	92	84
R^2	0.141	0.134	0.127	0.113
Adjusted R^2	0.023	0.039	0.066	0.056
Residual Std. Error	0.041 (df = 80)	0.041 (df = 82)	0.040 (df = 85)	0.038 (df = 78)
F Statistic	1.195 (df = 11; 80)	1.411 (df = 9; 82)	2.064* (df = 6; 85)	1.987* (df = 5; 78)
Note:				*p<0.1; **p<0.05; ***p<0.01

TABLE 16.11: Results of the step 2 regression on the entire sample. This table gives the result of the step 2 regression in which the green bond premium is explained by the characteristics of the bonds through specification (a) on the entire sample of 110 bonds

	Dependent variable: $\widehat{p_i}$
	Linear regression with White robust standard errors
Constant	−0.005
	(0.011)
Rating AA	−0.022**
	(0.010)
Rating A	−0.023*
	(0.014)
Rating BBB	−0.040
	(0.044)
Rating BB	−0.194***
	(0.010)
Non-rated	−0.011
	(0.018)
Sector Basic Materials	−0.011
	(0.018)
Sector Consumer, Non-cyclical	0.031
	(0.046)
Sector Financial	−0.007
	(0.012)
Sector Industrial	0.047
	(0.046)
Sector Utilities	0.037
	(0.034)
Currency AUD	−0.007
	(0.014)
Currency CAD	−0.005
	(0.013)
Currency CHF	0.020
	(0.015)
Currency CNY	0.059***
	(0.012)
Currency EUR	0.003
	(0.011)
Currency GBP	0.007
	(0.020)
Currency INR	0.060***
	(0.011)
Currency JPY	0.062**
	(0.024)
Currency RUB	−0.376***
	(0.011)
Currency SEK	0.011
	(0.011)
Currency TRY	0.084***
	(0.011)
Observations	110
R^2	0.606
Adjusted R^2	0.513
Residual Std. Error	0.040 (df = 88)
F Statistic	6.459*** (df = 21; 88)
Note:	*p<0.1; **p<0.05; ***p<0.01

Mithal, and Neis, 2005). However, contrary to the liquidity premium, which increases for low issue amounts (Longstaff, Mithal, and Neis, 2005), the green bond premium does not seem to be affected by low issue amounts. Moreover, Driessen, Nijman, and Simon (2016) find liquidity segmentation between long- and short-dated bonds, and Ejsing, Grother, and Grothe (2012) and Schuster and Uhrig-Hombourg (2012) show that the liquidity premium is greater in the short term. The green bond premium, in contrast, does not appear to be significantly impacted by the maturity of the bond.

Focusing on specification (b), we can express the green bond premia in absolute terms: they increase as the rating improves and are lower for financial bonds. For example, the yield of an AAA, AA, A and BBB EUR financial green bond is lower than that of an equivalent conventional bond by 0.9 bps, 3.2 bps, 3.2 bps and 4.9 bps, respectively. However, the yields of green and conventional AAA government-related bonds are in line (0 bp for EUR and -0.4bp for USD). As for the EUR (USD, resp.) utilities, although not significantly different from zero, the average premium is +1.2 bps (+0.8 bps, resp.) for A-green bonds and -0.5 bps (-0.9 bps, resp.) for BBB-green bonds.

These findings nuance several previous works that addressed this issue. We show that the yield differential between green and conventional bonds is negative for financial bonds–which are the most active corporate issuers–as suspected by Barclays (2015) and Ehlers and Packer (2017). Nevertheless, we substantially qualify the premium amount, of which the magnitude for A and AA bonds is closer to -3 bps than to -17 bps (Barclays, 2015) or -18 bps (Ehlers and Packer, 2017). Similar to HSBC (2016) and Climate Bonds Initiative (2017), we find evidence that this premium may be close to zero in several market segments, such as AAA government-related bonds or utilities. Lastly, we do not find evidence of a positive premium on USD-denominated bonds, as estimated by Karpf and Mandel (2018) (+7.8 bps).

In the final step, as a result of step 2, a green bond curve can be obtained from a conventional bond curve by applying the estimated green bond premium to the latter. This exercise is useful for investors as well as for issuers since few green bond benchmarks have been issued to date. Figure 16.4 presents the reconstituted green bond curve obtained by performing specification (b) as well as the conventional bond curve for eight different issuers. The quality of the fit achieved on the entire sample is satisfactory. However, the green bond curve does not always exactly intersect with the green bond market yields for three main reasons. Firstly, the green bond premia we calculate and explain here are long-term green premia, which reflect the average distortion since their inception. To obtain a closer fit, a short-term analysis would be more appropriate (see Section 16.6). Secondly, the low liquidity of several green bonds results in a yield that does not always reflect the actual yield on the reference date. Lastly, the greater the number of data available for estimating the green bond premium is, the closer the fit will be.

16.6 Robustness checks

In the first step of our robustness checks, we examine whether a negative premium may reflect the fact that the level of risk involved in a green bond is lower than in a conventional bond. We calculate the 10-day, 20-day and 30-day rolling annualized volatility during the period of interest in the case of both green and synthetic conventional bonds, following Equation (16.3) applied to the volatility, and take the difference between the members of each pair. We then estimate a step 1 regression adding the difference in volatility as an additional independent variable (Table 16.12). Using a robust standard errors estimation, we find no evidence that a difference in volatility is embedded in the yield differential

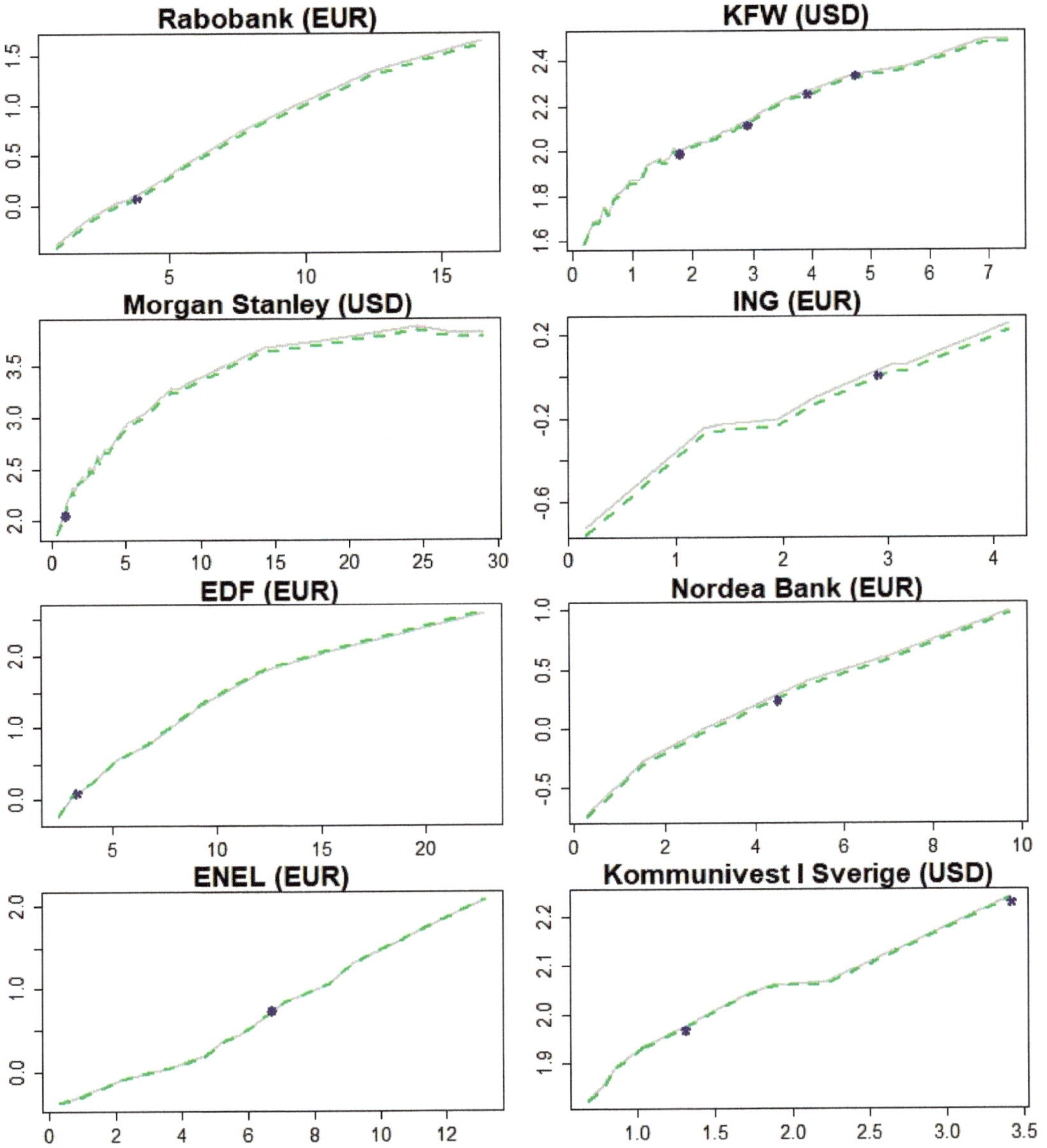

FIGURE 16.4: The green bond yield curves. This figure shows eight green bond curves (green dashed lines) reconstituted from conventional bond curves (grey solid lines) based on the parameters estimated in step 2 of regressions (b) performed on EUR and USD bonds. The market yields of the green bonds are also shown (blue stars).

between green and conventional bonds. This result indicates that the green bond premium should differ from a risk premium.

Another main issue is the question as to whether a green bond premium remains stable with time. We add a time fixed effect in the panel regression procedure. The estimated bid-ask spread parameter is found to be significant and almost equal to the parameter estimated above. Nevertheless, the individual time effect is significant during 24% of the 1162 days considered, which means that there might not be a durable daily time effect involved in the green bond premium.

TABLE 16.12: **Results of the step 1 regression with a control of the difference in volatility.** This table gives the results of the step 1 regression to which the difference in volatility between green and conventional bonds is added as an independent variable: $\Delta\tilde{y}_{i,t} = p_i + \beta\Delta\text{BA}_{i,t} + \Delta\text{Vol}_{i,t} + \epsilon_{i,t}$. Newey-West and Beck-Katz robust standard error tests are performed

	Dependent variable: $\Delta\tilde{y}_{i,t}$					
	Newey-West	Beck-Katz	Newey-West	Beck-Katz	Newey-West	Beck-Katz
ΔBA	−11.778***	−11.778***	−12.316***	−12.316***	−12.484***	−12.484***
	(3.178)	(3.861)	(3.330)	(3.989)	(3.459)	(4.129)
Δ 10-day volatility	−0.020	−0.020				
	(0.040)	(0.049)				
Δ 20-day volatility			0.037	0.037		
			(0.055)	(0.086)		
Δ 30-day volatility					0.017	0.017
					(0.060)	(0.119)
Note:						*p<0.1; **p<0.05; ***p<0.01

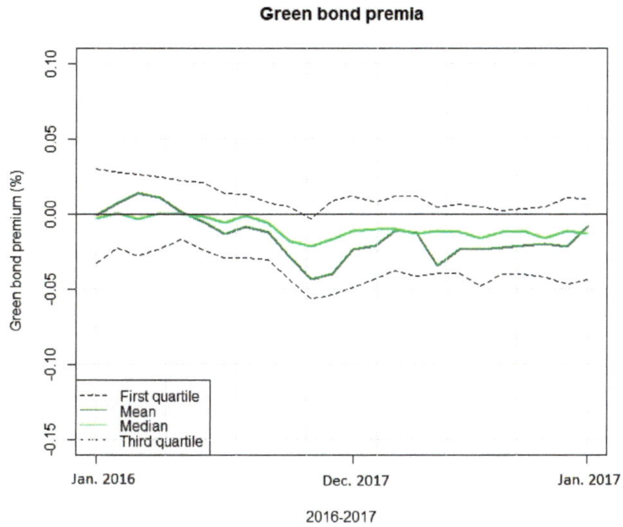

FIGURE 16.5: Green bond premium dynamics. This figure shows the evolution over time of the mean (light green solid line), the median (dark green solid line) and the quartiles (dashed blue lines) of the green bond premium during the years 2016 and 2017 based on the step 1 regression for the entire sample of green bonds.

However, upon applying the same regression procedure to the whole range of data on a monthly basis from January 2016 onward, we find the green bond premium to be variable,[21] although the mean and the median premia have become and remained negative since May 2016 (Figure 16.5), similar to what Karpf and Mandel (2018) reported. Moreover, interestingly, Delis, de Grieff, and Ongena (2018) find a similar result on bank loans: they show that, before 2015, bank did not price climate risk and, after 2015, a 2-bps average premium is charged to fossil fuel firms compared to non-fossil fuel firms.[22] We carry out the same analysis on each rating, sector (*Government* and *Financials*) and currency (EUR and USD) subgroups and find the same pattern for most of them with different amplitude ranges (Figure 16.6). It is worth noting that the robustness checks on a monthly basis are performed on rather small samples, and fewer bonds than in the main regression are therefore included. Thus, the information involved is somewhat different from that in the entire data history, which largely explains the discrepancies observed between the results.

A further potential concern is whether the green bond premium reflects a market risk premium over time. We therefore compare the daily returns of the time effects with three market indices' returns. Based on the S&P 500, the Eurostoxx 50 and the MSCI World indices, we first establish that the correlations between the index daily returns and the green bonds' time effects daily returns are low (10.9%, 7.8%, and 10.6%, respectively). In addition, to address the heteroscedasticity issue, we perform an OLS regression, with White robust standard errors,[23] to explain the daily returns of the green bond's time effects by the index daily returns (Table 16.13). Neither the S&P 500, the Eurostoxx 50, nor the MSCI

[21]As a comparison, Longstaff, Mithal, and Neis (2005), Favero, Pagano, and von Thadden (2010), and Huang and Huang (2012) show that the liquidity premium also varies over time.

[22]More precisely, Delis, de Grieff, and Ongena (2018) show that a one standard deviation increase in their measure of climate policy exposure induces a 2-bps increase of the loan rate.

[23]None of the Durbin Watson tests performed on the three specifications indicate any evidence of autocorrelation in the residuals. However, the hypothesis of heteroscedasticity is rejected.

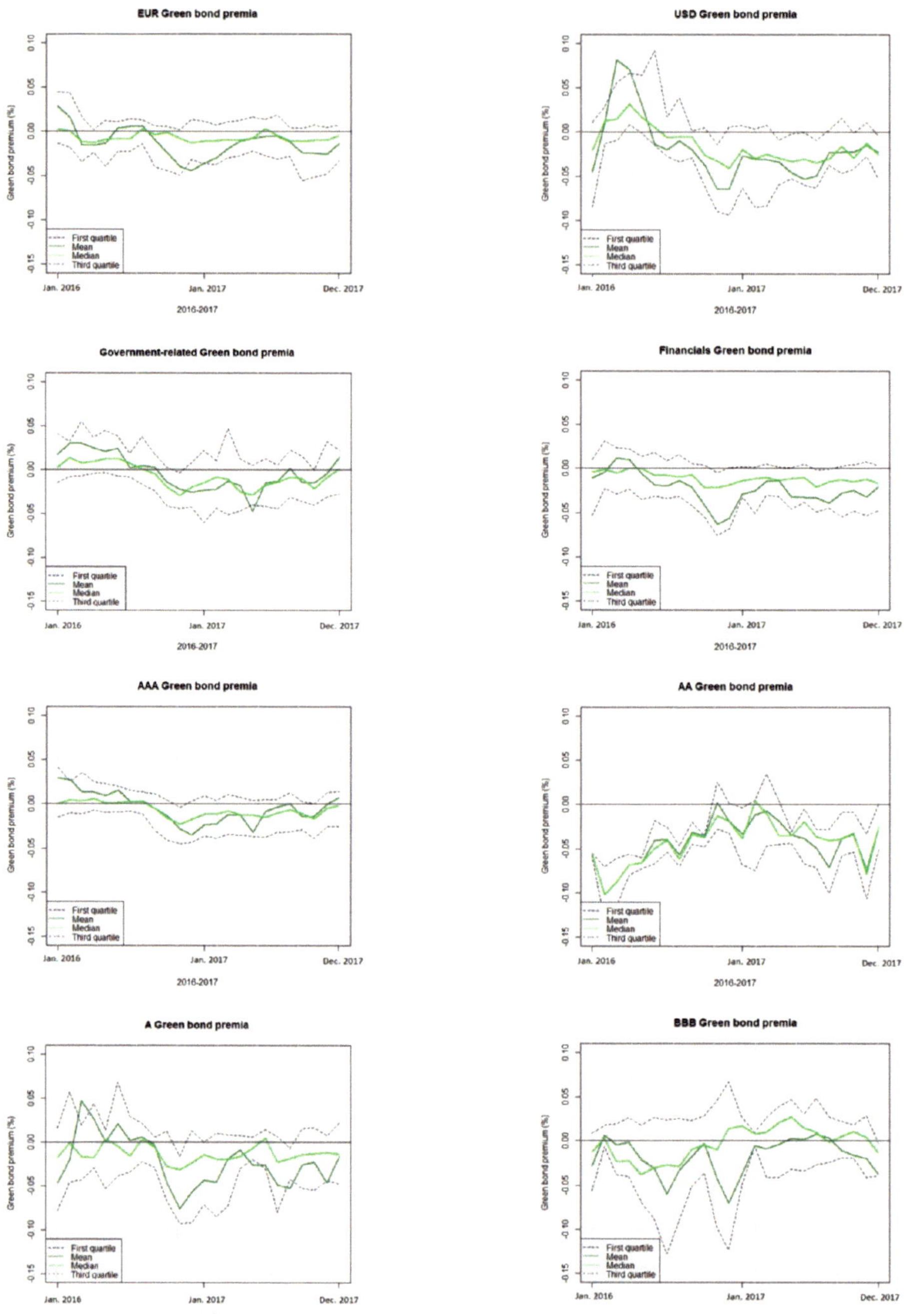

FIGURE 16.6: Green bond premium dynamics per group. These figures show the evolution over time of the mean (light green solid line), the median (dark green solid line) and the quartiles (dashed blue lines) of the green bond premia broken down by groups during the years 2016 and 2017 based on the step 1 regression for the entire sample of green bonds. The groups are as follows: (i) EUR, (ii) USD, (iii) Government-related, (iv) Financials, (v) AAA, (vi) AA, (vii) A, and (viii) BBB green bonds.

TABLE 16.13: Green premium and market returns. This table shows the regression of the daily returns of the time effects in the step 1 regression on the daily returns of several market indices

	Dependent variable: Time effects' returns		
	White robust std. err. estimation		
Constant	−0.818**	−0.764**	−0.802**
	(0.416)	(0.380)	(0.402)
S&P 500 returns	184.449		
	(133.206)		
Eurostoxx 50 returns		85.116	
		(60.130)	
MSCI World returns			203.006
			(135.363)
Note:	*p<0.1; **p<0.05; ***p<0.01		

World shows a significant effect. This analysis indicates that the time effect is not explained by a market risk premium and, hence, that the green bond premium does not reflect any market risk premium.

The quality of the matching method, as well as the interpolation or the extrapolation performed to obtain the synthetic conventional bond yield, must also be addressed. If CB1 and CB2 have significantly different levels of liquidity from that of the green bond, the first-step regression might not completely control for the residual liquidity. Furthermore, if the maturities of CB1 and CB2 differ greatly from that of the green bond, the yield of the synthetic conventional bond is liable to be over- or under-estimated. We therefore reproduce the matching method with more stringent liquidity constraints: we restrict the eligible conventional bonds to those (i) with an issue amount of less than twice the green bond's issue amount and greater than one-half of this amount and (ii) with an issue date that is, at most, two years earlier or two years later than the green bond's issue date. We also restrict the difference in maturity between CB1 and the green bond to a maximum of one year.[24] Comparing the estimated premia[25] of this sample to that of the same sample stemming from the matching constraints used in the general method, we find the descriptive statistics to be almost equal (Table 16.14). Moreover, after performing the step 2 regression, the estimated premia per subsample are very close for each of the two methods (Table 16.15). The minor difference is generally due to a poorer maturity matching with the second liquidity matching constraints. Therefore, in addition to restraining the obtained sample, requiring very stringent matching constraints can degrade the quality of the estimation.

Furthermore, we carry out a linear regression with White robust standard errors on the matched bond-day panel to explain the yield differential between GB and CB by the independent variables of specifications (a), (b), (c), (d), adding the liquidity differential

[24]Requiring the same restriction on CB2 leads to a total sample of only 30 matched pairs of bonds and, thus, to very small subsamples.

[25]The independent variable in step 1, ΔBA, is no longer significant with the second matching method, demonstrating that there is almost no more residual liquidity to be controlled.

TABLE 16.14: Descriptive statistics of more stringent matching criteria. This table gives the descriptive statistics of the estimated green bond premia through a step 1 regression on two different samples: a) the sample stemming from the matching criteria #2 and b) the sample stemming from the matching criteria #1 restricted to bonds in sample a). Matching criteria #1 require the conventional bonds to have (i) a maturity that is neither two years shorter nor two years longer than the green bond's maturity, (ii) an issue amount of less than four times the green bond's issue amount and greater than one-quarter of this amount, and (iii) an issue date that is at most six years earlier or six years later than the green bond's issue date. Matching criteria #2 require the conventional bonds to have (i) a maturity that is neither one (resp. two) year(s) lower nor one (resp. two) year(s) greater than the green bond's maturity for CB1 (resp. CB2), (ii) an issue amount of less than twice the green bond's issue amount and greater than one-half of this amount, and (iii) an issue date that is, at most, two years earlier or two years later than the green bond's issue date

	Green bond premia	
	Matching 2	*Matching 1 on M2's sample*
Min.	−0.130	−0.127
1st Quartile	−0.034	−0.039
Mean	−0.020	−0.018
Median	−0.012	−0.011
1st Quartile	0.003	0.001
Max.	0.079	0.079

control ΔBA.[26] For the sake of the comparison, we focus on the samples of 92 bonds ((a), (b) and (c)) and 84 bonds ((d)) used in the step 2 regression. The results (Table 16.16) confirm the necessity of controlling for the residual liquidity, although the effect is weakly significant for specifications (c) and (d). Moreover, the estimated effects are very close to that of the general method with less than 1 bp difference. However, the findings are slightly biased by overweighting the effect of bonds with the longest history. Furthermore, all of the estimated parameters are significant, which makes it difficult to discriminate between groups on the basis of the significance of their impact on the premium. Moreover, the R^2, approximately 5%, is less satisfactory than that of the second step in our general method.

It may also be interesting to contextualize our results with those of an OLS regression with White robust standard errors of the yield of green and conventional bonds on their characteristics. We apply specifications (a) and (b) on the sample consisting of the green and the closest conventional bonds (CB1),[27] using BA (instead of ΔBA) as a control for liquidity and adding a firm fixed effect as well as a dummy variable controlling for green bonds. Likewise, we find a significant negative premium that ranges from -0.6 bp to -0.9 bp (Table 16.17). However, as in the previous case, this method overweights premia for which a long price history is available.

Finally, the representativeness of the green premium estimated in our sample is addressed. Figure 16.7 compares the distribution of bonds in our sample with that of the global sample by rating and sector, which are the two factors that significantly influence the green premium. To assess goodness of fit, we perform a Chi-squared test on the

[26]It is worth noting that, as in our two-step regression, the better the matching, the more accurate the estimations.

[27]The same method could be applied to non-matched bonds, but the results would be much less accurate and would not be comparable with those of the main method presented in this paper.

TABLE 16.15: Estimated premia using more stringent matching criteria, broken down by rating and sector. This table gives the estimated average EUR and USD premia through a step 2 (b) regression using bonds stemming from matching criteria #1 and matching criteria #2, both restricted to the same largest common sample. The premia are broken down by ratings and sector. Matching criteria #1 require the conventional bonds to have (i) a maturity that is neither two years shorter nor two years longer than the green bond's maturity, (ii) an issue amount of less than four times the green bond's issue amount and greater than one-quarter of this amount, and (iii) an issue date that is at most six years earlier or six years later than the green bond's issue date. Matching criteria #2 require the conventional bonds to have (i) a maturity that is neither one (resp. two) year(s) shorter nor one (resp. two) year(s) longer than the green bond's maturity for CB1 (resp. CB2), (ii) an issue amount of less than twice the green bond's issue amount and greater than one-half of this amount, and (iii) an issue date that is at most two years earlier or two years later than the green bond's issue date

Matching criteria		*EUR*		*USD*	
		Govt	*Financials*	*Govt*	*Financials*
1	AAA	−0.00	−0.02	−0.02	−0.04
2	AAA	−0.02	−0.03	−0.03	−0.04
1	AA		−0.02		−0.04
2	AA		−0.01		−0.01
1	A		−0.02		−0.04
2	A		−0.03		−0.04
1	BBB		−0.02		−0.04
2	BBB		−0.03		−0.04

distributions of investment-grade bonds and on three of the four most represented sectors (*Government*, *Financials*, and *Utilities*) which account for 78% of the total sample of green bonds. With P-values of 21.3% and 19.9%, respectively, we find that our green premium estimate should be reasonably representative of the overall sample for investment-grade bonds in the considered sectors. Moreover, to estimate a premium over a broader scope, we use a matching method between each green bond and one conventional bond with less restrictive criteria requiring the same issuer, currency and coupon type; we also impose a maximum maturity difference of four years and an issue amount ratio between one-quarter and four. We therefore perform a cross-sectional regression on the 179 matched pairs, accounting for 40% of the global amount of green bonds issued, controlling for all the different characteristics of the bonds. The amount of the estimated negative premium (Table 16.18) is found to be of a similar magnitude to that which we find with our main method. Finally, we test the robustness of the result by restricting our estimate to subsamples. By carrying out 10,000 draws with and without replacement of 40, 60 and 80 pairs among the 110 studied, we observe that more than 99% of the estimated premia are negative in the six different cases.

TABLE 16.16: Results of a step 2 regression performed on the difference in the yield while controlling for the difference in liquidity. This table gives the results of step 2 regressions performed on the bond-day sample in which we explain the yield differential by a proxy of the difference in liquidity ΔBA and the bonds' characteristics of specifications (a),(b), (c), and (d). The yield differential and ΔBA are expressed as percentages. The rating is a qualitative variable, of which the four modalities are AAA (reference modality), AA, A and BBB. Maturity is the maturity of the bond expressed in years on December 31, 2017. The issue amount is the amount of green bonds issued expressed in USD billions. Sector is a qualitative variable, of which the three modalities are Government (reference modality), Financials and Utilities. We also consider Rating × Sector cross effects. Currency is a qualitative variable, of which the four modalities are USD (reference modality), AUD, EUR, and SEK

	Dependent variable: $\Delta\tilde{y}_{i,t}$			
	Linear regressions with White robust standard errors			
	(a)	(b)	(c)	(d)
Constant	0.001 (0.001)	-0.006^{***} (0.001)	-0.002^{***} (0.001)	-0.009^{***} (0.001)
ΔBA	-1.378^{***} (0.533)	-1.399^{***} (0.521)	-0.939^{*} (0.500)	0.578 (0.460)
Rating AA	-0.014^{***} (0.001)	-0.019^{***} (0.001)	-0.021^{***} (0.001)	
Rating A	-0.028^{***} (0.001)	-0.030^{***} (0.001)	-0.031^{***} (0.001)	
Rating BBB	-0.026^{***} (0.003)	-0.035^{***} (0.003)	-0.038^{***} (0.003)	
Non-rated	-0.006^{***} (0.001)	-0.009^{***} (0.001)	-0.002^{**} (0.001)	
Sector Financial	-0.010^{***} (0.001)	-0.009^{***} (0.001)	-0.006^{***} (0.001)	
Sector Utilities	0.027^{***} (0.002)	0.034^{***} (0.002)	0.041^{***} (0.002)	
AA × Government				-0.014^{***} (0.002)
AAA × Financial				0.006^{***} (0.001)
AA × Financial				-0.019^{***} (0.001)
A × Financial				-0.033^{***} (0.001)
NR × Financial				-0.002^{**} (0.001)
Currency AUD	0.003^{***} (0.001)	-0.0003 (0.001)		
Currency EUR	0.012^{***} (0.001)	0.009^{***} (0.001)		
Currency SEK	0.028^{***} (0.002)	0.014^{***} (0.001)		
Maturity	-0.002^{***} (0.0002)			
log(Issue Amount) (bn USD)	0.007^{***} (0.001)			
Observations	33,127	33,127	33,127	28,682
R^2	0.059	0.053	0.049	0.046
Adjusted R^2	0.059	0.052	0.049	0.045
Residual Std. Error	0.071 (df = 33114)	0.071 (df = 33116)	0.071 (df = 33119)	0.071 (df = 28675)
F Statistic	173.178^{***} (df = 12; 33114)	183.831^{***} (df = 10; 33116)	243.608^{***} (df = 7; 33119)	228.418^{***} (df = 6; 28675)
Note:				*p<0.1; **p<0.05; ***p<0.01

TABLE 16.17: Results of an OLS regression of the yields on the characteristics of green and conventional bonds. This table provides the results of an OLS regression with White standard errors performed on the yields of the green and the closest conventional bonds (CB1). Following specifications (a) and (b), the yields are explained by the characteristics of the bonds (rating, sector, currency, maturity) and a control for liquidity (bid-ask spread, BA), to which a dummy variable for green bonds and a firm fixed effect are added. The issue amount is not included in this regression since the bid-ask spread is used to control for bonds' liquidity

	Dependent variable: Bonds' yields	
	(a)	(b)
Constant	1.633*** (0.010)	1.748*** (0.015)
Green	−0.006** (0.002)	−0.009*** (0.003)
BA	80.880*** (4.009)	134.779*** (5.169)
Maturity	0.082*** (0.001)	
Rating control	Yes	Yes
Sector control	Yes	Yes
Currency control	Yes	Yes
Firm control	Yes	Yes
Observations	66,254	66,254
R^2	0.905	0.890
Adjusted R^2	0.905	0.890
Residual Std. Error	0.312 (df = 66198)	0.336 (df = 66199)
F Statistic	11,515.150*** (df = 55; 66198)	9,928.532*** (df = 54; 66199)
Note:		*p<0.1; **p<0.05; ***p<0.01

16.7 Discussion

The -2-bps average green bond yield premium (1.5% of the average yield in the sample) indicates the yield that investors are willing to give up to fund green investments rather than conventional investments with strictly equal risk. We find evidence of a low impact of investors' pro-environmental motives on bond prices. This statistically significant effect is consistent with existing theoretical works. Fama and French (2007) demonstrate that when

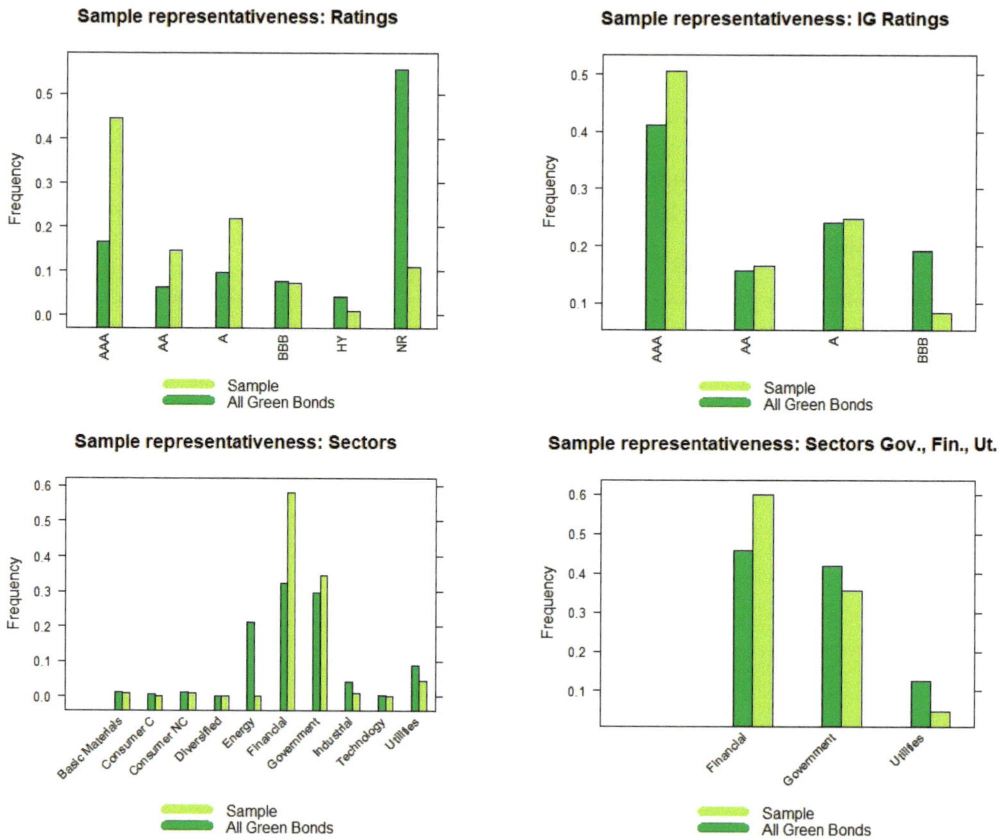

FIGURE 16.7: Analysis of the representativeness of the matched sample. This figure shows the distribution, by sector and rating, of green bonds in the matched sample (110 bonds) compared to the distribution of those in the global universe (1065 bonds). The right-hand figures correspond to the left-hand figures, and the comparison is focused on investment-grade bonds for the top figure and the Financial, Government, and Utilities sectors for the bottom figure.

a group of investors has a *taste* for a certain type of assets, equilibrium prices shift and the capital asset pricing model (CAPM) fails to explain asset returns. Focusing on equity, Heinkel, Kraus, and Zechner (2001) show that, by excluding polluting assets from their portfolio, *green investors* drive up the cost of capital of polluting companies. We also relate our result to the empirical finding that investors' pro-social and pro-environmental inclinations increase inflows to socially responsible investments (Hong and Kacperczyk, 2009; Riedl and Smeets, 2017; Hartzmark and Sussman, 2018), of which the psychological origin can be altruism (Brodback, Guenster, and Mezger, 2018) or social pressure (DellaVigna, List, and Malmendier, 2012). However, we show that, in contrast to the effects on the volume of financial flows, the impact on prices is very limited. In this respect, our findings suggest that the lower cost of debt for companies with good environmental performance should be predominantly related to a lower level of financial risk, through intangible asset creation[28]

[28]Intangible assets may refer to an improvement in the company's reputation, the attraction of new customers or a greater loyalty of employees toward the company.

TABLE 16.18: Results of an OLS regression of the yields on the characteristics of green and conventional bonds matched with less stringent criteria. This table provides the results of an OLS regression with White standard errors performed on the yields of the green and the matched conventional bonds with less stringent criteria than the main matching method. We require that both bonds have the same issuer, currency, coupon type, a maximum maturity difference of four years and an issue amount ratio of between one-quarter and four. The yields are explained by the characteristics of the bonds (rating, sector, currency, maturity, collateral, coupon type, bullet/callable structure) and the price percentage bid-ask spread as control for liquidity, to which a dummy variable for green bonds is added. The 179 bond pairs are studied over the same time period as the main regression: from July 18, 2013 to December 31, 2017

	Dependent variable: Bonds' yields
Constant	3.284***
	(0.049)
Green	−0.042***
	(0.003)
Maturity	0.099***
	(0.001)
Bid-Ask	21.952***
	(0.560)
Rating control	OK
Sector control	OK
Currency control	OK
Collateral control	OK
Coupon type control	OK
Bullet/Callable control	OK
Observations	138,272
R^2	0.901
Adjusted R^2	0.901
Residual Std. Error	0.649 (df = 138226)
F Statistic	27,975.340*** (df = 45; 138226)
Note:	*p<0.1; **p<0.05; ***p<0.01

(Porter and van der Linde, 1995; Hart, 1995; Jones, 1995; Ambec and Lanoie, 2008; Flammer, 2015) and better risk management and mitigation (Ambec and Lanoie, 2008; Bauer and Hann, 2014), rather than investors' non-pecuniary preferences.

A negative yield differential of 2 bps for green bonds has several implications for the different types of market participants. Regarding investors, the amount of this premium should not constitute a sufficient differential likely to discourage them from investing in green bonds. Becker and Ivashina (2015) study the arbitrage of insurers between investment-grade U.S. corporate bonds with the same rating but different yields, controlling for duration and liquidity, between 2004 and 2010. In particular, they show that a positive differential of 100 bps leads to a reallocation of between 3.6% and 7.4% of insurance companies' holdings on the primary market and 0% to 2.5% on the secondary market. Given the amounts highlighted by this article in a similar framework, a -2-bps premium should therefore not constitute a disincentive to invest in green bonds. Moreover, although this premium is low, it demonstrates investors' appetite for green bond issues and thus highlights the opportunity for issuers to broaden their bondholder base by issuing green bonds, as suggested by I4CE (2016). This premium is also consistent with the results of Flammer (2018), who finds that green bond issuances induce an increase in ownership by long-term and green investors. Finally, from the supervisory authority perspective, this result addresses the concern about the appearance of a bubble on green assets raised by the Dutch Central Bank (De Nederlandsche Bank, 2017): while the amount of this premium indicates investors' preference for green bonds, it does not yet reveal any substantial pricing discrepancy between green and conventional bonds.

The opportunity to increase the issuance of green bonds, which still accounted for 1.3% of the outstanding global debt in 2017,[29] is not only supported by the results of this paper but also consistent with political ambitions and the recommendations of financial players. Policymakers can play a crucial role by providing green project developers and investors with a clearer legal framework to unlock the full potential of the green bond market. Indeed, as recommended by the EU High-Level Group on Sustainable Finance (EU High-Level Expert Group (HLEG), 2018), the European Commission set a roadmap on March 8, 2018, to establish a common taxonomy (*EU Classification System*) for sustainable finance and to create EU labels for green financial products based on this classification (European Commission, 2018). These actions will notably help establish a precisely defined framework for green bond requirements and should streamline the approval process to increase the flow of low-carbon projects.[30]

16.8 Conclusion

In this paper, we use green bonds as an instrument to identify the effect of non-pecuniary motives, specifically pro-environmental preferences, on bond market prices. We analyze the yield of green bonds compared to that of equivalent synthetic non-green bonds through a matching method for bonds issued from July 2013 to December 2017. We identify the effect of pro-environmental preferences through a *green bond premium*, which is defined as the yield differential between a green bond and its counterfactual conventional bond after controlling for their difference in liquidity. We evidence a significant, albeit low, premium related to

[29]According to the Bank for International Settlements, the total outstanding debt worldwide amounted to USD 23,580 billion in the third quarter of 2017: https://www.bis.org/statistics/c1.pdf

[30]In 2016, green bonds accounted for only 17% of the USD 694 billion climate-aligned bonds universe (Climate Bonds Initiative, 2016) that gathers numerous potential candidates for a green bond label.

investors' pro-environmental preferences in the bond market. This result highlights the opportunity for issuers to benefit from an expansion of their bondholder base through this asset class, especially for low-rated and financial bonds. However, at this stage, the premium is still low enough not to demonstrate any substantial valuation discrepancy between green and conventional bonds or to dissuade investors from supporting the development of the green bond market.

The main limitation of this study arises from the quality of the data. Since bonds–and *a fortiori* corporate bonds–are not frequently traded, a bond yield does not accurately reflect the fair value of the bond in some cases. Further research along these lines could focus on pursuing the following two main objectives. An empirical study could be performed to assess whether the use of proceeds has a differentiating impact on the premium. This study could also be extended to social impact bonds, once this market is sufficiently mature, to analyze the impact of pro-social preferences on bond prices.

References

Abudy, M.M. and Raviv, A., 2016, How Much Can Illiquidity Affect Corporate Debt Yield Spread?, *Journal of Financial Stability* 25, 58–69.

Alexander, G., A. Edwards, and M. Ferri, 2000, The determinants of trading volume of high-yield corporate bonds, *Journal of Financial Markets* 3, 177–204.

Ambec, S., and P. Lanoie, 2008, Does It Pay to Be Green? A Systematic Overview, *Academy of Management Perspectives* 22, 45–62.

Amihud, Y., 2002, Illiquidity and stock returns: cross-section and time-series effects, *Journal of Financial Markets* 5, 31–56.

Baker, M., D. Bergstresser, G. Serafeim, and J. Wurgler, 2018, Financing the Response to Climate Change: The Pricing and Ownership of U.S. Green Bonds, *Working Paper*.

Bao, J., J. Pan, and J. Wang, 2011, The illiquidity of corporate bonds, *Journal of Finance* 66, 911–946.

Barclays, 2015, The cost of being green, *Credit Reseach*.

Bauer, R., and D. Hann, 2014, Corporate environmental management and credit risk, *Working Paper*.

Bauer, R., K. Koedijk, and R. Otten, 2005, International evidence on ethical mutual fund performance and investment style, *Journal of Banking and Finance* 29, 1751–1767.

Bauer, R., and P. Smeets, 2015, Social identification and investment decisions, *Journal of Economic Behavior and Organization* 117, 121–134.

Beber, A., M. W. Brandt, and K. A. Kavajecz, 2009, Flight-to-quality or flight-to-liquidity? Evidence from the euro-area bond market, *Review of Financial Studies* 22, 925–957.

Beck, N., and J. Katz, 1995, What to do (and not to do) with time-series cross-section data in comparative politics, *American Political Science Review* 89, 634–647.

Becker, B., and V. Ivashina, 2015, Reaching for yield in the bond market, *Journal of Finance* 70, 1863–1901.

Benabou, R., and J. Tirole, 2010, Individual and corporate social responsibility, *Economica* 77, 1–19.

Bloomberg, 2017, Investors are willing to pay a "green" premium, *Bloomberg New Energy Finance report*.

Brodback, D., N. Guenster, and D. Mezger, 2018, Altruism and Egoism in Investment Decisions, *Working Paper*.

Chava, S., 2014, Environmental externalities and cost of capital, *Management Science* 60, 2223–2247.

Chen, L., D. A. Lesmond, and J. Wei, 2007, Corporate yield spreads and bond liquidity, *Journal of Finance* 62, 119–149.

Climate Bonds Initiative, 2016, Bonds and Climate Change: The state of the market in 2016, `https://www.climatebonds.net/files/files/CBI%20State%20of%20the%20Market%202016%20A4.pdf`.

Climate Bonds Initiative, 2017, Green Bonds Highlights 2016, `https://www.climatebonds.net/files/files/2016%20GB%20Market%20Roundup.pdf`.

de Jong, F., and J. Driessen, 2012, Liquidity risk premia in corporate bond markets, *Quaterly Journal of Finance* 2, 1–34.

De Nederlandsche Bank, 2017, Waterprooof? An exploration of climate-related risks for the Dutch financial sector, *Report*.

Delis, M., K. de Grieff, and S. Ongena, 2018, Being Stranded on the Carbon Bubble? Climate Policy Risk and the Pricing of Bank Loans, *Swiss Finance Institute Research Paper Series* 18.

DellaVigna, S., J. List, and U. Malmendier, 2012, Testing for Altruism and Social Pressure in Charitable Giving , *Quarterly Journal of Economics* 127, 1–56.

Derwall, J., N. Guenster, R. Bauer, and K. Koedijk, 2005, The eco-efficiency premium puzzle, *Financial Analysts Journal* 61, 51–63.

Dhaliwal, D., O. Li, A. Tsang, and Y. Yang, 2011, Voluntary nonfinancial disclosure and the cost of equity capital: the initiation of corporate social responsibility reporting, *The Accounting Review* 86, 59–100.

Dick-Nielsen, J., P. Feldhütter, and D. Lando, 2012, Corporate bond liquidity before and after the onset of the subprime crisis, *Journal of Financial Economics* 103, 471–492.

Dixon, R., 2010, A framework for monitoring the performance impact on a global equity portfolio, *Mercer (15 June)*.

Driessen, J., T. Nijman, and Z. Simon, 2016, Much ado about nothing: A study of differential pricing and liquidity of short and long term bonds, *Working Paper*.

Ehlers, T., and F. Packer, 2017, Green bond finance and certification, *Bank for International Settlements Quarterly Review*.

Ejsing, J., M. Grother, and O. Grothe, 2012, Liquidity and Credit Risk Premia in Government Bond Yields, *European Central Bank Working Paper*.

El Ghoul, S., Guedhami, O., Kwok, C.C. and Mishra, D.R., 2011, Does corporate social responsibility affect the cost of capital?, *Journal of Banking and Finance* 35, 2388–2406.

Elton, E., M. Gruber, D. Agrawal, and C. Mann, 2004, Factors affecting the valuation of corporate bonds, *Journal of Banking and Finance* 28, 2747–2767.

EU High-Level Expert Group (HLEG), 2018, Final report 2018 by the high-level expert group on sustainable finance.

European Commission, 2018, Action plan: Financing sustainable growth, Accessed: 2024-03-25.

Fama, E., and K. French, 2007, Disagreement, tastes, and asset prices , *Journal of Financial Economics* 83, 667–689.

Favero, C., M. Pagano, and E.-L. von Thadden, 2010, How Does Liquidity Affect Government Bond Yields?, *Journal of Financial and Quantitative Analysis* 45, 107–134.

Financial Stability Board, 2018, Global shadow banking monitoring, report 2017, *Report*.

Flammer, C., 2015, Does corporate social responsibility lead to superior financial performance? A regression discontinuity approach, *Management Science* 61, 2549–2568.

Flammer, C., 2018, Corporate Green Bonds, *Working Paper*.

Fong, K. Y. L., C. W. Holden, and C. A. Trzcinka, 2017, What are the best liquidity proxies for global research, *Review of Finance* 21, 1355–1401.

Ge, W., and M. Liu, 2015, Corporate social responsibility and the cost of corporate bonds, *Journal of Accounting and Public Policy* 34, 597–624.

Ghouma, H., H. Ben-Nasr, and R. Yan, 2018, Corporate governance and cost of debt financing: Empirical evidence from Canada, *The Quarterly Review of Economics and Finance* 67, 138–148.

Goldreich, D., B. Hanke, and P. Nath, 2005, The Price of Future Liquidity: Time-Varying Liquidity in the U.S. Treasury Market, *Review of Finance* 9, 1–32.

Goldstein, H., 2003, *Multilevel Statistical Models*, third edition (New York: Wiley).

Goss, A., and G. Roberts, 2011, The impact of corporate social responsibility on the cost of bank loans, *Journal of Banking and Finance* 35, 1794–1810.

Hachenberg, B., and D. Schiereck, 2018, Are green bonds priced differently from conventional bonds? , *Journal of Asset Management* 1–13.

Han, S., and H. Zhou, 2016, Effects of liquidity on the nondefault component of corporate yield spreads: Evidence from intraday transactions data, *Quaterly Journal of Finance* 6, 1–49.

Hart, S., 1995, A natural-resource-based view of the firm, *Academy of Management Review* 20, 986–1014.

Hartzmark, S., and A. Sussman, 2018, Do Investors Value Sustainability? A Natural Experiment Examining Ranking and Fund Flows, *Working Paper* 565.

Hasan, I., C. Hoi, and H. Zhang, 2017, Social capital and debt contracting: Evidence from bank loans and public bonds, *Journal of Financial and Quantitative Analysis* 52, 1017–1047.

Heinkel, R., A. Kraus, and J. Zechner, 2001, The effect of green investment on corporate behaviour, *Journal of Financial and Quantitative Analysis* 36, 377–389.

Helwege, J., J.-Z. Huang, and Y. Wang, 2014, Liquidity effects in corporate bond spreads, *Journal of Banking and Finance* 45, 105–116.

Hong, H., and M. Kacperczyk, 2009, The price of sin: the effects of social norms on markets, *Journal of Financial Economics* 93, 15–36.

Houweling, P., A. Mentink, and T. Vorst, 2005, Comparing possible proxies of corporate bond liquidity, *Journal of Banking and Finance* 29, 1331–1358.

HSBC, 2016, Green Bonds 2.0, *Fixed Income Credit Report*.

Huang, J.-Z., and M. Huang, 2012, How much of the corporate-treasury yield spread is due to credit risk?, *Review of Asset Pricing Studies* 2, 153–202.

I4CE, 2016, Beyond transparency: unlocking the full potential of green bonds, *Institue for Climate Economics*.

Jiraporn, P., N. Jiraporn, A. Boeprasert, and K. Chang, 2014, Does corporate social responsibility (CSR) improve credit ratings? Evidence from geographic identification, *Financial Management* 43, 505–531.

Jones, T., 1995, Instrumental stakeholder theory: A synthesis of ethics and economics, *Academy of Management Review* 20, 404–437.

Karpf, A., and A. Mandel, 2018, The changing value of the 'green' label on the US municipal bond market, *Nature Climate Change* 8, 161–165.

Kempf, A., and P. Osthoff, 2007, The effect of socially responsible investing on portfolio performance, *European Financial Management* 13, 908–922.

Klock, M., S. Mansi, and W. Maxwell, 2005, Does corporate governance matter to bondholders, *Journal of Financial and Quantitative Analysis* 40, 693–717.

Konar, S., and M. Cohen, 2001, Does the market value environmental performance?, *Review of Economics and Statistics* 83, 281–289.

Kreander, N., R. Gray, D. Power, and C. Sinclair, 2005, Evaluating the performance of ethical and non-ethical funds: A matched pair analysis, *Journal of Business Finance and Accounting* 32, 1465–1493.

Krüger, P., 2015, Corporate goodness and shareholder wealth, *Journal of Financial Economics* 115, 304–329.

Longstaff, F. A., S. Mithal, and E. Neis, 2005, Corporate yield spreads: Default risk or liquidity? New evidence from the credit default swap market, *Journal of Finance* 60, 2213–2253.

Luo, X., and C. Bhattacharya, 2009, The debate over doing good: Corporate social performance, strategic marketing levers, and firm-idiosyncratic risk, *Journal of Marketing* 73, 198–213.

Magnanelli, B. S., and M. F. Izzo, 2017, Corporate social performance and cost of debt: the relationship, *Social Responsibility Journal* 13, 250–265.

Menz, K., 2010, Corporate social responsibility: is it rewarded by the corporate bond market? A critical note, *Journal of Business Ethics* 96, 117–134.

Merton, R., 1973, Theory of rational option pricing, *Bell Journal of Economics and Management Science* 4, 141–183.

Nilsson, J., 2008, Investment with a conscience: Examining the impact of pro-social attitudes and perceived financial performance on socially responsible investment behavior, *Journal of Business Ethics* 83, 307–325.

OECD, 2017a, Investing in climate, investing in growth, *OECD Publishing, Paris*.

OECD, 2017b, Green bonds: Mobilising bond markets for a low-carbon transition, *OECD Publishing, Paris*.

Oikonomou, I., C. Brooks, and S. Pavelin, 2014, The effects of corporate social performance on the cost of corporate debt and credit ratings, *The Financial Review* 49, 49–75.

Pasquariello, P., and C. Vega, 2009, The on-the-run liquidity phenomenon, *Journal of Financial Economics* 92, 1–24.

Porter, M., and C. van der Linde, 1995, Toward a new conception of the environment-competitiveness relationship , *Journal of Economic Perspective* 9, 97–118.

Renneboog, L., J. Ter Horst, and C. Zhang, 2008, The price of ethics and stakeholder governance: The performance of socially responsible mutual funds, *Journal of Corporate Finance* 14, 302–322.

Riedl, A., and P. Smeets, 2017, Why do investors hold socially responsible mutual funds?, *Journal of Finance* 72, 2505–2550.

Roll, R., 1984, A simple implicit measure of the effective bid-ask spread in an efficient market, *Journal of Finance* 39, 1127–39.

Schuster, P., and M. Uhrig-Hombourg, 2012, The Term Structure of Bond Market Liquidity Conditional on the Economic Environment: An Analysis of Government Guaranteed Bonds, *Working Paper*.

Semenova, N., and L. Hassel, 2008, Industry risk moderates the relation between environmental and financial performance, *Working paper, Sustainable Investment Research Platform*.

Sharfman, M., and C. Fernando, 2008, Environmental risk management and the cost of capital, *Strategic Management Journal* 29, 569–592.

Statman, M., and D. Glushkov, 2009, The wages of social responsibility, *Financial Analysts Journal* 65, 33–46.

Stellner, C., C. Klein, and B. Zwergel, 2015, Corporate social responsibility and Eurozone corporate bonds: The moderating role of country sustainability, *Journal of Banking and Finance* 59, 538–549.

Su, L., Y. Zhang, and J. Wei, 2016, A practical test for strict exogeneity in linear panel data models with fixed effects, *Economics Letters* 147, 27–31.

Sun, W., and K. Cui, 2014, Linking corporate social responsibility to firm default risk, *European Management Journal* 32, 275–287.

Wulandari, F., D. Schäfer, A. Stephan, and C. Sun, 2018, The impact of liquidity risk on the yield spread of green bonds, *Finance Research Letters,* in press.

Chapter 17

Stochastic carbon regulation in continuous time

René Aïd
University Paris-Dauphine, France

Sara Biagini
LUISS University, Italy

We deal with optimal dynamic carbon emission regulation of a set of firms. While the results are essentially taken from Aïd and Biagini (2023), the focus here is more on the methodology and its framing in the applied mathematics literature. Firms face idiosyncratic, as well as common, economic shocks on emissions, and they have linear quadratic abatement costs. Firms can trade allowances so as to minimize total expected costs, which arise from abatement, trading and terminal penalty. They may also receive free allowances from the regulator to aid in compliance. Using variational methods, we first exhibit in closed-form the market equilibrium as function of the regulator's allocation. We then solve the Stackelberg game between the regulator and the firms. The result is a closed-form expression of the optimal dynamic allocation policies that allow a desired expected emission reduction. In the presence of market frictions, optimal expected allocations are unique and the optimal allowances price is constant as well as abatement efforts and trading rates.

17.1 Introduction

Since its inception in 2005, the European Union Emission Trading System has been a major innovative tool to manage carbon emissions and help EU member states reach their agreed reduction targets. Former examples of cap-and-trade mechanism to reduce pollution include the successful cap-and-trade market for sulfur dioxide and nitrogen oxides in the US (Title IV of Clean Air Act Amendments, 1990). However, the striking novelty of the EU carbon market is the dimension: more than 30 billion euros of value gathering more than 15 thousand stationary installations across 30 countries. The seminal paper by Montgomery (1972) on market for licenses has found here a spectacular illustration of the idea that market mechanisms can be efficiently developed to achieve pollution reduction. Nevertheless, after 15 years, it is clear that the EU ETS is facing some issues.

Figure 17.1 (Left) provides the price of allowances from the beginning of the carbon market in 2005 to October 2024 and the cumulative difference between total verified emissions per year and total allowances (source: EEA of EU Emission Trading System) whereas (Right) provides the allowances and the emissions per year. First, note that the annual emissions rate that was 2 billion tons per year in 2005 has reached 1 billion tons in 2023.

DOI: 10.1201/9781032636252-17

This division by two of the emissions rate owes part to the economic recession, part to the development of renewable energy and the closing of coal fired plants across Europe and also, to the EU ETS mechanism (see Dechezleprêtre, Nachtigall, and Venmans (2023) and the references therein). Second, it shows that the market price of carbon is highly sensitive to the relation between supply (allowances) and demand (emissions). The 2008 financial crisis led to a large surplus that lasted until 2013 and led to a depressed market price, which reached less than 5 €/tonCO_2. During this period, emissions were reduced, not because of firms abatement efforts, but because the world was experiencing a major recession. This phenomenon led the EU to design the Market Stability Reserve mechanism (MSR), to reduce market price volatility and over supply (EU Decision 2015/1814 of October 6th, 2015 on EU Directive 2003/87). In a nutshell, the MSR regulates the potential market imbalances either by backloading allowances to the future or providing more allowances through auctions during the current phase and the next ones. This mechanism makes the carbon market regulation a dynamic process.

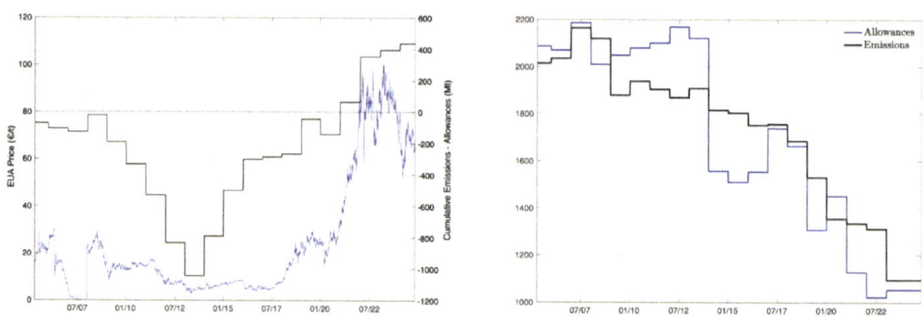

FIGURE 17.1: (Left) EUA price in €/tonCO_2 (left axis) compared to the difference between verified emissions and allocations in MtonCO_2 (right axis), (Right) Allowances and emissions. source: Eikon Reuters for EUA prices and European Environment Agency EU Emission Trading System data viewer.

Thus, the MSR mechanism implements the idea that contingent regulation should be preferred to fixed ones. This concept has been vastly supported in the literature, since Weitzman (1974) seminal paper on regulation under uncertainty. From there, an intense research activity has focused on dynamic regulation in different settings. Hahn (1989) and more recently Hepburn (2006) both provide surveys on regulation through prices or quantities as well as a framing of the admissible tools to regulators.

In our specific case of carbon market, there exists an extensive literature which analyzes the carbon emissions reduction problem under its many different aspects. The optimality of banking permits from one period to the next has been studied in Rubin (1996), Schennach (2000), Chaton, Creti, and Peluchon (2015), Lintunen and Kuusela (2018) and Kuusela and Lintunen (2020). Mechanism designs are proposed in Carmona et al. (2010) in a discrete time model to reduce electricity producers windfall profits. More recently, Meunier, Montero, and Ponssard (2018) deal with optimal allocation schemes indexed on the output of firms. An important stream of the literature on diffusive pollution deals with imperfect competition and strategic interaction among polluters. Requate (1993), Von der Fehr (1993) and recently Anand and Giraud-Carier (2020) develop static models of imperfect competition which point out the firms capacity to increase their profit with emissions regulation.

The focus in the present paper is continuous time regulation of the carbon emissions with dynamic allowances allocation. We regard carbon reduction as a stochastic Stackelberg game. Firms are the Followers, while the regulator is the Leader.

The market model for firms is a continuous time stochastic model, inspired by the Kollenberg and Taschini (2016, 2019) carbon market model. Firms experience individual emissions growth rate, plus idiosyncratic as well as common economic shocks. Abatement costs are heterogeneous and linear-quadratic. Although there are significant uncertainties on the marginal abatement costs of carbon reduction (see Gillingham and Stock (2018) for an introduction on the topic), we make the assumption that at the time scale of carbon market emissions phase, less than ten years, these costs are known and constant. Firms can trade the carbon emissions allowances provided by the regulator, and emissions market imperfection is taken into account by an impact on the price of carbon. The key feature is that each firm is endowed with a bank account, as in the cited papers by Kollenberg and Taschini. The bank account position is the result of the allocation received, the traded permits minus the realised emissions. Further, firms face a terminal quadratic penalty on their bank position at the end of the regulated period - this is a strong incentive to achieve emission reduction. There are quite large uncertainties on the damage cost function induced by carbon emissions (see Hsiang and Kopp's and Auffhammer's papers for a thorough introduction to the uncertainties in carbon emissions damage function). Nevertheless, the convex terminal penalty can be interpreted as a proxy of the future damage costs induced by non-compliant emissions.

Then, the Leader (authority) comes into play. The objective of the regulator is to achieve a given expected carbon emission reduction while minimizing the social cost. The latter is the expected total cost from abatement, trading and terminal penalty. Available dynamic allocation processes are quite general, as they are chosen in the space of (sufficiently integrable) semimartingales. In our context, such processes are the sum of a bounded variation part, with possible jumps, and of a (Brownian) martingale.

As far as methodology is concerned, carbon emissions equilibrium price is usually tackled using forward-backward stochastic differential equations as illustrated in Carmona, Fehr, and Hinz (2009); Carmona et al. (2010); Carmona and Hinz (2011); Carmona et al. (2013) and more recently in Chassagneux, Chotai, and Muûls (2017) book. Further, stochastic dynamic Stackelberg games are typically solved using Bellman's principle and coupled Hamilton Jacobi Bellman equations. On this, the reader is referred to Bensoussan, Chen, and Sethi (2014) for a survey on the problem and illustrations in the linear quadratic case.

Instead, our approach relies on the classic stochastic maximum principle[1]. Since the objective functions in the present framework are linear quadratic, they are Fréchet differentiable over the space of square integrable processes. Moreover, this convex optimization is unconstrained. Such regularity allows for a first order expansion, which leads to open loop expressions for the equilibrium price and optimal controls for both regulator and firms. For a given allocation, the optimal abatement effort for firms equates the marginal abatement cost to a martingale, namely the (conditional expectation (C.E.) process generated by the) marginal penalty as expected. The martingale property of the optimal abatement effort holds irrespectively of the specific allocation, which is a rather surprising fact. This is the content of Theorem 6.

In the subsequent Theorem 7 we obtain in closed form a unique market equilibrium, reached at the trading rates that clear the market. The corresponding equilibrium price $\hat{P}$ is a martingale given by the conditional expectation of the (average of) marginal penalties. Its dynamics are driven by the C.E. of allocations minus emission shocks: the higher the allocations, the lower the market price, and conversely for shocks. Further, we show that the same methodology can be applied to different situations, like risk-averse firms or alternative

[1]The interested reader can consult e.g. Peng (1990) (and the references there listed, which include pioneering works by Bismut). In a financial setup, Duffie's utility maximization via stochastic gradient methods is also based on the first order expansion of the objective function, see Duffie (2001).

terminal penalty function. These results are obtained in great generality, with no Markovianity assumption whatsoever on the controls, or on the form of the equilibrium price. This should be contrasted with Kollenberg and Taschini (2016), where the authors find the market equilibrium under such hypotheses and only for a specific, absolutely continuous allocation mimicking the MSR mechanism.

Then, in Theorem 8 we characterize (and exhibit) the optimal dynamic policies for the regulator. To our knowledge, this is the main novelty of our approach. We show that offsetting shocks in the emissions is indeed the optimal choice for the regulator. As a consequence of an optimal allocation policy then, the market equilibrium price is constant, together with the abatement efforts. In the absence of market frictions, the optimal policy is not unique, since the explanatory process is in fact the C.E. of the cumulative allocation A_T over the regulatory period $[0, T]$. And, clearly, a cumulative allocation A_T can be given to firms during the regulatory period in many ways. A simple example of optimal dynamic policy consists in, first, debiting the firms accounts by the level of the desired emission reduction, i.e., the regulator fixes a *reduction target* per firm. Then, the regulator credits them at the business-as-usual emission rate. By acting in this way, the regulator *hedges* or *protects* firms against heavy adjustment costs due to economic shocks that can't be controlled by the firms. In the presence of market frictions, the carbon equilibrium price is still constant as well as the abatement efforts. But, now the optimal allocations are unique and more surprisingly, there is no more trading. Indeed, because the dynamic allocations of allowances is a costless mechanism, it allows the regulator to circumvent the use of an imperfect costly exchange mechanism. This result owes to the perfect knowledge of the regulator of every shock affecting firms.

Compared to a simple pure tax mechanism, optimal stochastic dynamic allocation shares the same abatement cost compared to a pure proportional tax system. But they present different terminal penalty costs: we show that as long as the penalty in the cap-and-trade mechanism is large enough it outperforms the tax system. Threatening firms of heavy penalty is enough to ensure reductions efforts and when they do so, there is little to no penalty to collect. Further, we develop a comparison with a static allocation mechanism as implemented in the first phases of the EU ETS. It turns out that the dynamic allocation outperforms static ones only in the presence of adjustment costs, business cycles and, in particular, of common shocks. If any of these three features is absent, there is no benefit from the implementation of a dynamic allocation mechanism.

The paper is organized as follows. Section 17.2 illustrates the methodology. Section 17.3 describes the underlying model and formalizes the firms and regulator's problems, thus introducing the associated Stackelberg game. Section 17.3.1 deals with the firm individual optimization problem for a fixed allocation process and an exogenous market price process, in the presence of transaction costs. Section 17.3.2 provides the market equilibrium price and firms controls. Section 17.3.3 solves the regulator's problem and Section 17.4 concludes by summarizing the results and suggesting directions for further developments. Finally, the Appendix contains technical results and parts of the proofs of the main Theorems.

17.2 Optimization via the stochastic gradient method

17.2.1 Stochastic first order conditions

Gradient annihilation is one of the main pillars in convex optimization. Indeed, if a convex function $\psi : U \to \mathbb{R}$ with U open in $\mathbb{R}^n$ is differentiable at a point x_0 and the gradient $\nabla\psi(x_0) = 0_{\mathbb{R}^n}$, then x_0 is a global minimum for f on U.

The same result holds in a more abstract setting.[2] Consider a normed space $(\mathbb{E}, \|\cdot\|_{\mathbb{E}})$ and fix an open set $\mathbb{O}$. The concept of differentiability carries over in the following way. A map $\psi : \mathbb{O} \to \mathbb{R}$ is said to be Fréchet differentiable at the point $e_0 \in \mathbb{O}$ if:

$$\lim_{e \to e_0} \frac{|\psi(e) - \psi(e_0) - \phi(e - e_0)|}{\|e - e_0\|_{\mathbb{E}}} = 0$$

for some linear, continuous functional $\phi : \mathbb{E} \to \mathbb{R}$. In case, ϕ is called the Fréchet differential of ψ at e_0, which we also write as ϕ_{e_0}. If ψ is convex on $\mathbb{O}$, differentiable at e_0 and ϕ_{e_0} is null, then e_0 is a global minimum for ψ.

Spaces of square integrable processes In the present paper we deal with linear quadratic problems, so the focus is on a specific space $\mathbb{E}$ of square integrable functions, that is the space of $d\mathbb{P} \otimes dt$ - square integrable, adapted processes X on a given stochastic basis. Such space is here denoted by $\mathcal{L}^2_T$, and it is endowed with the scalar product:

$$\langle X, Y\rangle = \mathbb{E}\left[\int_0^T X_t Y_t \, dt\right].$$

The scalar product induces a complete norm:

$$\|X\|_2 := \left(\mathbb{E}\left[\int_0^T X_t^2 \, dt\right]\right)^{\frac{1}{2}},$$

which turns $\mathcal{L}^2_T$ in a Hilbert space. The Riesz Representation Theorem ensures that every continuous, linear functional ϕ on $\mathcal{L}^2_T$ can be represented via the scalar product with a suitable element $X^\phi \in \mathcal{L}^2_T$, namely:

$$\phi(X) = \mathbb{E}\left[\int_0^T X_t X_t^\phi dt\right], \text{ for all } X \in \mathcal{L}^2_T.$$

Therefore, given a differentiable map $\psi : \mathcal{L}^2_T \to \mathbb{R}$, its differential ϕ_{X^0} at any point X^0 can be written as:

$$\phi_{X^0}(X) = \mathbb{E}\left[\int_0^T X_t X_t^{(\phi_{X^0})} \, dt\right] \text{ for all } X \in \mathcal{L}^2_T,$$

with a representing process $X^{(\phi_{X^0})} \in \mathcal{L}^2_T$, called *the stochastic gradient* of ψ at X^0. Hence, if in addition ψ is convex and for some X^* the differential $\phi_{X^*} = 0$, or alternatively the gradient verifies: $X^{\phi_{X^*}} = 0$, then X^* is a global minimum for ψ. The First Order Condition, FOC, becomes then necessary and sufficient for identifying the optimizer of smooth functions. As usual, the optimizer is unique in case ψ is strictly convex. Let us see an easy example first.

Example 17.2.1 *Consider on $\mathcal{L}^2_T$ the real-valued, strictly convex map:*

$$\psi(X) = \mathbb{E}\left[\int_0^T h_t X_t + \frac{1}{2\eta} X_t^2 dt\right]$$

[2] A standard reference book is Aliprantis and Border (2007).

with $h \in \mathcal{L}^2_T, \eta \in \mathbb{R}_+$. *It is easy to show that, for every* X^0, *the differential of* ψ *is:*

$$\phi_{X^0}(X) = \mathbb{E}\left[\int_0^T X_t \left(h_t + \frac{1}{\eta}X^0_t\right) dt\right],$$

where in between parentheses we have isolated the gradient process. The minimizer is the unique process at which the gradient is zero:

$$h_t + \frac{1}{\eta}X^*_t = 0, \quad d\mathbb{P} \otimes dt - a.e.$$

The above FOC admits the solution $X^*_t = -\eta h_t$.

Remark 17.2.1 *It is common to work with convex maps* ψ *defined on the whole* $\mathbb{E}$, *which are not necessarily finite everywhere. A minimal requirement is that* $\psi(e_0) \in \mathbb{R}$ *for some* $e_0 \in \mathbb{E}$, *and that it never takes the value* $-\infty$: *that is,* ψ *is required to be* proper. *In case, let*

$$\mathbb{D} := \{e \mid \psi(e) \in \mathbb{R}\}$$

denote the proper domain of ψ. *The gradient method can still be applied, in the following version. Suppose that the interior of* $\mathbb{D}$ *is not empty:* $int(\mathbb{D}) \neq \emptyset$ *and that* ψ *is differentiable in some open set* $\mathbb{O} \subseteq int(\mathbb{D})$. *If for some* $e^* \in \mathbb{O}$ *the differential* $\phi_{e^*} = 0$, *then* e^* *is a global minimum of* ψ. *In case* ψ *is strictly convex,* e^* *is the unique minimizer.*

In the following, for illustrative purposes we place the FOC methodology next to stochastic control and duality methods in the resolution of the Merton's utility maximization problem.

17.2.2 First order conditions in portfolio selection

This Subsection is slightly more delicate from a functional analysis view point. Indeed, when portfolio selection is made according to the mean-variance criterion, it is classically formulated on $\mathcal{L}^2_T$. But the optimization domain depends on the utility function, so that other Banach spaces may be the natural choice when we leave a quadratic utility setup. Here, we use the FOC method to solve the Merton utility maximization problem, formulated on a suitable Banach space. Let's start.

An investor would like to maximize their utility from terminal wealth at a given horizon T, by investing in two given assets, a stock and a bond. The stock S evolves as a Geometric Brownian Motion on the Wiener space generated by a Brownian motion W:

$$dS_t = \mu S_t dt + \sigma S_t dW_t \quad t \in [0, T].$$

We work with discounted quantities, so that the bond B is constant: $B_t = 1$ on $[0, T]$. Starting with x as initial endowment, a self-financing portfolio is determined by the number of shares H_t held at time t, with associated wealth process:

$$X_t = x + \int_0^t H_s dS_s \quad t \in [0, T].$$

By a change of variable, we pass from the number of shares H to the wealth w invested in the stock, namely

$$w_t := H_t S_t,$$

and the portfolio value X rewrites as $X_t = X_t(w) = x + \int_0^t w_s(\mu ds + \sigma dW_s)$. To fix the ideas, suppose that agent maximizes the expected exponential utility from their terminal wealth $X_T = X_T(w)$:

$$u(w) := \mathbb{E}\left[-e^{-X_T(w)}\right],$$

on a suitable domain of wealth processes, which are:

$$\mathcal{L}_T^\infty := \left\{ w \text{ adapted } \mid \sup_{t\in[0,T]} |w_t| \leq c \text{ for some constant } c \right\}$$

that is the Banach space of bounded, adapted processes, endowed with the supnorm: $\|w\|_\infty := \sup_t \|w_t\|_\infty$. By flipping a sign, we aim to solve

$$v(0,x) := \min_{w\in\mathcal{L}_T^\infty} -u(w) = \min_{w\in\mathcal{L}_T^\infty} \mathbb{E}\left[e^{-X_T(w)}\right].$$

To be consistent with the notation of the previous Subsection, let $\psi := -u$. Exponentials may pose integrability problems, but the next Lemma comes to help.

Lemma 2 *The function ψ: $\psi(w) = \mathbb{E}\left[e^{-X_T(w)}\right], w \in \mathcal{L}_T^\infty$ has the following properties.*

1. *It is real valued and convex.*
2. *It is differentiable at any point w^0, with differential map given by:*

$$\phi_{w^0}(v) = \mathbb{E}\left[-e^{-X_T(w^0)} \int_0^T v_t(\mu dt + \sigma dW_t)\right].$$

3. *It admits a unique minimizer w^*, the constant $w_t^* = \frac{\mu}{\sigma^2}$. The minimum is:*

$$v(0,x) = e^{-x-\frac{\mu^2}{2\sigma^2}T}.$$

Proof.

1. Convexity is clear, so we only have to show finiteness. The next inequality relies only on boundedness of w:

$$\psi(w) = \mathbb{E}\left[e^{-X_T(w)}\right] \leq e^{|x|+|\mu|\|w\|_\infty T}\,\mathbb{E}\left[e^{-\int_0^T \frac{\sigma^2}{2}w_t^2 dt - \int_0^T \sigma w_t dW_t}\right].$$

$$\psi(w) = \mathbb{E}\left[e^{-X_T(w)}\right] \leq e^{|x|+|\mu|\|w\|_\infty T}\,\mathbb{E}\left[e^{-\int_0^T \frac{\sigma^2}{2}w_t^2 dt - \int_0^T \sigma w_t dW_t}\right].$$

If we show that the last expectation is finite, we conclude the proof. Consider the process $Y_t(w) := e^{-\int_0^t \frac{\sigma^2}{2}w_s^2 ds - \int_0^t \sigma w_s dW_s}$, which by construction is a local martingale. Since it nonnegative, it is also a supermartingale (see e.g., Cohen and Elliott, 2015, Lemma 5.6.8) and then $\mathbb{E}[Y_T(w)] \leq 1$. Alternatively one can verify the Novikov condition for the Doleans exponential Y to be a martingale (Karatzas and Shreve, 2000, Proposition 3.5.12), and get $\mathbb{E}[Y_T(w)] = 1$.

2. Differentiability at any w^0 follows from a Taylor expansion. Indeed we have:

$$\begin{aligned}\psi(w^0+v) - \psi(w^0) &= \mathbb{E}\left[e^{-X_T(w^0)}\left(e^{-\int_0^T v_t(\mu dt+\sigma dW_t)} - 1\right)\right] \\ &\sim \mathbb{E}\left[-e^{-X_T(w^0)} \int_0^T v_t(\mu dt + \sigma dW_t)\right] \\ &= \phi_{w^0}(v).\end{aligned}$$

3. Annihilating the differential map ϕ_{w^0} corresponds to finding a martingale measure for the stock. Indeed, the differential map at w^0 is null if $\phi_{w^0}(v) = 0$ for all v, which means that the probability $\mathbb{Q}$ with density:

$$\frac{d\mathbb{Q}}{d\mathbb{P}} := \frac{e^{-X_T(w^0)}}{\mathbb{E}\left[e^{-X_T(w^0)}\right]} \tag{17.1}$$

gives zero expectation to all stochastic integrals wrt to the process $\tilde{W}_t = \frac{\mu}{\sigma}t + W_t$:

$$\mathbb{E}^{\mathbb{Q}}\left[\int_0^T v_t\, d\tilde{W}_t\right] = 0 \quad \text{for all } v.$$

In particular, consider all the vs of the type: $v_t = I_A I_{]s_1,s_2]}(t)$ with arbitrary $A \in \mathcal{F}_{s_1}$ and $s_1 < s_2$ in $[0,T]$. The condition (17.1) reads:

$$\mathbb{E}^{\mathbb{Q}}\left[\int_0^T I_A I_{]s_1,s_2]}(t)(\mu dt + \sigma dW_t)\right] = 0$$

which is equivalent to:

$$\mathbb{E}^{\mathbb{Q}}\left[I_A \tilde{W}_{s_2}\right] = \mathbb{E}^{\mathbb{Q}}\left[I_A \tilde{W}_{s_1}\right]$$

Letting A and s_1, s_2 vary, $\tilde{W}$ must be a martingale under $\mathbb{Q}$.
By the Girsanov theorem then, such $\mathbb{Q}$ is unique and it is given by:

$$\frac{d\mathbb{Q}}{d\mathbb{P}} = e^{-\frac{\mu}{\sigma}W_T - \frac{\mu^2}{2\sigma^2}T}. \tag{17.2}$$

The latter expression, together with (17.1), gives a unique, constant solution w^*: $w^*_t = \frac{\mu}{\sigma^2}$. Note that S is also a martingale under $\mathbb{Q}$ by the Novikov condition. [3] None

For completeness, we *briefly* sketch other two solution methods, namely stochastic control and duality. For a detailed presentation, we refer to the books by Pham (2009) on stochastic control and to Duffie (2001) for the convex duality approach. For simplicity we take as space of admissible strategies $\mathcal{L}^2_T$, since it is not necessary anymore that the proper domain of ψ has non empty interior as in the FOC method.

1. Stochastic control/dynamic programming. We first make guess for the value function $v(t,x)$ of the dynamic version of problem:

$$v(t,x) = \min_{w \in \mathcal{L}^2_T} \mathbb{E}\left[e^{x + \int_t^T w_s(\mu ds + \sigma dW_s)}\right],$$

starting at a generic $t \in [0,T]$ with a generic initial wealth x. An educated choice is:

$$v(t,x) := a(t)e^{-x} - 1$$

with a deterministic and such that $a(T) = 1$. The corresponding Hamilton-Jacobi-Bellman equation is:

$$\inf_{w \in \mathbb{R}} \left\{ v_t + \mu v_x w + \frac{1}{2}\sigma^2 v_{xx} w^2 \right\} = 0$$

Substituting the candidate value function in the equation, after simple calculations one gets $a(t) = e^{-\frac{\mu}{2\sigma^2}(T-t)}$ and the optimal wealth is $w^*_t = const = \frac{\mu}{\sigma^2}$.

[3]The martingale measure, or better the density process $\mathbb{E}_t[\frac{d\mathbb{Q}}{d\mathbb{P}}] := Z_t$, is an element of the dual of $\mathcal{L}^\infty_T$, with duality induced by the scalar product $\langle Z, w\rangle = \mathbb{E}[\int_0^T Z_t w_t\, dt]$. This dual space contains the adapted processes which belong to L^1 for all t, like Z, but its full description is outside the scopes of the present chapter.

2. Convex Duality. The basic idea is to associate to $v(0, x)$ a problem which is easier to solve. This so-called dual problem in the present case admits a unique optimizer, which is a scalar multiple λ of the unique martingale measure density for S given by (17.2). Now, the optimality conditions give that the terminal utility is exactly:

$$e^{-X_T^*} = \lambda \frac{d\mathbb{Q}}{d\mathbb{P}}$$

This relation has an important financial interpretation. By rewriting it as $\frac{d\mathbb{Q}}{d\mathbb{P}} = \frac{1}{\lambda}(-e^{-X_T^*})'$, it is an instance of pricing by marginal utility (calculated at the optimal terminal wealth). Then,

$$X_T^* = -\ln \lambda + \frac{\mu}{\sigma} W_T + \frac{1}{2}\frac{\mu^2}{\sigma^2} T. \tag{17.3}$$

To find w^*, we recall that the wealth processes are all martingales under $\mathbb{Q}$, and the filtration generated by W is the same as that generated by the $\mathbb{Q}$-Brownian motion $W_t^{\mathbb{Q}} = W_t - \frac{\mu}{\sigma} t$. So, by the Predictable Representation Property, we can always write:

$$X_T^* = x + \int_0^T w_s^* \sigma dW_s^{\mathbb{Q}},$$

for a unique integrand w^*. Since the constant $w_t^* = \frac{\mu}{\sigma^2}$ does the job, we conclude the proof.

17.3 Stochastic regulation of emissions

The regulation of carbon allowances is quite complex. In particular, it occurs over several periods, and in addition allowances can be banked from one period to the other. We abstract from these features and focus on a single period of T years at the end of which compliance is assessed, as in Carmona et al. (2010, 2013), Kollenberg and Taschini (2016, 2019), Fell and Fell and Morgenstern (2010).

A regulator (the Leader) wishes to reduce the pollution produced by a set of N firms (Followers) over a period $[0, T]$. To this end, she dynamically allocates carbon permits to the firms. Given the received allocation, each firm minimizes its costs from reduction, trade and terminal penalty – till the system reaches a market equilibrium. Then, the regulator minimizes optimal social cost over possible allocations. The stochastic model formalization goes as follows.

Consider a filtered probability space $(\Omega, (\mathcal{F}_t)_{0 \leq t \leq T}, \mathbb{P})$, in which the filtration is the augmented Brownian filtration generated by a standard $N+1$ dimensional Brownian motion $(\widetilde{W}^0, \widetilde{W}^1, \ldots, \widetilde{W}^N)$. Fix further correlation factors $(k_1, \ldots, k_N)$ and let

$$W_t^i = \sqrt{1 - k_i^2}\, \widetilde{W}_t^i + k_i \widetilde{W}_t^0. \tag{17.4}$$

In particular, the correlation between W^i, W^j is $r_{ij} := k_i k_j$. The index notation we use is self-explanatory, in the sense that Y_t^j denotes a process at time t, while $Y_j, y_j, y(t)$ refer to constants or deterministic functions.

The Business As Usual (BAU) cumulative emissions dynamics for firm i are assumed to grow at a constant average rate $\mu_i > 0$ and are prone to Gaussian economic shocks with standard deviation rate $\sigma_i \geq 0$. Thus, at initial time, the cumulative emissions to be observed at time T are simply given by a Gaussian distribution with mean $\mu_i T$ and variance $\sigma_i^2 T$. Thus, the realized cumulative emissions up to time t can be modeled as a Brownian motion with drift:

$$E_t^i = \mu_i t + \sigma_i W_t^i, \quad 0 \leq t \leq T. \tag{17.5}$$

This relation states that compared to the expected average trend $\mu_i t$ the firm i has experienced economic shocks that lead to a cumulative emissions of E_t^i. If $W_t^i < 0$ (resp. $W_t^i > 0$), it means that cumulative emissions are below (resp. above) the expected trend. Furthermore, in the noise decomposition (17.4), we see that the emission of firm i is affected by its own idiosyncratic noise $d\widetilde{W}^i$ and by the common economic business cycle $d\widetilde{W}^0$. The modeling of cumulative emissions with the dynamics above should be understood as a decomposition between a deterministic trend and the economic shocks that affect them. Without carbon capture, cumulative emissions are always increasing. But what matters here is if they grow faster or slower than a given deterministic trend.

In the BAU case, the expected total emission over the period T is $\mathbb{E}[E_T] = N\bar{\mu}T$ where $E_t = \sum_{i=1}^N E_t^i$ and $N\bar{\mu} = \sum_{i=1}^N \mu_i$. The firms can control their emissions, e.g., by changing production technology. Under abatement effort rate α_t^i, the firm i emission becomes:

$$E_t^{i,\alpha^i} = \int_0^t (\mu_i - \alpha_s^i) ds + \sigma_i W_t^i.$$

Thus, the firm controls their emission trend, which increases at a rate $\mu_i - \alpha_t^i$, while the volatility is uncontrolled. The regulation wishes to reduce the expected emissions to

$$\rho T N \bar{\mu}, \quad 0 < \rho < 1,$$

namely, to achieve $1 - \rho$ percent reduction of the BAU case. The regulator has several instruments at her disposal (taxes, quotas) but she wishes to implement a dynamic cap–and–trade system working in this way. At $t = 0$ she opens a bank account X^i for each firm and provides incentives in the form of permits. The process $\tilde{A}^i$ is defined as the cumulative permits process, namely $\tilde{A}_t^i$ is the sum of all the permits given up to time t to firm i. So, the i-th bank follows the dynamics:

$$dX_t^i = \beta_t^i dt + d\tilde{A}_t^i - dE_t^{i,\alpha^i}, \quad dE^{i,\alpha} = (\mu_i - \alpha_t^i)dt + \sigma_i dW_t^i. \tag{17.6}$$

In the above, β^i is the trading rate in the allowances market. Since one allowance corresponds to one ton of pollutant, the emissions, trade and banks are measured in tons.

Assumptions on Followers and Leader's controls. The firms controls are square integrable wrt $d\mathbb{P} \otimes dt$, namely they belong to $\mathcal{L}_T^2 := L^2(\Omega \times [0,T], (\mathcal{F}_t)_t, \mathbb{P} \otimes dt)$. The allocation process $\tilde{A}^i$ is a square integrable semimartingale, that is $\mathbb{E}[(\tilde{A}_t^i)^2] < \infty$ for all t, decomposable into a square integrable finite variation part F and a square integrable stochastic integral:

$$\tilde{A}_t^i = F_t^i + \sum_{j=0}^N \int_0^t \tilde{b}_s^j d\widetilde{W}_s^j,$$

for some appropriate adapted integrands $\tilde{b}^j$. *We do not require that the finite variation part F of $\tilde{A}^i$ is absolutely continuous wrt dt.* Therefore, the regulator is free e.g., to allocate (credit or debit) permits at discrete instants, which can be fixed or stopping times, and/or at a rate

$\tilde{a}$. In addition, the (stochastic) Lebesgue Decomposition Theorem allows the decomposition of F into the sum of an absolutely continuous part with respect to the Lebesgue measure, and of singular part $\tilde{S}^i$. So,

$$\tilde{A}^i_t = \tilde{S}^i_t + \int_0^t \tilde{a}^i_s ds + \sum_{j=0}^{N} \int_0^t \tilde{b}^j_s d\widetilde{W}^j_s. \tag{17.7}$$

Each of the three addenda can be null. As already mentioned, the singular $\tilde{S}^i$ can be e.g., a pure jump part. Namely, a weighted sum of Dirac deltas along a targeted (stochastic) time grid, in which the regulator provides/cancels allowances in a discrete way:

$$\tilde{S}^i_t = q^i_1 I_{[t_1,T]}(t) + q^i_2 I_{[t_2,T]}(t) \dots$$

where the dates t_h form an increasing sequence of stopping times and $q^i_{t_h} \in \mathcal{F}_{t_h}$. This covers the case in which there is an initial allocation. In fact, it is enough take $t_1 = 0$ in the grid and $X^i_0 = \tilde{S}^i_0 = q^i_1$.

An important point is the interpretation of the sign of the various $\tilde{A}^i$. When the sign of the allocation is positive, the regulator is really giving permits to the firm. When the sign is negative, the regulator is placing a penalty on the firm bank account. This penalty may be different for every firm, depending on their induced efforts and various cost parameters. However, as soon as the permits across the system, $\tilde{A} = \sum_i \tilde{A}^i$, verify:

$$\tilde{A}_T > 0, \tag{17.8}$$

the regulator is indeed going to distribute permits contracts in $(0, T]$. This ensures the presence of a market of permits, which then are traded in $(0, T]$ according to the equilibrium induced by the specific allocation strategy. We will illustrate this aspect when analyzing the regulator's optimal policy in the conclusive step of the Stackelberg regulator-firms game in Section 17.3.3.

Given an allocation scheme, the bank X^i of firm i gives at each t the net cumulative position on $[0, t]$ of the firm in terms of emissions, abatement, allowances trading and allowances endowment by the regulator. A positive economic shock to the emissions induces a decrease in the bank accounts, while an increase in the allocations makes the accounts grow. The bank dynamics can be rewritten as

$$dX^i_t = dA^i_t + \left(\alpha^i_t + \beta^i_t\right)dt - \sigma_i dW^i_t. \tag{17.9}$$

in which A is the net allocation process over the trend

$$A^i_t = \tilde{A}^i_t - \mu_i t.$$

Since our results on the optimal regulation depend only on A^i, *the trend rate μ_i does not necessarily have to be constant.* Here, we set it constant for ease of presentation. For future use, note that

$$\mathbb{E}\left[A^i_T\right] = \mathbb{E}\left[\tilde{S}^i_T\right] + \mathbb{E}\left[\int_0^T (\tilde{a}^i - \mu_i)dt\right].$$

For a given market price of permits P and a given (net) allowance scheme A^i, the firm i aims to solve its cost minimization problem:

$$\inf_{\alpha^i,\beta^i} J^i(\alpha^i, \beta^i) := \inf_{\alpha^i,\beta^i} \mathbb{E}\left[\int_0^T \left(c_i(\alpha^i_t) + P_t\beta^i_t + \frac{1}{2\nu}(\beta^i_t)^2\right)dt + \lambda(X^i_T)^2\right], \tag{17.10}$$

in which three terms of cost appear: the cost from the intertemporal abatement c_i, the cost of trade and the terminal penalty on the bank.

The abatement cost function c_i is supposed to be quadratic

$$c_i(\alpha) = h_i\alpha + \frac{1}{2\eta_i}\alpha^2 \quad h_i, \eta_i > 0 \ ,$$

so to consider both the linear and the adjustment costs, the latter proportional to the square of the abatement rate. This choice is in line with the literature of carbon emission reduction (see Gollier (2020) and reference therein). Further, the linear-quadratic form captures the non-decreasing feature of marginal abatement cost. From an investment point of view, it means that there is some irreversibility in the abatement decision. The higher the values of η_i, the higher the flexibility of the abatement process and thus the higher the reversibility of the decision.

On trading costs, we take into account a price impact effect as in the original Kyle (1985) model , with constant market depth parameter $\nu > 0$. In 2022 the value of the bid–ask on the December contract was around 2 c€/t for a quoted carbon price around 30 €/t[4], which makes the transaction cost less than 0.06%. For the simpler case of regulation without frictions, we refer to Aïd and Biagini (2023).
The term $\lambda(X_T^i)^2$, with $\lambda > 0$ equal for all firms, is the terminal monetary penalty on the bank accounts set by the regulator. The firm is going to pay both if its bank is above or below the compliance zero level. It is a regularized version of the terminal (cap) penalty function applied in practice, which is zero if the firm is compliant and linear otherwise.

The objective of the regulator is to design net dynamic allocation schemes $\mathbf{A} = (A^1, \ldots, A^N)$ to reduce expected emission, while minimizing social cost. This is defined as the sum of the firms individual costs, so that

$$\inf_{\mathbf{A}\in\mathcal{S}^{\mathbf{N}}} \mathbb{E}\Big[\sum_{i=1}^{N}\int_0^T \Big(c_i(\alpha_t^i) + \beta_t^i P_t + \frac{1}{2}\frac{(\beta_t^i)^2}{\nu}\Big)dt + \lambda(X_T^i)^2\Big], \quad \mathbb{E}\Big[\sum_{i=1}^{N} E_T^{i,\alpha^i}\Big] = \rho T N\bar{\mu}. \tag{17.11}$$

in which $\mathcal{S}^N$ is the N-dimensional product space of square integrable semimartingales on our filtered space.

The problem of the regulator and firms falls in the category of dynamic stochastic Stackelberg games. In fact, the regulator aims at minimizing social cost from optimal firms reduction and trade. Thus, they act as a Leader while firms act as Followers (see Bensoussan, Chen, and Sethi (2014) for a survey).

In Section 17.3.1, we find the solution of the single firm problem, while Sections 17.3.2 and 17.3.3 tackle the market equilibrium and the optimal dynamic regulation.

17.3.1 The firm optimal response

The focus here is on the single firm cost minimization, for a given exogenous allowances price P and a net allocation scheme A^i. In Section 17.3.2 however, the price process will be the endogenous market equilibrium price. We now define

$$M_t^i := \mathbb{E}_t\Big[A_T^i\Big], \quad R_t^i := \mathbb{E}_t\Big[A_T^i - A_t^i\Big], \quad g_i(t) := \frac{2\lambda\eta_i}{1 + 2\lambda(\eta_i + \nu)(T-t)}, \tag{17.12}$$

in which:

[4]Source: Thomson-Reuters Refinitiv quotations of the EU EUA December contract.

- M^i is the martingale closed by A^i_T. It gives, at time t, the (conditional) expectation of the firm endowment A^i_T on $[0,T]$, net over trend. *Note that different intertemporal allocations, which share the same cumulative value on the regulatory horizon* $[0,T]$, *give rise to the same* M^i. From the definition of A^i, M^i, in particular we have:

$$M^i_0 = \mathbb{E}\big[A^i_T\big] = \mathbb{E}\big[\tilde{A}^i_T - \mu_i T\big].$$

- the process R^i gives, at time t, the conditional expectation of the *residual* net allocation on $[t,T]$.

As anticipated in the Introduction, the firm minimization problem can be tackled by means of a first order Fréchet expansion. In our case, the functional $J^i : \mathcal{L}^2_T \times \mathcal{L}^2_T \to \mathbb{R}$ is linear-quadratic:

$$J^i(\alpha^i, \beta^i) = \mathbb{E}\Big[\int_0^T \Big(c_i(\alpha^i_t) + P_t\beta^i_t + \frac{1}{2\nu}(\beta^i_t)^2\Big)dt + \lambda(X^i_T)^2\Big],$$

coercive in the controls, strictly convex and smooth. Therefore, as detailed in the beginning of Section 2 and Example 17.2.1, the unique optimal solution is found via the FOCs.

Theorem 6 *For a given cumulative, net allocation scheme* A^i *and for a given, exogenous, allowances price* $P \in \mathcal{L}^2_T$, *the i-th firm cost minimization problem* (17.10) *has a unique solution* $(\hat{\alpha}^i, \hat{\beta}^i)$ *in* $\mathcal{L}^2_T \times \mathcal{L}^2_T$.
(i) *The optimal abatement* $\hat{\alpha}^i$ *has the following dynamics:*

$$\begin{cases} \hat{\alpha}^i_t &= -g_i(t)\Big(d(M^i_t - \sigma_i W^i_t) + d\mathbb{E}_t\Big[\int_0^T \nu(h_i - P_s)ds\Big]\Big) \\ \hat{\alpha}^i_0 &= -g_i(0)\Big(\frac{1}{2\lambda}h_i + M^i_0 + \mathbb{E}\Big[\int_0^T \nu(h_i - P_t)dt\Big]\Big), \end{cases} \tag{17.13}$$

and is therefore a martingale.
(ii) *The optimal trade* $\hat{\beta}^i$ *is given by:*

$$\hat{\beta}^i_t = \nu\Big(h_i + \frac{\hat{\alpha}^i_t}{\eta_i} - P_t\Big). \tag{17.14}$$

(iii) *Both optimal controls can be rewritten as:*

$$\hat{\alpha}^i_t = \hat{\alpha}^i_t(\hat{X}^i_t) = -g_i(t)\Big(\frac{h_i}{2\lambda} + \hat{X}^i_t + R^i_t + \mathbb{E}_t\Big[\int_t^T \nu(h_i - P_s)ds\Big]\Big), \tag{17.15}$$

$$\hat{\beta}^i_t = \hat{\beta}^i_t(\hat{X}^i_t) = \nu\Big(h_i + \frac{\hat{\alpha}^i_t(\hat{X}^i_t)}{\eta_i} - P_t\Big). \tag{17.16}$$

Proof. The proof is given in Section 17.5.1. We anticipate the FOCs written on the gradient of J^i, as they will be explicitly referred to in the rest of the paper. They are

$$h_i + \frac{\alpha^i_t}{\eta_i} + 2\lambda\mathbb{E}_t\big[X^i_T\big] = 0, \tag{17.17}$$

$$P_t + \frac{\beta^i_t}{\nu} + 2\lambda\mathbb{E}_t\big[X^i_T\big] = 0. \tag{17.18}$$

□

Let us comment on these findings.

(i) The FOCs can be also seen as

$$c'(\alpha_t^i) = -2\lambda \mathbb{E}_t\big[X_T^i\big], \quad \beta_t^i = \nu\big(c'(\alpha_t^i) - P_t\big).$$

The marginal abatement cost is equal to the opposite of the (C.E. of the) marginal penalty, and so is the marginal cost of trading, $P_t + \beta_t^i/\nu$. And, consistently with economic intuition, the firm buys (resp. sells) if its marginal abatement cost is higher (resp. lower) than the market price.

(ii) The optimal abatement $\hat{\alpha}^i$ is a martingale. In fact, it is a stochastic integral, with a bounded, deterministic integrand $-g_i$, of *three explanatory martingales:* M, namely the C.E. of cumulative net allocation over $[0,T]$; the C.E. of the integrated price; and the emission noise $\sigma_i W^i$. A fortiori it is not the full intertemporal structure of the net allocation A^i which matters, as it appears in (17.7). The key quantity here is M.
The firm compares the dynamics of the expectation of what will be given during the whole regulatory period, A_T^i, to noise and to integrated price, and then makes the decision on effort. If there is a positive economic shock dW_t^i everything else being equal, the firm effort increases. It decreases if dM_t^i is positive, i.e., if the firm anticipates an increase in total expected net allocation.
The integrand g_i in the martingale representation for $\hat{\alpha}^i$ depends on the following parameters: the individual firm adjustment cost of abatement η_i, the common penalty coefficient λ and market depth coefficient ν.
When written in the form (17.15), $\hat{\alpha}^i$ depends not only on the bank state, but also on the C.E. of net residual allocation R^i at time t, $R_t^i = \mathbb{E}_t[A_T^i - A_t^i]$, and on the C.E. of the (residual) integrated price.
Finally, we remark that martingality of $\hat{\alpha}^i$ would be preserved if the abatement first order cost h_i became a martingale.

(iii) The optimal trade $\hat{\beta}^i$ is not a martingale, unless P is a martingale as well. However, this will be the case at equilibrium (see Section 17.3.2).

Remark 17.3.1 (Absence of frictions) *Absence of market frictions in the carbon market is a common assumption (see Kollenberg and Taschini (2016) or Carmona et al. (2010) and the references within). We can treat the optimal response in a market without frictions similarly to the frictions case in Theorem 6. The problem is however well posed only if P is a martingale. Under martingality of P, the complete proof with the resulting FOCs are obtained from* (17.17) *and* (17.18) *with $\nu = \infty$:*

$$h_i + \frac{\alpha_t^i}{\eta_i} + 2\lambda \mathbb{E}_t\big[X_T^i\big] = 0, \tag{17.19}$$

$$P_t + 2\lambda \mathbb{E}_t[X_T^i] = 0. \tag{17.20}$$

As economic intuition suggests, since P is a martingale now, each firm equates the marginal cost of abatement to the market price P_t. Also, the market price is equal to the conditional expectation of the marginal penalty.

17.3.2 Market equilibrium

Recall that the noises W^i in the firms activity have a quite general dependence structure, as described at the beginning of Section 17.3.1. For a fixed net allocation policy of the regulator $\mathbf{A} = (A^1, \ldots, A^N) \in (\mathcal{L}_T^2)^N$, we find a unique open loop equilibrium. Define the positive, deterministic function

$$\pi_i(t) := \frac{g_i(t)}{\eta_i}\Big(1 - \frac{\nu(T-t)}{N}\sum_{k=1}^{N}\frac{g_k(t)}{\eta_k}\Big)^{-1}, \tag{17.21}$$

where g_i is from (17.12). Market equilibrium consists in finding a price $\hat{P}$ that satisfies the market clearing condition:

$$\sum_{i=1}^{N}\hat{\beta}_t^i(\hat{P}) = 0, \quad \forall t \in [0,T]. \tag{17.22}$$

in which the $\hat{\beta}^i$ are given from the system (17.14). The resulting $\hat{P}$ is then called an *equilibrium price*, and it is described in the following Theorem.

Theorem 7 *For a given net cumulative allocation* $\mathbf{A}$ *the following hold.*

(i) *The equilibrium price* $\hat{P}$ *is given by:*

$$d\hat{P}_t = -\frac{1}{N}\sum_{i=1}^{N}\pi_i(t)\big(dM_t^i - \sigma_i dW_t^i\big), \quad \hat{P}_0 = \frac{1}{N}\sum_{i=1}^{N}\pi_i(0)\Big(\eta_i h_i T - M_0^i\Big). \tag{17.23}$$

and is therefore a martingale.

(ii) $\hat{P}$ *can also be written as*

$$\hat{P}_t = \frac{1}{N}\sum_{i=1}^{N}\pi_i(t)\Big(\eta_i h_i(T-t) - (\hat{X}_t^i + R_t^i)\Big). \tag{17.24}$$

(iii) *The optimal controls* $\hat{\alpha}^i, \hat{\beta}^i$ *are martingales, with dynamics:*

$$d\hat{\alpha}_t^i = -g_i(t)\Big[d(M_t^i - \sigma_i W_t^i) - \nu(T-t)d\hat{P}_t\Big], \tag{17.25}$$

$$\hat{\alpha}_0^i = -g_i(0)\Big[h_i\Big(\frac{1}{2\lambda} + \nu T\Big) + M_0^i - \nu T\hat{P}_0\Big], \tag{17.26}$$

$$d\hat{\beta}_t^i = \nu d(\hat{P}_t - h_i - \frac{1}{\eta_i}\hat{\alpha}_t^i), \quad \hat{\beta}_0^i = \nu\Big(h_i + \frac{\hat{\alpha}_0^i}{\eta_i} - \hat{P}_0\Big). \tag{17.27}$$

(iv) *In addition, the optimal controls admits a feedback form representation:*

$$\hat{\alpha}_t^i(\hat{X}_t^i, R_t^i) = g_i(t)\Big[\nu(T-t)(\hat{P}_t - h_i) - \Big(\frac{h_i}{2\lambda} + \hat{X}_t^i + R_t^i\Big)\Big], \tag{17.28}$$

$$\hat{\beta}_t^i(\hat{X}_t^i, R_t^i) = \nu(\hat{P}_t - \frac{\hat{\alpha}_t^i}{\eta_i} - h_i).$$

Proof. See appendix 17.5.2. □

Let us comment on the above results.

(i) The explanatory processes in (17.23) for the equilibrium price $\hat{P}$ are the martingales $(M^i - \sigma_i W^i)$, $i = 1, \ldots, N$. These martingales are better interpreted if each of them is seen as the C.E. process associated to the terminal variable $M_T^i - \sigma_i W_T^i = A_T^i - \sigma_i W_T^i$, namely the bank account balance that firm i will have at T in the BAU scenario. If these martingales experience a positive shock, the price decreases as it should. A positive shock would mean a positive shock to the bank accounts, which in turn reduces the hedging demand of allowances.

(ii) When the adjustment costs η_i are equal, the deterministic coefficients π_i become equal and can be factorized out in (17.24). The equilibrium price $\hat{P}$ in this case depends then *only on the aggregate quantities*

$$Z_t := \sum_{i=1}^{N}(M_t^i - \sigma_i W_t^i) \quad \text{and} \quad N\bar{h} := \sum_{i=1}^{N} h_i. \tag{17.29}$$

The martingale Z is in fact the C.E. process associated to the aggregate bank terminal balance in the BAU scenario, since

$$Z_T := \left(\sum_{i}^{N} A_T^i - \sum_{i=1}^{N} \sigma_i W_T^i\right).$$

This simplification also occurs in the case with no market frictions, for which we refer to Aïd and Biagini (2023).

(iii) The optimal efforts $\hat{\alpha}^i$ are obtained from (17.13) exploiting the martingality of $\hat{P}$. The trade $\hat{\beta}^i$ keeps its structure as in (17.14).

17.3.3 Optimal dynamic regulation

The regulator faces the social cost minimization problem in (17.11) when the firms play optimally, namely

$$\inf_{\mathbf{A}\in\mathcal{S}^N} \mathbb{E}\Big[\sum_{i=1}^{N}\int_0^T \Big(h_i\hat{\alpha}_t^i + \frac{(\hat{\alpha}_t^i)^2}{2\eta_i} + \frac{1}{2\nu}(\hat{\beta}_t^i)^2\Big)dt + \lambda(\hat{X}_T^i)^2\Big], \quad \mathbb{E}\big[E_T^{\hat{\alpha}}\big] = \rho T N\bar{\mu}. \tag{17.30}$$

In the above, $\mathcal{S}^N$ is the N-dimensional product space of square integrable semimartingales, $\hat{\alpha} = (\hat{\alpha}^1, \ldots, \hat{\alpha}^N)$ is the optimal effort of the system given the allocation vector $\mathbf{A} = (A^1, \ldots, A^N)$, while $E^{\hat{\alpha}}$ is the system optimal emission (sum of the individual emissions). From the equilibrium results, we can and do pass from $\mathbf{A}$ to the martingales $\mathbf{M}_t = (M_t^1, \ldots, M_t^N)$, with $M^i = \mathbb{E}_t[A_T^i]$ for all i as before. We denote the space of N-dimensional, square integrable martingales by $\mathcal{M}^N$. Now, from (17.30) rewrites as:

$$\mathbb{E}\big[E_T^{\hat{\alpha}}\big] = T\Big(N\bar{\mu} + \sum_{i=1}^{N} g_i(0)\Big[h_i\Big(\frac{1}{2\lambda} + \nu T\Big) + M_0^i - \nu T\hat{P}_0\Big]\Big).$$

So, the reduction constraint on the expected emissions amounts to *a constraint on the initial equilibrium price*

$$\hat{P}_0 = \frac{\bar{\mu}(1-\rho)}{\nu T\bar{g}_0} + \frac{1}{N}\sum_{i=1}^{N}\frac{g_i(0)}{\nu T\bar{g}_0}\Big[h_i\Big(\frac{1}{2\lambda} + \nu T\Big) + M_0^i\Big], \quad \bar{g}_0 := \frac{1}{N}\sum_{i=1}^{N} g_i(0). \tag{17.31}$$

The relation (17.31) gives the necessary average carbon price required to emerge from the market if the regulator wishes to reduce the average carbon emission rate by a factor $1-\rho$ as per (17.30). Note that it captures all the features of the system. Equating the constraint on $\hat{P}_0$ in (17.31) with the expression of $\hat{P}_0$ in (17.23), we obtain a hyperplane constraint on $\mathbf{M}_0 = (M_0^1, \ldots, M_0^N)$:

$$\frac{1}{N}\sum_{i=1}^{N}\Big(\frac{g_i(0)}{\nu T\bar{g}_0} + \pi_i(0)\Big)M_0^i = \ell(\rho), \tag{17.32}$$

$$\text{with} \quad \ell(\rho) := \frac{1}{N}\sum_{i=1}^{N}\eta_i h_i \pi_i(0) - \frac{(1-\rho)\bar{\mu}}{\nu T\bar{g}_0} - \frac{1+2\lambda\nu T}{2\lambda\nu T\bar{g}_0}\frac{1}{N}\sum_{i=1}^{N} h_i g_i(0). \tag{17.33}$$

We denote this hyperplane by $\mathcal{H}_\rho$.

Theorem 8 *The social cost minimization problem* (17.30) *can be recast as a minimization over* $\mathcal{M}^N$*:*

$$\inf_{\mathbf{M}\in\mathcal{M}^N} \mathbb{E}\Big[\sum_{i=1}^N \int_0^T \Big(h_i\hat\alpha_t^i + \frac{(\hat\alpha_t^i)^2}{2\eta_i}\Big) + \frac{1}{2\nu}(\hat\beta_t^i)^2\Big)dt + \lambda(\hat X_T^i)^2\Big], \quad \mathbf{M}_0 \in \mathcal{H}_\rho. \tag{17.34}$$

This is a consequence of the results on the firm optimizers in Theorem (6), in which the regulator's control appears only through M^i*. The solution* $\mathbf{M}^*$ *is unique, and it is found as follows.*

(1) *The dynamics of* $\mathbf{M}^*$ *are:*
$$dM^{i,*} = \sigma_i dW^i \quad \textit{for all } i,$$

(2) *The initial* $\mathbf{M}_0^* \in \mathcal{H}_\rho$ *is the minimizer of the scalar* N*-variate function* s *over the hyperplane* $\mathcal{H}_\rho$ *:*

$$s(\mathbf{M_0}) := \sum_{i=1}^N \left(T\Big(h_i\hat\alpha_0^i + \frac{(\hat\alpha_0^i)^2}{2\eta_i} + \frac{1}{2\nu}(\hat\beta_0^i)^2\Big) + \lambda(M_0^i + (\hat\alpha_0^i + \hat\beta_0^i)T)^2\right),$$

where $\hat\alpha_0^i, \hat\beta_0^i$ *are the initial controls from Theorem 6. Uniqueness follows by strict convexity of* s *and linearity of the constraint.*

(3) *Since* $dM^i = \sigma dW^i$*, for all* i*, the price is constant:*

$$\hat P_t = \hat P_0 = \frac{1}{N}\sum_i \pi_i(0)\left(\eta_i h_i T - M_0^{i,*}\right) \tag{17.35}$$

(4) *The firms optimal abatements are unique and constant*

$$\hat\alpha_t^i = \hat\alpha_0^i = -g_i(0)\left(\frac{1}{2\lambda}h_i + M_0^{i,*} + \nu(h_i - \hat P_0)T\right).$$

(5) *The firms optimal trading rates are also unique, and constant. The optimal rates are:*

$$\hat\beta_t^i = \hat\beta_0^i = \nu(h_i + \frac{\hat\alpha_0^i}{\eta_i} - \hat P_0) \tag{17.36}$$

Proof. The cost function (17.34) is convex in all the arguments, $\hat\alpha^i, \hat\beta^i, M^i, \hat X^i$. A combination of Jensen inequality and martingality of the controls $\hat\alpha^i, \hat\beta^i$, and of M^i for $i = 1, \dots, N$ then gives:

$$\begin{aligned}&\mathbb{E}\Big[\sum_{i=1}^N \int_0^T \Big(h_i\hat\alpha_t^i + \frac{(\hat\alpha_t^i)^2}{2\eta_i}\Big) + \frac{1}{2\nu}(\beta_t^i)^2\Big)dt + \lambda(\hat X_T^i)^2\Big]\\ &\geq \sum_{i=1}^N T\Big[\Big(h_i\hat\alpha_0^i + \frac{(\alpha_0^i)^2}{2\eta_i} + \frac{1}{2\nu}(\beta_0^i)^2\Big) + \lambda\mathbb{E}[X_T^i]^2\Big]\end{aligned} \tag{17.37}$$

So, we look for the optimizers among those $\mathbf{M}$ which induce constant firms controls and constant X^i. From (17.23) and (17.25), the abatement control is constant if and only if:

$$d(M^i - \sigma^i W^i) - \nu(T-t)d\hat P = 0, \text{ for all } i.$$

Moreover, the associated price is constant. The above displayed equation then becomes

$$d(M^i - \sigma^i W^i) = 0, \text{ for all } i.$$

To obtain the optimal initial constants $M_0^{i,*}$, we need to minimize the right hand side in (17.37) over the hyperplane constraint $\mathbf{M}_0 \in \mathcal{H}_\rho$. Substituting $\mathbb{E}[X_T^i]$ in there gives the function s in item (2) of the statement of the Theorem. The firms controls easily follows. □

Let's comment on above results. As a consequence of the optimal dynamic regulation, the market price is constant. Indeed, the level of carbon emission reduction fixes the average level of required effort, and thus the average required price. But price fluctuations induce variations of effort, which in turn produce irreversible cost because of the inflexibility of the system. By annihilating price changes, the regulator avoids firms expensive stop-and-goes. Within the optimal policy class described in Theorem 8 item 3), one easily obtains an optimal intertemporal allocation:

$$\tilde{A}_t^i = M_0^{i,*} + \sigma_i W_t^i + \mu_i t, \tag{17.38}$$

In total, the regulator allocates $M_0^{i,*} + \sigma_i W_T^i + \mu_i T$ permits (with sign) to each firm. They compensate each firm for the BAU cumulative emissions, modulo the term $M_0^{i,*}$.

Let us interpret the sign of the allocation first. The emissions drift is of higher order with respect to the volatility (2 ton/year and 0.2 ton/year$^{\frac{1}{2}}$, see EU ETS emission data in Section 6.4). As a consequence, the probability that in $(0, T]$ the BAU in our model becomes negative is small given the real estimated parameters[5].

Therefore the actual number of permits in the market is (with high probability) nonnegative, a necessary condition for the existence of a market of permits in $(0, T]$. However, the initial endowment in permits $M_0^{i,*}$ can be positive or negative and it should be interpreted as an incentive when positive, and as an initial penalty on the i-th firm when negative. Alternatively said, each firm may have an assist to compliance, or can have a stronger abatement requirement than others.

Moreover, since the optimal $\hat{\alpha}_t^i = \alpha_0^i$ and $\hat{\beta}_t^i = \hat{\beta}_0^i$ from item 4) and 5) in the previous Theorem are constant, the bank accounts have deterministic, linear dynamics:

$$\hat{X}_t^i = M_0^{i,*} + (\hat{\alpha}_0^i + \hat{\beta}_0^i)t \qquad i = 1, \ldots, N.$$

17.4 Conclusion and some ongoing research

We find the optimal dynamic allocation processes that achieve a target expected emission reduction of carbon emissions. In a market with frictions, they are unique, they induce constant abatement efforts, constant carbon price and they avoid trading. Even if the regulator is not pursuing any price control mechanism, the desired reduction goal naturally leads to that conclusion. This result is consequence of the perfect knowledge of the regulator on the different economic shocks that affect the firms.

[5]This can be quantified. The distribution of the running minimum of a Brownian motion with drift is known, see e.g., Karatzas and Shreve (2000, Section 3.3.5). The probability that it crosses level $-\epsilon$, with $\epsilon > 0$, is a decreasing function of $\frac{\mu}{\sigma}$.

There are many interesting related problems which can be addressed within a similar framework. For instance, in the working paper Aïd et al. (2025) the authors work in a similar mathematical model, but the regulator has to balance social concerns with environmental goals. This is indeed a key issue, pollution restrictions induce an increase in inflation and the cited paper provides a simple, yet effective, relation between price of carbon and inflation. In the preprint Biagini, Biffis, and Salazedeh Nobari (2024), the different decay behavior of the various gases is taken into account. The dedicated markets reach an equilibrium, on top of which the regulator imposes a temperature increase constraint. The resulting prices may be considered in the formation on environmental equivalence indicators of the various gases, in a more accurate way than the Global Warming Potential. Finally, in Biagini (2024) there is an analysis of significant regulatory policies and their impact on the firms system. There, the firms have to comply in every scenario at the regulation date and not merely in expectation. This is the mathematical translation of the carbon-neutrality or net-zero requirement, and the optimization problem is solved in an ETS setup first and then without allowances.

17.5 Proofs and computations

The following results are needed in the proofs of the main results.

Lemma 3 *Let h be a process in $\mathcal{L}^2_T$ and $Y \in L^2(\Omega, \mathcal{F}_T, P)$. Then*

$$\mathbb{E}\Big[Y\int_0^T h_t dt\Big] = \mathbb{E}\Big[\int_0^T dt\, h_t\, \mathbb{E}_t[Y]\Big].$$

Namely, the scalar product of Y and the pathwise integral $\int_0^T h_t dt$ equals the scalar product in $\mathcal{L}^2_T$ of h and the martingale process closed by Y, namely $(E_t[Y])_t$.

Proof. We have:

$$\mathbb{E}\Big[Y\int_0^T h_t dt\Big] = \mathbb{E}\Big[\int_0^T Y h_t\, dt\Big] = \int_0^T dt\,\mathbb{E}[Y h_t] = \int_0^T dt\,\mathbb{E}[h_t\,\mathbb{E}_t[Y]] = \mathbb{E}\Big[\int_0^T dt\, h_t\,\mathbb{E}_t[Y]\Big]$$

where the equalities from the second onwards follow from Fubini Theorem and from the properties of conditional expectation. An alternative is to apply integration by parts to the product of the martingale $(\mathbb{E}_t[Y])_t$ and the bounded variation process $\int_0^{\cdot} h_t dt$. □

The proof of the next Lemma is left as an exercise.

Lemma 4 *Let $\alpha \in \mathcal{L}^2_T$ be a martingale. Then, consider the $\mathcal{L}^2_T$-martingale M closed by the integral $\int_0^T \alpha_s ds$, $M_t := \mathbb{E}_t\Big[\int_0^T \alpha_s ds\Big]$. Then,*

$$M_t = \int_0^t \alpha_s ds + (T-t)\alpha_t$$

so that $dM_t = (T-t)d\alpha_t$.

17.5.1 Proof of Theorem 6

For ease of notation, we drop the dependence on i of the coefficients and of the controls. We split the cost function J in two parts, running cost C and terminal cost F:

$$J(\alpha,\beta)=\mathbb{E}\Big[\int_0^T\Big(h\alpha_t+\frac{\alpha_t^2}{2\eta}+P_t\beta_t+\frac{1}{2\nu}\beta_t^2\Big)dt\Big]+\mathbb{E}[\lambda X_T^2]=C(\alpha,\beta)+F(\alpha,\beta).$$

From the structure it is apparent that both C and F are differentiable. The differential of the running cost $C(\alpha,\beta):=\mathbb{E}\Big[\int_0^T\Big(h\alpha_t+\frac{\alpha_t^2}{2\eta}+P_t\beta_t+\frac{1}{2\nu}\beta_t^2\Big)dt\Big]$ is given by differentiation inside the integral

$$D_\alpha C_t=h+\frac{\alpha_t}{\eta},\quad D_\beta C_t=P_t+\frac{\beta_t}{\nu}.$$

Let us denote the terminal penalty by $F(\alpha,\beta):=\mathbb{E}[\lambda X_T^2]$. The bank X_T equals: $X_T=A_T+\int_0^T(\alpha_t+\beta_t)dt-\sigma W_T$. Now:

$$F(\alpha+\epsilon,\beta+\delta)-F(\alpha,\beta)\sim 2\lambda\,\mathbb{E}\left[X_T\int_0^T\epsilon_t dt+X_T\int_0^T\delta_t dt\right],$$

for any couple ϵ,δ in $\mathcal{L}_T^2$. A couple of applications of the Fubini Theorem gives:

$$\mathbb{E}\left[X_T\int_0^T\epsilon_t dt\right]=\int_0^T dt\,\mathbb{E}[X_T\epsilon_t]$$

$$=\int_0^T dt\,\mathbb{E}[\,\mathbb{E}_t[X_T]\epsilon_t]=\mathbb{E}\left[\int_0^T\mathbb{E}_t[X_T]\epsilon_t\,dt\right],$$

and similarly for the term in δ. We derive then that the gradient of F is $(2\mathbb{E}_t[X_T],2\mathbb{E}_t[X_T])$. Putting things together, the differential of the cost function J is:

$$D_\alpha J=h+\frac{\alpha_t}{\eta}+2\lambda\mathbb{E}_t\big[X_T\big],\quad D_\beta J=P_t+\frac{\beta_t}{\nu}+2\lambda\mathbb{E}_t\big[X_T\big]\tag{17.1}$$

The FOC equations for the generic agent i become

$$D_\alpha J=0,\quad D_\beta J=0.\tag{17.2}$$

Subtracting the two equations above,

$$\beta_t=\nu(h+\frac{\alpha_t}{\eta}-P_t),$$

which can be substituted into the first equation:

$$h+\frac{\alpha_t}{\eta}+2\lambda\mathbb{E}_t\left[A_T+\int_0^T(\alpha_t+\nu(h+\frac{\alpha_t}{\eta}-P_t))dt\right]-2\lambda\sigma W_t=0,\ \forall t\geq 0.\tag{17.3}$$

Therefore the optimal abatement (if it exists) $\hat{\alpha}$ is a martingale, and it is found as the solution of the equation:

$$\hat{\alpha}_t=-\eta\left(h+2\lambda M_t+2\lambda\mathbb{E}_t\Big[\int_0^T((1+\frac{\nu}{\eta})\hat{\alpha}_t+\nu(h-P_t))dt\Big]-2\lambda\sigma W_t\right)\tag{17.4}$$

In particular, using Lemma 4 and the definition of g, M in (17.12), the initial value of $\hat{\alpha}$ is:

$$\hat{\alpha}_0 = -g(0)\Big(\frac{1}{2\lambda}h + M_0 + \nu\mathbb{E}\Big[\int_0^T (h - P_t)dt\Big]\Big). \tag{17.5}$$

To solve for $\hat{\alpha}$, we rewrite (17.4) in differential version:

$$d\hat{\alpha}_t = -\eta\Big\{2\lambda dM_t + 2\lambda\Big(1 + \frac{\nu}{\eta}\Big)d\mathbb{E}_t\Big[\int_0^T \hat{\alpha}_s ds\Big] + 2\lambda\nu d\mathbb{E}_t\Big[\int_0^T (h - P_s)ds\Big] - 2\lambda\sigma dW_t\Big\}$$

By Lemma 4, $d\mathbb{E}_t\Big[\int_0^T \hat{\alpha}_s ds\Big] = (T - t)d\hat{\alpha}_t$ and thus

$$(1 + 2\lambda(\eta + \nu)(T - t))d\hat{\alpha}_t = -2\eta\lambda\left(dM_t + \nu d\mathbb{E}_t\Big[\int_0^T (h - P_s)ds\Big] - \sigma dW_t\right)$$

or

$$d\hat{\alpha}_t = -g(t)\Big(dM_t + \nu d\mathbb{E}_t\Big[\int_0^T (h - P_s)ds\Big] - \sigma dW_t\Big), \tag{17.6}$$

where the term between parentheses is (the differential of) a square integrable martingale. The Cauchy problem given by the SDE and the initial condition (A.6) uniquely identifies $\hat{\alpha}$, since the dynamics of P and a are exogenous here. The optimal trade $\hat{\beta}_t$ is then

$$\hat{\beta}_t = \nu\Big(h + \frac{\hat{\alpha}_t}{\eta} - P_t\Big). \tag{17.7}$$

Note that the optimal trade is a martingale of and only if P is a martingale. To conclude, we rewrite the optimal couple as a function of the state $\hat{X}$. Starting again from the first FOC,

$$h + \frac{\hat{\alpha}_t}{\eta} + 2\lambda\mathbb{E}_t\big[\hat{X}_T\big] = 0$$

and rewrite

$$h + \frac{\hat{\alpha}_t}{\eta} + 2\lambda\hat{X}_t + 2\lambda\mathbb{E}_t\Big[A_T - A_t + \int_t^T \Big(a_s + \hat{\alpha}_s + \nu\big(h + \frac{\hat{\alpha}_s}{\eta} - P_s\big)\Big)ds\Big] = 0$$

An application of Lemma 4 leads to

$$h + \frac{\hat{\alpha}_t}{\eta} + 2\lambda\Big(1 + \frac{\nu}{\eta}\Big)(T - t)\hat{\alpha}_t + 2\lambda\hat{X}_t + 2\lambda R_t + 2\lambda\mathbb{E}_t\Big[\int_t^T \nu(h - P_s)ds\Big] = 0$$

Finally, $\hat{\alpha}$ in feedback form reads as

$$\hat{\alpha}_t = -g(t)\Big(\frac{h}{2\lambda} + \hat{X}_t + R_t + \nu\mathbb{E}_t\Big[\int_t^T (h - P_s)ds\Big]\Big)$$

and this concludes the proof, since $M_0 = R_0$. □

17.5.2 Proof of Theorem 7

(i) Summing up over all i the relations (17.14) and using market clearing condition,

$$NP_t = \sum_{i=1}^{N} h_i + \frac{\alpha_t^i}{\eta_i}. \tag{17.8}$$

From Theorem 6 (i),

$$d\alpha_t^i = -g_i(t)\Big(dM_t^i - \sigma_i dW_t^i - \nu d\mathbb{E}_t\Big[\int_0^T P_s ds\Big]\Big). \tag{17.9}$$

Thus,

$$NdP_t = -\sum_{i=1}^{N} \frac{g_i(t)}{\eta_i}\Big(dM_t^i - \nu d\mathbb{E}_t\Big[\int_0^T P_s ds\Big] - \sigma_i dW_t^i\Big),$$

which shows that P must be a martingale. Hence, by Lemma 4,

$$\Big(N - \sum_i \frac{g_i(t)\nu(T-t)}{\eta_i}\Big)dP_t = -\sum_{i=1}^{N} \frac{g_i(t)}{\eta_i}\Big(dM_t^i - \sigma_i dW_t^i\Big),$$

which gives the SDE for the equilibrium price $\hat{P}$ in (17.23).

For the initial condition of the equilibrium price dynamics, using the first FOC in (17.3) written for $t = 0$ and substituting β_0^i as a function of α_0^i and P_0, we find that

$$\alpha_0^i = -g_i(0)\Big(x_0^i + M_0^i + h_i\Big(\frac{1}{2\lambda} + \nu T\Big) - \nu T P_0\Big). \tag{17.10}$$

Then substituting the value above in (17.8), we get after tedious computations:

$$P_0 = \frac{1}{N}\sum_{i=1}^{N} \pi_i(0)\Big(\eta_i h_i T - M_0^i\Big). \tag{17.11}$$

(ii) The equilibrium price $\hat{P}$ is given by the market clearing condition

$$\sum_{i=1}^{N} \hat{\beta}^i(\hat{P}(a), a) = 0.$$

According to Theorem 6 (ii), we deduce that:

$$\hat{P}_t = \frac{1}{N}\sum_{i=1}^{N} h_i + \frac{\hat{\alpha}_t^i}{\eta_i}, \tag{17.12}$$

from which we deduce that the equilibrium price is a martingale. Recalling that

$$\hat{\alpha}^i(\hat{X}_t^i, P) = -g_i(t)\Big(\frac{1}{2\lambda}h_i + \hat{X}_t^i + R_t^i + \nu(T-t)h_i - \mathbb{E}_t\Big[\int_0^T P_s ds\Big]\Big),$$

and using the fact that $\hat{P}$ is a martingale, we get that the equilibrium price is given by:

$$\Big[1 - \frac{1}{N}\sum_{i=1}^{N} \frac{\nu g_i(i)(T-t)}{\eta_i}\Big]\hat{P}_t = \frac{1}{N}\sum_{i=1}^{N} h_i - \frac{g_i(t)}{\eta_i}\Big(\frac{1}{2\lambda}h_i + \hat{X}_t^i + R_t^i + \nu(T-t)h_i\Big). \tag{17.13}$$

Hence,

$$\Big[1-\frac{1}{N}\sum_{i=1}^{N}\frac{\nu g_i(i)(T-t)}{\eta_i}\Big]\hat{P}_t=\frac{1}{N}\sum_{i=1}^{N}h_i\{1-\frac{g_i(t)}{\eta_i}(\frac{1}{2\lambda}+\nu(T-t))\}-\frac{g_i(t)}{\eta_i}\Big(\hat{X}_t^i+R_t^i\Big). \tag{17.14}$$

Further,

$$1-\frac{g_i(t)}{\eta_i}(\frac{1}{2\lambda}+\nu(T-t))\}=g_i(t)(T-t), \tag{17.15}$$

from which

$$\Big[1-\frac{1}{N}\sum_{i=1}^{N}\frac{\nu g_i(i)(T-t)}{\eta_i}\Big]\hat{P}_t=\frac{1}{N}\sum_{i=1}^{N}\frac{g_i(t)}{\eta_i}\Big(\eta_i h_i(T-t)-\hat{X}_t^i-R_t^i\Big).$$

Hence,

$$\hat{P}_t=\frac{1}{N}\sum_{i=1}^{N}\pi_i(t)\Big(\eta_i h_i(T-t)-\hat{X}_t^i-R_t^i\Big).$$

(iii) Using relation (17.9) and the expression of the dynamics of the equilibrium price, the relations are immediate. (iv) Direct consequence of the initial condition on the controls and on the price. □

References

Aliprantis, C., and K. Border, 2007, *Infinite Dimensional Analysis: A Hitchhiker's Guide*, third edition (Springer).

Anand, K. S., and F. C. Giraud-Carier, 2020, Pollution regulation of competitive markets, *Management Science* 66, 4193–4206.

Auffhammer, M., 2018, Quantifying economic damages from climate change, *Journal of Economic Perspectives* 32, 33–52.

Aïd, R., M. Arduca, S. Biagini, and L. Taschini, 2025, Emission impossible, Preprint.

Aïd, R., and S. Biagini, 2023, Optimal dynamic regulation of carbon emissions market, *Mathematical Finance* 33, 80–115.

Bensoussan, A., S. Chen, and S. P. Sethi, 2014, Feedback stackelberg solutions of infinite-horizon stochastic differential games, in F. El Ouardighi, and K. Kogan, eds., *Models and Methods in Economics and Management Science*, volume 198 of *International Series in Operations Research & Management Science* (Springer).

Biagini, S., 2024, Carbon neutrality and net-zero regulation, Preprint, available at `https://papers.ssrn.com/sol3/papers.cfm?abstract_id=4883544`.

Biagini, S., E. Biffis, and K. Salazedeh Nobari, 2024, Short-lived gases, carbon markets and climate risk mitigation, Preprint.

Carmona, R., F. Delarue, G.-E. Espinosa, and N. Touzi, 2013, Singular forward-backward stochastic differential equations and emissions derivatives, *Annals of Applied Probability* 23, 1086–1128.

Carmona, R., M. Fehr, and J. Hinz, 2009, Optimal stochastic control and carbon price formation, *SIAM Journal on Control and Optimization* 48, 2168–2190.

Carmona, R., M. Fehr, J. Hinz, and A. Porchet, 2010, Market design for emission trading schemes, *SIAM Review* 52, 403–452.

Carmona, R., and M. Hinz, 2011, Risk-neutral models for emission allowance prices and option valuation, *Management Science* 57, 1453–1468.

Chassagneux, J.-F., H. Chotai, and M. Muûls, 2017, *A Forward-Backward SDEs Approach to Pricing in Carbon Markets*, SpringerBriefs in Economics (Springer).

Chaton, C., A. Creti, and B. Peluchon, 2015, Banking and back-loading emission permits, *Energy Policy* 82, 332–341.

Cohen, S. N., and R. J. Elliott, 2015, *Stochastic Calculus and Applications* (Birkhauser).

Dechezleprêtre, A., D. Nachtigall, and F. Venmans, 2023, The joint impact of the european union emissions trading system on carbon emissions and economic performance, *Journal of Environmental Economics and Management* 118, 102758.

Duffie, D., 2001, *Dynamic Asset Pricing Theory*, third edition (Princeton University Press).

Fell, H., and R. D. Morgenstern, 2010, Alternative approaches to cost containment in a cap-and-trade system, *Environmental and Resource Economics* 47, 275–297.

Gillingham, K., and J. H. Stock, 2018, Quantifying economic damages from climate change, *Journal of Economic Perspectives* 32, 53–72.

Gollier, C., 2020, The cost–efficiency carbon pricing puzzle, Toulouse School of Economics Working Paper.

Hahn, R. W., 1989, Economic prescriptions for environmental problems: How the patient followed the doctor's orders, *Journal of Economic Perspectives* 3, 95–114.

Hepburn, C., 2006, Regulation by prices, quantities, or both: A review of instrument choice, *Oxford Review of Economic Policy* 22, 226–247.

Hsiang, S., and R. E. Kopp, 2018, An economist's guide to climate change science, *Journal of Economic Perspectives* 32, 3–32.

Karatzas, I., and S. E. Shreve, 2000, *Brownian Motion and Stochastic Calculus, Second Edition* (Springer).

Kollenberg, S., and L. Taschini, 2016, Emissions trading systems with cap adjustments, *Journal of Environmental Economics and Management* 80, 20–36.

Kollenberg, S., and L. Taschini, 2019, Dynamic supply adjustment and banking under uncertainty in an emission trading scheme: The market stability reserve, *European Economic Review* 118, 213–236.

Kuusela, O.-P., and J. Lintunen, 2020, A cap-and-trade commitment policy with allowance banking, *Environmental and Resource Economics* 75, 421–455.

Kyle, A. S., 1985, Continuous auctions and insider trading, *Econometrica* 53, 1315–1336.

Lintunen, J., and O.-P. Kuusela, 2018, Business cycles and emission trading with banking, *European Economic Review* 101, 397–417.

Meunier, G., J.-P. Montero, and J.-P. Ponssard, 2018, Output-based allocations in pollution markets with uncertainty and self-selection, *Journal of Environmental Economics and Management* 92, 832–851.

Montgomery, W. D., 1972, Markets in licences and efficient pollution control programs, *Journal of Economic Theory* 5, 395–418.

Peng, S., 1990, A general stochastic maximum principle for optimal control problems, *SIAM Journal of Control and Optimization* 28, 966–979.

Pham, H., 2009, *Continuous-Time Stochastic Control and Optimization with Financial Applications* (Springer).

Requate, T., 1993, Pollution control in a cournot duopoly via taxes and permits, *Journal of Economics* 58, 255–291.

Rubin, J. D., 1996, A model of intertemporal emission trading, banking, and borrowing, *Journal of Environmental Economics and Management* 31, 269–286.

Schennach, S., 2000, The economics of pollution permit banking in the context of title iv of the 1990 clean air act amendments, *Journal of Environmental Economics and Management* 40, 189–210.

Von der Fehr, N., 1993, Tradable emission rights and strategic interaction, *Environmental Resource Economics* 3, 129–151.

Weitzman, M. L., 1974, Prices vs. quantities, *Review of Economic Studies* 41, 683–691.

Chapter 18

Insurance products and sustainability

Enrico Biffis

Imperial College Business School, UK

Pietro Hughes

Astanor Ventures, UK

This chapter explores the role of insurance in advancing sustainability, particularly in the context of climate risk and environmental uncertainty. It highlights how innovative insurance mechanisms especially parametric insurance can be integrated into bundled finance solutions to support technology adoption and enhance financial inclusion in developing economies. A theoretical model is reviewed to demonstrate how such insurance can reduce collateral requirements and borrowing costs for smallholder farmers under both ex-ante and ex-post moral hazard. The chapter then turns to carbon markets and nature-based solutions (NbS), identifying wildfire as a growing threat to the permanence of carbon offsets, especially in forestry projects. It discusses the limitations of buffer pool-based risk management and presents alternative approaches, such as insurance and equivalence mechanisms, to improve resilience and credibility. Novel findings for insurance mechanisms applied to California Carbon Offsets (CCOs) are presented using a catastrophe modeling framework to estimate wildfire-related invalidation risk. The analysis reveals that insurance can offer more accurate and spatially differentiated protection than static buffer contributions. The chapter concludes by emphasizing the importance of integrating improved risk modeling and forward-looking data into climate finance, with a focus on transparency, cost-effectiveness, and long-term sustainability.

* Section 18.2 is based on research funded by Climate-KIC. Section 18.3 draws on research carried out by Pietro Hughes at the Centre for Environmental Policy at Imperial College London, and by Enrico Biffis at the Singapore Green Finance Centre.

Disclaimer. The views expressed in this chapter are those of the authors alone. Pietro Hughes would like to clarify that this research was conducted independently of Astanor Ventures.

18.1 Introduction

Sustainability has become a cornerstone of global economic and environmental policy, driven by the need to address the mounting challenges of climate change, resource depletion, and social inequality. Insurance, traditionally a financial risk management tool, is increasingly recognized for its potential to support sustainability efforts. By providing financial protection against environmental risks, insurance products can incentivize sustainable practices and investments, particularly in sectors vulnerable to climate impacts

DOI: 10.1201/9781032636252-18

(Surminski, Bouwer, and Linnerooth-Bayer, 2016; Birkmann, McMillan, and Tangwanichagapong, 2023).

Sustainability refers to the ability to meet the needs of the present without compromising the ability of future generations to meet their own needs (World Commission on Environment and Development, 1987). In the context of climate risk, sustainability involves managing and mitigating the impacts of climate change on natural and human systems. This includes reducing greenhouse gas emissions, protecting ecosystems, and ensuring that communities and economies are resilient to climate-related shocks (IPCC, 2022).

Insurance plays a critical role in managing climate risks by providing financial protection against losses resulting from extreme weather events, natural disasters, and other climate-related hazards. By spreading risk across a large pools of policyholders, insurance helps to reduce the financial burden on individuals, businesses, and governments, enabling them to recover more quickly from climate-related events. Integrating sustainability into insurance products involves designing policies that not only provide financial protection but also promote sustainable practices (Surminski, Bouwer, and Linnerooth-Bayer, 2016; Tellman and Eakin, 2022). This can include offering lower premiums for policyholders who adopt sustainable technologies, developing insurance products that support climate resilience, and ensuring that insurance products themselves are sustainable in their design and implementation (Shee and Turvey, 2012; Biffis et al., 2022).

Several trends are emerging at the intersection of insurance and sustainability. These include the growing use of parametric insurance in developed and developing markets (McIntosh, Sarris, and Papadopoulos, 2013; Carter, Janzen, and Stoeffler, 2014; Hill et al., 2019), the integration of satellite data and machine learning into risk assessment (Biffis and Chavez, 2017; van Wees et al., 2022; Li et al., 2023), and the development of insurance products that specifically address climate-related risks, such as the exposure of carbon sequestration vehicles to natural hazards (Swiss Re, 2019; Howden Group, 2022). This chapter draws on recent contributions and case studies to illustrate how innovative insurance products can contribute to sustainable development. We first examine the role of parametric insurance in bundled finance solutions promoting financial inclusion and climate resilient technology adoption. We then introduce insurance mechanisms to protect carbon offsets against the risks posed by wildfires.

Although parametric insurance has received considerable attention in the literature (Doherty and Schlesinger, 1990; Clarke, 2016; Bryan, 2019; Chen et al., 2024), its use in the context of bundled finance solutions, which are crucial for climate risk mitigation and adaptation, is still under exploration and often characterized by diverse normative and empirical predictions (Giné and Yang, 2009; Dercon and Christiaensen, 2011; Shee and Turvey, 2012; McIntosh, Sarris, and Papadopoulos, 2013; Carter, Janzen, and Stoeffler, 2014; Hill et al., 2019). We therefore review some novel findings presented in Biffis et al. (2022), which were applied in the context of de-risking agricultural value chains in sub-Saharan Africa. These findings shed light on how insurance mechanisms can play broader roles than standalone contingent capital provision, particularly when the wider spillovers across the ecosystem of value chain stakeholders are properly accounted for. When considering the area of carbon offset insurance, we note that if a forest protected under a carbon offset program is destroyed by wildfire, the carbon stored in the trees is released back into the atmosphere, thereby invalidating the offset. This creates a natural demand for insurance coverage against the risk of wildfire-related offset invalidation, a demand that is beginning to be met by solutions offered by some market participants (see Financial Times (2024); Gallagher Re (2024); Practical ESG (2024); CFC Underwriting (2024); Howden Group (2022); AXA XL (2022); MSCI (2022); Aviva (2024), for example). Although the insurance mechanism itself is standard, the nature of the exposure and its linkage to both voluntary and compliance carbon markets are not. We address a gap in the literature by providing an overview of these insurance solutions and a numerical analysis of the California Cap-and-Trade system.

18.2 Parametric insurance and technology adoption

Parametric insurance is a type of insurance that pays out when a specific event occurs, such as a hurricane or drought, rather than being based on the actual loss incurred by a policyholder. This type of insurance is particularly useful in situations where losses are costly to quantify, such as in agriculture, where the impact of a weather event on crop yields can vary widely by location, and carrying out granular loss assessment can be overly expensive. The role of parametric insurance in promoting technology adoption is well illustrated in the agricultural sector in developing countries. Smallholder farmers often face significant barriers to adopting new technologies due to limited access to credit and high financial risks. Parametric insurance, when bundled with financial products like loans, can help mitigate these risks by providing farmers with a safety net against adverse weather events.

Case Study: Derisking agricultural value chains in developing economies. A study focused on Tanzania (Biffis et al., 2022) demonstrates how parametric insurance, combined with input loans, can lower the barriers to technology adoption. By reducing the collateral required for loans and lowering interest rates, parametric insurance makes it more feasible for farmers to invest in more productive agricultural technologies, leading to higher yields and improved livelihoods. On the credit supply side, lenders are willing to lower collateral requirements and interest rates, conditional on parametric insurance being bundled with input loans, as the insurance mechanisms de-risks their loan portfolio. As such, parametric insurance plays the role of a credit enhancement device, mitigating credit rationing in situations in which farmers have limited credit history and/or collateral. The study addresses longstanding questions about how to make parametric insurance work for smallholder farmers (Meyer, Hazell, and Varangis, 2017) and provides new insights by examining it within broader financial and institutional contexts, as opposed to on a standalone basis.

18.2.1 Parametric insurance as a credit-enhancement device

We now discuss the model of Biffis et al. (2022), which builds on the seminal contributions of Jaffee and Russell (1976) and Stiglitz and Weiss (1981), linking credit rationing to information asymmetry. It is also closely related to the approach of Holmstrom and Tirole (1997), in that it considers the limitations of collateral that can be pledged by borrowers in a setting characterized by moral hazard. The literature on parametric insurance has traditionally focused on the efficiency losses induced by basis risk such as imperfect risk sharing and dampened insurance demand (Clarke, 2016) without necessarily acknowledging the potential efficiency gains that may arise from an imperfect insurance contract alleviating other frictions. In contrast, Biffis et al. (2022) clarifies the mechanism through which parametric insurance can help overcome credit rationing, providing theoretical grounding for several initiatives currently underway in East Africa. The model considers two settings: one with ex-ante moral hazard, where the farmer may divert borrowed funds toward unintended uses; and one with ex-post moral hazard, where the funds are used to acquire and implement a production technology as intended, but output is private information, allowing the farmer to strategically default by taking advantage of parallel market opportunities.

Irrespective of the setting, the structure of the model is single-period with two dates, 0 and 1, during which risk-neutral agents farmers and banks operate. The risk-free interest rate is set to zero for simplicity. There is a continuum of farmers. At time 0, each farmer can engage in a project by adopting a technology that requires an investment of capital I, yielding a random income $\widetilde{R}$ at time 1. The income takes the value R with probability $1-p$ (i.e., the project succeeds), and 0 with probability $p \in [0,1]$ (i.e., the project fails). Farmers have no liquid resources to finance the project and therefore require a loan to cover the capital investment I. Banks agree to extend credit provided the expected profit is no lower than $\eta \geq 0$, where $\eta = 0$ corresponds to perfect competition and $\eta > 0$ to monopolistic competition. At time 1, farmers must repay the principal I plus interest Ir, where r is the interest rate charged by the lender. At time 0, each farmer owns an illiquid asset valued at $A \geq 0$, which may be used as collateral but cannot be liquidated to finance the project. The value of A varies across farmers. To mitigate the risk of default, the bank requires a collateral amount C to be posted at time 0, with the understanding that it will be returned at time 1 if the loan is repaid in full. Farmers also have access to other resources, denoted by $\omega \geq 0$, generated from activities in the informal sector. These resources are not verifiable, cannot be used as collateral, but can be used to pay insurance premiums when required.

When the farmer experiences production losses, the project is deemed unsuccessful. These losses are driven by an exogenous risk factor, which may be interpreted as a weather shock and is modeled, for simplicity, as a binary random variable $\widetilde{y}$ taking values in $0,1$, with $\mathbb{P}(\widetilde{y}=1) = q \in (0,1)$. Equivalently, the events $\widetilde{y}=0$ and $\widetilde{y}=1$ represent favorable and adverse weather conditions, respectively. Note that $\widetilde{y}$ may be either farmer-specific or correlated across farmers, thereby capturing both idiosyncratic and systematic risks. In the event of a loss, the farmer defaults and the lender seizes the collateral $\widetilde{C}$.

The probabilistic relationship between the weather index $\widetilde{y}$ and income $\widetilde{R}$ is characterized by

$$\mathbb{P}\left(\widetilde{R}=0 \mid \widetilde{y}=1\right) = f+\varepsilon$$

$$\mathbb{P}\left(\widetilde{R}=0 \mid \widetilde{y}=0\right) = f,$$

where we assume that $\varepsilon > 0$, $0 < f < 1$, and $f+\varepsilon < 1$. In words, the occurrence of a weather shock increases the probability of project failure by ε. When $\varepsilon = 0$, the weather index is uninformative about income; conversely, the case $f=0$ and $\varepsilon = 1$ implies perfect correlation between the weather shock and income. As $\widetilde{y}=1$ with probability q and $\widetilde{y}=0$ with probability $1-q$, the unconditional probability of failure is given by $p = f+q\varepsilon$. See Table 18.1.

TABLE 18.1: Joint and marginal distributions of the project's outcome and the weather shock

		$\widetilde{y}$		
		0	1	
$\widetilde{R}$	R	$(1-q)(1-f)$	$q(1-f-\varepsilon)$	$1-p$
	0	$(1-q)f$	$q(f+\varepsilon)$	p
		$1-q$	q	1

18.2.2 Bundling parametric insurance

We consider a parametric insurance contract that pays an amount D to the lender when $\widetilde{y} = 1$, in which case the farmer is exempted from repaying the loan. The contract pays nothing when $\widetilde{y} = 0$. The loan agreement stipulates that the farmer recovers her collateral if the loan is repaid either by the farmer herself, when $\widetilde{R} = R$ and $\widetilde{y} = 0$, or by the insurer, when $\widetilde{y} = 1$ (regardless of whether $\widetilde{R} = 0$ or R). These terms are explicitly included in the contract between the lender and the farmer, meaning that the loan and insurance components are *bundled into a single agreement.* We assume that the actuarially fair insurance premium, P, is paid by the farmer to the bank, which then transfers it to the insurer. Since $\widetilde{y} = 1$ with probability q, the fair premium is given by $P = q(1 + r)I$. The case where the premium includes a risk loading, as well as potential subsidies from public funds, is discussed in Biffis et al. (2022), Section 2.7.2. The main message is that the higher the risk loading, the lower the attractiveness of technology adoption potentially requiring subsidies when insurance is prohibitively expensive.

Parametric insurance is imperfect: with probability $(1 - q)f$, it fails to provide a payout despite the farmer defaulting. Conversely, with probability $q(1 - f - \varepsilon)$, it pays out even if no loss occurs. We define **basis risk** as the probability α that the farmer's project fails but the insurance does not pay out. Formally,

$$\alpha = \Pr(\widetilde{y} = 0, \widetilde{R} = 0) = f(1 - q) < p,$$

which depends on the parameters f and q, with ε satisfying $p = f + q\varepsilon$. For a given failure probability p, a higher f and lower q (with $\varepsilon = \frac{p-f}{q}$) both lead to greater basis risk.

18.2.3 Ex-post moral hazard

With ex-post moral hazard, we consider the case in which credit is used as intended, but output is private information to the farmer, who may side-sell production in parallel markets and strategically default. Since verifying output is costly, the lender may prefer the posting of collateral to incentivize loan repayment. A lack of collateral can therefore result in financial exclusion and prevent the adoption of the target technology. The main result of Biffis et al. (2022) is that bundling insurance with the input loan reduces both collateral requirements and borrowing rates for farmers, while increasing the bank's expected profit.

In particular, let us denote by F the cost incurred by the farmer when deciding to report a low income and go bankrupt to avoid reimbursing the loan. The cost F may reflect a non-pecuniary cost akin to the cost of bankruptcy in Diamond (1984), or a moral/reputational cost of cheating. The following explicit expression is obtained for the collateral threshold $\underline{C}$ allowing credit to be extended by the bank:

$$\underline{C} = I + \eta - (1 - \alpha)F \leq C^N, \tag{18.1}$$

where C^N is the collateral required in the case of no insurance, with

$$C^N = I + \eta - (1 - p)F.$$

Furthermore, the interest rate charged by the lender is given by

$$r^I = \frac{\eta + \alpha F}{I} < r^N,$$

which is strictly lower than the one charged in the absence of parametric insurance, $r^N = \frac{\eta + pF}{I}$.

Finally, the parametric insurance contract increases the bank's expected profit, as the lower payoff, C, is obtained with probability $\alpha < p$, whereas the higher payoff, D, is achieved with probability $1 - \alpha > 1 - p$ (see Biffis et al. (2022), Sections 2.2–2.3, for details).

18.2.4 Ex-ante moral hazard

When credit is extended to fund technology adoption, but the use of financial resources is private information to the borrower, funds may be diverted for private benefit, thereby lowering the probability of debt repayment. We let each farmer's success probability, $p(e)$, depend on the farmer's privately observable effort choice, $e \in \{b, g\}$. This effort level should be understood as a measure of the quantity and quality of actions undertaken to maximize the likelihood that the crop harvest aligns with the production potential associated with the target technology.

Exerting a low level of effort is assumed to yield a private benefit $B > 0$. After receiving funds of amount I from the bank, the farmer's cash flow is $R > 0$ with probability $1 - p(e)$ and 0 with probability $p(e)$. This implies that both the exogenous (weather) shock and the farmer's effort jointly determine the distribution of the project's terminal payoff and, therefore, the conditional probability of default, which is given by:

$$\begin{aligned}\mathbb{P}(\tilde{R} = 0 \mid \tilde{y} = 1, e) &= f(e) + \varepsilon, \\ \mathbb{P}(\tilde{R} = 0 \mid \tilde{y} = 0, e) &= f(e),\end{aligned}$$

where $\tilde{R}$ denotes the random project outcome, and we assume $\varepsilon > 0$, $0 < f(g) < f(b)$, and $f(b) + \varepsilon < 1$. In words, for any effort level $e \in \{b, g\}$, the occurrence of the weather shock increases the probability of failure by ε. Conversely, for any realization of the weather shock, choosing $e = g$ reduces the default probability by $f(b) - f(g)$.

To incentivize the farmer to use funds productively, the lender requires the posting of collateral, which may not be available to a smallholder farmer. Parametric insurance bundled with input loans has two main effects (see Biffis et al. (2022), Section 2.5). First, the farmer's incentive compatibility constraint is relaxed, as insurance lowers the farmer's liability in the state of the world where she is able to repay the loan. This stands in contrast to the traditional shirking effect of insurance under moral hazard: bundling insurance with a loan incentivizes the farmer to exert effort, as her share of output increases in the case of success. Second, the introduction of parametric insurance lowers the bank's expected cost for any given behavior of the farmer, since the bank receives D with higher probability $1-(1-q)f(g) > 1-p(g)$; that is, it receives the indemnity from the insurer when the index takes the value 1. As in the ex-post moral hazard case, the parametric insurance contract therefore reduces both the collateral and the interest rate required by the bank.

18.2.5 Discussion

Joint ex-ante and ex-post moral hazard. When it comes to deterring both ex-ante and ex-post moral hazard simultaneously, the feasible reduction in collateral delivered by parametric insurance is constrained by the minimum and maximum possible reductions achievable when each type of moral hazard is addressed independently. In particular, Biffis et al. (2022) shows that the reduction in collateral requirements is not cumulative across sources of moral hazard.

Who is the direct recipient of insurance payouts? In the model outlined above, insurance payouts are assumed to be paid directly to the bank. The question of whether the borrower, rather than the lender, should be the recipient of the payout is a matter of ongoing debate among development economists and practitioners (Giné and Yang, 2009; Dercon and Christiaensen, 2011; Hill, Hoddinott, and Kumar, 2013; Carter et al., 2015). Some argue that channeling the payouts through farmers carries the risk of diverting the insurance compensation from its primary function securing loan repayment while others emphasize that this approach would enable farmers to smooth consumption over time in response to income fluctuations. In the ex-post moral hazard case, preventing the farmer from diverting the insurance payout does not require an additional incentive constraint, and the model's conclusions remain unchanged, regardless of who receives the payout. Matters differ in the ex-ante moral hazard case. Here, the need to prevent the farmer from retaining the insurance payment improperly introduces an additional incentive constraint arising from ex-post moral hazard. This leads to a setting in which both types of moral hazard co-exist, thereby dampening the potential reduction in collateral requirements. Ignoring the income-smoothing argument, we may therefore conclude that having the insurance payout go directly to the bank promotes financial inclusion and technology adoption.

Insurance premiums. In the model outlined above, insurance premiums are paid by farmers using resources generated from informal activity. These resources are not verifiable and therefore cannot be included in the collateral required by the bank, but they can be used to cover the insurance premium. Insurance is assumed to be actuarially fairly priced, and the parameterization implies that farmers can afford this payment based on the project's expected profit. However, as is common in insurance markets, premiums may include a loading factor, which would reduce the farmer's expected profit and potentially compromise the affordability of the insurance. Making technology adoption attractive may therefore require subsidizing insurance premiums when costs are too high. Biffis et al. (2022) shows that insurance subsidization is welfare-improving whenever the cost of public funds is not too large.

Basis risk. In the case of ex-post moral hazard, a smaller basis risk does not modify the farmer's truth-telling condition. It just allows the bank to be more completely covered by insurance and hence require a lower collateral amount. Things are different in the case of ex-ante moral hazard. On the one hand, the bank's participation constraint becomes more restrictive when $C < D$. On the other hand, parametric insurance affects farmers' incentives through two channels. Firstly, the farmer does not lose her collateral C when $\widetilde{R} = 0$ and $\widetilde{y} = 1$, while she would lose it in the absence of insurance. This first effect decreases the farmer's incentive to choose effort level $e = g$, as it corresponds to protection in the case of failure. Secondly, when $\widetilde{R} = R$ and $\widetilde{y} = 1$, the insurer pays D to the bank thus relieving the farmer from debt repayment, which provides an incentive to choose $e = g$, since the expected gain of the farmer in case of success is now increased. When $C < D$, the incentive effect of the reward D in case of success is larger than the disincentive of not losing C in case of failure, and consequently parametric insurance exerts a favorable effect on farmer's incentives. As a result, in both cases of ex-post and ex-ante moral hazard, a lower basis risk α reduces the required collateral, but the effect is amplified in the case of ex-ante moral hazard.

Basis risk. In the case of ex-post moral hazard, a smaller basis risk does not modify the farmer's truth-telling condition. It simply allows the bank to be more fully covered by insurance, and hence to require a lower collateral amount. Things are different in the case of ex-ante moral hazard. On the one hand, the bank's participation constraint becomes more restrictive when $C < D$. On the other hand, parametric insurance affects the farmer's incentives through two distinct channels. First, the farmer does not lose her collateral C when $\widetilde{R} = 0$ and $\widetilde{y} = 1$, whereas she would lose it in the absence of insurance. This effect reduces the incentive to choose effort level $e = g$, as it provides protection in the event of failure. Second, when $\widetilde{R} = R$ and $\widetilde{y} = 1$, the insurer pays D to the bank, thereby relieving the farmer from debt repayment. This increases the expected payoff for the farmer in the case of success and thus strengthens the incentive to choose $e = g$. When $C < D$, the incentive effect of the reward D in the success state outweighs the disincentive of not losing C in the failure state. Consequently, parametric insurance exerts a net positive effect on the farmer's effort incentives. As a result, in both ex-post and ex-ante moral hazard settings, a lower basis risk α reduces the required collateral, but the effect is amplified in the ex-ante moral hazard case.

Implications for broader bundled finance solutions. The trade-offs discussed in the context of the model outlined above Biffis et al. (2022) show that conventional predictions may be overturned once broader positive spillovers are taken into account within complex value chains. This opens the door to engineering de-risking structures in which suitably designed insurance components form only part of the solution, yet unlock significant value by reducing risk for a wide range of stakeholders. These mechanisms often work in conjunction with concessional lending, credit guarantees, and equity investors (e.g., first-loss tranche holders). An example of such structures is presented in Figure 18.1, which focuses on nature preservation and restoration, but the approach is equally applicable to financing for decarbonization (transition finance) and to climate risk de-risking through technology adoption (adaptation finance).

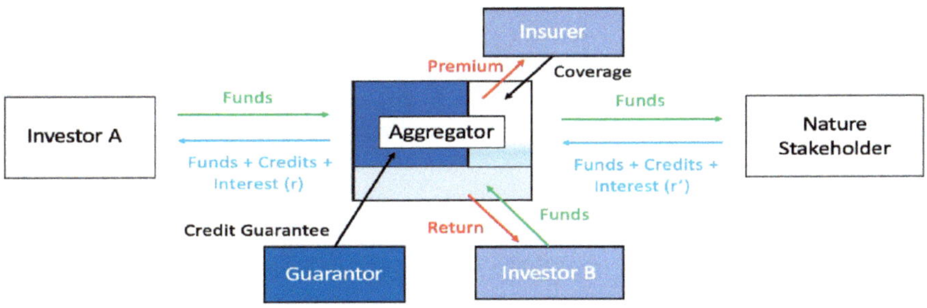

FIGURE 18.1: Example of investment structure aimed at nature conservation/restoration via a combination of concessional lending, insurance provision, credit guarantees, and first-loss equity investment.

The investment structure illustrated in Figure 18.1 highlights the potential for bundled finance to support nature preservation and restoration. Such initiatives frequently rely on nature-based solutions (NbS), which underpin many of the forestry projects generating carbon offsets in today's voluntary and compliance markets. However, the permanence of these offsets central to their credibility is increasingly threatened by climate-driven hazards such as wildfire. This raises important questions, which are addressed in the next section, about how risks to NbS can be more effectively managed, and where insurance can play a complementary role to traditional risk mitigation mechanisms like buffer pools.

18.3 Insurance for carbon offsets

18.3.1 The role of carbon markets in sustainability

Carbon markets are a key tool in the global effort to reduce greenhouse gas emissions and mitigate climate change. By putting a price on carbon, these markets create financial incentives for businesses and governments to reduce emissions and invest in cleaner technologies (Döbbeling-Hildebrandt et al., 2024). An emerging body of empirical literature offers mixed evidence on whether a carbon risk premium is shaping the valuation of publicly listed firms (Bolton and Kacperczyk, 2023; Aswani, Raghunandan, and Rajgopal, 2024; Zhang, 2025).

Carbon offsets, which represent emission reductions or removals elsewhere, are a critical component of carbon markets. In these markets, entities may trade either offsets representing emissions reductions outside the entity's boundary (see Figure 18.2) or allowances, which function as permits to emit. Beyond these markets, some jurisdictions impose carbon taxes as regulatory penalties for emissions.

Today, carbon pricing is neither sufficiently widespread nor sufficiently high to deliver on global temperature goals. Offsets currently cover 1.9% of global emissions (Ritchie and Roser, 2024), yet rarely exceed \$20/tCO_2e (Allied Offsets, 2025). Between allowances and carbon taxes, coverage extends to 24.0% of global emissions, with an estimated average price of only \$34/tCO_2e (World Bank, 2024). By 2030, if global efforts are to align with the Paris Agreement, a carbon price in the range of \$50–100/tCO_2e will be necessary (Stiglitz et al., 2017). Despite the promise of carbon pricing, substantial progress is still required in various strategic areas, including harmonization of emissions reporting standards (Biagini, Biffis, and Nobari, 2024; Cenci and Biffis, 2025) and carbon accounting (Comello, Reichelstein, and Reichelstein, 2023; Reichelstein, 2024).

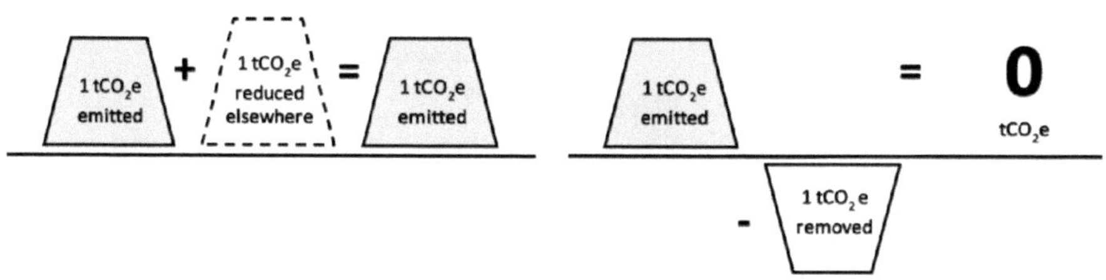

FIGURE 18.2: Carbon offsets: emissions reduction (left) versus emissions removal (right).

18.3.2 Wildfire risks to carbon offsets and the need for insurance

In carbon markets, the practice of offsetting has become closely linked with NbS, also referred to as nature climate solutions (NCS). These are defined as "conservation and management actions that reduce [...] emissions from ecosystems and harness their potential to store carbon" (Griscom et al., 2017), typically focused on the protection, management, and/or restoration of forests, mangroves, or peatlands (Seddon et al., 2020; Cheng et al., 2022). Project developers oversee the implementation of NbS, quantifying the carbon preserved and/or sequestered in vegetative biomass. Each tons of carbon is then packaged as an offset, which may be acquired by corporate actors to support carbon neutrality claims (Trouwloon et al., 2023). Such activities typically take place within the voluntary carbon market (VCM) a poorly regulated arena often likened to the "wild west" (Dhanda and Murphy, 2011; Temple-West, 2022), where NbS projects predominate (Figure 2).

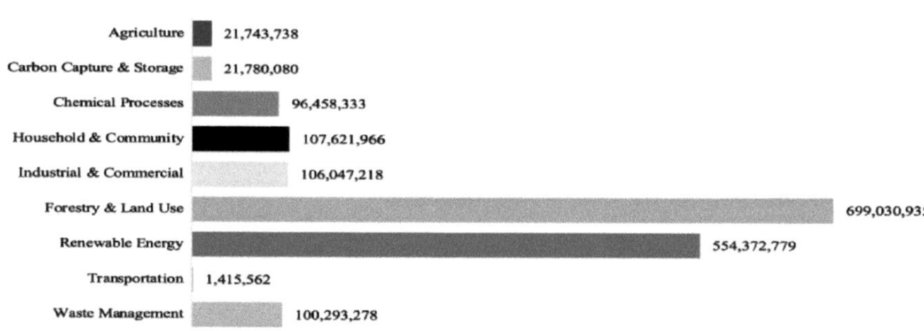

FIGURE 18.3: Carbon offsets: voluntary carbon market issuance by category. Figures represent 57.3% of issuance in 2022 (World Bank, 2023).

NbS are advantageous in terms of both practicality and cost. With respect to the former, NbS are readily deployable, requiring no extraction of raw materials, engineering, or manufacturing (Seddon et al., 2020). With respect to the latter, a widely accepted benchmark for cost-effective climate mitigation is \$100/tCO_2e (IPCC, 2022). Ecosystems can be protected, managed, and/or restored to mitigate 15.2–19.1% of global emissions at less than \$100/tCO_2e (Girardin et al., 2021), and up to 50.0% of this potential at less than \$20/tCO_2e (IPCC, 2022). Together, ease of deployment and cost make agriculture, forestry, and land use the sector with the highest mitigation potential through 2030 (IPCC, 2022).

Given the promise of carbon pricing and NbS, but the limited oversight in the voluntary carbon market (VCM) to fully capitalize on this potential, industry groups have moved to self-regulate. In 2023, the Integrity Council for the Voluntary Carbon Market (IC-VCM) published the *Core Carbon Principles*, an influential guide to best practices in carbon offsetting. Among these principles, the IC-VCM highlights the importance of permanence, stating that "emission reductions or removals [...] shall be permanent or, where there is a risk of reversal, there shall be measures in place to address those risks and compensate reversals" (Integrity Council for the Voluntary Carbon Market, 2023). Quantitatively, 'permanent' offsets are expected to withstand events with a return period greater than 1,000 years (Prado and MacDowell, 2023). In the case of NbS, this is often a challenge: reversal events, such as wildfire, can return stored carbon to the atmosphere. In an analysis of forestry projects in the VCM, wildfire risk is projected to increase by 37.0–76.0% through 2080 (Ballard et al., 2023). Measures are therefore needed to address such risks and compensate for reversals. Though still in its early stages, insurance is one emerging solution.

We next contextualize our discussion of insurance within the California Cap-and-Trade (CCT) system, before exploring de-risking instruments in more detail in the following section.

Case Study: California Cap-and-Trade (CCT). The CCT is a regional, regulated market in which a carbon price is applied through two instruments: offsets, known as California Carbon Offsets (CCOs), and allowances, known as California Carbon Allowances (CCAs). CCOs are integrated primarily for cost-containment purposes: they are permissible for up to 4% of a regulated entity's emissions between 2021–2025 and up to 6% between 2026–2030 (Figure 18.4), while typically trading at a 37.4% discount relative to CCAs (ICE, 2023; CME, 2023). The CCT's integration of CCOs has become a source of contention. CCOs are issued from projects across the United States, with forestry projects accounting for 81.3% of all credits (California Air Resources Board, 2023a). CCOs are intended to withstand 1-in-100-year adverse events (California Air Resources Board, 2015). To address risks of reversal, between 9–19% of credits are set aside in a *buffer pool*, with 2–4% reserved specifically for wildfire-related losses (California Air Resources Board, 2021a,b). Given recent trends, these safeguards appear insufficient. Between 2013 and 2022, wildfires invalidated over 95% of the dedicated wildfire buffer (Badgley, 2022), and the termination of even a single project due to wildfire could invalidate up to 50% of the entire buffer pool (Badgley and Cullenward, 2022). As administrator, the California Air Resources Board (CARB) has responded to such criticisms. It argues that, even if the specific 2–4% wildfire allocation is exceeded, CCOs remain protected by the full 9–19% buffer (California Air Resources Board, 2021a,b). However, in a more recent acknowledgment, California Air Resources Board (2022) conceded that wildfire risk may justify a review of buffer pool contributions. This concession opens the door to exploring alternative risk management mechanisms.

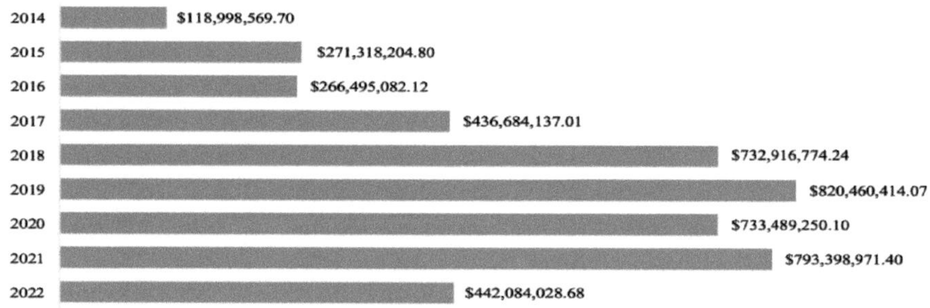

FIGURE 18.4: Carbon offsets: market valuation of transactions in the California Cap-and-Trade (CCT) system, as per California Air Resources Board (2023b).

Mechanism	Contribution (%)		Methodology
	Total	Wildfire	
American Carbon Registry (ACR, 2023)	8.8-50.5	3-12	Spatially explicit.
California Air Resources Board (CARB, 2015)	9-19	2-4	2% with fuel management, and 4% without. Not spatially explicit.
Climate Action Reserve (CAR, n.d., 2019, 2023)	15-20	1.4-7	Applies for wildfire, disease or insects. Not spatially explicit.
Gold Standard (GS, 2023)	20	n/a	Applies for all risk categories.
Verra (2019, 2023a, 2023b)	10-60	0.25-30	Spatially explicit.

FIGURE 18.5: Buffer pool contributions, with description of methodology for calculation. Data represents 59.7% of voluntary carbon market issuance in 2022 (World Bank (2023)). For the American Carbon Registry, the contribution assumes a project size ranging between 2,023ha and 4,047ha.

18.3.3 Risk management strategies for carbon offsets

Across carbon markets, prevailing risk management practices are centered on the use of buffer pools. Rather than being traded, a percentage of each year's offset issuance is set aside to cover potential reversals, and is used as a source of replacement credits. As shown in Figure 18.5, there is currently no consensus on best practices, as methodologies and contribution rates differ across standards. The Climate Action Reserve and Gold Standard do not mandate specific contributions for wildfire risk, while the American Carbon Registry and Verra calibrate their buffer requirements using spatially explicit but retrospective wildfire data (Figure 18.5). Depending on the mechanism, developers may be required to contribute between 8.8% and 60% of total offsets, and between 0% and 30% specifically for wildfire risk (Figure 18.5). In comparison, under the California Cap-and-Trade (CCT) program, the corresponding figures are 9.0–19% for total contributions and 2–4% for wildfire-specific coverage.

In the CCT, contributions to the buffer pool appear inadequate, particularly given that wildfire risk is expected to increase substantially in the future. Across California, higher temperatures and increasing aridity may extend the wildfire season by as much as 61% (Dong et al., 2022), while the risk to forestry projects may increase more than fourfold by 2080–2099 (Anderegg et al., 2022). These projections underscore shortcomings in existing risk management. In particular, CCT contributions do not incorporate spatial heterogeneity despite this being considered best practice (see Figure 18.5). Moreover, buffer pool contributions in the CCT are lower than those required under several independent mechanisms

(Figure 18.5), despite the sharp projected rise in wildfire risk (Anderegg et al., 2022; Ballard et al., 2023). Accounting for both spatial heterogeneity and future risk appears essential, yet existing risk management protocols fail to do so (California Air Resources Board, 2015).

Alternatives to the buffer pool include approaches based on equivalence and insurance (Whitmore and Preston Aragonès, 2022). Regarding equivalence, the literature defines the number of impermanent offsets required to equal one permanent offset. This is grounded in the concept of ton-year accounting, where 1 tCO_2e is valued according to the amount of atmospheric removal or avoidance over a specified time horizon (Fearnside, Lashof, and Moura-Costa, 2000; Galik et al., 2022). This framework can also be expressed in economic terms by incorporating the social cost of carbon and a discount rate (Swinfield et al., 2023; Groom and Venmans, 2023). For example, Swinfield et al. (2023), using a 3.0% discount rate and a social cost of carbon derived from Dietz and Venmans (2019), estimate that for a forestry project lasting 40 years in a wildfire-prone ecosystem, 0.26 impermanent offsets are equivalent to one permanent offset. A key strength of this approach lies in its conceptual clarity and ability to account for both spatial and temporal variation. However, a key drawback is methodological: there remains little consensus on critical assumptions such as discount rates and time horizons, which complicates real-world implementation (Swinfield et al., 2023; Galik et al., 2022).

Regarding insurance, a policy is purchased alongside an impermanent offset, paying out upon invalidation. This idea was originally proposed by Subak (2003) and Wong and Dutschke (2003), who extended insurance logic from forestry timber to forestry carbon. At the time, obstacles included insurers' aversion to long-term, niche products and concerns over offset price volatility. However, neither concern is as constraining today. With rapid growth in carbon markets (Taskforce on Scaling Voluntary Carbon Markets, 2021) and a proliferation of carbon-linked futures contracts (Coffman and Lockley, 2017; Spilker and Nugent, 2022), both academic and commercial interest in carbon offset insurance has revived (CISL, 2022; Biffis et al., 2023). For example, Lv and Shi (2023) models insurable loss in terms of wildfire-induced carbon emissions; He and Ren (2023) estimates that a 1.0% increase in insurance subsidy corresponds to a 0.1% increase in carbon preserved per hectare; Biffis et al. (2023); Biffis, Brandi, and Prentice (2024) consider multidimensional forestry project selection as a way to bundling and tranching forestry-backed assets to constrain wildfire risk exposure. In the private sector, Descartes Underwriting (2021) has introduced a parametric insurance product for offsets invalidated by wildfire. However, to date, neither academic nor grey literature has published explicit pricing calculations for such insurance. In this chapter, we present some of these calculations for the first time.

18.3.4 Overview of the insurance mechanism

We consider an annual insurance policy purchased alongside a California Carbon Offset (CCO), which pays out in the event of invalidation due to wildfire. In designing the insurance mechanism, we draw on long-standing theory from the forestry timber sector (Brunette and Couture, 2023). Wildfire is treated as a natural catastrophe, and a catastrophe modeling framework is used to understand hazard occurrence and loss propagation. This framework typically comprises three components: a hazard module, a vulnerability module, and a financial module (Figure 18.6). In this context, wildfire is represented in the hazard module, forestry projects in the vulnerability module, and CCOs in the financial module.

The insurance mechanism is outlined in equations (18.1)–(18.4) below for a given target location or bundle of locations. For ease of exposition, quantities are cumulated over a unitary time period and payouts are assumed to occur at the end of the period. The quantity of loss is represented by the random variable Q, which depends on the severity of a wildfire (vulnerability module), conditional on wildfire occurrence. Under standard

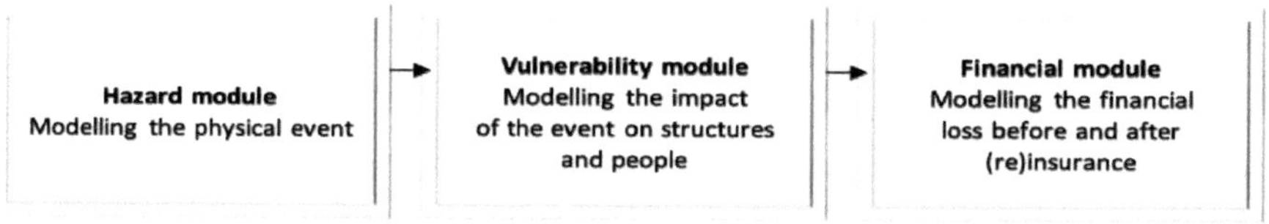

FIGURE 18.6: Components of a natural catastrophe model. Diagram based on Parodi (2023).

modeling assumptions, the expected loss is obtained as the product of wildfire frequency ϕ and expected severity:

$$E\left[Q\right] = \phi \times E\left[S\right]. \tag{18.1}$$

Severity is assumed to follow a multiplicative model combining fuel load (FL), burned area (BA), and combustion completeness (CC) (e.g., Magerl et al. (2022)). A simplified model that abstracts away from dependencies between severity drivers takes the form:

$$E\left[S\right] = E\left[FL\right] \times E\left[BA\right] \times E\left[CC\right] = l \times b \times c. \tag{18.2}$$

Key terminology in wildfire modeling.

- **Fuel load:** the dry biomass per unit area (e.g., Alonso-Rego et al. (2020)), equivalent to carbon stock.
- **Burned area:** the portion of the landscape affected by fire (e.g., Lutz et al. (2011)).
- **Combustion completeness:** the fraction of available biomass or fuel consumed (e.g., Balde et al. (2023)).

The financial module uses the information above to estimate expected financial losses associated with wildfire occurrence in a given area by multiplying the quantity of loss by its unit cost C. The insurance pricing functional is then commonly obtained by applying a multiplicative risk loading $\lambda \geq 0$, yielding the representation

$$P = (1 + \lambda) \times E\left[Q \times C\right]. \tag{18.3}$$

The risk loading reflects several considerations, including the insurer's cost of capital and geographical risk concentration. Consequently, the pricing functional is typically derived over a portfolio of locations, rather than being location-specific. Furthermore, it can be nonlinear in the quantity of loss.

In line with equation (18.2), a widely used simplification delivers the factorization:

$$E\left[Q\right] \times C_0, \tag{18.4}$$

where C_0 denotes the forward price of the CCO over the relevant horizon. While various generalizations and modeling choices are possible, we focus on the simplest possible framework here.

To parameterize the hazard module, we use data on carbon stock and wildfire emissions at 0.28 km^2 resolution and monthly intervals between 2002–2020 (van Wees et al., 2022). For the vulnerability module, we rely on 2020 forestry biome data (Friedl and Sulla-Menashe, 2019) and shapefiles for forestry projects (California Air Resources Board, 2023a). CCO forward prices (ICE, 2023) are used to parameterize the financial module. Insurance prices across locations are computed using equations (18.3)-(18.4) under the assumption of a zero risk loading for illustration purposes. Variation in insurance prices across locations highlights how pricing can reflect differing bundles of risk exposures.

The hazard module employs a negative binomial model to estimate wildfire occurrence. A Poisson process was also tested but delivered inferior performance consistent with wildfire risk literature (Plucinski et al., 2013; Becker, Woolford, and Dean, 2022) criticizing earlier reliance on the Poisson distribution (Cunningham and Martell, 1973). For the vulnerability module, the proportional loss of carbon stock in equation (18.3) is modeled by examining distributions and transformations of burned area BA for a given combustion completeness level CC (Becker, Woolford, and Dean, 2022). A simple Beta distribution performs better than a truncated Pareto distribution used, for example, by Schoenberg, Peng, and Woods (2003) and Li and Banerjee (2021). We consider insurance coverage for events causing losses greater than 50% of carbon stock. The corresponding unloaded insurance price (i.e., assuming a zero risk loading) is computed using equation (18.4), with forward prices used for C_0.

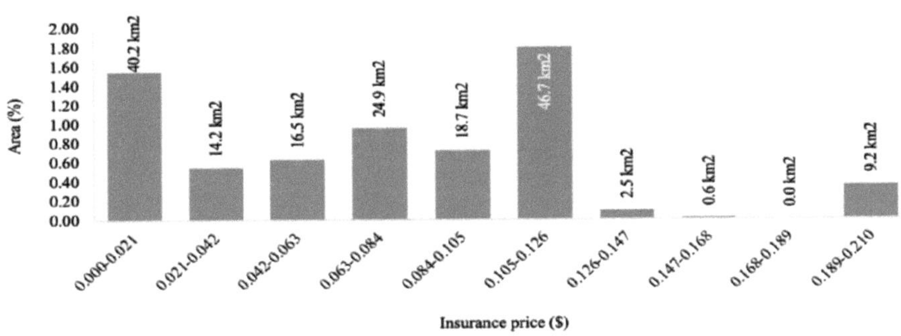

FIGURE 18.7: Insurance price and coverage area across forestry in California. 73.4% of area experienced no wildfire.

As illustrated in Figures 18.7–18.8, two main results are of interest. First, for forestry projects in California, the maximum insurance price is \$0.21/tCO_2e, which is below the implicit safety premium from buffer pool contributions (\$0.74/tCO_2e). However, across California, 12.4% of forestry project area exceeds this safety premium (Figure 18.8), highlighting shortcomings in current risk management practices. The CCT imposes no spatial restrictions on project siting, yet buffer pool contributions remain fixed at 2–4% of issuance, despite significant spatial heterogeneity in wildfire risk. In contrast, other mechanisms in the VCM account for spatial heterogeneity (Figure 18.5), and the CCT could follow suit. Second, the spatial distribution of wildfire risk aligns with existing literature. As shown in Figures 18.9–18.10, the highest probability of invalidation occurs around Los Angeles, followed by scattered risk concentrations in central and northern California. Prior studies find that wildfire frequency peaks near Los Angeles, while burned area is greatest in central and northern regions (Li and Banerjee, 2021). Burn severity is moderate in the former and high in the latter (Xu et al., 2022). These differences suggest that losses are driven by frequent, less severe human-ignited fires in southern California and by infrequent but more severe wildfires in the north, caused by both natural and anthropogenic ignition sources (Li

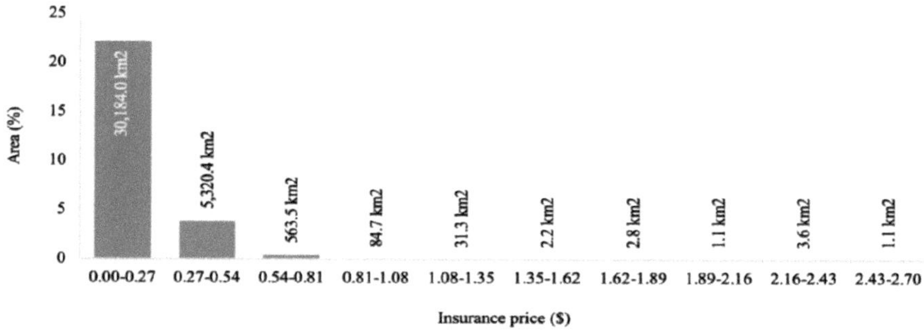

FIGURE 18.8: Insurance price and coverage area across forestry projects under the California Cap-and-Trade (CCT) program. The data reflect only forestry projects supporting carbon credits. Notably, 93.3% of project locations have experienced no wildfire to date.

FIGURE 18.9: Map of forestry projects in California under the California Cap-and-Trade (CCT) scheme.

and Banerjee, 2021; Xu et al., 2022). These dynamics have clear implications for pricing insurance across different exposure bundles.

Several additional considerations support the use of insurance as a risk management tool. First, as demonstrated, spatial heterogeneity should be integrated into risk models. Temporal dynamics such as projected increases in wildfire risk could also be incorporated using machine learning techniques (Anderegg et al., 2022; Ballard et al., 2023; Li et al., 2023). The insurance industry is particularly well-positioned to handle these complexities due to its long-standing experience with computationally intensive risk modeling (Ellili et al., 2023). Second, in designing the CCO insurance policy, we propose monetary payouts rather than like-for-like credit replacements (as done in the CCT; California Air Resources Board (2015)). Insurance is better suited to deliver monetary indemnity, which offers buyers greater optionality when replacing invalidated credits. This flexibility allows entities to select higher-quality offsets, such as NbS with lower reversal risk or engineered carbon removal solutions with greater permanence. In turn, this can improve standards in risk management, incentivize better project design, and strengthen market integrity overall (Axelsson et al., 2024; Herbert et al., 2022).

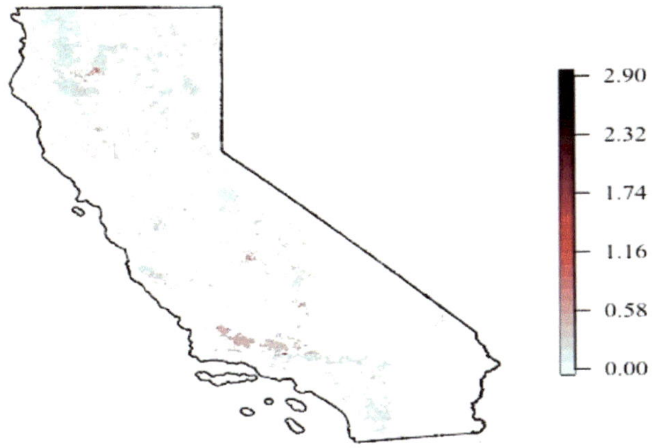

FIGURE 18.10: Map of benchmark insurance price for invalidation of a California Carbon Offset (CCO) to wildfire.

18.4 Conclusion

One of the main challenges in developing sustainable insurance products is the availability and quality of data. Accurate risk assessment requires high-resolution information on weather patterns, land use, and environmental conditions, as well as robust modeling techniques capable of capturing the complex interactions between these variables. Without reliable data and validated models, it is difficult to design risk transfer instruments that are both fair and financially viable.

Despite these challenges, the insurance sector is entering a period of unprecedented opportunity for innovation. Advances in satellite remote sensing, machine learning, and climate modeling are creating new possibilities for understanding and managing environmental risks. These technologies enable the development of more accurate, affordable, and timely insurance products - particularly for vulnerable sectors such as agriculture and forestry, which are disproportionately exposed to climate-related shocks.

The integration of insurance with other forms of financial support through *bundled finance solutions* allows insurance to play a key role in facilitating the adoption of more climate-resilient technologies. By combining parametric insurance with tools such as concessional lending, credit guarantees, and performance-based equity, stakeholders can unlock investment for nature-based solutions, climate adaptation projects, and sustainable infrastructure. These bundled instruments help reduce the cost of capital, crowd in private finance, and align risk mitigation with long-term development objectives.

In parallel, the emergence of smart contracts and blockchain technology presents new opportunities to increase transparency, automate claims settlement, and reduce administrative costs. Blockchain-based platforms can securely record insurance contracts and payout conditions, while smart contracts can execute payments automatically when predefined triggers such as weather index thresholds are met. This reduces transaction costs, enhances trust in insurance systems, and improves accessibility, particularly in remote or underserved regions.

Finally, policymakers have a critical role to play in enabling these innovations. This includes establishing clear and adaptive regulatory frameworks that support technology adoption and innovation in insurance markets, promoting data-sharing standards, and ensuring that the design of financial instruments aligns with broader sustainability goals. Financial incentives, technical assistance, and risk-sharing mechanisms can further accelerate the deployment of sustainable insurance solutions ensuring that those most vulnerable to climate impacts have access to the protection and investment they need to build resilience.

References

Allied Offsets, 2025, Prices and VCM activity, Available at: `https://alliedoffsets.com/pricing-activity/` (Accessed on: 15.05.25).

Alonso-Rego, C., S. Arellano-Pérez, C. Cabo, C. Ordoñez, J. Álvarez-González, R. Díaz-Varela, and A. Ruiz-González, 2020, Estimating fuel loads and structural characteristics of shrub communities by using terrestrial laser scanning, *Remote Sensing* 12, 3704.

Anderegg, W., O. Chegwidden, G. Badgley, A. Trugman, D. Cullenward, J. Abatzoglou, J. Hicke, J. Freeman, and J. Hamman, 2022, Future climate risks from stress, insects and fire across US forests, *Ecology Letters* 25, 1510–1520.

Aswani, J., A. Raghunandan, and S. Rajgopal, 2024, Are carbon emissions associated with stock returns?, *Review of Finance* 28, 75–106.

Aviva, 2024, Aviva zero - our new carbon-conscious car insurance, Available at: `https://zero.aviva.co.uk` (Accessed on: 31.12.24).

AXA XL, 2022, Enabling the offset; what role can insurance play in offsetting emissions?, Available at: `https://axaxl.com/fast-fast-forward/articles/enabling-the-offset-what-role-can-insurance-play-in-offsetting-emissions` (Accessed on: 31.12.24).

Axelsson, K., A. Wagner, I. Johnstone, M. Allen, B. Caldecott, N. Eyre, S. Fankhauser, T. Hale, C. Hepburn, C. Hickey, R. Khosla, S. Lezak, E. Mitchell-Larson, Y. Malhi, N. Seddon, A. Smith, and S. Smith, 2024, The oxford principles for net zero aligned carbon offsetting (revised 2024, Available at: `https://www.smithschool.ox.ac.uk/sites/default/files/2024-02/Oxford-Principles-for-Net-Zero-Aligned-Carbon-Offsetting-revised-2024.pdf` (Accessed on: 21.12.24).

Badgley, G., and D. Cullenward, 2022, California forest carbon buffer pool update, Available at: `https://carbonplan.org/blog/buffer-analysis-update` (Accessed on: 15.08.23).

Badgley Grayson, Chay Freya, Chegwidden Oriana S., Hamman Joseph J., Freeman Jeremy, Cullenward Danny., 2022, California's forest carbon offsets buffer pool is severely undercapitalized, *Frontiers in Forests and Global Change* 5, 930426.

Balde, B., C. Vega-Garcia, P. Gelabert, A. Ameztegui, and M. Rodrigues, 2023, The relationship between fire severity and burning efficiency for estimating wildfire emissions in mediterranean forests, *Journal of Forestry Research*.

Ballard, T., M. Cooper, C. Lowrie, and G. Erinjippurath, 2023, Widespread increases in future wildfire risk to global forest carbon offset projects revealed by explainable AI, Available at: `https://doi.org/10.48550/arXiv.2305.02397` (Accessed on: 15.08.23).

Becker, D., D. Woolford, and C. Dean, 2022, Assessing dependence between frequency and severity through shared random effects, *Public Library of Science One* 17, e0271904.

Biagini, S., E. Biffis, and K. Nobari, 2024, Short-lived gases, carbon markets and climate risk mitigation, Technical report, Imperial College Business School, available at SSRN 4997380.

Biffis, E., G. Brandi, L. Lee, and A. Snavely, 2023, Forestry-backed assets design, Technical report, Singapore Green Finance Centre.

Biffis, E., G. Brandi, and I. Prentice, 2024, Wildfire risk and carbon-based asset design, Technical report, Imperial College Business School and Imperial Life Sciences.

Biffis, E., and E. Chavez, 2017, Satellite data and machine learning for weather risk management and food security, *Risk Analysis* 37, 1508–1521.

Biffis, E., E. Chavez, A. Louaas, and P. Picard, 2022, Parametric insurance and technology adoption in developing countries, *The Geneva Risk and Insurance Review* 1–38.

Birkmann, J., J. McMillan, and S. Tangwanichagapong, 2023, Climate extremes call for resilience and adaptation, not just repair, *Nature*.

Bolton, P., and M. Kacperczyk, 2023, Global pricing of carbon-transition risk, *The Journal of Finance* 78, 3677–3754.

Brunette, M., and S. Couture, 2023, Forest insurance for natural events: An overview by economists, *Forests* 14, 289.

Bryan, G., 2019, Ambiguity aversion decreases demand for insurance: Evidence from field experiments in Africa, *Journal of Economic Behavior & Organization* 166, 369–382.

California Air Resources Board, 2015, Compliance offset protocol. U.S. forest projects, Available at: `https://ww2.arb.ca.gov/sites/default/files/cap-and-trade/protocols/usforest/forestprotocol2015.pdf` (Accessed on: 15.08.23).

California Air Resources Board, 2021a, California's compliance offset programme, Available at: `https://ww2.arb.ca.gov/sites/default/files/2021-10/nc-forest_offset_faq_20211027.pdf` (Accessed on: 14.08.23).

California Air Resources Board, 2021b, CARB responses to questions from ProPublica on California's Forest Offset Protocol, Available at: `https://ww2.arb.ca.gov/sites/default/files/2021-04/nc-carb-response-to-propublica-forest-questions.pdf` (Accessed on: 14.08.23).

California Air Resources Board, 2022, U.S. forest offset workshop presentations, November 2022, Available at: `https://ww2.arb.ca.gov/resources/documents/us-forest-offset-workshop-presentations-november-2022` (Accessed on: 14.08.23).

California Air Resources Board, 2023a, ARB offset credit issuance, Available at: `https://ww2.arb.ca.gov/our-work/programs/compliance-offset-program` (Accessed on: 14.08.23).

California Air Resources Board, 2023b, Summary of market transfers report, Available at: `https://ww2.arb.ca.gov/our-work/programs/cap-and-trade-program/program-data/summary-market-transfers-report` (Accessed on: 14.08.23).

Carter, M., A. de Janvry, E. Sadoulet, and A. Sarris, 2015, Index-based weather insurance for developing countries: A review of evidence and a set of propositions for up-scaling, *Revue d'Economie du Développement* 23, 5–57.

Carter, M., S. Janzen, and Q. Stoeffler, 2014, Can insurance help manage climate risk and food insecurity? Evidence from recent experiments in East Africa, *Global Environmental Change* 28, 217–226.

Cenci, S., and E. Biffis, 2025, Lack of harmonisation of greenhouse gases reporting standards and the methane emissions gap, *Nature Communications* 16, 1537.

CFC Underwriting, 2024, What is carbon insurance? guide and benefits, Available at: `https://www.cfc.com/en-gb/knowledge/resources/articles/2024/09/what-is-carbon-insurance-and-why-does-it-matter/` (Accessed on: 31.12.24).

Chen, Z., Y. Lu, J. Zhang, and W. Zhu, 2024, Managing weather risk with a neural network-based index insurance, *Management Science* 70, 4306–4327.

Cheng, S., S. Costedoat, E. Sterling, C. Chamberlain, A. Jagadish, P. Lichtenthal, A. Nowakowski, A. Taylor, J. Tinsman, S. Canty, et al., 2022, What evidence exists on the links between natural climate solutions and climate change mitigation outcomes in subtropical and tropical terrestrial regions? a systematic map protocol, *Environmental evidence* 11, 15.

CISL, 2022, Why nature matters: Nature-related risks and opportunities for insurance underwriting, University of Cambridge Institute for Sustainability Leadership.

Clarke, D., 2016, A theory of basis risk and insurance demand, *The Geneva Papers on Risk and Insurance* 41, 524–546.

CME, 2023, CBL California carbon allowance vintage-specific 2023, Chicago Mercantile Exchange. Available at: `https://www.cmegroup.com/markets/energy/emissions/cbl-california-carbon-allowance-vintage-specific-2023.html#venue=globex` (Accessed on: 14.08.23).

Coffman, D., and A. Lockley, 2017, Carbon dioxide removal and the futures market, *Environmental Research Letters* 12, 015003.

Comello, S., J. Reichelstein, and S. Reichelstein, 2023, Corporate carbon reporting: Improving transparency and accountability, *One Earth* 6, 803–810.

Cunningham, A., and D. Martell, 1973, A stochastic model for occurrence of man-caused forest fires, *Canadian Journal of Forest Fires* 3, 282–287.

Dercon, S., and L. Christiaensen, 2011, Consumption risk, technology adoption, and poverty traps: Evidence from Ethiopia, *Journal of Development Economics* 96, 159–173.

Descartes Underwriting, 2021, Parametric insurance & carbon credit risk, Available at: `https://descartesunderwriting.com/wildfire/` (Accessed on: 15.08.23).

Dhanda, K., and P. Murphy, 2011, The new wild west is green: Carbon offsets markets, transactions, and providers, *Academy of Management Perspectives* 25, 37–49.

Diamond, D., 1984, Financial intermediation and delegated monitoring, *The Review of Economic Studies* 51, 393–414.

Dietz, S., and F. Venmans, 2019, Cumulative carbon emissions and economic policy: In search of general principles, *Journal of Environmental Economics & Management* 96, 108–129.

Döbbeling-Hildebrandt, N., K. Miersch, T. Khanna, M. Bachelet, S. Bruns, M. Callaghan, O. Edenhofer, C. Flachsland, P. Forster, M. Kalkuhl, et al., 2024, Systematic review and meta-analysis of ex-post evaluations on the effectiveness of carbon pricing, *Nature Communications* 15, 4147.

Doherty, N., and H. Schlesinger, 1990, Rational insurance purchasing: Consideration of contract nonperformance, *The Quarterly Journal of Economics* 105, 243–253.

Dong, C., A. Williams, J. Abatzoglou, K. Lin, G. Okin, T. Gillespie, D. Long, Y.-H. Lin, A. Hall, and G. MacDonald, 2022, The season for large fires in southern California is projected to lengthen in a changing climate, *Communications Earth & Environment* 3, 22.

Ellili, N., H. Nobanee, L. Alsaiari, H. Shanti, B. Hillebrand, N. Hassanain, and L. Elfout, 2023, The applications of big data in the insurance industry: A bibliometric and systematic review of relevant literature, *The Journal of Finance and Data Science* 100102.

Fearnside, P., D. Lashof, and P. Moura-Costa, 2000, Accounting for time in mitigating global warming through land-use change and forestry, *Mitigation and Adaptation Strategies for Global Change* 5, 239–270.

Financial Times, 2024, Why the troubled carbon offset market is taking cover, Available at: `https://www.ft.com/content/0269694a-569f-44dc-ba3d-bf12db3f1466` (Accessed on: 31.12.24).

Friedl, M., and D. Sulla-Menashe, 2019, Mcd12q1 v006 modis/terra+aqua land cover type yearly l3 global 500 m sin grid.

Galik, C., J. Baker, A. Daigneault, and G. Latta, 2022, Crediting temporary forest carbon: Retrospective and empirical perspectives on accounting options, *Frontiers in Forests and Global Change* 5, 933020.

Gallagher Re, 2024, Insuring the transition: Safeguarding carbon credits, Available at: `https://www.ajg.com/gallagherre/news-and-insights/2024/november/insuring-the-transitions-safeguarding-carbon-credits/` (Accessed on: 31.12.24).

Giné, X., and D. Yang, 2009, Insurance, credit, and technology adoption: Field experimental evidence from Malawi, *Journal of Development Economics* 89, 1–11.

Girardin, C., S. Jenkins, N. Seddon, M. Allen, S. Lewis, C. Wheeler, B. Griscom, and Y. Malhi, 2021, Nature-based solutions can help cool the planet – if we act now, *Nature* 593, 191–194.

Griscom, B., J. Adams, P. Ellis, R. Houghton, G. Lomax, D. Miteva, W. Schlesinger, D. Shoch, J. Siikamäki, P. Smith, et al., 2017, Natural climate solutions, *Proceedings of the National Academy of Sciences of the USA* 114, 11645–11650.

Groom, B., and F. Venmans, 2023, The social value of offsets, *Nature* 619, 768–773.

He, Y., and Y. Ren, 2023, Can carbon sink insurance and financial subsidies improve the carbon sequestration capacity of forestry, *Journal of Cleaner Production* 397, 136618.

Herbert, C., B. Haya, S. Stephens, and V. Butsic, 2022, Managing nature-based solutions in fire-prone ecosystems: Competing management objectives in California forests evaluated at a landscape scale, *Frontiers in Forests and Global Change* 5, 957189.

Hill, R., J. Hoddinott, and N. Kumar, 2013, Adoption of weather index insurance: Learning from willingness to pay among a panel of households in rural Ethiopia, *Agricultural Economics* 44, 385–398.

Hill, R., J. Hoddinott, N. Kumar, and T. Mekasha, 2019, Ex ante and ex post effects of hybrid index insurance in Ethiopia, *Journal of Development Economics* 136, 123–135.

Holmstrom, B., and J. Tirole, 1997, Financial intermediation, loanable funds, and the real sector, *The Quarterly Journal of Economics* 112, 663–691.

Howden Group, 2022, Howden launches 'World-First' voluntary carbon credit insurance product, Available at: `https://www.howdengroup.com/news-and-insights/Howden-launches-World-First-voluntary-carbon-credit-insurance-product-to-help-scale-the-market` (Accessed on: 31.12.24).

ICE, 2023, California carbon offset futures, Intercontinental Exchange. Available at: `https://www.ice.com/products/71544060/California-Carbon-Offset-Futures` (Accessed on: 14.08.23).

Integrity Council for the Voluntary Carbon Market, 2023, The core carbon principles, Available at: `https://icvcm.org/the-core-carbon-principles/` (Accessed on: 14.08.23).

IPCC, 2022, *Climate Change 2022: Mitigation of Climate Change* (Cambridge University Press, Cambridge, UK), Intergovernmental Panel on Climate Change. Contribution of Working Group III to the Sixth Assessment Report.

Jaffee, D., and T. Russell, 1976, Imperfect information, uncertainty, and credit rationing, *The Quarterly Journal of Economics* 90, 651–666.

Li, F., Q. Zhu, W. Riley, L. Zhao, L. Xu, K. Yuan, M. Chen, H. Wu, Z. Gui, J. Gong, et al., 2023, AttentionFire v1.0: Interpretable machine learning fire model for burned-area predictions over tropics, *Geoscientific Model Development* 16, 869–884.

Li, S., and T. Banerjee, 2021, Spatial and temporal patterns of wildfires in California from 2000 to 2019, *Scientific Reports* 11, 8779.

Lutz, J., C. Key, C. Kolden, J. Kane, and J. Van Wagtendonk, 2011, Fire frequency, area burned, and severity: A quantitative approach to defining a normal fire year, *Fire Ecology* 7, 51–65.

Lv, H., and A. Shi, 2023, Setting the forestry carbon sink insurance compensation standard for reducing losses from forest fires: An empirical study from forest fires in a sample of 15 provinces, *Frontiers in Environmental Science* 11, 1112525.

Magerl, A., S. Gingrich, S. Matej, G. Cunfer, M. Forrest, C. Lauk, S. Schlaffer, F. Weidinger, C. Yuskiw, and K. Erb, 2022, Dynamics of forest carbon stocks, fires, and harvest under changing climatic conditions in the U.S. during the 20th century, Available at: `https://doi.org/10.1002/essoar.10511935.1` (Accessed on: 15.08.23).

McIntosh, C., A. Sarris, and F. Papadopoulos, 2013, Productivity, credit, risk, and the demand for weather index insurance in smallholder agriculture in Ethiopia, *Agricultural Economics* 44, 399–417.

Meyer, R., P. Hazell, and P. Varangis, 2017, Unlocking smallholder credit: Does credit-linked agricultural insurance work?, Technical report, Global Index Insurance Facility, The World Bank Group.

MSCI, 2022, Insurance joins list of options to de-risk carbon credits, Available at: `https://www.msci.com/www/blog-posts/insurance-joins-list-of-options/04718132808` (Accessed on: 31.12.24).

Parodi, P., 2023, *Pricing in General Insurance* (Chapman & Hall, New York, NY, USA).

Plucinski, M., W. McCaw, J. Gould, and B. Wotton, 2013, Predicting the number of daily human-caused bushfires to assist suppression planning in south-west Australia, *International Journal of Wildland Fire* 23, 520–531.

Practical ESG, 2024, Insuring carbon offsets: What CSOs should know, Available at: `https://practicalesg.com/2024/10/insuring-carbon-offsets-what-csos-should-know/` (Accessed on: 31.12.24).

Prado, A., and N. MacDowell, 2023, The cost of permanent carbon dioxide removal, *Joule* 7, 700–712.

Reichelstein, S., 2024, Corporate carbon accounting: balance sheets and flow statements, *Review of Accounting Studies* 29, 2125–2156.

Ritchie, H., and M. Roser, 2024, CO_2 emissions, *Our World in Data* Available at: `https://ourworldindata.org/co2-emissions` (Accessed on: 15.05.25).

Schoenberg, F., R. Peng, and J. Woods, 2003, On the distribution of wildfire sizes, *Environmetrics* 14, 583–592.

Seddon, N., A. Chausson, P. Berry, C. Girardin, A. Smith, and B. Turner, 2020, Understanding the value and limits of nature-based solutions to climate change and other global challenges, *Philosophical Transactions of the Royal Society B* 375, 20190120.

Shee, A., and C. Turvey, 2012, Collateral-free lending with risk-contingent credit for agricultural development: Indemnifying loans against pulse crop price risk in India, *Agricultural Economics* 43, 561–574.

Spilker, G., and N. Nugent, 2022, Voluntary carbon market derivatives: Growth, innovation & usage, *Borsa Istanbul Review* 22, S109–S118.

Stiglitz, J., N. Stern, M. Duan, O. Edenhofer, G. Giraud, G. Heal, E. La Rovere, A. Morris, E. Moyer, M. Pangestu, et al., 2017, Report of the high-level commission on carbon prices.

Stiglitz, J., and A. Weiss, 1981, Credit rationing in markets with imperfect information, *The American Economic Review* 71, 393–410.

Subak, S., 2003, Replacing carbon lost from forests: an assessment of insurance, reserves and expiring credits, *Climate Policy* 3, 107–122.

Surminski, S., L. Bouwer, and J. Linnerooth-Bayer, 2016, How insurance can support climate resilience, *Nature Climate Change* 6, 333–334.

Swinfield, T., A. Balmford, D. Coomes, A. Madhavapeddy, and S. Keshav, 2023, Realising the social value of impermanent carbon credits, Available at: `https://doi.org/10.33774/coe-2023-5v931-v5` (Accessed on: 14.08.23).

Swiss Re, 2019, Moving from carbon offsets to carbon removal, Available at: `https://www.swissre.com/dam/jcr:18fe31fb-f03e-4a71-a396-0be5b25fe229/2019_sustainability_report_swissre_sr19.pdf` (Accessed on: 23.12.24).

Taskforce on Scaling Voluntary Carbon Markets, 2021, Final report, Available at: https://www.iif.com/Portals/1/Files/TSVCM_Report.pdf (Accessed on: 15.08.23).

Tellman, B., and H. Eakin, 2022, Risk management alone fails to limit the impact of extreme climate events, *Nature*.

Temple-West, P., 2022, Critics take aim at 'wild west' carbon offset market, Available at: https://www.ft.com/content/9b02fcf7-9e04-4b71-ad14-251552d5a78e (Accessed on: 14.08.23).

Trouwloon, D., C. Streck, T. Chagas, and G. Martinus, 2023, Understanding the use of carbon credits by companies: A review of the defining elements of corporate climate claims, *Global Challenges* 7, 2200158.

van Wees, D., G. van der Werf, J. Randerson, B. Rogers, Y. Chen, S. Veraverbeke, L. Giglio, and D. Morton, 2022, Global biomass burning fuel consumption and emissions at 500m spatial resolution based on the global fire emissions database (gfed), *Geoscientific Model Development* 15, 8411–8437.

Whitmore, A., and M. Preston Aragonés, 2022, Addressing differences in permanence of carbon dioxide removal, Available at: https://network.bellona.org/content/uploads/sites/3/2022/04/Addressing-differences-in-permanence-of-Carbon-Dioxide-Removal.pdf (Accessed on: 15.08.23).

Wong, J., and M. Dutschke, 2003, Can permanence be insured? consideration of some technical and practical issues of insuring carbon credits from afforestation and reforestation, Available at: https://papers.ssrn.com/sol3/papers.cfm?abstract_id=419702 (Accessed on: 15.08.23).

World Bank, 2023, State and trends of carbon pricing 2023.

World Bank, 2024, State and trends of carbon pricing 2024.

World Commission on Environment and Development, 1987, *Our Common Future* (Oxford University Press, Oxford, UK), Also known as the Brundtland Report.

Xu, Q., A. L. Westerling, A. Notohamiprodjo, C. Wiedinmyer, J. Picotte, S. Parks, M. Hurteau, M. Marlier, C. Kolden, J. Sam, et al., 2022, Wildfire burn severity and emissions inventory: an example implementation over California, *Environmental Research Letters* 17, 085008.

Zhang, S., 2025, Carbon returns across the globe, *The Journal of Finance* 80, 615–645.

Index